Proteinase and Peptidase Inhibition

Proteinase and Peptidase Inhibition

Recent potential targets for drug development

Edited by

H. John Smith

and

Claire Simons

Welsh School of Pharmacy, Cardiff University, UK

London and New York

First published 2002
by Taylor & Francis
11 New Fetter Lane, London EC4P 4EE

Simultaneously published in the USA and Canada
by Taylor & Francis Inc,
29 West 35th Street, New York, NY 10001

Taylor & Francis is an imprint of the Taylor & Francis Group

Typeset in Goudy by
Integra Software Services Pvt. Ltd, Pondicherry, India
Printed and bound in Great Britain by
TJ International, Padstow, Cornwall

British Library Cataloguing in Publication Data
A catalogue record for this book is available
from the British Library

Library of Congress Cataloguing in Publication Data
A catalogue record has been requested

ISBN 0–415–27349–8

Contents

Color plates

All color artworks are reproduced in black and white in the maintext.

Contributors

Dr Paul J. Ala
The Althexix Company
1365 Main Street
Waltham MA 02451
USA

Professor Ralf Bartenschlager
Institute for Virology
Joahannes-Gutenberg University of Mainz
Obere Zahlbacher Str
67, 55131 Mainz
Germany

Dr Markus Bürgle
Wilex Biotechnology GmbH
München

Dr Chong-Hwan Chang
Bristol-Myers Squibb Company
P.O. Box 80353
Route 141 and Henry Clay Road
Wilmington
DE 19880
USA

Professor Olivier Coux
CRBM-CNRS
IFR 24
1919 route de Mende
34293 Montpellier cedex 5
France

Dr Christine E. Dabrowski
SmithKline Beecham Pharmaceuticals
1250 South Collegeville Road
P.O. Box 5089
Collegeville
PA 19426–0989
USA

Dr Hans-Ulrich Demuth
Probiodrug Research Ltd.
Weinbergweg 22/Biozentrum
D-06120 Halle
Germany

Dr Peter S. Dragovich
Agouron Pharmaceuticals Inc
3565 general Atomics Court
San Diego
CA 92121
USA

Dr Philip D. Edwards
Medicinal Chemistry Department
AstraZeneca Pharmaceuticals
P.O. Box 15437
1800 Concord Pike
Wilmington
Delaware 19850-5437
USA

Professor S.M. Ellerbroek
Departments of Obstetrics & Gynecology
and Cell & Molecular Biology
Northwestern University Medical School
303 E. Chicago Ave.
Tarry 4-751
Chicago, IL 60611
USA

Dr Bernd Gerhartz
Probiodrug Research Ltd.
Weinbergweg 22/Biozentrum
D-06120 Halle
Germany

Dr Michael Groll
Max-Planck Institute für Biochemie
Am Klopferspitz 18A
D-82152 Martinsried
Germany

Dr Fiona Grueninger
Pharma Division
PRPN
Bldg. 69/333
F. Hoffmann-La Roche Ltd
CH-4070
Basel
Switzerland

Dr Jörg Hauptmann
Klinikum der Universität Jena
Zentrum für Vaskuläre Biologie und Medizin
Institut für Biochemie
Nordhäuser Str. 78
D-99089
Erfurt
Germany

Dr Nigel M. Hooper
School of Biochemistry and Molecular Biology
University of Leeds
Leeds LS2 9JT
UK

Dr Gerda Huber
Pharma Division
PRPN
Bldg. 69/333
F. Hoffmann-La Roche Ltd
CH-4070
Basel
Switzerland

Dr Richard L. Jarvest
SmithKline Beecham Pharmaceuticals
New Frontiers Science Park
Third Avenue
Harlow
Essex CM19 5AW
UK

Dr Horst Kessler
Institut für Organische Chemie und Biochemie
Technische Universität München

Dr Daniel Kirchhofer
Departments of Protein Engineering and Cardiovascular Research
Genentech Inc.
1 DNA Way
South San Francisco
CA 94080
USA

Dr Jan-Oliver Koch
Institute for Virology
Joahannes-Gutenberg University of Mainz
Obere Zahlbacher Str
67, 55131 Mainz
Germany

Dr Joel A. Krauser
School of Chemistry and Biochemistry
Georgia Institute of Technology
315 Ferst Drive
Atlanta
GA 30332-0400
USA

Dr Achim Krüger
Clinical Research Group
Frauenklinik
Technische Universität München
Isomaniger Strasse 22
D-81669 München
Germany

Dr Robert A. Lazarus
Departments of Protein Engineering and Cardiovascular Research
Genentech Inc.
1 DNA Way
South San Francisco
CA 94080
USA

Dr Viktor Magdolen
Clinical Research Group
Frauenklinik
Technische Universität München
Isomaniger Strasse 22
D-81669 München
Germany

Dr Pari Malherbe
Pharma Division
PRPN
Bldg. 69/333
F. Hoffmann-La Roche Ltd.
CH-4070
Basel
Switzerland

Dr Andrè J. Niestroj
Probiodrug Research Ltd.
Weinbergweg 22/Biozentrum
D-06120 Halle
Germany

Professor James C. Powers
School of Chemistry and Biochemistry
Georgia Institute of Technology
315 Ferst Drive
Atlanta
GA 30332-0400
USA

Professor Philip J. Rosenthal
Bldg 30
Room 408
Sand Fransisco General Hospital
1001 Potrero Avenue
San Francisco
CA 94110
USA

Dr Brian Rukamp
School of Chemistry and Biochemistry
Georgia Institute of Technology
315 Ferst Drive
Atlanta
GA 30332-0400
USA

Dr Wolfgang Schmalix
Wilex Biotechnology GmbH
München

Professor Manfred Schmitt
Clinical Research Group
Frauenklinik
Technische Universität München
Isomaniger Strasse 22
D-81669 München
Germany

Professor Andrea Scozzafava
Laboratorio di Chimica Inorganica e Bioinorganica
Università degli Studi di Firenze
Via G. Capponi 7
50121 Firenze
Italy

Dr Stefan Sperl
Clinical Research Group
Frauenklinik
Technische Universität München
Isomaniger Strasse 22
D-81669 München
Germany

Professor M.S. Stack
Departments of Obstetrics & Gynecology and Cell & Molecular Biology
Northwestern University Medical School
303 E. Chicago Ave.
Tarry 4-751
Chicago, IL 60611
USA

Dr Torsten Steinmetzer
Klinikum der Universität Jena
Zentrum für Vaskuläre Biologie und Medizin
Institut für Biochemie
Nordhäuser Str. 78
D-99089
Erfurt
Germany

Dr Jörg Stürzebecher
Klinikum der Universität Jena
Zentrum für Vaskuläre Biologie und Medizin
Institut für Biochemie
Nordhäuser Str. 78
D-99089
Erfurt
Germany

Professor Claudiu T. Supuran
Laboratorio di Chimica Inorganica e Bioinorganica
Università degli Studi di Firenze
Via G. Capponi 7
50121 Firenze
Italy

Professor Allen Taylor
USDA Human Nutrition Research Centre on Ageing
Tufts University
711 Washington St
Boston MA 02111
USA

Dr Jason Warner
USDA Human Nutrition Research Centre on Ageing
Tufts University
711 Washington St
Boston MA 02111
USA

Dr Stephen E. Webber
Agouron Pharmaceuticals Inc
3565 general Atomics Court
San Diego
CA 92121
USA

Dr Olaf G. Wilhelm
Wilex Biotechnology GmbH
München

Dr Y. Wu
Departments of Obstetrics & Gynecology and Cell & Molecular Biology
Northwestern University Medical School
303 E. Chicago Ave.
Tarry 4-751
Chicago, IL 60611
USA

Abbreviations

Aβ	amyloid-β-protein
ACE	angiotensin converting enzyme
AD	Alzheimer's disease
ANP	α-human atrial natriuretic peptide
APs	aminopeptidases
βAPP	β-amyloid precursor protein
APPI	amyloid β-protein precursor inhibitor
APTT	activated partial thromboplastin tissue
ARDS	adult acute respiratory distress syndrome
ATIII	antithrombin III
BACE-1	beta-site APP cleaving enzyme
BPTI	bovine pancreatic trypsin inhibitor
BrAAP	branched chain amino-acid preferring
CA	carbonic anhydrase
Cbz	carbobenzoxy
ChC	*Clostridium histolyticum* collagenase
CMK	chloromethyl ketones
CMV	cytomegalovirus
COPD	chronic pulmonary destructive disease
EBV	Epstein-Barr virus
ECM	extracellular matrix
EST	expression sequence tag
FVIIa, IXa, Xa	factors VIIa, IXa, Xa
HHV	human herpes virus
HIV	human immunodeficiency virus
HMW-uPA	high moleular weight uPA
HNE	human neutrophil elastase
HRV 3C	human rhinovirus 3C
Hsp	heat shock protein
HSV	herpes simplex virus
LAP	leucine aminopeptidase
LMWH	low molecular weight heparin
MAP	methionine aminopeptidase
MHC	major histocompatability complex
MMP	matrix metalloproteinase
MMP-1	collagenase 1

MMP-2	gelatinase
MMP-3	stromelysin 1
MMP-7	matrilysin
MMP-8	collagenase 2
MMP-9	progelatinase B
MMP-10	stromeylsin 2
MMP-11	stromelysin 3
MMP-12	macrophage elastase
MMP-13	procollagenase 3
MMP-18	collagenase 4
MMPIs	matrix metalloproteinase inhibitors
NEP	neprilysin
NF-χB	inflammatory transcription factor
PAI-1	serpin PA inhibitor
PAR-1	proteinase-activated receptor 1
PKC	protein kinase C
PPE	porcine pancreatic elastase
ProMMP-3	prostromelysin 1
RT	reverse transcriptase
SNAAP	small neutral amino acid preferring
SREBP	sterol regulatory element-binding protein
suPA-R	soluble uPA-R
TACE	tumor necrosis factor-α converting enzyme
TAP	tick anticoagulant peptide
TF	tissue factor
TFPI	tissue factor pathway inhibitor
TIMP-2	tissue inhibitor of metalloproteinase-2
TMFK	trimethylfluoroketone
TNF-α	tumor necrosis factor-α
tPA	tissue plasminogen activator
TPPII	tripeptidyl peptidase II
Ub	ubiquitin
uPA	urokinase-type plasminogen activator
uPA-R	receptor for uPA
VZV	varicella zoster virus

Preface

Proteinases, initially identified as participants in mammalian food digestion in the intestinal tract, have more recently been recognised as of paramount importance as essential components for the functioning of normal body processes as well as unwelcome participants in diseases.

Proteinases have roles in normal physiological functions such as protein degradation and homeostasis, ("house-keeping"), protein and enzyme maturation, peptide hormone clearance after response, the blood clotting cascade, blood pressure control, antigen removal etc. Undesirable aspects of their activity are reflected in their role in tumor cell migration and tissue invasion (metastasis), maturation of plaque proteins in Alzheimer's disease, inflammatory processes such as rheumatoid arthritis, osteoporosis, viral infections (herpes, shingles, HIV, hepatitis, common cold etc.), protozoal diseases (malaria, leishmaniasis, trypanosomiasis etc.) and bacterial pathogenicity (gingivitis, gangrene, tetanus, botulism, anthrax etc.).

Targeting of a protease with an inhibitor can block or reduce its normal physiologically-accepted role as in blood pressure control and the blood clotting cascade, with a desired clinical treatment outcome for hypertension or thromboembolism. Alternatively removal of a non-physiological role, which has caused a disease condition, could be directed at such situations as tumor metastasis in tumor spread, viral maturation in HIV, plaque formation in Alzheimer's disease, inflammation in rheumatoid arthritis and osteoporosis.

Identification and characterization of the target enzyme in a cellular milieu is much easier nowadays owing to technological advances in molecular biology and protein separation techniques, leading to automated sequencing procedures and larger scale production of pure enzymes by recombinant DNA technology owing to modern molecular biology. Rational design strategy of an inhibitor for the nominated target enzyme intended for drug development is now well established based on a knowledge of the mechanism of action of the enzyme and its substrate specificity, although in many instances previously discovered lead compounds are available. Further refinement, especially in recent times, involves molecular modelling of inhibitors from a knowledge of the enzyme's 3-dimensional structure or that of a related enzyme, which further suggests suitable lead compounds for *in vitro* biochemical evaluation. However, the path from "bench to market place" is tortuous and the majority of drug candidates fall for not meeting the requirements for oral bioavailability, acceptable pharmacokinetics (metabolism, distribution, excretion), specificity for the target enzyme in the presence of related enzymes and an overall clean toxicological spectrum in further biochemical and pharmacological tests and clinical trials. Even then its clinical use may be curtailed by the development of resistance and/or mutation of the enzyme target.

This book considers, in detail, many proteases as new or existing inhibition targets for the development of drugs as clinical agents in situations where either none exist, or lead molecules have not been fully developed or resistance has developed towards drugs with other types of action at the molecular level. Hopefully it will spur on development of treatments for many common diseases where proteases have an important role for which clinical agents are not commercially available.

H. John Smith
Claire Simons
Cardiff
November, 2001

Chapter 1

Enzyme classes and mechanisms

Bernd Gerhartz, Andrè J. Niestroj and Hans-Ulrich Demuth

Proteases are an emerging class of enzymes. In the following chapter an introduction is given to the classification and terminology of these enzymes. Subdivided on the basis of the nucleophile, the chapter provides an overview on the variety of catalytic mechanisms operated by proteases.

1.1 INTRODUCTION

Research on proteolytic enzymes has a long history. As early as the late eighteenth century, it had been shown that specific proteins were responsible for the degradation of other proteins. Enzymes such as pepsin or trypsin which were among the first to be studied, are therefore also among the best characterized enzymes. During the last century, the number of enzymes known that demonstrate proteolytic activity has increased exponentially and we have gained a broad knowledge about the different physiological contributions made by this class of enzymes. In parallel with the increasing knowledge on the subject, a need for a systematic classification appeared. In this chapter, an introduction is given on terminology and classification of peptidases followed by an overview on the different catalytic mechanisms of these classes of enzymes.

1.2 TERMINOLOGY

In chemical terms, the enzymatic cleavage of peptide bonds is considered as hydrolysis, usually called proteolysis. The enzymes responsible for the catalysis of proteolysis have been named "proteases", a term that originated in the nineteenth century German literature on physiological chemistry. Due to the distribution of two different systems of nomenclature, confusion appeared in the terminology of proteolytic enzymes (Scheme 1.1). Today, the NC-IUBMB (Nomenclature Committee of the International Union of Biochemistry and Molecular Biology) has made recommendations which should help in overcoming this confusion (NC-IUBMB 1992). The EC list recommends use of the term "peptidase" for any enzymes that hydrolyze peptide bonds. Synonymous for peptidase, it is still possible to use the term "protease". Peptidases are further divided into "exopeptidases", acting only near a terminus of a polypeptide chain, and "endopeptidases", acting internally in polypeptide chains. The term "proteinase" used previously has been replaced by

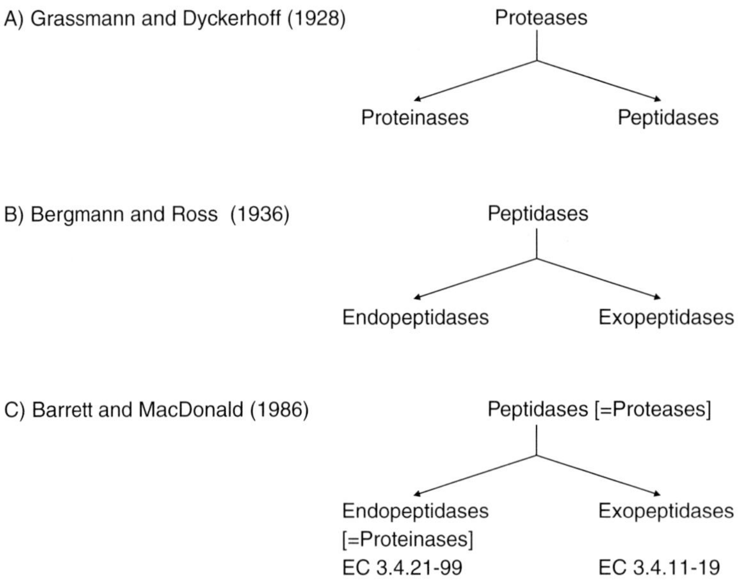

Scheme 1.1 Schemes of nomenclature. The conflict between the older schemes A and B is replaced by the unified scheme C.

Table 1.1 NC-IUBMB definition for subclassifications of peptidases

Subclasses	*Activity*
Exopeptidases	Cleave near a terminus of peptides or proteins
Aminopeptidases	Remove a single amino acid from the free N-terminus
Dipeptidyl peptidases	Remove a dipeptide from the free N-terminus
Tripeptidyl peptidases	Remove a tripeptide from the free N-terminus
Carboxypeptidases	Remove a single amino acid from the C-terminus
Peptidyl dipeptidases	Remove a dipeptide from the C-terminus
Dipeptidases	Cleave dipeptides
Omega peptidases	Remove terminal residues that are substituted, cyclized or linked by isopeptide bonds
Endopeptidases	Cleave internally in peptides or proteins
Oligopeptidases	Cleave preferentially on substrates smaller than proteins

"endopeptidase" for consistency. In addition, the EC list specifies different subtypes of exopeptidases and endopeptidases (Table 1.1).

To define a common nomenclature on the interaction of a substrate with a peptidase, the system of Berger and Schechter (1976) has become generally accepted and used (Scheme 1.2). This system is based on a schematic interaction of amino acid residues of the substrate with specific binding subsites located on the enzyme. By convention, the subsites on the protease are called S (for subsites) and the substrate amino acid residues are called P (for peptide). The numbering of the residues is given from the scissile bond.

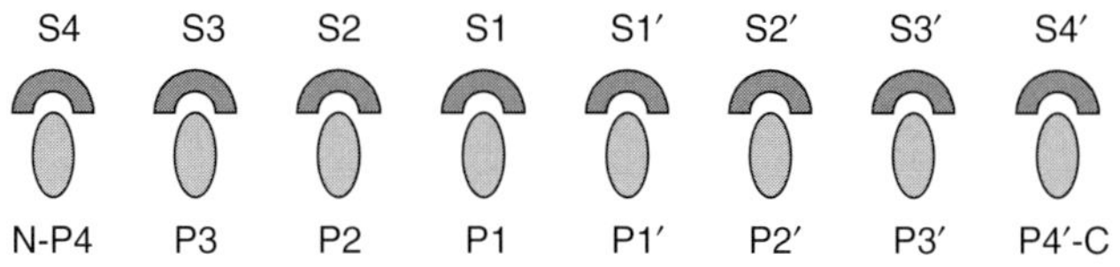

Scheme 1.2 Terminology of specificity subsites of proteases and the complementary features of the substrate (adapted from Berger and Schechter, 1976).

1.3 CLASSIFICATION

Early approaches, classifying peptidases according to molecular size, charge or substrate specificity, failed by the exponential increase of newly discovered enzymes. Observing that peptidases acted through different catalytic mechanisms, Hartley (1960) put forward a more rational system based on the catalytic mechanism of hydrolysis. This concept of distinguishing four groups of enzymes named "serine", "cysteine", "aspartic" and "metallo" peptidases still remains valid, although it has been supplemented by a new class called "threonine" peptidases. The catalytic nucleophile in serine-, threonine- and cysteine-type peptidases is the hydroxyl group of the active site serine, threonine, and the sulfhydryl group of the active site cysteine, respectively. In aspartic type peptidases, the nucleophilic water molecule is directly bound by two aspartic acid residues. Metallopeptidases contain one or two metal ions (in most cases zinc but also cobalt, nickel or manganese) that are usually bound by three amino acids. The nucleophile is a water molecule, as in aspartic peptidases, positioned and activated by the active site metal ions.

Increasing knowledge about structure and catalytic mechanism made a more detailed system of classification necessary. Recently, the subclassification of peptidase clans has been introduced (Barrett and Rawlings 1995; http://www.bi.bbsrc.ac.uk/Merops/Merops.htm). A clan is a group of families with indications of an evolutionary relationship without significant similarity in sequence. Members of a clan share an identical order of catalytic-site residues and similar tertiary folds. However, one clan can even include families from different catalytic classes. For example in clan PA, serine as well as cysteine peptidases are grouped together because current data indicate an exchange of the active site amino acid by keeping the tertiary fold conserved.

1.4 CATALYTIC MECHANISM

The active sites of peptidases are typically characterized by the nucleophile and its assistant groups, an oxyanion hole and specificity pockets. The divergence of substrate specificity is given by the structure of the specificity pockets. In some cases, like the exosite of thrombin even secondary binding sites have been reported to contribute substrate affinity. This chapter is far too short to give an overview of the inventiveness that nature used to create substrate specificity. Here we would rather focus on the divergence of hydrolytic mechanisms which in principle, can be allocated to the respective nucleophile. Nevertheless, in clans PA and PB a similar catalytic mechanism and tertiary fold is paired with different nucleophiles.

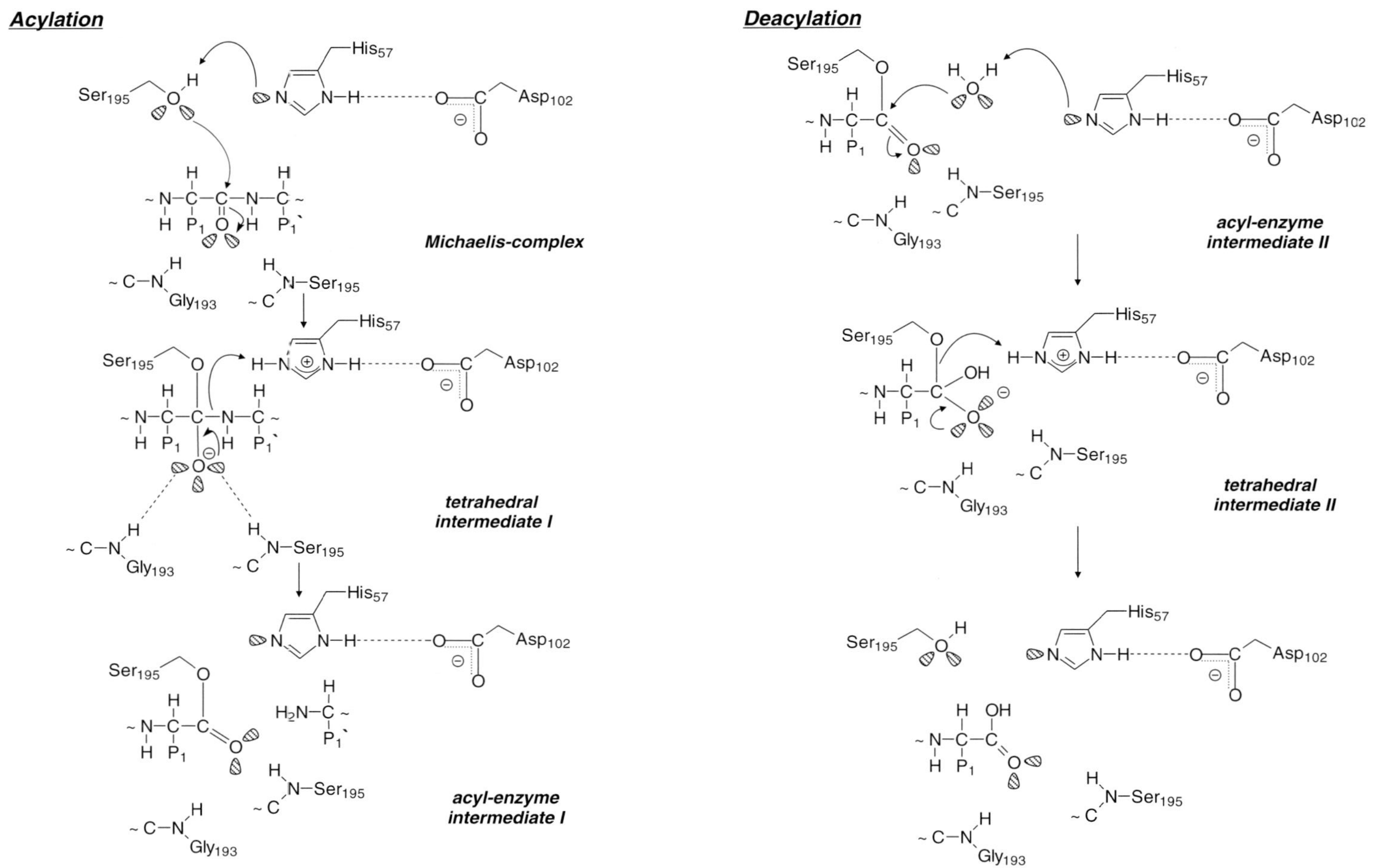

Scheme 1.3 Catalytic mechanism proposed for serine peptidases with a catalytic triad consisting of Ser, His, Asp (Chymotrypsin numbering, adapted from Beynon and Bond (1989)).

1.4.1 A serine residue as a nucleophile

1.4.1.1 Catalytic triad consisting of Ser, His, Asp

The classical mechanism of serine peptidases is based on the catalytic triad of serine, histidine and aspartate. This mechanism is used by serine peptidases of the clan PA (catalytic residues in the order His, Asp, Ser), clan SB (order Asp, His, Ser), and clan SC (order Ser, Asp, His). In all cases, the chemical mechanism of hydrolysis is identical. After formation of a Michaelis complex, the carbonyl carbon atom of the scissile bond is attacked by the active site Ser (Scheme 1.3). The nucleophilicity of the serine is enhanced by an adjacent histidine functioning as a general base catalyst. The formation of an acyl enzyme complex is achieved through a tetrahedral intermediate which collapses to yield the acyl enzyme and the cleavage of the peptide bond in parallel. Deacylation occurs via the same mechanistic steps. In this case, the nucleophilic attack is performed by a bound water molecule resulting in the release of the peptide and restoration of the Ser-hydroxyl of the enzyme. The mechanism requires a binding site for the oxyanion of the tetrahedral intermediate. This site is formed for example in the chymotrypsin family by the backbone amides of Ser195 and Gly193, in the subtilisin family by one amide nitrogen bond and the Asn155 side chain, and in serine carboxypeptidases by the backbone amides of Tyr147 and Gly53. The catalytic contribution of the third member in the catalytic triad, the conserved Asp, has been controversial. The early suggestion that the Asp accepts a proton to become uncharged in the transition state has been opposed by newer experimental and theoretical data (Kossiakoff and Spencer 1981; Warshel *et al* 1989). Further suggested roles of the Asp include stabilization of the imidazole orientation (Rogers and Bruice 1974) as well as stabilization of the local structure around the active site (Lau and Bruice 1999). In addition, it has been proposed that the hydrogen bond between Asp and His is a "low-barrier hydrogen bond" which accounts for the transition state stabilization (Frey *et al* 1994). However, this model has also been questioned (Ash *et al* 1997). Due to the lack of conservation of the position of the Asp, it has been suggested that instead of a triad, two catalytic dyads, a Ser/His and a Ser/Asp are at hand (Liao *et al* 1992).

1.4.1.2 Catalytic triad consisting of Ser, His, His

A slight modification of the classical catalytic triad can be observed in clan SH peptidases. This clan comprises just one family, the family 21 of cytomegalovirus assemblin. Besides a different fold, the aspartic acid of the catalytic triad is replaced by another histidine (Qiu *et al* 1996; Shieh *et al* 1996; Tong *et al* 1996). In human cytomegalovirus protease, Ser132 acts as the catalytic nucleophile and is in the vicinity of His63 and His157 (Tong *et al* 1996). Superimposing the structures of cytomegalovirus proteinase, chymotrypsin or papain reveals a general overlap of the position of the third members in the catalytic triad, His, Asp, and Asn residues (Tong *et al* 1996). It has been speculated that Asp65 may act as a proton acceptor of His157 resulting in a catalytic tetrad for the cytomegalovirus proteinase, where the second His would act as an extra component in a 'relay' proton transfer mechanism (Qiu *et al* 1996). However, Asp65 is not found in the structure of herpes simplex virus protease or varicella-zoster virus protease indicating that such a transfer mechanism is not a general feature in this clan (Hoog *et al* 1997; Qiu *et al* 1997). The simplex virus protease inhibitor complex revealed an oxyanion hole defined by a water molecule and the amide nitrogen of Arg165 (Hoog *et al* 1997).

1.4.1.3 Catalytic dyad consisting of Ser, Lys

After the catalytic triad had become almost a dogma for serine peptidases, one of the first tasks after discovering a new enzyme of this class was the assignment of active site Ser, His and Asp. However, cases occurred where an active site Ser could be identified but His and Asp could not be found to complete the triad. All evidences to date point towards a catalytic dyad in these cases. Surprisingly, a Lys completes the dyad instead of a His (Strynadka *et al* 1992). There are two clans of peptidases using the Ser/Lys dyad mechanism, clan SE and SF. Whereas the catalytic dyad occurs in clan SE peptidases within the

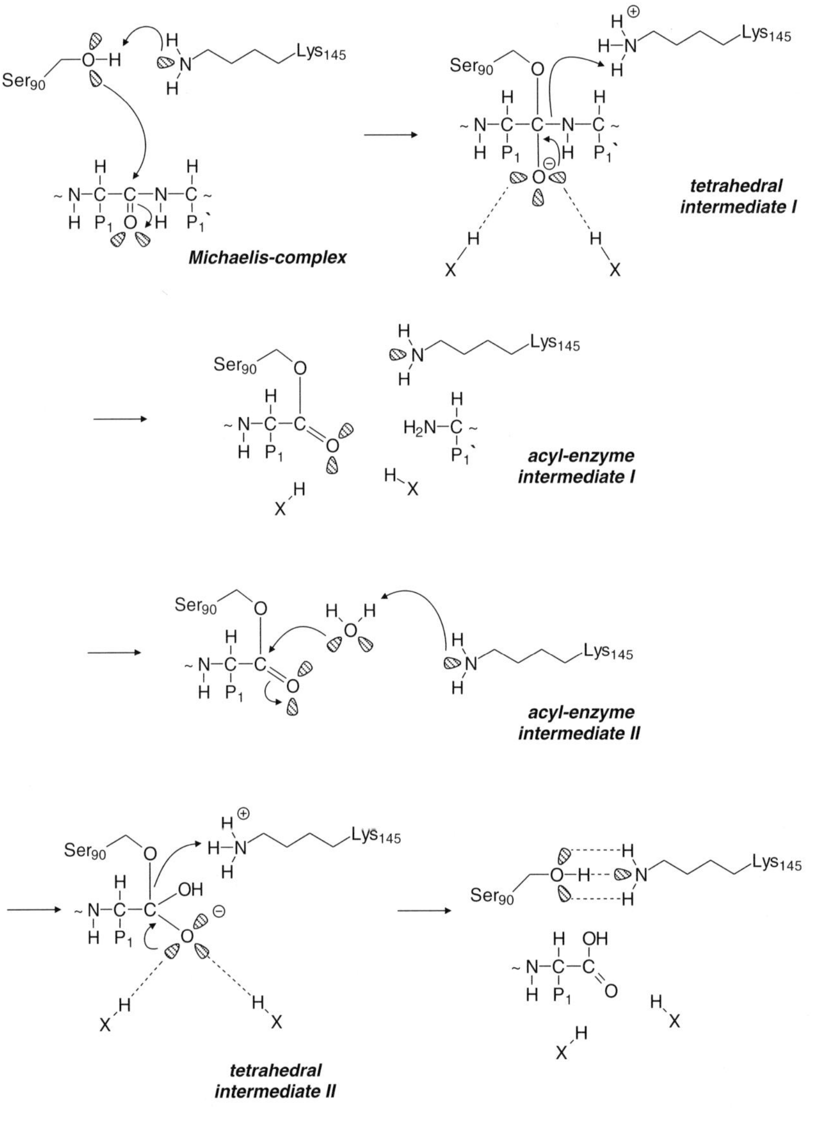

Scheme 1.4 Catalytic mechanism proposed for serine peptidases with a catalytic dyad consisting of Ser, Lys (*E. coli* leader peptidase numbering, adapted from Paetzel and Dalbey (1997)).

motif Ser-Xaa-Ybb-Lys, the catalytic residues are more widely spaced in clan SF peptidases (Paetzel and Dalbey 1997). In some enzymes of the clan SF the catalytic lysine was found to be replaced by a His (van Dijl *et al* 1992).

Paetzel and Dalbey (1997) proposed a mechanism for *Escherichia coli* leader peptidase based on site-directed mutagenesis and the structure of *E. coli* UmuD′protein (Scheme 1.4). Although the catalytic mechanism is far from being fully understood, it seems to be clear that the deprotonated ε-lysine has to act as a general base to abstract the proton from the hydroxyl group of the serine residue, making it nucleophilic enough to attack the scissile peptide bond. For this to be possible the enzyme must provide an environment for lysine in which its pK_a would be depressed. The tetrahedral intermediate I has to be stabilized by a yet unidentified oxyanion hole. Recently, in clan SE peptidases such an oxyanion hole has been identified (Kelly and Kuzin 1995).

1.4.2 A cysteine residue as a nucleophile

1.4.2.1 Catalytic triad consisting of Cys, His, Asn/Asp

The archetype of the cysteine peptidases is papain, a plant enzyme which is grouped in clan CA and provides the family C1 with its name. Most of the literature dealing with the enzyme mechanism of clan CA peptidases is based on papain. The general catalytic mechanism is similar to serine peptidases with a replacement of serine by cysteine. After the formation of a Michaelis complex an acyl enzyme intermediate is formed (Scheme 1.5). Through deacylation the cleaved peptide is released. In analogy with serine peptidases, it is widely believed that both acylation and deacylation steps have a tetrahedral intermediate. Still unclear is the acquisition of catalytic competence. Generally accepted is the status of the active site residues, Cys and His. The equilibrium between these residues lies in favour of the thiolate-imidazolium form (Lewis *et al* 1981). It has been proposed that the occurrence of this ion pair causes the attack at the scissile bond by its high nucleophilicity (Lewis *et al* 1981; Ménard *et al* 1995). However, doubt was cast on this interpretation by the observation that in some cases the thiolate-imidazolium ion pair is already at hand under conditions where the enzyme remains inactive (Pinitglang *et al* 1997). In the same report, the authors claim that the catalytic competence requires an additional protonic dissociation and suggest the Glu50 cluster (papain) as the source. The task of the third member in the catalytic triad is believed to be similar to the Asp in serine peptidase catalysis.

1.4.2.2 Catalytic triad consisting of Cys, His, Glu

In Clan CF and clan CE, the third member of the catalytic triad is replaced by a Glu. In spite of a different structure and catalytic residue order (Cys, His, Glu), pyroglutamyl peptidase I (clan CF, family 15) displays an almost conventional catalytic triad (Odagaki *et al* 1999; Singleton *et al* 1999). In addition, the active site residues of adenovirus (clan CE, family 5) proteinase were reported to be classically positioned in the novel tertiary structure of the enzyme (Mangel *et al* 1997). In continuation, the hydrolysis is mechanistically almost identical to the one known for the papain family. Even the Gln participating in the formation of the oxyanion was found to be identically positioned. This similarity of the catalytic site architecture reveals strong evidence that the Glu residue in the catalytic triad is used to stabilize the protonated His.

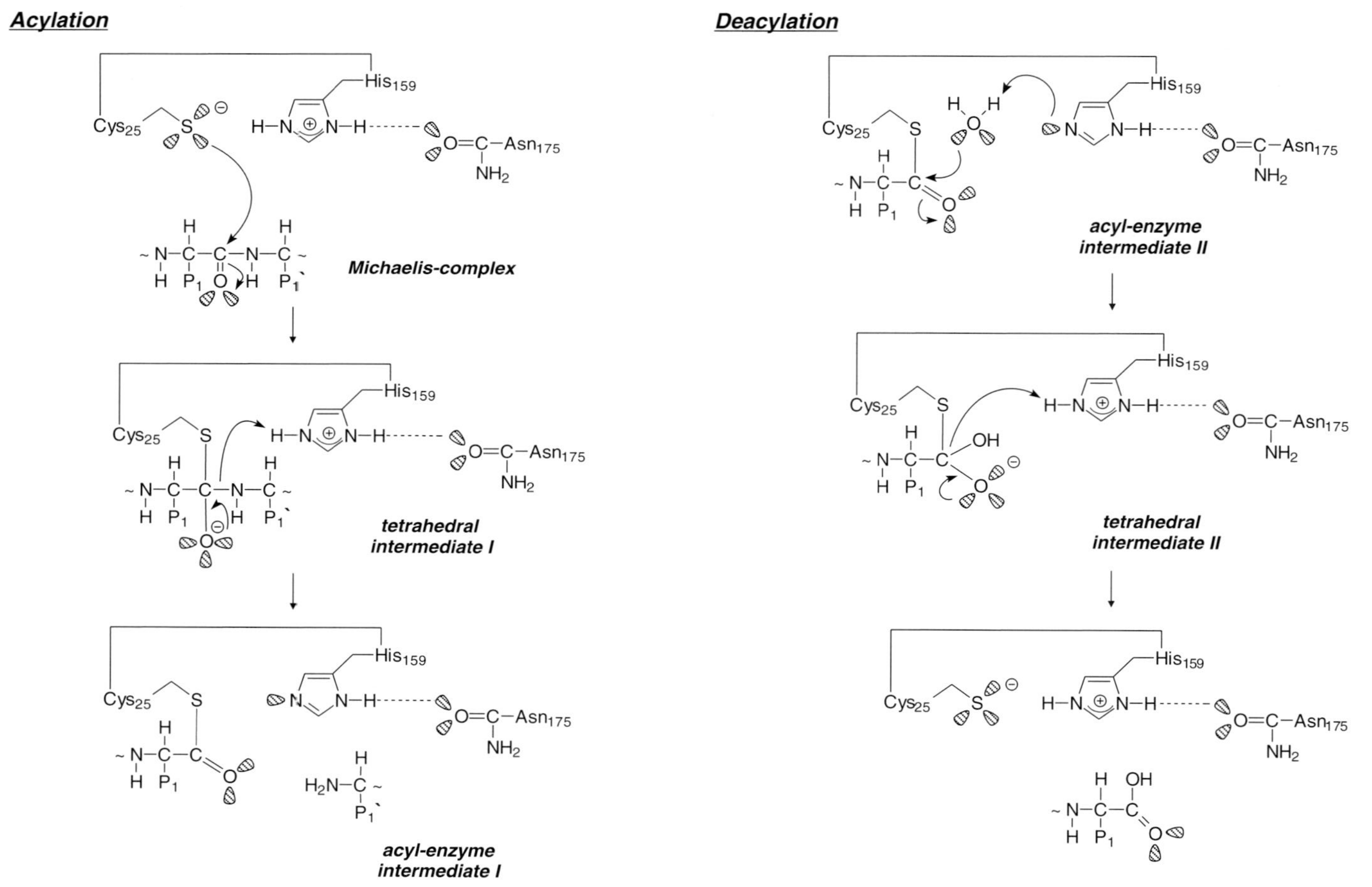

Scheme 1.5 Catalytic mechanism proposed for cysteine peptidases with a catalytic triad consisting of Cys, His, Asn/Asp (Papain numbering, adapted from Beynon and Bond (1989)).

1.4.2.3 Catalytic dyad consisting of Cys, His

Peptidases possessing a catalytic dyad in the order His, Cys are grouped together in clan CD. This clan contains at the present time four families of peptidases, the caspases, gingipains, clostripains and legumains. All CD clan peptidases have a high specificity for the amino acid in the P_1 position. X-ray studies of different caspases revealed that there is no amino acid in close proximity to the catalytic Cys and His to serve as the third member in a catalytic triad (Walker *et al* 1994). Recently, a catalytic mechanism for caspase 1 has been proposed (Scheme 1.6) (Brady *et al* 1999). Remarkably, in that mechanism the distance between the catalytic Cys and His is too far for a deprotonation of the Cys by the His to occur. Uniquely compared to the serine or the papain mechanism the substrate needs to fit between both

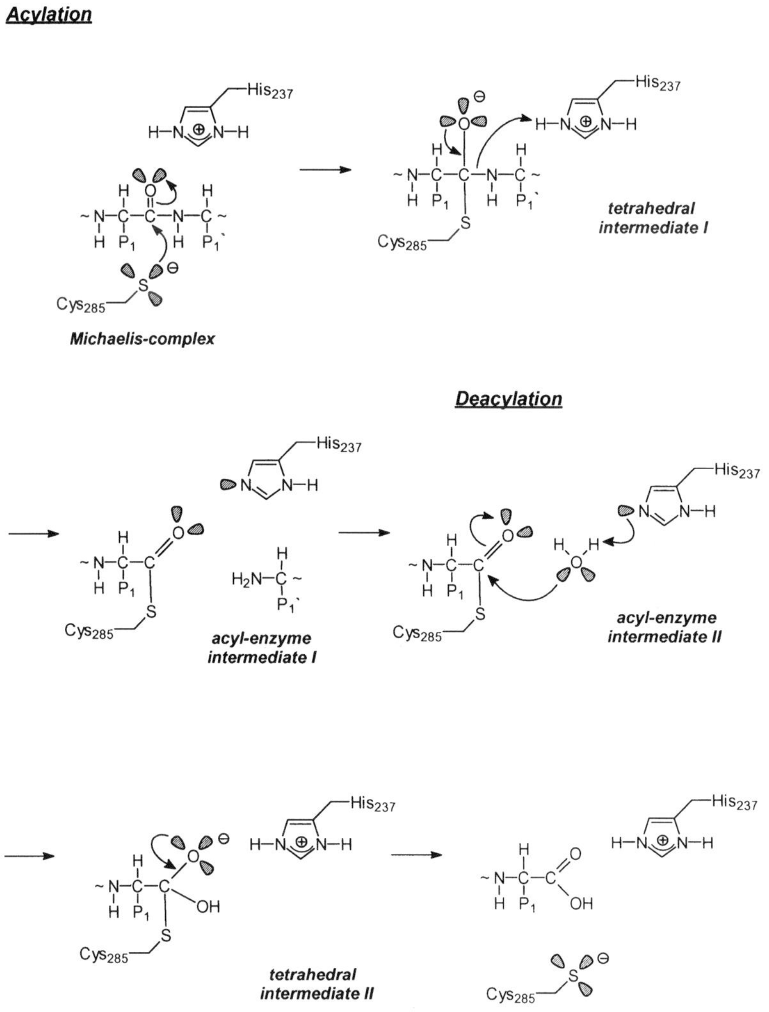

Scheme 1.6 Catalytic mechanism proposed for cysteine peptidases with a catalytic dyad consisting of Cys, His (Caspase 1 numbering, adapted from Brady *et al* (1999)).

active site residues. The catalytic His is protonated and polarizes the scissile carbonyl function. A still open question in this mechanism is the origin of the proton for the deacylation process. The His is very poorly positioned; instead the authors suggest a water molecule, held in place by a Gly, to serve as a proton donor.

1.4.2.4 Catalytic dyad consisting of Cys, Cys

For the members of the family C17, ER-60 protease and ER-72 protease a new clan, clan CG, has been introduced. By site-directed mutagenesis, it has been proposed that the C-terminal cysteines from the two CGHC motifs serve as the catalytic residues (Urade *et al* 1997). Nevertheless, this novel active site architecture has to be confirmed by further experiments and the novelty of the mechanism remains to be investigated.

1.4.2.5 Catalytic residue is Cys

The autoprocessing domain of hedgehog protein (clan CH, family C46) represents, together with self-splicing proteins, a total different mechanism of peptide bond cleavage. Hedgehog proteins are synthesized as precursors and undergo an intramolecular processing yielding a new C-terminal fragment and a new N-terminal fragment with a cholesterol moiety covalently attached to its carboxyl terminus. The active site cysteine, Cys258 acts as the nucleophile attacking the carbonyl group of the preceding amino acid residue (Scheme 1.7) (Hall *et al* 1997; Perler 1998); this results in the formation of a thioester linkage. The release of the N-terminal fragment is catalyzed by the nucleophilic attack of the 3β hydroxyl group of the cholesterol moiety. Other nucleophiles, like dithiothreitol

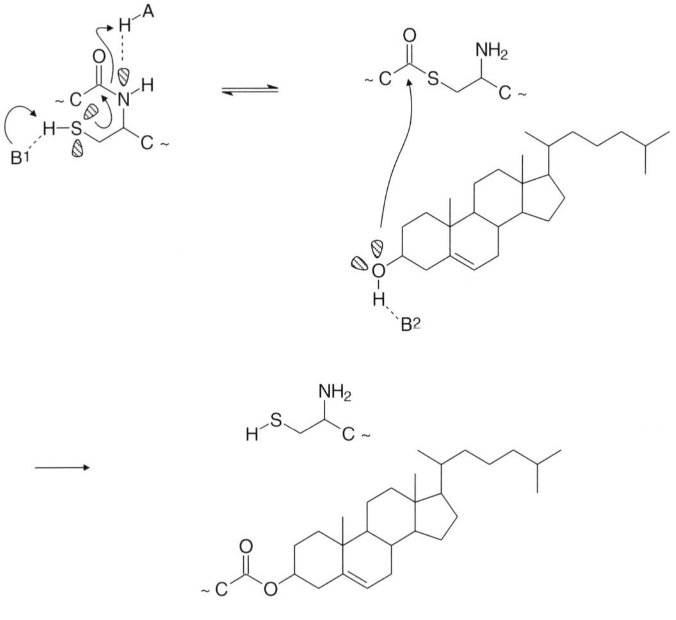

Scheme 1.7 Catalytic mechanism proposed for cysteine peptidases with a catalytic residue of Cys (Adapted from Hall *et al* (1997)).

can easily replace the cholesterol. A proposed catalytic triad of Ser, Thr and His could never be confirmed by any structural data. The task of the discussed His is more likely stabilization of the negative charge on the carbonyl oxygen of the preceding amino acid (Gly257), donation of a proton to the free α-amino group of Cys258 and maintenance of an appropriate orientation of reaction components. A stabilization of the reactive conformation is most likely the function of Thr326. The nucleophilic attack of the cholesterol is probably assisted by an aspartic acid residue which serves as a general base to deprotonate the 3β hydroxyl group. A similar mechanism is found in self splicing proteins (Perler 1998). These proteins provide an example of mechanistic class overlapping, the primary nucleophile was found to be Ser, Cys or Thr.

1.4.3 A serine or cysteine residue as a nucleophile

In clan PA, there are now peptidases grouped together which have either a serine or a cysteine as nucleophile. The catalytic mechanism is the classical one for serine peptidases as discussed in detail above. The only difference between serine and cysteine peptidases in this clan is the choice of the nucleophile, all members have a chymotrypsin-like fold.

1.4.4 A serine, cysteine or threonine residue as a nucleophile

The recently discovered catalytic mechanism of threonine peptidases involves a unique nucleophile (Seemuller *et al* 1995). The common feature of the new mechanism, on the other hand, is the use of the side chain from the amino-terminal residue, incorporated in a β-sheet, as the nucleophile in the attack at the carbonyl carbon. Therefore the name of Ntn (N-terminal nucleophile) hydrolases has been suggested (Brannigan *et al* 1995). The nucleophile is in the case of the proteasome, a threonine, in the case of glutamine PRPP amidotransferase, a cysteine, and in the case of penicillin acylase, a serine. Four threonine families, two cysteine families and one serine family are grouped together in this clan, clan PB. The catalytic mechanism of Ntn hydrolases using a single amino acid catalytic center has been described for penicillin acylase (Scheme 1.8) (Duggleby *et al* 1995). The nucleophilic attack is provided by SerB1 of the heterodimer. Due to the lack of an adjacent His, the only possibility of a serving base in close proximity is the α-amino group of the active residue. Nevertheless, there is evidence that a bridging water molecule may mediate the basic character of the α-amino group. This hypothesis is supported by a hydrogen bonding network located around the catalytic site. As in classical mechanisms, the tetrahedral intermediate is stabilized by an oxyanion hole, and the deacylation is accomplished by a water molecule.

1.4.5 A water molecule as a nucleophile

1.4.5.1 Water bound by two aspartic acid residues

It is commonly believed that the hydrolysis of aspartic peptidases proceeds by a general acid-base catalysis mechanism. Because of the great therapeutic interest in HIV-1 aspartic protease, the catalytic mechanism of this enzyme is the best understood in this mechanistic class. In HIV-1 aspartic protease, only one of the two Asp residues of the active site is protonated in the enzyme substrate complex (Scheme 1.9) (Silva *et al* 1996). One of the

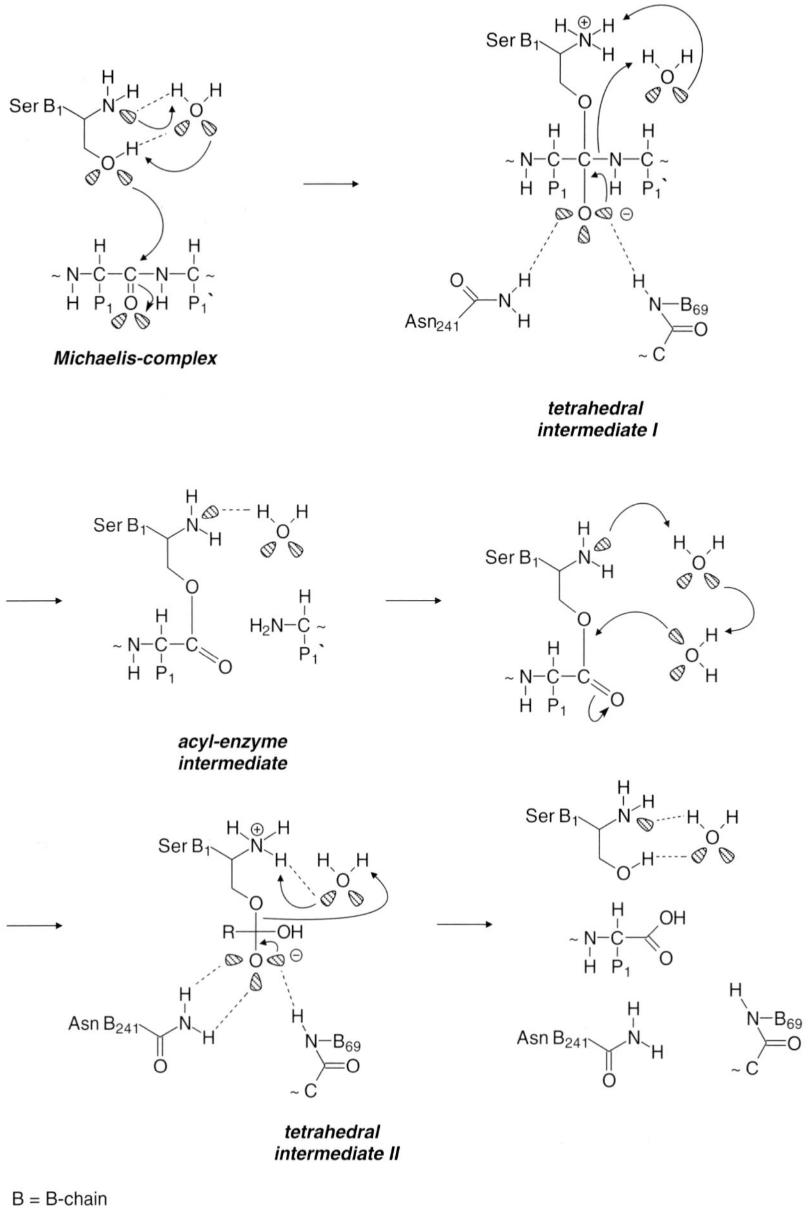

Scheme 1.8 Catalytic mechanism proposed for the clan PB (Penicillin acylase numbering, adapted from Duggleby *et al* (1995)).

Asp residues acts as a general base that activates the water molecule. After an asymmetric binding of the substrate, the nucleophilic attack of the water molecule results in the formation of a geminal diol and the amide nitrogen changes its hybridization from sp^2 to sp^3. Subsequent protonation of the sp^3 nitrogen together with proton release from the diol

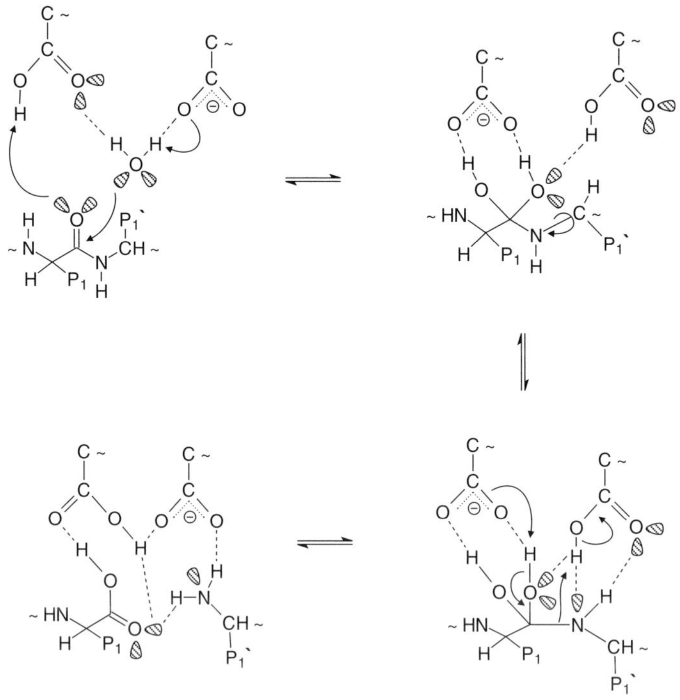

Scheme 1.9 Catalytic mechanism proposed for aspartic peptidases (Adapted from Silva *et al* (1996)).

leads to peptide bond breakage. The *trans* conformation of the scissile bond only permits a nucleophilic attack at the carbonyl due to steric interactions. Therefore the binding of the hydrated peptide has to be optimized by switching into a *gauche* conformation. The *anti-gauche* transition is calculated to be a fast step in the catalytic process. After the conformational change, the simultaneous proton donation from one hydroxyl to the charged Asp and from the second Asp to the nitrogen, results in peptide bond rupture. This second proton exchange requires, in addition, rotation of the Asp. Thus, the catalysis of aspartic proteases is another example of a concerted mechanism.

1.4.5.2 *Water bound by a single metal ion*

Only a single metal ion is involved in the catalytic mechanism of clan MA, MB, MC, MD and clan ME. For some time, the catalytic mechanism of thermolysin has stood as a prototype for this class of metallopeptidases. However, recently a divergent catalytic mechanism for thermolysin proteolysis has been proposed. So far, this novel mechanism has only been shown for thermolysin and the earlier proposed mechanism may still be valid for most other metallopeptidases. Therefore, we report here both the proposed mechanisms (using carboxypeptidase A as an example for the former mechanism).

Studies on thermolysin and carboxypeptidase A have suggested a general mechanism of a promoted water pathway for zinc peptidases (Scheme 1.10) (Christianson and Lipscomb

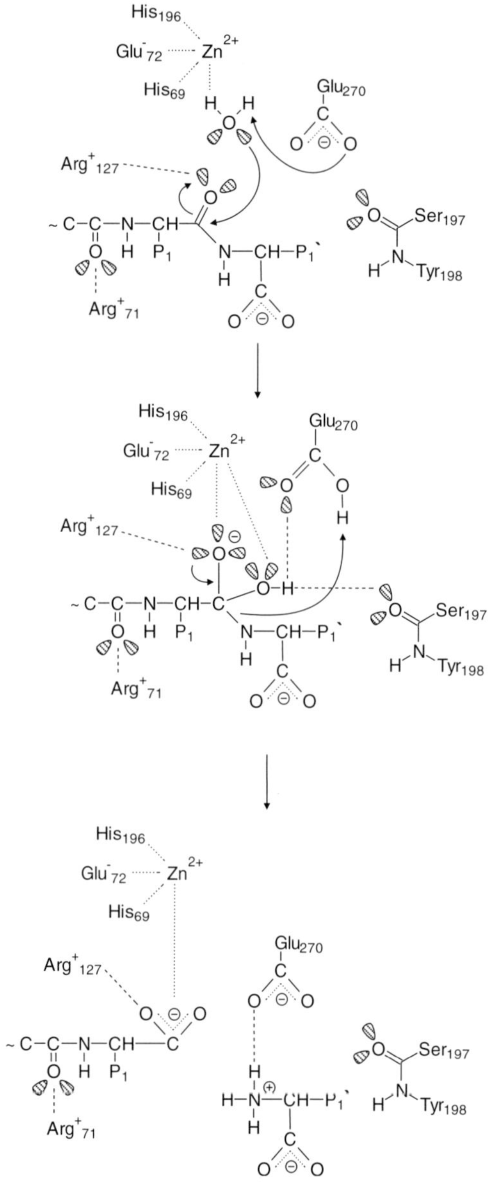

Scheme 1.10 One of the proposed catalytic mechanism for metallopeptidases with a water molecule bound to a single metal ion (Carboxypeptidase A numbering, adapted from Christianson and Lipscomb (1989)).

1989; Matthews 1989). The zinc ion has a tetrahedral coordination between two His, a Glu and a water molecule and the nucleophilicity of the water molecule is assisted by Glu270 out of the HELTH sequence. After nucleophilic attack by the water molecule a tetrahedral intermediate is formed which is decomposed by the transfer of the proton accepted by Glu270 to the leaving nitrogen. The release of the product may be facilitated by unfavorable electrostatic interactions between Glu270 and the products. Stabilization of the oxyanion is provided by the Zn^{2+} and an arginine.

However, studies with azoformyl dipeptide substrates revealed a major revision of this interpretation, at least for the catalytic mechanism of thermolysin (Scheme 1.11) (Mock and Stanford, 1996). In the first step, the activated water is replaced by the substrate and then the unprotonated His231 assists the nucleophilic water to attack the scissile peptide bond. Subsequent conversion into the products through a Zn^{2+}-ligated tetrahedral adduct is catalyzed by the protonated imidazole. Asp226 is, as in a catalytic triad, involved in the

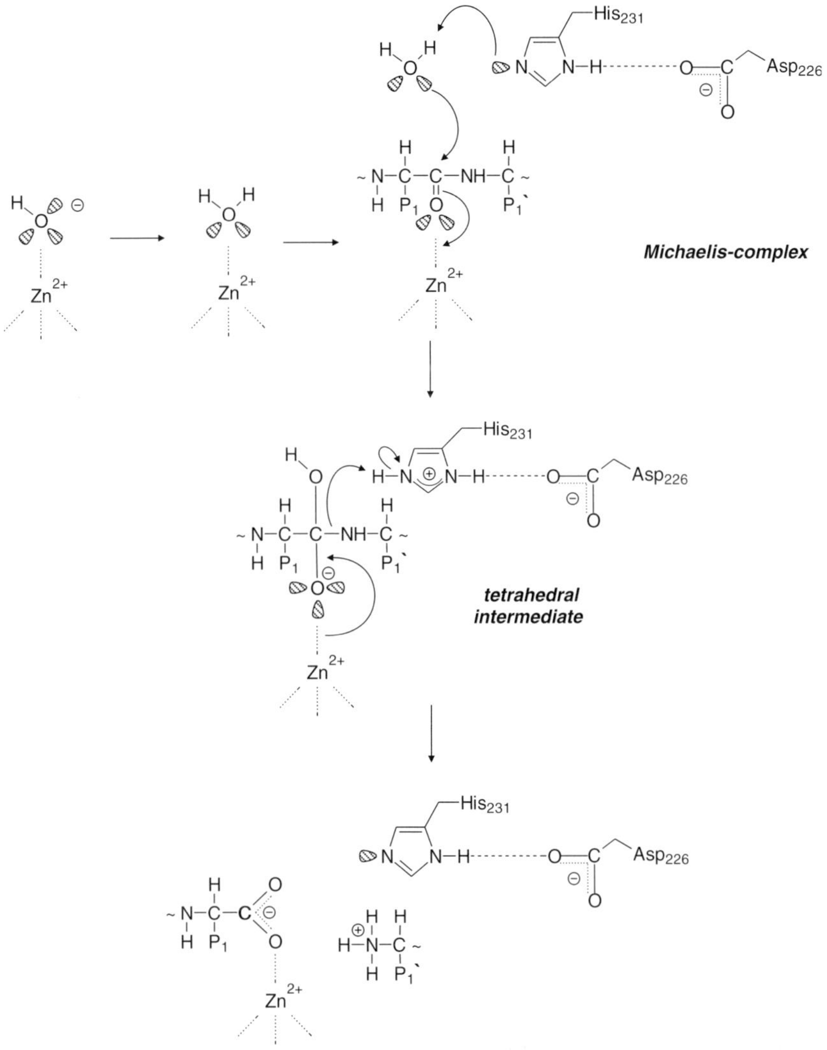

Scheme 1.11 Recently proposed catalytic mechanism for metallopeptidases with a water molecule bound to a single metal ion (Thermolysin numbering, adapted from Mock and Standford (1996)).

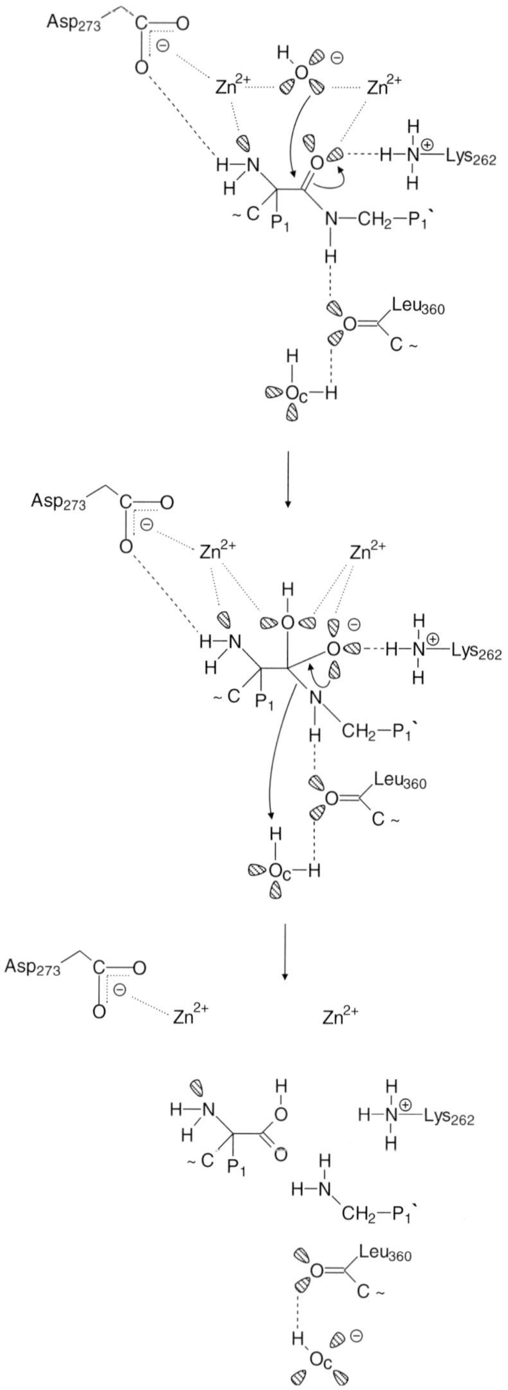

Scheme 1.12 Proposed catalytic mechanism for metallopeptidases with a water molecule bound by two metal ions (L-leucine aminopeptidase numbering, adapted from Strater and Lipscomb (1995)).

correct orientation of the imidazole ring. Mutation of Glu143 suppresses the activity of thermolysin totally (Glu270 in carboxypeptidase A). The authors suggest that the essential requirement for Glu143 arises from the anionic side chain carboxylate group serving as a negatively charged counter-ion providing electrostatic stabilization for electron defined reaction intermediates that are generated in proximity during the catalytic cycle.

1.4.5.3 Water bound by two metal ions

Four clans are presently known to contain two metal ions in their catalytic center. Clan MF and clan MH contain two zinc ions, clan MG contains two cobalt or manganese ions and clan MJ contains two nickel ions. One of the striking differences between double and single metal ion peptidases is the pyramidal coordination of the metal ions in the double metal ion peptidases (Strater and Lipscomb 1995; Tahirov *et al* 1998). The nucleophile is a hydroxide ion bridging the two metal ions symmetrically. This oxygen atom also bridges the metal ions in the transition state. As an example of a two metal ion, catalytic mechanism that of L-leucine aminopeptidase (Strater and Lipscomb 1995) is described. It is important to note, however, that this mechanism is not a general mechanism for this class of metallopeptidase. In Scheme 1.12, the nucleophilic hydroxide ion is bridged by the two Zn^{2+}-ions. Additionally, the Zn^{2+}-ions help to bind the substrate in the correct position by increasing the coordination number of both Zn^{2+}-ions. The nucleophilic attack provided by the hydroxide ion molecule results in a tetrahedral gem-diolate transition state. The rupture of the peptide bond is facilitated by the water molecule C which protonates the leaving group.

1.5 CONCLUSION

Considering the chemistry of the discussed peptidase-mediated mechanisms of peptide bond hydrolyses, we can distinguish between "covalent catalysis" and "non-covalent catalysis". Seriner, threonine and cysteine peptidases form transition states and covalent intermediates between themselves and their substrates. In contrast, substrates are bound throughout the whole catalytic cycle in a non-covalent manner to the active site of aspartic- and metallopeptidases.

Accordingly, chemistry evolved during the last 30 years has taken advantage of the different mechanistic classes to inhibit or modify proteolytic activity. Thus, either covalent binding inhibitors or transition-state analogues ("non-covalent catalyzing" enzymes) proved to be the inhibitors of choice for drug development.

Peptidase inhibitors suitable for treatment of a disease have to exhibit a high target enzyme specificity in an environment where a variety of peptidases are present. Therefore, the different catalytic mechanisms enable the design of class specific inhibitors as described in the following chapters of this book.

Moreover, subtle differences among sub-classes as in serine peptidases between the trypsin family and the prolyl oligopeptidase family (different rate-determining steps during catalysis due to substrate activation and substrate-assisted catalysis) or as in the cysteine peptidases between the papain family and caspase family (activation, reactivity and positioning of the active site nucleophile) will be of future guidance to improve the specificity of peptidase inhibitors so resulting in useful drugs.

REFERENCES

Ash, E.L., Sudmeier, J.L., De Fabo, E.C. and Bachovchin, W.W. (1997) A low-barrier hydrogen bond in the catalytic triad of serine proteases? Theory versus experiment. *Science*, **278**, 1128–1132.

Barrett, A.J. and Rawlings, N.D. (1995) Families and clans of serine peptidases. *Archives of Biochemistry and Biophysics*, **318**, 247–250.

Berger, A. and Schechter, I. (1976) On the size of the active site in proteases. I. Papain. *Biochemical and Biophysical Research Communications*, **27**, 157–162.

Beynon, R.J. and Bond, J.S. (1989) *Proteolytic enzymes*. Oxford: IRL Press.

Brady, K.D., Giegel, D.A., Grinnell, C., Lunney, E., Talanian, R.V., Wong, W. *et al* (1999) A catalytic mechanism for caspase-1 and for bimodal inhibition of caspase-1 by activated aspartic ketones. *Bioorganic and Medicinal Chemistry*, **7**, 621–631.

Brannigan, J.A., Dodson, G., Duggleby, H.J., Moody, P.C., Smith, J.L., Tomchick, D.R. *et al* (1995) A protein catalytic framework with an N-terminal nucleophile is capable of self-activation. *Nature*, **378**, 416–419.

Christianson, D.W. and Lipscomb, W.N. (1989) Carboxypeptidase A. *Accounts of Chemical Research*, **22**, 62–69.

Duggleby, H.J., Tolley, S.P., Hill, C.P., Dodson, E.J., Dodson, G. and Moody, P.C. (1995) Penicillin acylase has a single-amino-acid catalytic centre. *Nature*, **373**, 264–268.

Frey, P.A., Whitt, S.A. and Tobin, J.B. (1994) A low-barrier hydrogen bond in the catalytic triad of serine proteases. *Science*, **264**, 1927–1930.

Hall, T.M., Porter, J.A., Young, K.E., Koonin, E.V., Beachy, P.A. and Leahy, D.J. (1997) Crystal structure of a Hedgehog autoprocessing domain: homology between Hedgehog and self-splicing proteins. *Cell*, **91**, 85–97.

Hartley, B.S. (1960) Proteolytic enzymes. *Annual Review of Biochemistry*, **29**, 45–72.

Hoog, S.S., Smith, W.W., Qiu, X., Janson, C.A., Hellmig, B., McQueney, M.S. *et al* (1997) Active site cavity of herpesvirus proteases revealed by the crystal structure of herpes simplex virus protease/inhibitor complex. *Biochemistry*, **36**, 14023–14029. *http://www.bi.bbsrc.ac.uk/Merops/Merops.htm*.

Kelly, J.A. and Kuzin, A.P. (1995) The refined crystallographic structure of a DD-peptidase penicillin-target enzyme at 1.6 A resolution. *Journal of Molecular Biology*, **254**, 223–236.

Kossiakoff, A.A. and Spencer, S.A. (1981) Direct determination of the protonation states of aspartic acid-102 and histidine-57 in the tetrahedral intermediate of the serine proteases: neutron structure of trypsin. *Biochemistry*, **20**, 6462–6474.

Lau, E.Y. and Bruice, T.C. (1999) Consequences of breaking the Asp–His hydrogen bond of the catalytic triad: effects on the structure and dynamics of the serine esterase cutinase. *Biophysical Journal*, **77**, 85–98.

Lewis, S.D., Johnson, F.A. and Shafer, J.A. (1981) Effect of cysteine-25 on the ionization of histidine-159 in papain as determined by proton nuclear magnetic resonance spectroscopy. Evidence for a his-159–Cys-25 ion pair and its possible role in catalysis. *Biochemistry*, **20**, 48–51.

Liao, D.I., Breddam, K., Sweet, R.M., Bullock, T. and Remington, S.J. (1992) Refined atomic model of wheat serine carboxypeptidase II at 2.2-A resolution. *Biochemistry*, **31**, 9796–9812.

Mangel, W.F., Toledo, D.L., Ding, J., Sweet, R.M. and McGrath, W.J. (1997) Temporal and spatial control of the adenovirus proteinase by both a peptide and the viral DNA. *Trends in Biochemical Science*, **22**, 393–398.

Matthews, B.W. (1989) Structural basis of the action of thermolysin and related zinc peptidases. *Accounts of Chemical Research*, **21**, 333–340.

Ménard, R., Plouffe, C., Laflamme, P., Vernet, T., Tessier, D.C., Thomas, D.Y. *et al* (1995) Modification of the electrostatic environment is tolerated in the oxyanion hole of the cysteine protease papain. *Biochemistry*, **34**, 464–471.

Mock, W.L. and Stanford, D.J. (1996) Arazoformyl dipeptides substrates for thermolysin. Confirmation of a reverse protonation catalytic mechanism. *Biochemistry*, **35**, 7369–7377.

NC-IUBMB (Nomenclature Committee of the International Union of Biochemistry and Molecular Biology) (1992) *Enzyme Nomenclature 1992. Recommendations of the Nomenclature Committee of the International Union of Biochemistry and Molecular Biology on the Nomenclature and Classification of Enzymes*. Orlando: Academic Press.

Odagaki, Y., Hayashi, A., Okada, K., Hirotsu, K., Kabashima, T., Ito, K. *et al* (1999) The crystal structure of pyroglutamyl peptidase I from Bacillus amyloliquefaciens reveals a new structure for a cysteine protease. *Structure*, **7**, 399–411.

Paetzel, M. and Dalbey, R.E. (1997) Catalytic hydroxyl/amine dyads within serine proteases. *Trends in Biochemical Science*, **22**, 28–31.

Perler, F.B. (1998) Protein splicing of inteins and hedgehog autoproteolysis: structure, function, and evolution. *Cell*, **92**, 1–4.

Pinitglang, S., Watts, A.B., Patel, M., Reid, J.D., Noble, M.A., Gul, S. *et al* (1997) A classical enzyme active center motif lacks catalytic competence until modulated electrostatically. *Biochemistry*, **36**, 9968–9982.

Qiu, X., Culp, J.S., DiLella, A.G., Hellmig, B., Hoog, S.S., Janson, C.A. *et al* (1996) Unique fold and active site in cytomegalovirus protease. *Nature*, **383**, 275–279.

Qiu, X., Janson, C.A., Culp, J.S., Richardson, S.B., Debouck, C., Smith, W.W. *et al* (1997) Crystal structure of varicella-zoster virus protease. *Proceedings of the National Academy of Sciences of the United States of America*, **94**, 2874–2879.

Rogers, G.A. and Bruice, T.C. (1974) Synthesis and evaluation of a model for the so-called "charge-relay" system of the serine esterases. *Journal of the American Chemical Society*, **96**, 2473–2481.

Seemuller, E., Lupas, A., Stock, D., Lowe, J., Huber, R. and Baumeister, W. (1995) Proteasome from Thermoplasma acidophilum: a threonine protease. *Science*, **268**, 579–582.

Shieh, H.S., Kurumbail, R.G., Stevens, A.M., Stegeman, R.A., Sturman, E.J., Pak, J.Y. *et al* (1996) Three-dimensional structure of human cytomegalovirus protease. *Nature*, **383**, 279–282.

Silva, A.M., Cachau, R.E., Sham, H.L. and Erickson, J.W. (1996) Inhibition and catalytic mechanism of HIV-1 aspartic protease. *Journal of Molecular Biology*, **255**, 321–346.

Singleton, M., Isupov, M. and Littlechild, J. (1999) X-ray structure of pyrrolidone carboxyl peptidase from the hyperthermophilic archaeon *Thermococcus litoralis*. *Structure*, **7**, 237–244.

Strater, N. and Lipscomb, W.N. (1995) Two-metal ion mechanism of bovine lens leucine aminopeptidase: active site solvent structure and binding mode of L-leucinal, a gem-diolate transition state analogue, by X-ray crystallography. *Biochemistry*, **34**, 14792–14800.

Strynadka, N.C., Adachi, H., Jensen, S.E., Johns, K., Sielecki, A., Betzel, C. (1992) Molecular structure of the acyl-enzyme intermediate in beta-lactam hydrolysis at 1.7 A resolution. *Nature*, **359**, 700–705.

Tahirov, T.H., Oki, H., Tsukihara, T., Ogasahara, K., Yutani, K., Ogata, K. *et al* (1998) Crystal structure of methionine aminopeptidase from hyperthermophile, *Pyrococcus furiosus*. *Journal of Molecular Biology*, **284**, 101–124.

Tong, L., Qian, C., Massariol, M.J., Bonneau, P.R., Cordingley, M.G. and Lagace, L. (1996) A new serine-protease fold revealed by the crystal structure of human cytomegalovirus protease. *Nature*, **383**, 272–275.

Urade, R., Oda, T., Ito, H., Moriyama, T., Utsumi, S. and Kito, M. (1997) Functions of characteristic Cys-Gly-His-Cys (CGHC) and Gln-Glu-Asp-Leu (QEDL) motifs of microsomal ER-60 protease. *Journal of Biochemistry (Tokyo)*, **122**, 834–842.

van Dijl, J.M., de Jong, A., Vehmaanpera, J., Venema, G. and Bron, S. (1992) Signal peptidase I of *Bacillus subtilis*: patterns of conserved amino acids in prokaryotic and eukaryotic type I signal peptidases. *EMBO Journal*, **11**, 2819–2828.

Walker, N.P., Talanian, R.V., Brady, K.D., Dang, L.C., Bump, N.J., Ferenz, C.R. *et al* (1994) Crystal structure of the cysteine protease interleukin-1 beta-converting enzyme: a (p20/p10)2 homodimer. *Cell*, **78**, 343–352.

Warshel, A., Naray-Szabo, G., Sussman, F. and Hwang, J.K. (1989) How do serine proteases really work? *Biochemistry*, **28**, 3629–3637.

Chapter 2

Regulatory mechanisms for proteinase activity

S.M. Ellerbroek, Y. Wu and M.S. Stack

Proteolysis, the hydrolysis of peptide bonds, is required for a multitude of developmental and physiologic events including fertilization, implantation, cell motility, prohormone conversion, digestion and wound healing. Both proteolytic enzymes and proteinacious inhibitors are prevalent in all biologic tissues and fluids. Tight control of proteinase activity is essential for maintenance of tissue integrity and homeostasis, as uncontrolled proteolysis has been implicated in a variety of pathologic events including ulceration, arthritis, and tumor invasion and metastasis. Multiple mechanisms have evolved for both highly localized and systemic control of proteinase activity. The most common of these are secretion of proteinases as inactive precursors, or zymogens, and the formation of enzyme–inhibitor complexes that abrogate proteinase activity. As the remainder of this volume is devoted to discussion of proteinase inhibition, this chapter concentrates primarily on two major alternative mechanisms for the control of proteinase activity; limited proteolysis and proteinase compartmentalization. Specific examples are provided from the proteinase literature to illustrate the role of these mechanisms in regulating the location, duration, and concentration of enzyme activity.

2.1 LIMITED PROTEOLYSIS FOR ZYMOGEN ACTIVATION

2.1.1 Zymogen activation

The vast majority of proteinases are expressed as zymogens, which require some form of limited proteolysis for full catalytic activity. *Zymogenicity*, the activity ratio of latent protease to processed enzyme, can vary significantly both within and between mechanistic classes. For example, while the serine protease trypsin is >10,000-fold more active than trypsinogen, fully processed tissue-type plasminogen activator (tPA) is estimated to be only 2–10-fold more active than its zymogen (Tachias and Madison 1996). In cases of low zymogenicity, a commonly occurring regulatory mechanism is exosite interaction between the enzyme and non-substrate regulatory proteins that influence enzyme catalytic efficiency (i.e. tPA/fibrin, as discussed in Section 2.2.3 below).

Proteolytic maturation of a zymogen influences enzyme structure and conformation through various processes, including the removal of an inhibitory domain, formation or stabilization of either the substrate binding pocket or enzyme active site and alterations in quaternary structure via processing-induced association into multimers. The presence of an inhibitory domain, which contains a highly conserved cysteine residue that coordinates the catalytically essential active site zinc ion, is common in matrix metalloproteinase (MMP), zymogens. MMP zymogen activation requires proteinase cleavage of the pro-domain, thereby destabilizing the cysteine–zinc interaction (Springman *et al* 1990). Following the initial proteolytic event, the zymogen undergoes further inter- or intra-molecular cleavage(s), resulting in removal of the pro-domain and generation of fully active enzyme. This mechanism is distinct from the classic serine proteinase zymogen activation mechanism, which is illustrated by maturation of chymotrypsinogen to π-chymotrypsin (Appel 1986). This process involves a single trypsin-catalyzed cleavage of the Arg_{15}–Ile_{16} bond to generate a two-chain proteinase held together by a pair of interchain disulfide bonds. Propeptide removal enables a series of discrete electrostatic interactions to occur that mediate specific conformational changes, resulting in formation of the substrate binding pocket and stabilization of the tetrahedral transition state that is critical for enzyme catalysis.

2.1.2 Zymogen activation cascades

Proteinase activity is frequently regulated through zymogen activation cascades, where an initial event will produce an active proteinase that processes a downstream zymogen. The initiating enzyme within a zymogen activation cascade is often a highly regulated enzyme, as proteolytic potential is amplified substantially by progression through the cascade. A common example of this concept is the well-described zymogen activation cascade of blood clotting, in which initiation via an intrinsic or extrinsic pathway triggers a proteolytic event that is amplified with every activation step, resulting in a rapid response to trauma (Newland 1987). Such a cascade is not limited to proteolysis of extracellular substrates. Upon stimulation of programmed cell death, the "initiator" caspases, caspase-8 and caspase-9, directly activate "executioner" caspase zymogens, which in turn carry out the multiple proteolytic events of apoptosis (Budihardjo *et al* 1999; Salvesen and Dixit 1999).

Zymogen activation cascades may also involve processing of pro-enzymes between mechanistic groups. Effective examples of this concept can be found in the zymogen activation reactions of extracellular matrix-degrading proteinases (Figure 2.1). Urinary-type plasminogen activator (uPA) is secreted as a single chain zymogen (designated scuPA) that undergoes limited pericellular proteolytic processing by serine proteinases including plasmin, cathepsins, mast cell tryptase and plasma kallikrein, resulting in formation of the two-chain active proteinase (as described above for chymotrypsin) (Andreasen *et al* 1997). Following scuPA activation to uPA, pericellular proteolytic potential can be enhanced via uPA-catalyzed activation of the plasma zymogen plasminogen to the broad-spectrum serine proteinase plasmin. Plasmin can initiate the activation of the metalloproteinase zymogen prostromelysin-1 (proMMP-3) (Figure 2.1), which then participates in activation of other proMMPs, including procollagenase-1 (MMP-1), promatrilysin (MMP-7), neutrophil procollagenase (MMP-8), progelatinase B (MMP-9) and procollagenase-3 (MMP-13) (Nagase *et al* 1990; Birkedal-Hansen *et al* 1993; Knauper *et al* 1996; Murphy and Gavrilovic 1999; Ramos-DeSimone *et al* 1999).

Although it is tempting to envision a potent metalloproteinase cascade initiated by plasmin activation of stromelysin(s), closer inspection suggests that characterization as a linear activation pathway akin to that of the blood clotting response is over-simplistic. Plasmin can directly cleave a multitude of metalloproteinase zymogen pro-domains; however, this limited proteolysis must be accompanied by an additional MMP-mediated processing event(s) prior to acquisition of full catalytic activity (Figure 2.2). Whether the MMP

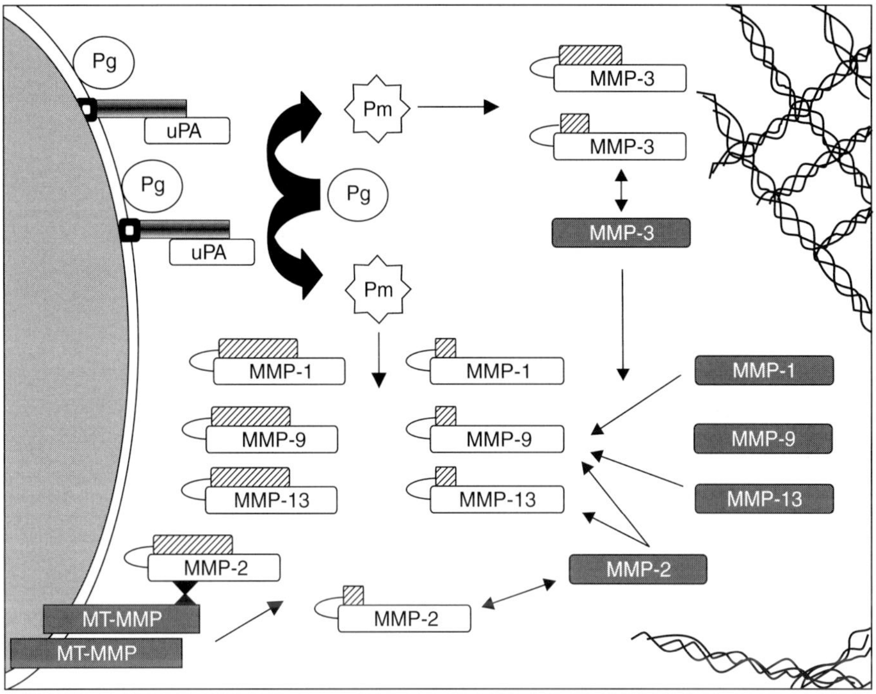

Figure 2.1 Interplay between pericellular serine- and metalloproteinase zymogen activation cascades. Single-chain urinary-type plasminogen activator (uPA) A-chain binds with high affinity to the GPI-anchored uPA receptor (uPAR) and is converted to the active, two-chain form. Activated uPA readily cleaves the cell surface-associated zymogen plasminogen to plasmin, which in turn can activate more single-chain uPA. Plasmin also initiates activation of multiple MMPs, including MMP-1,-3,-9, and -13, resulting in partial removal of the inhibitory propeptide. In the case of MMP-3, rapid autolytic activation ensures full enzymatic activity. MMP-3-mediated maturation of plasmin-modified MMPs may generate proteinases with greater specific activity ("superactive") than those produced through autolytic maturation (see Figure 2.2). Additionally, a number of transmembrane MMPs (MT-MMPs) can indirectly bind and process proMMP-2 at the cell surface. MT-MMP-catalyzed processing is followed by an autolytic cleavage resulting in fully active MMP-2 (Atkinson *et al* 1995), which can then process both MMP-9 and MMP-13. Additionally, MMP-1 and MMP-13 have been reported to activate MMP-9. Although not illustrated, serine protease-activated MMP-10 (stromelysin-2) also "superactivates" collagenases MMP-1 and MMP-8. Hatched box, pro-domain (inhibitory); white box, zymogen; gray box, mature enzyme; uPA, urinary-type plasminogen activator, Pg, plasminogen; Pm, plasmin; MMP, matrix metalloproteinase.

cleavage event(s) involves autolysis or another MMP appears to have important functional consequences with respect to catalytic efficiency and enzyme–inhibitor binding. For example, although stromelysin-1 cleavage of plasmin-modified collagenase-1 produces an active enzyme that differs in only a few amino acids from that generated from autolysis, stromelysin processed collagenase-1 is at least 60% more active (Figure 2.2) (Suzuki *et al* 1990). While stromelysin-1 can directly activate the collagenase-1 zymogen through a single cleavage at the Gln_{80}–Phe_{81} peptide bond, processing of this bond is 24,000-fold faster when

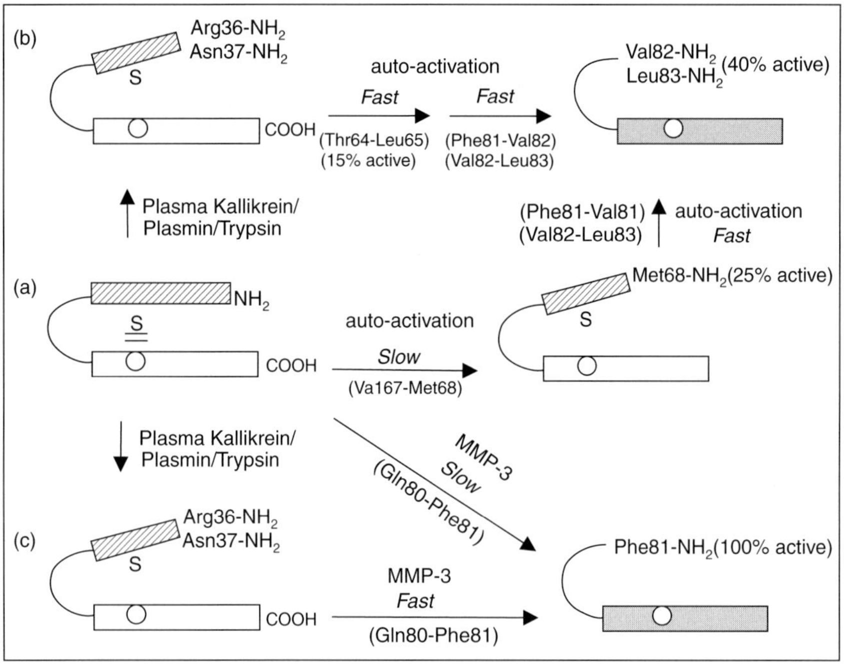

Figure 2.2 Superactivation of MMP-1. In solution, pro-MMP-1 (**a**) undergoes two autoproteolytic steps (one slow, one fast) to generate a processed enzyme with only 40% of its potential activity. Serine protease (plasmin, kallikrein, and trypsin) cleaved pro-MMP-1 (**b**) quickly undergoes intramolecular activation to produce an identically processed enzyme. MMP-3 can activate pro-MMP-1 (**a**) via a single cleavage of the Gln_{80}–Phe_{81} peptide bond to produce an enzyme that has 100% of its potential activity. While direct activation by MMP-3 is a slow event, prior cleavage of pro-MMP-1 (**c**) pro-domain by a serine protease enhances MMP-3 activation 24,000-fold. Hatched box, pro-domain (inhibitory); white box, zymogen; gray box, mature enzyme.

the substrate is plasmin-modified collagenase. Such "superactivation" also occurs for stromelysin-processed neutrophil collagenase, and this has also been correlated with a weaker binding potential of tissue inhibitor of matrix metalloproteinases-2 (TIMP-2), suggesting an additional mechanism for regulation of pericellular proteolytic activity (Knauper *et al* 1993; Reinemer *et al* 1994; Farr *et al* 1999). Thus, plasmin, and other extracellular serine proteases, can facilitate MMP activation through limited proteolysis and thereby contribute to accelerated matrix degradation during various physiological and pathological events.

2.2 COMPARTMENTALIZATION

2.2.1 Regulation of biochemical parameters for optimal activity

Proteinase trafficking and compartmentalization function to sequester enzymes into specific environments that promote optimal catalytic activity (via changes in pH, ion

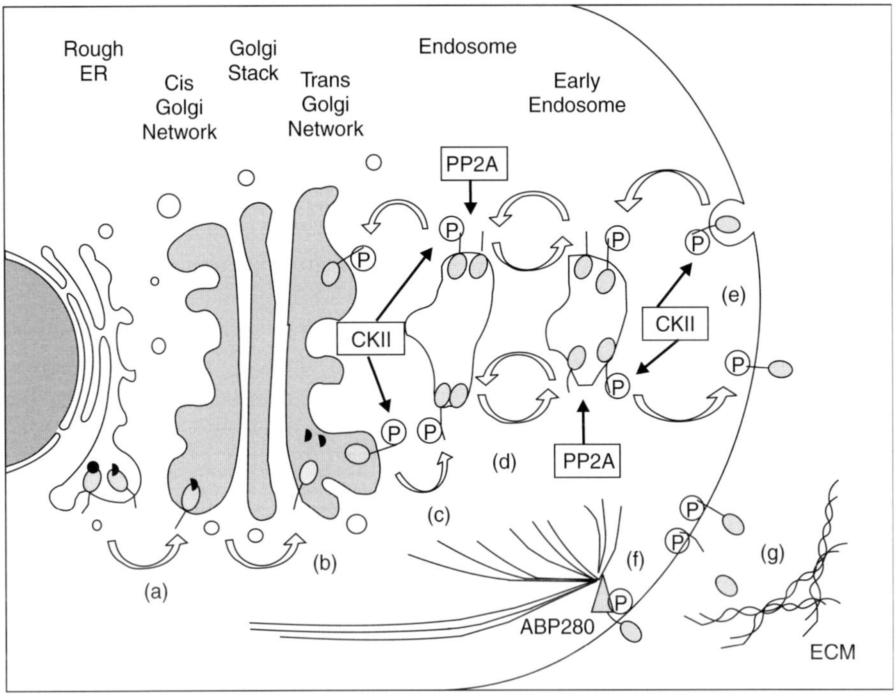

Figure 2.3 Furin activation and trafficking. Furin is synthesized as a zymogen containing an auto-inhibitory propeptide (●). Autolysis of the propeptide occurs in the endoplasmic reticulum (**a**); however, the cleaved propeptide (◗) remains associated with the enzyme and inhibits activity. A second cleavage of the propeptide in conjunction with a change to the acidic and Ca^{++}-rich environment of the trans-Golgi network (TGN) facilitates full release of the inhibitory propeptide (**b**). Furin phosphorylated by casein kinase II (CKII) cycles between the TGN and endosomes (**c**) (Molloy *et al* 1999). Upon dephosphorylation via specific protein phosphatase 2A isoforms, furin is transported from endosomes to the early endosomes (**d**), where re-phosphorylation by (CKII) provides the appropriate interaction site for shuttling to the cell surface (**e**). Cell surface furin can be tethered to the cytoskeleton-associated protein actin binding protein 280 (ABP280), thereby localizing the enzyme to specific regions of the plasma membrane (**f**). Furin shedding (**g**) has also been described and may lead to processing of ECM associated proteins.

composition, cofactors, substrate accessibility) and thereby ensure precise temporal and spatial control of proteolysis. This may involve restriction of proteinase expression or activation to specific tissues, cell types or subcellular organelles. For example, lysosomal cathepsins, which are most active at or near a pH of 5, require sequestration in acidic lysosomes for efficient proteolysis (Authier *et al* 1994). Similarly, calcium ions are required for calpain function, and recent structural studies (Hosfield *et al* 1999; Strobl *et al* 2000) demonstrate that calcium-free calpain does not contain a functional catalytic center, consistent with the hypothesis that calpain activity is regulated through regional calcium fluxes and/or membrane binding (Molinari and Carafoli 1997).

Trafficking of proteinases between subcellular compartments or from an intra- to an extra-cellular localization may function on multiple levels to regulate enzymatic activity. This is exemplified by the membrane-bound serine proteinase furin, which cleaves a large variety of precursor proteins at dibasic amino acid recognition sequences (R-X-K/R-R) and is regulated through spatial sequestration (Figure 2.3). The inhibitory propeptide domain of furin is removed through autolysis in the endoplasmic reticulum; however, this cleavage is not sufficient for activation as the propeptide remains associated with enzyme (Leduc *et al* 1992; Vey *et al* 1994). Trafficking of furin to the acidic and calcium-rich environment of the trans-Golgi network, where a second cleavage of the propeptide occurs, is required for release of this domain and acquisition of enzyme activity (Anderson *et al* 1997). In addition to modulating zymogen activation, proteinase trafficking also can regulate substrate accessibility. For example, trans-Golgi network-associated furin readily cleaves a multitude of proteins in the biosynthetic pathway, while cell surface-associated furin has been shown to hydrolyze extracellular substrates (Nakayama 1997; Molloy *et al* 1999). A truncated, soluble form of furin lacking a transmembrane domain has been reported, however, the functional significance of furin shedding is currently unclear (Vey *et al* 1994).

The biochemical events that regulate subcellular localization of furin have recently been described in a bi-cycling loop model (Figure 2.3) (Molloy *et al* 1999). An acidic cluster motif can be phosphorylated by casein kinase II, thereby promoting cycling of furin both from endosomes back to trans-Golgi network and from early endosomes to the cell surface (Jones *et al* 1995). Dephosphorylation of the same motif by protein phosphatase 2A isoforms mediates shuttling of furin between these two cycling loops (Molloy *et al* 1998). Lastly, a cytoskeleton-tethering signal provides an additional step of regulation of furin trafficking and activity (Liu *et al* 1997).

2.2.2 Receptor binding

Binding of proteinases to cell surface receptors can localize enzymatic activity to sites of cell–cell or cell–substratum contact and influence both the rate of zymogen activation and the accessibility of the enzyme to inactivation by proteinase inhibitors. In many cases, these interactions are thought to be essential for the control of cellular motility and migration during development, neurite outgrowth, angiogenesis, tumor invasion and metastasis. The serine proteinase uPA (Figure 2.4A) binds to a glycosyl phosphatidyl inositol (GPI)-anchored receptor (designated uPAR) on the cell surface (Ploug *et al* 1991). Receptor-bound uPA can efficiently activate cell-bound plasminogen, and activity is neutralized by the serpin PA inhibitor-1 (PAI-1) (Andreasen *et al* 1997). Recent studies have demonstrated a physical interaction between uPAR and integrins, heterodimeric transmembrane proteins that mediate cell-matrix contact (Aguirre Ghiso *et al* 1999; Carriero *et al* 1999; Wei *et al* 1999). These data suggest that cell surface localization may regulate activity in part by properly positioning pericellular proteinases for cleavage of extracellular matrix substrates.

Transmembrane proteinases can bind soluble proteinases or contribute directly to pericellular proteolytic modification of cell surface and extracellular substrates. Membrane type 1-MMP (MT1-MMP) (Figure 2.4B) is an integral membrane protein with a short (19 amino acid residues) cytoplasmic tail that functions in the activation of pro-gelatinase A (proMMP-2) zymogen (Sato *et al* 1994). Detailed biochemical studies have demonstrated

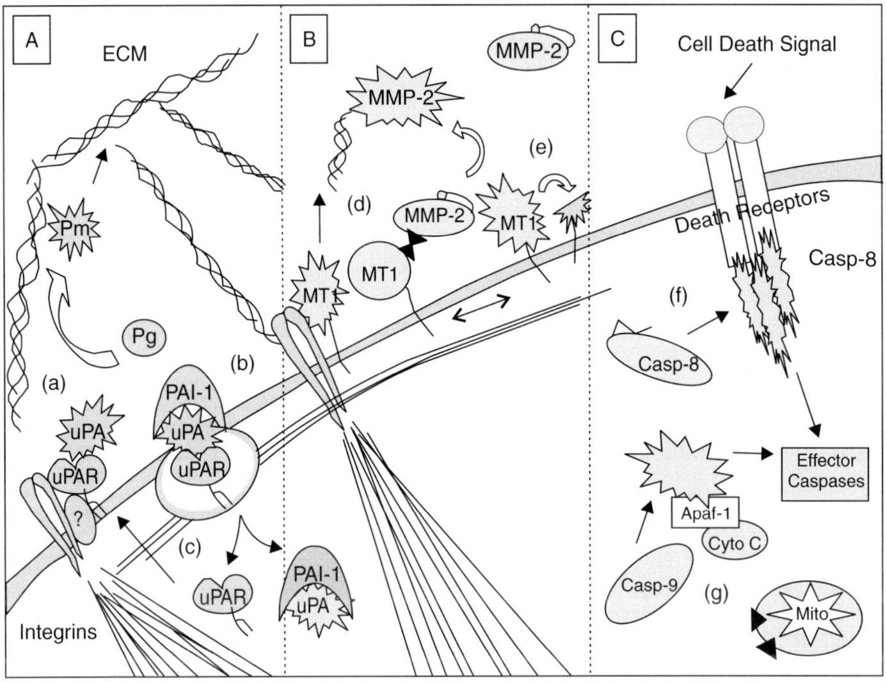

Figure 2.4 Regulation of proteinase activity by receptor binding and cofactor interaction. **A**. uPA mediates the activation of plasminogen (Pg) to plasmin (Pm), which can degrade or process a broad spectrum of extracellular matrix (ECM) proteins (**a**). Binding of uPA to the cell surface via association with the GPI-anchored uPA receptor (uPAR) localizes plasminogen activation to the peri-cellular environment. uPAR may physically associate with cell-surface integrins, either directly or via an accessory protein, localizing serine protease activity to sites of cell/matrix interactions (**b**). Plasminogen activator inhibitor type-1 (PAI-1) readily binds and inhibits uPAR-anchored uPA, thereby promoting internalization of the complex from the cell surface (**c**). Following internalization, dissociated uPAR can recycle back to the cell surface. **B**. The transmembrane proteinase MT1-MMP readily forms a ternary complex with tissue inhibitor of metalloproteinases type-2 (TIMP-2) and progelatinase A (proMMP-2) (**d**). Progelatinase A in the ternary complex is a substrate for a neighboring MT1-MMP, resulting in pro-peptide cleavage, additional autolytic processing and release of the gelatinase. Progelatinase A activation by MT1-MMP likely involves an induced clustering mechanism mediated through the short MT1-MMP cytoplasmic tails (arrows). This may be facilitated via interaction with other trans-membrane proteins co-localized with MT1-MMP in membrane protrusions and focal contacts. In the absence of substrate, MT1-MMP undergoes auto-lytic inactivation (**e**). **C**. The initiation of proteolytic cascades during apoptosis is regulated through the activation of the initiator caspases, caspase-8 and caspase-9. Ligand-stimulated clustering of cell death receptors induces autoactivation of caspase-8 zymogens bound to receptor cytoplasmic domains (**f**). Caspase-9 activity requires an activation complex of Apaf-1 and cytochrome C released from mitochondria (Mito) (**g**). The requirement of cytochrome C for caspase-9 activity ensures that the initiation of apoptosis is connected to mitochondrial integrity. Once activated, initiator caspases proteolytically activate executioner caspases, such as caspase-3, which then carry out the multiple proteolytic events of apoptosis.

that interaction between MT1-MMP and the proteinase inhibitor tissue inhibitor of metalloproteinases-2 (TIMP-2) generates a cellular binding site for pro-gelatinase A binding and subsequent activation (Strongin *et al* 1995; Butler *et al* 1998). This is a novel example of a mechanism by which a protein with inhibitory activity (TIMP-2) is also essential for zymogen activation. In addition to zymogen activation, the soluble catalytic domain of MT1-MMP has been shown to cleave extracellular matrix proteins such as laminins, fibronectin, vitronectin, and collagens, suggesting that the full length transmembrane proteinase may contribute directly to extracellular matrix processing events (Pei and Weiss 1996; d'Ortho *et al* 1997).

Receptor-mediated zymogen activation is not limited to extracellular proteinases. As indicated above, cysteine proteinases in the caspase family participate in zymogen activation cascades that result in induction of apoptosis (Figure 2.4C). Activation of the initiator caspases is more stringently regulated than those downstream and appears to involve autolysis induced by protein–protein interactions that alter local enzyme concentration. Interaction of the caspase-8 zymogen with activated cell death receptors results in proximity-induced autolytic processing (Salvesen and Dixit 1999). Similarly, caspase-9 activation depends on the association of the zymogen with apaf-1 and cytochrome C released from the mitochondria following cytotoxic stress (Li *et al* 1997).

2.2.3 Compartmentalization via non-substrate protein–protein interactions

Protein–protein interactions may function to temporally compartmentalize proteinases, thereby modulating zymogen activation, enzyme–inhibitor binding or substrate accessibility. A well-described example occurs during fibrinolysis, which is regulated via localization of proteinases to the surface of the fibrin clot (Collen and Lijnen 1992; Lijnen *et al* 1994). The serine proteinase tissue-type PA (tPA) is a very inefficient plasminogen activator in solution phase. However, tPA is targeted to the fibrin clot surface via exosite interactions involving the kringle and finger domains of its regulatory heavy chain, leading to a high affinity, non-substrate interaction (de Vries *et al* 1989). Stimulation of tPA activity by fibrin results from the formation of a ternary complex involving tPA, fibrin and plasminogen. This interaction enhances the catalytic efficiency of plasminogen activation up to 3 orders of magnitude by two mechanisms. First, the proximity of tPA to the substrate plasminogen is enhanced through fibrin binding, thus increasing the likelihood of enzyme–substrate contact. Secondly, fibrin-bound plasminogen undergoes a dramatic conformational change that exposes the tPA cleavage site, thereby facilitating activation (Mangel *et al* 1990). Furthermore, the reaction product plasmin is now positioned in close contact to its physiologic substrate fibrin. Similar interactions occur on the basement membrane, whereupon tPA and plasminogen interact with type IV collagen to enhance the catalytic efficiency of plasminogen activation (Stack *et al* 1990).

Protein–protein interactions may also regulate enzyme stability and thereby function to maintain a proteinase in a catalytically competent form. The trypsin-like serine proteinase tryptase is stored fully active in the cytoplasmic granules of mast cells and active enzyme is secreted following mast cell degranulation (Irani *et al* 1986). Tryptase is remarkably resistant to inhibition and is not susceptible to inactivation by any known plasma proteinase inhibitors (Smith *et al* 1984). However, recent data suggest that tryptase activity is regulated primarily by oligomerization, as the monomeric enzyme has negligible catalytic activity while the heparin-stabilized tetramer is catalytically competent (Schwartz *et al* 1981; Schechter *et al* 1995; Addington and Johnson 1996). Thus, enzymatic activity may be regulated by removal of heparin, resulting in dissociation of the tetramer into inactive monomers.

2.3 LIMITED PROTEOLYSIS FOR STRUCTURAL MODIFICATION AND PROTEINASE CLEARANCE

2.3.1 Proteinase clearance

One of the least-explored aspects of proteinase regulation is the clearance of enzymes from tissues and cells. Protease degradation can be regulated through various mechanisms including proteolysis, ubiquination, and endocytosis. Clearance of uPA from the pericellular environment involves an endocytotic mechanism of removal via interaction of PAI- 1-complexed uPA/uPAR with the α2-macroglobulin receptor/low-density lipoprotein receptor-related protein (α2-MR/LRP) (Strickland *et al* 1994). Both the uPA zymogen and active uncomplexed enzyme have lower affinities than uPA/PAI-1 complexes for the receptor and therefore are not internalized upon uPAR binding. Following endocytosis, uPAR may be recycled to the cell surface to resume its dual role in proteinase regulation by cell surface localization and enzyme–inhibitor complex clearance (Figure 2.4a) (Andreasen *et al* 1997).

In vitro, the majority of proteases display an appreciable amount of intramolecular proteolysis, suggesting that localized autolysis may function as a mechanism for proteinase clearance under conditions of substrate depletion. In support of this hypothesis, recent studies demonstrate that MT1-MMP undergoes autoproteolysis to species that lack catalytic activity in the absence of substrate (pro-gelatinase A) or inhibitor (TIMP-2) (Stanton *et al* 1998; Ellerbroek *et al* 1999). As this enzyme depends on cell surface clustering for effective activation of gelatinase A (MMP-2) and hydrolysis of matrix proteins, it is likely that proteolytic activity will be down-regulated rapidly in the absence of substrate (Figure 2.4b).

2.3.2 Removal of structural/functional domains

Protease degradation often removes domains that are essential for protease localization and/or activity. For instance, the first autoproteolytic cleavage of gelatinase A releases the large carboxyl-terminal hemopexin domain from the enzyme (Bergmann *et al* 1995). The potential physiological significance of hemopexin domain loss can be predicted based on its function as a high affinity binding site for the inhibitor TIMP-2 and indispensable role in heparin, fibronectin, and cell surface binding (Murphy and Knauper 1997). Although the release of this regulatory region from the catalytic domain does not appear to alter the kinetics of synthetic peptide substrate cleavage (Willenbrock *et al* 1993), it is difficult to envision a biologically relevant role for the gelatinase in the absence of the hemopexin domain. However, it should be noted that the truncated proteinase retains the type I collagen-binding domain and has been reported to exist in tissues (Brooks *et al* 1998).

2.3.3 Exposure of biologically active domains

Protein processing can result in formation of products with biological activity distinct from the precursor protein, and the proteinases themselves are no exception. Active enzymes such as uPA are also proteinase substrates, and processing of uPA can release the receptor-binding amino terminal fragment (ATF) from the serine proteinase domain (Marcotte *et al* 1992; Rabbani *et al* 1992; Fishman *et al* 1999). The ATF retains the ability to bind to uPAR and influence cellular proliferation and migration via proteinase-independent mechanisms, demonstrating that the influence of uPA on cell behavior is not abolished upon loss of cell surface proteolytic activity (Rabbani *et al* 1992; Aguirre Ghiso *et al* 1999; Fishman *et al* 1999; Wei *et al* 1999).

Another notable example is the conversion of the pro-angiogenic proteinase plasmin(ogen) into anti-angiogenic polypeptides, designated angiostatin, that inhibit endothelial cell proliferation and/or migration (O'Reilly *et al* 1994). Plasmin(ogen) is comprised of a serine protease domain coupled to 5 kringle domains that participate in ligand binding

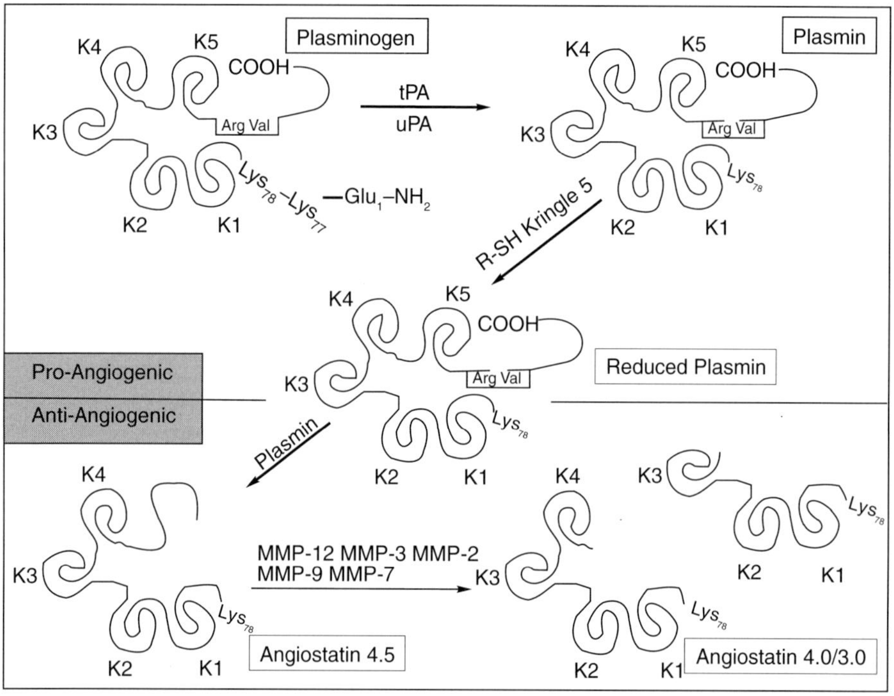

Figure 2.5 Conversion of plasminogen to angiostatin. Plasminogen is activated through cleavage of the Arg_{560}–Val_{561} peptide bond producing plasmin, a two-chain proteinase containing two inter-chain disulfide bonds. Following maturation through autolysis of the Lys_{78}–Lys_{77} peptide bond, plasmin functions as a pro-angiogenic proteinase. Reduction of two disulfide bonds in kringle 5 via a plasmin-specific reductase facilitates an autolytic cleavage within kringle 5, leading to loss of the catalytic domain. The N-terminal product, comprised of kringles 1–4 and a portion of kringle 5 (designated angiostatin 4.5), is functionally distinct from pro-angiogenic plasmin and has anti-angiogenic properties *in vitro* and *in vivo*. Angiostatin 4.5 is a substrate for a variety of MMPs, generating smaller kringle-containing angiostatin species. Differences in biological activity among angiostatin species and identification of the predominant *in vivo* reaction product(s) are areas of current investigation. K – kringle; tPA – tissue-type plasminogen activator; uPA – urinary-type plasminogen activator; MMP – matrix metalloproteinase.

(Figure 2.5) (Castellino 1995). Following plasminogen activation, reduction of the Cys_{461}–Cys_{540} and Cys_{511}–Cys_{535} disulfide bonds within kringle 5 results in a partially reduced plasmin which functions as a substrate for limited proteolysis to generate the angiostatin product(s) (Stathakis *et al* 1997). Cleavage can be autolytic resulting in generation of a large angiostatin fragment containing 4.5 kringles (Gately *et al* 1997; Stathakis *et al* 1997). Alternatively, reduced plasmin may be further processed by a variety of proteinases including gelatinases A and B, matrilysin, and macrophage metalloelastase to produce smaller angiostatin products containing 3–4 kringles (Dong *et al* 1997; Patterson and Sang 1997; Lijnen *et al* 1998). The functional differences among angiostatin species are currently unclear and the fate of the liberated plasmin catalytic domain has not been evaluated. However, limited proteolysis of plasmin to angiostatin provides a mechanism whereby proteinase processing generates a product with a biologically distinct function, thereby providing a high fidelity system of homeostasis.

REFERENCES

Addington, A.K. and Johnson, D.A. (1996) Inactivation of human lung tryptase: evidence for a re-activatable tetrameric intermediate and active monomers. *Biochemistry*, **35**, 13511–13518.

Aguirre Ghiso, J.A., Kovalski, K. and Ossowski, L. (1999) Tumor dormancy induced by down-regulation of urokinase receptor in human carcinoma involves integrin and MAPK signaling. *Journal of Cell Biology*, **147**, 89–104.

Anderson, E.D., VanSlyke, J.K., Thulin, C.D., Jean, F. and Thomas, G. (1997) Activation of the furin endoprotease is a multiple-step process: requirements for acidification and internal propeptide cleavage. *EMBO Journal*, **16**, 1508–1518.

Andreasen, P.A., Kjoller, L., Christensen, L. and Duffy, M.J. (1997) The urokinase-type plasminogen activator system in cancer metastasis: a review. *International Journal of Cancer*, **72**, 1–22.

Appel, W. (1986) Chymotrypsin: molecular and catalytic properties. *Clinical Biochemistry*, **19**, 317–322.

Atkinson, S.J., Crabbe, T., Cowell, S., Ward, R.V., Butler, M.J., Sato, H. *et al* (1995) Intermolecular autolytic cleavage can contribute to the activation of progelatinase A by cell membranes. *Journal of Biological Chemistry*, **270**, 30479–30485.

Authier, F., Posner, B.I. and Bergeron, J.J.M. (1994) In *Cellular Proteolytic Systems*, edited by A.J. Ciechanover, pp. 89–113. New York: Wiley-Liss.

Bergmann, U., Tuuttila, A., Stetler-Stevenson, W.G. and Tryggvason, K. (1995) Autolytic activation of recombinant human 72 kilodalton type IV collagenase. *Biochemistry*, **34**, 2819–2825.

Birkedal-Hansen, H., Moore, W.G., Bodden, M.K., Windsor, L.J., Birkedal-Hansen, B. and DeCarlo, A. *et al* (1993) Matrix metalloproteinases: a review. *Critical Reviews in Oral Biology and Medicine*, **4**, 197–250.

Brooks, P.C., Silletti, S., von Schalscha, T.L., Friedlander, M. and Cheresh, D.A. (1998) Disruption of angiogenesis by PEX, a noncatalytic metalloproteinase fragment with integrin binding activity. *Cell*, **92**, 391–400.

Budihardjo, I., Oliver, H., Lutter, M., Luo, X. and Wang, X. (1999) Biochemical pathways of caspase activation during apoptosis. *Annual Review of Cell Developmental Biology*, **15**, 269–290.

Butler, G.S., Butler, M.J., Atkinson, S.J., Will, H., Tamura, T., van Westrum, S.S. *et al* (1998) The TIMP2 membrane type 1 metalloproteinase "receptor" regulates the concentration and efficient activation of progelatinase A. A kinetic study. *Journal of Biological Chemistry*, **273**, 871–880.

Carriero, M.V., Del Vecchio, S., Capozzoli, M., Franco, P., Fontana, L., Zannetti, A. *et al* (1999) Urokinase receptor interacts with alpha(v)beta5 vitronectin receptor, promoting urokinase-dependent cell migration in breast cancer. *Cancer Research*, **59**, 5307–5314.

Castellino, F.J. (1995) Plasminogen. In *Molecular Basis of Thrombosis and Hemostasis*, edited by K.A. High and H.R. Roberts, pp. 517–544. New York: Marcel Dekker, Inc.

Collen, D. and Lijnen, H.R. (1992) Fibrin-specific fibrinolysis. *Annals of the New York Academy of Sciences*, **667**, 259–271.

de Vries, C., Veerman, H. and Pannekoek, H. (1989) Identification of the domains of tissue-type plasminogen activator involved in the augmented binding to fibrin after limited digestion with plasmin. *Journal of Biological Chemistry*, **264**, 12604–12610.

Dong, Z., Kumar, R., Yang, X. and Fidler, I.J. (1997) Macrophage-derived metalloelastase is responsible for the generation of angiostatin in Lewis lung carcinoma. *Cell*, **88**, 801–810.

d'Ortho, M.P., Will, H., Atkinson, S., Butler, G., Messent, A., Gavrilovic, J. *et al* (1997) Membrane-type matrix metalloproteinases 1 and 2 exhibit broad-spectrum proteolytic capacities comparable to many matrix metalloproteinases. *European Journal of Biochemistry*, **250**, 751–757.

Ellerbroek, S.M., Fishman, D.A., Kearns, A.S., Bafetti, L.M. and Stack, M.S. (1999) Ovarian carcinoma regulation of matrix metalloproteinase-2 and membrane type 1 matrix metalloproteinase through beta1 integrin. *Cancer Research*, **59**, 1635–1641.

Farr, M., Pieper, M., Calvete, J. and Tschesche, H. (1999) The N-terminus of collagenase MMP-8 determines superactivity and inhibition: a relation of structure and function analyzed by biomolecular interaction analysis. *Biochemistry*, **38**, 7332–7338.

Fishman, D.A., Kearns, A., Larsh, S., Enghild, J.J. and Stack, M.S. (1999) Autocrine regulation of growth stimulation in human epithelial ovarian carcinoma by serine-proteinase-catalysed release of the urinary-type-plasminogen-activator N-terminal fragment. *Biochemical Journal*, **341**, 765–769.

Gately, S., Twardowski, P., Stack, M.S., Cundiff, D.L., Grella, D., Castellino, F.J. *et al* (1997) The mechanism of cancer-mediated conversion of plasminogen to the angiogenesis inhibitor angiostatin. *Proceedings of the National Academy of Sciences of the U.S.A.*, **94**, 10868–10872.

Hosfield, C.M., Elce, J.S., Davies, P.L. and Jia, Z. (1999) Crystal structure of calpain reveals the structural basis for Ca(2+)-dependent protease activity and a novel mode of enzyme activation. *EMBO Journal*, **18**, 6880–6889.

Irani, A.A., Schechter, N.M., Craig, S.S., DeBlois, G. and Schwartz, L.B. (1986) Two types of human mast cells that have distinct neutral protease compositions. *Proceedings of the National Academy of Sciences of the U.S.A.*, **83**, 4464–4468.

Jones, B.G., Thomas, L., Molloy, S.S., Thulin, C.D., Fry, M.D., Walsh, K.A. *et al* (1995) Intracellular trafficking of furin is modulated by the phosphorylation state of a casein kinase II site in its cytoplasmic tail. *EMBO Journal*, **14**, 5869–5883.

Knauper, V., Wilhelm, S.M., Seperack, P.K., DeClerck, Y.A., Langley, K.E., Osthues, A. *et al* (1993) Direct activation of human neutrophil procollagenase by recombinant stromelysin. *Biochemical Journal*, **295**, 581–586.

Knauper, V., Will, H., Lopez-Otin, C., Smith, B., Atkinson, S.J., Stanton, H. *et al* (1996) Cellular mechanisms for human procollagenase-3 (MMP-13) activation. Evidence that MT1-MMP (MMP-14) and gelatinase a (MMP-2) are able to generate active enzyme. *Journal of Biological Chemistry*, **271**, 17124–17131.

Leduc, R., Molloy, S.S., Thorne, B.A. and Thomas, G. (1992) Activation of human furin precursor processing endoprotease occurs by an intramolecular autoproteolytic cleavage. *Journal of Biological Chemistry*, **267**, 14304–14308.

Li, P., Nijhawan, D., Budihardjo, I., Srinivasula, S.M., Ahmad, M., Alnemri, E.S. *et al* (1997) Cytochrome c and dATP-dependent formation of Apaf-1/caspase-9 complex initiates an apoptotic protease cascade. *Cell*, **91**, 479–489.

Lijnen, H.R., Bachmann, F., Collen, D., Ellis, V., Pannekoek, H., Rijken, D.C. *et al* (1994) Mechanisms of plasminogen activation. *Journal of Internal Medicine*, **236**, 415–424.

Lijnen, H.R., Ugwu, F., Bini, A. and Collen, D. (1998) Generation of an angiostatin-like fragment from plasminogen by stromelysin-1 (MMP-3). *Biochemistry*, **37**, 4699–4702.

Liu, G., Thomas, L., Warren, R.A., Enns, C.A., Cunningham, C.C., Hartwig, J.H. *et al* (1997) Cytoskeletal protein ABP-280 directs the intracellular trafficking of furin and modulates proprotein processing in the endocytic pathway. *Journal of Cell Biology*, **139**, 1719–1733.

Mangel, W.F., Lin, B.H. and Ramakrishnan, V. (1990) Characterization of an extremely large, ligand-induced conformational change in plasminogen. *Science*, **248**, 69–73.

Marcotte, P.A., Kozan, I.M., Dorwin, S.A. and Ryan, J.M. (1992) The matrix metalloproteinase pump-1 catalyzes formation of low molecular weight (pro)urokinase in cultures of normal human kidney cells. *Journal of Biological Chemistry*, **267**, 13803–13806.

Molinari, M. and Carafoli, E. (1997) Calpain: a cytosolic proteinase active at the membranes. *The Journal of Membrane Biology*, **156**, 1–8.

Molloy, S.S., Anderson, E.D., Jean, F. and Thomas, G. (1999) Bi-cycling the furin pathway: from TGN localization to pathogen activation and embryogenesis. *Trends in Cellular Biology*, **9**, 28–35.

Molloy, S.S., Thomas, L., Kamibayashi, C., Mumby, M.C. and Thomas, G. (1998) Regulation of endosome sorting by a specific PP2A isoform. *Journal of Cell Biology*, **142**, 1399–1411.

Murphy, G. and Gavrilovic, J. (1999) Proteolysis and cell migration: creating a path? *Current Opnion in Cell Biology*, **11**, 614–621.

Murphy, G. and Knauper, V. (1997) Relating matrix metalloproteinase structure to function: why the "hemopexin" domain? *Matrix Biology*, **15**, 511–518.

Nagase, H., Enghild, J.J., Suzuki, K. and Salvesen, G. (1990) Stepwise activation mechanisms of the precursor of matrix metalloproteinase 3 (stromelysin) by proteinases and (4-aminophenyl)mercuric acetate. *Biochemistry*, **29**, 5783–5789.

Nakayama, K. (1997) Furin: a mammalian subtilisin/Kex2p-like endoprotease involved in processing of a wide variety of precursor proteins. *Biochemical Journal*, **327**, 625–635.

Newland, J.R. (1987) Blood coagulation: a review. *American Journal of Obstetrics and Gynecology*, **156**, 1420–1422.

O'Reilly, M.S., Holmgren, L., Shing, Y., Chen, C., Rosenthal, R.A., Moses, M. *et al* (1994) Angiostatin: a novel angiogenesis inhibitor that mediates the suppression of metastases by a Lewis lung carcinoma. *Cell*, **79**, 315–328.

Patterson, B.C. and Sang, Q.A. (1997) Angiostatin-converting enzyme activities of human matrilysin (MMP-7) and gelatinase B/type IV collagenase (MMP-9). *Journal of Biological Chemistry*, **272**, 28823–28825.

Pei, D. and Weiss, S.J. (1996) Transmembrane-deletion mutants of the membrane-type matrix metalloproteinase-1 process progelatinase A and express intrinsic matrix-degrading activity. *Journal of Biological Chemistry*, **271**, 9135–9140.

Ploug, M., Ronne, E., Behrendt, N., Jensen, A.L., Blasi, F. and Dano, K. (1991) Cellular receptor for urokinase plasminogen activator. Carboxyl-terminal processing and membrane anchoring by glycosyl-phosphatidylinositol. *Journal of Biological Chemistry*, **266**, 1926–1933.

Rabbani, S.A., Mazar, A.P., Bernier, S.M., Haq, M., Bolivar, I., Henkin, J. *et al* (1992) Structural requirements for the growth factor activity of the amino-terminal domain of urokinase. *Journal of Biological Chemistry*, **267**, 14151–14156.

Ramos-DeSimone, N., Hahn-Dantona, E., Sipley, J., Nagase, H., French, D.L. and Quigley, J.P. (1999) Activation of matrix metalloproteinase-9 (MMP-9) via a converging plasmin/stromelysin-1 cascade enhances tumor cell invasion. *Journal of Biological Chemistry*, **274**, 13066–13076.

Reinemer, P., Grams, F., Huber, R., Kleine, T., Schnierer, S. and Piper, M. *et al* (1994) Structural implications for the role of the N terminus in the 'superactivation' of collagenases. A crystallographic study. *FEBS Letters*, **338**, 227–233.

Salvesen, G.S. and Dixit, V.M. (1999) Caspase activation: the induced-proximity model. *Proceedings of the National Academy of Sciences of the U.S.A.*, **96**, 10964–10967.

Sato, H., Takino, T., Okada, Y., Cao, J., Shinagawa, A., Yamamoto, E. *et al* (1994) A matrix metalloproteinase expressed on the surface of invasive tumour cells. *Nature*, **370**, 61–65.

Schechter, N.M., Eng, G.Y., Selwood, T. and McCaslin, D.R. (1995) Structural changes associated with the spontaneous inactivation of the serine proteinase human tryptase. *Biochemistry*, **34**, 10628–10638.

Schwartz, L.B., Lewis, R.A. and Austen, K.F. (1981) Tryptase from human pulmonary mast cells. Purification and characterization. *Journal of Biological Chemistry*, **256**, 11939–11943.

Smith, T.J., Hougland, M.W. and Johnson, D.A. (1984) Human lung tryptase. Purification and characterization. *Journal of Biological Chemistry*, **259**, 11046–11051.

Springman, E.B., Angleton, E.L., Birkedal-Hansen, H. and Van Wart, H.E. (1990) Multiple modes of activation of latent human fibroblast collagenase: evidence for the role of a Cys73 active-site zinc complex in latency and a "cysteine switch" mechanism for activation. *Proceedings of the National Academy of Sciences of the U.S.A.*, **87**, 364–368.

Stack, M.S., Gonzalez-Gronow, M. and Pizzo, S.V. (1990) Regulation of plasminogen activation by components of the extracellular matrix. *Biochemistry*, **29**, 4966–4970.

Stanton, H., Gavrilovic, J., Atkinson, S.J., d'Ortho, M.P., Yamada, K.M., Zardi, L. *et al* (1998) The activation of ProMMP-2 (gelatinase A) by HT1080 fibrosarcoma cells is promoted by culture on a fibronectin substrate and is concomitant with an increase in processing of MT1-MMP (MMP-14) to a 45 kDa form. *Journal of Cell Science*, **111**, 2789–2798.

Stathakis, P., Fitzgerald, M., Matthias, L.J., Chesterman, C.N. and Hogg, P.J. (1997) Generation of angiostatin by reduction and proteolysis of plasmin. Catalysis by a plasmin reductase secreted by cultured cells. *Journal of Biological Chemistry*, **272**, 20641–20645.

Strickland, D.K., Kounnas, M.Z., Williams, S.E. and Argraves, W.S. (1994) LDL receptor-related protein (LRP): a multiligand receptor. *Fibrinolysis*, **8**, 204–215.

Strobl, S., Fernandez-Catalan, C., Braun, M., Huber, R., Masumoto, H., Nakagawa, K. *et al* (2000) The crystal structure of calcium-free human m-calpain suggests an electrostatic switch mechanism for activation by calcium. *Proceedings of the National Academy of Sciences of the U.S.A.*, **97**, 588–592.

Strongin, A.Y., Collier, I., Bannikov, G., Marmer, B.L., Grant, G.A. and Goldberg, G.I. (1995) Mechanism of cell surface activation of 72-kDa type IV collagenase. Isolation of the activated form of the membrane metalloprotease. *Journal of Biological Chemistry*, **270**, 5331–5338.

Suzuki, K., Enghild, J.J., Morodomi, T., Salvesen, G. and Nagase, H. (1990) Mechanisms of activation of tissue procollagenase by matrix metalloproteinase 3 (stromelysin). *Biochemistry*, **29**, 10261–10270.

Tachias, K. and Madison, E.L. (1996) Converting tissue-type plasminogen activator into a zymogen. *Journal of Biological Chemistry*, **271**, 28749–28752.

Vey, M., Schafer, W., Berghofer, S., Klenk, H.D. and Garten, W. (1994) Maturation of the trans-Golgi network protease furin: compartmentalization of propeptide removal, substrate cleavage, and COOH-terminal truncation. *Journal of Cell Biology*, **127**, 1829–1842.

Wei, Y., Yang, X., Liu, Q., Wilkins, J.A. and Chapman, H.A. (1999) A role for caveolin and the urokinase receptor in integrin-mediated adhesion and signaling. *Journal of Cell Biology*, **144**, 1285–1294.

Willenbrock, F., Crabbe, T., Slocombe, P.M., Sutton, C.W., Docherty, A.J., Cockett, M.I. *et al* (1993) The activity of the tissue inhibitors of metalloproteinases is regulated by C-terminal domain interactions: a kinetic analysis of the inhibition of gelatinase A. *Biochemistry*, **32**, 4330–4337.

Chapter 3

Matrix metalloproteinases (MMPs)

Claudiu T. Supuran and Andrea Scozzafava

A review of the MMPs and their inhibitors is presented. The 20 vertebrate enzymes isolated up to now degrade all components of extracellular matrix (ECM) being thus involved in important physiological and physiopathological events, such as embryonic development, blastocyst implantation, nerve growth, ovulation, morphogenesis, angiogenesis, tissue resorption and remodeling (such as wound healing), bone remodeling, apoptosis, cancer invasion and metastasis, arthritis, atherosclerosis, aneurysm, breakdown of blood–brain barrier, periodontal disease, skin ulceration, corneal ulceration, gastric ulcer, and liver fibrosis among others. The structure, catalytic and inhibition mechanisms of these metalloproteinases are well understood, thus allowing for the design of inhibitors with clinical significance. Several classes of such inhibitors are known, such as carboxylates, hydroxamates, 1,3,4-thiadiazoles or phosphorus based derivatives. Many of them show high affinities (in the nanomolar range) for all or only some specific MMPs. Bacterial collagenase inhibitors structurally related to the hydroxamate MMP inhibitors are also discussed. Clinical trials of some potent MMP inhibitors showed them to be valuable candidates for the development of drugs useful in the treatment of tumors, metastasis, rheumatoid- and osteoarthritis, as well as other diseases characterized by extensive ECM degradation and remodeling.

3.1 INTRODUCTION

The extracellular matrix (ECM) plays a critical role for the structure and integrity of various tissue types in higher vertebrates (Tschesche 1995; Whittaker *et al* 1999). ECM turnover is involved in important physiological and physiopathological events, such as embryonic development, blastocyst implantation, nerve growth, ovulation, morphogenesis, angiogenesis, tissue resorption and remodeling (such as wound healing), bone remodeling, apoptosis, cancer invasion and metastasis, arthritis, atherosclerosis, aneurysm, breakdown of blood–brain barrier, periodontal disease, skin ulceration, corneal ulceration, gastric ulcer, and liver fibrosis among others (Dioszegi *et al* 1995; Nagase 1997; Bottomley *et al* 1998; Johnson *et al* 1998; Nagase and Woessner 1999; Whittaker *et al* 1999). The matrix metalloproteinases (MMPs), a family of zinc-containing endopeptidases (also called matrixins), are thought to play a central role in the above mentioned processes (Nagase 1997; Brandstetter *et al* 1998; Johnson *et al* 1998; Nagase and Woessner 1999; Whittaker *et al* 1999).

At least 20 members of this enzyme family, sharing significant sequence homology, have been reported (Table 3.1) (Barrett *et al* 1998; Nagase and Woessner 1999). They can be

Table 3.1 Vertebrate MMPs, their molecular weights, substrates and preferred scissile amide bonds

Protein	*MMP*	*MW* (kDa)	*Principal substrate*(s)	*Preferred scissile amide bond*(s)
Collagenase 1	MMP-1	52	fibrillar and nonfibrillar collagens (types I, II, III, VI and X), gelatins	Gly-Ile
Gelatinase A	MMP-2	72	basement membrane and nonfibrillar collagens (types IV, V, VII, X), fibronectin, elastin	Ala-Met
Stromelysin 1	MMP-3	57	proteoglycan, laminin, fibronectin, collagen (types III, IV, V, IX); gelatins; pro-MMP-1	Gly-Leu
Matrilysin	MMP-7	28	fibronectins, gelatins, proteoglycan	Ala-Ile
Collagenase 2	MMP-8	64	fibrillar collagens (types I, II, III)	Gly-Leu; Gly-Ile
Gelatinase B	MMP-9	92	basement membrane collagens (types IV, V), gelatins	Gly-Ile; Gly-Leu
Stromelysin 2	MMP-10	54	fibronectins, collagen (types III, IV), gelatins, pro-MMP-1	Gly-Leu
Stromelysin 3	MMP-11	45	serpin	Ala-Met
Macrophage elastase	MMP-12	53	elastin	Ala-Leu; Tyr-Leu
Collagenase 3	MMP-13	51.5	fibrillar collagens (types I, II, III), gelatins	Gly-Ile
MT1-MMP	MMP-14	66	pro-72 kDa gelatinase	not determined
MT2-MMP	MMP-15	61	not determined	not determined
MT3-MMP	MMP-16	55	pro-72 kDa gelatinase	not determined
MT4-MMP	MMP-17	58	not determined	Ala-Gly
Collagenase 4 (*Xenopus*)	MMP-18	53	not determined	Gly-Ile
RASI 1	MMP-19	?	gelatin	not determined
Enamelysin	MMP-20	?	amelogenin (dentine), gelatin	not determined
XMMP (*Xenopus*)	MMP-21	?	not determined	not determined
CMMP (chicken)	MMP-22	?	not determined	not determined
(No trivial name)	MMP-23	?	not determined	not determined

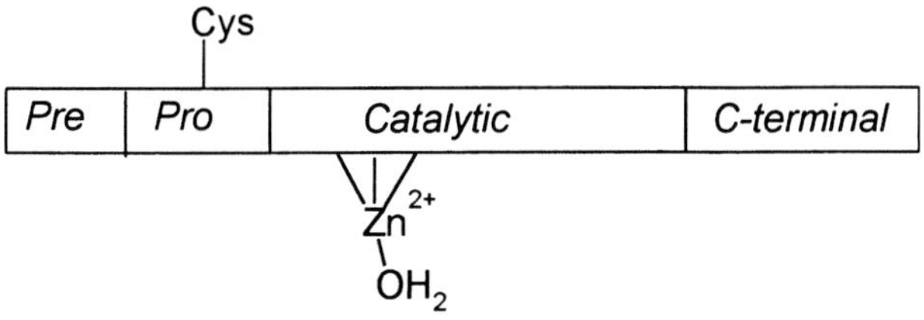

Figure 3.1 Schematic representation of the structure of a typical MMP enzyme. *Pre* represents the N-terminal signal peptide, *Pro* the propeptide sequence, followed by the catalytic domain and the C-terminal domain.

subdivided (considering the macromolecular substrate requirements) into: (1) collagenases (MMP-1, -8, -13 and -18); (2) gelatinases (MMP-2 and -9); (3) stromelysins (MMP-3, -10 and -11); and (4) membrane-type MMPs (MT-MMPs) (MMP-14, -15, -16 and -17). Recently, some new members of the family have been found, but little is known for the moment regarding their properties, substrate specificity, and inhibition (Table 3.1) (Nagase and Woessner 1999).

MMPs possess a modular structure (Grams *et al* 1995a,b; Barrett *et al* 1998; Nagase and Woessner 1999) consisting of: (1) an N-terminal signal peptide sequence (*Pre*, see Figure 3.1); (2) a propeptide sequence (*Pro* in Figure 3.1) which has the role of conferring latency to the enzyme. In fact this domain contains a conserved Cys residue which is coordinated to the catalytic Zn(II) ion, inhibiting in this way the autolysis of these highly active enzymes. MMPs require the removal of this pro-domain to acquire catalytic activity; (3) the catalytic domain (of about 170 amino acid residues), which contains a highly conserved zinc binding motif, consisting of three histidine residues and a conserved glutamate, important in catalysis. The Zn(II) binding motif is HEXXHXXGXXH (where X can be any amino acid residue). This is in fact characteristic for many other metalloproteases belonging to the M7, M10, M11 and M12 families (Barrett *et al* 1998). Stromelysin 3 and the MT-MMPs also contain a furin recognition sequence between the propeptide sequence and the catalytic domain; (4) a variable, C-terminal domain. In matrylisin this domain is missing, whereas for other MMPs (such as the collagenases) it is essential for the recognition of macromolecular substrates. MT-MMPs also contain a transmembrane region within the C-terminal domain, which serves to anchor the enzyme to the cell membrane, whereas the N-terminal part of the molecule protrudes into the extracellular space; (5) several metal ions, playing different functions. All MMPs contain two Zn(II) and two–three Ca(II) ions. One of the zinc ions, coordinated by the histidines belonging to the binding motif mentioned above, is critical for catalysis, since the water coordinated to it, as the fourth ligand in a quasi tetrahedral geometry of Zn(II), acts as the nucleophile during the proteolytic process (see Section 3.2). The other zinc ion and the calcium ions have a structural role, probably in stabilizing the enzyme from autocleavage (Barrett *et al* 1998; Nagase and Woessner 1999).

In general, MMPs are secreted as zymogens which are inactive, latent pro-enzymes (Johnson *et al* 1998). These proforms need activation in order to give fully active proteases. Extracellular activation is generally a two-step process: an initial cleavage by an activator protease to an exposed susceptible loop in the propeptide domain (the so-called "bait" region), leading to the destabilization of the propeptide binding interactions and disrupting the coordination of the conserved Cys residue to Zn(II); this is then followed by a final cleavage, usually assisted by another MMP, with release of the amino terminus of the mature enzyme (Springman *et al* 1990).

Due to their ubiquitous spread in many tissues, where they play critical physiological functions, MMPs have recently become interesting targets for drug design, in the search for novel types of anticancer, anti-arthritis or other pharmacological agents useful in the management of osteoporosis, restenosis, aortic aneurysm, glomerulonephritis, or multiple sclerosis among others (Babine and Bender 1997; Bottomley *et al* 1998; Whittaker *et al* 1999). Some of the recent developments in the field will be reviewed in this chapter.

3.2 CATALYTIC AND INHIBITION MECHANISM OF MMPs

3.2.1 Catalytic mechanism

MMPs and the carbonic anhydrases (CAs) possess very similar zinc coordination spheres within their catalytic sites, consisting of a Zn(II) ion coordinated by three histidines, with the fourth ligand being a water molecule/hydroxide ion, which is the nucleophile intervening in the catalytic cycle of both enzymes (Figure 3.2) (Lovejoy *et al* 1994; Supuran 1994; Grams *et al* 1995a,b; Supuran and Scozzafava 2000). The main structural difference between these two types of enzymes concerns the residues with which the zinc-bound water molecule interacts: in CAs, the zinc-bound water forms a hydrogen bond with the hydroxylic moiety of Thr 199 (human isozyme CA II numbering), which in turn is hydrogen-bonded to the carboxylate of Glu 106, leading thus to a dramatic enhancement in nucleophilicity of the water molecule (Supuran 1994; Briganti *et al* 1999). CAs catalyze with great efficiency the hydration of CO_2 to bicarbonate, ester hydrolysis as well as some other related reactions, but they are devoid of protease activity (Supuran *et al* 1997; Supuran and Scozzafava 2000). In the case of MMPs, the zinc-bound water molecule interacts with the carboxylate moiety of the conserved glutamate (Glu 198 in MMP-8), probably forming two hydrogen bonds with it (Lovejoy *et al* 1994; Grams *et al* 1995a,b).

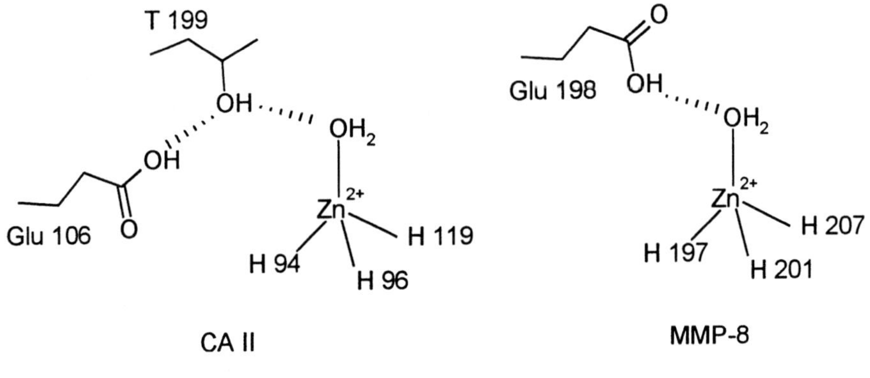

Figure 3.2 Coordination of the Zn(II) ion within the active sites of carbonic anhydrase isozyme II (CA II) and MMP-8.

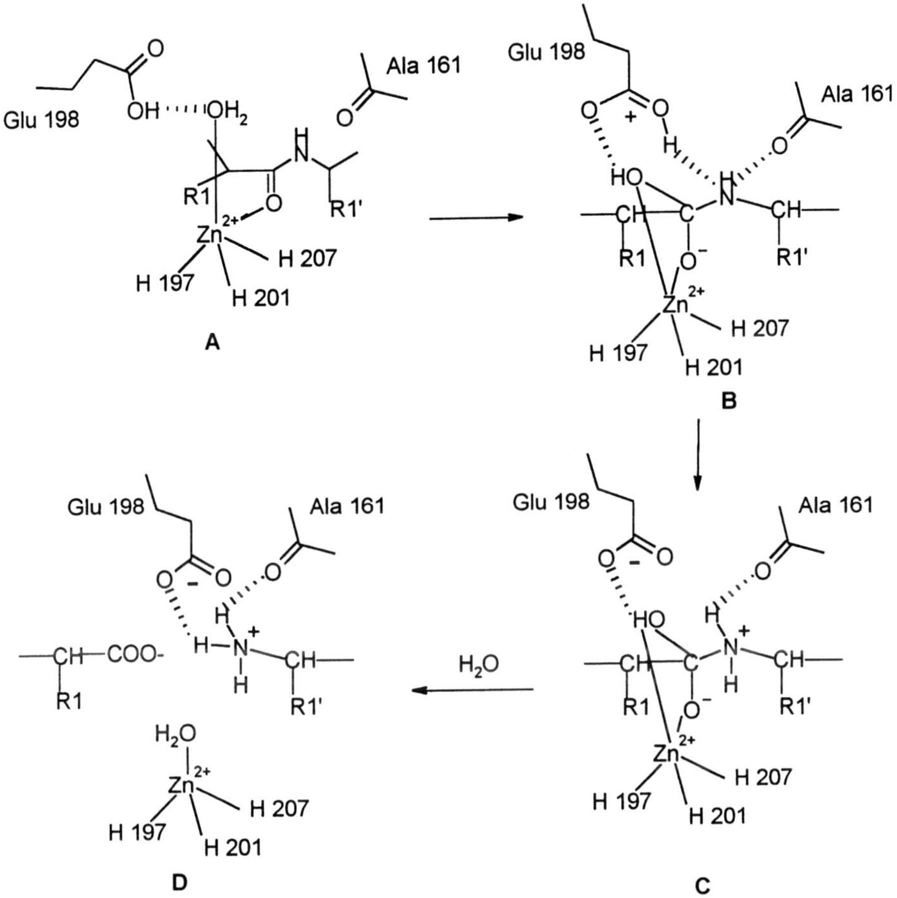

Figure 3.3 Catalytic mechanism of MMPs (exemplified for one of the best studied cases, MMP-8), adapted from Lovejoy *et al* (1994).

Thus, a very effective nucleophile is formed again, which will attack the amide scissile bond (see later in the text).

The proteolytic mechanism of MMPs involves the binding of the substrate with its scissile carbonyl moiety weakly coordinated to the catalytic Zn(II) ion (Figure 3.3A), followed by nucleophilic attack of the zinc-bound (and glutamate hydrogen-bonded) water molecule (Figure 3.3B) on this carbon atom. The water molecule donates a proton to the carboxylate moiety of Glu 198, which transfers it to the nitrogen atom of the scissile amide bond (Figure 3.3C). Then, the Glu 198 residue shuttles the second remaining proton of the water to the nitrogen of the scissile amide bond, resulting in peptide bond cleavage (Figure 3.3D). During these processes, the Zn(II) ion stabilizes the developing negative charge on the carbon atom of the scissile amide bond, whereas a conserved alanine residue (Ala 161 in MMP-8) helps to stabilize the positive charge at the nitrogen atom of the scissile amide (Lovejoy *et al* 1994; Grams *et al* 1995a,b).

Thus, the main difference between the enzymatic mechanisms of CAs and MMPs is that the nucleophilic adduct is just the reaction product in the case of CAs (bicarbonate), whereas the nucleophilic adduct is a reaction intermediate in the case of the MMPs.

3.2.2 Inhibition mechanism

As for other metallo-enzymes, inhibition of MMPs is correlated with binding of the inhibitor molecule to the catalytic metal ion, with or without substitution of the metal-bound water molecule. Thus, MMP inhibitors (MMPIs) must contain a zinc-binding function attached to a framework that interacts with the binding regions of the protease (Pavlovsky *et al* 1999; Whittaker *et al* 1999). The usual MMPIs of peptidic nature generally belong to the so-called "right-hand-side" inhibitors in reference to the convention of drawing the peptide linkages of a peptide substrate on the right side of the constituent residues (Babine and Bender 1997; Whittaker *et al* 1999). Depending on the zinc-binding function contained in their molecule, MMPIs belong to several chemical classes, such as carboxylates, hydroxamates, thiols, phosphorus-based inhibitors, sulfodiimines, etc. The strongest inhibitors are generally the hydroxamates; 5–10-fold less potent are the "reverse hydroxamates" (of the type HCON(OH)-R), while thiols are 20–50 times less potent, and carboxylates/phosphonates 100–2000 times less potent (Babine and Bender 1997). Many of these inhibitors were derived by replacing the scissile peptide bond with such a zinc-binding function (eventually followed by a methylene moiety) in such a way that the zinc-binding moiety is available for coordination to the catalytic Zn(II) ion.

The interaction of the catalytic domain of several MMPs with some inhibitors has been recently investigated by means of X-ray crystallography, NMR and homology modeling (Figures 3.4 and 3.5) (Grams *et al* 1995a,b; Babine and Bender 1997; Brandstetter *et al* 1998; Graff von Roedern *et al* 1998; Jacobsen *et al* 1999; Kiyama *et al* 1999; Pavlovsky *et al* 1999).

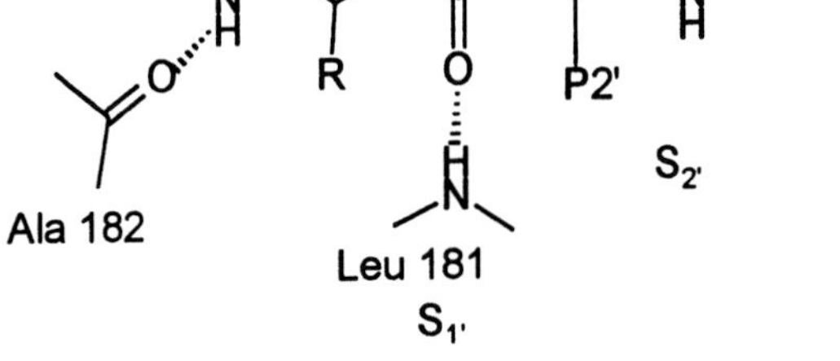

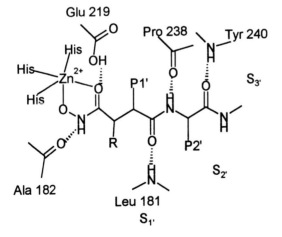

Figure 3.4 Binding of a hydroxamate inhibitor to MMP-7, as determined by X-ray crystallography. The Zn(II)-ligand and hydrogen bond interactions in the enzyme-inhibitor adduct are shown (adapted from Grams *et al* (1995a)).

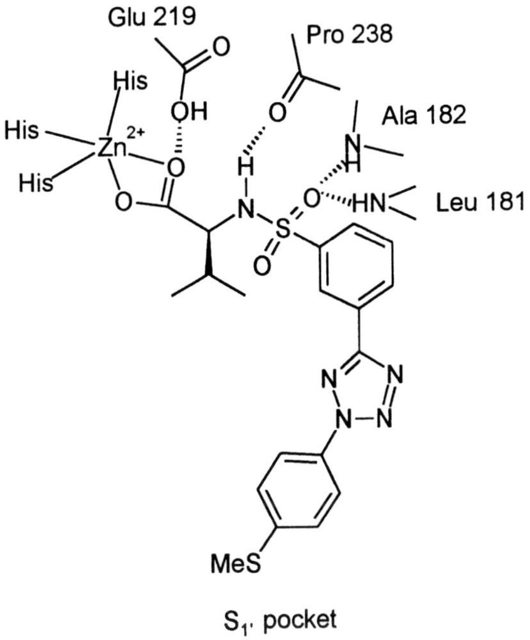

Figure 3.5 Proposed binding of a carboxylate inhibitor to MMP-9, as determined by molecular modeling. The Zn(II)-ligand and hydrogen bond interactions between enzyme and inhibitor are shown (adapted from Kiyama *et al* (1999)).

As seen in the above figures, hydroxamates bind bidentately to the catalytic Zn(II) ion of the enzyme, which in this way acquires a distorted trigonal bipyramidal geometry, as shown by X-ray crystallography (Grams *et al* 1995a,b; Brandstetter *et al* 1998; Graff von Roedern *et al* 1998). The hydroxamate anion forms a short and strong hydrogen bond with the carboxylate moiety of Glu 219 that is oriented towards the unprimed binding regions. The NH hydroxamate also forms a hydrogen bond with the carbonyl oxygen of Ala 182 (Grams *et al* 1995a,b; Brandstetter *et al* 1998; Graff von Roedern *et al* 1998). Thus, several strong interactions are achieved at the zinc site without any significant unfavorable contacts. In the case of the sulfonylated carboxylate inhibitor shown in Figure 3.5, the carboxylate is also coordinated to the Zn(II) ion (in mono- or bidentate fashion, but most probably in monodentate form) and is hydrogen-bonded to the carboxylate moiety of Glu 219, similarly to the hydroxamate-type inhibitors. One of the oxygens belonging to the SO_2 moiety participates in two hydrogen bonds with the main chain amide nitrogens of Leu 181 and Ala 182, whereas the sulfonyl substituent makes extensive hydrophobic contacts with the $S_{1'}$ site (Kiyama *et al* 1999). Such data allows the classification of the MMPs into two main types, depending on their $S_{1'}$ pocket: (1) the deep pocket enzymes (such as MMP-2, MMP-3, MMP-8, MMP-9 and MMP-13), possessing a relatively big $S_{1'}$ pocket; and (2) the shallower pocket enzymes (MMP-1, MMP-7 and MMP-11 among others) which possess a somehow smaller specificity $S_{1'}$ pocket due to its partial occlusion by bulkier amino acid residues, such as that in position 193 (MMP-8 numbering) which from Leu in MMP-8, becomes Arg in MMP-1, Tyr in MMP-7 and Gln in MMP-11 (Whittaker *et al* 1999).

The $S_{2'}$ and $S_{3'}$ subsites are also important for the binding of inhibitors as well as for the specificity of such inhibitors towards different MMPs. This $S_{2'}$ subsite is a solvent-exposed cleft with a general preference for hydrophobic $P_{2'}$ residues, in both substrates and MMPIs. The $S_{3'}$ subsite on the other hand is a relatively ill-defined, solvent exposed region (Grams *et al* 1995a,b; Brandstetter *et al* 1998; Graff von Roedern *et al* 1998; Pavlovsky *et al* 1999).

NMR structural studies of the catalytic domain of MMP-1 seem to indicate that substantial structural changes occur in the active site cleft after binding of inhibitors (Jacobsen *et al* 1999).

3.3 MMP INHIBITORS (MMPIs)

As for many other proteases, the main approach for the identification of synthetic, potent MMPIs was the substrate-based design of peptide-like compounds, derived from information of the amino acid sequence at the cleavage site (Babine and Bender 1997). Both right-hand-side, as well as left-hand-side inhibitors were investigated initially, but as the compounds of the first type acted as much stronger inhibitors (as compared with the other type), they were subsequently the most investigated for different types of pharmacological applications (Whittaker *et al* 1999). Thus, mainly this type of MMPIs will be discussed in detail here, although a few left-hand inhibitors important for drug design are also mentioned.

The tissue inhibitors of metalloproteinases (TIMPs), which are glycoproteins of molecular weight of 21–30 kDa (consisting of 184–194 amino acid residues), are also not dealt with in the present chapter. These endogenous inhibitors react in an 1:1 stoichiometry with MMPs, so regulating their proteolytic activities *in vivo* (Murphy and Willenbrock 1995; Nagase and Woessner 1999). Four homologous TIMPs (TIMP-1–TIMP-4) have been identified to date (Nagase and Woessner 1999). Their conserved amino-terminal Cys residue coordinates to the catalytic Zn(II) ion of MMPs, bidentately, by means of the terminal H_2N moiety and carbonyl oxygen, inhibiting thus the strong proteolytic activity of these enzymes (Nagase and Woessner 1999).

3.3.1 Carboxylic acids

Although carboxylic acids are among the weakest MMPIs, interest in this class of compounds has been revived recently due to two reasons: (1) the report that some carboxylates act as relatively selective inhibitors for the deep pocket enzymes (MMP-2, MMP-3, MMP-8, MMP-9 and MMP-13), as compared with the analogous hydroxamates and; (2) the good oral bioavailability of such derivatives, and their lack of toxicity as compared to other MMP ligands (Whittaker *et al* 1999).

Some of the most active MMPIs of this class, such as (**3.1–3.4**), (**3.7**) and (**3.8–3.10**) possess elongated, relatively bulky and hydrophobic $P_{1'}$ moieties (such as phenethyl, substituted phenethyls, substituted biphenyls, 1,4-diphenyl-tetrazole, diphenylpiperidine, etc.) which make extensive hydrophobic contacts with the $S_{1'}$ subsite. Since the other MMPs possess a much smaller $S_{1'}$ pocket, these compounds generally show a good selectivity for MMP-2, MMP-3 and MMP-9 over MMP-1. Typically, their K_I values are >1000 nM against the latter enzyme, being of around 1.1–310 nM against the deep pocket

MMPs (Whittaker *et al* 1999). Compounds of types (**3.1–3.4**) showed interesting *in vivo* activities after injection into the murine pleural cavity, but generally showed little benefit after oral dosing, except for (**3.4**), which is a potent MMP-3 inhibitor with an oral bioavailability of 78% in the mouse and an ED_{50} of 4.7 mg/kg i.v., and 11 mg/kg p.o., in the arthritis model mentioned above (Whittaker *et al* 1999).

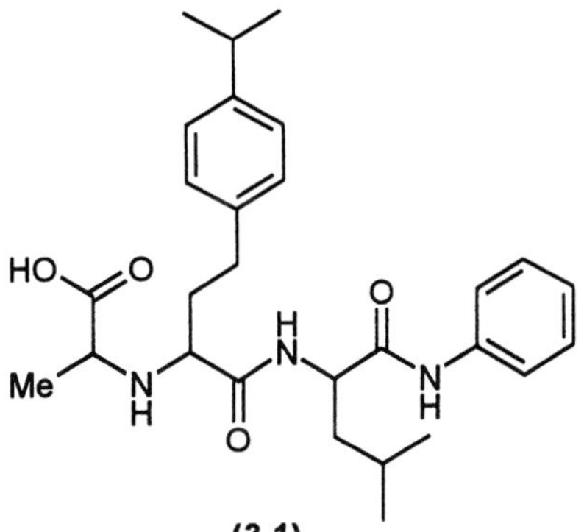

(3.1)

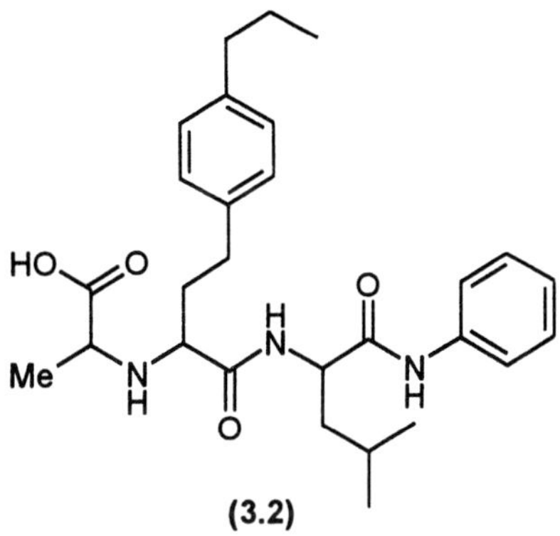

(3.2)

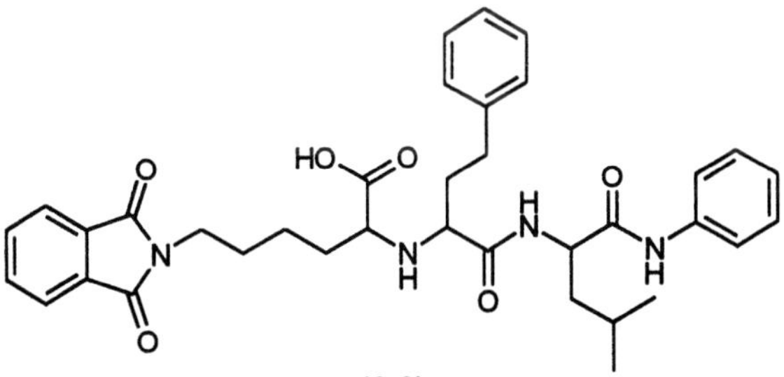

(3.3)

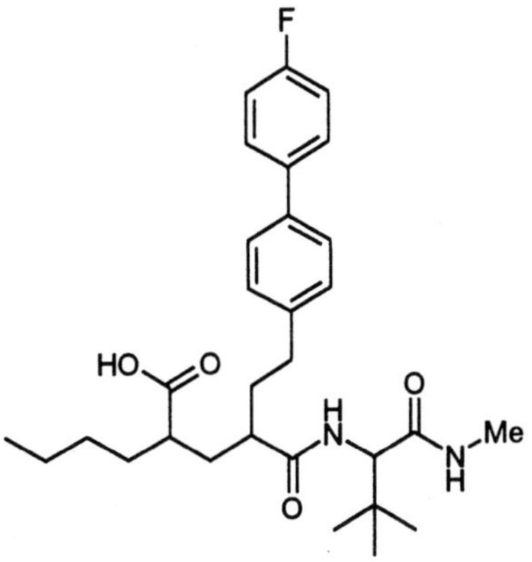

(3.4)

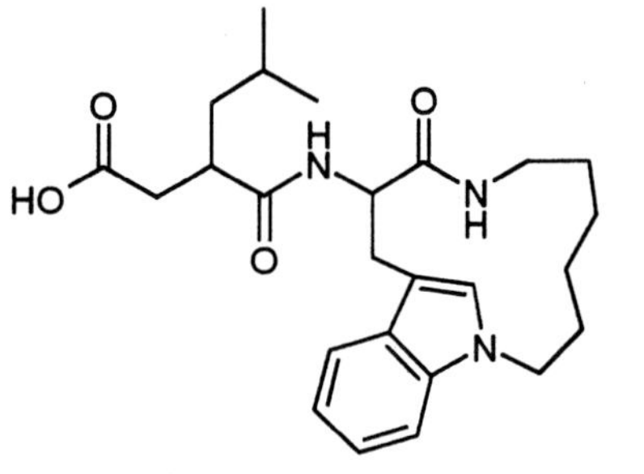

(3.5)

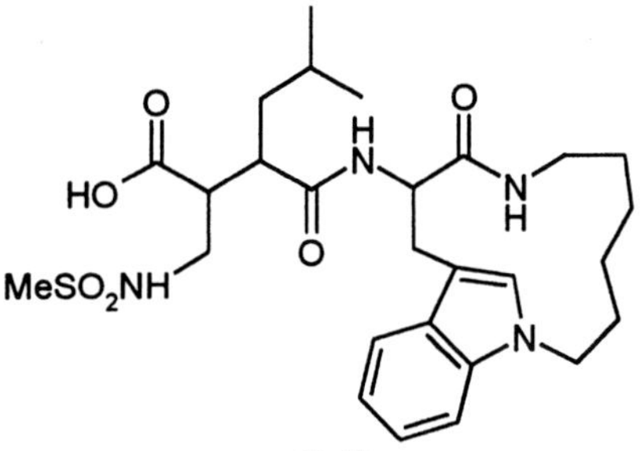

(3.6)

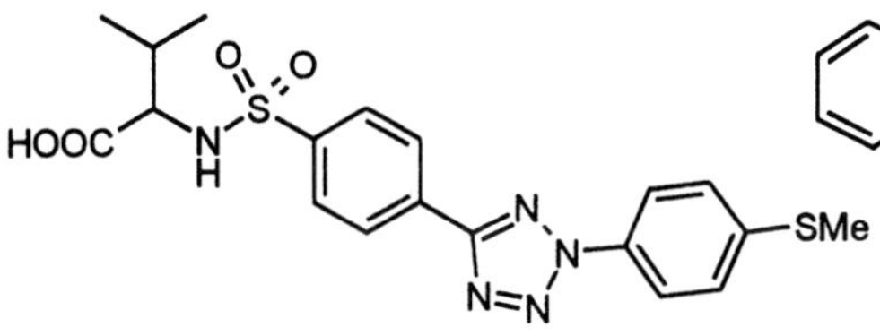

(3.7)

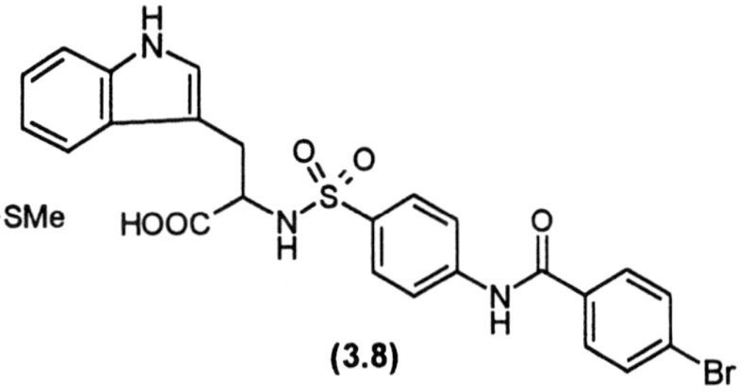

(3.8)

	K_I-s(nM)			
	MMP-1	MMP-2	MMP-3	MMP-9
(3.1)	>10000	130	180	
(3.2)	>10000	310	68	68
(3.3)	720	86	8	–
(3.4)	>10000	22	17	10
(3.7)	>1000	1.1	>1000	5.8
(3.8)	3200	1.5	12	6.5

(–) means that the compound was not tested.

Compounds (**3.5**) and (**3.6**) are important as they were studied in detail by X-ray crystallography, in order to examine the differences in binding with the corresponding hydroxamates, to MMP-7. It was thus established that several factors contribute to the loss of energy of 3–5 kcal/mol on the binding of carboxylates (**3.5**) and (**3.6**), compared with the corresponding hydroxamates: (1) the difference of acidity of the two binding functions. Thus, the binding of hydroxamates is pH independent in the pH range of 5–8, whereas carboxylates bind much tighter as the pH is lowered. There is also at least 3 pH units difference between the pK_a of the hydroxamic and carboxylic acid moieties (typically with a pK_a of 4.5–5.5 for the COOH, and 7.5–8.5 for the CONHOH moieties); (2) the loss of the hydrogen bond of the hydroxamic NH group to the carbonyl oxygen of Ala 182 (Babine and Bender 1997).

For the adducts of MMP-9 with the sulfonamide inhibitors of types (**3.7**) and (**3.8**), besides the Zn(II) ion coordination by means of the COOH moiety, and the extensive hydrophobic contacts of the $S_{1'}$ pocket with the bulky, elongated diphenyl-tetrazole/*p*-bromophenylcarboxamidophenyl moieties, the formation of two strong hydrogen bonds between one of the sulfonamido oxygen atoms and the backbone NH groups of Leu 181 and Ala 182, seems to be responsible for the strong inhibitory properties of these compounds. Thus, many sulfonyl-amino acid derivatives were recently investigated for their interaction with different MMPs (Kiyama *et al* 1999).

Some potent sulfonamide stromelysin (MMP-3) inhibitors of types (**3.9**) and (**3.10**) were recently reported and investigated by means of X-ray crystallography (Jacobsen *et al* 1999).

Inhibitors of types (**3.9**) and (**3.10**) bind within the active site of MMP-3 so that the bulky diphenylpiperidine moiety penetrates into the deep, predominantly hydrophobic $S_{1'}$ subsite.

The above-mentioned data provides the structural basis for explaining the selectivity of some of the carboxylate MMPIs for different MMPs, and also leads to the design of compounds with improved inhibition profiles (Whittaker *et al* 1999).

3.3.2 Hydroxamates

Hydroxamates are by far the most investigated class of MMPIs. Thousands of structural variants containing the CONHOH moiety have been synthesized and assayed as inhibitors of MMPs and other types of metallo-enzymes. Three main classes of such MMPIs have been reported: (1) the succinyl hydroxamates (and their derivatives); (2) the malonic acid-based inhibitors and; (3) the sulfonamide-based inhibitors.

3.3.2.1 *Succinyl hydroxamates*

Early studies in this field (Johnson *et al* 1987) showed that succinyl hydroxamic acid derivatives such as (**3.11**) act as much stronger MMPIs as compared to their homologues, derived from malonic or glutaric acid hydroxamates. Thus, most studies have been concentrated (until recently) on the succinyl hydroxamate derivatives. It was thus observed that the presence of a P_1 substituent (α to the hydroxamate moiety) in this type of

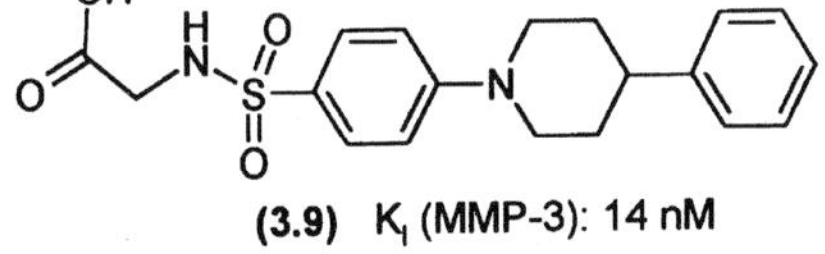

(3.9) K_I (MMP-3): 14 nM

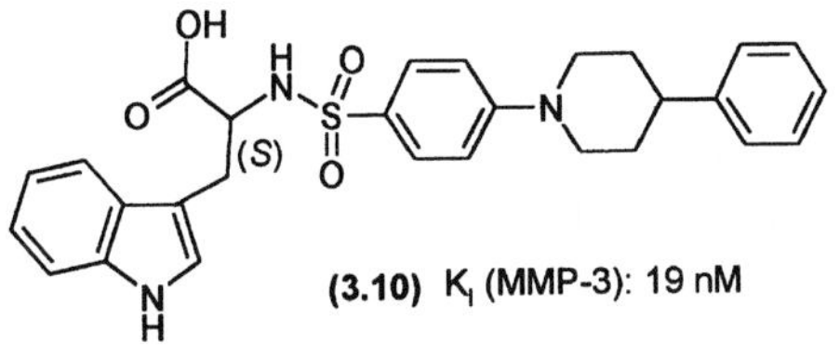

(3.10) K_I (MMP-3): 19 nM

compound confers broad spectrum activity against a variety of MMPs (Bottomley *et al* 1998; Whittaker *et al* 1999). Thus, two important MMPIs, batimastat (**3.12**) and marimastat (**3.13**) were discovered by scientists from the British Biotech Pharmaceuticals (Whittaker *et al* 1999). These are two of the compounds which have entered advanced clinical studies and might be soon licensed as the first drugs from the MMPIs family. Batimastat possesses a thienylthiomethylene α substituent, whereas in marimastat this is an OH group. These compounds showed very good *in vivo* activities in several disease models (see later), but batimastat is not orally bioavailable, whereas marimastat is, probably due to the increased water solubility induced by the presence of the hydrophilic OH moiety. Different compounds were then obtained considering marimastat as lead molecule, such as (**3.14–3.16**), and although they showed good MMP inhibitory properties, no pharmacological benefits as compared with the parent compound were seen (Whittaker *et al* 1999).

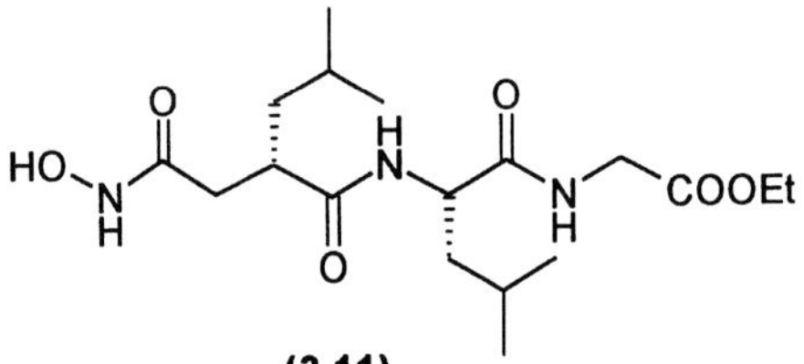

(3.11)

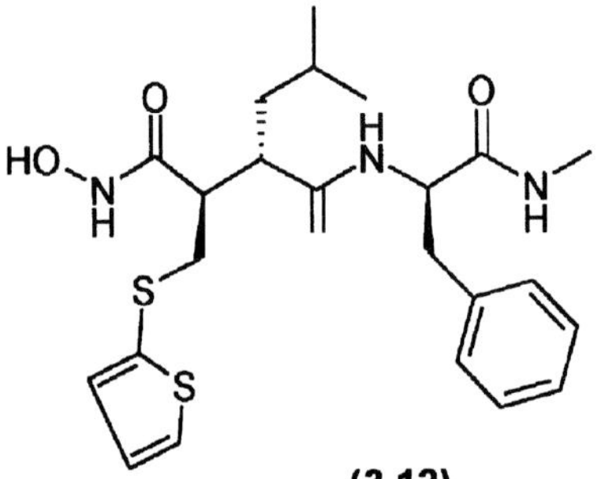

(3.12)

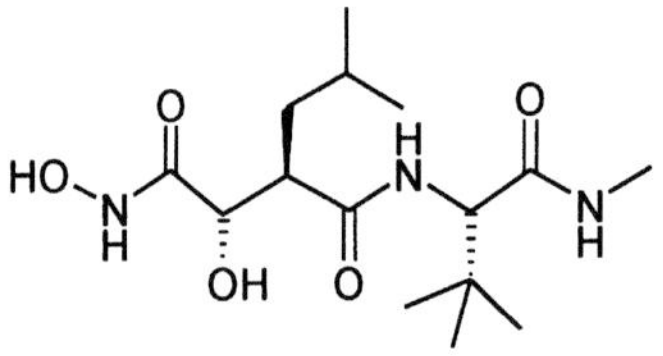

(3.13)

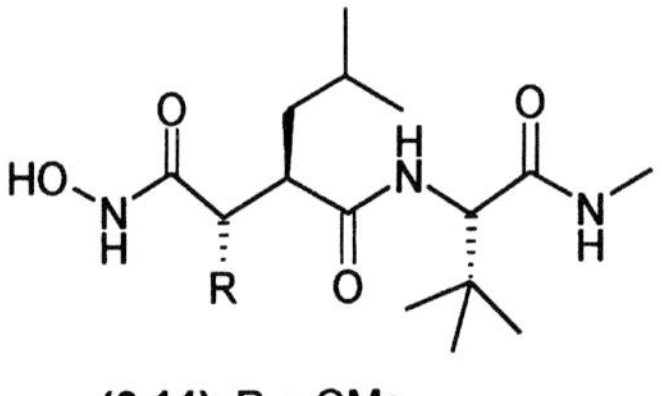

(3.14): R = OMe
(3.15): R = cyclopentyl
(3.16): R = thienyl-2-sulfonyl

	IC_{50}(nM)					
	MMP-1	MMP-2	MMP-3	MMP-8	MMP-9	MMP-14
(3.11)	40	–	–	–	–	–
(3.12)	10	4	20	10	1	3
(3.13)	5	6	200	2	3	1.8
(3.14)	6	30	40	–	–	–
(3.15)	4	3	30	20	9	–
(3.16)	2	20	30	–	7	–

(–) means that the compound was not tested.

Further developments in this field involved variations of the α substituent and the $P_{1'}$–$P_{3'}$ moieties in order to obtain stronger or/and more selective inhibitors. Thus, the synthesis and assay of a large number of succinyl hydroxamates led to the following SAR conclusions. (1) The presence of an α substituent leads to an increase in activity against MMP-1 and MMP-3. Certain substituents, together with truncation of the $P_{2'}/P_{3'}$ provide good MMP-1, MMP-8 and MMP-13 inhibitors; (2) the $P_{1'}$ moiety is the major determinant of activity and selectivity of MMPIs. Thus, small alkyl groups are preferred for good MMP-1 activity. Longer alkyl (C_9H_{19}, $C_{10}H_{21}$, etc.) and phenalkyl (phenethyl, phenylpropyl, phenethyloxybutyl, biphenyl-propyl, etc.) chains provide selectivity over MMP-1 and MMP-7, whereas charged and polar groups are not tolerated in this position. (3) In the $P_{2'}$ position, a wide range of substituents are tolerated, with preference for aromatic moieties. Steric bulk close to amides, such as in derivatives (**3.13–3.16**) is beneficial for the oral bioavailability of these compounds. It has also been observed that the α substituent and the $P_{2'}$ moiety can be incorporated in a cyclic structure; (4) also in the $P_{3'}$ position a wide range of substituents is tolerated, with aromatic moieties improving activity against MMP-3. Charged groups in this position may also affect the biliary excretion of these pharmacological agents, which is deleterious for their clinical usefulness. Some of the most active MMPIs designed taking account of the above-mentioned facts, of types (**3.17–3.21**) are shown below together with their inhibition data (Whittaker *et al* 1999).

Compound (**3.21**), Ro 32-3555 or Trocade, a very strong collagenase inhibitor, was selected for development as an anti-arthritis agent (Bottomley *et al* 1998). For this compound, a favorable balance between active-site interactions and solvation is probably optimally achieved, despite the removal of three hydrogen-bonding groups present in the succinyl hydroxamates possessing $P_{2'}$ moieties. The presence of the cyclic imide moiety in the P_1 position appears to be important for activity too, and might compensate for the loss of the hydrogen bonds mentioned above. Trocade selectively inhibits all three members of the collagenase family (MMP-1, MMP-8 and MMP-13) over MMP-2, MMP-3 and MMP-9. It was also shown that this compound possessed good oral bioavailability in rats, acting as a cartilage protecting agent and provided a promising therapy for osteo- and rheumatoid arthritis (see Section 3.5) (Bottomley *et al* 1998).

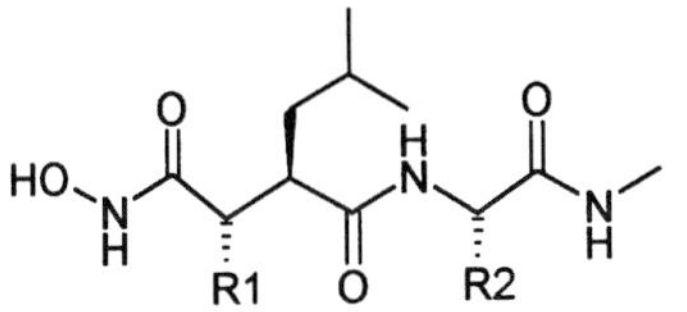

(3.17): R1 = H, R2 =

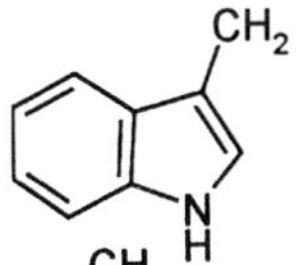

(3.18): R1 = Me; R2 =

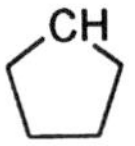

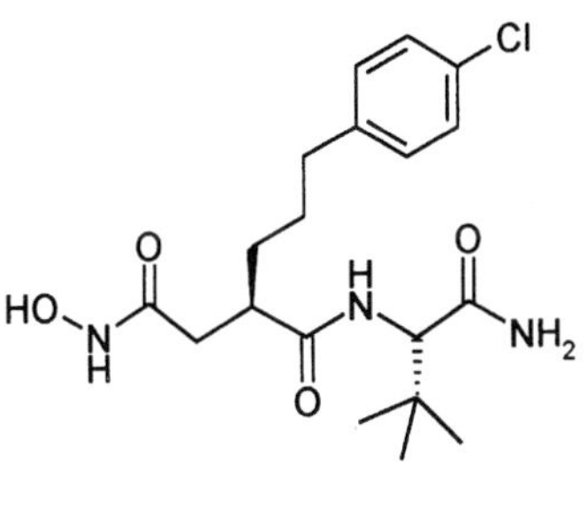

(3.19)

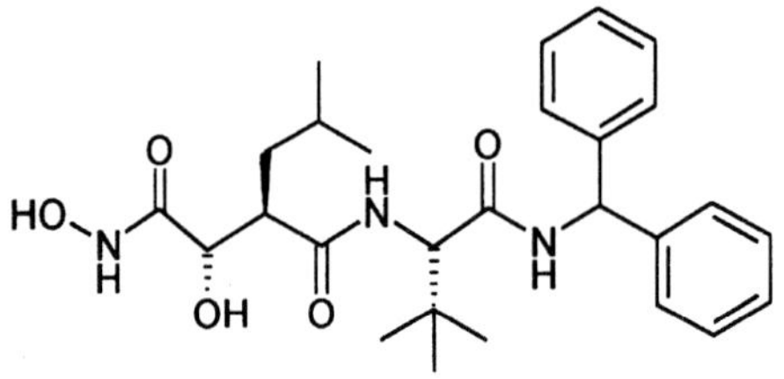

(3.20)

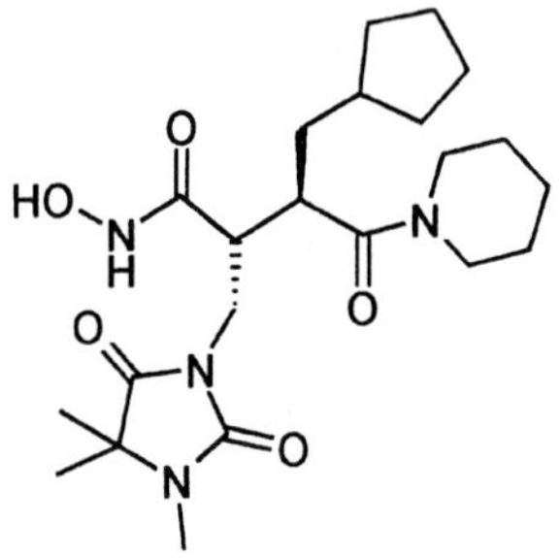

(3.21)

	K_I(nM)					
	MMP-1	MMP-2	MMP-3	MMP-8	MMP-9	MMP-14
(3.17)	0.4	0.39	26	0.18	0.57	–
(3.18)	5.4	8.4	2.3	–	5	2.3
(3.19)	122	0.04	11	–	0.17	–
(3.20)	600	3000	50	MMP-7:4	–	–
(3.21)	7	154	527	4	59	MMP-13:3

(–) means that the compound was not tested.

3.3.2.2 *Malonic acid-based hydroxamates*

X-ray crystallographic experiments showed that the malonic acid hydroxamate (**3.22**), HONHCOCH(iso-Bu)CO-L-Ala-Gly-NH_2, unexpectedly binds in a different manner than anticipated from its design and the binding of the "normal" hydroxamates (which interact in a substrate-like mode with the MMP active site) (Brandstetter *et al* 1998). The hydroxamate moiety of this inhibitor is bidentately coordinated to the catalytic Zn(II) ion but its isobutyl chain remains outside the $S_{1'}$ pocket presumably due to severe constraints imposed by the adjacent planar hydroxamate group. The C-terminal Ala-GlyNH_2 moiety adopts a bent conformation, being inserted into the $S_{1'}$ pocket. Thus, this type of inhibitor, with a nonsubstrate-like binding to the active site, represented a new interesting lead for obtaining malonic acid-based MMPIs by Moroder's and Tschesche's groups (Graff von Roedern *et al* 1998; Krumme *et al* 1998). Derivatives of types (**3.23–3.28**) showing enhanced MMP-8 inhibitory properties as compared with the lead (**3.22**) were thus prepared. Other structural variants of types (**3.29–3.31**) were also obtained and assayed

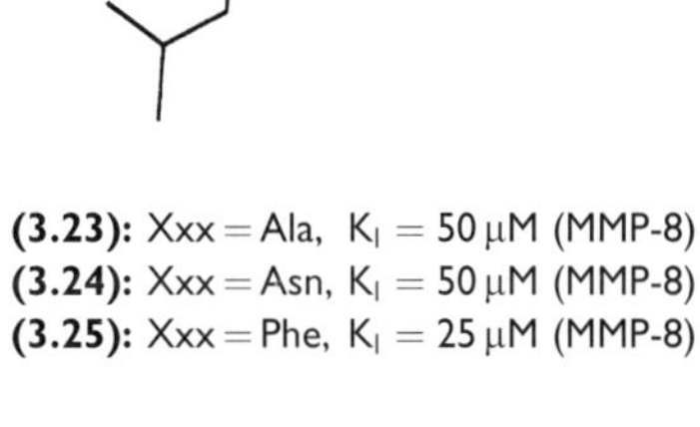

(3.23): Xxx = Ala, K_I = 50 μM (MMP-8)
(3.24): Xxx = Asn, K_I = 50 μM (MMP-8)
(3.25): Xxx = Phe, K_I = 25 μM (MMP-8)

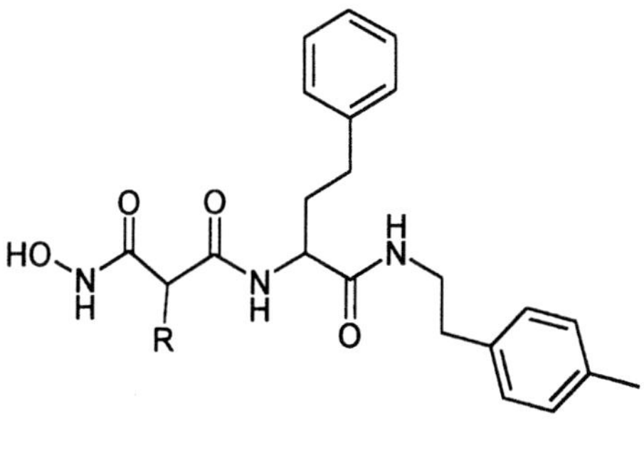

(3.26): R = i-Bu; K_I = 1.4 μM (MMP-8)
(3.27): R = CH_2Ph; K_I = 1.7 μM (MMP-8)
(3.28): R = OH; K_I = 1.6 μM (MMP-8)

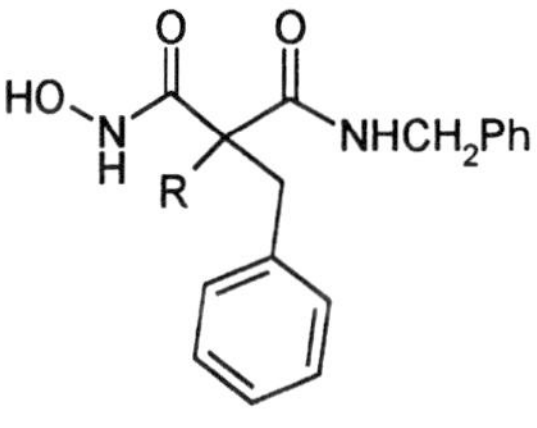

(3.29): R = NHAc; K_I = 2.3 μM (MMP-8)
(3.30): R = OH; K_I = 1.9 μM (MMP-8)

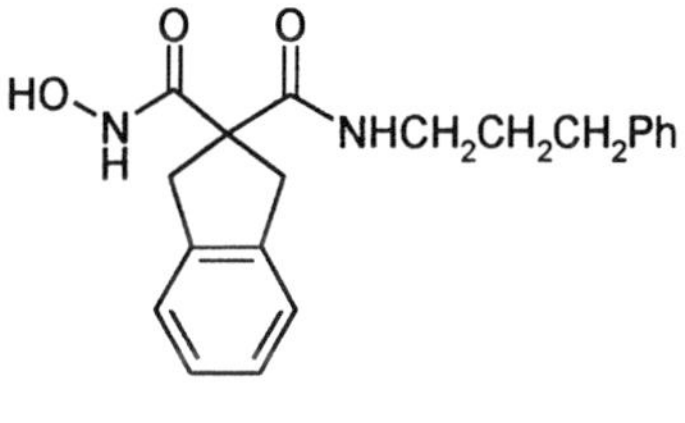

(3.31): K_I = 0.30 μM (MMP-8)

as collagenase inhibitors. Although this class of inhibitors is less potent than the succinyl hydroxamates discussed earlier, the design of novel and very active compounds of this type might lead in the future to effective MMPIs based on the malonic/amino malonic acids (the drug design process for this class of MMPIs has not yet been fully optimized) (Graff von Roedern *et al* 1998; Krumme *et al* 1998).

Some very effective MMP-9 inhibitors were recently reported by Tschesche's group (Krumme *et al* 1998). They contain the sequence -Pro-Leu-Ama(NHOH) – (where Ama represents aminomalonic acid) in the P_3, P_2 and P_1 positions, respectively, and a bulky benzyl-oxyphenyl moiety in the $P_{1'}$ subsite. Some of these compounds, of types (**3.32–3.34**) and their MMP inhibitory properties are shown below.

These inhibitors are characterized by a much higher affinity (>1000-fold) for MMP-9 over MMP-8, and might lead to the design of isozyme-specific MMPIs.

3.3.2.3 *Sulfonamide hydroxamates*

Sulfonylated amino acid hydroxamates were recently discovered to act as efficient MMPIs (MacPherson *et al* 1997; Whittaker *et al* 1999). The first compounds from this class to be developed for clinical trials, of types (**3.35**) (CGS 27023A) and (**3.36**) (CGS 25966), possess the following structural features: (1) an isopropyl substituent α to the hydroxamic acid moiety, considered to slow down metabolism of the zinc-binding function. It probably

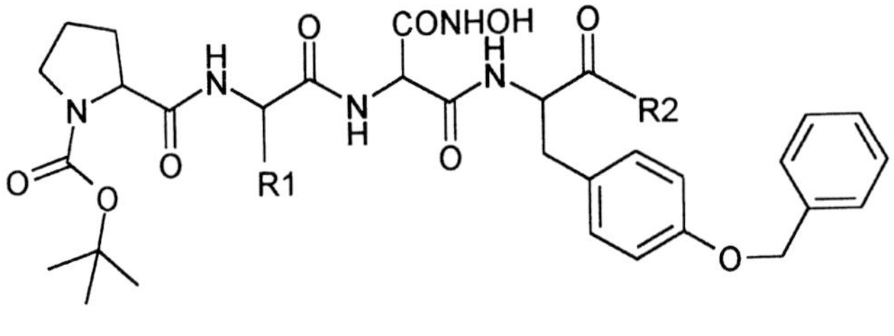

(3.32): R1 = Me; R2 = NHCH(Ph)Me
(3.33): R1 = i-Bu; R2 = NHCH(Ph)Me
(3.34): R1 = i-Bu; R2 = NHCH(c-C_6H_{11})Me

	K_I MMP-9 (nM)	MMP-8 (μM)
(3.32)	5	1.9
(3.33)	5	0.8
(3.34)	18	2.5

binds within the S_1 subsite; (2) a bulkier pyridylmethyl or benzyl moiety substituting the amino nitrogen atom and probably binding within the $S_{2'}$ pocket; (3) the arylsulfonyl group occupies (but does not fill!) the specificity $S_{1'}$ pocket (Whittaker *et al* 1999). Generally, for reasons poorly explained as yet, this group is 4-methoxybenzenesulfonyl in many of the clinically investigated derivatives. CGS 27023A is a potent inhibitor of MMP-12, an enzyme that seems to be implicated in the development of emphysema that results from chronic inhalation of cigarette smoke (Jeng *et al* 1998).

Further developments in this field involved changes of the groups substituting the α carbon atom, the length of the chain between the hydroxamate and amino groups (from one to two carbon atoms), the usual variations of the $P_{1'}-P_{3'}$ moieties, as well as replacing the arylsulfonamido moiety by an arylsulfone one (such as in (**3.39**)) (Whittaker *et al* 1999; Jeng *et al* 1998; Hanessian *et al* 1999). Some strong and relatively selective inhibitors of this type, such as (**3.37–3.39**), and their MMP inhibitory properties are shown below. Some of these inhibitors, such as (**3.36**), (**3.37**) or (**3.39**) (which are orally bioavailable) were chosen for further development as anti-metastasis (Whittaker *et al* 1999) or anti-arthritis drugs (Bottomley *et al* 1998) (see Section 3.5).

Recently, a large number of arylsulfonyl hydroxamates derived from glycine, L-alanine, L-valine, and L-leucine, possessing *N*-benzyl- or *N*-benzyl-substituted moieties, of the types (**3.40–3.43**), with nanomolar affinities for MMP-1, MMP-2, MMP-8 and MMP-9, were reported (Scozzafava and Supuran 2000c; Clare *et al* 2001). The most important parameters influencing activity in these classes of MMPIs are: (1) the bulky group substituting the amino acid moiety (with nitrobenzyl derivatives more active than chlorobenzyl derivatives, which in turn are more active than the unsubstituted benzyl derivatives); (2) the nature of the alkyl/arylsulfonyl moieties. Aromatic derivatives were generally much more active than the aliphatic ones, except for the perfluorobutyl and perfluorooctylsulfonyl compounds which showed very good inhibitory effects, similar to those of simple aromatic compounds incorporating substituted-phenyl moieties (such as *p*-methoxyphenyl, *p*-aminophenyl, *p*-halogenophenylsulfonyl, etc.). The most promising

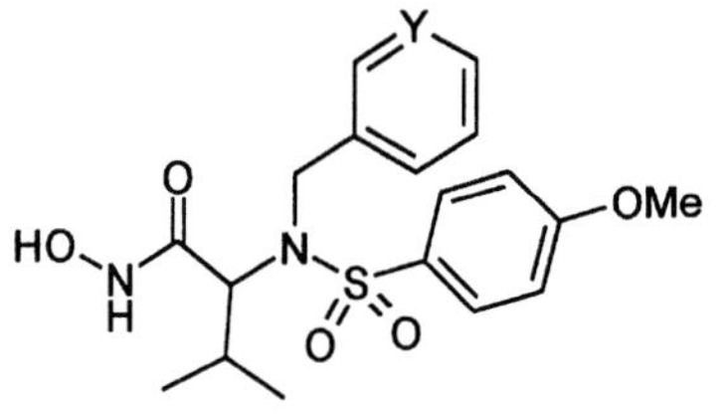

(3.35): Y = N
(3.36): Y = CH

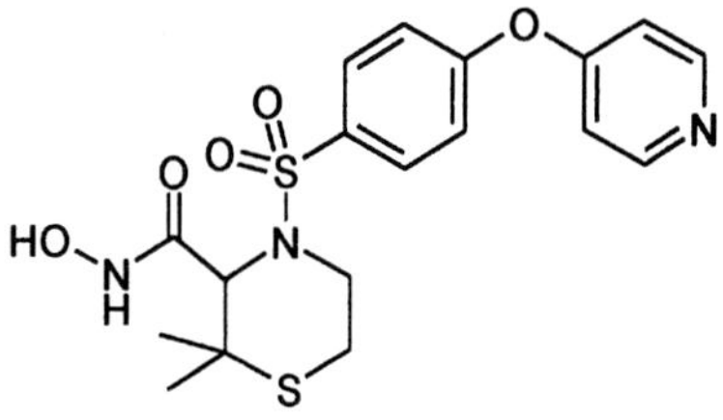

(3.37)

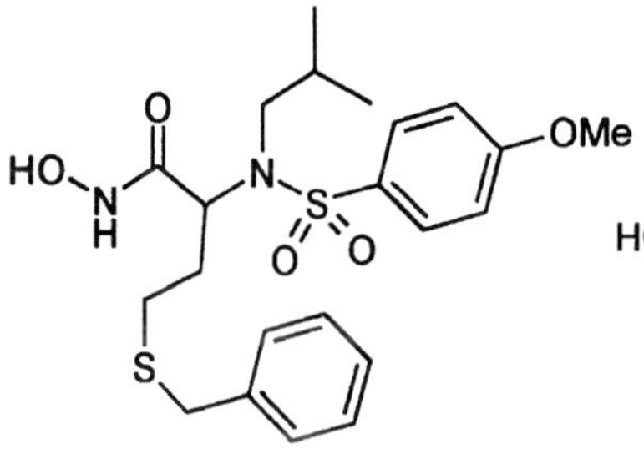

(3.38)

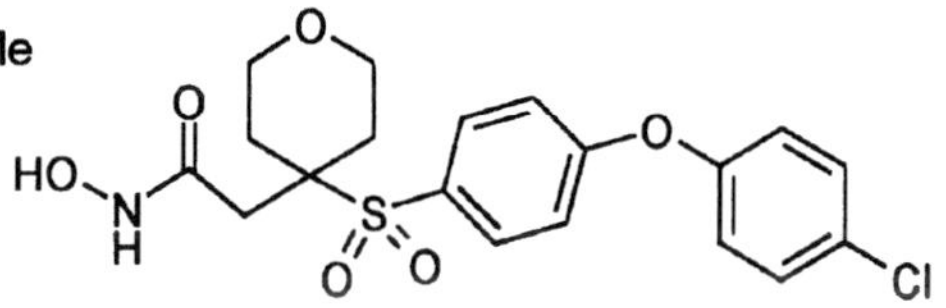

(3.39)

	K_I(nM)					
	MMP-1	MMP-2	MMP-3	MMP-7	MMP-9	MMP-13
(3.35)*	33	20	43	–	8	–
(3.36)	–	–	92	–	–	–
(3.37)	8	0.08	0.27	54	–	0.038
(3.38)	104	0.7	0.7	–	2.5	12
(3.39)	70	0.054	5.2	240	0.065	0.17

*100 nM of **35** produced a 80% inhibition of MMP-12 (Jeng *et al* 1998)
(–) means that the compound was not tested.

aromatic substitutions were those including perfluorophenylsulfonyl or 3-trifluoromethylbenzene-sulfonyl among others (Scozzafava and Supuran 2000c; Clare *et al* 2001).

3.3.3 Thiols, 1,3,4-thiadiazoles and phosphorus ligands

Although the mercapto group generally possesses good coordinating properties towards heavy metal ions, included Zn(II), the initially investigated thiol MMPIs (such as (**3.44**)) acted rather inefficiently when compared with the hydroxamates (Babine and Bender 1997; Bottomley *et al* 1998).

The SH moiety of such inhibitors is monodentately coordinated to the catalytic Zn(II) ion of the enzyme, as shown by X-ray crystallography on adducts of diverse MMPs with

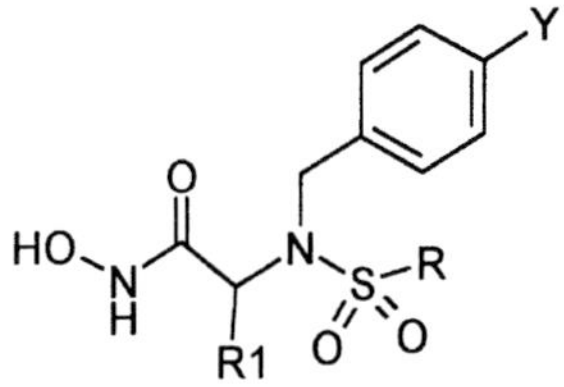

Y = H, Cl, NO_2
R = alkyl, aryl

(3.40): R1 = H (Gly derivatives)
(3.41): R1 = Me (Ala derivatives)
(3.42): R1 = i-Pr (Val derivatives)
(3.43): R1 = i-Bu (Leu derivatives)

	Y	R	R'	K_I (nM)			
				MMP-1	MMP-2	MMP-8	MMP-9
(3.40a)	H	n-C_4F_9	H	75	12	120	8.1
(3.40b)	H	n-C_8F_{17}	H	98	2.7	8.6	5.1
(3.40c)	H	C_6F_5	H	8.5	1.6	5.4	3.2
(3.40d)	NO_2	n-C_4F_9	H	62	1.5	2.4	2.0
(3.40e)	NO_2	n-C_8F_{17}	H	79	0.9	1.3	1.3
(3.40f)	NO_2	C_6F_5	H	3.0	0.7	0.1	0.6
(3.40g)	NO_2	3-$CF_3C_6H_4$	H	5.2	1.1	0.7	0.8
(3.41a)	NO_2	C_6F_5	Me	3.1	0.6	0.1	0.7
(3.42a)	NO_2	C_6F_5	i-Pr	3.0	0.5	0.2	0.6
(3.43a)	NO_2	C_6F_5	i-Bu	2.5	0.5	0.1	0.5

thiol inhibitors. The interesting discovery was that the binding geometry of the side-chains of thiol inhibitors was quite similar to that of the substrate-like hydroxamates (possessing the same substitution pattern), so that much SAR information could be derived considering the thoroughly investigated hydroxamate MMPIs (Grams *et al* 1995a,b). Some stronger inhibitors, such as (**3.45–3.47**) were then developed, which possessed moieties present in strong hydroxamate inhibitors (such as the Trocade imide moiety in (**3.46**) – D2163, for example – and inhibited different MMPs in the nanomolar range. D2163, recently entered clinical development as an anticancer drug (Whittaker *et al* 1999). It is a strong MMP-8 and MMP-13 inhibitor, and possesses oral activity in a rat model of cancer. Except for the simple thiols of type (**3.44**) and (**3.45**), other structural variants, such as acyl-thiols; hydroxy-thiols (compound (**3.47**)) or aminothiols were also investigated (Grams *et al* 1995a,b).

Recently, a novel type of selective stromelysin (MMP-3) inhibitor, derived from 5-amino-2-mercapto-1,3,4-thiadiazole, was reported and its binding to the enzyme investigated by means of X-ray crystallography (Finzel *et al* 1998; Jacobsen *et al* 1999). Compounds such as (**3.48**) and (**3.49**) are coordinated monodentately to the Zn(II) ion of the enzyme through the endocyclic sulfur atom whereas the remainder of the ligand extends into the S_1–S_3 subsites of the enzyme. The fluorine-containing inhibitor (**3.49**) binds particularly strongly probably due to a strong, coplanar interaction between the perfluorophenyl ring and the aromatic ring of Tyr 155 of the enzyme active site. This class of

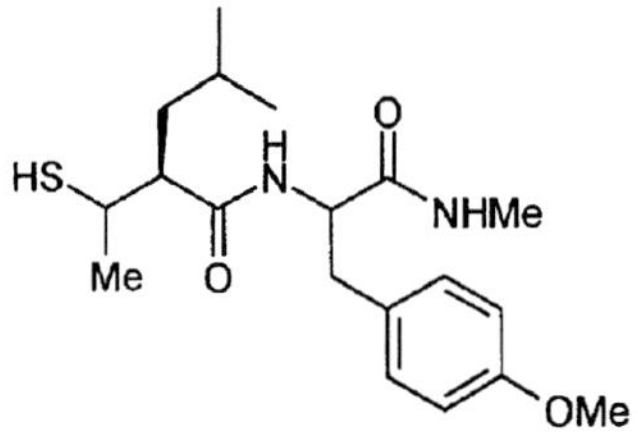

(3.44)

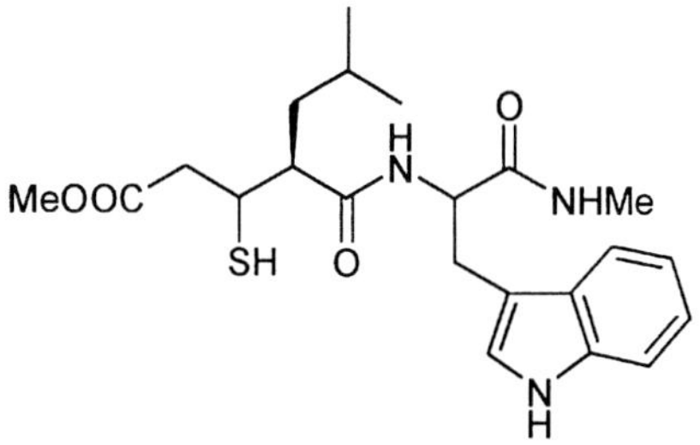

(3.45)

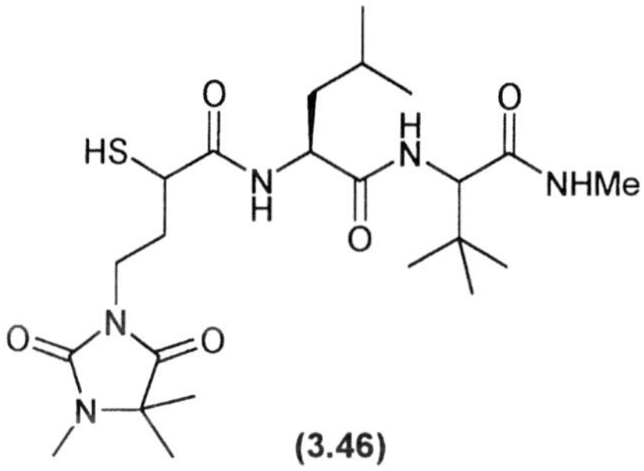

(3.46)

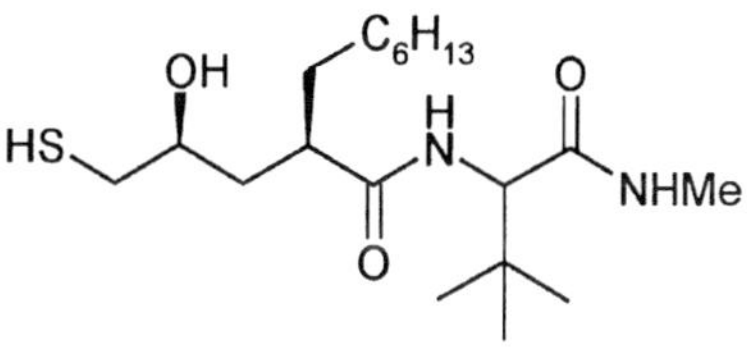

(3.47)

	K_i(nM)					
	MMP-1	MMP-2	MMP-3	MMP-8	MMP-9	MMP-13
(3.44)	220	–	–	–	–	–
(3.45)	2.5	–	–	–	–	–
(3.46)	25	41	157	10	25	4
(3.47)	5	–	9	–	0.14	–

(–) means that the compound was not tested.

inhibitor, which does not inhibit MMP-1 and weakly inhibits MMP-2, has not yet been optimized, and it is probable that important future developments will emerge, for obtaining selective and stronger MMP-3 inhibitors (Jacobsen *et al* 1999).

Another class of thoroughly investigated MMPIs is constituted by the phosphorus-based ligands (Bottomley *et al* 1998). In contrast to other inhibitors discussed here, the phosphorus-based derivatives act as transition-state mimics, and incorporate both left-hand-side as well as right-hand-side moieties in their molecules. Phosphinic acids and phosphonic acids, such as compounds (**3.50–3.53**) were among the most investigated MMPIs, leading in some cases to powerful inhibitors, though so far none seem effective enough to have reached the clinic (Whittaker *et al* 1999).

Phosphorus-based ligands coordinate to the catalytic Zn(II) ion by means of one, two or three oxygen atoms (phosphorus oxygen atoms), in mono- or bidentate fashion (Whittaker *et al* 1999). The phthalamidobutyl group in derivative (**3.50**) was shown to bind in the S_1–S_3 subsites of MMP-3, whereas its phenethyl moiety was bound in the $S_{1'}$ pocket.

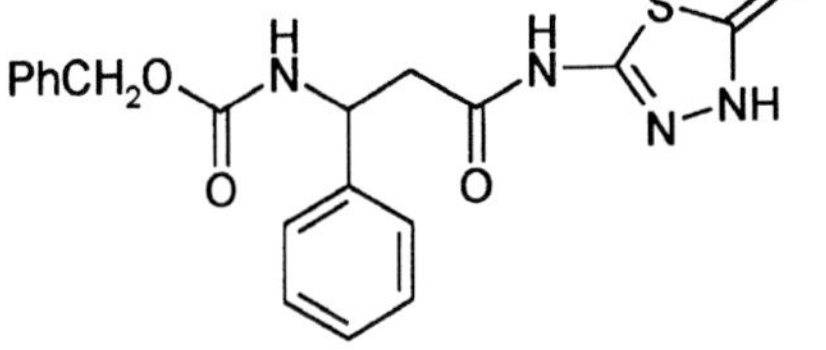

(3.48)

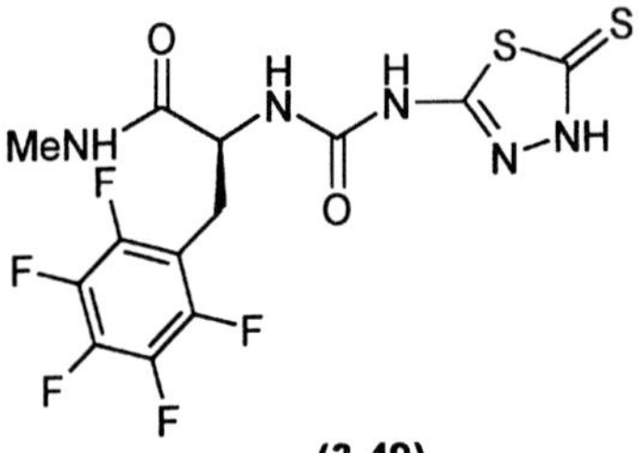

(3.49)

		K_I(nM)	
	MMP-1	MMP-2	MMP-3
(3.48)	inactive	49500	310
(3.49)	inactive	3000	18

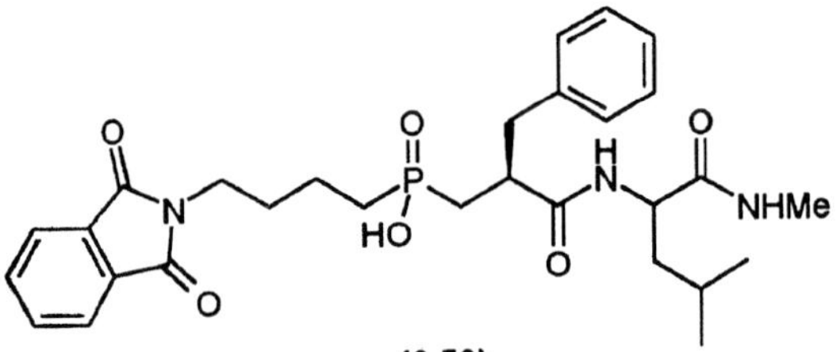

(3.50)

This compound inhibited MMP-2 and MMP-3, but was ineffective against MMP-1. The introduction of the arylthiomethylene phosphinic acid as the zinc-binding function, such as in compound (**3.53**) led to improved efficiency in the inhibition of MMP-1, MMP-2 and MMP-8, whereas the affinity for MMP-3 remained low. Other structural variants included phosphonic acids of types (**3.51**) and (**3.52**), which showed relatively good inhibitory properties against MMP-1 (Whittaker *et al* 1999).

An under investigated class of MMPIs is represented by the sulfodiimines (Babine and Bender 1997), such as (**3.54**). This inhibitor is coordinated in a monodentate fashion to the zinc ion with an NH = moiety bound to Zn(II) (a 2.0 Å bond) (the so-called "outside" NH moiety, which is the opposite one to the isobutyl group) and the other, "inner" imino group (on the same side of the molecule as the isobutyl group) bonding to Glu 219. Being monodentate ligands, sulfodiimines are generally weaker inhibitors as compared to the hydroxamates.

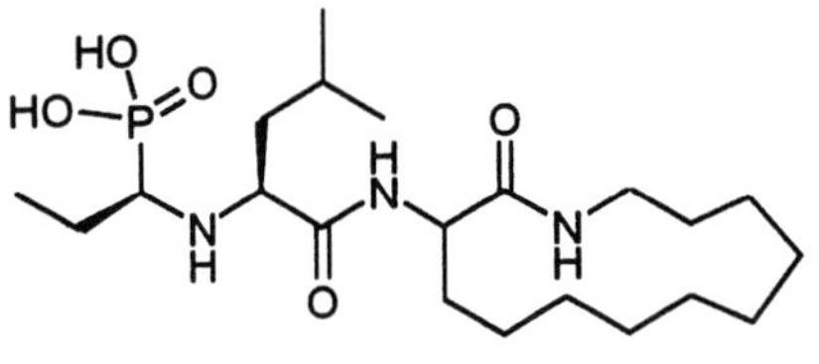

(3.51)

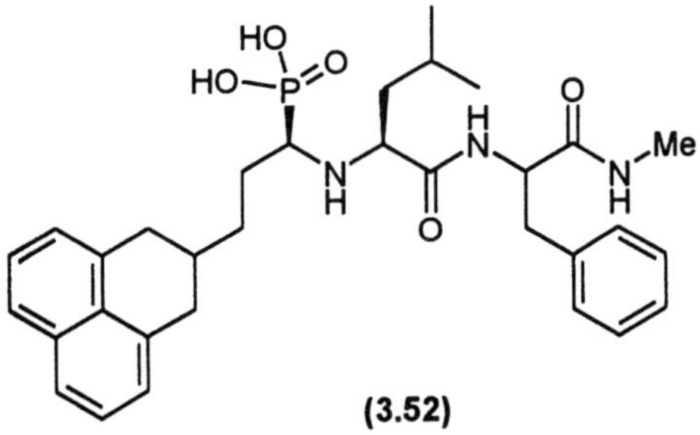

(3.52)

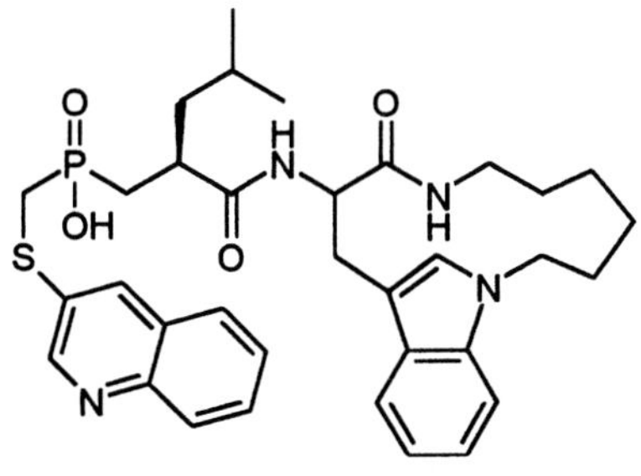

(3.53)

	K_I(nM)				
	MMP-1	MMP-2	MMP-3	MMP-7	MMP-8
(3.50)	>10000	20	1.4	–	–
(3.51)	180	–	–	–	–
(3.52)	20	–	–	–	–
(3.53)	5	14	806	78	4

(–) means that the compound was not tested.

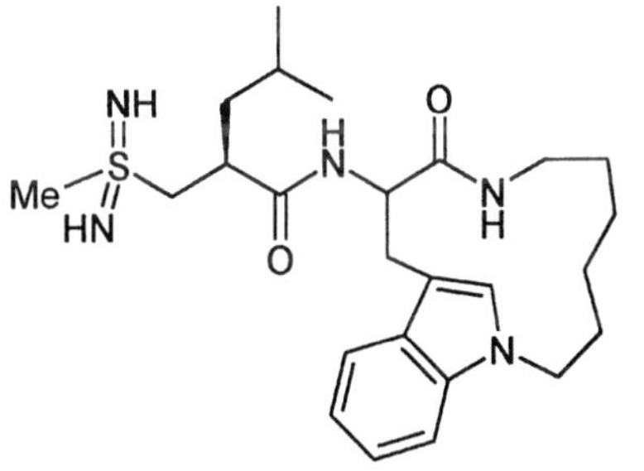

(3.54)

It is clear from all this data that, except for the hydroxamates which were extensively investigated, the other classes of MMPIs have generally not been optimized, and further synthetic, crystallographic and modeling work is necessary for a better understanding of the factors governing SAR in these important classes of putative pharmacological agents.

3.4 BACTERIAL COLLAGENASE INHIBITORS

As shown in the preceding sections, the MMPIs were extensively studied over a period of 15 years in order to discover pharmacological applications. The same situation is not true for the inhibitors of other enzymes that degrade ECM, such as the bacterial collagenases (for instance the enzyme isolated from *Clostridium histolyticum*), which have been much less investigated (Van Wart 1998). This collagenase (EC 3.4.24.3) is a 116 kDa protein belonging to the M9 metalloproteinase family, which is able to hydrolyze triple helical regions of collagen under physiological conditions, as well as an entire range of synthetic peptide substrates. In fact the crude homogenate of *Clostridium histolyticum*, which contains several distinct collagenase isozymes, is the most efficient system known for the degradation of connective tissue, being also involved in the pathogenicity of this and related clostridia, such as C. *perfringens*, which causes human gas gangrene and food poisoning among other diseases. Typically, these bacteria (and their collagenases) cause so much damage and so quickly, that antibiotics are ineffective. Thus, development of inhibitors against these collagenases should be a welcome development (Scozzafava and Supuran 2000a–c; Scozzafava *et al* 2000; Supuran *et al* 2000; Clare *et al* 2001).

Similar to the vertebrate MMPs, *Clostridium histolyticum* collagenase (ChC) incorporates the conserved HEXXH zinc-binding motif, which in this specific case is His 415-EXXH, with the two histidines (His 415 and His 419) acting as Zn(II) ligands, whereas the third ligand seems to be Glu 447, and a water molecule/hydroxide ion acts as nucleophile in the hydrolytic scission. Similarly to the MMPs, ChC is also a multiunit protein, consisting of four segments, S1, S2a, S2b and S3, with S1 incorporating the catalytic domain (Van Wart 1998). Although the two types of collagenases mentioned above (the MMPs type and the bacterial ChC) are relatively different, it is generally considered that their mechanism of action for the hydrolysis of proteins and synthetic substrates is relatively similar (Van Wart 1998). These enzymes, as yet, have not been successfully crystallized for X-ray studies.

Thus, it was hypothesized that amino acid hydroxamates and some of their derivatives which strongly inhibit MMPs would also act as potent ChC inhibitors (Supuran *et al* 2000). Our interest in this type of enzyme inhibitor is related to the development of pharmacological agents for the treatment of bacterial corneal keratitis, a condition leading

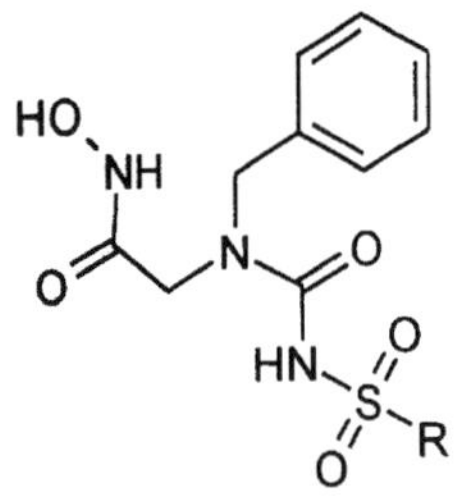

(3.55–3.58)

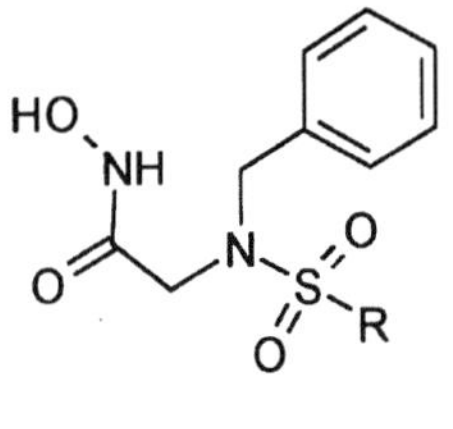

(3.59–3.60)

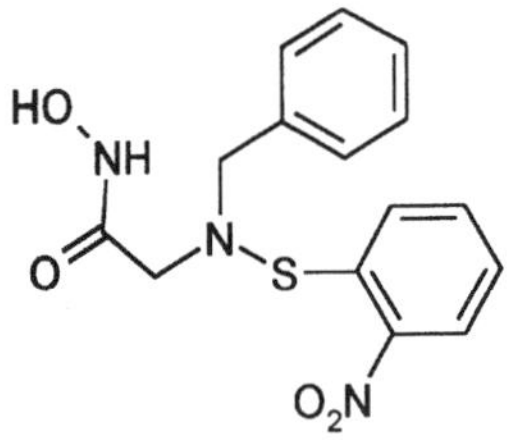

(3.61)

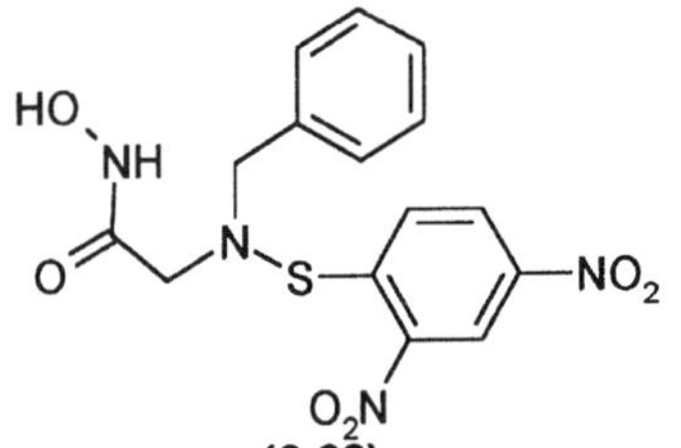

(3.62)

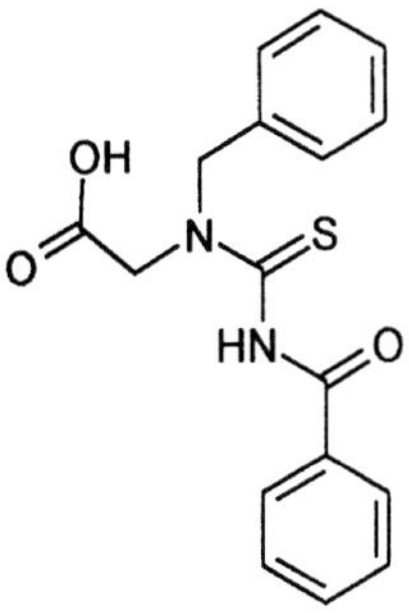

(3.63)

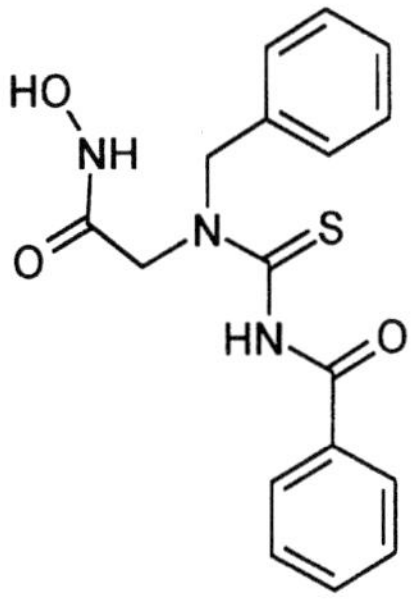

(3.64)

	R	K_I(nM) (against ChC)
(3.55)	4-F-C_6H_4	18
(3.56)	4-Cl-C_6H_4	15
(3.57)	4-Me-C_6H_4	15
(3.58)	2-Me-C_6H_4	14
(3.59)	3-CF_3-C_6H_4	6
(3.60)	C_6F_5	6
(3.61)	–	12
(3.62)	–	10
(3.63)	–	500
(3.64)	–	5

(–) means that R is not present.

to serious complications for which efficient cures are not foreseeable (Scozzafava and Supuran 2000a–c). It was in fact reported that collagen shields applied to the cornea of patients with bacterial keratitis degrade rapidly, due to the collagenases secreted by the pathogen bacterial species, but these shields also protect to some extent the corneal collagen degradation and thus the ocular surface. The use of such a shield impregnated with an inhibitor specific for the collagen-degrading bacteria would thus have a double benefit for the patient: (1) the collagenase inhibitor would kill or impair the growth of bacteria present on the cornea, improving and accelerating healing of the keratitis; (2) the protective collagen shield would possess an augmented stability, as its degradation by the secreted collagenases would be delayed, promoting/accelerating in this way the healing of the wound. We have thus developed a series of sulfonylated, sulfenylated or arylsulfonylureido-derivatized amino acid hydroxamates structurally related to compounds of type (**3.40–3.43**), some of which proved to possess nanomolar affinity for the type II ChC (the most abundant and active isozyme) (Scozzafava and Supuran 2000a–c; Supuran *et al* 2000; Clare *et al* 2001). Some of the most active inhibitors and their K_I data are shown (Scozzafava and Supuran 2000a–c).

As seen from the above data, both arylsulfonyl-, arylsulfenyl- as well as arylsulfonylureido-derivatives possess good inhibitory properties. Obviously, carboxylates were much less inhibitory as compared to the analogous hydroxamates (compare (**3.63**) and (**3.64**)). The proposed binding mode for one of these hydroxamate inhibitors within the ChC active site is shown schematically in Figure 3.6.

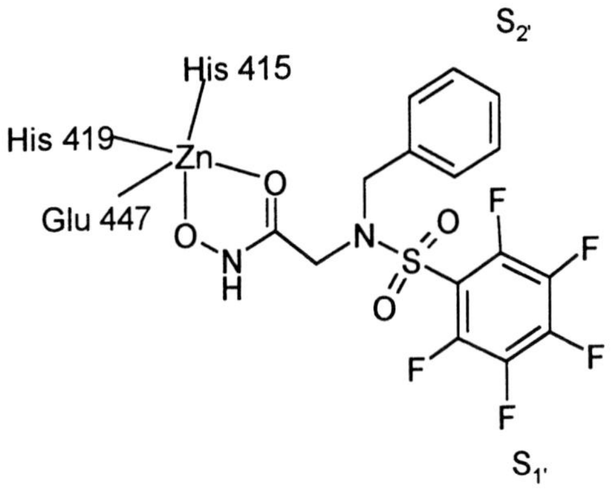

Figure 3.6 Proposed schematic binding of a sulfonyl-glycine hydroxamate inhibitor within the ChC active site (adapted from Scozzafava and Supuran (2000c)).

3.5 CLINICAL DEVELOPMENTS

MMPIs were investigated recently in several animal models of human disease, mainly cancer and arthritis, and promising pharmacological effects have been observed in many cases. Thus, as the controlled degradation of ECM is crucial for the growth, invasive capacity, metastasis and angiogenesis in human tumors, inhibition of some of the enzymes involved, such as the MMPs, can lead to the introduction of novel anticancer therapies based on inhibitors of these proteases (Bottomley *et al* 1998; Johnson *et al* 1998; Whittaker *et al* 1999).

One of the most investigated derivatives is batimastat (**3.12**) (Whittaker *et al* 1999). It has been shown that in a rat mammary carcinoma model, (**3.12**) effectively suppressed the micrometastatic disease, leading to a significantly fewer number of lung metastasis but distant lymph node metastasis was seen both in the batimastat-treated as well as the control animals. Both batimastat as well as the related derivative (**3.19**) were also shown to inhibit the local invasive growth of many types of human carcinomas, in xenograft models of the disease. The Agouron compound (**3.37**), a deep pocket MMP inhibitor, has a range of pharmacological activities in animals, inhibiting tumour growth in models of human glioma, human colon carcinoma, Lewis lung carcinoma and human non-small cell lung carcinoma (Babine and Bender 1997; Whittaker *et al* 1999).

Some MMPIs have also been used in combination therapy with classical cytotoxic chemotherapeutic agents with interesting results. It also seems that some MMPIs, such as batimastat, are able to suppress the development of human tumors by nearly 50% (Whittaker *et al* 1999).

Another promising field for the application of MMPIs-based drugs is in the treatment of arthritis. Both rheumatoid- as well as osteoarthritis are characterized by the loss of normal joint function due to the destruction of the articular cartilage, a process mediated by MMPs (Bottomley *et al* 1998). In both diseases, cartilage destruction involves loss of aggrecan and collagen II, the two major structural components of this tissue. Experimental studies in animal models have shown that MMPIs are able to prevent the breakdown of these two cartilage components. Thus, batimastat given intraperitoneally from the onset of symptoms (in an arthritis-like syndrome induced in rodents as an animal model of the human disease) significantly reduced paw edema bone degradation and cartilage breakdown (Bottomley *et al* 1998). The sulfonamide hydroxamate **3.35** was also shown to inhibit cartilage proteoglycan loss in the rabbit, following injection of MMP-3 into the knee joint (Whittaker *et al* 1999). But one of the most attractive candidates for development as an anti-arthritis drug is Trocade (**3.21**) (Bottomley *et al* 1998). This compound not only exhibited selectivity for collagenases over the other MMPs but also possessed a greatly improved water solubility and oral bioavailability (up to 41%). This compound successfully inhibited degradation of collagen in several animal models of the disease. Trocade is currently in phase II trials as a cartilage protective agent and promises to be an important breakthrough in the field of osteo- and rheumatoid arthritis treatment.

Some of the MMPIs discussed in this chapter, such as batimastat (**3.12**), were also shown to be beneficial in the treatment of restenosis (a complication of balloon catheter angioplasty), aortic aneurysm, glomerulonephritis and stroke (Whittaker *et al* 1999).

Except for batimastat (**3.12**), marimastat (**3.13**) and Trocade (**3.21**), other compounds which are in clinical trials (mainly as anticancer drugs) include: (**3.17**), (**3.19**), (**3.35**) and (**3.46**) among others (Whittaker *et al* 1999). Thus, MMPIs may constitute an important new class of therapeutic agents for the treatment of diseases characterized by extensive ECM degradation and remodeling. As the field is relatively new, future developments in the design, synthesis and clinical evaluation of novel types of MMPIs constitutes an important task for many chemists and pharmacologists working towards the development of new types of such pharmacological agents.

REFERENCES

Babine, R.E. and Bender, S.L. (1997) Molecular recognition of protein–ligand complexes: Applications to drug design. *Chemical Reviews*, **97**, 1359–1472.

Barrett, A.J., Rawlings, N.D. and Woessner, J.F. (Eds) (1998) *Handbook of proteolytic enzymes*, Academic Press, London (CD-ROM), and references cited therein.

Bottomley, K.M., Johnson, W.H. and Walter, D.S. (1998) Matrix metalloproteinase inhibitors in arthritis. *Journal of Enzyme Inhibition*, **13**, 79–102.

Brandstetter, H., Engh, R.A., Graf von Roedern, E., Moroder, L., Huber, R., Bode, W. *et al* (1998) Structure of malonic acid-based inhibitors bound to human neutrophil collagenase. A new binding mode explains apparently anomalous data. *Protein Science*, **7**, 1303–1309.

Briganti, F., Mangani, S., Scozzafava, A., Vernaglione, G. and Supuran, C.T. (1999) Carbonic anhydrase catalyzes cyanamide hydration to urea: Is it mimicking the physiological reaction? *Journal of Biological and Inorganic Chemistry*, **4**, 528–536.

Clare, B.W., Scozzafava, A. and Supuran, C.T. (2001) Protease inhibitors. Part 10. Synthesis and QSAR study of bacterial collagenase inhibitors incorporating N-2-nitrobenzylsulfonyl alanine hydroxamate moieties. *Journal of Medicinal Chemistry*, **44**, 2253–2258.

Dioszegi, M., Cannon, P. and Van Wart, H.E. (1995) Vertebrate collagenases. *Methods in Enzymology*, **248**, 413–431.

Finzel, B.C., Baldwin, E.T., Bryant, G.L., Hess, G.F., Wilks, J.W., Trepod, C.M. *et al* (1998) Structural characterization of nonpeptidic thiadiazole inhibitors of matrix metalloproteinases reveal the basis for stromelysin selectivity. *Protein Science*, **7**, 2118–2126.

Graff von Roedern, E., Grams, F., Brandstetter, H. and Moroder, L. (1998) Design and synthesis of malonic acid-based inhibitors of human neutrophil collagenase (MMP-8). *Journal of Medicinal Chemistry*, **41**, 339–345.

Grams, F., Crimmin, M., Hinnes, L., Huxley, P., Pieper, M., Tschesche, H. and Bode, W. (1995a) Structure determination and analysis of human neutrophil collagenase complexed with a hydroxamate inhibitor. *Biochemistry*, **34**, 14012–14020.

Grams, F., Reinemer, P., Powers, J.C., Kleine, T., Pieper, M., Tschesche, H. *et al* (1995b) X-ray structures of human neutrophil collagenase complexed with hydroxamate and peptide thiols inhibitors. Implications for substrate binding and rational drug design. *European Journal of Biochemistry*, **228**, 830–841.

Hanessian, S., Bouzbouz, S., Boudon, A., Tucker, G.C. and Peyroulan, D. (1999) Picking the S_1, $S_{1'}$ and $S_{2'}$ pockets of matrix metalloproteinases. A niche for potent acyclic sulfonamide inhibitors. *Bioorganic and Medicinal Chemistry Letters*, **9**, 1691–1696.

Jacobsen, E.J., Mitchell, M.A., Hendges, S.K., Belonga, K.L., Skaletzky, L.L., Stelzer, L.S. *et al* (1999) Synthesis of a series of stromelysin-selective thiadiazole urea matrix metalloproteinase inhibitors. *Journal of Medicinal Chemistry*, **42**, 1525–1536.

Jeng, A.Y., Chou, M. and Parker, D.T. (1998) Sulfonamide-based hydroxamic acids as potent inhibitors of mouse macrophage metalloelastase. *Bioorganic and Medicinal Chemistry Letters*, **8**, 897–902.

Johnson, L.L., Dyer, R. and Hupe, D.J. (1998) Matrix metalloproteinases. *Current Opinion in Chemical Biology*, **2**, 466–471.

Johnson, W.H., Roberts, N.A. and Borkakoti, N. (1987) Collagenase inhibitors: their design and potential therapeutic use. *Journal of Enzyme Inhibition*, **2**, 1–22.

Kiyama, R., Tamura, Y., Watanabe, F., Tsuzuki, H., Ohtani, M. and Yodo, M. (1999) Homology modeling of gelatinase catalytic domains and docking simulations of novel sulfonamide inhibitors. *Journal of Medicinal Chemistry*, **42**, 1723–1738.

Krumme, D., Wenzel, H. and Tschesche, H. (1998) Hydroxamate derivatives of substrate analogous peptides containing aminomalonic acid are potent inhibitors of matrix metalloproteinases. *FEBS Letters*, **436**, 209–212.

Lovejoy, B., Hassell, A.M., Luther, M.A., Weigl, D. and Jordan, S.R. (1994) Crystal structures of recombinant 19-kDa human fibroblast collagenase complexed to itself. *Biochemistry*, **33**, 8207–8217.

MacPherson, L.J., Bayburt, E.K., Capparelli, M.P., Caroll, B.J., Goldstein, R., Justice, M.R. *et al* (1997) Discovery of CGS 27023A, a non-peptidic, potent, and orally active stromelysin inhibitor that blocks cartilage degradation in rabbits. *Journal of Medicinal Chemistry*, **40**, 2525–2532.

Murphy, G. and Willenbrock, F. (1995) Tissue inhibitors of matrix metalloendopeptidases. *Methods in Enzymology*, **248**, 496–510.

Nagase, H. (1997) Activation mechanisms of matrix metalloproteinases. *Biological Chemistry*, **378**, 151–160.

Nagase, H. and Woessner, J.F. Jr. (1999) Matrix metalloproteinases. *Journal of Biological Chemistry*, **274**, 21491–21494.

Pavlovsky, A.G., Williams, M.G., Ye, Q.Z., Ortwine, D.F., Purchase, C.F., White, A.D. *et al* (1999) X-ray structure of human stromelysin catalytic domain complexed with nonpeptide inhibitors: Implications for inhibitor selectivity. *Protein Science*, **8**, 1455–1462.

Scozzafava, A. and Supuran, C.T. (2000a) Protease inhibitors. Part 5. Alkyl/arylsulfonyl- and arylsulfonylureido-/arylureido-glycine hydroxamate inhibitors of *Clostridium histolyticum* collagenase. *European Journal of Medicinal Chemistry*, **35**, 299–307.

Scozzafava, A. and Supuran, C.T. (2000b) Protease inhibitors. Part 9. Synthesis of *Clostridium histolyticum* collagenase inhibitors incorporating sulfonyl-L-alanine hydroxamate moieties. *Bioorganic and Medicinal Chemistry Letters*, **10**, 499–502.

Scozzafava, A. and Supuran, C.T. (2000c) Protease inhibitors. Synthesis of potent matrix metalloproteinase and bacterial collagenase inhibitors incorporating N-4-nitrobenzylsulfonyl glycine hydroxamate moieties. *Journal of Medicinal Chemistry*, **43**, 1858–1865.

Scozzafava, A., Manole G., Ilies, M.A. and Supuran, C.T. (2000) Protease inhibitors. Part 12. Synthesis of potent matrix metalloproteinase and bacterial collagenase inhibitors incorporating sulfonylated *N*-4-nitrobenzyl-β-alanine hydroxamate moieties. *European Journal of Pharmaceutical Sciences*, **11**, 69–79.

Springman, E.B., Angleton, E.L., Birkedal-Hansen, H. and Van Wart, H.E. (1990) Multiple modes of activation of latent human fibroblast collagenase: evidence for the role of Cys 73 active-site zinc complex in latency and a "cysteine switch" mechanism for activation. *Proceedings of the National Academy of Sciences of the U.S.A*, **87**, 364–368.

Supuran, C.T. (1994) Carbonic anhydrase inhibitors. In "*Carbonic anhydrase and modulation of physiologic and pathologic processes in the organism*", Puscas, I. (Ed.), Helicon Press, Timisoara, pp. 29–130.

Supuran, C.T., Briganti, F., Mincione, G. and Scozzafava, A. (2000) Protease inhibitors: Synthesis of L-alanine hydroxamate sulfonylated derivatives as inhibitors of *Clostridium histolyticum* collagenase. *Journal of Enzyme Inhibition*, **15**, 111–128.

Supuran, C.T., Conroy, C.W. and Maren, T.H. (1997) Is cyanate a carbonic anhydrase substrate? *Proteins: Structure, Function and Genetics*, **27**, 272–278.

Supuran, C.T. and Scozzafava, A. (2000) Carbonic anhydrase inhibitors and their therapeutic potential. *Expert Opinion on Therapeutic Patents*, **10**, 575–600.

Tschesche, H. (1995) Human neutrophil collagenase. *Methods in Enzymology*, **248**, 431–449.

Van Wart, H.E. (1998) Clostridium collagenases. In *Handbook of proteolytic enzymes*, Barrett, A.J., Rawlings, N.D., Woessner, J.F. (Eds), Academic Press: London (CD-ROM), chapter 368.

Whittaker, M., Floyd, C.D., Brown, P. and Gearing, A.J.H. (1999) Design and therapeutic application of matrix metalloproteinase inhibitors. *Chemical Reviews*, **99**, 2735–2776.

Chapter 4

Proteasomes

Michael Groll and Olivier Coux

The Ubiquitin-proteasome pathway is an essential multienzymatic system in eukaryotic cells, playing a central role in intracellular proteolysis. It is particularly important for the regulated degradation of many critical proteins controlling a vast array of biological pathways, including proliferation, differentiation and inflammation. Therefore, proteasome inhibitors are attractive candidates as anti-tumoral or anti-inflammatory drugs.

There is a large effort, reflected by an increasing number of reports, to develop and study new molecules able to block proteasomal activities. This article reviews the current literature on this topic, and details the present knowledge on the mechanisms of action and biological effects of proteasome inhibitors.

4.1 INTRODUCTION

4.1.1 The ubiquitin-proteasome pathway

The Ubiquitin (Ub) and proteasome dependent proteolytic pathway is the major system for intracellular protein degradation in eukaryotes (Hershko and Ciechanover 1998). Its central components are ubiquitous and generally distributed in both the nucleus and the cytoplasm. This pathway plays a primary role in the degradation of the bulk of proteins in mammalian cells, as well as in the degradation of abnormal proteins, and thus produces most of the antigenic peptides presented to the immune system by the MHC (Major Histocompatibility Complex) class I molecules. Moreover, this pathway is involved in the turnover of membrane proteins, and is responsible for the regulated degradation of many critical proteins, including proteins important for the control of cell growth, cell differentiation or metabolic adaptation.

The Ub-proteasome pathway uses a complex enzymatic machinery to degrade proteins. It functions in two steps: first, a protein substrate is marked by covalent addition of a poly-Ub chain; second, the poly-ubiquitinated substrate is degraded by a 2,500 kDa proteolytic complex called the 26S proteasome.

The poly-ubiquitination reaction requires the action of three types of enzymes, which function sequentially to covalently attach Ub, via an isopeptidic linkage, to a lysine residue of the substrate or of the previous Ub in the chain (Hershko and Ciechanover 1998). The Ub-activating protein, E1, utilizes ATP to form a high-energy Ub-thiol ester, and then transfers the activated Ub to a second protein, E2 (Ub-carrier protein or UBC), forming an E2-Ub thiol ester. The Ub is then linked to the substrate in a reaction requiring a third component, E3 (or Ub-protein ligase). Cells contain many E2s (11 in yeast, more than 20 in humans for example), among which some show overlapping specificities. The E3s seem to provide most of the substrate specificity of the ubiquitination process, and their number appears to be very high (Deshaies 1999). The action of the ubiquitinating enzymes is countered by that of deubiquitinating enzymes, or isopeptidases, some of which are able to remove Ub from poly-Ub adducts. Although the function of most of these isopeptidases is unclear, some of them might be important in reversing the ubiquitination of specific proteins, and thus in preventing, possibly transiently, their degradation (Huang *et al* 1995).

Once ubiquitinated, proteins are usually rapidly degraded to small peptides by the 26S proteasome. This complex is an essential enzyme found in all eukaryotic cells (Coux *et al* 1996; Voges *et al* 1999). It is formed by a cylinder-shaped multimeric complex referred to as the 20S proteasome (core particle), capped at each end by another multimeric component called the 19S complex (regulatory particle) or PA700. The 20S proteasome contains the proteolytic activities (see below), and the 19S complex contributes multiple functions to the 26S proteasome, including a subunit able to bind poly-Ub chains *in vitro* (Deveraux *et al* 1994), an isopeptidase that catalyzes the release of free Ub (Kam *et al* 1997), and six essential ATPase subunits (Rubin *et al* 1998). These ATPases, despite their sequence similarity, are not functionally redundant and are thought to collectively assume multiple roles within the complex. They are most likely involved in the unfolding of the substrates and in their translocation into the 20S proteasome (Braun *et al* 1999). They are also critical for the activation of the proteolytic activities of the 20S proteasome upon binding of the 19S complex.

In addition to the 19S complex to which it associates to form the 26S proteasome, the 20S proteasome can separately interact with another complex called PA28 or 11S regulator (DeMartino and Slaughter 1999). PA28 (200 kDa) is a multimer of two homologous 28 kDa subunits, which activates the peptidase activities of the 20S proteasome upon binding to its poles. The exact function of PA28 is still unclear, but it is important for antigen processing *in vivo* (Groettrup *et al* 1996). Interestingly, hybrid molecules made of one 20S proteasome bound to one 19S complex and one PA28 have been found in cell extracts (Hendil *et al* 1998), and could represent a significant fraction of total proteasomes, at least in some cells (Tanahashi *et al* 2000).

4.1.2 Biological and therapeutic interest of proteasome inhibitors

Since regulated degradation of specific proteins is necessary for a large range of cellular processes important in particular for cell integrity, proliferation and differentiation, any dysfunctioning of the degradation machinery can lead to aberrant expression of proteins and consequent deleterious effects for the cell or the organism (Schwartz and Ciechanover 1999). Therefore, there is considerable interest in being able to manipulate the Ub-proteasome system in order to control the stability of important regulatory proteins.

A promising route for precise therapeutic intervention aiming at stabilizing or destabilizing a given substrate of the Ub-proteasome pathway is to interfere with the function(s) or substrate recognition elements of its specific E3 Ub-protein ligase, since these factors provide most of the selectivity in the pathway. However, because of its central role in intracellular proteolysis, the proteasome is also a potential target, in particular for treatments of pathologies associated with excessive degradation of one or more proteins

(Lee and Goldberg 1998b). For example, proteasome inhibitors have anti-inflammatory effects by blocking activation of the transcription factor NF-κB (Grisham *et al* 1999). They could be used as anti-cancer drugs since they can block degradation of negative regulators of the cell cycle or induce apoptosis in certain cell types (Adams *et al* 1999). Finally, proteasome inhibitors could also be used to prevent proteasome-dependent muscle wasting associated with many pathologies, including certain types of cancer (Mitch and Goldberg 1996). Several laboratories and companies are currently developing specific proteasome inhibitors for these purposes, and some of these compounds are already in clinical trials (Lightcap *et al* 2000). Despite their toxic effects at high doses due to the pleiotropic role of proteasomes, certain inhibitors appear to elicit significant antitumoral effects at a dose well tolerated by animals (Orlowski *et al* 1998; Teicher *et al* 1999).

4.2 STRUCTURE AND MECHANISM OF ACTION OF THE 20S PROTEASOME

The 26S proteasome is a sophisticated multimeric proteolytic machine, in which several enzymatic (proteolytic, ATPasic, de-ubiquitinating) activities work together to degrade proteins. To date however, only its 20S proteolytic sub-component is well understood at the molecular level, owing to the extensive enzymatic and structural data that have been accumulated over the years.

4.2.1 Multiple peptidase activities of the 20S proteasome

In view of its pleiotropic functions, it seems logical that the proteasome evolved to hydrolyze most of its numerous substrates into peptides of relatively small size, since large fragments might retain biological activity. Indeed, recent studies have shown that most of the peptides produced by the proteasome are shorter than 10 residues (Kisselev *et al* 1999b). The requirement for extensive hydrolysis of substrates explains why this complex contains several proteolytically active sites.

Initial studies on mammalian 20S proteasome, using synthetic model peptides as substrates, established that at least three different active sites are present in the complex. Their activities can be distinguished by their different kinetics, pH optima and inhibitor sensitivities. By comparison with substrate specificities of known proteases, the three peptidase activities of the 20S proteasome, cleaving substrates after large hydrophobic, basic or acidic residues, were designated as "chymotrypsin-like", "trypsin-like", and "peptidylglutamyl-peptide hydrolyzing" (PGPH) activities, respectively (Orlowski 1990). Recently, the latter has also been named "post-acidic" or "caspase-like" activity (Kisselev *et al* 1999a).

It must be emphasized that the terms used to describe these peptidase activities do not exactly reflect the true nature of proteasome active sites, i.e. that the proteasome can not be understood as the simple integration within the same complex of a "chymotrypsin-like", a "trypsin-like" and a "caspase-like" enzymes. Instead, as structural analyses clearly showed, the proteasome active sites are sequestered into a proteolytic chamber (see below), and their different specificities reflect more a physical constraint on the peptide substrate due to the local structure around each active site, rather than a true preference for certain P1 residues on the substrate. In line with this idea, it has been shown that proteasomes can cleave a protein substrate at almost every peptide bond (Wenzel *et al* 1994), that substrate residues other than P1 influence degradation (Cardozo *et al* 1994; Bogyo *et al* 1998), and that neighboring subunits can interfere with the functions of the catalytic subunits (Heinemeyer *et al* 1997). In addition, detailed analyses demonstrated that certain active sites have overlapping specificities (Dick *et al* 1998).

As the three peptidase activities of the 20S proteasome can be easily probed with specific fluorogenic peptides, they are very often monitored as a measure of proteasomal activity *in vitro*, and most of the studies on proteasome inhibitors were pursued using these activities as targets. However, it should be noted that the serine protease inhibitor DCI (3,4-dichloroisocoumarin), that efficiently blocks these three activities against peptide substrates, does not inhibit degradation of protein substrates by purified 20S proteasome. In fact, this compound even stimulates *in vitro* protein degradation by the proteasome, an observation still not understood that was the basis for the suggestion that the 20S proteasome contains two additional peptidase activities: one named BrAAP (for "branched chain amino acid preferring"), preferentially hydrolyzing peptide bonds on the carboxyl side of branched chain amino acids, and the other named SNAAP (for "small neutral amino acid preferring"), cleaving peptide bonds between small neutral amino acids (Orlowski *et al* 1993; Cardozo *et al* 1999).

4.2.2 Structural features of the 20S proteasome

Since different types of classical protease inhibitors were active against 20S proteasomes, the nature of the active sites within this complex remained obscure for several years. Eventually, a better understanding of its mechanisms of action came from structural and mutational analyses of the simpler 20S proteasome of the archaebacteria *Thermoplasma acidophilum* (Löwe *et al* 1995; Seemuller *et al* 1995). This work has since been completed by extensive studies on yeast proteasome (Groll *et al* 1997; Heinemeyer *et al* 1997). A detailed view of proteasome organization can be found in several recent reviews (Bochtler *et al* 1999; Voges *et al* 1999). Briefly, the eukaryotic 20S proteasome is a cylinder-shaped complex composed

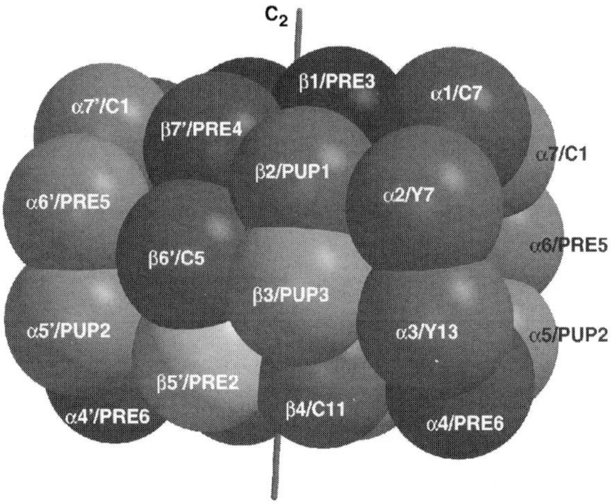

Figure 4.1 Topology of the 28 subunits of the yeast 20S proteasome drawn as spheres. (See *Color plate 1*)

of four stacked rings (Figure 4.1). It possesses 14 different subunits (28 per complex) ranging in size between 20 and 35 kDa, which can all be classified, based on their sequences, into two groups (α and β), each containing seven families. The two outer rings of the 20S proteasome contain only α subunits, and its two inner ones only β subunits, with 7 subunits per ring ($\alpha_7\beta_7\beta_7\alpha_7$). The active subunits are all of the β-type, and the catalytic sites are enclosed within a central cavity defined by the two β rings (Groll *et al* 1997; Löwe *et al* 1995).

4.2.2.1 Active sites of the 20S proteasome

Although in the *Thermoplasma* proteasome the 14 β-type subunits are identical and all form an active site (Löwe *et al* 1995), the eukaryotic proteasome has seven different β-type subunits (Figure 4.1), among which only three (β1/Pre3, β2/Pup1, β5/Pre2) are active in degradation of model peptides (Groll *et al* 1997; Heinemeyer *et al* 1997). Due to its symmetrical structure, the eukaryotic proteasome thus possesses six active sites (Figure 4.2). One important feature of these active subunits, is their expression in the form of a proprotein that is cleaved near the N-terminus by autoprocessing during proteasome assembly, thus generating a N-terminal threonine that is responsible for peptidase activity (Seemuller *et al* 1996; Ditzel *et al* 1998). Therefore, the protease belongs to the family of Ntn (N-terminal nucleophile) hydrolases (Brannigan *et al* 1995).

Mutational studies allowed the determination of the role of each β-subunit for activity. In eukaryotes, three distinct pairs of active sites can be distinguished, each of them associated with a specific subunit: β5/Pre2 is responsible for cleavage after hydrophobic residues, β2/Pup1 for cleavage after basic residues, and β1/Pre3 for cleavage after acidic residues (Heinemeyer *et al* 1997; Dick *et al* 1998). However, the specificity of the active sites is not absolute, since β1/Pre3 and β5/Pre2 additionally can cleave after some hydrophobic and small neutral amino acids, respectively (Dick *et al* 1998). In addition, the

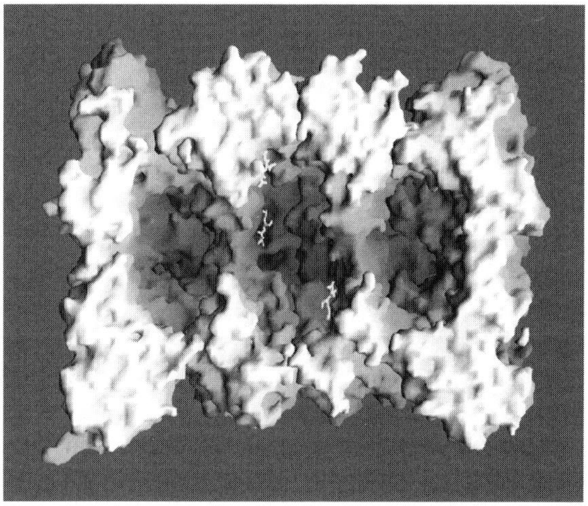

Figure 4.2 Surface view of the yeast 20S proteasome crystallized in the presence of calpain inhibitor I, clipped along the cylindrical axis. The inhibitor molecules are shown as space filling models in yellow. The sealed α-ports at both ends of the yeast proteasome and a few narrow side windows can be seen. (*See Color plate 2*)

neighboring subunits play an important role for activity, probably because they help to shape the catalytic centers (Heinemeyer *et al* 1997). Another important fact is that, although the proteasome is essential for life, none of its peptidase activities is absolutely required, at least in yeast, as long as the mutation does not prevent complex formation. Even double mutants having only one functional active site (either the site responsible for "chymotrypsin-like" or the site for "trypsin-like" activity) are viable (Heinemeyer *et al* 1997). This clearly shows some redundancy in the proteolytic active sites of the proteasome, although the sites are not functionally equivalent. Indeed, analysis of the phenotypes of several mutants indicates a possible hierarchy among the active sites, the "chymotrypsin-like" activity being the most important for cell growth, and the post-acidic the least important (Jager *et al* 1999). However, it is possible that the stronger phenotype of the defect in the chymotrypsin-like activity reflects a defect in auto-processing of the β subunits and thus a defect in proteasome assembly rather than a true impairment in degradation of certain type of substrates (Heinemeyer *et al* 1997). It was proposed recently, that there is an ordered functioning of the different active sites during the degradation of a protein by the proteasome (Kisselev *et al* 1999a). According to this model, the chymotrypsin-like sites are responsible for the first cuts within the substrate, while their occupancy activates the other sites that then hydrolyse the products into smaller peptides. Such a feature might contribute to the stronger phenotype of "chymotrypsin-like" site mutants. In any case, it is interesting to note that there is a correlation between the results obtained with mutants and those obtained with pharmacological compounds (see below): the best proteasome inhibitors to date are the inhibitors which react directly with the β5 subunit responsible for the "chymotrypsin-like" activity. Whether this general trend will hold true in the future remains to be seen.

Structural analyses of the 20S proteasome, together with mutational studies, allowed a better understanding of the intimate organization of its active sites. The catalytic system for each active site is formed in the *T. acidophilum* proteasome by Thr1, Glu17 and Lys33. Close to Thr1 are residues Ser129, Ser169 and Asp166, which seem to be required for structural integrity of the site, but may additionally be involved in catalysis (Seemuller *et al* 1996; Groll *et al* 1999). These residues are invariant in the active yeast-proteasome subunits. In addition, the electron density of the crystal structure from the yeast proteasome showed a fully occupied solvent molecule in close proximity to all three catalytically active subunits near Thr1(Oγ and N), Ser129(Oγ and N) and Gly47(N). This was not apparent in the *Thermoplasma* proteasome electron density map at lower resolution, but has been seen in penicillin acylase, a member of the Ntn-hydrolase family (Duggleby *et al* 1995). In general, Thr1(N) is hydrogen-bonded via the water molecule to the carbonyl oxygen of residue 168 and to Ser129(Oγ) and Ser169(Oγ), whereas Thr1(Oγ) is hydrogen-bonded to Lys33(Nζ). Lys33(Nζ) makes three hydrogen bonds with Asp17(Oγ2), Thr1(Oγ) and residue 19(O) (Figure 4.3). The pattern of hydrogen bonds suggests that both Asp17 and Lys33 are charged. The positive charge on Lys33 may shift the intrinsic pKa of the water molecule and of the amino and hydroxyl-group of Thr1, enhancing their nucleophilicity. Thr1(N) is most probably the proton acceptor when Thr1(Oγ) adds to an electrophilic center, whereas the water molecule participates in proton shuttling. This is well illustrated by the structure of the 20S complex with inhibitors which make covalent bonds to the Thr1(Oγ) (Figure 4.5, see later). Thr1(N) is in complex with the water molecule and is therefore ideally positioned to serve as a proton shuttle from Thr1(Oγ) to the carbonyl oxygen atom of the inhibitor for this process (Groll *et al* 1999; Loidl *et al* 1999).

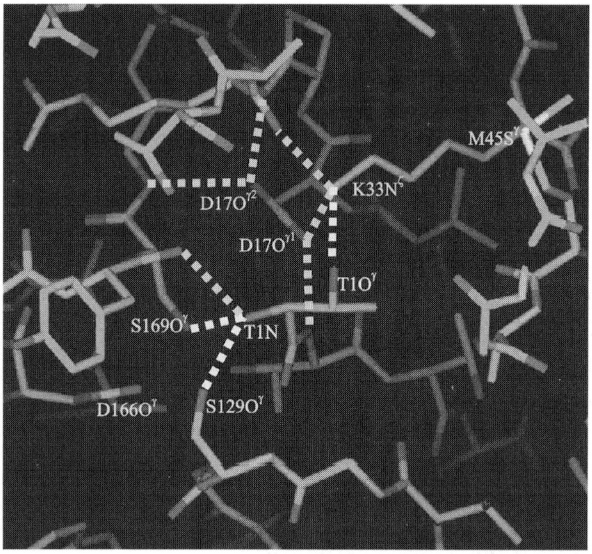

Figure 4.3 The nucleophilicity of threonine 1 in the active site of the 20S proteasome shown for the subunit β5/Pre2. Hydrogen-bonds of the Thr1(Oγ) to Lys33(Nε) and of the N-terminus to Ser129(Oγ) and Ser169(Oγ) are shown as yellow dotted lines. (*See Color plate 3*)

The proteasome subunit β1, with Arg45 at the base of S1, is well suited for P1 glutamate residues (Figure 4.4). This is in agreement with mutational analyses that associated it with the post acidic activity of the proteasome (Enenkel *et al* 1994). The subunit β2 has a glycine at residue 45 and consequently has a spacious S1 pocket confined at the bottom by Glu53 (Figure 4.4), explaining why it is mainly responsible for the "tryptic-like" activity (Dick *et al* 1998). In the case of the subunit β5, Met45 makes the pocket hydrophobic, so that this subunit mainly has "chymotrypsin-like" activity (Figure 4.4).

Mutations in β4 and β7 affect the chymotrypsin and post acidic activity, respectively (Heinemeyer *et al* 1997). These subunits are inactive but adjacent to the β5 and β1 subunits in both rings (Figure 4.1). The exchange in β4 of the internal Ser136 by the bulky phenylalanine residue disrupts the β-trans-β contact between β4 and β5 and may distort the adjacent Thr1 site. In the case of the subunit β1/Pre3 the C-terminal residues from β7 form extensive contacts with the active site in β1/Pre3 of the opposite ring, so that the deletion of the last 15 amino acids of β7 results in an inactive subunit β1 (Groll *et al* 1997). These data clearly show that non-active subunits influence proteolysis and illustrate the tremendous interdependence of the whole structure of the eukaryotic 20S proteasome.

4.2.2.2 *Access to the central catalytic cavity*

An obvious function of the structural organization of the 20S proteasome is to isolate its proteolytic compartment from the cellular components, preventing unwanted degradation of endogenous proteins and probably favoring processive degradation of substrates by restricting dissociation of partially digested polypeptides (Akopian *et al* 1997). However, it

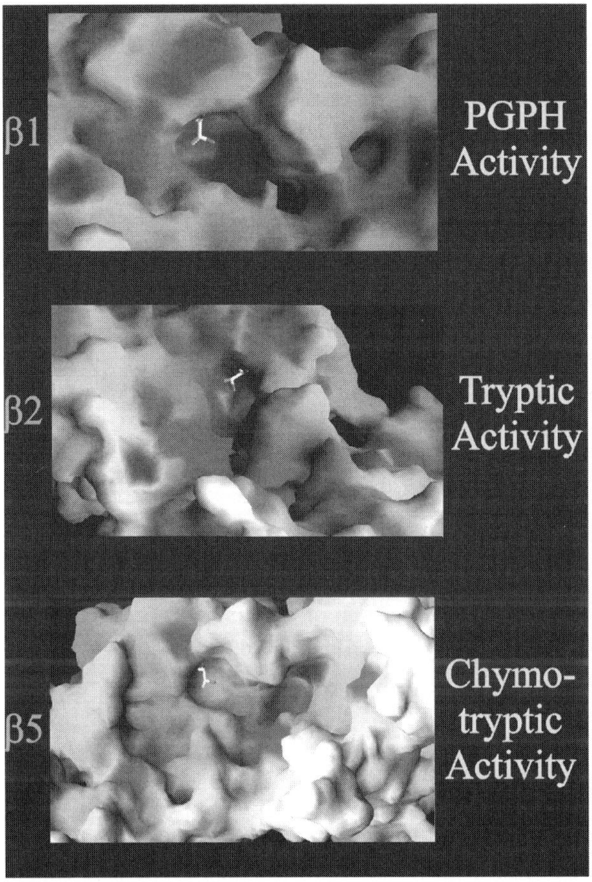

Figure 4.4 Surface representation of the three active sites in the yeast 20S proteasome. Each picture shows the nucleophilic Thl in sticks, the basic residues in blue, the acidic residues in red, and the hydrophobic residues in white. (*See Color plate 4*)

also imposes strong constraints on the access of substrates into the proteolytic chamber, and on the release of degradation products. In the *Thermoplasma* proteasome, two entry ports of ~13Å diameter, constricted by an annulus made by turn-forming segments of the seven identical α-subunits, can be visualized at the ends of the cylindrical particle (Löwe *et al* 1995). However, the 12 N-terminal residues of each α-subunit are disordered in the crystal of the *Thermoplasma* proteasome and thus cannot be positioned. By contrast, the hydrolytic chamber of the yeast 20S proteasome seems quite inaccessible (Groll *et al* 1997): the N-termini of the α subunits are projected into the ports seen in *Thermoplasma*, and fill them completely in several layers with tightly interdigitating side chains (Figure 4.2). There is thus no possible access to the interior of the particle without substantial structural rearrangement, excepting several narrow "side windows" (Figure 4.2) located at the interface of the α- and β-rings. These are coated with polar and charged residue side chains and lead to the active site threonines. It seems unlikely that these narrow side windows could allow access for the substrates. Instead, since it has been suggested that products will be released only when they become small enough to diffuse out of the complex (Kisselev *et al* 1999b), these small openings could participate in the release of peptides resulting from protein degradation.

Substrate entrance thus most likely occurs through the α-rings of the cylinder. Experimental evidence supporting this model includes the observations that proteins must be unfolded to access the catalytic chamber and that substrates that cannot be degraded anymore when linked to 2 nm nanogold beads, since the beads prevent their access to the catalytic chamber of the proteasome, accumulate at both extremities of the complex (Wenzel and Baumeister 1995). In eukaryotes, a mechanism must thus exist to open the ends of the 20S proteasome. Most likely, this function is carried out by the 19S complex and PA28 that interact with the α-rings of the 20S proteasome. It is believed that both regulatory complexes activate peptide degradation of the 20S proteasome by altering the arrangement of the α-rings to enhance access of substrates (DeMartino and Slaughter 1999).

4.3 INHIBITORS OF THE PROTEASOME

4.3.1 Endogenous inhibitors of the 20S proteasome

Soon after the characterization of the 20S proteasome, systematic screens allowed the discovery of several endogenous molecules able to inhibit its proteolytic activities *in vitro*. To date however, since these proteins have been poorly studied, their mechanism of action and their role(s) *in vivo* remain obscure.

The first endogenous proteasome inhibitor to be identified is a labile 240 kDa hexamer of a single 40 kDa subunit (Murakami and Etlinger 1986). This complex was able to block the degradation of protein substrates *in vitro* by the 20S proteasome and by calpain, but had no effect on other proteases like trypsin, chymotrypsin or papain (Murakami and Etlinger 1986). It was later suggested that this inhibitor corresponds to CF-2 (Driscoll *et al* 1992), one of the 3 factors necessary to reconstitute the 26S proteasome from ATP-depleted reticulocytes (Ganoth *et al* 1988), and that its 40 kDa subunit was present within the 26S proteasome in an ubiquitinated form (Li and Etlinger 1992). Finally, this inhibitor has been reported to be identical to δ-aminolevulinic acid dehydratase (ALAD), the second enzyme in the pathway of heme synthesis (Guo *et al* 1994), but the presence of this protein has not been confirmed in highly purified 26S proteasomes of different species. A 200 kDa complex of a single subunit of 50 kDa has been reported to inhibit protein degradation by the 20S proteasome, and also certain of its peptidase activities (Li *et al* 1991), but has not been characterized further. An interesting molecule, termed PI31, has been shown to stably bind and to inhibit the 20S proteasome. Recent data suggest that it acts as a monomer, and that its physiological role might be to alter the function or the binding of PA28 or of the 19S complex (McCutchen-Maloney *et al* 2000). Finally, the heat-shock/chaperone proteins HSP90 and α-crystallin also were shown to inhibit certain proteasome peptidase activities *in vitro* (Tsubuki *et al* 1994; Conconi *et al* 1998).

Clearly, more research is required to understand the mechanisms of action, the exact role(s) and the significance of the inhibitory activities of these molecules. An attractive hypothesis is that these inhibitors, like the regulatory proteins PA28 and 19S complex, act by binding to the α-rings at the poles of the 20S proteasome, but with an opposite effect resulting in the prevention of substrate entry.

4.3.2 Pharmacological compounds

After the discovery of the 20S proteasome, its mode of action was first analysed with non-specific protease inhibitors. In general however, these agents are no longer used to study the functions of the proteasome since several molecules, having greater (but not absolute) specificity against proteasome, are now available (see Table 4.1 for a partial list). These inhibitors have greatly facilitated testing for involvement of the proteasome in biological processes *in vivo* as well as *in vitro*.

Initially, cell-penetrating peptide aldehydes, such as ALLN, MG115, MG132 and PSI have been developed. More recently, new compounds were synthesized with increased potency and selectivity for the proteasome (Adams *et al* 1998; Elofsson *et al* 1999). In addition, based on the crystal structure of the yeast proteasome, molecular modeling can now be used to engineer improved inhibitors (Loidl *et al* 1999).

In addition to the synthetic molecules especially developed as proteasome inhibitors, a variety of compounds have been described that also inhibit these complexes. Some are of natural origin and were discovered through their biological effects on cells before it was recognized that their main target is the proteasome. Lactacystin (Fenteany *et al* 1995), eponemycin (Meng *et al* 1999a) and epoxomicin (Meng *et al* 1999b) are such molecules. Other compounds were initially aimed at targeting (or thought to target) other proteins before they were found to also inhibit the proteasome. For example, it has been reported that the HIV-1 protease inhibitor Ritonavir and the caspase-1 inhibitor Ac-YVADal inhibit the "chymotrypsin-like" and the post acidic activities of the proteasome, respectively (Andre *et al* 1998; Kisselev *et al* 1999a). Likewise, the *Streptomyces* metabolite leptomycin B, an inhibitor of CRM1-dependent nucleocytoplasmic protein export, was found to inhibit additionally the post acidic activity of the proteasome *in vitro*, and to stabilize several well-known proteasome substrates in living cells (M. Kroll, personal communication). Based on these documented cases, it seems reasonable to assume that, in fact, many compounds small enough to enter the proteolytic chamber of the proteasome might react with certain residues in this chamber and alter proteasomal activity.

In the following sections, the current knowledge will be reviewed on the mechanisms of action and biological effects of some proteasome inhibitors. An overview of the inhibitors shown to be active *in vivo* is presented in Table 4.1. Several difficulties hinder a quantitative summary. First, data sets from different laboratories are not totally consistent, reporting for example divergent Ki values. This is probably due to the long known extreme sensitivity of proteasomal peptidase activities to experimental conditions, as well as to the diversity of the sources, purification procedures and activity assays used by different groups. Second, most *in vitro* inhibition experiments with purified components were performed using the 20S proteasome, and the results cannot be simply extrapolated to the 26S proteasome which is the predominant active form *in vivo*. Indeed, in the few cases where inhibitors were tested simultaneously on both forms of the proteasome, their effects on both complexes were different: for example, peptide aldehydes inhibit preferentially the 20S proteasome *in vitro* (Rock *et al* 1994), but the opposite seems true for lactacystin (Craiu *et al* 1997). Third, when proteins, not peptides, are used as substrates, higher concentrations of inhibitors are necessary to inhibit proteasomal activity (Craiu

Table 4.1 Proteasome inhibitors active *in vivo*

Name(s)	*Inhibitor of*			*Properties (indicative working conc.)*	*Other known protease targets*	*Selected references*
	Chym.	*Tryp.*	*PGPH*			
Peptide aldehydes						
ALLN (acetyl-leu-leu-norleucinal, also called calpain inhibitor I)	++	+	+	Reversible (50 μM)	Cathepsin B, Calpains	Vinitsky *et al* 1992; Rock *et al* 1994
Z-LLF-CHO (Cbz-leu-leu-phenylalaninal)	+++	+	–	Reversible (50 μM)		Vinitsky *et al* 1994; Orlowski *et al* 1998
MG132(Cbz-leu-leu-leucinal)	+++	++	++	Reversible (50 μM)	Cathepsin B, Calpains	Palombella *et al* 1994; Bogyo *et al* 1997
PSI (Cbz-Ile-Glu(O-t-Bu)-Ala-leucinal)	+++			Reversible (50 μM)		Figueiredo-Pereira *et al* 1995; Drexler *et al* 2000
MG115 (Cbz-leu-leu-norvalinal)	++		+	Reversible (50 μM)	Cathepsin B, Calpains	Rock *et al* 1994
CEP1612	+++	–		Reversible (10 μM)	Cathepsin B	Harding *et al* 1995
Other specific inhibitors						
Lactacystin (β-lactone)	+++	++	+	Irreversible (10 μM)	Cathepsin A	Fenteany *et al* 1995; Fenteany and Schreiber 1998
NLVS (NIP-leu-leu-leu-vinyl sulfone)	+++	++	++	Irreversible (50 μM)	Cathepsin B, Calpains	Bogyo *et al* 1997
Glyotoxin	+++	++	+	Reversible (1 μM)		Kroll *et al* 1999
Lovastatin (β-lactone ring form)	++			Irreversible (40 μM)		Rao *et al* 1999
Eponemycin	+++	+	+++	Irreversible (2 μM)		Meng *et al* 1999a
Epoxomicin	+++	++	+	Irreversible (1 μM)		Meng *et al* 1999b
Peptide boronate proteasome inhibitors	+++			Reversible (1–10 μM)		Grisham *et al* 1999; Adams *et al* 1999

Abbreviations: Cbz: benzyloxycarbonyl; Chym, Tryp, PGPH: “chymotrypsin-like”, “trypsin-like”, post-acidic activities of the proteasome, respectively.

et al 1997). It is thus recommended that care should be taken when interpreting data obtained with proteasome inhibitors. This is especially true with experiments done *in vivo*, for which cell permeability and inhibitor stability are an additional source of variability likely to interfere with the results. A solution is to compare, when possible, the effects of different proteasome inhibitors on the process being studied.

4.3.3 Mechanisms of action of proteasome inhibitors

With a multifunctional enzyme like the proteasome, it is difficult to obtain clear results concerning the effects of its inhibitors. However, structure analyses of the yeast 20S proteasome in complex with several inhibitors, as well as experiments performed with radiolabeled irreversible inhibitors, show that most proteasome inhibitors covalently bind in the active site.

N-Acetyl-Leu-Leu-Norleucinal (ALLN, also called Calpain inhibitor 1), has been widely used to study proteasome function *in vivo*, despite its lack of specificity (Rock *et al* 1994; Vinitsky *et al* 1994). This inhibitor binds reversibly to the N-terminal threonine of the active subunits and abolishes the "chymotrypsin-like" and, to a lesser extent, the "trypsin-like" and post-acidic activities of the proteasome. The crystal structure of the proteasome in complex with ALLN shows the inhibitor covalently bound to Thr1(Oγ) of all active subunits, as a hemiacetal. It adopts a β-conformation and fills a gap between strands to which it is hydrogen bonded, generating an antiparallel β-sheet structure. The norleucine-side chain projects into the S1 pocket which opens sidewise towards a tunnel leading to the particle surface, whereas the leucine side chain at P2 is not stabilized. The leucine side chain at P3 is in contact with the amino acids of the adjacent β-type subunit and is therefore fixed (Figure 4.5).

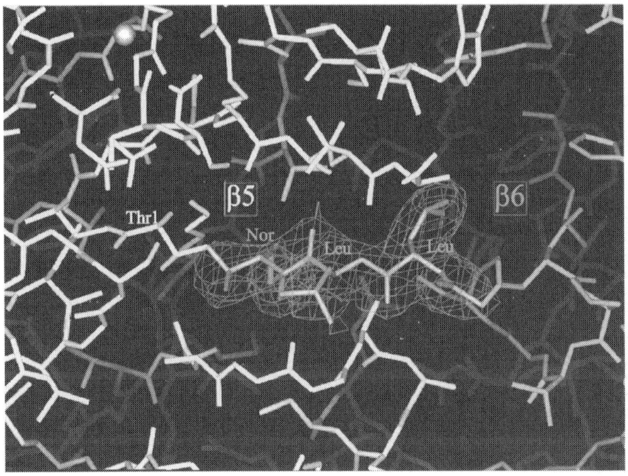

Figure 4.5 Calpain inhibitor I binding and S1–S3 pocket of the subunit β5/Pre2. The inhibitor is shown with the electron density map (contoured from 1σ on) with 2Fo–Fc coefficients after two-fold averaging as orange sticks. The β5/Pre2 and β6/C5 subunit are shown as yellow and blue sticks respectively, the magnesium ion as a gray ball. (*See Color plate 5*)

Lactacystin is a natural product from *Streptomyces* that was discovered by its ability to induce neurite outgrowth in a murine neuroblastoma cell line. Incubation of cells in the presence of radioactive lactacystin resulted mainly in the labeling of the X/MB1 (β5) subunit (Fenteany *et al* 1995), although lactacystin can be found bound to all active β-type subunits of the proteasome in certain conditions (Craiu *et al* 1997). Similar results have been obtained with the peptide vinyl sulfone NLVS (Bogyo *et al* 1997) and other irreversible proteasome inhibitors (Meng *et al* 1999a,b), suggesting that the mode of action of all irreversible inhibitors of the proteasome is to covalently bind the N-terminal threonine of the active β-subunits. Lactacystin effectively and irreversibly inhibits the "chymotrypsin-like" activity of the proteasome. It also blocks the "trypsin-like" and the post-acidic activities with however progressively lower efficiencies (Fenteany *et al* 1995; Craiu *et al* 1997). In aqueous solutions at pH 8, lactacystin is spontaneously hydrolyzed into *clasto*-lactacystin β-lactone which is in fact the active compound inhibiting the proteasome (Dick *et al* 1997). The crystal structure of the complex between lactacystin and yeast proteasome shows the molecule covalently bound only to subunit β5/Pre2 (Groll *et al* 1997), in accord with the observed chemical modification of subunit X/MB1 of the mammalian proteasome. Thereby, lactacystin displays a host of hydrogen bonds with protein main chain atoms (Figure 4.6). The irreversible inhibition by lactacystin of the active site of the proteasome is due to the formation of an ester bond with the N-terminal threonine. In principle, this ester bond and almost every hydrogen bonding interaction with lactacystin could be made in all active subunits. However, subunit β1/Pre3 and β2/Pup1 do not form a covalent complex in the crystal of the yeast proteasome, as there is a major difference in their S1 pocket in comparison to subunit β5/Pre2 (Figure 4.4). As a consequence, the dimethyl side chain of lactacystin could be bound analogously to a valine or a leucine side chain, only in the S1 pocket of β5/Pre2 (Figure 4.6). Met45, which

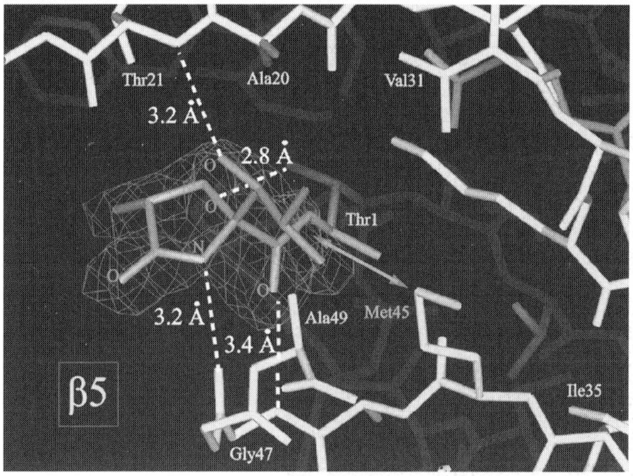

Figure 4.6 β5/Pre2 with the covalently bound inhibitor lactacystin. The inhibitor is surrounded via the electron density map (contoured from 1σ on) with 2Fo–Fc coefficients after two-fold averaging. The ester bond between lactacystin and the active site threonine of β5/Pre2 is formed via an acetylation of the Thr1(Oγ) as a result of the β-lactone ring opening. Hydrogen-bonds to the β5-backbone are shown as dashed lines, the isopropyl-β5/Met45 interaction as a blue arrow. (*See Color plate 6*)

is mainly responsible for the "chymotryptic-like" activity, closely interacts with the branched side chain of lactacystin and has a major role for the selective inhibition. The S1 pocket of β5 consists additionally of several acidic amino acids contributed by subunit β6, that may allow binding of substrates with basic residues, consistent with the observation that lactacystin inhibits both the "chymotrypsin-like" and, to a lesser extent, the "trypsin-like" activity against chromogenic substrates (Fenteany *et al* 1995; Groll *et al* 1997). These data gave impetus to the first structure based design efforts for inhibitor development (Loidl *et al* 1999).

Recently, it was shown that the α′, β′-epoxyketone peptide natural product epoxomicin potently and irreversibly inhibits the catalytic activity of the proteasome (Meng *et al* 1999b). Unlike most other proteasome inhibitors, epoxomicin is highly specific for the proteasome and does not inhibit other proteases. The crystal structure of epoxomicin bound to the 20S proteasome (Figure 4.7) reveals the molecular basis for selectivity of α′, β′-epoxyketone inhibitors (Groll *et al* 2000). The complex showed an unexpected morpholino ring formation between the amino terminal threonine and epoxomicin, providing the first insights into the unique specificity of epoxomicin for the proteasome. The morpholino derivative formation is most likely a two step process. First, activation of the Thr1(Oγ) is believed to occur by its N-terminal amino group acting as a base either directly or via a neighboring water molecule. Subsequent nucleophilic attack of the Thr1(Oγ) on the carbonyl of the epoxyketone pharmacophore would produce a hemiacetal, as is observed in the proteasome: ALLN complex. The formation of the hemiacetal facilitates the second step in the formation of the morpholino adduct. In this cyclization, the threonine 1N opens the epoxide ring via an intramolecular displacement with consequent inversion of the C2 carbon (Groll *et al* 2000, see Figure 4.7). The observed selectivity of epoxomicin for the proteasome is rationalized by the requirement for both a N-terminal amino group and side chain nucleophile for adduct formation with the epoxyketone pharmacophore. These observations explain the high selectivity of epoxomicin against Ntn-hydrolases.

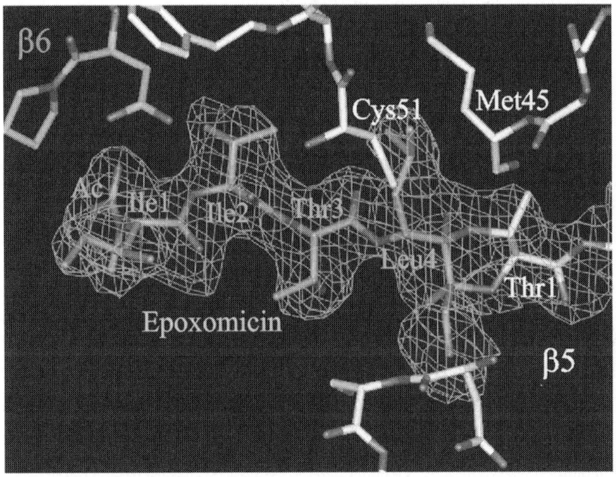

Figure 4.7 View of the electron density map of the epoxomicin adduct at β5/Pre2. Epoxomicin is covalently bound to Thr1, resulting in the formation of a morpholino derivative, and the extended substrate binding site is composed of β5/Pre2 and β6. (*See Color plate* 7)

The human immunodeficiency virus protease inhibitor Ritonavir, used successfully in AIDS therapy, has been found recently to also inhibit the chymotrypsin-like activity of the human 20S proteasome (Andre *et al* 1998). It was not possible to obtain any electron density for Ritonavir by analyzing the yeast proteasome–inhibitor complex crystallographically (Groll, unpublished data). However, a structural model in which Ritonavir interacts with the yeast subunit β5/Pre2 (the subunit responsible for the "chymotrypsin-like" activity that is homologous to the mammalian subunit LMP7 or X), has been proposed (Schmidtke *et al* 1999).

4.3.4 Biological effects of proteasome inhibitors

The availability of proteasome inhibitors with increasing specificity and potency has generated a large collection of data documenting the critical roles of the proteasome and of the Ub pathway in many biological processes. Indeed, these inhibitors enabled the identification, through their stabilization, of numerous proteasome substrates. These results clearly establish that, beyond the necessary housekeeping functions of intracellular proteolysis, precise regulated degradation of key proteins is an irreversible mechanism largely used in cells to switch off specific pathways. Therefore, when applied to cells, proteasome inhibitors elicit diverse biological effects, depending on the processes that are the most affected, which in turn is a function of several parameters, including cell type and proliferation status, nature and dose of the inhibitor, and time of exposure (see for example Lin *et al* 1998). It must be kept in mind however that most proteasome inhibitors are not exclusively specific for this complex, and that the biological effect(s) of inhibitor treatment can be due in part to inhibition of other proteases. For example, peptide aldehydes and peptide vinyl sulfones also inhibit, at certain concentrations, intracellular cysteine proteases such as cathepsins and calpains. Likewise, although lactacystin was initially thought to be highly specific for the proteasome, it was shown later to additionally inhibit cathepsin A (Ostrowska *et al* 1997).

As can be expected from the Ub-proteasome dependent degradation of numerous proteins regulating cell-cycle (G1 and mitotic cyclins, CDK inhibitors, p53 etc.), cell proliferation is profoundly affected by proteasome inhibitors. In fact, several of these compounds have been identified via their effect on cell growth. Proteasome inhibition can cause cell-cycle arrest at various stages: there are reports of arrest at the G1/S boundary, as well as at the G2/M transition (Sherwood *et al* 1993; Katagiri *et al* 1995; Wojcik *et al* 1996).

Probably in part because proteasome inhibitors can stabilize both positive and negative regulators of cell growth and thereby activate conflicting signaling pathways, they often trigger apoptosis (Orlowski 1999). However, in some cases, proteasome inhibitors protect cells against apoptosis (Grimm *et al* 1996), indicating a role for proteasome in certain forms of cell death (Sadoul *et al* 1996). The pro- and anti-apoptotic effects of proteasome inhibitors seem thus to be cell specific, probably because the stabilization of many proteins critical for cellular growth, homeostasis and defense, including p53 and NF-κB, has a differential impact according to the cellular context. In general, rapidly-dividing cells are more sensitive to the pro-apoptotic effects of proteasome inhibitors than non-dividing ones (Drexler 1997; Lopes *et al* 1997). For example, SV40-transformed fibroblasts, but not normal fibroblasts, are susceptible to inhibitor-induced apoptosis (An *et al* 1998), and 340-fold higher concentrations of PSI are necessary to induce apoptosis in quiescent primary endothelial cells, as compared to proliferating cells (Drexler *et al* 2000). Interestingly, proteasome inhibitor-induced apoptosis seems to be mediated through activation of the c-Jun N-terminal kinase, JNK1 (Meriin *et al* 1998).

On the other hand, at non pro-apoptotic concentrations, proteasome inhibitors can protect cells against apoptosis induced by other factors (Lin *et al* 1998). An interesting link between protection by proteasome inhibitors and accumulation of heat-shock proteins (hsp) has been pointed out. Inhibitor treatment induces production of the major hsps (Bush *et al* 1997), that in turn increases cell resistance to various stresses and to apoptosis (Lee and Goldberg 1998a; Meriin *et al* 1998). Indeed, blocking Hsp70 production with antisense oligonucleotides has been shown to potentiate apoptosis induced by proteasome inhibitors (Robertson *et al* 1999). Concomitant to induction of hsps, proteasome inhibitors confer thermotolerance to the cells (Bush *et al* 1997). However, work with yeast suggested that thermotolerance induced by these inhibitors also requires the induction of the thermoprotectant disaccharide trehalose (Lee and Goldberg 1998a).

In addition to their effects on cell growth or death, proteasome inhibitors affect other biological processes that also depend on the Ub-proteasome system for proper regulation, including differentiation, inflammatory and immune responses. Their effect on cell differentiation is illustrated by the fact that MG132 and lactacystin have been characterized as compounds that promote neurite outgrowth in PC12 and Neuro-2a cells (Omura *et al* 1991T; subuki *et al* 1993), and by the fact that several proteasome inhibitors can block myoblast to myotube fusion during muscle differentiation (Kim *et al* 1998). Due to the prominent role of the transcription factor NF-κB in inflammatory response, proteasome inhibitors may be used as anti-inflammatory agents since they strongly stabilize its inhibitor IκBα (Palombella *et al* 1994; Traenckner *et al* 1994). Finally, since the proteasome is the main producer of antigenic peptides (Rock and Goldberg 1999) and also is directly involved in the control of immune response, proteasome inhibitors present interesting immuno-suppressive or immuno-modulating properties (Groettrup and Schmidtke 1999).

Inhibitors proved to be very useful in giving a better understanding of how proteasomes contribute to total intracellular proteolysis. They were for example instrumental in showing that the Ub-proteasome pathway is directly involved in the increase of proteolysis responsible for muscle wasting in many pathologies (Attaix *et al* 1994). They can also help in understanding how the different intracellular proteolytic pathways cooperate to degrade proteins. For example, as mentioned above, studies with proteasome inhibitors have shown that these complexes are important for apoptosis, at least in certain conditions, in addition to the caspases. More recently, an unexpected observation suggested that another protease can, at least partially, take over some of the proteasome's functions within the cell. After exposure of cultured EL-4 mouse lymphoma cells to proteasome inhibitors, although most of them died, some continued to grow and became resistant to the inhibitor. Analysis of the inhibitor-resistant cells suggested that a new protease, different from the proteasome, was induced (Glas *et al* 1998). It has been proposed that this protease could be the giant protein TPPII (tripeptidyl peptidase II) (Geier *et al* 1999), but another report, based on work with fission yeast, suggested that it could be another protease that the authors named "multicorn" (Osmulski and Gaczynska 1998).

Using proteasome inhibitors, it is possible to determine where proteasome-dependent proteolysis normally occurs. Interestingly, undegraded proteins seem to aggregate in a perinuclear region (Wojcik 1997) that appears to contain proteasomes, ubiquitin and chaperone proteins. As this region also contains γ-tubulin and corresponds to the centrosome (Wigley *et al* 1999), it has been suggested that the centrosome could be a center regulating proteasome function (Fabunmi *et al* 2000).

4.4 PERSPECTIVES

It has now become clear that proteasomes play an important role in most intracellular processes, especially those of an "irreversible" character such as cell-cycle, differentiation, apoptosis and signal transduction. Proteasome inhibition is thus a promising avenue to retard or block degradation of specific proteins to correct diverse pathologies. The prominent role of proteasomes in proliferation, through degradation of cell-cycle regulators, and in inflammation, through activation of NF-κB, suggests the potential to exploit proteasome inhibitors as anti-tumoral, pro-apoptotic or anti-inflammatory agents. As mentioned above, encouraging anti-inflammatory or anti-tumoral effects have already been obtained with certain inhibitors. In addition, proteasome inhibitors have a promising future for the treatment of muscle wasting (cachexia) due to excessive proteolysis. However, due to the pleiotropic role of the Ub-proteasome system, the toxic side effects of these compounds may strongly limit their potential. Obviously, an important route for progress will be to improve targeting proteasome inhibitors to the appropriate cells, but, at this point, it seems difficult to use these inhibitors as drugs to correct abnormal degradation of specific proteins. A possible solution could be to design inhibitors that block or retard degradation of only a specific set of substrates, and thus decrease the toxicity of the inhibitors. It can be imagined that molecules able to interfere with only one of the proteolytic sites of the proteasome, leaving the others able to function normally, could have such properties. But this goal is more likely to be attained by targeting the regulatory particle of the 26S proteasome. Indeed, this complex possesses other activities (ATPases, isopeptidase) that could possibly play a substrate-specific role. In addition, it contains several subunits that have no clear function, and that could be involved in substrate recognition. Another field that remains to be explored is the domain of endogenous inhibitors of proteasomes. As mentioned above, very little work has been done in this respect, and it is possible that new molecules will be discovered that could be specifically activated to modulate proteasome-dependent turn-over of specific proteins.

ACKNOWLEDGEMENTS

The authors wish to thank their colleagues, especially Dr. Marc Piechaczyk and Dr. Richard A. Engh, for critical reading of the manuscript. Prof. Robert Huber is acknowledged for his support to MG. The work of OC is supported by the French CNRS, by the "Association pour la Recherche sur le Cancer", and the "Fondation pour la Recherche Médicale".

REFERENCES

Adams, J., Behnke, M., Chen, S., Cruickshank, A.A., Dick, L.R., Grenier, L. *et al* (1998) Potent and selective inhibitors of the proteasome: dipeptidyl boronic acids. *Bioorganic & Medicinal Chemistry Letters*, 8, 333–338.

Adams, J., Palombella, V.J., Sausville, E.A., Johnson, J., Destree, A., Lazarus, D.D. *et al* (1999) Proteasome inhibitors: a novel class of potent and effective antitumor agents. *Cancer Research*, **59**, 2615–2622.

Akopian, T.N., Kisselev, A.F. and Goldberg, A.L. (1997) Processive degradation of proteins and other catalytic properties of the proteasome from Thermoplasma acidophilum. *Journal of Biological Chemistry*, **272**, 1791–1798.

An, B., Goldfarb, R.H., Siman, R. and Dou, Q.P. (1998) Novel dipeptidyl proteasome inhibitors overcome Bcl-2 protective function and selectively accumulate the cyclin-dependent kinase inhibitor p27 and induce apoptosis in transformed, but not normal, human fibroblasts. *Cell Death and Differentiation*, **5**, 1062–1075.

Andre, P., Groettrup, M., Klenerman, P., de Giuli, R., Booth, B.L., Jr., Cerundolo, V. *et al* (1998) An inhibitor of HIV-1 protease modulates proteasome activity, antigen presentation, and T cell responses. *Proceedings of the National Academy of Sciences of the U.S.A.*, **95**, 13120–13124.

Attaix, D., Taillandier, D., Temparis, S., Larbaud, D., Aurousseau, E., Combaret, L. *et al* (1994) Regulation of ATP-ubiquitin-dependent proteolysis in muscle wasting. *Reproduction, Nutrition, Development*, **34**, 583–597.

Bochtler, M., Ditzel, L., Groll, M., Hartmann, C. and Huber, R. (1999) The proteasome. *Annual Review of Biophysics and Biomolecular Structure*, **28**, 295–317.

Bogyo, M., McMaster, J.S., Gaczynska, M., Tortorella, D., Goldberg, A.L. and Ploegh, H. (1997) Covalent modification of the active site threonine of proteasomal beta subunits and the *Escherichia coli* homolog HslV by a new class of inhibitors. *Proceedings of the National Academy of Sciences of the U.S.A.*, **94**, 6629–6634.

Bogyo, M., Shin, S., McMaster, J.S. and Ploegh, H.L. (1998) Substrate binding and sequence preference of the proteasome revealed by active-site-directed affinity probes. *Chemistry & Biology*, **5**, 307–320.

Brannigan, J.A., Dodson, G., Duggleby, H.J., Moody, P.C.E., Smith, J.L., Tomchick, D.R. *et al* (1995) A protein catalytic framework with an N-terminal nucleophile is capable of self-activation. *Nature*, **378**, 416–419.

Braun, B.C., Glickman, M., Kraft, R., Dahlmann, B., Kloetzel, P.M., Finley, D. *et al* (1999) The base of the proteasome regulatory particle exhibits chaperone-like activity. *Nature Cell Biology*, **1**, 221–226.

Bush, K.T., Goldberg, A.L. and Nigam, S.K. (1997) Proteasome inhibition leads to a heat-shock response, induction of endoplasmic reticulum chaperones, and thermotolerance. *Journal of Biological Chemistry*, **272**, 9086–9092.

Cardozo, C., Michaud, C. and Orlowski, M. (1999) Components of the bovine pituitary multicatalytic proteinase complex (proteasome) cleaving bonds after hydrophobic residues. *Biochemistry*, **38**, 9768–9777.

Cardozo, C., Vinitsky, A., Michaud, C. and Orlowski, M. (1994) Evidence that the nature of amino acid residues in the P-3 position directs substrates to distinct catalytic sites of the pituitary multicatalytic proteinase complex (proteasome). *Biochemistry*, **33**, 6483–6489.

Conconi, M., Petropoulos, I., Emod, I., Turlin, E., Biville, F. and Friguet, B. (1998) Protection from oxidative inactivation of the 20S proteasome by heat-shock protein 90. *Biochemical Journal*, **333**, 407–415.

Coux, O., Tanaka, K. and Goldberg, A.L. (1996) Structure and functions of the 20S and 26S proteasomes. *Annual Review of Biochemistry*, **65**, 801–847.

Craiu, A., Gaczynska, M., Akopian, T., Gramm, C.F., Fenteany, G., Goldberg, A.L. *et al* (1997) Lactacystin and clasto-lactacystin beta-lactone modify multiple proteasome beta-subunits and inhibit intracellular protein degradation and major histocompatibility complex class I antigen presentation. *Journal of Biological Chemistry*, **272**, 13437–13445.

DeMartino, G.N. and Slaughter, C.A. (1999) The proteasome, a novel protease regulated by multiple mechanisms. *Journal of Biological Chemistry*, **274**, 22123–22126.

Deshaies, R.J. (1999) SCF and cullin/ring H2-based ubiquitin ligases. *Annual Review of Cell Developmental Biology*, **15**, 435–467.

Deveraux, Q., Ustrell, V., Pickart, C. and Rechsteiner, M. (1994) A 26 S protease subunit that binds ubiquitin conjugates. *Journal of Biological Chemistry*, **269**, 7059–7061.

Dick, L.R., Cruikshank, A.A., Destree, A.T., Grenier, L., McCormack, T.A., Melandri, F.D. *et al* (1997) Mechanistic studies on the inactivation of the proteasome by lactacystin in cultured cells. *Journal of Biological Chemistry*, **272**, 182–188.

Dick, T.P., Nussbaum, A.K., Deeg, M., Heinemeyer, W., Groll, M., Schirle, M. *et al* (1998) Contribution of proteasomal beta-subunits to the cleavage of peptide substrates analyzed with yeast mutants. *Journal of Biological Chemistry*, **273**, 25637–25646.

Ditzel, L., Huber, R., Mann, K., Heinemeyer, W., Wolf, D.H. and Groll, M. (1998) Conformational constraints for protein self-cleavage in the proteasome. *Journal of Molecular Biology*, **279**, 1187–1191.

Drexler, H.C., Risau, W. and Konerding, M.A. (2000) Inhibition of proteasome function induces programmed cell death in proliferating endothelial cells. *FASEB Journal*, **14**, 65–77.

Drexler, H.C.A. (1997) Activation of the cell death program by inhibition of proteasome function. *Proceedings of the National Academy of Sciences of the U.S.A.*, **94**, 855–860.

Driscoll, J., Frydman, J. and Goldberg, A.L. (1992) An ATP-stabilized inhibitor of the proteasome is a component of the 1500-kDa ubiquitin conjugate-degrading complex. *Proceedings of the National Academy of Sciences of the U.S.A.*, **89**, 4986–4990.

Duggleby, H.J., Tolley, S.P., Hill, C.P., Dodson, E.J., Dodson, G. and Moody, P.C. (1995) Penicillin acylase has a single-amino-acid catalytic centre. *Nature*, **373**, 264–268.

Elofsson, M., Splittgerber, U., Myung, J., Mohan, R. and Crews, C.M. (1999) Towards subunit-specific proteasome inhibitors: synthesis and evaluation of peptide alpha′,beta′-epoxyketones. *Chemistry & Biology*, **6**, 811–822.

Enenkel, C., Lehmann, H., Kipper, J., Guckel, R., Hilt, W. and Wolf, D.H. (1994) PRE3, highly homologous to the human major histocompatibility complex-linked LMP2 (ring12) gene, codes for a yeast proteasome subunit necessary for the peptidylglutamyl-peptide hydrolyzing activity. *FEBS Letters*, **341**, 193–196.

Fabunmi, R.P., Wigley, W.C., Thomas, P.J. and DeMartino, G.N. (2000) Activity and regulation of the centrosome-associated proteasome. *Journal of Biological Chemistry*, **275**, 409–413.

Fenteany, G., Standaert, R.F., Lane, W.S., Choi, S., Corey, E.J. and Schreiber, S.L. (1995) Inhibition of proteasome activities and subunit-specific amino-terminal threonine modification by lactacystin. *Science*, **268**, 726–731.

Ganoth, D., Leshinsky, E., Eytan, E. and Hershko, A. (1988) A multicomponent system that degrades proteins conjugated to ubiquitin. Resolution of factors and evidence for ATP-dependent complex formation. *Journal of Biological Chemistry*, **263**, 12412–12419.

Geier, E., Pfeifer, G., Wilm, M., Lucchiari-Hartz, M., Baumeister, W., Eichmann, K. *et al* (1999) A giant protease with potential to substitute for some functions of the proteasome. *Science*, **283**, 978–981.

Glas, R., Bogyo, M., McMaster, J.S., Gaczynska, M. and Ploegh, H.L. (1998) A proteolytic system that compensates for loss of proteasome function. *Nature*, **392**, 618–622.

Grimm, L.M., Goldberg, A.L., Poirier, G.G., Schwartz, L.M. and Osborne, B.A. (1996) Proteasomes play an essential role in thymocyte apoptosis. *EMBO Journal*, **15**, 3835–3844.

Grisham, M.B., Palombella, V.J., Elliott, P.J., Conner, E.M., Brand, S., Wong, H.L. *et al* (1999) Inhibition of NF-kappa B activation *in vitro* and *in vivo*: role of 26S proteasome. *Methods in Enzymology*, **300**, 345–363.

Groettrup, M. and Schmidtke, G. (1999) Selective proteasome inhibitors: modulators of antigen presentation. *Drug Discovery Today*, **4**, 63–71.

Groettrup, M., Soza, A., Eggers, M., Kuehn, L., Dick, T.P., Schild, H. *et al* (1996) A role for the proteasome regulator PA28-alpha in antigen presentation. *Nature*, **381**, 166–168.

Groll, M., Ditzel, L., Lowe, J., Stock, D., Bochtler, M., Bartunik, H.D. *et al* (1997) Structure of 20S proteasome from yeast at 2.4 Å resolution. *Nature*, **386**, 463–471.

Groll, M., Heinemeyer, W., Jager, S., Ullrich, T., Bochtler, M., Wolf, D.H. *et al* (1999) The catalytic sites of 20S proteasomes and their role in subunit maturation: A mutational and crystallographic study. *Proceedings of the National Academy of Sciences of the U.S.A.*, **96**, 10976–10983.

Groll, M., Kim, K.B., Kairies, N., Huber, R. and Crews, C.M. (2000) Crystal structure of epoxomicin: 20S proteasome reveals a molecular basis for selectivity of α', β'-epoxyketone proteasome inhibitors. *Journal of the American Chemical Society*, **122**, 1237–1238.

Guo, G.G., Gu, M. and Etlinger, J.D. (1994) 240-kDa proteasome inhibitor (CF-2) is identical to delta-aminolevulinic acid dehydratase. *Journal of Biological Chemistry*, **269**, 12399–12402.

Heinemeyer, W., Fischer, M., Krimmer, T., Stachon, U. and Wolf, D.H. (1997) The active sites of the eukaryotic 20 S proteasome and their involvement in subunit precursor processing. *Journal of Biological Chemistry*, **272**, 25200–25209.

Hendil, K.B., Khan, S. and Tanaka, K. (1998) Simultaneous binding of PA28 and PA700 activators to 20S proteasomes. *Biochemical Journal*, **332**, 749–754.

Hershko, A. and Ciechanover, A. (1998) The ubiquitin system. *Annual Review of Biochemistry*, **67**, 425–479.

Huang, Y.Z., Baker, R.T. and Fischervize, J.A. (1995) Control of cell fate by a deubiquitinating enzyme encoded by the fat facets gene. *Science*, **270**, 1828–1831.

Jager, S., Groll, M., Huber, R., Wolf, D.H. and Heinemeyer, W. (1999) Proteasome beta-type subunits: unequal roles of propeptides in core particle maturation and a hierarchy of active site function. *Journal of Molecular Biology*, **291**, 997–1013.

Kam, Y.A., Xu, W., Demartino, G.N. and Cohen, R.E. (1997) Editing of ubiquitin conjugates by an isopeptidase in the 26S proteasome. *Nature*, **385**, 737–740.

Katagiri, M., Hayashi, M., Matsuzaki, K., Tanaka, H. and Omura, S. (1995) The neuritogenesis inducer lactacystin arrests cell cycle at both G0/G1 and G2 phases in neuro 2a cells. *Journal of Antibiotics*, **48**, 344–346.

Kim, S.S., Rhee, S., Lee, K.H., Kim, J.H., Kim, H.S., Kang, M.S. *et al* (1998) Inhibitors of the proteasome block the myogenic differentiation of rat L6 myoblasts. *FEBS Letters*, **433**, 47–50.

Kisselev, A.F., Akopian, T.N., Castillo, V. and Goldberg, A.L. (1999a) Proteasome active sites allosterically regulate each other, suggesting a cyclical bite-chew mechanism for protein breakdown. *Molecular Cell*, **4**, 395–402.

Kisselev, A.F., Akopian, T.N., Woo, K.M. and Goldberg, A.L. (1999b) The sizes of peptides generated from protein by mammalian 26 and 20 S proteasomes. Implications for understanding the degradative mechanism and antigen presentation. *Journal of Biological Chemistry*, **274**, 3363–3371.

Lee, D.H. and Goldberg, A.L. (1998a) Proteasome inhibitors cause induction of heat shock proteins and trehalose, which together confer thermotolerance in *Saccharomyces cerevisiae*. *Molecular & Cellular Biology*, **18**, 30–38.

Lee, D.H. and Goldberg, A.L. (1998b) Proteasome inhibitors: valuable new tools for cell biologists. *Trends in Cell Biology*, **8**, 397–403.

Li, X.C., Gu, M.Z. and Etlinger, J.D. (1991) Isolation and characterization of a novel endogenous inhibitor of the proteasome. *Biochemistry*, **30**, 9709–9715.

Li, X.S. and Etlinger, J.D. (1992) Ubiquitinated proteasome inhibitor is a component of the 26S proteasome complex. *Biochemistry*, **31**, 11964–11967.

Lightcap, E.S., McCormack, T.A., Pien, C.S., Chau, V., Adams, J. and Elliott, P.J. (2000) Proteasome inhibition measurements: clinical application. *Clinical Chemistry*, **46**, 673–683.

Lin, K.I., Baraban, J.M. and Ratan, R.R. (1998) Inhibition versus induction of apoptosis by proteasome inhibitors depends on concentration. *Cell Death & Differentiation*, **5**, 577–583.

Loidl, G., Groll, M., Musiol, H.J., Ditzel, L., Huber, R. and Moroder, L. (1999) Bifunctional inhibitors of the trypsin-like activity of eukaryotic proteasomes. *Chemistry & Biology*, **6**, 197–204.

Lopes, U.G., Erhardt, P., Yao, R.J. and Cooper, G.M. (1997) P53-dependent induction of apoptosis by proteasome inhibitors. *Journal of Biological Chemistry*, **272**, 12893–12896.

Löwe, J., Stock, D., Jap, B., Zwickl, P., Baumeister, W. and Huber, R. (1995) Crystal structure of the 20S proteasome from the archaeon *T. acidophilum* at 3.4 Å resolution. *Science*, **268**, 533–539.

McCutchen-Maloney, S.L., Matsuda, K., Shimbara, N., Binns, D.D., Tanaka, K., Slaughter, C.A. *et al* (2000) cDNA cloning, expression, and functional characterization of PI31, a proline-rich inhibitor of the proteasome. *Journal of Biological Chemistry*, **275**, 18557–18565.

Meng, L., Kwok, B.H., Sin, N. and Crews, C.M. (1999a) Eponemycin exerts its antitumor effect through the inhibition of proteasome function. *Cancer Research*, **59**, 2798–2801.

Meng, L., Mohan, R., Kwok, B.H., Elofsson, M., Sin, N. and Crews, C.M. (1999b) Epoxomicin, a potent and selective proteasome inhibitor, exhibits *in vivo* antiinflammatory activity. *Proceedings of the National Academy of Sciences of the U.S.A.*, **96**, 10403–10408.

Meriin, A.B., Gabai, V.L., Yaglom, J., Shifrin, V.I. and Sherman, M.Y. (1998) Proteasome inhibitors activate stress kinases and induce Hsp72. Diverse effects on apoptosis. *Journal of Biological Chemistry*, **273**, 6373–6379.

Mitch, W.E. and Goldberg, A.L. (1996) Mechanisms of disease – Mechanisms of muscle wasting – The role of the Ubiquitin-proteasome pathway. *New England Journal of Medicine*, **335**, 1897–1905.

Murakami, K. and Etlinger, J.D. (1986) Endogenous inhibitor of nonlysosomal high molecular weight protease and calcium-dependent protease. *Proceedings of the National Academy of Sciences of the U.S.A.*, **83**, 7588–7592.

Omura, S., Fujimoto, T., Otoguro, K., Matsuzaki, K., Moriguchi, R., Tanaka, H. *et al* (1991) Lactacystin, a novel microbial metabolite, induces neuritogenesis of neuroblastoma cells. *Journal of Antibiotics*, **44**, 113–116.

Orlowski, M. (1990) The Multicatalytic Proteinase Complex, a major extralysosomal proteolytic system. *Biochemistry*, **29**, 10289–10297.

Orlowski, M., Cardozo, C. and Michaud, C. (1993) Evidence for the presence of five distinct proteolytic components in the pituitary multicatalytic proteinase complex. Properties of two components cleaving bonds on the carboxyl side of branched chain and small neutral amino acids. *Biochemistry*, **32**, 1563–1572.

Orlowski, R.Z. (1999) The role of the ubiquitin-proteasome pathway in apoptosis. *Cell Death & Differentiation*, **6**, 303–313.

Orlowski, R.Z., Eswara, J.R., Lafond-Walker, A., Grever, M.R., Orlowski, M. and Dang, C.V. (1998) Tumor growth inhibition induced in a murine model of human Burkitt's lymphoma by a proteasome inhibitor. *Cancer Research*, **58**, 4342–4348.

Osmulski, P.A. and Gaczynska, M. (1998) A new large proteolytic complex distinct from the proteasome is present in the cytosol of fission yeast. *Current Biology*, **8**, 1023–1026.

Ostrowska, H., Wojcik, C., Omura, S. and Worowski, K. (1997) Lactacystin, a specific inhibitor of the proteasome, inhibits human platelet lysosomal cathepsin A-like enzyme. *Biochemical & Biophysical Research Communications*, **234**, 729–732.

Palombella, V.J., Rando, O.J., Goldberg, A.L. and Maniatis, T. (1994) The ubiquitin-proteasome pathway is required for processing the NF-kappa-B1 precursor protein and the activation of NF-kappa-B. *Cell*, **78**, 773–785.

Robertson, J.D., Datta, K., Biswal, S.S. and Kehrer, J.P. (1999) Heat-shock protein 70 antisense oligomers enhance proteasome inhibitor-induced apoptosis. *Biochemical Journal*, **344**, 477–485.

Rock, K.L. and Goldberg, A.L. (1999) Degradation of cell proteins and the generation of MHC class I-presented peptides. *Annual Review of Immunology*, **17**, 739–779.

Rock, K.L., Gramm, C., Rothstein, L., Clark, K., Stein, R., Dick, L. *et al* (1994) Inhibitors of the proteasome block the degradation of most cell proteins and the generation of peptides presented on MHC class 1 molecules. *Cell*, **78**, 761–771.

Rubin, D.M., Glickman, M.H., Larsen, C.N., Dhruvakumar, S. and Finley, D. (1998) Active site mutants in the six regulatory particle ATPases reveal multiple roles for ATP in the proteasome. *EMBO Journal*, **17**, 4909–4919.

Sadoul, R., Fernandez, P.A., Quiquerez, A.L., Martinou, I., Maki, M., Schroter, M. *et al* (1996) Involvement of the proteasome in the programmed cell death of NGF-deprived sympathetic neurons. *EMBO Journal*, **15**, 3845–3852.

Schmidtke, G., Holzhutter, H.G., Bogyo, M., Kairies, N., Groll, M., de Giuli, R. *et al* (1999) How an inhibitor of the HIV-I protease modulates proteasome activity. *Journal of Biological Chemistry*, **274**, 35734–33740.

Schwartz, A.L. and Ciechanover, A. (1999) The ubiquitin-proteasome pathway and pathogenesis of human diseases. *Annual Review of Medicine*, **50**, 57–74.

Seemuller, E., Lupas, A. and Baumeister, W. (1996) Autocatalytic processing of the 20S proteasome. *Nature*, **382**, 468–470.

Seemuller, E., Lupas, A., Stock, D., Lowe, J., Huber, R. and Baumeister, W. (1995) Proteasome from *Thermoplasma acidophilum* – a threonine protease. *Science*, **268**, 579–582.

Sherwood, S.W., Kung, A.L., Roitelman, J., Simoni, R.D. and Schimke, R.T. (1993). *In vivo* inhibition of cyclin B degradation and induction of cell-cycle arrest in mammalian cells by the neutral cysteine protease inhibitor N-acetylleucylleucylnorleucinal. *Proceedings of the National Academy of Sciences of the U.S.A.*, **90**, 3353–3357.

Tanahashi, N., Murakami, Y., Minami, Y., Shimbara, N., Hendil, K.B. and Tanaka, K. (2000) Hybrid proteasomes. Induction by interferon-gamma and contribution to ATP-dependent proteolysis. *Journal of Biological Chemistry*, **275**, 14336–14345.

Teicher, B.A., Ara, G., Herbst, R., Palombella, V.J. and Adams, J. (1999) The proteasome inhibitor PS-341 in cancer therapy. *Clinical Cancer Research*, **5**, 2638–2645.

Traenckner, E.B.M., Wilk, S. and Baeuerle, P.A. (1994) A proteasome inhibitor prevents activation of NF-kappa-B and stabilizes a newly phosphorylated form of I-kappa-B-alpha that is still bound to NF-kappa-B. *EMBO Journal*, **13**, 5433–5441.

Tsubuki, S., Kawasaki, H., Saito, Y., Miyashita, N., Inomata, M. and Kawashima, S. (1993) Purification and characterization of a Z-Leu-Leu-Leu-MCA degrading protease expected to regulate neurite formation: a novel catalytic activity in proteasome. *Biochemical & Biophysical Research Communications*, **196**, 1195–1201.

Tsubuki, S., Saito, Y. and Kawashima, S. (1994) Purification and characterization of an endogenous inhibitor specific to the Z-leu-leu-leu-MCA degrading activity in proteasome and its identification as heat-shock protein 90. *FEBS Letters*, **344**, 229–233.

Vinitsky, A., Cardozo, C., Sepp-Lorenzino, L., Michaud, C. and Orlowski, M. (1994) Inhibition of the proteolytic activity of the multicatalytic proteinase complex (proteasome) by substrate-related peptidyl aldehydes. *Journal of Biological Chemistry*, **269**, 29860–29866.

Voges, D., Zwickl, P. and Baumeister, W. (1999) The 26S proteasome: a molecular machine designed for controlled proteolysis. *Annual Review of Biochemistry*, **68**, 1015–1068.

Wenzel, T. and Baumeister, W. (1995) Conformational constraints in protein degradation by the 20S proteasome. *Nature Structural Biology*, **2**, 199–204.

Wenzel, T., Eckerskorn, C., Lottspeich, F. and Baumeister, W. (1994) Existence of a molecular ruler in proteasomes suggested by analysis of degradation products. *FEBS Letters*, **349**, 205–209.

Wigley, W.C., Fabunmi, R.P., Lee, M.G., Marino, C.R., Muallem, S., DeMartino, G.N. *et al* (1999) Dynamic association of proteasomal machinery with the centrosome. *Journal of Cell Biology*, **145**, 481–490.

Wojcik, C. (1997) An inhibitor of the chymotrypsin-like activity of the proteasome (PSI) induces similar morphological changes in various cell lines. *Folia Histochemica et Cytobiologica*, **35**, 211–214.

Wojcik, C., Schroeter, D., Stoehr, M., Wilk, S. and Paweletz, N. (1996). An inhibitor of the chymotrypsin-like activity of the multicatalytic proteinase complex (20S proteasome) induces arrest in G2-phase and metaphase in HeLa cells. *European Journal of Cell Biology*, **70**, 172–178.

Chapter 5

Cathepsins

Brian Rukamp and James C. Powers

Cysteine proteases enzymes, both inherent and virus-introduced ones, are found in plants, bacteria, protozoa, fungi, and mammals. They are involved in the hydrolysis of many different proteins, playing a major role in intracellular protein degradation and turnover. Cathepsins are known to have many biological roles in addition to their major role in intracellular proteolysis. These include bone remodeling, prohormone activation, and antigen processing. They are suspected to have roles in a variety of disease states including viral and parasitic diseases, rheumatoid arthritis, Alzheimer's disease, cancer metastasis, osteoporosis, and pulmonary emphysema. For these reasons cathepsins make attractive targets for inhibitor drug therapy.

5.1 INTRODUCTION

Cysteine proteases are found in plants, bacteria, protozoa, fungi, and mammals, both those that are inherent in the species and those introduced by infecting viruses (Barrett 1986). These enzymes participate in the hydrolysis of many different proteins and play a major role in intracellular protein degradation and turnover. The cathepsins are one group of cysteine proteases of particular interest. Originally, cathepsins were defined as proteases that were active at acidic pH but that were distinguished from pepsin (Willstätter and Bamann 1929). The majority of cathepsins are presently defined as intracellular proteases, mostly found in the lysosome, that are active at acidic pH, and belong to the C1 family of cysteine proteases in the papain superfamily (Rawlings and Barrett 1993, 1994; Berti and Storer 1995; Barrett *et al* 1998). The proposed evolutionary relationship of the cathepsins, based on that proposed by Santamaría in 1999 and expanded on by Sol-Church in 2000, is shown in Figure 5.1 (Santamaría *et al* 1999; Sol-Church *et al* 2000c). Other cathepsins include the serine protease cathepsins A, G, and R, the aspartic protease cathepsins D and E, and the metalloprotease cathepsin III (Otto and Schirmeister 1997). This chapter will only discuss cathepsins which are cysteine proteases.

Cathepsins are usually characterized as lysosomal enzymes due to the presence of potential N-glycosylation sites, the presence of signal sequences, usually having a pH optima at lower pH's, and lack of stability in the physiological pH range (Wang *et al* 1998). In addition, they can be divided into two functional groups: (1) extensively expressed, intracellular housekeeping proteases responsible for the general lysosomal breakdown of proteins (cathepsins B, C, F, H, L, M (rabbit liver), O, and X); and (2) cathepsins significantly restricted in distribution to certain tissues and/or assigned specific functions

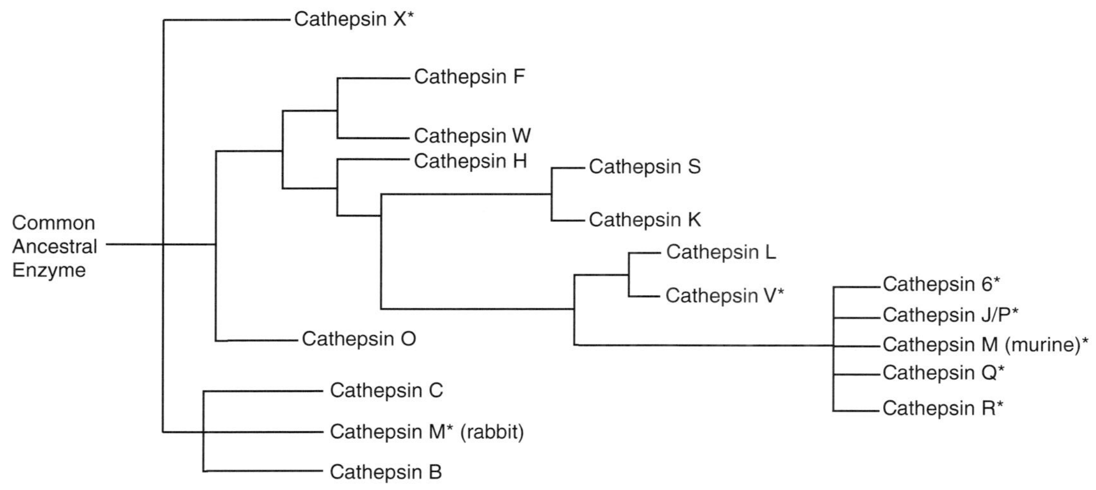

* probable position

Figure 5.1 Evolutionary relationship of cathepsins.

(cathepsins 6, J/P, K, M (murine placental), Q, R, S, V, and W) (Wang *et al* 1998; Sol-Church *et al* 2000c).

Cathepsins are known to have many biological roles in addition to their major role in intracellular proteolysis. These include bone remodeling (Tezuka *et al* 1994), prohormone activation (Krieger and Hook 1991), and antigen processing (Riese 1996). They are suspected to have roles in a variety of disease states including viral and parasitic diseases (Korant *et al* 1979), rheumatoid arthritis (Mort *et al* 1984), Alzheimer's disease (Golde *et al* 1992), cancer metastasis (Sloane 1990), osteoporosis (Delaisse *et al* 1991), and pulmonary emphysema (Mason *et al* 1986). For these reasons, cathepsins make attractive targets for inhibitor drug therapy.

5.2 MECHANISM AND ACTIVE SITE

Cathepsins, similar to all cysteine proteases, contain the highly conserved catalytic triad of Cys-25, His-159, and Asn-175 (papain numbering system) (Berti and Storer 1995). It has been speculated that the Cys-25 and His-159 residues of the active site may not be capable of full hydrolysis on their own, and that the Asn-175 helps to align the His residue, through H-bonding, to help stabilize and aid in the hydrolytic activity of the catalytic pair (Kirschke and Wiederanders 1994). The normal state of these residues in the catalytic pocket appears to be a negatively charged thiolate residue in conjunction with a protonated His residue (Lewis *et al* 1976; Polgar and Halasz 1982). The highly nucleophilic cysteine thiolate hydrolyzes proteins through an acyl thioester intermediate formed upon attack on the peptide bond (Fink and Angelides 1976). The hydrolysis of a substrate specific to the cathepsin begins via an attack by the enzyme's active site cysteine on the amide scissile bond forming a covalent tetrahedral adduct. The substrate is protonated immediately before or during the attack (Plapp 1982; Dufour *et al* 1995). Release of the amide portion of the substrate results in the acyl group of the substrate being bound in the active site in the form of a thioester. Finally, an attack by a water molecule on the carbonyl and the subsequent "reverse" reaction releases the newly cleaved peptide, returning the enzyme to its native state. This type of peptide degradation is reminiscent of that of serine proteases, making inhibitor distinction between the two types of enzymes challenging (Govardhan and Abeles 1996). However, cysteine proteases, unlike serine proteases, have a protonated His residue in the active site that is able to donate a proton to the scissile bond in conjunction with the peptide cleavage (Walsh 1979; Fersht 1985). This mechanistic difference is the reason that some inhibitor types are able to distinguish between these two classes of enzymes (Rich 1986). Crystal structures of substrate-like inhibitors complexed with papain show that the nucleophilic attack by the thiolate on the carbonyl takes place from the *si* face, while other cysteine proteases attack the *re* face (Babine and Bender 1997).

A highly conserved region called the oxyanion hole, composed of Gln-19 and the backbone hydrogen donated from Cys-25, is used to stabilize the tetrahedral intermediate during substrate hydrolysis (Ménard *et al* 1991; Ménard *et al* 1995). In addition, mature cathepsins contain a conserved Trp-177 immediately adjacent to the active site cysteine residue in the S1′ binding pocket, and both a Gly-67 and Gly-68 located in the C-terminal region (Berti and Storer 1995).

The extended substrate binding site of cysteine proteases are frequently able to recognize peptide residues from the S4 subsite to the S3′ subsite (Schechter and Berger

1967). The primary specificity determinant of cathepsins is the P2-S2 subsite interaction (Baker and Drenth 1987; Gour-Salin *et al* 1994). While the other subsites are shallow clefts on the surface of the enzymes, the S2 subsite forms an actual pocket (Kamphuis *et al* 1985; Musil *et al* 1991; McGrath *et al* 1995, 1997). The depth, size, and residues lining this pocket account for the specificity differences found for substrates and inhibitors between the varying cathepsins. The S′ subsites are quite similar in most cathepsins and the majority of differences are seen in the S subsites (Schechter and Berger 1967). Although commonly having only endo- or exo-peptidase activity, several cathepsins exhibit both types of activity. Depending on the P2-P1 sequence of the substrates, one activity generally dominates over the other (Nägler *et al* 1999a). The subsite specificities of many of the cathepsins are shown in Table 5.1.

The hunt for selective inhibitors for cathepsins has been difficult due to their broad substrate specificity and their common active site structure (Nägler *et al* 1999a). Generally, known inhibitors of cysteine proteases occupy only one-half of the enzyme binding site and usually contain a reactive functional group or warhead (Demuth 1990). Most of the inhibitors designed have had a peptide sequence that can be recognized by the desired enzyme as well as an electrophilic warhead that can interact with the thiolate of the active site.

5.3 THE ENZYMES

Cathepsin B (EC 3.4.22.1), also known as cathepsin B1, is a lysosomal cysteine dipeptidyl carboxypeptidase (Aronson and Barrett 1978) that possesses some endopeptidase activity (McDonald and Ellis 1975). It is suspected that the primary role of cathepsin B is its dipeptidyl activity (Illy *et al* 1997). The enzymatic dipeptidyl carboxypeptidase activity seems due to two His residues, present on an "occluding loop" (Musil *et al* 1991), which line up well with and interact with the carboxy end of substrates beyond the S2′ subsite (Turk *et al* 1995; Illy *et al* 1997; Nägler *et al* 1997; Schaschke *et al* 1998). The favorable interactions of these His residues allow the acceptance of a wide range of substrates in the normally dominating S2 subsite, giving the enzyme broad specificity (Nägler *et al* 1999a). The enzyme is widely distributed throughout the body and in many tissues, although it is rarely found outside the cells in non-pathological tissue, possibly due to its instability in neutral and high pH's (Barrett 1973).

Cathepsin C (EC 3.4.14.1), also known as dipeptidyl peptidase I (DPPI), cathepsin J (Liao and Lenney 1984), and dipeptidyl transferase, is an aminoprotease of the papain family (McDonald and Schwabe 1977; Nikawa *et al* 1992). It has been suggested that the enzyme results from convergent evolution rather than diverging from a common cathepsin ancestor (Rao *et al* 1997). Unlike many other cathepsins which have a dominating S2 subsite for substrate specificity, cathepsin C can accept a larger variety of residues at that position (Nägler *et al* 1999a). The enzyme is believed to be involved in the regulation of the plasminogen-plasmin system and activation of granule-associated proteases of cytotoxic lymphocytes, myeloid and mast cells, in addition to its general housekeeping role (Horn *et al* 2000).

Cathepsin F is a new member of the papain protease family first obtained as a cDNA clone of γgt10-skeletal muscle (Wang *et al* 1998). Only a partial propeptide sequence was originally isolated, leading to the incorrect assumption that this enzyme lacked a signal sequence (Wang *et al* 1998). More recently, the enzyme was found to have an uncharacteristically large 251 residue sequence belonging to the propeptide (Santamaría *et al* 1999). Cathepsin F is thought to be a lysosomal enzyme, being expressed liberally

Table 5.1 Cathepsin subsite specificity

Cathepsin	*S3 subsite*	*S2 subsite*	*S1 subsite*	*S1′ subsite*	*S2′ subsite*
B	large hydrophobic[a]	basic, large hydrophobic, Phe[b,c,e]	small or Arg[f,g]	hydrophobic[f,g]	
C			wide range, Arg[g,r]	no Pro, Lys, or Arg[f]	no Pro, Lys, or Arg[f]
F		Leu and Phe[h]			
H			large hydrophobic group, free amine[q]		
K		small hydrophobic, Leu or Val[d,i]	hydrophobic, Arg or Lys[d,i]	aliphatic or aromatic[p]	
L	Leu or Phe[s]	non-polar, bulky, hydrophobic[j]	non-polar, bulky, hydrophobic[j]	long/aliphatic sidechains with or without charges[k]	
M[u]	no known substrate specificity data				
O	no known substrate specificity data				
S		small, non-polar[m]	aromatic or small non-polar[m,n]	basic residue[l]	
V			aromatic and non-aromatic[o]		
W	no known substrate specificity data				
X		hydrophobic, Phe[t]	Arg[t]	hydrophobic, Trp[t]	
6, J/P, M, R, Q (Murine placental)	no known substrate specificity data				

[a] Yasuma *et al* 1998; [b] Green and Shaw 1981; [c] Xing *et al* 1998; [d] Brömme *et al* 1996a; [e] Nägler *et al* 1997; [f] Otto and Schirmeister 1997; [g] Nägler *et al* 1999a,b; [h] Wang *et al* 1998; [i] Bossard *et al* 1996; [j] Fujishima *et al* 1997; [k] Gour-Salin *et al* 1994; [l] Brömme *et al* 1994; [m] Brömme *et al* 1989b; [n] Shaw *et al* 1993; [o] Brömme *et al* 1999; [p] Thompson *et al* 1997; [q] McGrath 1999; [r] McDonald *et al* 1969; [s] Brinker *et al* 2000; [t] Klemencic *et al* 2000; [u] Isolated from rabbit liver, believed to be homologous to cathepsin B based on tryptic cleavage, Erickson-Viitanen *et al* 1985.

throughout the body and suggesting an essential housekeeping function in cells (Wang *et al* 1998). The abnormally high expression of the enzyme in the brain and the testis suggests a possible role in brain processes and in fertilization, respectively (Wang *et al* 1998; Santamaría *et al* 1999).

Cathepsin H (EC 3.4.22.16), also known as cathepsin B3 and cathepsin BA, is a long known, but little studied lysosomal papain-like protease. The enzyme was originally isolated from rat liver (Kirschke *et al* 1976). Cathepsin H appears to have both aminopeptidase activity (Koga *et al* 1992), uncommon to cathepsins, and weak endopeptidase activity (Schwartz and Barrett 1980; Kirschke and Shaw 1981). Cathepsin H is thought to play its most prominant role in general protein degradation, although it has been suspected to also play a part in a number of pathological states, including tumor metastasis (Tsushima *et al* 1991).

Cathepsin K (EC 3.4.22.38), also called cathepsin O2 (Brömme and Okamoto 1995), cathepsin O (Shi *et al* 1995), cathepsin OC2 (Tezuka *et al* 1994), and cathepsin X (Li *et al* 1995), is a relatively new cysteine endoprotease of the papain-like superfamily (Tezuka *et al* 1994; Shi *et al* 1995). This enzyme was originally isolated from rabbit osteoclasts (Tezuka *et al* 1994). This was followed by isolation from human osteoclasts and ovaries (Brömme and Okamoto 1995), and porcine thyroid epithelial cells (Tebel *et al* 2000). Cathepsin K, which has an overall positive charge (McGrath 1999), possibly due to the acid environment where it is active, has been found to be nearly exclusively expressed in osteoclasts, the cells responsible for bone resorption (Tezuka *et al* 1994; Brömme *et al* 1996b). This enzyme represents a novel molecular target in the treatment of diseases associated with bone loss, like osteoporosis, bone malformation, like Paget's disease and pycnodysostosis (Gelb *et al* 1996), and certain types of arthritis (Bossard *et al* 1996; Marquis *et al* 1998) and cancers (Littlewood-Evans *et al* 1997).

Cathepsin L (EC 3.4.22.15) is a cysteine endopeptidase originally isolated from rat liver lysosomes (Kirschke *et al* 1977) and is thought to be involved in bone resorption (Kakegawa *et al* 1993), glomerulonephritis, arthritis, and cancer metastasis (Sheahan *et al* 1989; Baricos *et al* 1991; Bohley and Seglen 1992). The enzyme is also believed to function in thyroid hormone liberation (Brix *et al* 1996). Cathepsin L is of particular interest because it has the highest activity of all the cathepsins, as well as a higher specific activity in degradation of proteins than many of the collagen and lysosomal proteases (Barrett and Kirschke 1981; Kirschke *et al* 1982; Wang *et al* 1998).

Cathepsin M is a long known, but little studied cysteine protease. It was first isolated from rabbit liver and is associated with the lysosomal membrane (Pontremoli *et al* 1982). The enzyme is believed to be involved in modification and inactivation of fructose 1,6-bisphosphate aldolase (EC 4.1.2.13), fructose 1,6-bisphosphatase (EC 3.1.3.11) (Pontremoli *et al* 1982), and other cytosolic enzymes (Erickson-Viitanen *et al* 1985). Studies have shown that cathepsin M has a primary structure which is homologous to cathepsin B (Erickson-Viitanen *et al* 1985). However, its substrate specificity is similar to cathepsin H, so cathepsin M does not cleave many of the synthetic substrates specific for cathepsin B (Erickson-Viitanen *et al* 1985). In addition, the enzyme is sensitive to leupeptin, and is more stable at alkaline pH's than cathepsin H (Kirschke *et al* 1977). Cathepsin M is primarily a carboxypeptidase, but does seem to have some aminopeptidase activity (Erickson-Viitanen *et al* 1985). It also seems to be more reactive toward native proteins than cathepsin B (Erickson-Viitanen *et al* 1985).

Cathepsin O, different than the cathepsin K mistakenly labeled as cathepsin O (see above), is a relatively new cysteine protease isolated by cDNA encoding from a breast carcinoma cDNA library (Velasco *et al* 1994). The enzyme has been found in many different tissue types, suggesting that it plays a role in normal cellular protein degradation and turnover (Velasco *et al* 1994).

Cathepsin S (EC 3.4.22.27) is a highly active cysteine endopeptidase originally purified from bovine lymph nodes and spleen (Turnsek *et al* 1975). The enzyme was the first known tissue specific cysteine protease (Turnsek *et al* 1975; Kirschke *et al* 1986) and is highly expressed in lymphatic tissue (Kirshke *et al* 1989), macrophages (Shi *et al* 1992), brain (Petanceska and Devi 1992), spleen, lymph nodes, peripheral leukocytes (Kirschke and Wiederanders 1994), B-cells and dendritic cells (Chapman *et al* 1994; Reddy *et al* 1995; Linnevers *et al* 1997), and thyroid (Petanceska and Devi 1992). It is much less abundantly distributed throughout other tissues and organs in comparison to cathepsins L, B, and H (Brömme and McGrath 1996; Otto and Schirmeister 1997). The enzyme is active in and highly stable at slightly acidic, neutral, and alkaline pH's, but little is known about the actual function of cathepsin S (Kirschke *et al* 1986; Kirshke *et al* 1989). However, the locations of expression and its stability suggest a role for cathepsin S in inflammation and antigen processing (Petanceska and Devi 1992; Brömme and McGrath 1996). The enzyme appears to interact with MHC-II molecules, a class of membrane-bound proteins encoded by the major histocompatibility complex (MHC) that function as antigen-presenting markers which allow the immune system to distinguish between immune system cells and other cells. Cathepsin S plays a major role in MHC-II mediated antigen processing by preparing the peptide binding site of the MHC-II molecules, making the enzyme an attractive target for drug therapy against diseases that have hyper-immune responses (Riese 1996; Riese *et al* 1998).

Cathepsin V, also known as cathepsin L2 (Santamaría *et al* 1998a), is a newly discovered cysteine peptidase specifically expressed in the thymus and testis (Brömme *et al* 1999) and corneal epithelium. Oddly enough, detectable levels were not found in other immune-related organs such as fetal liver, appendix, lymph nodes, and bone marrow (Brömme *et al* 1999). Cathepsin V has also been found in colon carcinomas (Schilling and Ahlquist 1999), suggesting that it may play a role in cancer progression, and its detection may be able to be used as a marker (Somoza *et al* 2000). The expression of cathepsin V in antigen-presenting cells and organs is very similar to that of cathepsin S, and in turn is thought to play a potential role in the MHC class II antigen processing as well as potential degradation roles (Brömme *et al* 1999).

Cathepsin W is a relatively new lymphocyte-specific cysteine protease (Linnevers *et al* 1997; Wex *et al* 1988). The enzyme is specifically expressed in lymphatic tissues and high levels are expressed in T-lymphocytes, especially the $CD8^+$ cells (Linnevers *et al* 1997). This suggests a specific function in or regulation of the cytolytic activity of T-cells (Linnevers *et al* 1997). The enzyme is also detected extensively throughout the tissues of the immune system (Linnevers *et al* 1997). An uncommon feature associated with this enzyme is a 21 amino acid insertion between the active site histidine and asparagine residues with no known function as well as an eight amino acid C-terminal expansion (Linnevers *et al* 1997). Due to these additions, it has been speculated that cathepsin W belongs to its own subgroup of the papain superfamily (Linnevers *et al* 1997).

Cathepsin X, also called cathepsin Z, is different than cathepsin K (mistakenly labeled as cathepsin X, above). The enzyme is a relatively new cysteine protease found expressed in varying levels throughout the body, suggesting its function may be general housekeeping (Nägler and Ménard 1998; Santamaría *et al* 1998b). Cathepsin X was originally discovered through the human EST (expression sequence tag) database, and isolated through PCR amplification of human ovary (Nägler and Ménard 1998) and human brain cDNA libraries (Santamaría *et al* 1998b). Several unique features of this cathepsin suggest that it may be in a subfamily all its own. Other than several highly conserved residues, cathepsin X has very little homology with any other cathepsin (Nägler and Ménard 1998). Unlike the rest of the cathepsins, it contains a very short propeptide region, about 38 residues (or 41; Santamaría *et al* 1998b), compared to propeptide amino acid lengths of about 60 (cathepsin B) to more than 200 (cathepsin F) (Nägler and Ménard 1998). The propeptide also contains no secondary structure (Sivaraman *et al* 2000) and neither of the conserved regions found in the cathepsin B or cathepsin L propeptides (Santamaría *et al* 1998b). The conserved regions in the propeptides are usually used to classify a cathepsin into either the cathepsin B or cathepsin L subfamily. Autolytic cleavage of cathepsin X's proregion is also not seen, under appropriate conditions, like it is with other cathepsins (Nägler *et al* 1999b). This is likely due to a Cys–Cys bond between the active site residue and a cysteine on the proregion (Nägler *et al* 1999b). Cathepsin X has a three amino acid residue insertion, termed the "mini-loop", in the highly conserved region between the glutamine and cysteine on the primed side of the active site (Nägler and Ménard 1998; Santamaría *et al* 1998b; Guncar *et al* 2000; Sivaraman *et al* 2000). The mini-loop has a histidine residue which blocks access to the S2′ position. This allows interaction with free carboxylates on the C-terminus of substrates and appears to give cathepsin X mono-peptidyl carboxypeptidase activity, which is unique among cathepsins (Nägler *et al* 1999b; Guncar *et al* 2000; Klemencic *et al* 2000; Therrien *et al* 2001). In addition, cathepsin X is suspected to have some dipeptidyl carboxypeptidase activity because the histidine residue in the mini-loop can twist into a conformation resembling one of the histidines in cathepsin B's occluding loop (Guncar *et al* 2000; Klemencic *et al* 2000; Therrien *et al* 2001). However, because of the mini-loop, exopeptidase activity is severely restricted, compared to cathepsin B (Nägler *et al* 1999b), and there appears to be limited or no endopeptidase activity (Guncar *et al* 2000; Klemencic *et al* 2000; Therrien *et al* 2001). It is speculated that cathepsin X may be involved in Alzheimer's disease due to the enzyme's ability to cleave β-amyloid peptides, which can accumulate and form amyloid plaques (Guncar *et al* 2000).

A series of tissue specific cathepsin proteases, including cathepsin-6 (Nakajima *et al* 2000); cathepsin J/P (Sol-Church *et al* 1999; Tisljar *et al* 1999), cathepsin M (different than the rabbit liver enzyme listed above) (Sol-Church *et al* 2000c), cathepsin Q (Sol-Church *et al* 2000b), and cathepsin R (Sol-Church *et al* 2000a), have recently been found in labyrinthine trophoblasts of murine embryo placenta, but not in any tissue-type of adult rodents. Sol-Church and colleagues have suggested that this may be a new subfamily of cathepsins, expressed only during embryonic development (Sol-Church *et al* 2000a). However, no human equivalent has yet been found (Nakajima *et al* 2000). All of these enzymes were originally isolated by the use of a murine EST (expression sequence tag) database, followed by cloning cDNA from mouse or rat placenta. They are all very similar in structure and homology, and seem to have a likeness to cathepsin L. In addition, the location of the murine cathepsin J/P gene is near the location of the cathepsin L gene, suggesting a common ancestral gene or possibly a duplication of genes for this group of cathepsins (Tisljar *et al* 1999). Although the function and catalytic properties of these enzymes are not yet known, an examination of the amino acid residues surrounding the subsites of the active site vary between the enzymes, suggesting different substrate specificities (Sol-Church *et al* 2000b). Suggested roles for the placental cathepsins include embryo implantation, placental development and function (Tisljar *et al* 1999), fetal nutrition (Sol-Church *et al* 1999), immunological modulation, and processing of secretory protein factors (Nakajima *et al* 2000).

5.4 REVERSIBLE INHIBITORS

Most reversible cathepsin inhibitors are transition-state analogs containing electrophilic carbonyl groups and include peptidyl aldehydes, trifluoromethyl ketones, methyl ketones, α-ketoacids, α-ketoesters, α-ketoamides, α-keto-β-aldehydes, and diketones. The hydrolysis of substrates by cysteine proteases goes through a tetrahedral intermediate before the respective components are released. Transition-state inhibitors bind in the active site to form a hemithioketal adduct, similar to the tetrahedral intermediate. The mechanism of interaction of electrophilic carbonyl containing inhibitors is shown in Figure 5.2. The highly nucleophilic cysteine thiolate attacks the carbonyl of the inhibitor, forming a tetrahedral intermediate that is stabilized by the enzyme's oxyanion hole.

The peptidyl nitriles are another type of reversible inhibitor for cysteine proteases, but differ from the electrophilic carbonyl inhibitors due to the formation of an isothioamide derivative in the active site. The mechanism of the peptidyl nitriles can be found in Figure 5.3. Inhibition occurs by formation of an E•I complex (measured by K_I) followed by a

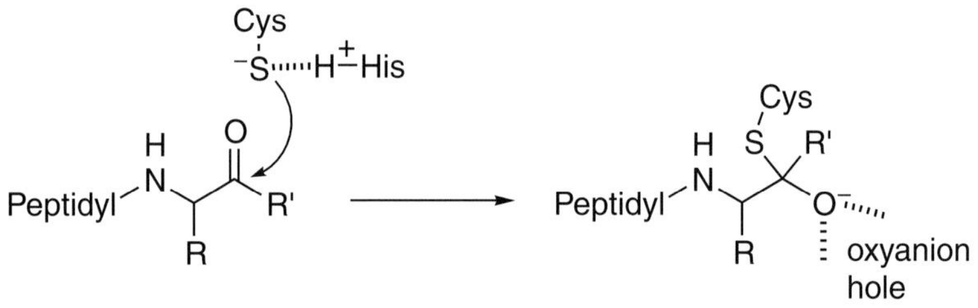

Figure 5.2 Mechanism of reversible transition state inhibitors of cathepsins. The enzyme's active site cysteine residue nucleophilically attacks the carbonyl of the inhibitor, forming a tetrahedral intermediate that is stabilized by the enzyme's oxyanion hole. For aldehydes, R′ = H; for trifluoromethyl ketones, R′ = CF_3; for methyl ketones, R′ = CH_3; for α-ketoacid inhibitors, R′ = CO_2H; for α-ketoester inhibitors, R′ = CO_2R''; for α-ketoamide inhibitors, R′ = CONHR″ or $CONH_2$; for α-keto-β-aldehydes, R′ = CHO; and for diketones, R′ = CO-alkyl.

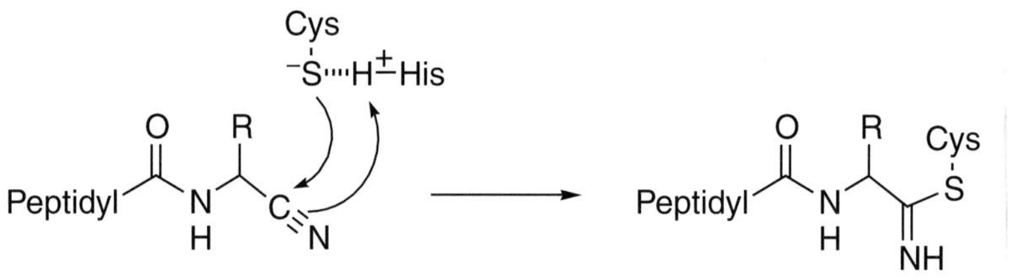

Figure 5.3 Mechanism of inhibition of cathepsins by peptidyl nitriles. The active site cysteine nucleophilically attacks the nitrile carbon, while the nitrogen simultaneously is protonated by the histidine residue.

nucleophilic attack of the cysteine on the nitrile containing carbon in conjunction with the protonation of the nitrogen to form an isothioamide (Moon *et al* 1986). The isothioamide cannot be hydrolyzed, thus stopping the catalytic activity in the enzyme (Moon *et al* 1986; Shaw 1990).

5.4.1 Peptidyl aldehydes

Peptide aldehydes are inhibitors of both cysteine and serine proteases, but are usually more potent with cysteine proteases. Although the inhibition is not completely selective between the two types of enzymes, increased selectivity can be obtained by varying the amino acid residues in the P1 and P2 positions to match the enzyme's preferred substrates (Otto and Schirmeister 1997).

Peptidyl aldehyde inhibitors exhibit "slow binding" due to the low concentration of free aldehyde actually present in solution, about 22% with most aldehydes (Schultz *et al* 1989). The remainder is the hydrated form. Other problems include poor *in vivo* activity due to both their low cell permeability (Mehdi *et al* 1988; Tsujinaka *et al* 1988) and their ease of degradation (Shaw 1990). Several examples of peptidyl aldehyde inhibitors are shown in Table 5.2.

5.4.2 Peptidyl trifluoromethyl and methyl ketones

Peptidyl trifluoromethyl and methyl ketones are inhibitors of both cysteine and serine proteases, but favor serine proteases (Liang and Abeles 1987; Brady *et al* 1989). It was expected that these inhibitors would have increased metabolic stability compared to peptidyl aldehyde inhibitors (Otto and Schirmeister 1997). Trifluoromethyl ketones are almost completely hydrated and, compared to the corresponding peptide aldehydes, are

Table 5.2 Peptidyl aldehyde inhibitors

Enzyme	*Inhibitor*	IC_{50} (nM)	*Reference*
Cathepsin B	2-NapSO_2-Ile-Trp-H[a]	23	Yasuma *et al* 1998
	Bz-Ile-Trp-H	52	Yasuma *et al* 1998
	1-NapSO_2-Phe-Trp-H	30	Yasuma *et al* 1998
	1-NapSO_2-Val-Trp-H	65	Yasuma *et al* 1998
	1-NapSO_2-Ile-Leu-H	100	Yasuma *et al* 1998
	Cbz-Phe-Leu-H	17.7	Woo *et al* 1995
	Cbz-Phe-Ala-H	45.1	Woo *et al* 1995
	Cbz-Phe-Phe-H	69.7	Woo *et al* 1995
Cathepsin L	1-NapSO_2-Ile-Trp-H	1.9	Yasuma *et al* 1998
	1-NapSO_2-Val-Trp-H	0.97	Yasuma *et al* 1998
	1-NapSO_2-Ile-Trp-H	1.9	Yasuma *et al* 1998
	1-NapSO_2-Ile-Val-H	0.74	Yasuma *et al* 1998
	1-NapSO_2-Ile-Leu-H	0.95	Yasuma *et al* 1998
	1-NapSO_2-Ile-Phe-H	0.95	Yasuma *et al* 1998
	Cbz-Phe-Phe-H	0.74	Woo *et al* 1995
	Cbz-Phe-Leu-H	0.78	Woo *et al* 1995
	Cbz-Phe-Tyr-H	0.85	Woo *et al* 1995

[a] Nap = naphthyl.

weak inhibitors of cysteine proteases (Babine and Bender 1997). The poor inhibition capabilities of the peptidyl trifluoromethyl and methyl ketones may also be due to steric hinderance (Lienhard and Jencks 1966). Some examples of these types of inhibitors for cathepsin B are Cbz-Phe-Ala-CH_3 ($K_I = 31\,\mu M$) and Cbz-Phe-Ala-CF_3 ($K_I = 300-470\,\mu M$) (Smith *et al* 1988a).

5.4.3 Diamino ketone inhibitors and related compounds

A new class of reversible and irreversible inhibitors, selective and specific for cathepsin K, has been developed by a group at SmithKline Beecham Pharmaceuticals (now known as GlaxoSmithKline) (Yamashita *et al* 1997). The inhibitors work by forming a thiohemiketal between the active site cysteine and the carbonyl of the ketone, and span both the S and S′ subsites. Replacement of the carbonyl with an alcohol led to a three order of magnitude drop in cathepsin K inhibition, demonstrating the importance of the carbonyl moiety (Yamashita *et al* 1997). The inhibitors were designed using the poorly electrophilic 1,3-diamino-2-propanone scaffold. The scaffold design originated from an overlay of leupeptin (Ac-Leu-Leu-Arg-H) and the peptide aldehyde, Cbz-Leu-Leu-Leu-H (**5.1** in Figure 5.4). X-ray crystal structures of the two inhibitors binding to papain show leupeptin binding in the S subsites (Schroder *et al* 1993) and the leupeptin analog bound to the S′ subsites. Because binding to one half of the enzyme's subsites often results in the poor selectivity between similar enzymes, it was speculated that a combination of these two peptidyl aldehyde compounds spanning the catalytic subsites would increase selectivity for a particular target enzyme. This combination molecule was modified to be specific for cathepsin K and an example is compound (**5.2**) in Figure 5.4. The P1 amino acid side chains were removed and the peptide chain was shortened by one amino acid on each end of the molecule. Compound (**5.2**) was found to inhibit papain poorly ($K_I = >10\,\mu M$), but was a very potent and selective for cathepsin K ($K_I = 22\,nM$) (Yamashita *et al* 1997). X-ray crystallography showed that the inhibitor bound as expected (Yamashita *et al* 1997). Compound (**5.2**) was also found to be a poor inhibitor for cathepsin L ($K_I = 0.34\,\mu M$), cathepsin B ($K_I = 1.3\,\mu M$), cathepsin S ($K_I = 0.89\,\mu M$), trypsin, and chymotrypsin ($K_I > 50\,\mu M$ for both) (Yamashita *et al* 1997).

Diamino ketones are transition-state inhibitors and lack reactive groups. Thus, there is little potential for immunological or antigenic side effects (Amos *et al* 1985), giving these compounds considerable therapeutic potential. The inhibitors are stable toward simple thiols in neutral, acidic, and basic conditions, suggesting the highly nucleophilic cysteine and the molecular interactions between cathepsin K and the inhibitor are necessary for inhibition.

Development of this class of cathepsin K inhibitors has resulted in a large number of inhibitors with related structures (Figure 5.4). The first of these was the diacylcarbohydrazide derivative (**5.3**), which was a much more potent inhibitor of cathepsin K than the original 1,3-diamino-2-propanone derivatives (Thompson *et al* 1997, 1998). This inhibitor also irreversibly inactivated cathepsin B ($k_{obs}/[I] = 1.3 \times 10^3\,M^{-1}s^{-1}$), cathepsin L ($k_{obs}/[I] = 5.8 \times 10^4\,M^{-1}s^{-1}$), and cathepsin S ($K_I = 11\,nM$). These compounds irreversibly inactivate the active site cysteine of cathepsin K by forming a thiocarbazoic ester, as determined by mass spectrometry (Bossard *et al* 1999). The acyl enzyme is hydrolyzed slowly ($10^{-4}\,s^{-1}$), reactivating the enzyme. The slow hydrolysis may be a result of lowered electrophilicity of the carbonyl group compared to other thioesters (Gupton *et al* 1984). This

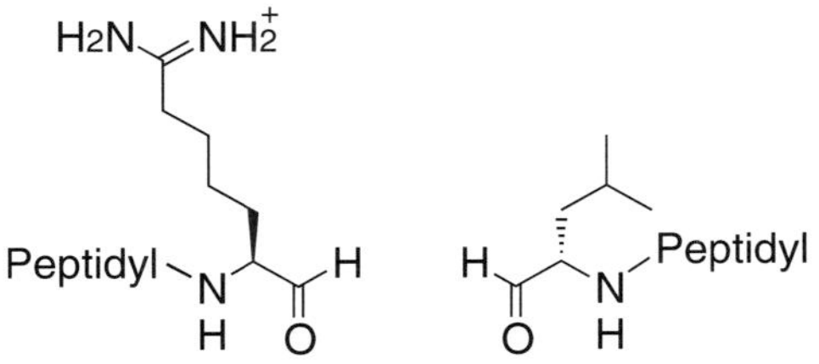

5.1

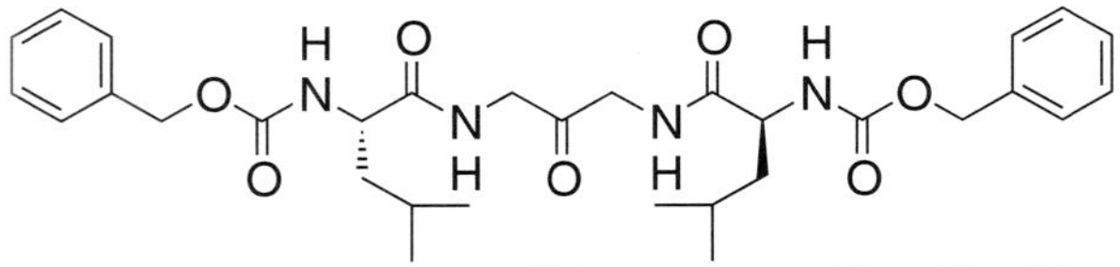

5.2 $K_I = 22$ nM

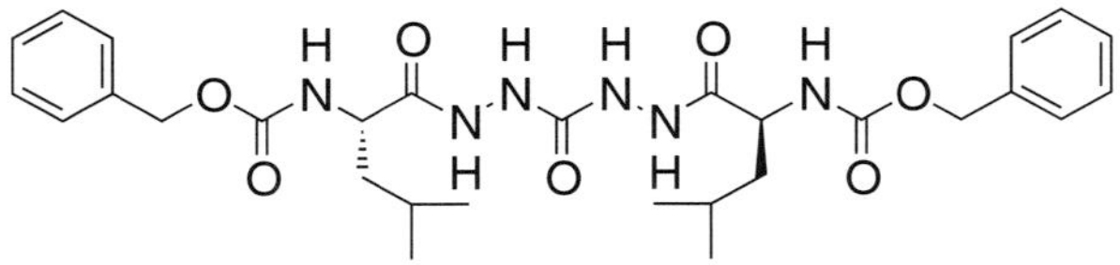

5.3 $k_{obs}/[I] = 3.1 \times 10^6\ M^{-1}s^{-1}$
$K_I = 0.7$ nM

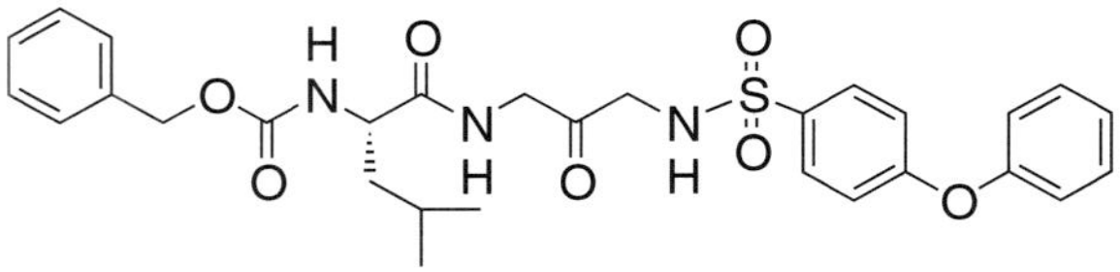

5.4 $K_I = 1.8$ nM

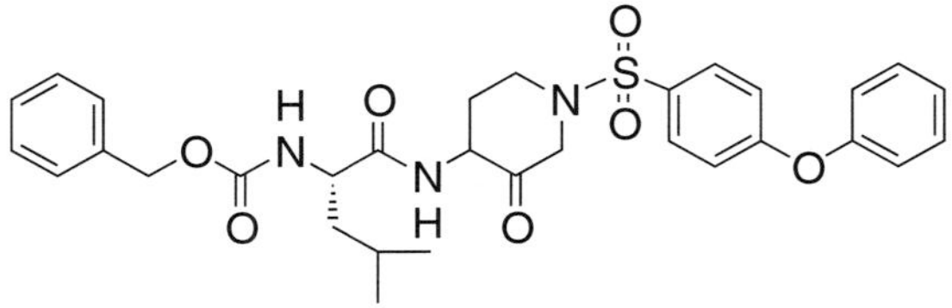

5.5 $K_I = 0.5$ nM

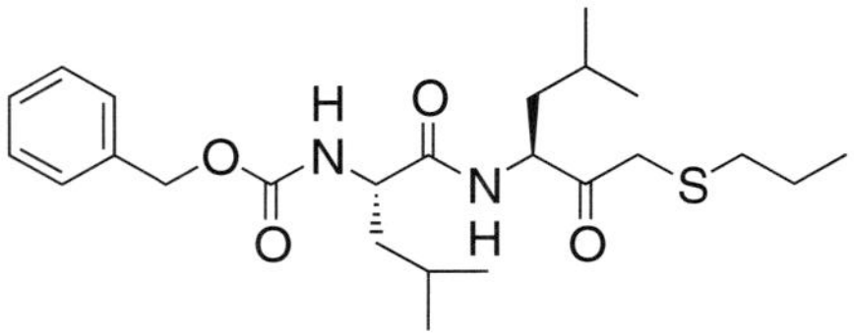

5.6 $k_{obs}/[I] = 2.2 \times 10^4\ M^{-1}s^{-1}$

Figure 5.4 Diamino ketones and related compounds. The design of diamino ketones involved the combination of two peptide aldehydes (**5.1**) to give ketone (**5.2**) which spans both S and S′ subsites. Related inhibitors for Cathepsin K are compounds (**5.3**) through (**5.6**).

inhibitor is related in concept to the azapeptide inhibitors, discussed in section 5.5.6. Other derivatives include the 4-phenoxylphenyl sulfonamide analog of 1, 3-di-amino-2-propanone (**5.4**) (Yamashita *et al* 1997), the cyclic 1,3-diamino ketone analog (**5.5**) (Marquis *et al* 1998), and the thiolmethylketone (**5.6**) (Marquis *et al* 1999). Phenoxymethyl and thiomethyl ketones are irreversible cathepsin K inhibitors and alkylate the enzyme in the same manner as halomethyl and acyloxymethyl ketones (Section 5.5.1).

5.4.4 Acyclic and cyclic alkoxyketones

Another new class of inhibitor, developed by a group at SmithKline Beecham Pharmaceuticals, has been designed with strong inhibitory activity for cathepsin K (Marquis *et al* 1999, 2001; Fenwick *et al* 2001a). According to crystal structures, these compounds inhibit through the formation of a hemithioketal with the active site cysteine, similar to the mechanistic pathway presented in Figure 5.2 (Marquis *et al* 2001). The acyclic compounds appear to bind in the primed side of the active site, while the cyclic analogs bind in the unprimed side (Marquis *et al* 2001). A major drawback for both acyclic and cyclic alkoxyketones is epimerization at the α-amino ketone (Marquis *et al* 2001).

Acyclic alkoxymethylketones, which contain an α-heteroatom that increases reactivity of the carbonyl toward nucleophilic attack by the cysteine active site residue, were found to be potent inhibitors for cathepsin K (**5.7** in Figure 5.5) (Marquis *et al* 1999). In order to improve both the *in vivo* and *in vitro* activity of these inhibitors, the reactive moiety was cyclized by linking the alkoxymethyl to the α-carbon adjacent to the reactive carbonyl (**5.8** in Figure 5.5) (Fenwick *et al* 2001a). The initial attempts led to a decrease in activity for the cyclic analogs, but this was due to a mixture of diastereomers. A series of cyclic analogs were synthesized by solid-phase with a variety of substituents on the N-terminus of the inhibitors (**5.9** and **5.10** in Figure 5.5) (Fenwick *et al* 2001a). To further refine their work, the group developed a way to synthesize the compounds with known *R* or *S* stereochemistry at the ring attachment (**5.10** in Figure 5.5) (Fenwick *et al* 2001b). Several of the *S* stereoisomers had approximately 40-fold more inhibition activity than the equivalent *R* isomer.

The importance of the five membered ring was studied by its replacement with a six membered ring. Although the diastereomer stereochemistry was not elucidated, the results of the six membered ring analogs were comparable to those of the five membered rings (**5.11** in Figure 5.5). The authors suggest that using the six membered ring rather than the five membered ring might help stabilize the epimerization observed in these compounds (Fenwick *et al* 2001b). The heteroatom of the ring was also replaced and/or moved to see what role it played in inhibition potency. It was found that, in most cases, an oxygen is preferred in the ring, and shifting the heteroatom in the ring resulted in a loss of potency (Marquis *et al* 2001). However, when a nitrogen with a substituent was used in the ring (pyrrolidinone and piperidinone derivatives), the inhibitor potency was improved dramatically (**5.12** in Figure 5.5) (Marquis *et al* 2001).

The cyclic alkoxyketones (**5.12** in Figure 5.5) were also tested against cathepsin B (for R, $R' = H; n = 1, K_I = >1000\,nM$ and for $n = 2, K_I = 440\,nM$), cathepsin S (for R, $R' = H; n = 1, K_I = 90\,nM$ and for $n = 2, K_I = 8.0\,nM$), and cathepsin L (for R, $R' = H; n = 1, K_I = 39\,nM$ and for $n = 2, K_I = 16\,nM$). Nearly all of the analogs were

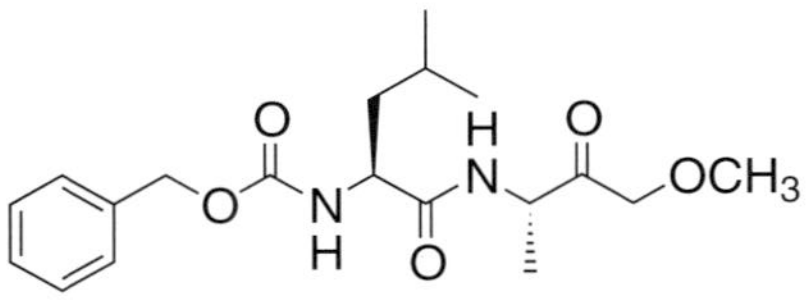

K_I = 60 nM

5.7

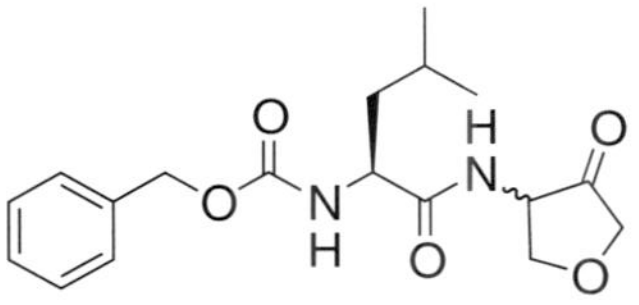

K_I = 140 nM

5.8

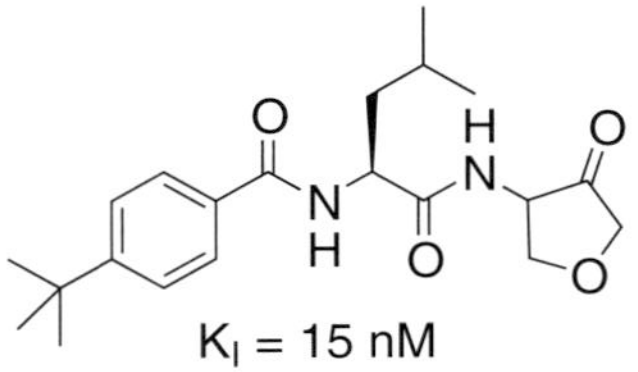

K_I = 15 nM

5.9

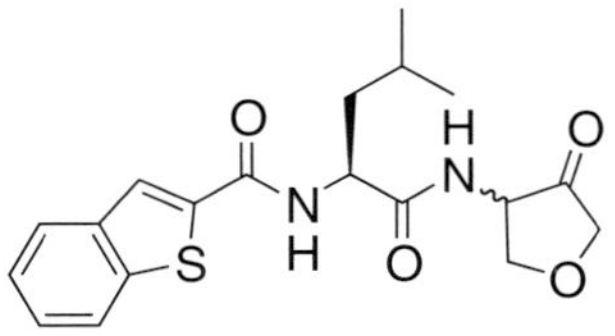

Diastereomeric Mixture, K_I = 11 nM
S isomer, K_I = 7 nM
R isomer, K_I = 68 nM

5.10

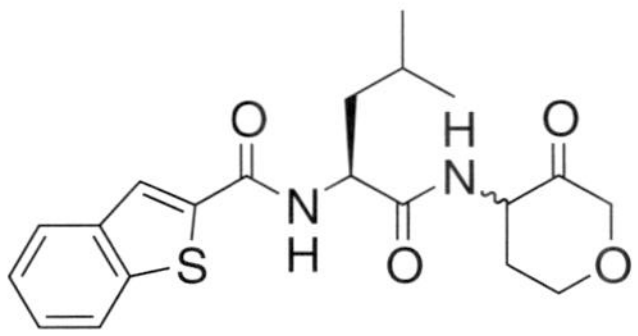

K_I = 8 nM (Diastereomer-1)
K_I = 70 nM (Diastereomer-2)

5.11

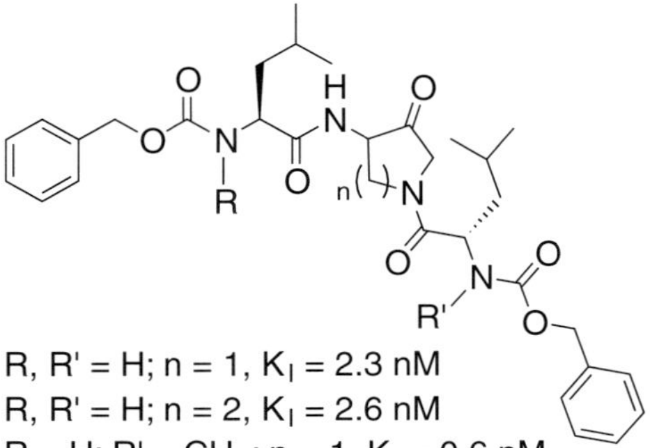

For R, R' = H; n = 1, K_I = 2.3 nM
For R, R' = H; n = 2, K_I = 2.6 nM
For R = H; R' = CH_3; n = 1, K_I = 0.6 nM

5.12

Figure 5.5 Acyclic and cyclic alkoxyketone inhibitors of Cathepsin K. Acyclic alkoxyketones (**5.7**) were found to be good inhibitors of cathepsin K and are the basis for cyclic alkoxyketones (**5.8**). Poorer inhibition by cyclic analogs led to the synthesis of a set of N-terminal derivatives and single isomers (**5.9**) and (**5.10**). Analogs containing a reactive six membered ring (**5.11**) and pyrrolidinone and piperidinone acylatated analogs (**5.12**) are also represented.

more selective for cathepsin K over the other three enzymes, but in some cases, these differences were negligible (Marquis *et al* 2001).

5.4.5 Peptidyl α-ketoacids, esters, amides, β-aldehydes, and diketones

The α-dicarbonyl inhibitors were originally developed as an alternative electrophilic carbonyl inhibitor for serine and cysteine proteases. It was hoped that by having more electron withdrawing groups, the α-dicarbonyl compounds would have increased inhibitory potency. The initial α-dicarbonyl compounds were poorly soluble in aqueous solutions and very hydrophobic, but newer compounds containing more polar groups have increased solubility (Walker *et al* 1993; Harbeson *et al* 1994). The α-dicarbonyl compounds are usually better cysteine protease inhibitors than trifluoromethyl ketones and methyl ketones (Hu and Abeles 1990; Li *et al* 1993). However, with the exception of α-keto-β-aldehydes, the α-dicarbonyl inhibitors are generally not as potent as the peptidyl aldehydes (Hu and Abeles 1990; Li *et al* 1993; Walker *et al* 1993). Some of these inhibitors have poor *in vivo* efficacy due to their low membrane permeability (Li *et al* 1993). Several examples of the α-dicarbonyl inhibitors can be seen in Table 5.3.

The α-keto-β-aldehydes, also called peptide glyoxals, are a new addition to the transition state inhibitor class. They contain a glyoxal moiety at the C-terminus that can react with the active site residue of cysteine and serine proteases (Walker *et al* 1993). Like other α-dicarbonyl compounds, these inhibitors are more potent with cysteine proteases. In addition, the α-keto-β-aldehydes are up to 10-fold more potent than the peptidyl aldehyde analogs (Walker *et al* 1993). Walker *et al* suggest that this increased potency comes from the formation of the hydrated aldehyde, which gives the inhibitor the ability to form positive hydrogen-bonding interactions in the primed side of the active site, similar to vinyl sulfones (see Section 5.5.5) (Lynas *et al* 2000; Walker *et al* 2000).

Table 5.3 Peptidyl α-ketoacid, ester, amide, β-aldehydes, and diketone inhibitors

Enzyme	*Inhibitor*	K_I (nM)	*Reference*
Cathepsin B	Cbz-Phe-Gly-$CO_2CH_2CO_2Et$	200	Hu and Abeles 1990
	Cbz-Gly-Phe-Gly-CO_2-*n*-Bu	1300	Hu and Abeles 1990
	Cbz-Phe-Gly-CO_2-*n*-Bu	1500	Hu and Abeles 1990
	Cbz-Phe-Gly-CO_2H	2000	Hu and Abeles 1990
	Cbz-Leu-Phe-CO_2H	4500	Li *et al* 1993
	Cbz-Leu-Abu-CO_2H[a]	1500	Li *et al* 1993
	Cbz-Phe-Gly-CO-NHEt	4000	Hu and Abeles 1990
	Cbz-Phe-Ala-CHO	76.8	Walker *et al* 1993
	Cbz-Phe-Arg-CHO	40	Lynas *et al* 2000
Cathepsin L	Cbz-Phe-Arg-CHO	90	Lynas *et al* 2000
	Cbz-Phe-Tyr(OtBu)-CHO	0.6	Lynas *et al* 2000
Cathepsin S	Cbz-Phe-Ala-CHO	23.2	Walker *et al* 2000
	Cbz-Phe-Tyr(OtBu)-CHO	2.23	Walker *et al* 2000
	Cbz-Phe-Arg-CHO	2.4	Walker *et al* 2000
	Cbz-Phe-Leu-CHO	0.185	Walker *et al* 2000

[a] Abu = α-aminobutanoic acid.

The α-keto-β-aldehydes, shown in Table 5.3, were first tested on cathepsin B (Walker *et al* 1993). They were found to be approximately 3-fold more potent than the corresponding peptidyl aldehyde analogs (Walker *et al* 1993). More recently, these inhibitors have been designed with amino acid sequences more specific for cathepsin L (Lynas *et al* 2000) and cathepsin S (Walker *et al* 2000). Although the inhibitors worked well for cathepsins B, L, and S, they were highly selective for their target cathepsin (Lynas *et al* 2000; Walker *et al* 2000). The sequence used for cathepsin S inhibitors was inspired by the fluoroketone, Cbz-Phe-Ala-CH_2F (see Section 5.5.1) (Brömme *et al* 1989b).

5.4.6 Peptidyl nitriles

Peptidyl nitriles are widely studied (Lowe and Yuthavong 1971) selective inhibitors of cysteine proteases with very poor inhibitory activity toward serine proteases (Westerik and Wolfenden 1972; Thompson *et al* 1986). The inhibition specificity of peptide nitriles for cysteine proteases is due to the higher nucleophilic character of the cysteine thiolate compared to the catalytic residue of serine proteases. It has also been suggested that the isothioamide triagonal structure more closely resembles the substrate transition state preferred by cysteine protease compared to that of serine proteases (Dufour *et al* 1995). Little inhibitor data exists for cathepsins other than cathepsin C (Gly-NH-CH(CH_2Ph)-CN, $K_I = 1.1\,\mu M$; Moon *et al* 1986) and papain (CH_3O-CO-Phe-NH-CH_2CN, $K_I = 1.8\,\mu M$; Lowe and Yuthavong 1971). In spite of their higher selectivity, peptide nitriles are poor inhibitors compared to the corresponding peptidyl aldehydes (Otto and Schirmeister 1997).

5.4.7 Oligopeptides as inhibitors

Oligopeptides can be used as inhibitors if the sequences of amino acid residues are optimized to interact with the proteases of interest. They are designed to bind with the enzyme noncovalently (Eichler and Houghten 1993; Brinker *et al* 2000; Horn *et al* 2000). This class differs from most inhibitors, in that it does not have substrate-like sequences and lacks reactive groups. The oligopeptide inhibitors are generally synthesized as a large library, through peptide combinatorial chemistry, in which one amino acid residue is changed until an optimal one is found, while the rest are held constant. This continues until the whole sequence of amino acid residues is optimal for the enzyme being considered.

Horn and colleagues, the first to use this technique for mammalian cysteine proteases, have synthesized several of these substrates for cathepsin C (Horn *et al* 2000). Surprisingly, they found that a series of arginine residues gave the best inhibition of all the amino acid sequences tested (acyl-$(Arg)_{10}$-NH_2, $IC_{50} = 35\,\mu M$). In addition, they found that the best inhibitors were those that had a free amino terminus (Arg_8-NH_2, $IC_{50} = 29\,\mu M$), and were at least eight but no more than ten amino acid residues long (Arg_{10}-NH_2, $IC_{50} = 18\,\mu M$). The group also tested the equivalent D-amino acid residues of the best sequences and found little or no inhibition of cathepsin C activity. The oligopeptides were tested against cathepsin B, cathepsin H, papain, and ficin, but no loss of catalytic activity was observed (Horn *et al* 2000), suggesting that these inhibitors are specific for cathepsin C. The authors suggest that, by using the best amino acid sequences as scaffolds, irreversible, efficient, and specific inhibitors for cathepsin C could be generated with the addition of reactive groups, like vinyl sulfonyl or epoxy moieties.

Brinker *et al* designed a set of pentapeptide inhibitors to test the extended amino acid preference of the active site of cathepsin L (Brinker *et al* 2000). They found that pentapeptides containing aromatic, bulky, hydrophobic amino acid residues, like leucine, and positively charged residues, like arginine, were preferred (Leu-Leu-Leu-Thr-Arg-NH_2, $IC_{50} = 0.5\,\mu M$). Nonproteinogenic amino acids were also tested, but only two gave positive results, *p*-nitrophenylalanine and homophenylalanine. All other amino acids that were tested, both proteinogenic and nonproteinogenic, dramatically reduced the effect of inhibition of cathepsin L by the pentapeptides. One of the best pentapeptides, Arg-Lys-Leu-Leu-Trp-NH_2 ($IC_{50} = 0.6\,\mu M$, $K_I = 130\,nM$), resembled part of the propeptide sequence of cathepsin L, which also effectively blocks the enzyme's function. Altering the pentapeptide sequence, either by truncation, blocking the N-terminus, using D-amino acids, reversing the direction of the preferred sequence, or changing the P3 from Leu or Phe, dramatically reduced or eliminated inhibition activity.

5.5 IRREVERSIBLE INHIBITORS

5.5.1 Peptidyl halomethyl, acyloxymethyl, and diazomethyl ketones

Irreversible inhibitors for cathepsins come in several chemical types. One group consists of halomethyl, acyloxymethyl, and diazomethyl ketones, which possess a carbonyl and a leaving group. The mechanism of inactivation by these inhibitors is shown in Figure 5.6. Inactivation of enzymes by peptidyl halomethyl, acyloxymethyl, and diazomethyl ketones follows one of the two routes. The first is direct nucleophilic attack by the active site cysteine, displacing the leaving group from the substituted carbon (Figure 5.6, route a) (Barrett 1986). The second possible route is the formation of a reversible tetrahedral intermediate, followed by rearrangement to a sulfonium ion three membered ring with the loss of the halogen, carboxylate ion, or nitrogen, and finally ring opening to yield the alkylated thiol (Figure 5.6, route b) (Barrett 1986). In cysteine proteases, the leaving group is protonated by the active site histidine (Rauber *et al* 1986).

Peptidyl halomethyl ketones, which include the highly chemically reactive α-chloromethyl ketones and the less reactive α-fluoromethyl ketones, interact with both cysteine and serine proteases (Rauber *et al* 1986; Albeck *et al* 1996). Peptidyl chloromethyl ketones are chemically reactive toward simple thiols and non-proteolytic enzymes and can have toxic side effects due to their high reactivity (Rauber *et al* 1986; Babine and Bender 1997). These inhibitors are not suitable as drugs because of their high toxicity (Shaw 1990).

Peptidyl fluoromethyl ketones have less chemical activity with other molecules, especially simple thiols, when compared to chloromethyl ketones, and thus have greater selectivity toward cysteine proteases (Rauber *et al* 1986; Babine and Bender 1997). The fluoride is resistant to being substituted in the enzyme complex and first forms a stable tetrahedral adduct with the active site residue, making a tight enzyme–inhibitor complex which then undergoes alkylation (Rauber *et al* 1986). Although their inhibition of cathepsins is slower than the corresponding chloromethyl ketone, they have much tighter binding (Rauber *et al* 1986). Several examples of this inhibitor type can be seen in Table 5.4 for cathepsins B, L, and S. An interesting fact is that despite substrate analysis showing that cathepsin S prefers a small amino acid in the P2 position, the best inhibitor has a Phe

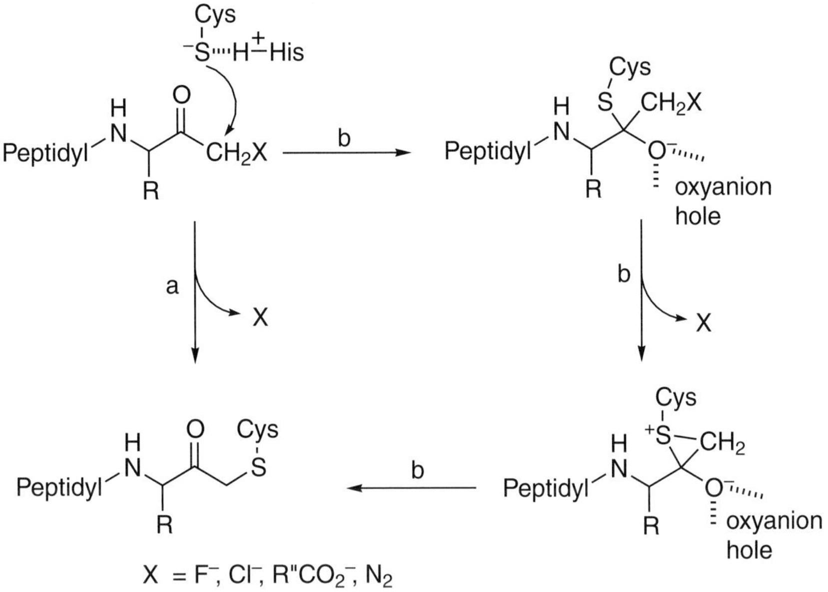

Figure 5.6 Mechanism of inhibition of cathepsins by halomethyl ketones, acyloxymethyl ketones, and diazomethyl ketones. The enzyme's active site cysteine nucleophilically attacks the substituted carbon, displacing the leaving group and inhibiting the enzyme (route a). An alternate mechanism (route b) involves the cysteine attacking the carbonyl of the inhibitor, forming a tetrahedral intermediate, which cyclizes to release the leaving group, followed by rearrangement to give the inhibited enzyme.

Table 5.4 Peptidyl halomethyl ketones inhibitors

Enzyme	*Inhibitor*	$k_{obs}/[I]$ ($M^{-1}s^{-1}$)[a]	k_2/K_I ($M^{-1}s^{-1}$)	*Reference*
Cathepsin B	Cbz-CH_2-CH_2-CO-Phe-Ala-CH_2F	21 000		Ahmed *et al* 1992
	Cbz-Phe-Ala-CH_2F	12 300		Ahmed *et al* 1992
	Ph-CH_2-NH-$COCH_2$-CH_2-CO-Phe-Ala-CH_2F	3 900		Ahmed *et al* 1992
	Cbz-Phe-Phe-CH_2Cl		9 040	Rauber *et al* 1986
	Cbz-Phe-Phe-CH_2F		3 920	Rauber *et al* 1986
	Cbz-Phe-Ala-CH_2F		54 500	Rauber *et al* 1986
	Pro-Phe-Arg-CH_2Cl	5 100 000		Barrett *et al* 1982
	Cbz-Leu-Gly-CH_2Cl		520 000	Graybill *et al* 1992
Cathepsin L	Cbz-Tyr-Ala-CH_2F	7 400 000		Angliker *et al* 1992
	Cbz-Leu-Leu-Tyr-CH_2F	680 000		Angliker *et al* 1992
	Cbz-Leu-Tyr-CH_2F	12 000		Angliker *et al* 1992
	Pro-Phe-Arg-CH_2Cl	30 000 000		Kirschke and Barrett 1987
	Leu-Leu-Lys-CH_2Cl		21 500 000	Shaw 1990
Cathepsin S	Cbz-Phe-Ala-CH_2F		2 780 000	Brömme *et al* 1989b

[a] apparent second order rate constant.

at this site (Brömme *et al* 1989b; Walker *et al* 2000). This suggests that enzymes may be inhibited by amino acid sequences contrary to their amino acid substrate specificity (Walker *et al* 2000).

The acyloxymethyl ketone inhibitors are very similar to the halomethyl ketones, but have a reduced chemical activity due to a larger leaving group which is only weakly nucleophilic (Honn *et al* 1982; Smith *et al* 1988b). Acyloxymethyl ketones vary greatly because a wide variety of leaving groups can be used that span the S′ subsites of cysteine proteases (Brömme *et al* 1994). This allows these inhibitors to be highly selective and have controlled reactivity toward cysteine proteases (Brömme *et al* 1994). Acyloxymethyl ketones are very selective toward cysteine proteases and can be made chemically inert toward most non-enzymatic nucleophiles (Smith *et al* 1988b). Several inhibitors of this type are shown in Table 5.5 for cathepsins B, L, and S.

Peptidyl diazomethyl ketones are based on the antibiotic azaserine, which alkylates the thiol group of an aminotransferase involved in purine synthesis (Buchanan 1973; Shaw 1984). They are selective inhibitors of cysteine proteases, but not serine or metalloproteases even with complimentary designs for those enzymes (Watanabe *et al* 1979; Green and Shaw 1981; Kirschke and Shaw 1981). Although they do not inhibit other proteases, peptidyl diazomethyl ketones can serve as substrates (Green and Shaw 1981). They are unreactive toward simple thiols and that, in combination with their selectivity toward cysteine proteases, make them a useful class of inhibitors (Green and Shaw 1981; Crawford *et al* 1988). The selectivity of peptidyl diazomethyl ketones can be controlled by altering the peptide portion of the inhibitor, which has led to inhibitors designed for cathepsin B, cathepsin C, as well as many other cysteine proteases (Green and Shaw 1981). Inhibitors of this type are shown in Table 5.6 for cathepsins B, C, H, L, and S.

A drawback of the peptidyl diazomethyl ketone inhibitors is the difficulty of their synthesis due to their instability in acids (Green and Shaw 1981). In addition, some believe that they will not make good drugs because of this instability, the reactivity of the diazomethyl group, and possible toxicity (Page 1990). However, a study with mice that used 10 times more inhibitor than needed to cure a fatal streptococcal infection had no toxic effects (Bjorck *et al* 1989).

The inhibitors work better at lower pH values due to the need to protonate the diazoketone (Shaw 1990). The easy alkylation of sulfides to sulfonium cations versus the highly energetic and unfavorable oxonium ion, along with the unfavorable enzyme-inhibitor complex geometry in serine proteases, is the reason why these inhibitors are cysteine protease specific (Shaw 1984; Albeck *et al* 1996).

5.5.2 Peptidyl epoxysuccinyl inhibitors

Peptidyl epoxysuccinyl inhibitors contain a reactive three-membered ring that opens when attacked by a good nucleophile (Yabe *et al* 1988; Varughese *et al* 1989). The nucleophilic opening of the ring is displayed in Figure 5.7. The molecules inhibit by forming a thioether bond with the enzyme via a nucleophilic attack at C-2 or C-3 on the ring by the active site cysteine residue, inhibiting the enzyme.

Peptidyl epoxysuccinyl inhibitors are based on the design of the natural inhibitor of cysteine proteases isolated from *Aspergillus juponicus*, E-64 (Hanada *et al* 1978). Unlike many other microbial inhibitors, epoxysuccinyl peptides and their derivatives are potent and specific irreversible inhibitors of cysteine proteases (Hanada *et al* 1978; Barrett *et al*

Table 5.5 Acyloxymethyl ketone inhibitors

Enzyme	*Inhibitor*	$k_2(s^{-1})$	$K_I(\mu M)$	$k_2/K_I(M^{-1}s^{-1})$	*Reference*
Cathepsin B	Cbz-Val-Lys-CH_2-OCOPh(2,6-$(CF_3)_2$)			2 000 000	Pliura *et al* 1992
	Cbz-Phe-Lys-CH_2-OCOPh(2,6-$(CF_3)_2$)			2 000 000	Pliura *et al* 1992
	Cbz-Leu-Leu-CH_2-OCOPh(2,6-$(CF_3)_2$)			270 000	Pliura *et al* 1992
Cathepsin L	Cbz-Phe-Cys(Bzl)-CH_2-OCOPh(2,6-$(CF_3)_2$)	0.015 ± 0.002	0.0014 ± 0.0004	10 700 000	Brömme *et al* 1994
	Cbz-Phe-Ser(Bzl)-CH_2-OCOPh(2,6-$(CF_3)_2$)	0.003 ± 0.0012	0.0007 ± 0.0003	4 290 000	Brömme *et al* 1994
	Cbz-Phe-Ala-CH_2-OCOPh(2,6-$(CF_3)_2$)	0.083 ± 0.037	0.25 ± 0.12	332 000	Brömme *et al* 1994
Cathepsin S	Cbz-Phe-Cys(Bzl)-CH_2-OCOPh(2,6-$(CF_3)_2$)			1 550 000	Brömme *et al* 1994
	Cbz-Phe-Ala-CH_2-OCOPh(2,6-Cl_2)	0.096 ± 0.001	0.14 ± 0.02	686 000	Brömme *et al* 1994
	Cbz-Phe-Ala-CH_2-OCOPh(2,6-$(CF_3)_2$)	0.040 ± 0.009	0.12 ± 0.04	364 000	Brömme *et al* 1994

Table 5.6 Peptidyl diazomethyl ketone inhibitors

Enzyme	*Inhibitor*	$k_{obs}/[I]$ ($M^{-1}s^{-1}$)	k_2/K_I ($M^{-1}s^{-1}$)	*Reference*
Cathepsin B	Cbz-Phe-Ala-CHN_2	630		Kirschke and Shaw 1981
	Cbz-Ala-Phe-Ala-CHN_2	1 180		Green and Shaw 1981
	Cbz-Phe-Gly-CHN_2	702		Green and Shaw 1981
Cathepsin C	Gly-Phe-CHN_2	17 300		Green and Shaw 1981
	Cbz-Phe-Ala-CHN_2	18		Green and Shaw 1981
	Cbz-Phe-Gly-Phe-CHN_2	17		Green and Shaw 1981
Cathepsin H	Cbz-Phe-Ala-CHN_2	0.6		Kirschke and Shaw 1981
	Ser(Bzl)-CHN_2	2 600		Shaw *et al* 1993
	Gly-Phe-CHN_2	6 700		Green and Shaw 1981
Cathepsin L	Cbz-Phe-Phe-CHN_2	660 000		Kirschke and Shaw 1981
	Cbz-Phe-Ala-CHN_2	620 000		Kirschke and Shaw 1981
	DNP-Ahx-Gly-Phe-Ala-CHN_2[a,b]		2 160 000	Hawthorne *et al* 1998
Cathepsin S	Cbz-Leu-Leu-Nle-CHN_2[c]	9 200 000		Shaw *et al* 1993
	Cbz-Val-Val-Nle-CHN_2	4 600 000		Shaw *et al* 1993
	Cbz-Leu-Leu-Tyr-CHN_2	2 100 000		Shaw *et al* 1993

[a] DNP = 2,4-dinitrophenyl; [b] Ahx = aminohexanoic; [c] Nle = norleucine.

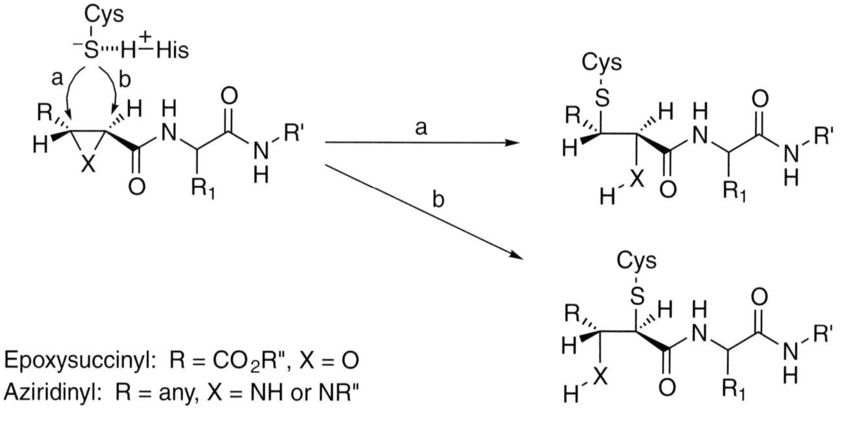

Figure 5.7 Mechanism of inhibition of cathepsins by epoxysuccinyl and aziridinyl peptides. The active site cysteine nucleophilically attacks either the C-2 or C-3, opening the reactive three membered ring and alkylating the enzyme.

1982). All active peptidyl epoxysuccinyl inhibitors have the *trans*-configuration (Otto and Schirmeister 1997).

Based on kinetic studies of esters of epoxysuccinic acid, it was predicted that E-64 would bind with its amino acid residues in the S′ subsites (Barrett *et al* 1982). A crystal structure of papain inhibited by E-64 shows, however, that the molecule binds with its amino acid residues in the S subsites, resulting in the peptide backbone binding in the direction opposite to that of peptide substrates, as seen in Figure 5.8 (Varughese *et al* 1989; Matsumoto *et al* 1999). In contrast, the E-64 relatives CA-074 (Yamamoto *et al* 1997; Matsumoto *et al* 1999) and CA-030 (Turk *et al* 1995; Matsumoto *et al* 1999) bind to

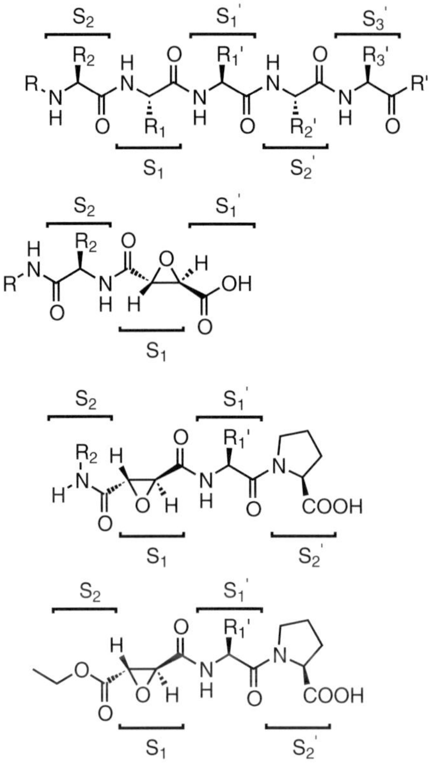

Substrate Binding

E-64
$R = (CH_2)_4NHC(NH_2)^+{}_2$
$R_2 = CH_2CH(CH_3)_2$

CA-074
$R_2 = CH_2CH_2CH_3$
$R_1' = CH(CH_3)CH_2CH_3$

CA-030
$R_1' = CH(CH_3)CH_2CH_3$

Figure 5.8 Binding modes of peptidyl epoxysuccinyl inhibitors. The top structure shows the binding mode of substrates with cathepsins. The second structure shows E-64 in the reverse binding mode which was observed in a crystal structure bound to papain. Structures three and four show CA-074 and CA-030 binding in the same direction as cathepsin substrates.

cathepsin B with the peptide backbone in the same direction as a peptide substrate, probably due to favorable interactions with the carboxyl oxygens of the terminal proline and the two His residues of the occluding loop.

Generally, epoxysuccinyl derivatives, including E-64, with the *S*,*S*-configuration of the oxirane ring are more potent than the corresponding *R*,*R*-configuration (Barrett *et al* 1982; Tamai *et al* 1987). This has led to the nearly exclusive study of inhibitors with the *S*,*S*-configuration. However, recent research has shown that some inhibitors in the *R*,*R*-configuration also have high activity (Schaschke *et al* 1997). Peptidyl epoxysuccinyl derivatives can bind in either direction with their amino acid residues in either the S subsite (*S*,*S*-configuration binding) or the S′ subsite (*R*,*R*-configuration binding) (Barrett *et al* 1982; Musil *et al* 1991).

Knowledge of how the epoxysuccinate derivatives bind is important for determining whether it is the C-2 or C-3 carbon of the oxirane ring that is attacked, whether it is important to have the *S*,*S*- or *R*,*R*-configuration designed into the inhibitor, and which subsite (S or S′) individual amino acid residues will use. For example, epoxysuccinyl derivatives with amino acid residues that can interact with the two protonated His residues (110 and 111) of the occluding loop of cathepsin B will likely bind in the S′ binding region (Buttle *et al* 1992; Gour-Salin *et al* 1993). Because of possible directional changes, the subsites at which the amino acid residues of an epoxysuccinyl inhibitor bind must be identified prior to determining the preferences of the S and S′ subsites of cysteine proteases. Examples of this type of inhibitor can be seen in Table 5.7.

One advantage of peptidyl epoxysuccinyl inhibitors is their stability under physiological conditions toward simple thiols (Barrett *et al* 1982). In addition, although they have limited selectivity toward different cysteine proteases, their reactivity toward cysteine proteases and not other proteases, along with their chemical unreactivity, make this class of inhibitors useful as pharmaceutical agents (Hanada *et al* 1978; Barrett *et al* 1982). Esterification of the carboxylate function and replacement of the charged residues with uncharged alkyl groups can increase cell permeability. The esters are less potent than free acids (100–1000 fold less), but the rapid hydrolysis of these esters into their active forms makes them encouraging as prodrugs (Noda *et al* 1981). The major drawback of the peptidyl epoxysuccinyl derivatives is that they can covalently bind to proteins other than cysteine proteases (Fukushima *et al* 1990). Overall, the epoxysuccinyl inhibitors appear to be one of the most promising inhibitor classes for drug design.

5.5.3 Peptidyl aziridinyl inhibitors

This new inhibitor class contains a reactive aziridine-2,3-dicarboxylic acid or aziridine-2-carboxylic acid moiety, termed Azi, attached to a peptide segment (Morodor *et al* 1992; Schirmeister 1999a, 1999b). Aziridinyl peptide inhibitors are aza analogs of epoxysuccinyl peptide inhibitors and are susceptible to ring opening by nucleophiles like their epoxysuccinyl counterparts (Figure 5.7) (Morodor *et al* 1992; Schirmeister 1999b). However, unlike the epoxysuccinyl analogs, substituents can be placed at the heteroatom of the aziridine ring (Schirmeister 1999b). Activation of the aziridines toward ring opening by nucleophilic attack results either by protonation or acylation of the heteroatom of the ring (Schirmeister 1999b).

There are three types of aziridine inhibitors of cysteine proteases (Schirmeister 1999a, 1999b). Type I are N-acylated aziridines with the Azi on the C-terminus of the molecule, type II are N-unsubstituted aziridines with the Azi at the N-terminus of the molecule, and type III are N-acylated derivatives with the Azi moiety within the molecule (Schirmeister 1999b). With few exceptions, type II are the most powerful of the peptidyl aziridine inhibitors against cathepsins, particularly at low pH values, and are comparable to their epoxysuccinyl peptide analogs in potency at approximately pH 4 (Otto and Schirmeister 1997, Schirmeister 1999b). This increased inhibitory action is due to the protonation of the aziridine nitrogen which, when interacting with the oxyanion hole of cysteine proteases, allows for energetically favorable H-bonding (Schirmeister 1999b). In addition, those type II and type III with a carboxylic acid at one or both ends of the inhibitor have even greater potency toward cathepsins (Schirmeister and Perics 2000). Type III have been found to increase in selectivity and potency, particularly for cathepsin B, if they are synthesized as bispeptidyl derivatives, with Boc-Phe linked to the aziridinyl nitrogen (Schirmeister and Perics 2000). The increased selectivity for cathepsin B is thought to be due to favorable interactions with Boc-Phe in the S2 subsite of the enzyme, and the free carboxylic acid at the C-terminus hydrogen bonding with the enzyme's occluding loop (Schirmeister and Perics 2000).

Table 5.7 Peptidyl epoxysuccinyl inhibitors

Enzyme	*Inhibitor*	$k_{obs}/[I]$ ($M^{-1}s^{-1}$)	k_2/K_I ($M^{-1}s^{-1}$)	*Reference*
Cathepsin B	HO-Eps-Leu-NH-$(CH_2)_7$-NH_2[b]	339 000		Barrett *et al* 1982
	HO-Eps-Leu-NH-$(CH_2)_2$-$CH(CH_3)_2$	298 000		Barrett *et al* 1982
	HO-Eps-Leu-NH-$(CH_2)_4$-NH-Cbz	175 000		Barrett *et al* 1982
	EtO-(2*R*, 3*R*)-*t*Eps-Leu-Pro-OH		567 000	Schaschke *et al* 1997
	EtO-(2*R*, 3*R*)-*t*Eps-Leu-Arg-OH		291 000	Schaschke *et al* 1997
	Agm ← Orn ← (2*S*, 3*S*)-*t*Eps-Leu-Pro-OH[a,c,d]		197 000	Schaschke *et al* 1997
	MeO-Gly ← Gly ← Leu ← (2*R*, 3*R*)-*t*Eps-Leu-Pro-OH		215 000	Schaschke *et al* 1998
	MeO-Gly ← Gly ← Leu ← (2*S*, 3*S*)-*t*Eps-Leu-Pro-OH		1 520 000	Schaschke *et al* 1998
Cathepsin C	HO-Eps-Leu-NH-$(CH_2)_2$-$CH(CH_3)_2$		153 000	Nikawa *et al* 1992
Cathepsin H	HO-(2*S*, 3*S*)-*t*Eps-Leu-Agm	4 000		Barrett *et al* 1982
	HO-Eps-Leu-NH-$(CH_2)_4$-NH_2	3 080		Barrett *et al* 1982
	HO-Eps-Leu-NH-$(CH_2)_7$-NH_2	2 070		Barrett *et al* 1982
Cathepsin L	HO-Eps-Leu-NH-$(CH_2)_4$-NH-Cbz	231 000		Barrett *et al* 1982
	HO-Eps-Leu-NH-$(CH_2)_2$-$CH(CH_3)_2$	206 000		Barrett *et al* 1982
	HO-Eps-Leu-NH-$(CH_2)_7$-NH_2	143 000		Barrett *et al* 1982
	Bzl-O-Leu ← Eps-OH		791 000	Gour-Salin *et al* 1993
	Bzl-NH-Phe ← Eps-OH		27 800 000	Gour-Salin *et al* 1993
	Bzl-NH-Leu ← Eps-OH		424 000	Gour-Salin *et al* 1993
	OH-(2*S*, 3*S*)-*t*Eps-Leu-Agm, E-64		43 800	Schaschke *et al* 1998
	HO-(2*R*, 3*R*)-*t*Eps-Leu-Agm		4 930	Schaschke *et al* 1998
Cathepsin S	Bzl-O-Leu ← Eps-OH		171 000	Gour-Salin *et al* 1993
	Bzl-NH-Leu ← Eps-OH		542 000	Gour-Salin *et al* 1993
	Bzl-NH-Phe ← Eps-OH		501 000	Gour-Salin *et al* 1993

[a] Agm = agmatine (1-amino-4-guanidino-butane); [b] Eps = epoxysuccinyl; [c] Orn = ornithine; [d] ← = peptide chain runs right to left.

Aziridines have been tested against several types of proteases, including serine, aspartate, and metalloproteases, but were found to selectively inhibit only cysteine proteases (Morodor *et al* 1992; Schirmeister 1999b). In fact, the aziridine-2-carboxylates and aziridine-2,3-dicarboxylates are hydrolyzed by serine proteases (Bucciarelli *et al* 1993a, 1993b). The peptidyl aziridines inhibit cathepsin B and cathepsin L in an irreversible manner, while inhibition of cathepsin H is not time dependent (Schirmeister 1999b). Examples of these inhibitors for cathepsins B and L are shown in Table 5.8.

Although both the aziridine and the epoxysuccinyl may have similar chemical reactivity, differences exist between the classes of inhibitor. Ready protonation of the nitrogen of type II aziridines is one difference between aziridines and their epoxysuccinyl analogs (Schirmeister 1999b). A second difference is the H-bonding abilities of the two classes. Aziridines are H-bond donors while the epoxysuccinyl inhibitors are H-bond acceptors (Schirmeister 1999b). These differences suggest that the two classes of inhibitor have different binding modes and possibly vary in their interactions with the enzymes (Schirmeister 1999b). Lastly, unlike most epoxysuccinyl inhibitors, the *R,R*-configuration of the aziridine ring is preferred for inhibition in both types II and III aziridine inhibitors, while the type I aziridine inhibitor prefers the *S,S*-configuration (Schirmeister 1999a).

Table 5.8 Aziridinyl peptide inhibitors

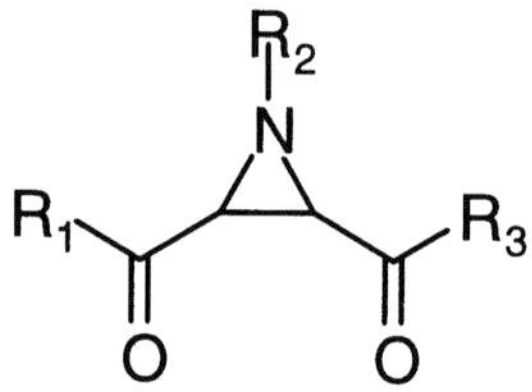

Enzyme	R_1	R_2	R_3	*Configuration*	k_2/K_I ($M^{-1}s^{-1}$)	*Reference*
Cathepsin B	OEt	Boc-Phe	Leu-Pro-OBzl	*R,R*	32	Schirmeister 1999a
	OEt	H	Leu-OBzl	*R,R*	27	Schirmeister 1999a
	OEt	Boc-Phe	OH	*S,S* + *R,R*	21	Schirmeister 1999b
	OEt	Boc-Phe	Leu-Pro-OH	*S,S*	114	Schirmeister and Perics 2000
	OEt	Boc-Phe	Leu-Pro-OH	*R,R*	109	Schirmeister and Perics 2000
Cathepsin L	OEt	H	Leu-OBzl	*R,R*	271	Schirmeister 1999a
	OEt	Boc-Phe	Leu-Pro-OBzl	*R,R*	98	Schirmeister 1999a
	OEt	Boc-Leu-Gly	OBzl	*S,S* + *R,R*	54	Schirmeister 1999b
	OH	Boc-Phe	OH	*S,S* + *R,R*	635	Schirmeister and Perics 2000
	OH	Boc-Leu-Gly	OEt	*S,S* + *R,R*	183	Schirmeister and Perics 2000

5.5.4 Peptidyl epoxide inhibitors

Peptidyl epoxide inhibitors are a relatively new family of mechanism-based inhibitors for cysteine proteases that are stable to neutral and basic conditions (Pocker *et al* 1988). The mechanism involves opening of the epoxide ring, formation of a thioether bond with the enzyme via a nucleophilic attack at C-2 or C-3 on the ring by the active site cysteine residue, resulting in inhibition of the enzyme (Figure 5.9) (Albeck and Kliper 1997). A variety of structures have been reported with epoxides attached to or replacing the carboxyl group of amino acid or peptide derivatives. The epoxides were designed to take advantage of the initial/simultaneous protonation step displayed by cysteine proteases and become highly electrophilic when protonated (Albeck *et al* 1996). They are thought to form a Michaelis-type complex, aligning the epoxide near the active site (Albeck *et al* 1996). Inhibition of an enzyme with a radioactive epoxide inhibitor resulted in one molar equivalent of inhibitor being retained and loss of all enzymatic activity, indicating that the peptidyl epoxides form irreversible covalent complexes by alkylating the active site cysteine residue (Albeck and Estreicher 1997; Albeck and Kliper 1997). Peptidyl epoxides are ineffective toward serine proteases even when complimentary sequences to substrates of those enzymes were used (Albeck *et al* 1996; Albeck and Kliper 1997). They can be directed toward various cysteine proteases by varying the amino acid sequence of the epoxides (Albeck *et al* 1996; Albeck and Kliper 1997). The *threo*-configuration of peptidyl epoxides are not active toward cysteine proteases, but those with the *erythro*-configuration are (Albeck *et al* 1996). Although the actual inhibitory potency of current peptidyl epoxides is poor, it is likely that they can be improved. Several examples of this type of inhibitor can be seen in Table 5.9 for cathepsin B.

5.5.5 Peptidyl vinyl sulfone inhibitors and α, β-unsaturated carbonyl derivatives

Michael-acceptor derivatives are another class of irreversible inhibitors and include vinyl sulfone inhibitors and α, β-unsaturated carbonyl derivatives. The mechanism of inhibition

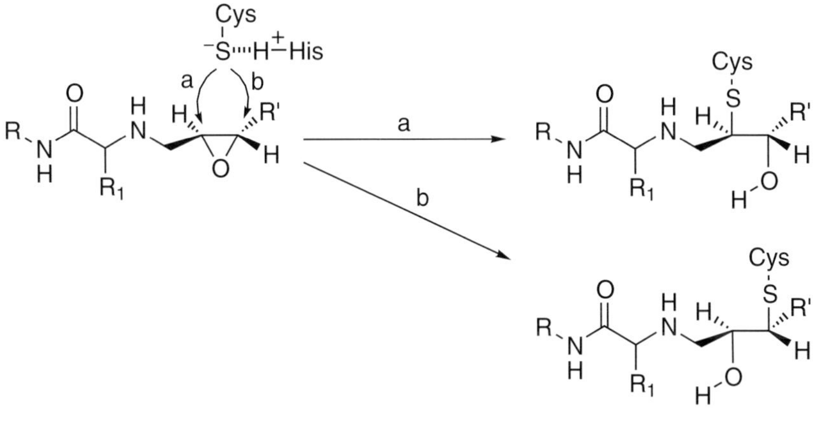

Figure 5.9 Mechanism of inhibition of cathepsins by peptidyl epoxides. The active site cysteine nucleophilically attacks either the C-2 or C-3, opening the reactive three membered ring and alkylating the enzyme.

Table 5.9 Peptidyl epoxide inhibitors

Enzyme	*Inhibitor*	$k_{obs}/[I]$ ($M^{-1}s^{-1}$)	$k_2(s^{-1})$	$K_I(M)$	k_2/K_I ($M^{-1}s^{-1}$)	*Reference*
Cathepsin B	Ac-Phe-NH-CH_2-Epo[a]	0.058				Giordano *et al* 1990
	Ph-CH = CH-CO_2-CH_2-Epo	0.022				Giordano *et al* 1990
	Ph-O-CH_2CO_2-NH-CH_2-Epo	0.018				Giordano *et al* 1990
	Cbz-Phe-Thr(Bzl)-Epo		0.0033	0.00001	333	Albeck *et al* 1996
	Cbz-NH-CH(CH_2Ph)-Epo		0.0015	0.00037	4.05	Albeck *et al* 1996
	Cbz-Phe-NH-CH(CH_3)-Epo		0.0012	0.0006	1.94	Albeck *et al* 1996

[a] Epo = Epoxide.

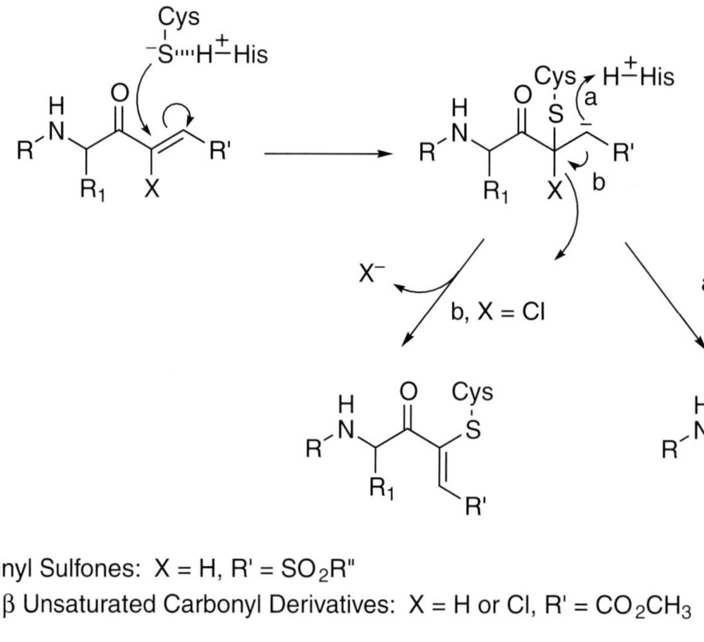

Figure 5.10 Mechanism of inhibition of cathepsins by vinyl sulfones and α, β-unsaturated carbonyl derivatives. The inhibition is believed to take place through a Michael addition where the active site cysteine nucleophilically attacks the β-carbon, alkylating the enzyme, followed by protonation of the α-carbon (a) or departure of the chlorine (b).

by vinyl sulfones and α, β-unsaturated carbonyl derivatives proceeds via a Michael addition by an attack at the β-carbon by the active site cysteine, followed by protonation of the α-carbon (Figure 5.10) (Hanzlik and Thompson 1984). Evidence of this mechanism was obtained using a halogenated α, β-unsaturated carbonyl derivative in which a Michael addition would release a chloride ion (Govardhan and Abeles 1996). A stoichiometric amount of the chloride ion was released (Govardhan and Abeles 1996). This class of inhibitors irreversibly inhibits cathepsins, but acts only as a competitive inhibitor of serine proteases (Palmer *et al* 1995).

Vinyl sulfones are a new and highly potent class of specific, irreversible cysteine protease inhibitors which contain a double bond activated by an electron withdrawing sulfone (Palmer *et al* 1995). Peptidyl vinyl sulfone inhibitors are stable, unreactive toward nucleophiles, and need the catalytic machinery of the cysteine proteases for activation (Palmer *et al* 1995). Vinyl sulfone inhibitors can be manipulated on both the P and P′ sides of the molecules, allowing selectivity and reactivity toward target enzymes to be controlled (Palmer *et al* 1995). Vinyl sulfones can also hydrogen bond with the glutamine side chain and protonated histidine in the active site, which would align the inhibitor for nucleophilic attack by the active site cysteine residue (Palmer *et al* 1995). Vinyl sulfone inhibitors have considerable potential for use as drugs, and examples of this class of inhibitor are shown in Table 5.10 for cathepsins B, L, S, and K.

Table 5.10 Peptidyl vinyl sulfones

Enzyme	*Inhibitor*	$k_{obs}/[I]$ ($M^{-1}s^{-1}$)	k_2/K_I ($M^{-1}s^{-1}$)	*Reference*
Cathepsin B	Mu-2Np-Hph-CH = CH-SO_2-CH_2CH_2Ph[a,b,c]	33 000		Palmer *et al* 1995
	Mu-Phe-Hph-CH = CH-SO_2-CH_2CH_2Ph	29 000		Palmer *et al* 1995
	Mu-Phe-Lys-CH = CH-SO_2-Ph•HBr	11 300		Palmer *et al* 1995
Cathepsin L	Mu-2Np-Hph-CH = CH-SO_2-CH_2CH_2Ph	2 240 000		Palmer *et al* 1995
	Ac-Leu-Leu-Met(O_2)-CH = CH-SO_2-Ph	1 500 000		Palmer *et al* 1995
	Ac-Leu-Leu-Nle-CH = CH-SO_2-Ph	930 000		Palmer *et al* 1995
Cathepsin S	Mu-2Np-Hph-CH = CH-SO_2-CH_2CH_2Ph	29 000 000		Palmer *et al* 1995
	Mu-Phe-Lys-CH = CH-SO_2-Ph•HBr	10 700 000		Palmer *et al* 1995
	Mu-Leu-Hph-CH = CH-SO_2-Ph		14 600 000	Brömme *et al* 1996a
	Mu-Met-Hph-CH = CH-SO_2-Ph		14 000 000	Brömme *et al* 1996a
Cathepsin K	Mu-Phe-Lys-CH = CH-SO_2-Ph•HBr	83 300		Palmer *et al* 1995
	Mu-Phe-Leu-Leu-Met(O_2)-CH = CH-SO_2-Ph	28 500		Palmer *et al* 1995
	Mu-Leu-Hph-CH = CH-SO_2-Ph		727 000	Brömme *et al* 1996a
	Mu-Met-Hph-CH = CH-SO_2-Ph		71 000	Brömme *et al* 1996a

[a] Mu = morpholine-CO-; [b] 2Np = 2-naphthylalanine; [c] Hph = homophenylalanine.

Another type of Michael-acceptor are the α, β-unsaturated carbonyl derivatives. The first compound tested was a fumarate derivative of E-64c, HO-Fum-Leu-$NH(CH_2)_2CH(CH_3)_2$ (DC-11, $k_2 = 625\,M^{-1}s^{-1}$ for cathepsin B, $11\,M^{-1}s^{-1}$ for cathepsin H, and $2272\,M^{-1}s^{-1}$ for cathepsin L) (Barrett *et al* 1982). However, Hanzlik *et al* (Hanzlik and Thompson 1984; Thompson *et al* 1986) was the first group to prepare a series of α, β-unsaturated carbonyl derivatives as inhibitors of papain (*trans* Ac-Phe-NH-CH_2-CH = CH-COOMe, $k_2 = 70\,M^{-1}s^{-1}$ and Ac-Phe-NH-CH_2-CH=CH_2, $k_2 = 9.2\,M^{-1}s^{-1}$) and cathepsin C (*trans* Gly-NH-CH(Bzl)-CH=CHCOOMe, $k_2 = 68\,M^{-1}s^{-1}$).

Both the *cis*- and *trans*-derivatives of α, β-unsaturated non-halogenated inhibitors are selective inhibitors of papain and cathepsin B (Govardhan and Abeles 1996). However, only the *trans*-configuration of the α, β-unsaturated halogen derivatives were found to be active inhibitors of papain and cathepsin B (Govardhan and Abeles 1996). Ester derivatives interact with the cathepsin B and papain more rapidly and readily than the carboxylate derivatives (Govardhan and Abeles 1996). In addition, the non-halogen unsaturated derivatives (*trans* Ac-Phe-NHCH_2-CH=CH-CO_2Me, $k_2/K_I = 20\,M^{-1}s^{-1}$ for cathepsin B) seem to inactivate much better than the halogenated anaglogs (*trans* Ac-Phe-NHCH_2-CCl=CH-CO_2Me, $k_2/K_I = 1.7\,M^{-1}s^{-1}$ for cathepsin B) (Govardhan and Abeles 1996). However, α, β-unsaturated inhibitors are not very potent and show little promise for future use as inhibitors for cysteine proteases.

5.5.6 Azapeptide inhibitors

Azapeptide esters were originally designed as active site titrants and inhibitors for serine proteases (Powers *et al* 1984), but were later found to inhibit cysteine proteases (Magrath and Abeles 1992). The mechanism of inhibition of cathepsins by azapeptides begins with a nucleophilic attack by the active site thiol on the azapeptide carbonyl, which forms a covalent, stoichiometric, acyl enzyme derivative (Figure 5.11). The acyl derivative formed between a cysteine protease and an aza peptide is a thiocarbazoic ester derivative, while a simple thiol ester is formed with a peptide substrate. As a result, the acyl carbonyl group has decreased electrophilicity, triagonal geometry, and hydrolyzes slowly, yielding irreversible inhibition (Xing and Hanzlik 1998). The inhibitory potency of azapeptides increases with more electronegative leaving groups (Xing and Hanzlik 1998). However, inhibition by azapeptides toward cysteine proteases is modest at best, including both the ester and the α-halomethyl ketone derivatives. Examples of this type of inhibitor are shown in Table 5.11 for cathepsin B.

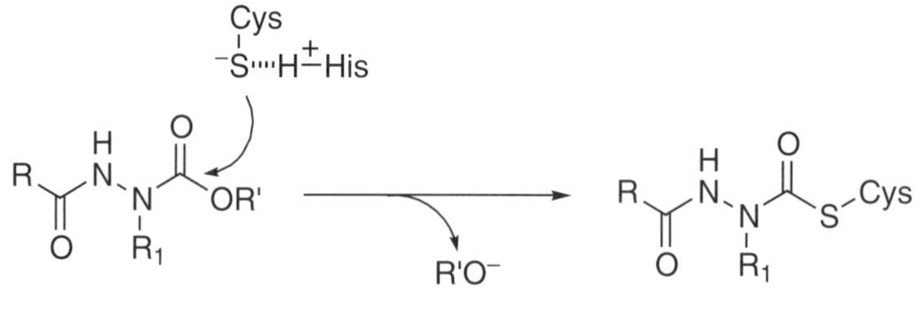

Figure 5.11 Mechanism of inhibition of cathepsins by azapeptides. The active site cysteine nucleophilically attacks the carbonyl, forming a tetrahedral intermediate, which decomposes to the acyl enzyme derivative.

Table 5.11 Azapeptide inhibitors

Enzyme	*Inhibitor*	k_2/K_I ($M^{-1}s^{-1}$)	*Reference*
Cathepsin B	Ac-Phe-NHNH-CO_2Ph	7730	Xing *et al* 1998
	Ac-Phe-NHNH-$CO_2CH_2CCl_3$	730	Xing *et al* 1998
	Ac-Phe-NHNH-CO_2CH_2Ph	310	Xing *et al* 1998

5.5.7 Peptidyl hydroxamates

Peptidyl hydroxamates or O-acylhydroxamates were originally designed to inhibit serine proteases, in particular dipeptidyl peptidase IV (Fischer *et al* 1983). However, their true usefulness was discovered when they were found to be much more effective inhibitors of cysteine proteases (Smith *et al* 1988a; Brömme *et al* 1989a). Peptidyl hydroxamates inhibit cysteine proteases by amidating the active site thiolate as shown in Figure 5.12. Experi-

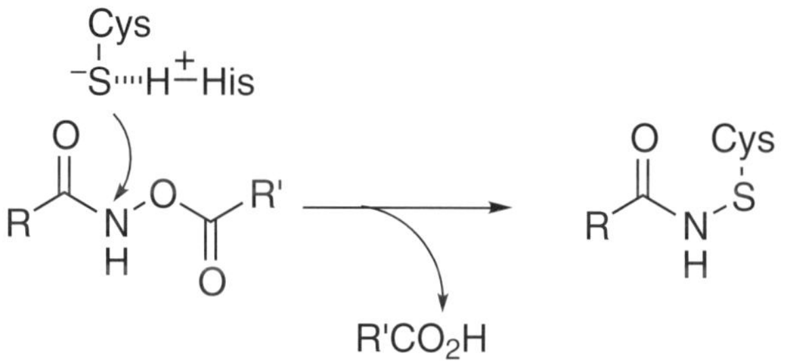

Figure 5.12 Mechanism of inhibition of cathepsins by peptidyl hydroxamates. The active site cysteine nucleophilically attacks the amide, displacing the leaving group and amidating the enzyme.

Table 5.12 Peptidyl Hydroxamates

Enzyme	*Inhibitor*	$k_{obs}/[I]$ ($M^{-1}s^{-1}$)	k_2/K_I ($M^{-1}s^{-1}$)	*Reference*
Cathepsin B	Cbz-Phe-Ala-NH-OMes[b]	640000		Shaw 1990
	Cbz-Phe-Gly-NH-OMes		580000	Smith *et al* 1988a
	Boc-Phe-Ala-NH-ONbz[c]		14000	Brömme *et al* 1989a
	Boc-Ala-Phe-Leu-NH-ONbz		12000	Brömme *et al* 1989a
Cathepsin H	Boc-Ala-Phe-Leu-NH-ONbz		32	Brömme *et al* 1989a
	Boc-Phe-Ala-NH-ONbz		21	Brömme *et al* 1989a
	Cbz-Phe-Phe-NH-OMa[a]		19	Brömme *et al* 1989a
Cathepsin L	Cbz-Phe-Phe-NH-OMa		1220000	Brömme *et al* 1989a
	Boc-Gly-Phe-Phe-NH-ONbz		800000	Brömme *et al* 1989a
	Boc-Ala-Phe-Leu-NH-ONbz		696000	Brömme *et al* 1989a
	Boc-Phe-Ala-NH-ONbz		437000	Brömme *et al* 1989a
Cathepsin S	Boc-Gly-Phe-Phe-NH-ONbz		267000	Brömme *et al* 1989a
	Boc-Ala-Phe-Leu-NH-ONbz		229000	Brömme *et al* 1989a
	Boc-Phe-Ala-NH-ONbz		42000	Brömme *et al* 1989a
	Cbz-Phe-Phe-NH-OMa		21000	Brömme *et al* 1989a

[a] Ma = methacroyl ($CH_2{=}C(CH_3)$-CO-); [b] Mes = mesitoyl; [c] Nbz = 4-nitrobenzoyl.

mental evidence supports a mechanism involving enzymatic attack on the amide in the enzyme inhibitor complex or attack on the carbonyl carbon followed by rearrangement to the amide, forming the sulfenamide (Robinson *et al* 1991). Several hydroxamate inhibitors can be seen in Table 5.12 for cathepsins B, H, L, and S.

In the absence of reducing agents normally used in assays for cysteine proteases, the inhibition pathway followed a different course. The thiol group of the enzyme active site is oxidized by the O-acylhydroxyl amine, possibly forming the sulfenic (Enz-SOH) or sulfinic acid (Enz-SO_2H) (Robinson *et al* 1991). This process is reversed by the addition of reducing agents and the enzyme's activity is completely recovered. In the presence of simple thiols, there is a catalytic, cyclic process by which the enzyme inhibitor complex is hydrolyzed, the thiol oxidized, and then reduced, followed by the final irreversible formation of the sulfenamide (Robinson *et al* 1991).

5.6 THERAPEUTIC PROPERTIES

Cathepsin inhibitors have demonstrated therapeutic benefit in disease models of rheumatoid arthritis, osteoporosis, and glomerulonephritis. Rheumatoid arthritis is a disease where the degradation of extracellular matrices of cartilage and bone lead to a loss of joint function. Cathepsin B has been implicated in this disease since high levels of the enzyme are found in the synovial lining tissues and fluids of arthritis patients (Mort *et al* 1984). Both peptidyl fluoromethyl ketones and peptidyl aldehydes have been successfully used in the treatment of rheumatoid arthritis animal models. Two fluoromethyl ketones studied by Ahmed *et al* (1992), Cbz-Phe-Ala-CH_2F and MeO-$COCH_2CH_2CO$-Phe-Ala-CH_2F, were found to reduce the severity of inflammation and the extent of cartilage and bone damage in adjuvant-induced arthritis in rats as evidenced by a dramatic reduction in focal ulcers and bone destruction. The peptide aldehyde Cbz-Leu-Leu-Leu-H, a cathepsin K inhibitor ($K_I = 1.4\,nM$), also reduced bone loss in the rat adjuvant-induced arthritis model (Votta *et al* 1997). This inhibitor is thought to slows the rate of bone resorption by cathepsin K.

Osteoporosis is a condition where bony tissue is lost, causing the bones to become brittle and to fracture easily. This disease results primarily from the activity of cathepsin K (Bossard *et al* 1996) and cathepsin L (Kakegawa *et al* 1993). Woo *et al* (1995) found that several peptide aldehyde cathepsin L inhibitors reduced the loss of radioactive Ca^{2+} in organ culture of chick calvaria by 50% and completely inhibited the formation of osteoclastic pits in bone cultures. For example, administration of Cbz-Phe-Tyr-H for 4 weeks in ovariectomyzed mice, an animal model of osteoporosis, resulted in a recovery of bone weight loss (Woo *et al* 1995). Similar results were observed with 2-NapSO_2-Ile-Trp-H ($IC_{50} = 1.9\,nM$, cathepsin L) (Yasuma *et al* 1998). Because of the structural similarities between cathepsin K and L, and the obvious role of cathepsin K in bone modification, Yasuma *et al* suggest the biological effects may be due to dual inhibition of cathepsin K and L.

The SmithKline Beecham cathepsin K inhibitors are likely to be marketed for osteoporosis in the near future and have been tested extensively in animals. A diacylhydrazide derivative of (**5.3**) (Figure 5.4), with a $-CH_2N(CH_3)_2$ substituent at the 4-position on the S′ benzyloxycarbonyl group ($k_{obs}/[I] = 5.3 \times 10^6\,M^{-1}s^{-1}$ for cathepsin K and $k_{obs}/[I] = 1.2 \times 10^5\,M^{-1}s^{-1}$ for cathepsin L) (Thompson *et al* 1997), was also used in a study of bone resorption in adult male thyroparathyroidectomized rats (Thompson *et al* 1988). The rats were fed a calcium-deficient diet for 24 hours and then injected with human parathyroid hormone, a hormone that promotes increased levels of calcium release from bones, or human parathyroid hormone plus the diacylhydrazide derivative. After 6 hours of treatment and monitoring, the rats given only the hormone, had a dramatic increase in blood calcium-ion levels (almost 50% increase). However, those rats given both the hormone and the diacylhydrazide derivative maintained near baseline levels of calcium over the entire time period. This suggests the inhibition of an enzyme, probably cathepsin K or cathepsin L in osteoclastic cells, which is active in bone resorption.

Glomerulonephritis is a disease that involves the degradation of blood capillaries that filter the blood in the kidneys and prevents the flow of plasma proteins into the urine. The prominent characteristic of this disease is the detection of protein in the urine, most notably fragments of the glomerular basement membrane (GBM) protein, which cathepsins B and L are suspected of degrading (Baricos *et al* 1988; Thomas and Davies 1989). When E-64 was administered to rats with glomerulonephritis, Barcios *et al* (1991) noticed that the animals had a dramatically reduced protein concentration in their urine. This led to the study of the E-64 derivative, *trans*-epoxysuccinyl-L-leucylamido-(3-methyl)butane (called Ep475 or E-64c), *in vivo* in rats (Baricos *et al* 1991). They discovered that Ep475 dramatically reduced the protein concentration in the animal's urine. The activity of cathepsin L was reduced by about two-fold and that of cathepsin B by about four fold compared to the anti-GBM antibody alone. However, the compound was not specific for cathepsin L or B, and so a peptidyl diazomethyl ketone, specific for cathepsin L (Kirschke *et al* 1988), was tested *in vivo* as well. The diazomethyl ketone, Cbz-Phe-Tyr(O*t*Bu)-CHN_2, was found to decrease the protein concentration in urine by about five-fold compared to the anti-GBM antibody alone, suggesting that cathepsin L is the specific proteinase involved in this disease.

Cathepsin B is important in muscular dystrophy, pulmonary emphysema, and tumor invasion, and effective inhibitors for it are highly sought after as therapeutics. Peptidyl acyloxymethyl ketones have been tested *in vivo* for their inhibitory efficiency toward cathepsin B. For example, Cbz-Phe-Lys-CH_2-OCO-Ph(2,4,6-Me_3)•HCl was found to be stable when administered subcutaneously to rats and was a potent inhibitor of cathepsin B with ED_{50} values of 2.4, 1.0, and 0.1 mg/kg in liver, skeletal muscle, and the heart, respectively (Wanger *et al* 1994), suggesting these compounds have the potential of becoming therapeutics. Peptidyl epoxysuccinyl compounds have also been tested *in vivo* for their effect on cathepsin B. E-64 and E-64c were found to be very effective inhibitors toward cathepsins B and L *in vivo*, but were not selective (Hashida *et al* 1980, 1982). Two E-64 derivatives, CA-030 and CA-074, were found to be potent and selective inhibitors of cathepsin B *in vitro* (Towatari *et al* 1991). CA-030 bound cathepsin B approximately two fold better than E-64c, and CA-074 approximately 4.5 fold better. However, *in vivo*, CA-030 was not selective for cathepsin B, possibly due to hydrolysis of its ester, but CA-074 is resistant to the hydrolysis and was highly selective *in vivo*. In fact, it bound to cathepsin B approximately 40,000-fold better than to cathepsin H and 120,000-fold better than cathepsin L.

5.7 SUMMARY

Cathepsins have been linked to a wide variety of diseases and disorders including rheumatoid arthritis, osteoporosis, glomerulonephritis, muscular dystrophy, pulmonary emphysema, tumor invasion, and many others. Numerous classes of compounds have been developed with the hopes of inhibiting the destruction caused by cathepsins. Fluoromethyl ketones, vinyl sulfones, peptidyl epoxysuccinyl derivatives, and diamino ketones have already found use in a number of animal models of disease states. The diamino ketone inhibitors of cathepsin K or L developed by SmithKline Beecham Pharmaceuticals show promise for treatment of osteoporosis. Synthetic inhibitors are also useful to study the role of cathepsins in biological systems and to identify their location. With the increasing number of disease states being associated with cathepsins, many additional inhibitors are likely to be developed for therapeutic applications.

REFERENCES

Ahmed, N., Martin, L., Watts, L., Palmer, J., Thornburg, L., Prior, J. and Esser, R. (1992) Peptidyl fluoromethyl ketones as inhibitors of cathepsin B. Implication for treatment of rheumatoid arthritis. *Biochemical Pharmacology*, **44**, 1201–1207.

Albeck, A., Fluss, S. and Persky, P. (1996) Peptidyl epoxides: novel selective inactivators of cysteine proteases. *Journal of the American Chemical Society*, **118**, 3591–3596.

Albeck, A. and Estreicher, G. (1997) Functionalized *erythro* N-protected α-amino epoxides. Stereocontrolled synthesis and biological activity. *Tetrahedron*, **53**, 5325–5338.

Albeck, A. and Kliper, S. (1997) Mechanism of cysteine protease inactivation by peptidyl epoxides. *Biochemical Journal*, **322**, 879–884.

Amos, H.E., Park, B.K. and Dixon, R. (1985). In *Immunotoxicology and Immunopharmacology*, Edited by J.H. Dean, pp. 207–288. New York: Raven Press.

Angliker, H., Anagli, J. and Shaw, E. (1992) Inactivation of calpain by peptidyl fluoromethyl ketones. *Journal of Medicinal Chemistry*, **35**, 216–220.

Aronson, N.N. and Barrett, A.J. (1978) The specificity of cathepsin B. Hydrolysis of glucagon at the C-terminus by peptidyldipeptidase mechanism. *Biochemical Journal*, **171**, 759–765.

Babine, R. and Bender, S. (1997) Molecular recognition of protein-ligand complexes: Applications to drug design. *Chemical Reviews*, **97**, 1359–1472.

Baker, E.N. and Drenth, J. (1987). In *Biological Macromolecules and Assemblies. Vol. 3. Active Sites of Enzymes*, Edited by F.A. Jurank and A. McPherson, pp. 313–368. New York: John Wiley and Sons.

Baricos, W.H., Zhou, Y., Mason, R.W. and Barrett, A.J. (1988) Human kidney cathepsins B and L. Characterization and potential role in degradation of glomerular basement membrane. *Biochemical Journal*, **252**, 301–304.

Baricos, W.H., Cortez, S.L., Le, Q.C., Wu, L.-T., Shaw, E., Hanada, K. and Shah, S.V. (1991) Evidence suggesting a role for cathepsin L in an experimental model of glomerulonephritis. *Archives of Biochemistry & Biophysics*, **288**, 468–472.

Barrett, A.J. (1973) Human cathepsin B1. Purification and some properties of the enzyme. *Biochemical Journal*, **131**, 809–822.

Barrett, A.J. and Kirschke, H. (1981) Cathepsin B, cathepsin H, and cathepsin L. *Methods in Enzymology*, **80**, 535–561.

Barrett, A., Kembhavi, A., Brown, M., Kirschke, H., Knight, C., Tamai, M. and Hanada, K. (1982) L-*trans*-Epoxysuccinyl-leucylamido(4-guanidino)butane (E-64) and its analogues as inhibitors of cysteine proteinases including cathepsins B, H and L. *Biochemical Journal*, **201**, 189–198.

Barrett, A. (1986). In *Proteinase Inhibitors*, Edited by A. Barrett and G. Salvesen, pp. 3–22. Amsterdam: Elsevier.

Barrett, A.J., Rawlings, N.D. and Woessner, J.F. (1998) Introduction: Cysteine peptidases and their clans. In *Handbook of Proteolytic Enzymes*, Edited by A.J. Barrett, N.D. Rawlings and J.F. Woessner, pp. 545. San Diego: Academic Press.

Berti, P. and Storer, A. (1995) Alignment/phylogeny of papain superfamily of cysteine proteases. *Journal of Molecular Biology*, **246**, 273–283.

Bjorck, L., Akesson, P., Bohus, M., Trojnar, J., Abrahamson, M., Olafsson, I. and Grubb, A. (1989) Bacterial-growth blocked by a synthetic peptide based on the structure of a human proteinase-inhibitor. *Nature*, **337**, 385–386.

Bohley, P. and Seglen, P.O. (1992) Proteases and proteolysis in the lysosome. *Experientia*, **48**, 151–157.

Bossard, M., Tomaszek, T., Thompson, S., Amegadzie, B., Hanning, C., Jones, C., Kurdyla, J., McNulty, D., Drake, F., Gowen, M. and Levy, M. (1996) Proteolytic activity of human osteoclast cathepsin K. Expression, purification, activation, and substrate identification. *Journal of Biological Chemistry*, **271**, 12517–12524.

Bossard, M.J., Tomaszek, T.A., Levy, M.A., Ijames, C.F., Huddleston, M.J., Briand, J., Thompson, S., Halpert, S., Veber, D.F., Carr, S.A., Meek, T.D. and Tew, D.G. (1999) Mechanism of inhibition of cathepsin K by potent, selective 1,5-diacylcarbohydrazides: A new class of mechanism-based inhibitors of thiol proteases. *Biochemistry*, **38**, 15893–15902.

Brady, K., Liang, T.-C. and Abeles, R.H. (1989) pH dependence of the inhibition of chymotrypsin by a peptidyl trifluoromethyl ketone. *Biochemistry*, **28**, 9066–9070.

Brinker, A., Weber, E., Stoll, D., Voigt, J., Müller, A., Sewald, N., Jung, G., Wiesmüller, K.-H. and Bohley, P. (2000) Highly potent inhibitors of human cathepsin L identified by screening combinatorial pentapeptide amide collections. *European Journal of Biochemistry*, **267**, 5085–5092.

Brix, K., Lemansky, P. and Herzog, V. (1996) Evidence for extracellularly acting cathepsins mediating thyroid hormone liberation in thyroid epithelial cells. *Endocrinology*, **137**, 1963–1974.

Brömme, D., Schierhorn, A., Kirschke, H., Wiederanders, B., Barth, A., Fittkau, S. and Demuth, H.-U. (1989a) Potent and selective inactivation of cysteine proteinases with *N*-peptidyl-*O*-acyl hydroxylamines. *Biochemical Journal*, **263**, 861–866.

Brömme, D., Steinert, A., Friebe, S., Fittkau, S., Wiederanders, B. and Kirschke, H. (1989b) The specificity of bovine spleen cathepsin S. A comparison with rat liver cathepsin L and B. *Biochemical Journal*, **264**, 475–481.

Brömme, D., Smith, R., Coles, P., Kirschke, H., Storer, A. and Krantz, A. (1994) Potent inactivation of cathepsins S and L by peptidyl (acyloxy)methyl ketones. *Biological Chemistry Hopper-Seyler*, **375**, 343–347.

Brömme, D. and Okamoto, K. (1995) Human cathepsin O2, a novel cysteine protease highly expressed in osteoclastomas and ovary molecular cloning, sequencing and tissue distribution. *Biological Chemistry Hopper-Seyler*, **376**, 379–384.

Brömme, D. and McGrath, M. (1996) High level expression and crystallization of recombinant human cathepsin S. *Protein Science*, **5**, 789–791.

Brömme, D., Klaus, J., Okamoto, K., Ranick, D. and Palmer, J. (1996a) Peptidyl vinyl sulphones: A new class of potent and selective cysteine protease inhibitors. S_2P_2 specificity of human cathepsin O2 in comparison with cathepsins S and L. *Biochemical Journal*, **315**, 85–89.

Brömme, D., Okamoto, K., Wang, B. and Biroc, S. (1996b) Human cathepsin O2, a matrix protein-degrading cysteine protease expressed in osteoclasts. *Journal of Biological Chemistry*, **271**, 2126–2132.

Brömme, D., Li, Z., Barnes, M. and Mehler, E. (1999) Human cathepsin V functional expression, tissue distribution, electrostatic surface potential, enzymatic characterization, and chromosomal localization. *Biochemistry*, **38**, 2377–2385.

Bucciarelli, M., Forni, A., Moretti, I., Prati, F. and Torre, G. (1993a) Candida-cylindracea lipase-catalyzed hydrolysis of methyl aziridine-2-carboxylates and aziridine-2,3-dicarboxylates. *Journal of the Chemical Society, Perkin Transactions* 1, 3041–3045.

Bucciarelli, M., Forni, A., Moretti, I., Prati, F. and Torre, G. (1993b) Enzymatic resolution of aziridine-carboxylates. *Tetrahedron: Asymmetry*, **4**, 903–906.

Buchanan, J. (1973) The amidotransferases. *Advances in Enzymology & Related Areas in Molecular Biology*, **39**, *91–183*.

Buttle, D.J., Murata, M., Knight, G.C. and Barrett, A.J. (1992) CA074 methyl ester: a proinhibitor for intracellular cathepsin B. *Archives of Biochemistry & Biophysics*, **299**, 377–380.

Chapman, H.A.J., Munger, J.S. and Shi, G.P. (1994) The role of thiol proteases in tissue injury and remodeling. *American Journal of Respiratory and Critical Care Medicine*, **150**, S155–S159.

Crawford, C., Mason, R., Wikstrom, P. and Shaw, E. (1988) The design of peptidyldiazomethane inhibitors to distinguish between the cysteine proteinases calpain II, cathepsin L and cathepsin B. *Biochemical Journal*, **253**, 751–758.

Delaisse, J.M., Ledent, P. and Vaes, G. (1991) Collagenolytic cysteine proteinases of bone tissue. Cathepsin B, (pro)cathepsin L and a cathepsin L-like 70 kDa proteinase. *Biochemical Journal*, **279**, 167–174.

Demuth, H.-U. (1990) Recent developments in inhibiting cysteine and serine proteases. *Journal of Enzyme Inhibition*, **3**, 249–278.

Dufour, É., Storer, A. and Ménard, R. (1995) Peptide aldehydes and nitriles as transition state analog inhibitors of cysteine proteases. *Biochemistry*, **34**, 9136–9143.

Eichler, J. and Houghten, R.A. (1993) Identification of substrate-analog trypsin inhibitors through the screening of synthetic peptide combinatorial libraries. *Biochemistry*, **32**, 11035–11041.

Erickson-Viitanen, S., Balestreri, E., McDermott, M.J., Horecker, B.L., Melloni, E. and Pontremoli, S. (1985) Purification and properties of rabbit liver cathepsin M and cathepsin B. *Archives of Biochemistry & Biophysics*, **243**, 46–61.

Fenwick, A.E., Garnier, B., Gribble, A.D., Ife, R.J., Rawlings, A.D. and Witherington, J. (2001a) Solid-phase synthesis of cyclic alkoxyketones, inhibitors of the cysteine protease cathepsin K. *Bioorganic & Medicinal Chemistry Letters*, **11**, 195–198.

Fenwick, A.E., Gribble, A.D., Ife, R.J., Stevens, N. and Witherington, J. (2001b) Diastereoselective synthesis, activity and chiral stability of cyclic alkoxyketone inhibitors of cathepsin K. *Bioorganic & Medicinal Chemistry Letters*, **11**, 199–202.

Fersht, A. (1985). In *Enzyme Structure and Mechansim, 2nd ed.*, pp. 405–426. New York: W.H. Freeman and Co.

Fink, A.L. and Angelides, K.J. (1976) Papain-catalyzed reactions at subzero temperatures. *Biochemistry*, **15**, 5287–5293.

Fischer, G., Demuth, H.-U. and Barth, A. (1983) N,O-diacylhydroxylamines as enzyme-activated inhibitors for serine proteases. *Pharmazie*, **38**, 249–250.

Fujishima, A., Imai, Y., Nomura, T., Fujisawa, Y., Yamamoto, Y. and Sugawara, T. (1997) The crystal structure of human cathepsin L complexed with E-64. *FEBS Letters*, **407**, 47–50.

Fukushima, K., Arai, M., Kohno, Y., Suwa, T. and Satoh, T. (1990) An epoxysuccinic acid derivative(loxistatin) – induced hepatic injury in rats and hamsters. *Toxicolology and Applied Pharmacology*, **105**, 1–12.

Gelb, B.D., Shi, G.P., Chapman, H.A. and Desnick, R.J. (1996) Pycnodysostosis, a lysosomal disease caused by cathepsin K deficiency. *Science*, **273**, 1236–1238.

Giordano, C., Gallina, C., Consalvi, V. and Scandurra, R. (1990) Irreversible inactivation of papain and cathepsin B by epoxidic substrate analogues. *European Journal of Medicinal Chemistry*, **25**, 479–487.

Golde, T.E., Estus, S., Younkin, L.H., Selkoe, D.J. and Younkin, S.G. (1992) Processing of the amyloid protein precursor to potentially amyloidogenic derivatives. *Science*, **255**, 728–730.

Gour-Salin, B., Lachance, P., Plouffe, C., Storer, A. and Ménard, R. (1993) Epoxysuccinyl dipeptides as selective inhibitors of cathepsin B. *Journal of Medicinal Chemistry*, **36**, 720–725.

Gour-Salin, B., Lachance, P., Magny, M., Plouffe, C., Ménard, R. and Storer, A. (1994) E64 [*trans*-epoxysucciyl-L-leucylamido-(4-guanidino)butane] analogues as inhibitors of cysteine proteinases: Investigation of S_2 subsite interactions. *Biochemical Journal*, **299**, 389–392.

Govardhan, C. and Abeles, R. (1996) Inactivation of cysteine proteases. *Archives of Biochemistry & Biophysics*, **330**, 110–114.

Graybill, T., Ross, M., Gauvin, B., Gregory, J., Harris, A., Ator, M., Rinker, J. and Dolle, R. (1992) Synthesis and evaluation of azapeptide-derived inhibitors of serine and cysteine proteases. *Bioorganic & Medicinal Chemistry Letters*, **2**, 1375–1380.

Green, D. and Shaw, E. (1981) Peptidyl diazomethyl ketones are specific inactivators of thiol proteinases. *Journal of Biological Chemistry*, **256**, 1923–1928.

Guncar, G., Klemencic, I., Turk, B., Turk, V., Karaoglanovic-Carmona, A., Juliano, L. and Turk, D. (2000) Crystal structure of cathepsin X: a flip-flop of the ring of His23 allows carboxy-monopeptidase and carboxy-dipeptidase activity of the protease. *Structure*, **8**, 305–313.

Gupton, B.F., Carroll, D.L., Tuhy, P.M., Kam, C.M. and Powers, J.C. (1984) Reaction of azapeptides with chymotrypsin-like enzymes. New inhibitors and active site titrants for chymotrypsin A alpha, subtilisin BPN′, subtilisin Carlsberg, and human leukocyte cathepsin G. *Journal of Biological Chemistry*, **259**, 4279–4287.

Hanada, K., Tamai, M., Yamagishe, M., Ohmura, S., Sawada, J. and Tanaka, I. (1978) Isolation and characterization of E-64, a new thiol protease inhibitor. *Agriculture and Biological Chemistry*, **42**, 523–528.

Hanzlik, R. and Thompson, S. (1984) Vinylogous amino acid esters: a new class of inactivators for thiol proteases. *Journal of Medicinal Chemistry*, **27**, 711–712.

Harbeson, S., Abelleira, S., Akiyama, A., Barrett, R.I., Carroll, R., Straub, J., Tkacz, J., Wu, C. and Musso, G. (1994) Stereospecific synthesis of peptidyl alpha-ketoamides as inhibitors of calpain. *Journal Medicinal Chemistry*, **37**, 2918–2929.

Hashida, S., Towatari, T., Kominami, E. and Katunuma, N. (1980) Inhibitions by E-64 derivatives of rat liver cathepsin B and cathepsin L *in vitro* and *in vivo*. *Journal of Biochemistry*, **88**, 1805–1811.

Hashida, S., Kominami, K. and Katunuma, N. (1982) Inhibitions of cathepsin B and cathepsin L by E-64 *in vivo*. II. Incorporation of [3H]E-64 into rat liver lysosomes *in vivo*. *Journal of Biochemistry*, **91**, 1373–1380.

Hawthorne, S., Pagano, M., Harriot, P., Halton, D. and Walker, B. (1998) The synthesis and utilization of 2,4-dinitrophenyl-labeled irreversible peptidyl diazomethyl ketone inhibitors. *Analytical Biochemistry*, **261**, 131–138.

Honn, K., Cavanaugh, P., Evens, C., Taylor, J. and Sloane, B. (1982) Tumor cell-platelet aggregation: induced by cathepsin B-like proteinase and inhibited by prostacyclin. *Science*, **217**, 540–542.

Horn, M., Pavlík, M., Dolecková, L., Baudys, M. and Mares, M. (2000) Arginine-based structures are specific inhibitors of cathepsin C: Application of peptide combinatorial libraries. *European Journal of Biochemistry*, **267**, 3330–3336.

Hu, L.-Y. and Abeles, R. (1990) Inhibition of cathepsin-B and papain by peptidyl, α-ketoesters, α-ketoamides, α-diketones, and α-ketoacids. *Archives of Biochemistry & Biophysics*, **281**, 271–274.

Illy, C., Quraish, O., Wang, J., Purisima, E., Vernet, T. and Mort, J. (1997) Role of occluding loop in cathepsin B activity. *Journal of Biological Chemistry*, **272**, 1197–1202.

Kakegawa, H., Nikawa, T., Tagami, K., Kamioka, H., Sumitani, K., Kawata, T., Drobnic-Kosorok, M., Lenarcic, B., Turk, V. and Katunuma, N. (1993) Participation of cathepsin L on bone resorption. *FEBS Letters*, **321**, 247–250.

Kamphuis, I.G., Drenth, J. and Baker, E.N. (1985) Thiol proteases. Comparative studies based on the high-resolution structures of papain and actinidin, and on amino acid sequence information for cathepsins B and H, and stem bromelain. *Journal of Molecular Biology*, **182**, 317–329.

Kirschke, H., Langner, J., Wiederanders, B., Ansorge, S., Bohley, P. and Broghammer, U. (1976) Intracellular protein breakdown. VII. Cathepsin L and H; two new proteinases from rat liver lysosomes. *Acta Biologica et Medica Germanica*, **35**, 285–299.

Kirschke, H., Langner, J., Wiederanders, B., Ansorge, S. and Bohley, P. (1977) Cathepsin L: A new proteinase from rat-liver lysosomes. *European Journal of Biochemistry*, **74**, 293–301.

Kirschke, H. and Shaw, E. (1981) Rapid inactivation of cathepsin L by Z-Phe-PheCHN$_2$ and Z-Phe-AlaCHN$_2$. *Biochemical & Biophysical Research Communications*, **101**, 454–458.

Kirschke, H., Kembhavi, A.A., Bohley, P. and Barrett, A.J. (1982) Action of rat liver cathepsin L on collagen and other substrates. *Biochemical Journal*, **201**, 367–372.

Kirschke, H., Schmidt, I. and Wiederanders, B. (1986) Cathepsin S: The cysteine proteinase from bovine lymphoid tissue is distinct from cathepsin L (EC 3.4.22.15). *Biochemical Journal*, **240**, 455–459.

Kirschke, H. and Barrett, A. (1987). In *Lysosomes: Their Roles in Protein Breakdown*, Edited by H. Glaumann and F.J. Ballard, pp. 193–238. London: Academic Press.

Kirschke, H., Wikstrom, P. and Shaw, E. (1988) Active center differences between cathepsins L and B: the S1 binding region. *FEBS Letters*, **228**, 128–130.

Kirshke, H., Wiederanders, B., Brömme, D. and Rinne, A. (1989) Cathepsin S from bovine spleen. *Biochemical Journal*, **264**, 467–473.

Kirschke, H. and Wiederanders, B. (1994) Cathepsin S and related lysosomal endopeptidases. *Methods in Enzymology*, **244**, 500–511.

Klemencic, I., Carmona, A.K., Cezari, M.H.S., Juliano, M.A., Juliano, L., Guncar, G., Turk, D., Krizaj, I., Turk, V. and Turk, B. (2000) Biochemical characterization of human cathpesin X revealed that the enzyme is an exopeptidase, acting as carboxymonopeptidase or carboxydipeptidase. *European Journal of Biochemistry*, **267**, 5404–5412.

Koga, H., Mori, N., Yamada, H., Nishimura, Y., Tokuda, K., Kato, K. and Imoto, T. (1992) Endo- and aminopeptidase activities of rat cathepsin H. *Chemical & Pharmaceutical Bulletin* **40**, 965–970.

Korant, B., Chow, N., Lively, M. and Powers, J. (1979) Virus-specified protease in poliovirus-infected HeLa cells. *Proceedings of National Academy of Sciences of the U.S.A.*, **76**, 2992–2995.

Krieger, T.J. and Hook, V.Y.H. (1991) Purification and characterization of a novel thiol protease involved in processing the enkephalin precursor. *Journal of Biological Chemistry*, **266**, 1106–1109.

Lewis, S.D., Johnson, F.A. and Shafer, J.A. (1976) Potentiometric determination of ionizations at the active site of papain. *Biochemistry*, **15**, 5009–5017.

Li, Z., Patil, G., Golubski, Z., Hori, H., Tehrani, K., Foreman, J., Eveleth, D., Bartus, R. and Powers, J. (1993) Peptide α-ketoester, α-ketoamide, and α-ketoacid inhibitors of calpains and other cysteine proteases. *Journal of Medicinal Chemistry*, **36**, 3472–3480.

Li, Y.-P., Alexander, M., Wucherpfennig, A., Chen, W., Yelick, P. and Stashenko, P. (1995) Cloning and complete coding sequence of a novel human cathepsin expressed in giant cells of osteoclastmas. *Journal of Bone & Mineral Research* **10**, 1197–1202.

Liang, T.-C. and Abeles, R.H. (1987) Complex of alpha-chymotrypsin and N-acetyl-L-leucyl-L-phenylalanyl trifluoromethyl ketone: structural studies with NMR spectroscopy. *Biochemistry*, **26**, 7603–7608.

Liao, J.C.R. and Lenney, J.F. (1984) Cathepsin J and K: High molecular weight cysteine proteinases from human tissues. *Biochemical & Biophysical Research Communications*, **124**, 909–916.

Lienhard, G. and Jencks, W. (1966) Thiol addition to the carbonyl group. Equilibria and kinetics. *Journal of the American Chemical Society*, **88**, 3982–3995.

Linnevers, C., Smeekens, S. and Brömme, D. (1997) Human cathepsin W, a putative cysteine proteases predominantly expressed in $CD8^+$ T-lymphocytes. *FEBS Letters*, **405**, 253–259.

Littlewood-Evans, A.J., Bilbe, G., Bowler, W.B., Farley, D., Wlodarski, B., Kokubo, T., Inaoka, T., Sloane, J., Evans, D.B. and Gallagher, J.A. (1997) The osteoclast-associated protease cathepsin K is expressed in human breast carcinoma. *Cancer Research*, **57**, 5386–5390.

Lowe, G. and Yuthavong, Y. (1971) Kinetic specificity in papain-catalysed hydrolyses. *Biochemical Journal*, **124**, 107–115.

Lynas, J.F., Hawthorne, S.J. and Walker, B. (2000) Development of peptidyl α-keto-β-aldehydes as new inhibitors of cathepsin L – Comparisons of potency and selectivity profiles with cathepsin B. *Bioorganic & Medicinal Chemistry Letters*, **10**, 1771–1773.

Magrath, J. and Abeles, R. (1992) Cysteine protease inhibition by azapeptide esters. *Journal of Medicinal Chemistry*, **35**, 4279–4283.

Marquis, R., Yamashita, D., Ru, Y., LoCastro, S., Oh, H.-J., Erhard, K., DesJarlais, R., Head, M., Smith, W., Zhao, B., Janson, C., Abdel-Meguid, S., Tomaszek, T., Levy, M. and Veber, D. (1998)

Conformationally constrained 1,3-diamino ketones: A series of potent inhibitors of the cysteine protease cathepsin K. *Journal of Medicinal Chemistry*, **41**, 3563–3567.

Marquis, R.W., Ru, Y., Yamashita, D.S., Oh, H.J., Yen, J., Thompson, S.K., Carr, T.J., Levy, M.A., Tomaszek, T.A., Ijames, C.F., Smith, W.W., Zhao, B., Janson, C.A., Abdel-Meguid, S.S., D'Alessio, K.J., McQueney, M.S. and Veber, D.F. (1999) Potent dipeptidylketone inhibitors of the cysteine protease cathepsin K. *Bioorganic & Medicinal Chemistry*, **7**, 581–588.

Marquis, R.W., Ru, Y., Zeng, J., Trout, R.E.L., LoCastro, S.M., Gribble, A.D., Witherington, J., Fenwick, A.E., Garnier, B., Tomaszek, T., Tew, D., Hemling, M.E., Quinn, C.J., Smith, W.W., Zhao, B., McQueney, M.S., Janson, C.A., D'Alessio, K. and Veber, D.F. (2001) Cyclic ketone inhibitors of cysteine protease cathepsin K. *Journal of Medicinal Chemistry*, **44**, 725–736.

Mason, R.W., Johnson, D.A., Barrett, A.J. and Chapman, H.A. (1986) Elastinolytic activity of human cathepsin L. *Biochemical Journal*, **233**, 925–927.

Matsumoto, K., Mizoue, K., Kitamura, K., Tse, W.-C., Huber, C.P. and Ishida, T. (1999) Structural basis of inhibition of cysteine proteases by E-64 and its derivatives. *Biopolymers*, **51**, 99–107.

McDonald, J.K., Zeitman, B.B., Reilly, T.J. and Ellis, S. (1969) New observations on the substrate specificity of cathepsin C (dipeptidylaminopeptidase I). Including the degradation of beta-corticotropin and other peptide hormones. *Journal of Biological Chemistry*, **244**, 2693–2709.

McDonald, J. and Ellis, S. (1975) On the substrate specificity of cathepsins B1 and B2 including a new fluorogenic substrate for cathepsin B1. *Life Science*, **17**, 1269–1276.

McDonald, J.K. and Schwabe, C. (1977). In *Proteinases in Mammalian Cells and Tissues*, Edited by A.J. Barrett, pp. 311–391. Amsterdam: North-Holland.

McGrath, M.E., Eakin, A.E., Engel, J.C., McKerrow, J.H., Craik, C.S. and Fletterick, R.J. (1995) The crystal structure of cruzain: a therapeutic target for Chagas' disease. *Journal of Molecular Biology*, **247**, 251–259.

McGrath, M.E., Klaus, J.L., Barnes, M.G. and Brömme, D. (1997) Crystal structure of human cathepsin K complexed with a potent inhibitor. *Nature Structural Biology*, **4**, 105–109.

McGrath, M.E. (1999) The lysosomal cysteine proteases. *Annual Review of Biophysics & Biophysical Structure*, **28**, 181–204.

Mehdi, S., Angelastro, M., Wiseman, J. and Bey, P. (1988) Inhibition of the proteolysis of rat erythrocyte-membrane proteins by a synthetic inhibitor of calpain. *Biochemical & Biophysical Research Communications*, **157**, 1117–1123.

Ménard, R., Carriére, J., Laflamme, P., Plouffe, C., Khouri, H.E., Vernet, T., Tessier, D.C., Thomas, D.Y. and Storer, A.C. (1991) Contribution of the glutamine 19 side chain to transition-state stabilization in the oxyanion hole of papain. *Biochemistry*, **30**, 8924–8928.

Ménard, R., Plouffe, C., Laflamme, P., Vernet, T., Tessier, D.C., Thomas, D.Y. and Storer, A.C. (1995) Modification of the electrostatic environment is tolerated in the oxyanion hole of the cysteine protease papain. *Biochemistry*, **34**, 464–471.

Moon, J., Coleman, R. and Hanzlik, R. (1986) Reversible covalent inhibition of papain by a peptide nitrile. ^{13}C NMR evidence for a thiolimidate ester adduct. *Journal of the American Chemical Society*, **108**, 1350–1351.

Morodor, L., Musiol, H.-J. and Scharf, R. (1992) Aziridine-2-carboxylic acid. A reactive amino acid unit for a new class of cysteine proteinase inhibitors. *FEBS Letters*, **299**, 51–53.

Mort, J.S., Recklies, A.D. and Poole, A.R. (1984) Extracellular presence of the lysosomal proteinase cathepsin B in rheumatoid synovium and its activity at neutral pH. *Arthritis & Rheumatism* **27**, 509–515.

Musil, D., Zucic, D., Turk, D., Engh, R.A., Mayr, I., Huber, R., Popovic, T., Turk, V., Towatari, T., Katunuma, N. and Bode, W. (1991) The refined 2.15 A X-ray crystal structure of human liver cathepsin B: the structural basis for its specificity. *EMBO Journal*, **10**, 2321–2330.

Nägler, D., Stoer, A., Portaro, F., Carmona, E., Juliano, L. and Ménard, R. (1997) Major increase in endopeptidase activity of human cathepsin B upon removal of occluding loop contacts. *Biochemistry*, **36**, 12608–12615.

Nägler, D.K. and Ménard, R. (1998) Human cathepsin X: A novel cysteine protease of the papain family with a very short proregion and unique insertions. *FEBS Letters*, **434**, 135–139.

Nägler, D., Tam, W., Storer, A., Krupa, J., Mort, J. and Ménard, R. (1999a) Interdependency of sequence and positional specificities for cysteine proteases of the papain family. *Biochemistry*, **38**, 4868–4874.

Nägler, D.K., Zhang, R., Tam, W., Sulea, T., Purisima, E.O. and Ménard, R. (1999b) Human cathepsin X: A cysteine protease with unique carboxypeptidase activity. *Biochemistry*, **38**, 12648–12654.

Nakajima, A., Kataoka, K., Takata, Y. and Huh, N.-H. (2000) Cathepsin-6, a novel cysteine proteinase showing homology with and co-localized expression with cathepsin J/P in the labyrinthine layer of mouse placenta. *Biochemical Journal*, **349**, 689–692.

Nikawa, T., Towatari, T. and Katunuma, N. (1992) Purification and characterization of cathepsin J from rat liver. *European Journal of Biochemistry*, **204**, 381–393.

Noda, T., Isogal, K., Katunuma, N., Tarumoto, Y. and Ohzeki, M. (1981) Effects on cathepsins B, H, and D in pectoral muscle of dystrophic chickens of *in vivo* administration of E-64-c (N[N-(L-3-transca rboxyoxirane-2-carbonyl)-L-leucyl]-3-methyl-butylamine). *Journal of Biochemistry*, **90**, 893–896.

Otto, H.-H. and Schirmeister, T. (1997) Cysteine proteases and their inhibitors. *Chemical Reviews*, **97**, 133–171.

Page, M. (1990) In *Comprehensive Medicinal Chemistry*, Edited by P. Sammes, pp. 61–87. Oxford: Pergamon Press.

Palmer, J., Rasnick, D., Laus, J. and Brömme, D. (1995) Vinyl sulfones as mechanism-based cysteine protease inhibitors. *Journal of Medicinal Chemistry*, **38**, 3193–3196.

Petanceska, S. and Devi, L. (1992) Sequence analysis, tissue distribution, and expression of rat cathepsin S. *Journal of Biological Chemistry*, **267**, 26038–26043.

Plapp, B.V. (1982) Application of affinity labeling for studying structure and function of enzymes. *Methods in Enzymology*, **87**, 469–499.

Pliura, D., Bonaventura, B., Smith, R., Coles, P. and Krantz, A. (1992) Comparative behavior of calpain and cathepsin B toward peptidyl acyloxymethyl ketones, sulphonium methyl ketones and other potential inhibitors of cysteine proteinases. *Biochemical Journal*, **288**, 759–762.

Pocker, Y., Ronald, B.P. and Anderson, K.W. (1988) A mechanistic characterization of the spontaneous ring-opening process of epoxides in aqueous-solution-kinetic and product studies. *Journal of the American Chemical Society*, **110**, 6492–6397.

Polgar, L. and Halasz, P. (1982) Current problems in mechanistic studies of serine and cysteine proteinases. *Biochemical Journal*, **207**, 1–10.

Pontremoli, S., Melloni, E., Salamino, F., Sparatore, B., Michetti, M. and Horecker, B.L. (1982) Cathepsin M: A lysosomal proteinase with aldolase-inactivating activity. *Archives of Biochemical & Biophysics*, **214**, 376–385.

Powers, J., Boone, R., Carroll, D., Gupton, B., Kam, C.-M., Nishino, N., Sakamoto, M. and Tuhy, P. (1984) Reaction of azapeptides with human leukocyte elastase and porcine pancreatic elastase. New inhibitors and active site titrants. *Journal of Biological Chemistry*, **259**, 4288–4294.

Rao, N., Rao, G. and Hoidal, J. (1997) Human dipeptidyl-peptidase I. *Journal of Biological Chemistry*, **272**, 10260–10265.

Rauber, P., Angliker, H., Walker, B. and Shaw, E. (1986) The synthesis of peptidyl fluoromethanes and their properties as inhibitors of serine proteinases and cysteine proteinases. *Biochemical Journal*, **239**, 633–640.

Rawlings, N.D. and Barrett, A.J. (1993) Evolutionary families of peptidases. *Biochemical Journal*, **290**, 205–218.

Rawlings, N.D. and Barrett, A.J. (1994) Families of cysteine peptidases. *Methods in Enzymology*, **244**, 461–486.

Reddy, V.Y., Zhang, Q.Y. and Weiss, S.J. (1995) Pericellular mobilization of the tissue-destructive cysteine proteinases, cathepsins B, L, and S, by human monocyte-derived macrophages. *Proceedings of the National Academy of Sciences of the U.S.A.*, **92**, 3849–3853.

Rich, D.H. (1986) Research monographs in cell and tissue physiology. In *Proteinase Inhibitors*, Edited by A.J. Barrett and G. Salvensen, pp. 153–178. Amsterdam: Elsevier.

Riese, R.J. (1996) Essential role for cathepsin S in MHC class II-associated invariant chain processing and peptide loading. *Immunity*, **4**, 357–366.

Riese, R.J., Mitchell, R.N., Villadangos, J.A., Karp, E.R., DeSanctis, G.T., Ploegh, H.L. and Chapman, H.A. (1998) Cathepsin S activity regulates antigen presentation and immunity. *The Journal of Clinical Investigation*, **101**, 2351–2363.

Robinson, V., Coles, P., Smith, R. and Krantz, A. (1991) Competing redox and inactivation processes in the inhibition of cysteine proteinases by peptidyl O-acylhydroxamates – C-13 and N-15 NMR evidence for a novel sulfenamide enzyme adduct. *Journal of the American Chemical Society*, **113**, 7760–7761.

Santamaría, I., Velasco, G., Cazorla, M., Fueyo, A., Campo, E. and López-Otín, C. (1998a) Cathepsin L2, a novel human cysteine proteinase produced by breast and colorectal carcinomas. *Cancer Research*, **58**, 1624–1630.

Santamaría, I., Velasco, G., Pendás, A.M., Fueyo, A. and López-Otín, C. (1998b) Cathepsin Z, a novel human cysteine proteinase with a short propeptide domain and a unique chromosomal location. *Journal of Biological Chemistry*, **273**, 16816–16823.

Santamaría, I., Velasco, G., Pendás, A., Paz, A. and López-Otín, C. (1999) Molecular cloning and structural and functional characterization of human cathepsin F, a new cysteine proteinase of papain family with a long propeptide domain. *Journal of Biological Chemistry*, **274**, 13800–13809.

Schaschke, N., Assfalg-Machleidt, I., Machleidt, W., Turk, D. and Moroder, L. (1997) E-64 analogues as inhibitors of cathepsin B. On the role of absolute configuration of the epoxysuccinyl group. *Bioorganic & Medicinal Chemistry*, **5**, 1789–1797.

Schaschke, N., Assfalg-Machleidt, I., Machleidt, W. and Moroder, L. (1998) Substrate/propeptide-derived endo-epoxysuccinyl peptides as highly potent and selective cathepsin B inhibitors. *FEBS Letters*, **421**, 80–82.

Schechter, I. and Berger, A. (1967) On the size of the active site in proteases. I. Papain. *Biochemical & Biophysical Research Communications*, **27**, 157–162.

Schilling, D.M. and Ahlquist, D.A. (1999) Cathepsin L2 overexpression in colorectal neoplasia. *Gastroenterology*, **116**, A370.

Schirmeister, T. (1999a) Inhibition of cysteine proteases by peptides containing aziridine-2,3-dicarboxylic acid building blocks. *Biopolymers*, **51**, 87–97.

Schirmeister, T. (1999b) New peptidic cysteine protease inhibitors derived from the electrophilic α-amino acid aziridine-2,3-dicarboxylic acid. *Journal of Medicinal Chemistry*, **42**, 560–572.

Schirmeister, T. and Perics, M. (2000) Aziridinyl peptides as inhibitors of cysteine proteases: Effect of free carboxylic acid function on inhibition. *Bioorganic & Medicinal Chemistry*, **8**, 1281–1291.

Schroder, E., Phillips, C., Garman, E., Harlos, K. and Crawford, C. (1993) X-ray crystallographic structure of a papain-leupeptin complex. *FEBS Letters*, **315**, 38–42.

Schultz, R., Varma-Nelson, R., Oritz, R., Kozlowski, K., Orawski, A., Pagast, P. and Frankfater, A. (1989) Active and inactive forms of the transition-state analog protease inhibitor leupeptin: explanation of the observed slow binding of leupeptin to cathepsin B and papain. *Journal of Biological Chemistry*, **264**, 1497–1507.

Schwartz, W. and Barrett, A. (1980) Human cathepsin H. *Biochemical Journal*, **191**, 487–497.

Shaw, E. (1984) The selective inactivation of thiol proteases *in vitro* and *in vivo*. *Journal of Protein Chemistry*, **3**, 109–120.

Shaw, E. (1990) Cysteinyl proteinases and their selective inactivation. In *Advances in Enzymology and Related Areas in Molecular Biology*, Edited by A. Meister, pp. 271–347. New York: J. Wiley & Sons.

Shaw, E., Mohanty, S., Colic, A., Stoka, V. and Turk, V. (1993) The affinity-labelling of cathepsin S with peptidyl diazomethyl ketones. Comparison with the inhibition of cathepsin L and calpain. *FEBS Letters*, **334**, 340–342.

Sheahan, K., Shuja, S. and Murnane, M.J. (1989) Cysteine protease activities and tumor development in human colorectal carcinoma. *Cancer Research*, **49**, 3809–3814.

Shi, G.-P., Munger, J.S., Meara, J.P., Rich, D.H. and Chapman, H.A. (1992) Molecular cloning and expression of human alveolar macrophage cathepsin S, an elastinolytic cysteine protease. *Journal of Biological Chemistry*, **267**, 7258–7262.

Shi, G.-P., Chapman, H., Bhairi, S., DeLeeuw, C., Reddy, V. and Weiss, S. (1995) Molecular cloning of human cathepsin O, a novel endoproteinase and homologue of rabbit OC2. *FEBS Letters*, **357**, 129–134.

Sivaraman, J., Nägler, D.K., Zhang, R., Ménard, R. and Cygler, M. (2000) Crystal structure of human procathepsin X: A cysteine protease with the proregion covalently linked to the active site cysteine. *Journal of Molecular Biology*, **295**, 939–951.

Sloane, B.F. (1990) Cathepsin B and cystatins: evidence for a role in cancer progression. *Seminars in Cancer Biology*, **1**, 137–152.

Smith, R., Coles, P., Spencer, R., Copp, L., Jones, C. and Krantz, A. (1988a) Peptidyl O-acyl hydroxymates: Potent new inactivators of cathepsin B. *Biochemical & Biophysical Research Communications*, **155**, 1201–1206.

Smith, R., Copp, L., Coles, P., Pauls, H., Robinson, V., Spencer, R., Heard, S. and Krantz, A. (1988b) New inhibitors of cysteine proteinases – peptidyl acyloxymethyl ketones and the quiescent nucleofuge strategy. *Journal of American Chemical Society*, **110**, 4429–4431.

Sol-Church, K., Frenk, J., Troeber, D. and Mason, R.W. (1999) Cathepsin P, a novel protease in mouse placenta. *Biochemical Journal*, **343**, 307–309.

Sol-Church, K., Frenck, J., Bertenshaw, G. and Mason, R.W. (2000a) Characterization of mouse cathepsin R, a new member of a family of placentally expressed cysteine proteases. *Biochimica et Biophysica Acta*, **1492**, 488–492.

Sol-Church, K., Frenck, J. and Mason, R.W. (2000b) Cathepsin Q, a novel lysosomal cysteine protease highly expressed in placenta. *Biochemical & Biophysical Research Communications*, **267**, 791–795.

Sol-Church, K., Frenck, J. and Mason, R.W. (2000c) Mouse cathepsin M, a placenta-specific lysosomal cysteine protease related to cathepsin L and P. *Biochimica et Biophysica Acta*, **1491**, 289–294.

Somoza, J.R., Zhan, H., Bowman, K.K., Yu, L., Mortara, K.D., Palmer, J.T., Clark, J.M. and McGrath, M.E. (2000) Crystal structure of human cathepsin V. *Biochemistry*, **39**, 12543–12551.

Tamai, M., Yokoo, C., Murata, M., Oguma, K., Sota, K., Sato, E. and Kanaoka, Y. (1987) Efficient synthetic method for ethyl (+)-(2S,3S)-3-[(S)-3-methyl-1-(3-methylbutylcarbamoyl)butylcarbamoyl]-2-oxiranecarboxylate (EST), a new inhibitor of cysteine proteinases. *Chemical & Pharmaceutical bulletin*, **35**, 1098–1104.

Tebel, C., Brömme, D., Herzog, V. and Brix, K. (2000) Cathepsin K in thyroid epithelial cells: sequence, localization and possible function in extracellular proteolysis of thyroglobulin. *Journal of Cell Science*, **113**, 4487–4498.

Tezuka, K., Tezuka, Y., Maejima, A., Sato, T., Nemoto, K., Kamioka, H., Hakeda, Y. and Kumegawa, M. (1994) Molecular cloning of possible cysteine proteinases predominantly expressed in osteoclasts. *Journal of Biological Chemistry*, **269**, 1106–1109.

Therrien, C., Lachance, P., Sulea, T., Purisima, E.O., Qi, H., Ziomek, E., Alvarez-Hernandez, A., Roush, W.R. and Ménard, R. (2001) Cathepsins X and B can be differentiated through their respective mono- and dipeptidyl carboxypeptidase activities. *Biochemistry*, **40**, 2702–2711.

Thomas, G.J. and Davies, M. (1989) The potential role of human kidney cortex cysteine proteinases in glomerular basement membrane degradation. *Biochimica et Biophysica Acta*, **990**, 246–253.

Thompson, D.D., Seedor, J.G., Fisher, J.E., Rosenblatt, M. and Rodan, G.A. (1988) Direct action of the parathyroid hormone-like human hypercalcemic factor on bone. *Proceedings of the National Academy of Sciences of the U.S.A.*, **85**, 5673–5677.

Thompson, S., Halbert, S., Bossard, M., Tomaszek, T., Levy, M., Zhao, B. *et al* (1997) Design of potent and selective human cathepsin K inhibitors that span the active site. *Proceedings of the Nationl Academy of Sciences of the U.S.A.*, **94**, 14249–14254.

Thompson, S.A., Andrews, P.R. and Hanzlik, R.P. (1986) Carboxyl-modified amino acids and peptides as protease inhibitors. *Journal of Medicinal Chemistry*, **29**, 104–111.

Thompson, S.K., Smith, W.W., Zhao, B., Halbert, S.M., Tomaszek, T.A., Tew, D.G., Levy, M.A., Janson, C.A., D'Alessio, K.J., McQueney, M.S., Kurdyla, J., Jones, C.S., DesJarlais, R.L., Abdel-Meguid, S.S. and Veer, D.F. (1998) Structure-based design of cathepsin K inhibitors containing a benzyloxy-substituted benzoyl peptidomimetic. *Journal of Medicinal Chemistry*, **41**, 3923–3927.

Tisljar, K., Deussing, J. and Peters, C. (1999) Cathepsin J, a novel murine cysteine protease of the papain family with placenta-restricted expression. *FEBS Letters*, **459**, 299–304.

Towatari, T., Nikawa, T., Murata, M., Yokoo, C., Tamai, M., Hanada, K. and Katunuma, N. (1991) Novel epoxysuccinyl peptides: A selective inhibitor of cathepsin B, *in vivo*. *FEBS Letters*, **280**, 311–315.

Tsujinaka, T.Y.K., Kambayashi, J., Sakon, M., Higuchi, N., Tanaka, T. and Mori, T. (1988) Synthesis of a new cell penetrating calpain inhibitor (calpeptin). *Biochemical & Biophysical Research Communications*, **153**, 1201–1208.

Tsushima, H., Ueki, A., Matsuoka, Y., Mihara, H. and Hopsu-Havu, V.K. (1991) Characterization of a cathpesin H-like enzyme from a human melanoma cell line. *International Journal of Cancer*, **48**, 726–732.

Turk, D., Podobnik, M., Popovic, T., Katunuma, N., Bode, W., Huber, R. and Turk, V. (1995) Crystal structure of cathepsin B inhibited with CA030 at 2.0-A resolution: A basis for the design of specific epoxysuccinyl inhibitors. *Biochemistry*, **34**, 4791–4797.

Turnsek, T., Kregar, I. and Lebez, D. (1975) Acid sulphydryl protease from calf lymph nodes. *Biochimica et Biophysica Acta*, **403**, 514–520.

Varughese, K., Ahmed, F., Carey, P., Hasnain, S., Huber, C. and Storer, A. (1989) Crystal structure of a papain-E-64 complex. *Biochemistry*, **28**, 1330–1332.

Velasco, G., Ferrando, A., Puente, X., Sánchez, L. and López-Otín, C. (1994) Human cathepsin O. *Journal of Biological Chemistry*, **269**, 27136–27142.

Votta, B.J., Levy, M.A., Badger, A., Bradbeer, J., Dodds, R.A., James, I.A., Thompson, S.T., Bossard, M.J., Carr, T., Connor, J.R., Tomaszek, T.A., Szewczuk, L., Drake, F.H., Veber, D.F. and Gowen, M. (1997) Peptide aldehyde inhibitors of cathepsin K inhibit bone resorption both *in vitro* and *in vivo*. *Journal of Bone and Mineral Research*, **12**, 1396–1406.

Walker, B., McCarthy, N., Healy, A., Ye, T. and McKervey, M.A. (1993) Peptide glyoxals: a novel class of inhibitor for serine and cysteine proteinases. *Biochemical Journal*, **293**, 321–323.

Walker, B., Lynas, J.F., Meighan, M.A. and Brömme, D. (2000) Evaluation of dipeptide α-keto-β-aldehydes as new inhibitors of cathepsin S. *Biochemical & Biophysical Research Communications*, **275**, 401–405.

Walsh, C. (1979). In *Enzymatic Reaction Mechanisms*, pp. 53–107. New York: W H. Freeman and Co.

Wang, B., Shi, G.-P., Yao, P., Li, Z., Chapman, H. and Brömme, D. (1998) Human cathepsin F. *Journal of Biological Chemistry*, **273**, 32000–32008.

Wanger, B.M., Smith, R.A., Coles, P.J., Copp, L.J., Ernest, M.J. and Krantz, A. (1994) *In vivo* inhibition of cathepsin B by peptidyl (acyloxy)methyl ketones. *Journal of Medicinal Chemistry*, **37**, 1833–1840.

Watanabe, H., Green, G.D.J. and Shaw, E. (1979) A comparison of the behavior of chymotrypsin and cathepsin B towards peptidyl diazomethyl ketones. *Biochemical & Biophysical Research Communications*, **89**, 1354–1360.

Westerik, J.O. and Wolfenden, R. (1972) Aldehydes as inhibitors of papain. *Journal of Biological Chemistry*, **247**, 8195–8197.

Wex, T., Levy, B., Smeekens, S.P., Ansorge, S., Desnick, R.J. and Brömme, D. (1988) Genomic structure, chromosomal localization, and expression of human cathepsin W. *Biochemical & Biophysical Research Communications*, **248**, 255–261.

Willstätter, R. and Bamann, E. (1929) Über die proteasen der magenschleimhaut. Erste abhandlung über die enzyme der leukocyten. *Hoppe-Seylers Zeitschrift fur Physiologische Chemie*, **180**, 127–143.

Woo, J.-T., Sigeizumi, S. and Yamaguchi, K. (1995) Peptidyl aldehyde derivatives as potent and selective inhibitors of cathepsin L. *Bioorganic & Medicinal Chemistry Letters*, **5**, 1501–1504.

Xing, R. and Hanzlik, R. (1998) Azapeptides as inhibitors and active site titrants for cysteine proteinases. *Journal of Medicinal Chemistry*, **41**, 1344–1351.

Xing, R., Addington, A. and Mason, R. (1998) Quantification of cathepsins B and L in cells. *Biochemical Journal*, **332**, 499–505.

Yabe, Y., Guillaume, D. and Rich, D. (1988) Irreversible inhibition of papain by epoxysuccinyl peptides. ^{13}C NMR characterization of the site of alkylation. *Journal of the American Chemical Society*, **110**, 4043–4044.

Yamamoto, A., Hara, T., Tomoo, K., Ishida, T., Fujii, T., Hata, Y., Murata, M. and Kitamura, K. (1997) Binding mode of CA074, a specific irreversible inhibitor, to bovine cathepsin B as determined by X-ray crystal analysis of the complex. *Journal of Biochemistry*, **121**, 974–977.

Yamashita, D.S., Smith, W.W., Zhao, B., Janson, C.A., Tomaszek, T.A., Bossard, M.J., Levy, M.A., Oh, H.-J., Carr, T.J., Thompson, S.K., Ijames, C.F., Carr, S.A., McQueney, M., D'Alessio, K.J., Amegadzie, B.Y., Hanning, C.R., Abdel-Meguid, S., DesJarlais, R.L., Gleason, J.G. and Veber, D.F. (1997) Stucture and Design of Potent and Selective Cathepsin K Inhibitors. *Journal of the American Chemical Society*, **119**, 11351–11352.

Yasuma, T., Oi, S., Choh, N., Nomura, T., Furuyama, N., Nishimura, A., Fujisawa, Y. and Sohda, T. (1998) Synthesis of peptide aldehyde derivatives as selective inhibitors of human cathepsin L and their inhibitory effect on bone resorption. *Journal of Medicinal Chemistry*, **41**, 4301–4308.

Chapter 6

Calpain

Joel A. Krauser and James C. Powers

Calpain is a ubiquitous cysteine protease that exists in two major forms, calpain I (μ-calpain) and calpain II (m-calpain) (Melloni and Pontremoli 1989; Sorimachi *et al* 1994). The enzyme is unique among cysteine proteases in that it has a calmodulin-like calcium binding domain, and must have Ca^{++} for activation. The nomenclature of the two calpain forms is derived from the μM and mM calcium concentrations required to activate calpain I and calpain II respectively. Calpain is involved in many biological roles and the initiation of multiple degenerative conditions including the pathogenesis of peripheral neuropathy, Alzheimer's disease and muscular dystrophy (Wang and Yuen 1994; Spencer *et al* 1995). Enhanced calpain activity has also been linked to cellular injury caused by physical damage observed in the case of ischemic stroke (Bartus *et al* 1995). Calpain inhibitors could be useful for elucidating the physiological role of calpains and could lead to the development of treatments for both acute and chronic neurologic diseases. This has stimulated extensive research into calpain's active site structure and mechanism. Several different reversible and irreversible calpain inhibitors are available for probing the biological function of calpain.

6.1 BIOLOGICAL ROLES

The activation of calpain requires calcium and results in removal of small peptidyl segments at the N-terminal ends of both subunits, thus converting the dormant 80 kD to the active 76 kD form. In addition, the 30 kD subunit is converted to the 21 kD form, which promotes the dissociation of the two subunits (Yoshizawa *et al* 1995a). It was previously thought that both the 80 kD and 30 kD subunits were required for catalytic activity; however, Suzuki and coworkers have demonstrated that full catalytic activity can be achieved with the 80 kD subunit alone (Yoshizawa *et al* 1995b). They showed that the purified and appropriately renatured 80 kD subunit is fully active and further speculate

Abbreviations: Standard three letter codes for amino acids are used throughout. Other abbreviations are as follows: A_{23187}, calcimycin; ABP, actin binding protein; Ac, acetyl; AMC, 7-amino-4-methylcoumarin; AMPA, amino-3-hydroxy-5-methyl-4-isoxazolepropionic acid; ATyr, azatyrosine; Azi, aziridinyl; β-MAPI, β-microbial alkaline protease inhibitor; Biot, biotinylated; Boc, *t*-butyloxycarbonyl; Bu, butyl; Bzl, benzyl; Cha, cyclohexylalanyl; Dns, dansyl; Eps, epoxysuccinate; Fmoc, 9-fluorenylmethoxycarbonyl; PB, 4-phenylbutanoyl; Nle, norleucine; P235, talin; *t*-BuAla, *t*-butylalanine; *t*-BuGly, *t*-butylglycine; TxB_2, thromboxane B_2; THP, tetrahydropyranyl; USF, urea soluble fraction; VS, vinyl sulfone; Z, benzyloxycarbonyl.

that the 30 kD subunit acts as a chaperonine, converting the 80 kD form, an inactive form, to an active form in the presence of calcium (Yoshizawa *et al* 1995a).

The calcium concentrations required to activate m- and μ-calpain for autolysis at maximal activity are 5–50 μM and 200–1000 μM respectively, whereas physiological Ca^{++} concentration ranges from 100 to 300 nM. Thus, factors and proteins that increase the sensitivity of calpain to calcium are also required for calpain activation. A Ca^{++} binding activator protein isolated from rat brain was found to form a 1:1 complex with calpain and enhance Ca^{++} binding about 10 fold without affecting the V_{max} for μ-calpain, but not m-calpain. Another activator, found in muscle, bound in a 1:1 manner to m-calpain and lowered the association constant (K_a) for Ca^{++} from 400 μM to 15 μM thus facilitating autolysis. The activator proteins appear to function when bound to membranes. Phospholipids also enhance μ-calpain sensitivity to Ca^{++} which results in autolysis.

Current dissociation and reassociation studies support a membrane activation mechanism for calpain (Suzuki and Sorimachi 1998). Calpain binds calcium at a low concentration, and then the protease translocates to the membrane where it undergoes a conformational change, not dissociation. The conformational change exposes hydrophobic surfaces, probably in region IV and VI, which are responsible for membrane binding. The dissociation occurs in the presence of activators, phospholipids and calcium as described in the previous paragraph. The phospholipids alter the large subunit and decrease the Ca^{++} concentration required for autolysis. This dissociation mechanism might facilitate the release of the large subunit from the membrane to the cytosol, where the dissociated large subunit hydrolyzes substrates.

Calpain hydrolyzes a large range of substrates and many of those substrates have hydrophilic residues near the enzyme's cleavage site (Donkor 2000). Several cellular based substrates for calpain include membrane receptors (Bi *et al* 1994), transcription factors (Hirai *et al* 1991), and G-proteins (signal transduction enzymes) (Greenwood and Jope 1994).

6.2 STRUCTURE, ACTIVE SITE AND MECHANISM

Calpain is a heterodimer made up of two subunits, 80 kD and 30 kD. The primary structure of the large subunit (80 kD) from rat calpain II contains 704 amino acids that are divided into four domains: residues 1–79, 80–319, 320–559 and 560–704. The roles of domain I (1–79) and domain III (320–559) are not yet fully understood. Domain I does appear to be involved in the N-terminal autolysis of calpain, which is associated with conversion to the fully active 76 kD polypeptide. Domain II (80–319), highly conserved between calpain I and calpain II, contains the cysteine protease activity. The next most conserved sequence is domain IV (560–704), which is the Ca^{++} binding domain. The small 30 kD subunit is divided into two domains (V and VI). Domain V is the hydrophobic glycine rich N-terminal region of the small 30 kD subunit.

X-ray structural studies on domain VI of the small subunit uncovered five EF hand motifs instead of the original four EF motifs predicted from the sequence (Blanchard *et al* 1997; Lin *et al* 1997). Four of the five EF motifs bound calcium while the most C-terminal EF finger does not appear to bind calcium. The fifth EF finger interacts with its counterpart in the small or large subunit of calpain, which results in either homo or hetero dimerization. This new model for association of calpain subunits was confirmed when ca. 8–10 amino acid residues were cleaved by carboxypeptidase, which resulted in the loss of calpain activity and the ability to dimerize (Imajoh *et al* 1987). A recent crystal structure of unactivated calpain II showed that the catalytic triad is not in an active configuration, suggesting that calcium binding induces the changes necessary for catalytic activity (Hosfield *et al* 1999).

Calpain is homologous to the papain family of cysteine proteases, including cathepsin B and papain, and likely has many features in common (Rawlings and Barrett 1994; Storer and Ménard 1994). Thus, the active site of calpain contains characteristic residues of cysteine proteases and includes cysteine (Cys 105), histidine (His 262) and asparagine (Asn 286). When a substrate is hydrolyzed, the active site cysteine (Cys 105) nucleophilically attacks the scissile peptide carbonyl group to form a tetrahedral intermediate, which collapses to release the first product (an amine) and then forms an acyl enzyme. The acyl enzyme is eventually hydrolyzed by water releasing the second product (carboxylic acid) and regenerating the active site cysteine.

6.3 TRANSITION STATE AND REVERSIBLE INHIBITORS

6.3.1 Aldehydes

The first peptide aldehyde protease inhibitors, leupeptin, chymostatin, elastinal, antipain and β-MAPI, were originally isolated from the *Streptomyces* species (Umezawa 1973). The aldehydes inhibit cysteine proteases by forming a tetrahedral thiohemiacetal adduct with the active site cysteine. The inhibitors were only moderately potent and exhibited low specificity for cysteine proteases such as papain or the cathepsins. The inhibitors also exhibited poor membrane permeability due to their polar C-terminal groups and sidechains. Therefore potent peptidic and non-peptidic aldehydes with improved membrane permeability and calpain specificity were developed for calpain I and calpain II (Tables 6.1 and 6.2).

Calpeptin (Z-Leu-Nle-H), one of the first selective calpain inhibitors, had an inhibitory potency approximately four times better than leupeptin for calpain I, 28 times better for calpain II and 45 times better for platelet calpain I (Table 6.1) (Tsujinaka *et al* 1988).

Table 6.1 Peptide aldehyde inhibitors of calpain

	Calpain I IC_{50} (μM)	*Calpain II* IC_{50} (μM)	*Reference*
Ac-Leu-Leu-Arg-H (leupeptin)	0.211[a], 1.8[b], 0.072[d]	0.938[c]	(Tsujinaka *et al* 1988)
Z-Leu-Nle-H (calpeptin)	0.052[a], 0.005[d]	0.034[c]	(Tsujinaka *et al* 1988)
Z-Leu-Met-H	0.434[a], 0.033[d]	0.025[c]	(Tsujinaka *et al* 1988)
PB-Leu-Phe-H	0.038[a]	0.078[c]	(Sasaki *et al* 1990; Woo *et al* 1995)
Z-*t*-BuGly-Leu-H	0.004[d]		(Iqbal *et al* 1997)
Z-Ser-Leu-H	0.530[d]		(Iqbal *et al* 1997)
Z-(N-Me)Leu-Leu-H	>10[d]		(Iqbal *et al* 1997)
Z-Leu-Leu-H	0.008[d]		(Iqbal *et al* 1997)
Z-Leu-Val-H	0.004[d]		(Iqbal *et al* 1997)
Z-Leu-Nle-H	0.005[d]		(Iqbal *et al* 1997)
Z-Leu-Tyr(Bzl)-H	0.007[d]		(Iqbal *et al* 1997)

[a]Porcine Erythrocyte Calpain; [b]Human Platelet Calpain; [c]Porcine Kidney Calpain; [d]Recombinant Human Calpain.

Calpeptin, unlike leupeptin which has a polar C-terminal group and an Arg side chain, also effectively penetrated cells. Thus, calpeptin prevented the degradation of actin binding protein (ABP) and talin (P235), natural calpain substrates, in A_{23187} stimulated platelets. Calpeptin also inhibited the generation of TxB_2 by direct inhibition of thromboxane (Tx) synthase, also a measure of cell penetration. Calpeptin analogues containing different N-terminal groups were studied and some were found to have comparable inhibitory potency and were actually more potent than calpeptin at preventing TxB_2 generation (Ariyoshi *et al* 1991).

Subsequently dipeptide, tripeptide and tetrapeptide inhibitors have been evaluated for calpain inhibitory activity (Table 6.1). Dipeptide aldehydes such as Z-Leu-Met-H and PB-Leu-Phe-H (Figure 6.1) were found to be effective calpain inhibitors, however, they are not selective for calpain I or calpain II, and they also inhibited cathepsins (Sasaki *et al* 1990; Woo *et al* 1995). The N-terminal protecting groups and the P site residues of calpain I dipeptide aldehydes have been systematically varied (Iqbal *et al* 1997). Many dipeptide aldehydes are as potent as their tripeptide and tetrapeptide analogues and some of the subsite requirements are shown in Table 6.1 and Figure 6.1. Different N-terminal blocking groups for dipeptide aldehyde inhibitors such as acyl, alkoxycarbonyl or alkanesulfonyl are tolerated very well by calpain I. In general, Leu was the best choice at the P2 position; however for some dipeptide aldehydes, *t*-BuGly at P2 was found to be more potent than Leu (i.e. Z-*t*-BuGly-Leu-H; $IC_{50} = 4\,nM$). Substitution of other amino acids at the P2 position led to decreased potency (i.e. Z-Ser-Leu-H; $IC_{50} = 530\,nM$). Calpain I tolerates many different substituents such as aliphatic, aromatic and basic amino acids at P1 and P3 in tripeptide aldehydes. At P2, aliphatic amino acids such as Nle, Nva and Ile decreased potency. Aromatic and bulky groups at P2 are even less potent (Iqbal *et al* 1997). The nomenclature for calpain subsites is based on Schechter & Berger (Schechter and Berger 1967), and is used to describe the individual residues of the inhibitors and the corresponding subsites of the enzyme.*

The P1 and P2 NH groups in dipeptide aldehyde inhibitors of calpain were once believed to be required for hydrogen bonding as illustrated by the N-methylation of the P1–P2 or P2–P3 peptide bonds, which greatly decreases binding affinity (Iqbal *et al* 1997). Comparison of Z-(N-Me)Leu-Leu-H ($IC_{50} > 10000\,nM$) and Z-Leu-Leu-H ($IC_{50} = 8\,nM$) illustrates the potential importance of H-bonding. The most potent peptide aldehyde

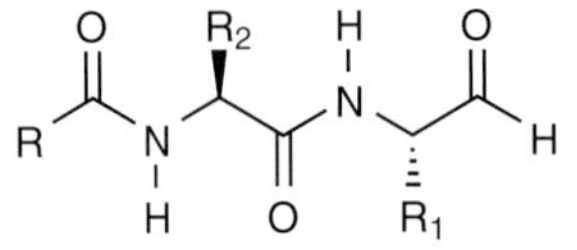

Figure 6.1 Structure of a dipeptide aldehyde calpain inhibitor. The order of reactivity with calpain I is: P1: Val ~ Tyr ~ Tyr(O-Bzl)> Leu ~ Abu ~ Phe ~ Cha > His; P2: *t*-BuGly > Leu > Val ~ Nle > Ile > *t*-BuAla; blocking groups (RCO-): Z ~ 4-Nitro-Z ~ Ts ~ Fmoc ~ (+)-Menthyloxy-CO > CH_3SO_2 > CH_3CO.

* This nomenclature assigns P to each amino acid residue of the peptide substrate and S for each enzyme subsite. The residues on the N-terminal side of the scissile bond are numbered P3, P2, P1 and the residues on the C-terminal side are numbered P1′, P2′, P3′. The scissile bond is located between the P1 or P1′ residues. The subsites within the protease complementary to substrate binding residues are numbered S3, S2, S1, S1′, S2′, S3′ respectively.

calpain I inhibitors found in this series were Z-Leu-Val-H, Z-Leu-Nle-H, Z-*t*-BuGly-Leu, and Z-Leu-Tyr(Bzl)-H.

The replacement of Leu at the P2 position with alkyl or arylsulfonyl-D-amino acids resulted in a series of potent calpain I inhibitors. The two most potent D-amino acid inhibitors are (**6.1**) and (**6.2**) (Figure 6.2), which have K_I values of 2 and 4 nM, respectively (Chatterjee *et al* 1998a). Tripathy *et al* (1998) proved that many P2 proline derivatives were potent in the nanomolar range. Thus, the presence of L-Leu or L-Val is not always a requirement for high affinity binding (Chatterjee *et al* 1998a; Tripathy *et al* 1998).

Nonpeptidic ketomethylene (-$COCH_2$) and carbamethylene (-CH_2CH_2) isosteres of dipeptide aldehydes revealed that the NH at the P2–P3 position is indeed not a strict requirement for potent binding activity as described above from the work of Iqbal *et al* (Table 6.2) (Iqbal *et al* 1997; Chatterjee *et al* 1998b). The P2–P3 ketomethylene isosteres such as (**6.3–6.5**) (Table 6.2) ($IC_{50} = 55$, 50 and 25 nM respectively) were quite potent (Chatterjee *et al* 1996a; Chatterjee *et al* 1998b). However, the P1–P2 ketomethylene isosteres such as (**6.6**) and (**6.7**) ($K_I = 45$ and 2.5 μM respectively) were poor inhibitors suggesting the importance of NH hydrogen bonding at the P1–P2 position (Angelastro *et al* 1999). In general, the NH at the P2 position of a dipeptide inhibitor can be effectively replaced by a CH_2 (ketomethylene isosteres) if there is an aromatic moiety present at the P3 position. Carbamethylene isostere (**6.8**) ($IC_{50} = 50$ nM) is 10 times more potent than isostere (**6.9**) ($IC_{50} = 500$ nM), indicating a strict stereochemical requirement at the pseudo P2 position (Chatterjee *et al* 1997b). Other favorably calpain I binding carbamethylene isosteres are the sulfoxynaphthyl derivative (**6.10**) and the thionaphthyl derivative (**6.11**) with IC_{50} values of 30 and 75 nM respectively (Table 6.2) (Chatterjee

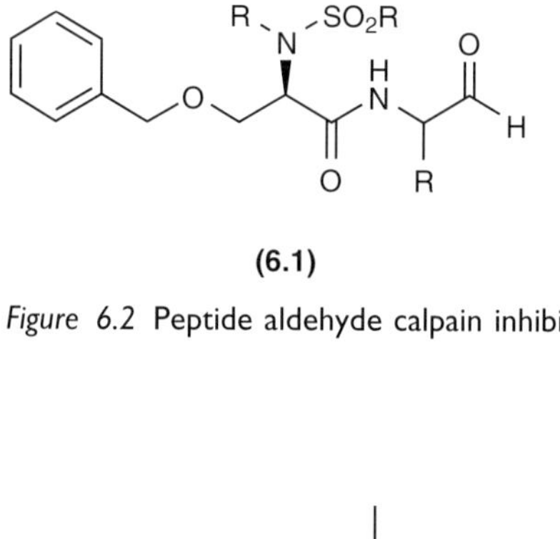

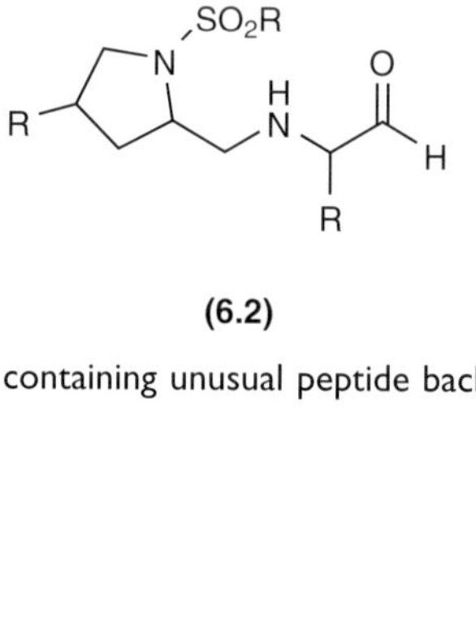

(6.1) **(6.2)**

Figure 6.2 Peptide aldehyde calpain inhibitors containing unusual peptide backbone structures.

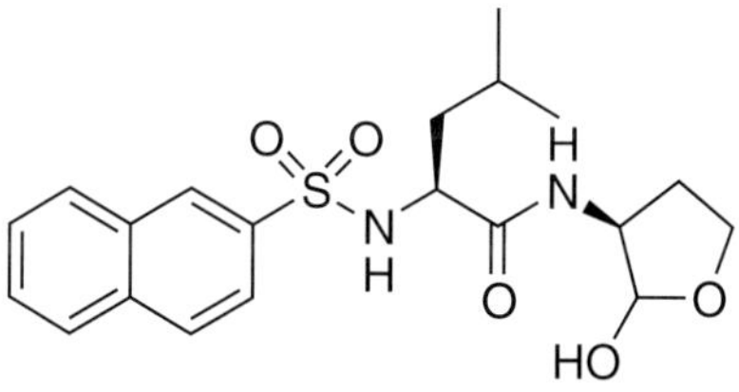

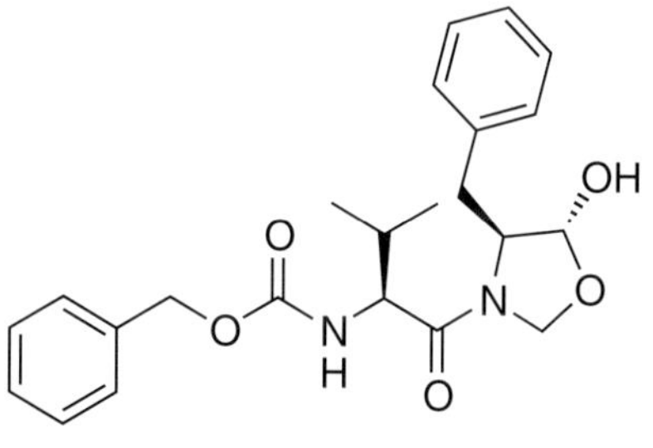

(6.14) **(6.15)**

Figure 6.3 Pro-drug forms of dipeptide aldehydes.

Table 6.2 Ketomethylene and carbamethylene isosteres of peptide aldehydes

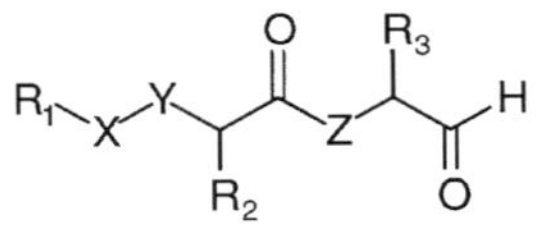

No.	R_1	R_2	R_3	X	Y	Z	Calpain I IC_{50}(nM)	Reference
(**6.3**)	$CH(CH_2CH_3)Ph$	*i*-Bu (S)	*i*-Bu	CO	CH_2	NH	55[b]	(Chatterjee *et al* 1998b)
(**6.4**)	$CH(Ph)_2$	*i*-Bu	*i*-Bu	CO	CH_2	NH	50[b]	(Chatterjee *et al* 1998b; Chatterjee *et al* 1996a)
(**6.5**)	Xanthen-9-yl	*i*-Bu	*i*-Bu	CO	CH_2	NH	25[b]	(Chatterjee *et al* 1998b; Chatterjee *et al* 1996a)
(**6.6**)	Z	*i*-Pr	Bzl	CO	NH	CH_2	45000[a]	(Angelastro *et al* 1999)
(**6.7**)	Z	*i*-Pr	Bzl	CO	NH	CH_2	2500[a]	(Angelastro *et al* 1999)
(**6.8**)	2-Sulfonylnaphthyl	*i*-Bu (R)	*i*-Bu	CH_2	CH_2	NH	50[b]	(Chatterjee *et al* 1998b; Chatterjee *et al* 1997b)
(**6.9**)	2-Sulfonylnaphthyl	*i*-Bu (S)	*i*-Bu	CH_2	CH_2	NH	500[b]	(Chatterjee *et al* 1998b; Chatterjee *et al* 1997b)
(**6.10**)	2-Sulfoxylnaphthyl	*i*-Bu	*i*-Bu	CH_2	CH_2	NH	30[b]	(Chatterjee *et al* 1998b; Chatterjee *et al* 1997b)
(**6.11**)	2-Thionaphthyl	*i*-Bu	*i*-Bu	CH_2	CH_2	NH	75[b]	(Chatterjee *et al* 1998b; Chatterjee *et al* 1997b)
(**6.12**)	1,2,3,4-Tetrahydroisoquinolyl	*i*-Bu (R)	Bzl	CO	CH_2	NH	28[b]	(Chatterjee *et al* 1998b)
(**6.13**)	1,2,3,4-Tetrahydroisoquinolyl	*i*-Bu (S)	Bzl	CO	CH_2	NH	1000[b]	(Chatterjee *et al* 1998b)

[a] Chicken Gizzard Calpain; [b] Recombinant Human Calpain.

et al 1998b). The 1,2,3,4-tetrahydroisoquinolyl derivatives exhibit a stereochemical preference for the *R*-configuration at the R_2 position as indicated by the respective 28 nM and 1 μM IC_{50} values for compounds (**6.12**) and (**6.13**).

Cyclic hemiacetals (**6.14**) and (**6.15**) (Figure 6.3) designed to circumvent the detrimental metabolic effects on aldehydes and enhance permeability through the blood-brain barrier are potent calpain inhibitors. Modified blocking group derivatives of (**6.14**) exhibit IC_{50} values that range from 30–110 nM *in vitro*. The dipeptide aldehyde derivative (**6.15**), which is a modified 5-hydroxy oxazolidine derivative, has an IC_{50} value of 33 nM (Wells and Bihovsky 1998).

6.3.2 α-Dicarbonyls

α-Dicarbonyl derivatives, originally developed as serine protease and aminopeptidase inhibitors, are some of the most effective calpain inhibitors. These transition state inhibitors (Figure 6.4) contain an electronegative functional group (amide, ester or ketone) adjacent to a ketonic carbonyl group which corresponds to the scissile peptide bond of a substrate. The electronegative functional group activates the adjacent carbonyl for nucleophilic attack to form a stable tetrahedral thiohemiacetal adduct with the active site cysteine. The α-dicarbonyl derivatives include α-diketones, α-ketoesters, α-ketoacids and α-ketoamides.

6.3.2.1 *α-Diketones*

Peptide α-diketones are reasonable inhibitors for calpain (Table 6.3). For example, Z-Val-Phe-$COCH_3$ is an effective inhibitor for both purified chicken gizzard calpain ($K_I = 0.7$ μM) and α-chymotrypsin ($K_I = 0.2$ μM) (Angelastro *et al* 1990). In general, α-diketones and α-ketoester analogues are equipotent; however, they are less potent than the corresponding aldehydes (Angelastro *et al* 1990).

6.3.2.2 *α-Ketoesters*

Potent dipeptide α-ketoester inhibitors are specific for calpain I and calpain II compared to cathepsin B and papain (Table 6.3) (Li *et al* 1993). The dipeptide α-ketoester inhibitors were also more potent inhibitors than simple N-protected amino acid α-ketoesters. For the Z-Leu-AA-$COCH_2CH_3$ derivatives, the K_I decreased in the following order: Nle (7.0 μM) < Abu (4.5) < 4-Cl-Phe (4.0) Phe (1.8) < Nva (1.4) and Met (1.0) with calpain I. Essentially the reverse order was observed for calpain II: Met (1.5 μM) < Nva (1.2) < Phe,

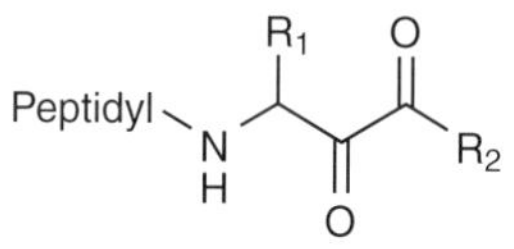

Figure 6.4 Transition state inhibitors of calpain containing α-dicarbonyl functional groups. R_1 = Various amino acid side chains; R_2 = NHR′ for α-ketoamides; OH for α-keto acids; OR′ for α-ketoesters and CH_2R' for α-diketones; Peptidyl refers to various acyl groups, amino acid residues and blocking groups.

Table 6.3 α-Dicarbonyl inhibitors of calpain

Inhibitor	*Calpain I*[a] $K_i(\mu M)$	*Calpain II*[b] $K_i(\mu M)$	*Reference*
Z-Phe-$COCH_3$		3[c]	(Angelastro *et al* 1990)
Z-Leu-Abu-COOEt	4.5	0.40	(Li *et al* 1993)
Ph_2CHCO-Leu-Abu-COOEt	0.1	0.2	(Li *et al* 1993)
Z-Leu-Leu-Abu-COOEt	1.8	2.6	(Li *et al* 1993)
2-NapCO-Leu-Leu-Abu-COOEt	1.3	0.09	(Li *et al* 1993)
Z-Leu-Abu-$CONH_2$	0.28	0.019	(Li *et al* 1993)
Z-Leu-Abu-CONH-CH_2-furyl	0.8	0.033	(Li *et al* 1996)
Z-Leu-Abu-CONH-2-pyridyl	0.64	0.017	(Li *et al* 1996)
Z-Leu-Phe-CONH-CH_2-quinolinyl	0.11	0.023	(Li *et al* 1996)
Z-Leu-Abu-CONH-CH_2-CHOH-C_6F_5	0.05	0.20	(Li *et al* 1996)
1-Napthyl-CO-Leu-Abu-CONH-Et	0.3	0.25	(Li *et al* 1996)
1-Isoquinolinyl-CO-Leu-Abu-CONH-Et	0.35	0.11	(Li *et al* 1996)
2-Quinolinyl-CO-Leu-Abu-CONH-Et	0.5	0.3	(Li *et al* 1996)
Z-Leu-Nva-CONH-CH_2-2-pyridinyl	0.019	0.12	(Li *et al* 1996)
Z-Leu-Abu-CONH-CH_2CH_2-Ph(4-OMe)	0.12	0.060	(Li *et al* 1996)
Z-Leu-Abu-CONH-CH_2-CHOH-Ph(3,4-$(OCH_2Ph)_2$)	0.48	0.67	(Li *et al* 1996)
Z-Leu-Abu-COOH	0.075	0.022	(Li *et al* 1993)
Z-Leu-Phe-COOH	0.0085	0.0057	(Li *et al* 1993)

[a] Human Neutrophil Calpain; [b] Rabbit Muscle Calpain; [c] Chicken Gizzard Calpain.

Abu and 4-Cl-Phe (1.8) < Nle (0.18). Dipeptide α-ketoesters in this series were more potent than simple N-protected amino α-ketoesters. Thus, inhibitory potency for calpain increases in the presence of a P2 residue (Li *et al* 1993).

The N-protecting group of dipeptide α-ketoesters was modified to improve inhibitory potency (Li *et al* 1993). In general, changing the N-protecting group from benzyloxycarbonyl resulted in significantly poorer inhibitors with one exception. The dipeptide, Ph_2CHCO-Leu-Abu-COOEt was the best inhibitor for calpain I and calpain II with K_I values 0.1 and 0.2 μM, respectively. This was a 45-fold improvement for calpain I and a two-fold improvement for calpain II over Z-Leu-Abu-COOEt. Extending the dipeptide to a tripeptide resulted in a three-fold improvement with calpain I as in the case of 2-NapCO- and Z-derivatives of Leu-Leu-Abu-COOEt. The tripeptide inhibitor 2-NapCO-Leu-Leu-Abu-COOEt was the best inhibitor for calpain II with a K_I value 0.09 μM.

Modification of the ester group functionality had no significant effect on the inhibitory potency (Li *et al* 1993). Changing the ester group to a butoxy group resulted in a slight improvement for calpain I but no significant change for calpain II. Also, when α-ketoesters were tested *in vivo*, they degraded rapidly, most likely due to cleavage by plasma esterases (Li *et al* 1993). Efforts were then directed toward the development of more stable α-ketoamides.

6.3.2.3 α-Ketoamides

Dipeptidyl α-ketoamides with the general structure R_1-L-Leu-D,L-AA-CONH- are effective inhibitors for calpain I and calpain II (Table 6.3) (Li *et al* 1996). The P1 position, R_1 protecting groups and R_2 substituents on the α-ketonitrogen were varied in an attempt

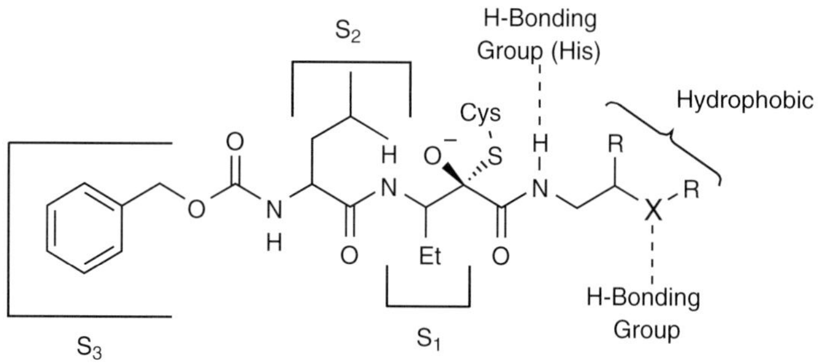

Figure 6.5 Proposed model for the binding of a α-ketoamide transition state inhibitor to the active site of calpain. The leucine side chain is shown interacting with the S2 binding pocket of calpain, the primary specificity determinant. Other important interactions are the S′ hydrophobic pocket and a preference for Z in the S3 pocket.

to find the most potent inhibitor. Figure 6.5 shows the proposed interactions of an α-ketoamide and calpain. In general, calpain II was more sensitive to these inhibitors than calpain I with a K_I typically in the 10–100 nm range. Some inhibitors were selective for calpain I and calpain II.

Calpain I inhibitory potency was not changed significantly (less than 5-fold) upon varying the substituent of the nitrogen atom on the α-ketoamide. The best calpain I inhibitors were Z-Leu-Abu-CONH-CH_2-2-pyridyl with a K_I value of 19 nm and the phenyl substituted derivatives Z-Leu-Abu-CONH-CH_2-CHOH-C_6F_5, $K_I = 50$ nM and Z-Leu-Abu-CONH-CH_2-CHOH-C_6H_4-3-OC_6H_4(3-CF_3), $K_I = 70$ nM. Modification of the N-terminal blocking group of R_1-L-Leu-D,L-AA-CONH-Et resulting in a large decrease in inhibitory potency. The best calpain I inhibitors were R_1 = 1-napthyl-CO, 1-isoquinolinyl and 2-quinolinyl, which had approximately the same potency as the benzyloxycarbonyl group (K_I = 0.30, 0.35, 0.50 and 0.21 μM respectively). Inhibitors possessing Abu in the P1 position had a slight advantage over compounds containing Phe at that position. Calpain I inhibitors with Nva in the P1 position were sometimes favored or disfavored. For example, Z-Leu-Nva-CONH-CH_2-2-pyridinyl is one of the best calpain I inhibitors with a K_I of 19 nM. However, Z-Leu-Nva-CH_2CHOH-Ph is a poor calpain I inhibitor with a K_I of 7.8 μM.

An attempt to increase the selectivity of α-ketoamides for calpain I over cathepsin B led to the design of a set of 2,3-methanoleucine stereoisomers (Donkor *et al* 2000). These compounds were based on the compound made by Li *et al*, Z-Leu-Phe-CONH-CH_2CH_2Ph (K_I = 52 nM), which was a potent inhibitor of calpain I (Li *et al* 1993). Nearly all of the 2,3-methanoleucine derivatives were more potent for cathepsin B than calpain I. However, one compound did emerge, Z-[*E*-(2*S*,3*S*)]-methanoLeu-Phe-CONH-CH_2CH_2Ph, that had moderate inhibitor activity for calpain I and was nine-fold more selective for calpain I (K_I = 0.75 μM) than cathepsin B (K_I = 6.78 μM) (Donkor *et al* 2000).

Most, but not all, α-ketoamides were better inhibitors of calpain II. Calpain II activity was not affected greatly by variation of the N-terminal blocking group. No blocking group derivatives gave better inhibitory potencies than the parent benzyloxycarbonyl group. Bulky derivatives such as Ph_2CHCO-Leu-Abu-CONH-Et with a K_I of 1.2 μM were poor inhibitors. Calpain II inhibitory potency was favored with Abu in the P1 position as compared to Phe and Nva. Replacement of Abu by Phe typically caused less than a two-fold change in inhibitory potency. Many potent calpain II inhibitors contain P1′ hydrophobic groups, (i.e. R_1-L-Leu-D,L-AA-CONH-R_2 with $R_2 = CHOH\text{-}CH_2\text{-}Ph$ or CH_2CH_2-Ph), which suggests the existence of a S1′ hydrophobic pocket. Calpain II activity was not improved when heteroatoms were introduced into the P1′ substituent of the α-ketoamide or the N-terminal blocking group. However, heterocyclic inhibitors such as Z-Leu-Abu-CONH-CH_2-2-furyl, Z-Leu-Abu-CONH-CH_2-2-pyridyl or Z-Leu-Phe-CH_2-quinolinyl were still effective calpain II inhibitors.

6.3.2.4 *α-Ketoacids*

Peptide α-ketoacids are very effective α-dicarbonyl inhibitors (Table 6.3). The best inhibitor for calpain I and calpain II was Z-Leu-Phe-COOH ($K_I = 8.5$ nM, calpain I) along with Z-Leu-Abu-COOH ($K_I = 75$ nM, calpain I). The α-ketoacids Z-Leu-Phe-COOH and Z-Leu-Abu-COOH were more potent calpain I inhibitors than the corresponding α-ketoamide and α-ketoester. However, Z-Leu-Abu-COOH was not a significantly more potent calpain II inhibitor than the analogous simple α-ketoamide. These α-ketoacids were poor inhibitors for cathepsin B.

Each class of α-dicarbonyl derivatives presents certain advantages and disadvantages. For instance, α-diketone inhibitors are reasonable inhibitors; however, they inhibit both calpain and α-chymotrypsin. The α-ketoacids are among the most potent calpain inhibitors; however, they have very poor membrane permeability. The α-ketoacid potency may be attributed to a Coulombic interaction between the carboxylic acid group and the histidine residue of the catalytic triad. The α-ketoester inhibitors are good inhibitors with good membrane permeability; however, they are not stable to esterases. The α-ketoamides possess good membrane permeability, stability and potency.

6.3.3 Nonpeptide quinolinecarboxamides

Nonpeptidic 1,4-dihydro-4-oxo-3-quinolinecarboxamides are reasonably potent and specific reversible inhibitors for erythrocyte human calpain I (Graybill *et al* 1995). Interestingly, the calpain ^{3}H-caseinolytic assay using multiple concentrations showed that the quinolinecarboxamide inhibitors bind to calpain at a separate site from the casein substrate. The most effective calpain I inhibitors were determined by the high-throughput screening of >500 quinolinecarboxamide derivatives (Table 6.4). The potency and selectivity of these inhibitors is dependent upon their R substituents. The R_1 position of quinolinecarboxamide derivatives exhibit a preference for the 4-hydroxyphenyl group. Derivatives (**6.16**) ($IC_{50} = 7\,\mu M$) and (**6.17**) ($IC_{50} = 20\,\mu M$) which have strong electron withdrawing groups are 4–10-fold less potent than (**6.18**). Substitution at the 3- or 5- or both positions of the phenyl, as in the case of (**6.19**), led to <10% inhibition at 50 μM. Derivatives (**6.20**) ($IC_{50} = 0.5\,\mu M$) and (**6.21**) with a 2-CH_3 ($IC_{50} = 0.6\,\mu M$) or 2-Cl substituent are much more potent inhibitors. The R_2 substituent of inhibitors such as (**6.18**) ($IC_{50} = 2\,\mu M$) exhibits a required intramolecular hydrogen bond between the β-ketoamide amine group and quinoline carbonyl, and extramolecular hydrogen bonding to the active site. Quinolinecarboxamide derivatives such as (**6.21**) or (**6.22**) with a 4-pyridinyl or 2-pyrazolyl ring were more potent than other aryl derivatives (none shown).

Table 6.4 Quinolinecarboxamide inhibitors of calpain

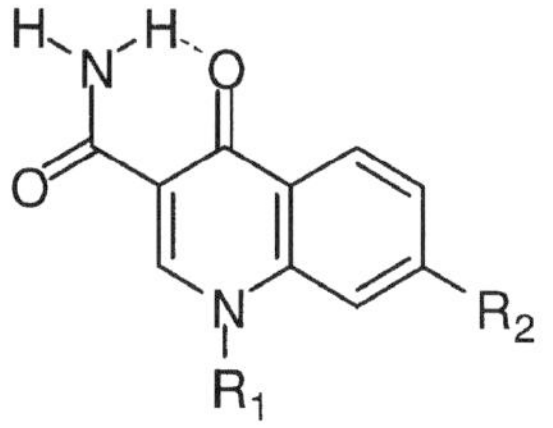

No.	R_1	R_2	*Calpain I*[a] IC_{50}(μM)	*Reference*
(**6.16**)	2-F-4-HOC_6H_3	4-pyridinyl	7	(Graybill *et al* 1995)
(**6.17**)	2-CF_3-4-HOC_6H_3	4-pyridinyl	20	(Graybill *et al* 1995)
(**6.18**)	4-HOC_6H_3	4-pyridinyl	2	(Graybill *et al* 1995)
(**6.19**)	3,5-CH_3-4-HOC_6H_3	4-pyridinyl	–	(Graybill *et al* 1995)
(**6.20**)	2-CH_3-4-HOC_6H_3	4-pyridinyl	0.5	(Graybill *et al* 1995)
(**6.21**)	2-Cl-4-HOC_6H_3	4-pyridinyl	0.6	(Graybill *et al* 1995)
(**6.22**)	2-Cl-4-HOC_6H_3	4-pyrazolyl	0.6	(Graybill *et al* 1995)

[a]Human Erythrocyte Calpain.

In summary, inhibitors (**6.20**), (**6.21**), and (**6.22**), are the most potent inhibitors (Graybill *et al* 1995).

6.3.4 Nonpeptide alpha-mercaptoacrylic acids

Alpha-mercaptoacrylic acid derivatives are uncompetitive inhibitors of calpain that block the calcium binding site, not the catalytic site. These inhibitors exhibit good cell permeability and neuroprotection but do not prevent irreversible calpain inactivation by E64, an active site reagent. These inhibitors require an unsaturated double bond and an unmodified carboxylic acid and sulfhydryl groups, which probably serve to chelate the calcium ion. The K_I of (**6.23**) (PD150606) with m- and μ-calpain is respectively 0.21 and 0.37 μM (Figure 6.6). Also, these compounds were selective for calpain relative to other proteases such as cathepsin B ($K_I = 127\,\mu$M), papain ($K_I > 500\,\mu$M) or trypsin ($K_I > 500\,\mu$M).

The inhibitor (**6.23**) also effectively penetrated cells and prevented the A_{23187} induced degradation of α-spectrin, which was proteolytically degraded into two fragments (150 and

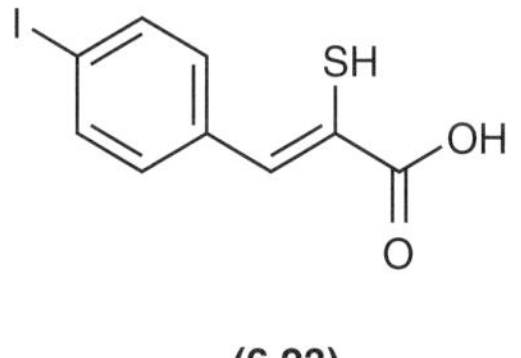

(6.23)

Figure 6.6 An alpha-mercaptoacrylic acid inhibitor (**6.23**).

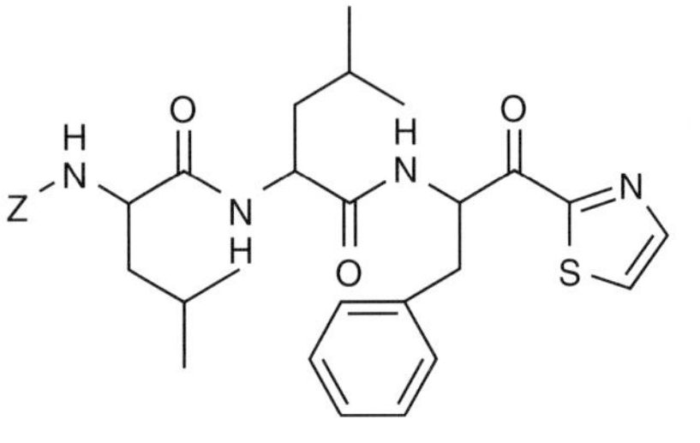

(6.24)

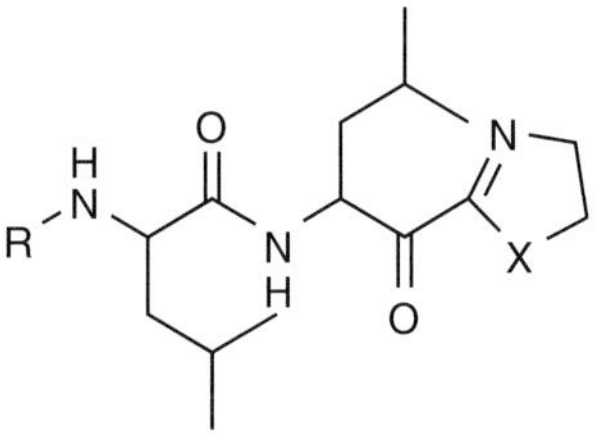

(6.25) R = Boc, X = N-H
(6.26) R = Z, X = N-H
(6.27) R = Z, X = N-$CH_2C_6H_5$

Figure 6.7 Heterocyclic transition-state inhibitors of calpain.

145 kDa). Inhibitor (**6.23**) (10 μM) completely blocked formation of the 145 kDa fragment in a dose-dependent manner. However, formation of the 150 kDa fragment was only partially blocked even at high inhibitor concentration, which could be due to the uncompetitive inhibition mechanism. Cell permeability of the alpha-mercaptoacrylic acid derivatives was also verified using a human neuroblastoma cell assay. Cells were incubated with the inhibitors and maitotoxin before adding Ser-Leu-Leu-Val-Tyr-AMC, a fluorogenic membrane permeable substrate. The inhibitor (**6.23**) diminished *in situ* substrate hydrolysis by ca. 75%.

6.3.5 Peptide heterocycles

Peptide heterocycles that mimic peptide α-ketoamides and α-ketoacids are inhibitors of calpain I (Figure 6.7). Tripeptide thiazole (**6.24**, 54% inhibition at 10 μM) and α-ketoimidazoles (**6.25**, 54% inhibition at 10 μM; **6.26**, 0% inhibition at 10 μM; **6.27**, 39% inhibition at 10 μM) exhibit potencies dependent upon their protecting groups. For example, the N-terminal Boc protected α-ketoimidazole derivative (**6.25**) is much more potent than its Z analogue. However, the protected imidazole group derivative (**6.27**) exhibits some potency whereas the unprotected imidazole group derivative (**6.26**) does not inhibit.

6.3.6 Phosphorus derivatives

Dipeptidyl phosphorus derivatives, which were also designed to mimic peptide α-ketoesters and α-ketoacids, are inhibitors of human calpain I. The phosphonate Z-Leu-Leu-$P(O)(OCH_3)_2$ and the phosphine oxides ($IC_{50} = 0.43$ μM), Z-Leu-Leu-$P(O)(Ph)_2$ ($IC_{50} = 0.35$ μM) and Z-Leu-Leu-$P(O)(C_6H_4$-*p*-$Cl)_2$ ($IC_{50} = 0.35$ μM) have inhibitory potencies that compare favorably to the α-ketoester analogue Z-Leu-Leu-CO_2Et ($IC_{50} = 0.60$ μM). The α-ketophosphine oxide Z-Leu-Leu-$P(O)(C_6H_4$-*p*-$Cl)_2$ ($IC_{50} = 0.35$ μM) was more potent than its α-ketoester analogue Z-Leu-Leu-CO_2Et ($IC_{50} = 0.60$ μM). The monomethyl α-ketophosphonate Z-Leu-Leu-$P(O)(OH)CH_3$ ($IC_{50} = 5.2$ μM) was suprisingly less potent than the dimethyl α-ketophosphonate Z-Leu-

Leu-P(O)(OCH_3)$_2$ (IC_{50} = 0.43 μM), which might be due to inhibitor geometry or the additional charge. The tetrahedral geometry of the α-ketophosphonate, unlike the planar trigonal geometry of the α-ketoacid, could place the anionic oxygen of the phosphonic acid in a non-optimum orientation with respect to the histidine of the catalytic triad (Hanada *et al* 1978).

6.4 IRREVERSIBLE INHIBITORS

6.4.1 Epoxysuccinates

The first epoxysuccinyl inhibitor, L-*trans*-epoxysuccinyl-leucylamido(4-guanidino)butane (E-64) was originally isolated and characterized from solid cultures of *Aspergillus japonicus* by Hanada *et al* (1978). E-64 and its derivatives have proven to be effective inhibitors for calpain as well as cathepsins (Figure 6.8). E-64 and its analogues irreversibly bind to cysteine proteases via formation of a covalent bond between the cysteine thiol and C-3 of the epoxysuccinyl unit. Many epoxysuccinyl inhibitors have been developed and assayed for inhibitory potency and cell permeability.

The potency of E-64 derivatives is affected both by stereochemistry and the substitution pattern on the epoxide. The L-isomer of E-64 ($k_2 = 7500\,M^{-1}s^{-1}$) inactivates calpain II about seven-fold better than the D-isomer ($k_2 = 1070\,M^{-1}s^{-1}$). Inactivation by HO-Eps-Leu-NH-[CH_2]$_4$-NH_2 and HO-Eps-Leu-NH-[CH_2]$_4$-NH-$COCH_3$ implies that the presence or absence of the free NH_2 group does not greatly affect binding. However, the presence of a Z group in HO-Eps-Leu-NH-[CH_2]$_4$-NH-Z ($k_2 = 23340\,M^{-1}s^{-1}$) drastically increases inhibition. Increasing the sidechain chain length from HO-Eps-Leu-NH-[CH_2]$_4$-NH_2 ($k_2 = 2790\,M^{-1}s^{-1}$) to HO-Eps-Leu-NH-[CH_2]$_7$-NH_2 ($k_2 = 4990\,M^{-1}s^{-1}$) increases the rates of inactivation (Parkes *et al* 1985). In general, epoxysuccinate inhibitors were found to inhibit other cysteine proteases more effectively than they inhibit calpain.

Calpain E-64 inhibitors containing either an ester or amide group were assayed in lysed platelets using a caseinolytic assay (Table 6.5) (Huang *et al* 1992). The E-64c ester and amide derivatives such as E-64d (4 μM), $ClCH_2CH_2O$-Eps-Leu-NH-[(CH_2)$_2$CH(CH_3)$_2$], CCl_3CH_2O-Eps-Leu-NH-[(CH_2)$_2$CH(CH_3)$_2$], CBr_3CH_2O-Eps-Leu-NH-[(CH_2)$_2$CH(CH_3)$_2$], CF_3CH_2O-Eps-Leu-NH-[(CH_2)$_2$CH(CH_3)$_2$] and Z-Leu-Nle-Eps-Leu-NH-[(CH_2)$_2$CH(CH_3)$_2$ are at least 20 times less potent than E-64c itself, suggesting that the acid E-64c (0.04 μM) is the active form. Therefore, inhibitor potency is affected little by variation of the ester group unless it is bulky, which then significantly decreases potency. The amide derivatives except for N-2-(5-hydroxy-1H-indol-3-yl)-Eps-Leu-NH-[(CH_2)$_2$CH(CH_3)$_2$] are poor inhibitors.

Calpain E-64 derivatives containing either an ester or amide group were assayed for their ability to enter intact platelets (Table 6.5) (Huang *et al* 1992). In the first assay, the inhibitors were incubated with the platelets, which were subsequently washed to remove free inhibitor and lysed. The E-64c derivatives had IC_{50} values that ranged from 0.3 to 6.0 μM, which suggest that all the ester and amide derivatives are membrane permeable. In the second assay, the inhibitors were incubated with platelets before addition of the ionophore A23187 and calcium, both of which activate endogenous calpain and degrade ABP and talin as measured by SDS-PAGE gel electrophoresis. All of the ester and amide inhibitors in Table 6.5 had IC_{50} values of 5–44 μM in this assay. The carboxylic acid E-64c exhibited no cell permeability, probably due to its polar group.

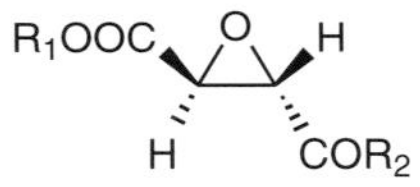

Peptide Epoxide

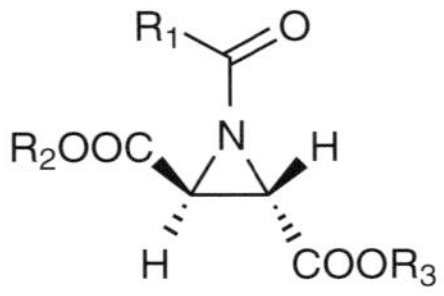

(6.28, 6.29) Peptide Aziridine

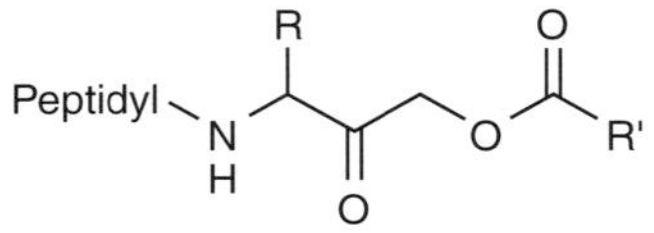

Peptide Acyloxymethyl Ketone

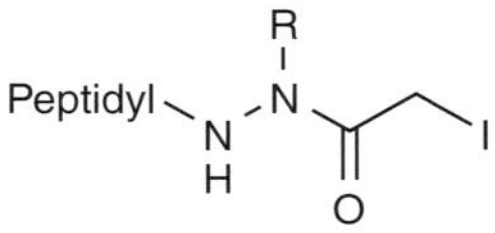

Azapeptide

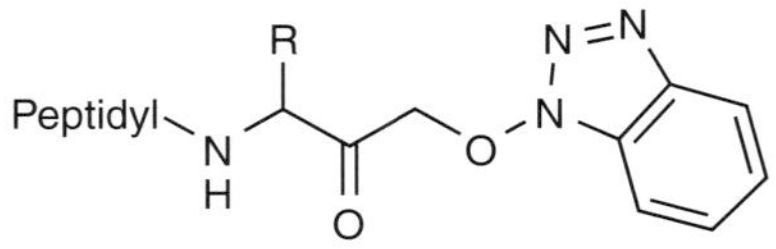

Peptide Benzotriazoloxymethyl Ketone

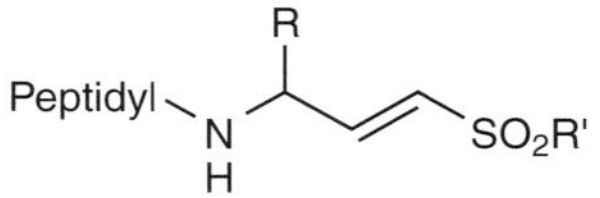

Peptide Vinyl Sulfone

(Peptidyl-AA-VS-R')

Figure 6.8 Irreversible inhibitors for calpain. Peptide epoxide inhibitors (top left) of calpain include E-64 analogues: $R_1 = H$, alkyl or halo-alkyl, R_2 = Leu-NH-4-guanidinobutane. Peptide aziridines (top right): R_1CO- = Boc-Phe-, R_2 = Et, and R_3 = Bzl for (**6.28**), and R_1CO- = Cbz-Ala-, $R_2 = R_3$ = Et for (**6.29**). Peptide acyloxymethyl ketones, azapeptides, and peptide benzotriazoloxymethyl ketones (middle row) are haloketone analogues. Peptide vinyl sulfones (bottom) are Michael acceptors.

Table 6.5 Epoxysuccinate inhibitors of calpain

Inhibitor	*Calpain I*[a] IC_{50}(μM)	*Calpain II*[b] $k_2(M^{-1}s^{-1})$	*Reference*
HO-Eps-Leu-NH-$[CH_2]_4$-NH-C=NH-NH_2 (E-64(L))		7500	(Parkes *et al* 1985)
HO-Eps-Leu-NH-$[CH_2]_4$-NH-C=NH-NH_2 (E-64(D))		1070	(Parkes *et al* 1985)
HO-Eps-Leu-NH-$[CH_2]_4$-NH-Z		23340	(Parkes *et al* 1985)
HO-Eps-Leu-NH-$[CH_2]_4$-NH_2		2790	(Parkes *et al* 1985)
HO-Eps-Leu-NH-$[CH_2]_4$-NH-$COCH_3$		3040	(Parkes *et al* 1985)
HO-Eps-Leu-NH-$[CH_2]_7$-NH_2		4990	(Parkes *et al* 1985)
HO-Eps-Leu-NH-$[(CH_2)_2CH(CH_3)_2]$ (E-64c)	0.04		(Huang *et al* 1992)
EtO-Eps-Leu-NH-$[(CH_2)_2CH(CH_3)_2]$ (E-64d)	4.0		(Huang *et al* 1992)
$ClCH_2CH_2O$-Eps-Leu-NH-$[(CH_2)_2CH(CH_3)_2]$	2		(Huang *et al* 1992)
FCH_2CH_2O-Eps-Leu-NH-$[(CH_2)_2CH(CH_3)_2]$	0.5		(Huang *et al* 1992)
ICH_2CH_2O-Eps-Leu-NH-$[(CH_2)_2CH(CH_3)_2]$	Substantial[c]		(Huang *et al* 1992)
$BrCH_2CH_2O$-Eps-Leu-NH-$[(CH_2)_2CH(CH_3)_2]$	2.0		(Huang *et al* 1992)
CCl_3CH_2O-Eps-Leu-NH-$[(CH_2)_2CH(CH_3)_2]$	0.7		(Huang *et al* 1992)
CBr_3CH_2O-Eps-Leu-NH-$[(CH_2)_2CH(CH_3)_2]$	Substantial[c]		(Huang *et al* 1992)
CF_3CH_2O-Eps-Leu-NH-$[(CH_2)_2CH(CH_3)_2]$	Substantial[c]		(Huang *et al* 1992)
Z-Leu-Nle-Eps-Leu-NH-$[(CH_2)_2CH(CH_3)_2]$	100		(Huang *et al* 1992)
N-2-(5-Hydroxy-H-indol-3-yl)-Eps-Leu-NH-$[(CH_2)_2CH(CH_3)_2]$	6		(Huang *et al* 1992)

[a] Human Platelet Calpain; [b] Chicken Gizzard Calpain; [c] substantial inhibition by 5–44 μM.

6.4.2 Aziridines

Aziridine inhibitors (Figure 6.8) are similar to epoxides and susceptible to ring opening by nucleophiles. Effective aziridine inhibitors of calpain have not yet been developed (Schirmeister 1999). The aziridine derivative Boc-Phe-(S,S)-(EtO)-Azi-Leu-OBzl (**6.28**) has a K_I of 19 μM and 42 μM for calpain I and II respectively and a K_I of 137 μM for cathepsin H. Inhibitor Z-Ala-(S,S)-Azi-$(OEt)_2$ (**6.29**) showed poor inhibition of calpain I and no inhibition with cathepsin H.

6.4.3 Acyloxymethyl ketones

Peptide acyloxymethyl ketone inhibitors (Figure 6.8) for calpain I are irreversible calpain inhibitors (Harris *et al* 1995). The S1 and S2 subsite prefers Phe or Tyr or Leu or Val respectively as illustrated by Z-Leu-Phe-CH_2-OCO-2,6-Cl_2-Ph ($k_2 = 11000\,M^{-1}s^{-1}$), Z-Val-Phe-$CH_2$-OCO-2,6-$Cl_2$-Ph ($k_2 = 11000\,M^{-1}s^{-1}$) and Z-Leu-Tyr-$CH_2$-OCO-2,6-$Cl_2$-Ph ($k_2 = 28000\,M^{-1}s^{-1}$) as compared to Z-Gly-Phe-CH_2-OCO-2,6-Cl_2-Ph ($k_2 = 700\,M^{-1}s^{-1}$), Z-Val-Gly-CH_2-OCO-2,6-Cl_2-Ph ($k_2 = 800\,M^{-1}s^{-1}$) and Z-Leu-Gly-CH_2-OCO-2,6-Cl_2-Ph ($k_2 = 6000\,M^{-1}s^{-1}$) respectively. In general, the subsite specificity for dipeptide acyloxymethyl ketones roughly parallels that observed with dipeptide aldehydes. The peptidyl acyloxymethyl ketone leaving group in the S′ subsite also affects the potency of the inhibitor. For example, the dipeptide Z-Leu-Gly-CH_2-OCO-(2,6-Cl_2-3-SO_2-morpholine)C_6H_2 has an inactivation rate of $23000\,M^{-1}s^{-1}$ even with a non-optimal P1 Gly residue. Thus, leaving groups of the acyloxymethyl ketones can override calpain's P1–P2 specificity preferences.

Some acyloxymethyl ketones inhibitors were specific for calpain I over cathepsin B and L (Harris *et al* 1995). The acyloxymethyl ketones, Z-Leu-Phe-CH_2-OCO-2,6-Cl_2-Ph ($k_{obs}/[I] = 11000\,M^{-1}s^{-1}$ and $250\,M^{-1}s^{-1}$) and Z-Leu-Phe-CH_2-OCO-(2,6-Cl_2-3-OCH_2CH_2-morpholine)C_6H_2 ($k_{obs}/[I] = 30000\,M^{-1}s^{-1}$, $80\,M^{-1}s^{-1}$), are calpain I and cathepsin B specific, respectively. Lastly, Z-(D)-Ala-Leu-Phe-CH_2-OCO-2,6-F_2-Ph ($k_{obs}/[I] = 31000\,M^{-1}s^{-1}$) is a selective potent calpain inhibitor relative to both cathepsin B ($k_{obs}/[I] = 100\,M^{-1}s^{-1}$) and cathepsin L ($k_{obs}/[I] = 300\,M^{-1}s^{-1}$).

6.4.4 Benzotriazoloxymethyl ketones

There is little data on the primed side of calpian's active site compared to what has been collected for calpain's unprimed amino acid subsites (Donkor 2000). A new set of analogues of methyl ketone inhibitors with *N*-hydroxy peptide coupling reagents as leaving groups has recently been designed for this purpose (Tripathy *et al* 2000). The compounds that inhibited calpain the best had an *N*-hydroxy benzotriazole as the leaving group, and are called benzotriazoloxymethyl ketones. A representative inhibitor of this type can be seen in Figure 6.8. One of the major drawbacks of this type of inhibitor is its low to moderate stability in aqueous solutions at a neutral pH, limiting its therapeutic benefits (Tripathy *et al* 2000).

It was suggested that either steric hindrance and/or leaving groups in configurations unable to be protonated by a residue in the active site of calpain I were the causes of the poor inhibitory activity of acyloxymethyl ketones (Pliura *et al* 1992). The idea behind the N-hydroxymethyl ketone inhibitors was to improve on the rate of inactivation of calpain by eliminating these problems (Tripathy *et al* 2000). A variety of leaving groups were chosen to determine which would be the least sterically intrusive, while at the same time being in a conformation able to be protonated and quickly released from the active site. The *N*-hydroxy benzotriazole (Peptidyl-AA-CH_2-OBt, HOBt = *N*-hydroxy benzotriazole) derivatives did well, while the *N*-hydroxy succinimide and other derivatives did much more poorly. Some of the best inhibitors for calpain were Z-Leu-Leu-Phe-CH_2-OBt ($k_{obs}/[I] = 524000\,M^{-1}s^{-1}$), Z-Leu-Phe-$CH_2$-OBt ($K_I = 35\,nM$, $k_{obs}/[I] = 320000\,M^{-1}s^{-1}$), and Z-Leu-Ile-$CH_2$-OBt ($k_{obs}/[I] = 175000\,M^{-1}s^{-1}$) (Tripathy *et al* 2000). Several of these inhibitors were also found to be good inhibitors for cathepsin B, but most had a higher degree of selectivity for calpain I (Tripathy *et al* 2000).

Tripathy *et al* (2000) believe that the conformation of the leaving group given by the N–O bond is critical for optimal inhibition of calpain. This speculation was supported by the dramatically reduced inhibition of calpain by an analogue containing the nitrogen of the benzotriazole linked directly to the methyl group of the ketone.

6.4.5 Halomethyl ketones

6.4.5.1 *Monofluoromethyl ketones*

Peptide fluoromethyl ketones are effective inhibitors for calpain I and calpain II. Peptide fluoromethyl ketones inhibit calpain and other cysteine proteases via nucleophilic displacement of fluorine by the active site cysteine residue. Peptide fluoromethyl ketone inhibitors for calpain I also exhibit good membrane permeability. Fluoromethyl ketone inhibitors also inhibit other cysteine proteases.

Potent and selective fluoromethyl ketone inhibitors of recombinant human calpain I were obtained through subsite mapping. In general, the fluoromethyl ketone inhibitors prefer Phe at the P1 position and Leu at the P2 position (Table 6.6). Interestingly, the

Table 6.6 Fluoromethyl ketone inhibitors of calpain

Inhibitor	*Calpain I*[a] $k_2(M^{-1}s^{-1})$	*Intact Cell*[b] $IC_{50}(\mu M)$	*Calpain II*[c] $k_2(M^{-1}s^{-1})$	*Reference*
Z-Leu-Tyr-CH_2F			17000	(Angliker *et al* 1992)
Z-Leu-Leu-Tyr-CH_2F			28900	(Angliker *et al* 1992)
Z-Leu-Phe-CH_2F	136300	0.2		(Chatterjee *et al* 1996b)
Z-Leu-Ser-CH_2F	21000			(Chatterjee *et al* 1996b)
Z-Leu-Ser(THP)-CH_2F	100000	1.3		(Chatterjee *et al* 1996b)
Boc-Leu-Phe-CH_2F	68600	10[d]		(Chatterjee *et al* 1996b)
1,2,3,4-Tetrahydroisoquin-2-yl-Leu-Phe-CH_2F	276000	0.2		(Chatterjee *et al* 1996b)
Ph$(CH_2)_2$CO-Leu-Phe-CH_2F	26600			(Chatterjee *et al* 1996b)
Morphilino-4-sulfonyl-Leu-Phe-CH_2F	67200	10		(Chatterjee *et al* 1996b)
Z-Leu-Leu-Phe-CH_2F	290000	0.1		(Chatterjee *et al* 1996b)
Benzylaminocarbonyl-Leu-PheCH_2F	67000	0.8		(Chatterjee *et al* 1996b)

[a]Human Recombinant Calpain; [b]Human Leukemic T cells; [c]Chicken Gizzard Calpain; [d]42% at 10 μM.

fluoromethyl ketone Z-Leu-Ser(THP)-CH_2F ($k_2 = 100000\,M^{-1}s^{-1}$) is about five times more potent than the unprotected Z-Leu-Ser-CH_2F ($k_2 = 21000\,M^{-1}s^{-1}$). The Z N-terminal protecting group was preferred over *t*-Boc, morpholinosulfonyl and benzyl carbamyl blocking groups. Modification of OCH_2 in the Z group to CH_2CH_2 led to decreased potency, which indicates a favorable binding interaction with the oxygen. The benzyl carbamyl group decreases inhibitory activity, probably due to the unfavorable interaction of the additional NH in the P3 region. The dipeptide inhibitor (1,2,3,4-tetrahydroisoquin-2-yl)carbonyl-Leu-Phe-CH_2F ($k_2 = 276000\,M^{-1}s^{-1}$), containing a tetrahydroquinolyl capping group with a strained benzyl urea motif, is the most potent inhibitor of calpain I.

Peptide fluoromethyl ketone inhibitors exhibited excellent selectivity for calpain I over cathepsins B and L as well as good cell permeability (Chatterjee *et al* 1996b; Chatterjee *et al* 1997a). For example, morphilino-4-sulfonyl-Leu-Phe-CH_2F was more than 670 times more potent with calpain I than cathepsin B. Also, Boc-Leu-Phe-CH_2F was about 690 times more potent for calpain I than cathepsin B and about 460 times more potent than cathepsin L. Cell permeability for the peptide fluoroketones was measured using an intact cell assay that monitored formation of spectrin breakdown products. The inhibitors, Z-Leu-Phe-CH_2F, Bzl-NHCO-Leu-Phe-CH_2F, 1,2,3,4-tetrahydroisoquin-2-yl-Leu-Phe-CH_2F, Z-Leu-Ser(THP)-CH_2F and Z-Leu-Leu-Phe-CH_2F, were cell permeable. In particular, the dipeptides 1,2,3,4-tetrahydroisoquin-2-yl-Leu-Phe-CH_2F and Z-Leu-Phe-CH_2F had equal IC_{50} values of 0.2 μM, and the tripeptide Z-Leu-Leu-Phe-CH_2F had an IC_{50} value of 0.1 μM. The Boc-Leu-Phe-CH_2F was not as permeable since its IC was only 42% at 10 μM in the cell assay (Chatterjee *et al* 1996b, 1997a).

Calpain II was inhibited by Z-Leu-Tyr-CH_2F with a second order rate of inactivation ($k_2 = 17000\,M^{-1}s^{-1}$) which is greater than the rate of inactivation by the diazomethyl ketone Z-Leu-Tyr-CH_2N_2 ($k_2 = 1470\,M^{-1}s^{-1}$). The tripeptide derivatives Z-Leu-Leu-Tyr-CH_2F and Z-Leu-Leu-Tyr-CH_2N_2, exhibited the converse relationship with k_2 values of 28900 and 230000 $M^{-1}s^{-1}$ respectively. The tripeptide Z-Leu-Leu-Tyr-CH_2F was also found to inhibit *in vitro* intact platelets much more effectively than the corresponding diazomethyl ketones, even though the diazomethane derivative had a much better rate constant in solution. The two tripeptides also inhibited cathepsin L ($k_2 = 6.8 \times 10^5$ and $1.5 \times 10^6\,M^{-1}s^{-1}$, respectively) more effectively than calpain II (Angliker *et al* 1992).

6.4.5.2 Difluoromethyl and trifluoromethyl ketones

Difluoro and trifluoroketones were ineffective inhibitors for calpain. For instance, the trifluoromethyl ketones, Z-Val-Phe-CF_3 ($K_I > 180$ μM) and Bz-Phe-CF_3 ($K_I > 1000$ μM), were both poor inhibitors of calpain. The difluoroketone, Bz-Phe-CF_2H ($K_I > 1000$ μM) was also a poor inhibitor of calpain. The poor K_I values of these inhibitors may be due to poor solubility in the assay medium (Angelastro *et al* 1990).

6.4.5.3 Chloromethyl ketones

Chloromethyl ketones inhibit calpain by the same nucleophilic displacement mechanism as the fluoromethyl ketones and are fairly selective inhibitors of calpain I and calpain II. The chloromethyl ketone Leu-Leu-Phe-CH_2Cl was a very potent inhibitor with an IC_{50} of 0.2 μM for calpain I and an IC_{50} of 0.19 μM for calpain II. The dansyl derivative Dns-Leu-Leu-CH_2Cl ($IC_{50} = 0.12$ μM and 0.18 μM for calpain I and II respectively) was the most selective when compared to papain ($IC_{50} = 12\,\mu M$) (Sasaki *et al* 1986). These inhibitors were not compared to the cathepsins, however, the chloromethyl ketone Z-Leu-Leu-Phe-CH_2Cl ($k_2/K_I = 100000\,M^{-1}s^{-1}$) was not very selective for calpain II and inhibited cathepsin B twice as fast ($k_2/K_I = 190000\,M^{-1}s^{-1}$). Also, the chloromethyl ketones, unlike diazomethyl ketones, are reactive against the thiol reducing agents used in most enzymatic assays of cysteine proteases (Pliura *et al* 1992).

6.4.6 Azapeptides

Azapeptide halomethyl ketones (Figure 6.8) are analogues of peptide halomethyl ketones, and are specific inhibitors for cysteine proteases such as cathepsin B. The serine protease chymotrysin was not inhibited (Giordano *et al* 1993). Azapeptide halomethyl ketone analogues are poor inhibitors for calpain I and II: Boc-Val-Lys(Z)-Leu-ATyr-CH_2I ($k_{obs}/[I] < 10\,M^{-1}s^{-1}$), Boc-Val-Lys(COOMe)-Leu-ATyr-CH_2I ($k_{obs}/[I] < 10\,M^{-1}s^{-1}$), Boc-Val-Lys(Tos)-Leu-ATyr-CH_2I ($k_{obs}/[I] < 10\,M^{-1}s^{-1}$) and Z-Leu-Leu-ATyr-$CH_2I$ ($k_{obs}/[I] = 20\,M^{-1}s^{-1}$).

6.4.7 Sulfonium methyl ketones

Sulfonium methyl ketones are haloketone analogues containing a sulfide leaving group. They are among the most potent calpain inhibitors known (Donkor 2000). Two sulfonium methyl ketone inhibitors have been made by Pliura *et al* and the most potent and selective calpain II inhibitor, Z-Leu-Leu-Phe-$CH_2S^+(CH_3)_2 \bullet Br^-$, has a $k_2/K_I >> 200000\,M^{-1}s^{-1}$ (Pliura *et al* 1992). The k_2/K_I for cathepsin B is only $1300\,M^{-1}s^{-1}$.

6.4.8 Diazomethyl ketones

Diazomethyl ketones inhibit calpain by a nucleophilic substitution reaction similar to halomethyl ketones. Potent and selective diazomethyl ketones of calpain II have been obtained by subsite mapping (Table 6.7) (Crawford *et al* 1988). The specificity of calpain II for diazomethyl ketones inhibitors is similar to the observed specificity with AMC substrates (Sasaki *et al* 1990). The N-terminal protecting group influences potency and the Z group is more favorable, compare Leu-Leu-Tyr-CHN_2 ($k_2 = 2300\,M^{-1}s^{-1}$),

Table 6.7 Diazomethyl ketone inhibitors of calpain

Inhibitor	*Calpain I[a]* $k_2(M^{-1}s^{-1})$	*Calpain II[b]* $k_2(M^{-1}s^{-1})$	*Reference*
Leu-Leu-Tyr-CHN_2		2300	(Crawford *et al* 1988)
Ac-Leu-Leu-Tyr-CHN_2		10570	(Crawford *et al* 1988)
Z-Leu-Leu-Tyr-CHN_2		230000	(Crawford *et al* 1988)
Boc-Val-Lys-(Z)-Leu-Tyr-CHN_2		20640	(Crawford *et al* 1988)
Biot-Aca-Leu-Tyr-CHN_2	2900		(Wikstrom *et al* 1993)
Biot-Aca-Leu-Leu-Tyr-CHN_2		240	(Wikstrom *et al* 1993)

[a]Human Platelet Calpain; [b]Chicken Gizzard Calpain.

Ac-Leu-Leu-Tyr-CHN_2 ($k_2 = 10570\,M^{-1}s^{-1}$) and Z-Leu-Leu-Tyr-$CHN_2$ ($k_2 = 230000\,M^{-1}s^{-1}$). The P2 site has a preference for Leu since Z-Phe-Phe-CHN_2 and Z-Phe-Ala-CHN_2 were ineffective inhibitors of calpain II ($k_2 < 10\,M^{-1}s^{-1}$) (Crawford *et al* 1988).

Tripeptidyl and tetrapeptidyl diazomethyl ketone inhibitors were more effective calpain II inhibitors than dipeptidyl inhibitors. The diazomethyl ketone inhibitors, Ac-Leu-Leu-Tyr-CHN_2 ($k_2 = 10570\,M^{-1}s^{-1}$) and Z-Leu-Leu-Tyr-$CHN_2$ ($k_2 = 230000\,M^{-1}s^{-1}$), were effective inhibitors of calpain II but inhibited cathepsin L more effectively. Interestingly, Boc-Val-Lys(Z)-Leu-Tyr-CHN_2 ($k_2 = 20640\,M^{-1}s^{-1}$) was the most specific inhibitor for calpain II and inhibited cathepsin L ca. 1.5-fold poorer (Crawford *et al* 1988).

Peptidyl diazomethyl ketone derivatives have been used as radio-labeled inhibitors that affinity label calpain (Shaw 1994). This method incorporates ^{125}I into a diazomethyl ketone, such as ^{125}I-labeled Z-Leu-Leu-Tyr(I)-CHN_2, which affinity labels calpain in platelets. Since the inhibitors are cell permeable, they can be used for *in vivo* studies for the identification of specific proteases so giving an insight into their function.

Biotinylated peptidyl diazomethyl ketone derivatives have been used to affinity label calpain (Wikstrom *et al* 1993). Biotin can be attached to the N-terminal group of calpain inhibitors since it is tolerant to various substitutions. The inhibitor Biot-Aca-Leu-Tyr-CHN_2($k_2 = 2900\,M^{-1}s^{-1}$) was an effective inhibitor while Biot-Aca-Leu-Leu-Tyr-CHN_2 ($k_2 = 240\,M^{-1}s^{-1}$) was less effective. This difference could be due to the hydrophobic aminocaproic acid spacer, which replaces a P3 Leu in preferred inhibitors.

6.4.9 Vinyl sulfones

Peptide vinyl sulfone inhibitors (Figure 6.8) are Michael acceptors that react irreversibly with the active site cysteine. Effective irreversible vinyl sulfone inhibitors for calpain I and II have not yet been developed (Palmer *et al* 1995). Current vinyl sulfone inhibitors are much more potent and specific for cathepsins than calpains. For example, the inhibitor Z-Leu-Leu-Tyr-VS-Ph ($k_2 = 24300\,M^{-1}s^{-1}$ and $6400\,M^{-1}s^{-1}$ for calpain I and II respectively) is more potent for cathepsin S ($k_2 = 280000\,M^{-1}s^{-1}$).

6.4.10 Cyclopropenone derivatives

Cyclopropenone derivatives inhibit rat brain calpain II with IC_{50} values of 0.5 to >200 μM (Ando *et al* 1999) (Table 6.8). They are not very specific and also inhibit other cysteine proteases such as the cathepsins. The mechanism of inhibition could involve either reaction of the active site cysteine residue with the cyclopropenone carbonyl group to yield a reversible tetrahedral adduct or nucleophilic attack (1,2 or 1,4 addition) on the cyclopropenone to form a covalent bond. Kinetic analysis indicates that the inhibition is reversible, thus favoring the formation of the tetrahedral adduct. There is no evidence for formation of a irreversible covalent bond.

The cyclopropenone (**6.38**) was the most specific inhibitor for calpain II with a IC_{50} of 2.10 μM for calpain II and 29.0 μM for cathepsin B. Cyclopropenyl derivatives that contain alkyl substituents at R_3 such as (**6.30**) ($IC_{50} > 200$ μM), had poorer IC_{50} values than (**6.31**) ($IC_{50} = 2.7$ μM), which contains an aryl substituent. Interestingly, the most potent calpain inhibitors, (**6.32**) ($IC_{50} = 0.5$ μM) and (**6.33**) ($IC_{50} = 0.81$ μM), had no substituents. Cyclopropenyl conformation also has a profound effect on activity. For example, inhibitor (**6.34**) ($IC_{50} = 3.25$ μM), which contains a phenyl group planar to the

Table 6.8 Cyclopropenone inhibitors of calpain

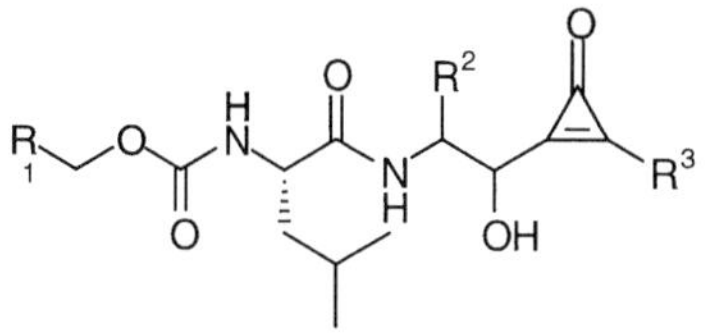

No.	R_1	R_2	R_3	*Hydroxyl (S):(R)*	*Calpain II*[a] *IC_{50}(μM)*	*Reference*
(6.30)	Cyclohexyl	Butyl	$CH_2CH(CH_3)_2$	6:4	>200	(Ando *et al* 1999)
(6.31)	Cyclohexyl	Butyl	Phenyl	7:3	2.7	(Ando *et al* 1999)
(6.32)	Cyclohexyl	Isobutyl	H	7:3	0.50	(Ando *et al* 1999)
(6.33)	Cyclohexyl	Butyl	H	7:2	0.81	(Ando *et al* 1999)
(6.34)	Cyclohexyl	Isobutyl	Phenyl	R	3.25	(Ando *et al* 1999)
(6.35)	Cyclohexyl	Butyl	2-Me-phenyl	R	29	(Ando *et al* 1999)
(6.36)	Cyclohexyl	Isobutyl	4-F-phenyl	S	1.51	(Ando *et al* 1999)
(6.37)	Cyclohexyl	Isobutyl	4-F-phenyl	R	7.0	(Ando *et al* 1999)
(6.38)	Phenyl	Isobutyl	Phenyl	S	0.6	(Ando *et al* 1999)

[a]Rat Brain Calpain.

cyclopropenone ring, was much more effective than inhibitor **(6.35)** ($IC_{50} = 29\,\mu M$), which is suggested to have a twisted conformation between the cyclopropenyl group and phenyl group. The bulky cyclopropenyl derivative **(6.30)** also has a poor IC_{50} value. The (*S*) hydroxyl enantiomers are typically two to four-fold more potent than the (*R*) hydroxyl enantiomers as illustrated by **(6.36)** and **(6.37)**. The authors suggest that the substituents and conformation may be changing the electronic nature of the cyclopropenone ring which could affect inhibitory potency.

6.5 THERAPEUTIC APPROACH

Several calpain inhibitors have demonstrated therapeutic potential in animal and cell based assays. Both peptide α-ketoamides and peptide aldehydes are effective in animal models of stroke and reduce the infarct size in these models. An alpha-mercaptoacrylic acid derivative exhibited *in vitro* neuroprotective effects in attenuation of hypoxic/hypoglycemic injury in cultured fetal rat cerebro-cortical neurons and excitotoxic injury in Purkinje cerebellar slices (Wang *et al* 1996). Peptide aldehydes and epoxides were shown to reduce calcium ionophore-induced cataracts in rat lenses (Sanderson *et al* 1996; Fukiage *et al* 1997).

Two dipeptide α-ketoamides (Li *et al* 1996) have a protective effect against ischemic brain damage in a rat model (Bartus *et al* 1994a,b, 1995). Focal ischemia was induced in rats using a variation of the middle cerebral artery occlusion method. One compound, Z-Leu-Abu-CONH-Et was perfused directly onto the infarcted cortical surface (Bartus *et al* 1994b). Alternately, Z-Leu-Abu-CONH-$(CH_2)_3$-4-morpholi nyl (AK295) was infused through the internal carotid artery (Bartus *et al* 1994a, 1995). After a delay, rats were sacrificed and the infarct volume was quantified. With Z-Leu-Abu-CONH-Et, a 75% reduction in infarct volume was achieved, while the maximum reduction obtained with AK295 was 32% using a different dosing protocol. A dose-dependent neuroprotective effect could be demonstrated in both cases. The morpholine compound (AK295) is also able to attenuate motor dysfunction in rats following a head injury (Saatman *et al* 1996). The significant improvement in behavioral outcome measurements suggests that AK295 could be used in the treatment of head injuries. Ketoamide inhibitors have the advantage of good membrane solubility (unlike the α-ketoacids) and excellent chemical stability to esterases (unlike the α-ketoesters) in animal models.

The peptide aldehyde inhibitor Z-Val-Phe-H was neuroprotective if the inhibitor was administered within six hours of the induced focal cerebral ischemia in a rat model. The inhibitor effectively reduced the infarct volume in Wistar rats over six hours after induction of focal ischemia using the modified middle cerebral artery occlusion method (Markgraf 1998). The Z-Val-Phe-H inhibitor was neuroprotective up to six hours whereas AK275 only exhibited significant neuroprotection up to 3 hours (Markgraf 1998).

The pharmacological profile of Z-Leu-Phe-H was optimized for blood brain barrier permeability as well as efficacious dose in the ischemia-reperfusion rodent model. Brain penetration of Z-Leu-Phe-H was determined by monitoring brain protease activity in the supernatant extracted from the homogenized supratentorial region of the brain. The aldehyde Z-Leu-Phe-H was very effective in penetrating the blood brain barrier.

An alpha-mercaptoacrylic acid derivative exhibited *in vitro* neuroprotective effects through attenuation of hypoxic/hypoglycemic injury to cultured cerbebral cortical neurons and excitotoxic injury to Purkinje cells in cerebral slices (Wang *et al* 1996). The inhibitor PD150606 (**6.23**) protected the glutamatergic neurons from hypoxic/hypoglycemic injury. Inhibitor (**6.23**) also protected cerebral Purkinje cells from AMPA, which is a highly toxic glutamate receptor agonist. Thus, alpha-mercaptoacrylic acid derivatives could prove useful in treatment of disorders involving calpain overactivation.

Cataracts formed by calcium ionophore stimulation were attenuated upon treatment with peptide aldehyde and epoxide inhibitors (Sanderson *et al* 1996). In an initial cell-free based assay, treatment of the rat lens with E-64, calpeptin, or Z-Val-Phe-H prevented formation of vimentin, filensin and spectrin breakdown products. In a different assay which involved pretreatment of the rat lens with an ionophore and vimentin, calpeptin (5 and 20 μM) had little effect on the prevention of the calpain breakdown products filensin and spectrin. However, calpeptin at 200 μM totally prevented filensin breakdown (IC_{50} 50 μM), which is one of the most sensitive calpain substrates. The aldehyde inhibitor Z-Leu-Phe-H was more potent than calpeptin with IC_{50} values less than 20 μM.

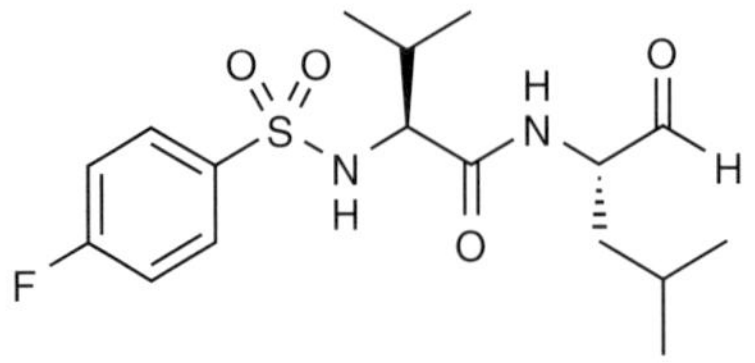

(6.39): SJA6017

Figure 6.9 Structure of SJA6017, an aldehyde calpain inhibitor (**6.39**).

In another study, the peptide aldehyde SJA6017 (**6.39**) ($IC_{50} = 0.08\,\mu M$) (Figure 6.9), E-64 ($IC_{50} = 2.31\,\mu M$) and leupeptin ($IC_{50} = 0.49\,\mu M$) prevented calcium ionophore induced cataract formation in cultured rat cells (Fukiage *et al* 1997). The aldehyde (**6.39**) prevented proteolysis of α-spectrin more effectively than leupeptin or E-64 since (**6.39**) is suggested to be more membrane permeable. The aldehyde (**6.39**) and E-64 prevented proteolysis of lens crystallins by 75% and 95%, respectively. Similarly, (**6.39**) and E-64 were the most effective at preventing calcium ionophore induced cataracts as measured by opacity density.

6.6 PERSPECTIVES

Research has linked calpain with many physiological disorders ranging from acute and chronic neurodegeneration to various disorders that involve other organs such as the eye or heart. Acute neurological disorders such as ischemic stroke, traumatic brain injury, and subarachnoid hemmorage and chronic neurodegenerative disorders such as Alzheimer's, Parkinson's, Huntington's disease, spinal cord injury and muscular dystrophy have been associated with calpain activity or activation. Diseases not associated with neurodegeneration include cardiac ischemia, cataract formation, thrombolytic platelet aggregation, restenosis, joint inflammation and arthritis.

Several investigators have focused on the use of calpain inhibitors in animal models of stroke. Calpain inhibition may extend the time-frame for treatment of stroke since calpain activation occurs downstream in the ischemic cascade. Prolonging the time-window for treatment of stroke while retaining efficacy would be ideal for neuroprotection. Current treatment of thrombolytic stroke with recombinant tissue plasminogen activator must occur within three hours after stroke onset. However, 40% of stroke victims do not reach medical attention during this time, therefore, time is crucial in treating ischemic stroke. Calpain inhibition has the potential for extending this therapeutic window.

Both academic and pharmaceutical have been very interested in calpain inhibition as a therapeutic strategy for the treatment of both acute and chronic neurological diseases. An ideal calpain inhibitor that was potent, bioavailable, specific and nontoxic would likely be quickly evaluated as a drug. Novel compounds that show neural protection in response to global ischemia have already been developed and tested in rodent stroke models. Calpain inhibitors have also been proposed as potential therapeutic agents against peripheral nerve injuries and other chronic neurological diseases associated with neuronal degeneration. In an era that focuses on increasing life expectancy, the significance of therapeutically targeting calpain is now more important than ever.

REFERENCES

Ando, R., Sakaki, T., Morinaka, Y., Takahashi, C., Tamao, Y., Yoshii, N., Katayama, S., Saito, K., Tokuyama, H., Isaka, M. and Nakamura, E. (1999) Cyclopropenone-containing cysteine proteinase inhibitors. Synthesis and enzyme inhibitory activities. *Bioorganic & Medicinal Chemistry*, **7**, 571–579.

Angelastro, M.R., Mehdi, S., Burkhart, J.P., Peet, N.P. and Bey, P. (1990) α-Diketone and α-ketoester of N-protected amino acids and peptides as novel inhibitors of cysteine of serine proteases. *Journal of Medicinal Chemistry*, **33**, 11–13.

Angelastro, M.R., Marquart, A.L., Mehdi, S., Koehl, J.R., Vaz, R.J., Bey, P. and Peet, N.P. (1999) The synthesis of ketomethylene pseudopeptide analogues of dipeptide aldehyde inhibitors of calpain. *Bioorganic & Medicinal Chemistry Letters*, **9**, 139–140.

Angliker, H., Anagli, J. and Shaw, E. (1992) Inactivation of calpain by peptidyl fluoromethyl ketones. *Journal of Medicinal Chemistry*, **35**, 216–220.

Ariyoshi, H., Shiba, E., Kambayashi, J., Sakon, M., Tsujinaka, T., Uemura, Y. and Mori, T. (1991) Characteristics of various synthetic peptide calpain inhibitors and their application for the analysis of platelet reaction. *Biochemistry International*, **23**, 1019–1033.

Bartus, R., Baker, K., Heiser, A., Sawyer, S., Dean, R., Akiyama, A., Elliot, P., Straub, J., Harbeson, S. and Powers, J. (1994a) Calpain inhibitor AK295 protects neurons from focal brain ischemia. *Stroke*, **25**, 2265–2270.

Bartus, R., Baker, K., Heiser, A., Sawyer, S., Dean, R., Elliot, P. and Straub, J. (1994b) Postischemic administration of AK275, a calpain inhibitor, provides substantial protection against focal ischemic brain damage. *Journal of Cerebral Blood Flow and Metabolism*, **14**, 537–544.

Bartus, R.T., Elliot, P.J., Hayward, N.J., Dean, R.L., Harbeson, S., Straub, J.A., Li, Z. and Powers, J.C. (1995) Calpain as a novel target for treating acute neurodegenerative disorders. *Neurological Research*, **17**, 249–258.

Bi, X., Tocco, G. and Baudry, M. (1994) Calpain-mediated regulation of AMPA receptors in adult rat brain. *Neuroreport*, **6**, 61–64.

Blanchard, H., Grochulski, P., Li, Y., Arthur, J., Davies, P., Elce, J. and Cygler, M. (1997) Structure of a calpain Ca^{2+}-binding domain reveals a novel EF-hand and Ca^{2+}-induced conformational changes. *Nature Structural Biology*, **4**, 532–538.

Chatterjee, S., Iqbal, M., Kauer, J.C., Mallamo, J.P., Senadhi, S., Mallya, S., Bozyczko-Coyne, D. and Siman, R. (1996a) Xanthene derived potent nonpeptidic inhibitors of recombinant human calpain I. *Bioorganic & Medicinal Chemistry Letters*, **6**, 1619–1622.

Chatterjee, S., Josef, K., Wells, G., Iqbal, M., Bihovsky, R., Mallamo, J.P., Ator, M.A., Bozyczko-Coyne, D., Mallya, S., Senadhi, S. and Siman, R. (1996b) Potent fluoromethyl ketone inhibitors of recombinant human calpain I. *Bioorganic & Medicinal Chemistry Letters*, **6**, 1237–1240.

Chatterjee, S., Ator, M.A., Bozyczko-Coyne, D., Josef, K., Wells, G., Tripathy, R., Iqbal, M., Bihovsky, R., Senadhi, S.E., Mallya, S., Okane, T., McKenna, B.A., Siman, R. and Mallamo, J.P. (1997a) Synthesis and biological activity of a series of potent fluoromethyl ketone inhibitors of recombinant human calpain I. *Journal of Medicinal Chemistry*, **40**, 3820–3828.

Chatterjee, S., Senadhi, S., Bozyczko-Coyne, D., Siman, R. and Mallamo, J.P. (1997b) Non-peptidic inhibitors of recombinant human calpain I. *Bioorganic & Medicinal Chemistry Letters*, **7**, 287–290.

Chatterjee, S., Gu, Z.Q., Dunn, D., Tao, M., Josef, K., Tripathy, R., Bihovsky, R., Senadhi, S.E., O'Kane, T.M., McKenna, B.A., Mallya, S., Ator, M.A., Bozyczko-Coyne, D., Siman, R. and Mallamo, J.P. (1998a) D-Amino acid containing high-affinity inhibitors of recombinant human calpain I. *Journal of Medicinal Chemistry*, **41**, 2663–2666.

Chatterjee, S., Iqbal, M., Mallya, S., Senadhi, S.E., O'Kane, T.M., McKenna, B.A., Bozyczko-Coyne, D., Kauer, J.C., Siman, R. and Mallamo, J.P. (1998b) Exploration of the importance of the P-2-P-3-NHCO-Moiety in a potent Di- or tripeptide inhibitor of calpain I: Insights into the development of nonpeptidic inhibitors of calpain I. *Bioorganic & Medicinal Chemistry*, **6**, 509–522.

Crawford, C., Mason, R.W., Wikstrom, P. and Shaw, E. (1988) The design of peptidyldiazomethane inhibitors to distinguish between the cysteine proteinases calpain-II, cathepsin-L and cathepsin-B. *Biochemical Journal*, **253**, 751–758.

Donkor, I.O. (2000) A survey of calpain inhibitors. *Current Medicinal Chemistry*, **7**, 1171–1188.

Donkor, I.O., Zheng, X. and Miller, D.D. (2000) Synthesis and calpain inhibitory activity of α-ketoamides with 2,3-methanoleucine stereoisomers at the P2 position. *Bioorganic & Medicinal Chemistry Letters*, **10**, 2497–2500.

Fukiage, C., Azuma, M., Nakamura, Y., Tamada, Y., Nakamura, M. and Shearer, T. (1997) SJA6017, a newly synthesized peptide aldehyde inhibitor of calpain: amelioration of cataract in cultured rat lenses. *Biochemical & Biophysics Acta*, **1361**, 304–312.

Giordano, C., Calabretta, R., Gallina, C., Consalvi, V., Scandurra, R., Noya, F.C. and Franchini, C. (1993) Synthesis and inhibiting activities of 1-peptidyl-2-haloacetyl hydrazines toward cathepsin-B and calpains. *European Journal of Medicinal Chemistry*, **28**, 297–311.

Graybill, T.L., Dolle, R.E., Osifo, I.K., Schmidt, S.J., Gregory, J.S., Harris, A.L. and Miller, M.S. (1995) Inhibition of human erythrocyte calpain-I by novel quinoline carboxamides. *Bioorganic & Medicinal Chemistry Letters*, **5**, 387–392.

Greenwood, A.F. and Jope, R.S. (1994) Brain G-protein proteolysis by calpain: enhancement by lithium. *Brain Research*, **636**, 320–326.

Hanada, K., Tamai, M., Yamagishi, M., Ohmura, S., Sawada, J. and Tanaka, I. (1978) Studies on thiol protease inhibitors. Part I. isolation and characterization of E-64, a new thiol protease inhibitor. *Agricultural & Biological Chemistry*, **42**, 523–528.

Harris, A.L., Gregory, J.S., Maycock, A.L., Graybill, T.L., Osifo, I.K., Schmidt, S.J. and Dolle, R.E. (1995) Characterization of a continuous fluorogenic assay for calpain I – kinetic evaluation of peptide aldehydes, halomethyl ketones and (acyloxy) methyl ketones as inhibitors of the enzyme. *Bioorganic & Medicinal Chemistry Letters*, **5**, 393–398.

Hirai, S., Kawasaki, H., Yaniv, M. and Suzuki, K. (1991) Degradation of transcription factors, c-Jun and c-Fos, by calpain. *FEBS Letters*, **287**, 57–61.

Hosfield, C.M., Elce, J.S., Davis, P.L. and Jia, Z. (1999) Crystal structure of calpain reveals the structural basis for Ca(2+)-dependent protease activity and a novel mode of enzyme activation. *EMBO Journal*, **18**, 6880–6889.

Huang, Z.Y., McGowan, E.B. and Detwiler, T.C. (1992) Ester and amide derivatives of E64c as inhibitors of platelet calpains. *Journal of Medicinal Chemistry*, **35**, 2048–2054.

Imajoh, S., Kawasaki, H. and Suzuki, K. (1987) The COOH-terminal E-F hand structure of calcium-activated neutral protease (CANP) is important for the association of subunits and resulting proteolytic activity. *Journal of Biochemistry*, **101**, 447–452.

Iqbal, M., Messina, P.A., Freed, B., Das, M., Chatterjee, S., Tripathy, R., Tao, M., Josef, K.A., Dembofsky, B., Dunn, D., Griffith, E., Siman, R., Senadhi, S.E., Biazzo, W., Bozyczko-Coyne, D., Meyer, S.L., Ator, M.A. and Bihovsky, R. (1997) Subsite requirements for peptide aldehyde inhibitors of human calpain I. *Bioorganic & Medicinal Chemistry Letters*, **7**, 539–544.

Li, Z.Z., Patil, G.S., Golubski, Z.E., Hori, H., Tehrani, K., Foreman, J.E., Eveleth, D.D., Bartus, R.T. and Powers, J.C. (1993) Peptide alpha-ketoester, alpha-ketoamide, and alpha-ketoacid inhibitors of calpains and other cysteine proteases. *Journal of Medicinal Chemistry*, **36**, 3472–3480.

Li, Z.Z., Ortega-Vilain, A.C., Patil, G.S., Chu, D.L., Foreman, J.E., Eveleth, D.D. and Powers, J.C. (1996) Novel peptidyl alpha-ketoamide inhibitors of calpains and other cysteine proteases. *Journal of Medicinal Chemistry*, **39**, 4089–4098.

Lin, G., Chattopadhyay, D., Maki, M., Wang, K., Carson, M., Jin, L., Yuen, P., Takano, E., Hatanaka, M., DeLucas, L. and Narayana, S. (1997) Crystal structure of calcium bound domain VI of calpain at 1.9 angstrom resolution and its role in enzyme assembly, regulation, and inhibitor binding. *Nature Structural Biology*, **4**, 539–547.

Markgraf (1998) Six-hour window of opportunity for calpain inhibition in focal cerebral ischemia in rats. *Stroke*, **29**, 152–158.

Melloni, E. and Pontremoli, S. (1989) The calpains. *Trends in Neuroscience*, **12**, 438–444.

Palmer, J.T., Rasnick, D., Klaus, J.L. and Bromme, D. (1995) Vinyl sulfones as mechanism-based cysteine protease inhibitors. *Journal of Medicinal Chemistry*, **38**, 3193–3196.

Parkes, C., Kembhavi, A.A. and Barrett, A.J. (1985) Calpain inhibition by peptide epoxides. *Biochemical Journal*, **230**, 509–516.

Pliura, D.H., Bonaventura, B.J., Smith, R.A., Coles, P.J. and Krantz, A. (1992) Comparative behavior of calpain and cathepsin-B toward peptidyl acyloxymethyl ketones, sulfonium

methyl ketones and other potential inhibitors of cysteine proteinases. *Biochemical Journal*, **288**, 759–762.
Rawlings, N.D. and Barrett, A.J. (1994) Families of cysteine peptidases. *Methods in Enzymology*, **244**, 461–486.
Saatman, K.E., Murai, H., Bartus, R.T., Smith, D.H., Hayward, N.J., Perri, B.R. and McIntosh, T.K. (1996) Calpain inhibitor AK295 attenuates motor and cognitive deficits following experimental brain injury in the rat. *Proceedings of the National Academy of Sciences of the U.S.A.*, **93**, 3428–3433.
Sanderson, J., Marcantonio, J. and Duncan, G. (1996) Calcium ionophore induced proteolysis and cataract: Inhibition by cell permeable calpain antagonists. *Biochemical & Biophysical Research Communications*, **218**, 893–901.
Sasaki, T., Kikuchi, T., Fukui, I. and Murachi, T. (1986) Inactivation of calpain I and calpain II by specificity-oriented tripeptidyl chloromethyl ketones. *Journal of Biochemistry*, **99**, 173–179.
Sasaki, T., Kishi, M., Saito, M., Tanaka, T., Higuchi, N., Kominami, E., Katunuma, N. and Murachi, T. (1990) Inhibitory effects of di- and tripeptidyl aldehydes on calpains and cathepsins. *Journal of Enzyme Inhibition*, **3**, 195–201.
Schechter, I. and Berger, A. (1967) On the size of the active site in protease. 1. Papain. *Biochemical & Biophysical Research Communications*, **27**, 157–162.
Schirmeister, T. (1999) New peptidic cysteine protease inhibitors derived from the electrophilic alpha-amino acid aziridine-2,3-dicarboxylic acid. *Journal of Medicinal Chemistry*, **42**, 560–572.
Shaw, E. (1994) Peptidyl diazomethanes as inhibitors of cysteine and serine proteinases. In *Proteolytic Enzymes: Serine and Cysteine Peptidases*, pp. 649–656. San Diego: Academic Press Inc.
Sorimachi, H., Saido, T.C. and Suzuki, K. (1994) New era of calpain research – discovery of tissue-specific calpains. *FEBS Letters*, **343**, 1–5.
Spencer, M.J., Croall, D.E. and Tidball, J.G. (1995) Calpains are activated in necrotic fibers from MDX dystrophic mice. *Journal of Biological Chemistry*, **270**, 10909–10914.
Storer, A. and Ménard, R. (1994) Catalytic mechanism in papain family of cysteine peptidases. *Methods in Enzymology*, **244**, 486–500.
Suzuki, K. and Sorimachi, H. (1998) A novel aspect of calpain activation. *FEBS Letters*, **433**, 1–4.
Tripathy, R., Gu, Z.Q., Dunn, D., Senadhi, S.E., Ator, M.A. and Chatterjee, S. (1998) P-2-Proline-derived inhibitors of calpain I. *Bioorganic & Medicinal Chemistry Letters*, **8**, 2647–2652.
Tripathy, R., Ator, M.A. and Mallamo, J.P. (2000) Calpain inhibitors based on the quiescent affinity label concept: high rates of calpain inactivation with leaving groups derived from N-hydroxy peptide coupling reagents. *Bioorganic & Medicinal Chemistry Letters*, **10**, 2315–2319.
Tsujinaka, T., Kajiwara, Y., Kambayashi, J., Sakon, M., Higuchi, N., Tanaka, T. and Mori, T. (1988) Synthesis of a new cell penetrating calpain inhibitor (calpeptin). *Biochemical & Biophysical Research Communications*, **153**, 1201–1208.
Umezawa, H. (1973) Chemistry of enzyme inhibitors of microbial origin. *Pure and Applied Chemistry*, **33**, 129–144.
Wang, K.K.W. and Yuen, P.W. (1994) Calpain inhibition – an overview of its therapeutic potential. *Trends in Pharmacological Sciences*, **15**, 412–419.
Wang, K., Nath, R., Posner, A., Raser, K., Buroker-Kilgore, M., Hajimohammadreza, I., Hatanaka, M., Maki, M., Caner, H., Collins, J., Fergus, A., Lee, K., Lunney, W., Hays, S. and Yuen, P. (1996) An alpha-mercaptoacrylic acid derivative is a selective nonpeptide cell-permeable calpain inhibitor and is neuroprotective. *Proceedings of the National Academy of Sciences of the U.S.A.*, **93**, 6687–6692.
Wells, G.J. and Bihovsky, R. (1998) Calpain inhibitors as potential treatment for stroke and other neurodegenerative diseases: Recent trends and developments. *Expert Opinion on Therapeutic Patents*, **8**, 1707–1727.

Wikstrom, P., Anagli, J., Angliker, H. and Shaw, E. (1993) Additional peptidyl diazomethyl ketones, including biotinyl derivatives, which affinity-label calpain and related cysteinyl proteinases. *Journal of Enzyme Inhibition*, **6**, 259–269.

Woo, J.T., Sigeizumi, S. and Yamaguchi, K. (1995) Peptidyl aldehyde derivatives as potent and selective inhibitors of cathepsin-L. *Bioorganic & Medicinal Chemistry Letters*, **5**, 1501–1504.

Yoshizawa, T.H.S., Tomioka, S., Ishiura, S. and Suzuki, K. (1995a) Calpain dissociates into subunits in the presence of calcium ions. *Biochemical & Biophysical Research Communications*, **208**, 376–383.

Yoshizawa, T.H.S., Tomioka, S., Ishiura, S. and Suzuki, K. (1995b) A catalytic subunit of calpain possesses full catalytic activity. *FEBS Letters*, **3588**, 101–103.

Chapter 7

Human neutrophil elastase inhibitors

Philip D. Edwards

Human neutrophil elastase is a particularly destructive serine protease and has been implicated in a variety of inflammatory diseases. With respect to the development of enzyme inhibitors, elastase has been the most intensely studied serine protease. With only a few exceptions, all of the low-molecular-weight inhibitors of elastase have been developed as a result of first principle, rational design approaches. This paper will review the strategies and tactics that have been applied to the development of potent inhibitors of elastase.

7.1 INTRODUCTION

The search for, and development of, inhibitors of human neutrophil elastase (HNE) has been the focus of intense effort in both academia and the pharmaceutical industry for over a quarter of a century. The basis of this interest was the seminal finding by Laurel and Erickson that individuals genetically deficient in α_1-protease inhibitor (α_1-antitrypsin, α_1PI) are susceptible to developing emphysema even if they do not smoke cigarettes (Laurell and Eriksson 1963; Eriksson 1965). Subsequently, it was determined that α_1PI is the most important endogenous inhibitor of HNE. Thus developed the Protease-Antiprotease Theory of emphysema (Aboussouan and Stoller 1999; Snider 1992).

HNE is one of the most destructive enzymes in the body. It has the ability to degrade a variety of proteins, including the structural proteins fibronectin, collagen and elastin (Bieth 1986). It is a key component of the body's inflammatory defenses, assisting the neutrophil in its migration to the site of inflammation and participating in the proteolytic degradation of invading microorganisms. In addition, HNE is involved in tissue remodeling and wound healing. Under homeostatic conditions, the destructive effects of HNE are limited to the microenvironment immediately surrounding the neutrophil by endogenous proteases inhibitors such as α_1PI and secreatory leukocyte protease inhibitor. As a consequence of chronic inflammation, however, the balance between HNE and these endogenous inhibitors can be shifted in favor of HNE, resulting in uncontrolled tissue destruction. The protease/antiprotease balance may also be upset by a decreased availability of α_1PI, either through inactivation by oxidants such as cigarette smoke, or as a result of a genetic inability to produce sufficient serum levels. HNE has been implicated in the promotion or exacerbation of a number of diseases including pancreatitis, acute respiratory distress syndrome, rheumatoid arthritis, atherosclerosis, pulmonary emphysema, ischemia reperfusion injury and cystic fibrosis (Bernstein *et al* 1994; Farley *et al* 1997).

As a means of developing therapies to treat these diseases, a number of strategies have been pursued to identify compounds that inhibit the proteolytic activity of HNE. This review is aimed at introducing the reader to the general approaches used for developing HNE inhibitors. A brief description of the properties and catalytic mechanism of HNE catalysis is followed by a discussion of the various classes of inhibitors that have been developed. The mechanism of inhibition and the resulting physical and biological properties that should be considered when developing a particular class of inhibitor are emphasized. Although recent, novel inhibitors are discussed, no attempt has been made to comprehensively catalogue the various structural classes of inhibitors since this information can be found in several recent reviews (Bernstein *et al* 1994; Edwards and Bernstein 1994; Hlasta and Pagani 1994; Powers *et al* 1996; Edwards and Veale 1997; Farley *et al* 1997; Metz and Peet 1999; Skiles and Jeng 1999). Rather, specific inhibitor classes are used to develop an historical background and to illustrate important concepts. Detailed mechanisms and kinetic constants are generally not specified. For more specific details the reader is directed to reviews by Powers *et al* (1996) and Edwards and Bernstein (1994) and the primary literature cited in this review. Attempts have been made to include only the most recent references, from which the reader can access the earlier literature. Finally, a discussion of the progress of HNE inhibitors in the clinic is included.

7.2 PROPERTIES OF HNE

Human neutrophil elastase (EC 3.4.21.37) is a 218 amino acid glycoprotein with a molecular weight of 33,000-Da that exists in several isoforms differing in their extent of glycosylation. All isoforms have been shown to have identical catalytic properties. HNE has a high isoelectric point due to the large number of arginine residues on its outer surface. The primary structure (Sinha *et al* 1987), X-ray crystal structure (Bode *et al* 1986), and gene sequence (Takahashi *et al* 1988a) for HNE have all been determined. While HNE can be found in small amounts in monocytes and mast cells, the bulk of HNE is found in the human neutrophil, where it is compartmentalized in the cytoplasmic azurophilic granules (Ohlsson and Olsson 1974). After release from these cellular stores, HNE assists the neutrophil in both its migration to the site of inflammation and in its attack on xenobiotic agents. HNE is classified as a member of the elastase sub-family of proteases based on its ability to cleave the structural protein elastin. Of more significance with respect to the development of synthetic inhibitors is the fact that HNE is a member of a large class of enzymes known as serine proteases.

7.3 CATALYTIC MECHANISM OF SERINE PROTEASES

What distinguishes HNE and the serine protease family from the metallo, thiol, and aspartate classes of proteases is the molecular machinery used to effect peptide bond hydrolysis (Kraut 1977). The active site of proteases consists of two domains: an extended binding site where non-covalent binding interactions occur, and a catalytic site where the covalent bond forming and bond breaking reactions take place. The key catalytic element of the serine proteases is the hydroxyl group of Ser-195. The mechanism of peptide bond hydrolysis by a serine protease is depicted in Figure 7.1. After complexation of the

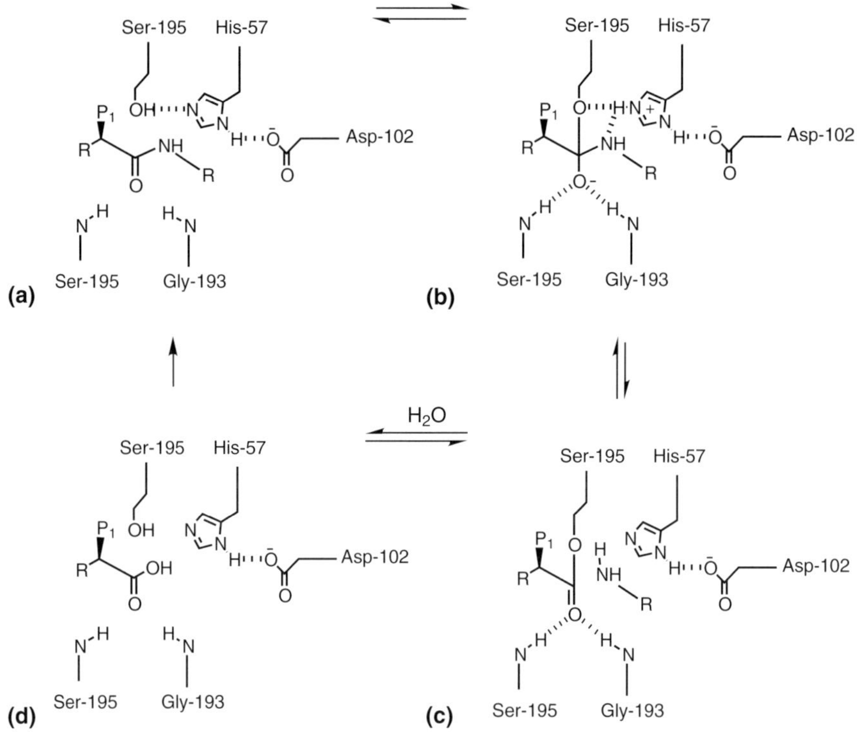

Figure 7.1 Schematic representation of the mechanism of peptide bond hydrolysis by serine proteases. (**a**) initial enzyme – substrate complex; (**b**) tetrahedral adduct with Ser-195; (**c**) acyl-enzyme plus *C*-terminal amine product; (**d**) regenerated catalytic triad plus *N*-terminal acid product.

substrate with the extended binding site of the enzyme (Figure 7.1a), the hydroxyl group of Ser-195 undergoes nucleophilic addition to the carbonyl carbon atom of the scissile peptide bond to form the tetrahedral intermediate (Figure 7.1b). The nucleophilicity of the serine hydroxyl group is increased through hydrogen bond formation with the imidazole of His-57, which itself is involved in a hydrogen-bonding interaction with Asp-102. These three amino acid residues are known as the "catalytic triad" and are responsible for proton transfer during peptide bond hydrolysis.

While the precise movement of protons between the residues of the catalytic triad remains controversial, the proton on the serine hydroxyl group is ultimately transferred via His-57 to the nitrogen atom of the eventual amine product, thereby facilitating collapse of the tetrahedral intermediate to the acyl-enzyme ester (Figure 7.1c) and departure of the C-terminal peptide fragment. The acyl-enzyme is hydrolyzed by the catalytic addition of water to afford the *N*-terminal carboxylic acid fragment of the peptide, along with the free enzyme which is ready to repeat the cycle. In addition to the catalytic triad, another key feature of the catalytic site of serine proteases is the "oxyanion hole" which is comprised of the NH groups of Ser-195 and Gly-193. These two groups participate in a hydrogen-bonding interaction with the carbonyl oxygen atom of the scissile bond, increasing its electrophilicity toward addition of the Ser-195 hydroxyl group, and stabilizing the oxyanion of the tetrahedral adduct.

In addition to the interactions within the catalytic site, the substrate also interacts with the enzyme through the amino acid residues extending to either side of the scissile bond. Those residues extending towards the *N*-terminus of the substrate are designated as P_1, P_2, P_3, ... etc., while those towards the C-terminus are designated as P'_1, P'_2, ... etc. The corresponding binding subsites on the enzyme are identified as S_1, S'_1, ... and so on (Schechter and Berger 1967). Binding outside of the catalytic site occurs through non-covalent interactions such as hydrogen bonds and hydrophobic forces. The most important of these interactions is that between the S_1 subsite and the P_1 residue. This interaction is the primary determinant of substrate specificity among the different serine proteases, and it is generally possible to alter the enzyme selectivity of a substrate or inhibitor by modifying the group which binds in the S_1 subsite.

Knowledge of the mode of binding of substrates and of the mechanism of peptide bond hydrolysis has facilitated the development of extremely potent inhibitors of HNE. While the ability to form a covalent attachment with Ser-195 forms the basis of most synthetic HNE inhibitors, the cumulative binding forces obtainable in the extended binding pocket can be extremely strong and hence sufficient to hold some inhibitors to the enzyme without covalent attachment to Ser-195. Indeed, many proteinaceous inhibitors of HNE do not form a covalent adduct, but rely solely on these non-covalent interactions. In contrast to proteinaceous inhibitors, low-molecular-weight inhibitors of HNE require the formation of a covalent adduct with Ser-195 to attain significant potency, since their small size limits the number of possible non-covalent interactions.

7.4 SUBSTRATE SPECIFICITY

While HNE has the ability to degrade a wide range of proteins, it hydrolyzes elastin very efficiently. Elastin is an insoluble structural protein found in skin, blood vessels, and lung. The primary sequence of elastin is rich in hydrophobic amino acid residues such as glycine, alanine, valine, and proline. Most of our knowledge of the substrate specificity of HNE is the result of the pioneering studies directed at mapping the active site of porcine pancreatic elastase (PPE) by Thompson and Blout (1970 and 1973), and Powers (Powers *et al* 1977a; Harper *et al* 1984). Since HNE has a relatively small S_1 subsite, it is more dependent than most serine proteases on the interactions in the extended binding pocket as a means of achieving sufficient binding of substrates and inhibitors, hence its relatively long active site (S_5–S'_3). The S_1 pocket of HNE prefers substrates with small hydrophobic residues such as valine and alanine at the P_1 position, while the S_2 subsite prefers proline, presumably due to the bend it imparts to the peptide backbone as observed in the X-ray crystal structures of complexes between small peptidic inhibitors with HNE (An-Zhi *et al* 1988; Navia *et al* 1989) and PPE (Takahashi *et al* 1988b). In addition, the P_2 proline helps to restrict the number of low energy conformations accessible to the unbound inhibitor, thereby "pre-organizing" its backbone into an enzyme-complementary conformation.

Substrate length plays a key role in determining the efficiency of amide bond hydrolysis catalyzed by HNE. Increasing the length of a peptide substrate such that it occupies the remote subsites S_5–S_2 results in increased catalytic efficiency. This same effect is observed with peptidic inhibitors, and a step jump in *in vitro* potency (e.g. 1,000-fold) is obtained

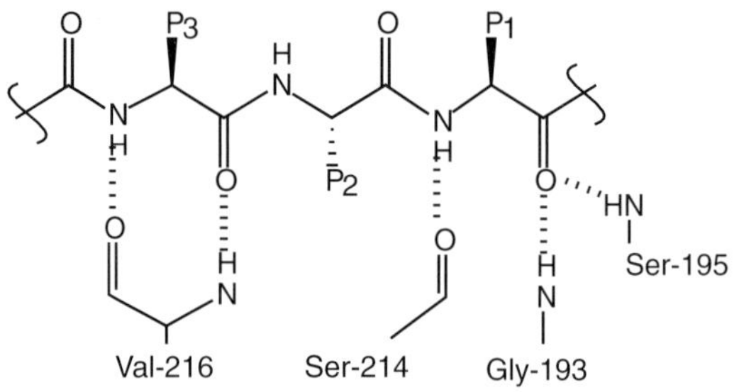

Figure 7.2 Representation of the hydrogen bonding network between a serine protease and a peptide ligand.

upon going from a di- to a tripeptidic inhibitor (Stein *et al* 1987). Particularly important are the P_3 to P_1 residues which bind to the enzyme's extended binding pocket in an anti-parallel, β-pleated sheet arrangement through three critical hydrogen bonds: the NH of the P_1 residue to the carbonyl of Ser-214; the carbonyl of P_3 to the NH of Val-216; and the NH of P_3 to the carbonyl of Val-216. The binding scheme depicted in Figure 7.2 has been observed in most of the X-ray crystal structures of complexes between peptidic inhibitors with PPE and HNE. The hydrogen bond between the P_3 NH and the carbonyl of Val-216 is a very important interaction, one which cannot be formed with dipeptides. The ability to form this hydrogen bond is the most likely explanation for the large increase in activity observed for tripeptide inhibitors relative to the corresponding dipeptides.

Limited studies of the S' subsites of HNE have indicated that hydrophobic residues are favored in the S'_1 position and that improved binding can be realized by occupying subsites out to S'_3. Few synthetic inhibitors have been designed that take advantage of interactions in the S' subsites. While this could be a result of the lack of important interactions in this region, it is more likely a function of the fact that occupation of this region is generally not possible, given the structural limitations of most of the synthetic inhibitors which have been developed.

7.5 IRREVERSIBLE INHIBITORS OF HNE

7.5.1 Alkylating agents

Alkylating agents are mechanistically the simplest class of serine protease inhibitors. Nucleophilic attack of the serine hydroxyl group on an electrophilic atom with displacement of a leaving group results in formation of an irreversible, covalent bond to the serine hydroxyl oxygen atom, permanently inactivating the enzyme. The relative reactivity of irreversible inhibitors cannot be analyzed by evaluation of a dissociation or inhibition constant (K_i) since the inhibitor is irreversibly bound in the final enzyme – inhibitor adduct, and is therefore not in equilibrium with unbound inhibitor. However, assay conditions can be chosen such that the ratio of the observed rate of enzyme inactivation (k_{obs}) to inhibitor concentration ([I]) will remain fairly linear over a range of inhibitor concentrations. The derived constant, k_{obs}/[I], can then be used to evaluate the relative reactivity of irreversible inhibitors in general. Since k_{obs}/[I] is related to the rate constant for inactivation, larger values correspond to more rapid inactivation.

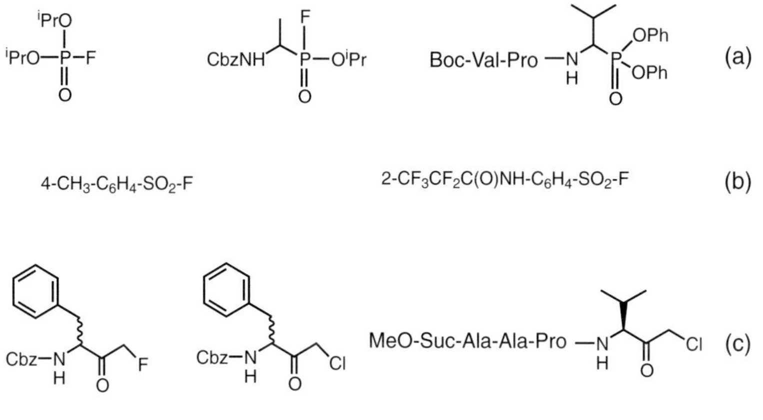

Figure 7.3 (a) Organophosphorus; (b) organosulfur; and (c) halomethyl ketone inhibitors of serine proteases.

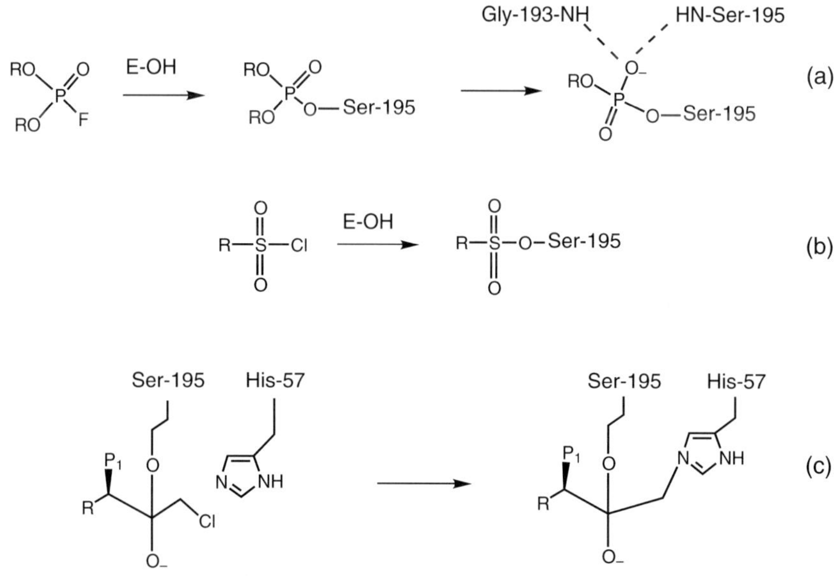

Figure 7.4 Mechanism of inhibition of serine proteases by (a) organophosphorus; (b) organosulfur; and (c) chloromethyl ketone inhibitors of serine proteases.

Some of the earliest inhibitors of serine proteases were organophosphorus and organosulfur based electrophiles (Figures 7.3 and 7.4, Lively and Powers 1978; Yoshimura *et al* 1982; Lamden and Bartlett 1983; Powers and Harper 1986; Oleksyszyn and Powers 1991). Simple dialkylfluorophosphonates and sulfonyl fluorides lack specificity and inhibit all serine proteases. However, they do not inhibit other classes of enzymes. Indeed, inhibition of an enzyme by either diisopropylfluorophosphonate (Figure 7.3a) or *p*-methylphenylsulfonyl fluoride (Figure 7.3b) serves as a diagnostic test to identify an enzyme as being a serine protease (Jansen *et al* 1949; Gold and Farney 1964). The X-ray crystal structures of the irreversible adducts between sulfonyl fluorides and a number of serine proteases, including chymotrypsin (Matthews *et al* 1967), subtilisin (Wright *et al* 1969), and PPE (Shotton and Watson 1970), have contributed greatly to our understanding of enzyme structure and the mechanism of catalysis. Fluorophosphonates tend to be very reactive, but hydrolytically unstable, whereas sulfonyl fluorides are less reactive and, consequently, more stable towards aqueous hydrolysis. However, by attaching appropriate groups to the sulfur or phosphorus atoms that will bind to the S_1 and subsequent binding pockets, both enzyme selectivity and hydrolytic stability can be imparted to this group of inhibitors. Moreover, occupation of the extended binding pocket activates the catalytic machinery and increases enzyme reactivity without increasing aqueous hydrolysis.

Peptidyl chloromethyl ketones (CMKs) have contributed greatly to our understanding of elastase-ligand interactions, especially subsite specificity and the effect of peptide length on the inhibition of HNE (Figure 7.3c, Powers 1977b). Upon binding to the enzyme, CMKs form a covalent, reversible hemi-ketal adduct between the ketone carbonyl carbon atom and the oxygen atom of the hydroxyl group of Ser-195. However, subsequent nucleophilic displacement of the α-chloro group by the nitrogen of the imidazole ring of His-57 results in irreversible alkylation of His-57 (Figure 7.4). In addition to chloromethyl ketones, iodo-, bromo- and fluoromethyl ketones have been studied. Although iodo- and bromomethyl ketones are more reactive inhibitors of serine proteases, they have not been evaluated extensively since their high reactivity renders them nonspecific alkylating agents. Fluoromethyl ketones, on the other hand, have been developed as effective inhibitors of a number of thiol and serine proteases (Rasnick 1985; Imperiali and Abeles 1986; Angliker *et al* 1988). Since they are less reactive than the corresponding chloromethyl ketones, they may offer therapeutic advantage in that nonspecific reactions leading to side effects might be minimized. Fluoromethyl ketones are more reactive towards thiol than serine proteases (Shaw *et al* 1986) and, therefore, most studies on this class of inhibitors have been conducted with thiol proteases.

7.5.2 Enzyme-activated heterocyclic inhibitors

A class of serine protease inhibitors that has been studied intensively over the years is enzyme-activated heterocyclic inhibitors. Also termed mechanism-based inhibitors, these compounds function similarly to halomethyl ketones, in that they form covalent linkages to both the Ser-195 and His-57. Unlike halomethyl ketones, however, a second, reactive functionality is not present in the parent compound. Rather, initial covalent bond formation between an electrophilic center in the heterocycle and a catalytic-site residue initiates a bond-breaking reaction within the heterocyclic ring which thereby unmasks an even more reactive functionality. The newly generated reactive species then forms a second, covalent bond with another catalytic residue (Figure 7.5a). This latter covalent bond is usually very stable, and results in the irreversible inhibition which is characteristic of this class of inhibitors. Because the action of the enzyme triggers the release of the latent functional group which leads to the enzyme's own irreversible inhibition, these compounds are often referred to as "suicide inactivators".

Depending upon the effectiveness of the second covalent bond-forming reaction, the initial mono-covalent adduct may be hydrolyzed affording active enzyme and hydrolyzed,

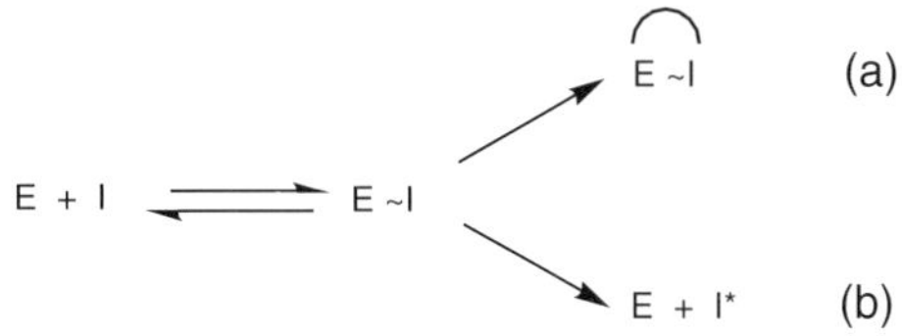

Figure 7.5 Generalized mechanism of inhibition of serine proteases by heterocyclic inhibitors: (a) pathway leading to irreversible inhibition; (b) pathway leading to reactivated enzyme and hydrolyzed inhibitor.

inactive inhibitor (Figure 7.5b). The ratio of inhibitor molecules which are hydrolyzed to those which irreversibly inactivate the enzyme has been used as a measure of the effectiveness of enzyme-activated inhibitors. Since these inhibitors require a catalytic trigger as well as a second, appropriately positioned active-site group, very good enzyme selectivity can be obtained, especially among different classes of enzymes. On the other hand, within a particular enzyme class such as serine proteases, adequate selectivity is often difficult to achieve for heterocyclic inhibitors since such compounds generally only interact with the catalytic residues and the S_1 subsite, which may not exert maximum discrimination unless more remote subsites are occupied.

Table 7.1 lists representative examples of some of the more actively investigated enzyme-activated serine protease inhibitors along with the proposed structure of the final enzyme-inhibitor complex. One of the earliest enzyme activated inhibitors were the halomethyl dihydrocoumarins (Bechét *et al* 1977). While at first glance, this inhibitor appears to be a variation of a chloromethyl ketone, the intermediate acylated chloromethyl phenyl moiety does not alkylate the active-site His-57. Rather, alkylation occurs via a reactive quinone methide generated by elimination of chloride (Figure 7.6).

The first enzyme-activated inhibitors of serine proteases which were eventually shown to be selective inhibitors of HNE were the halo enol lactones (**7.1** (Table 7.1), Katzenellenbogen *et al* 1992). Initial nucleophilic attack by Ser-195 forms an acyl-enzyme and opens the lactone ring to form an acyl halo enol ketone which tautomerizes to a halomethyl ketone. The halo ketone then alkylates His-57. A possible alternative mechanism for irreversible inhibition is the hydrolysis of the acyl halomethyl ketone to the unbound halomethyl ketoacid, which could then alkylate His-57 in an intermolecular fashion. Inhibitors operating by such a mechanism have been termed pseudo-suicide or paracatalytic inhibitors (Walsh 1982). This distinction is of more importance than just passing mechanistic interest, since, if the reactive intermediate escapes from the active site, it may function as a nonspecific alkylating agent, limiting the inhibitor's usefulness as a drug. A similar class of inhibitors are the ynenol lactones (**7.2**, Copp *et al* 1987). Following enzyme acylation and ring opening, the initially formed ynenolate rearranges to an allenone which is irreversibly captured by His-57.

One of the most thoroughly investigated heterocyclic inhibitors of serine proteases are the isocoumarins developed by Powers (**7.3**, Harper and Powers 1985). While isocoumarin itself is not an inhibitor, introduction of a chlorine atom into the 3-position affords inhibitors of a number of serine proteases, including HNE. Initial attachment to Ser-195 forms an acyl-enzyme and opens the ring to unmask a latent acid chloride which acylates His-57. Although for all intents and purposes the enzyme is irreversibly inhibited, this

Table 7.1 Heterocyclic inhibitors of HNE

Compound number	*Parent inhibitor*	*Reversibly bound inhibitor*	*Irreversibly bound inhibitor*
7.1	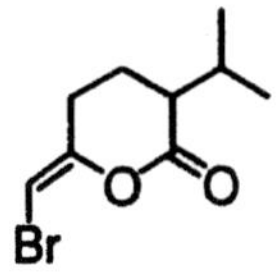	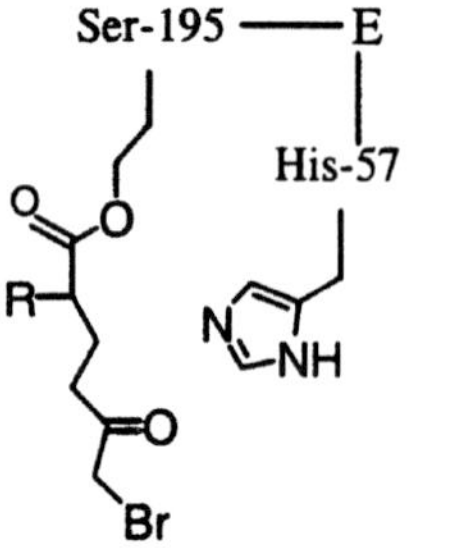	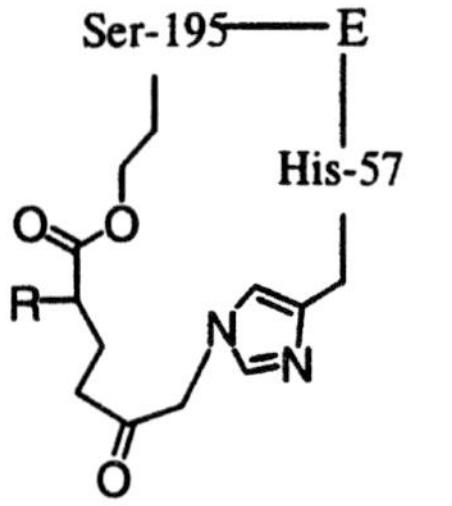
7.2	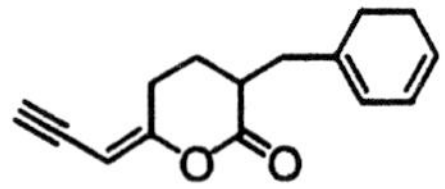	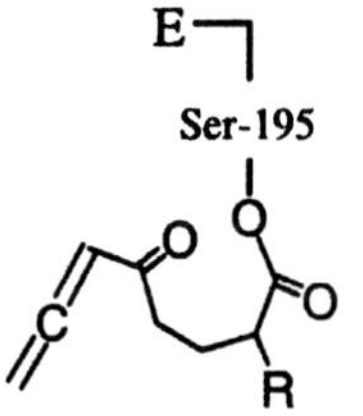	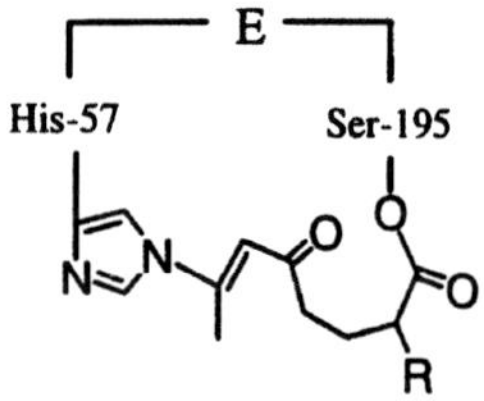
7.3	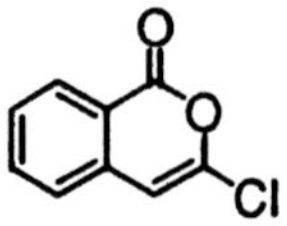	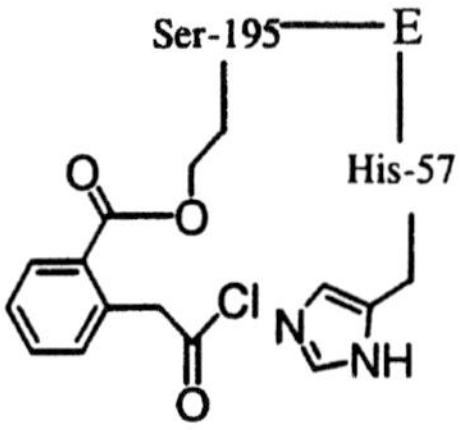	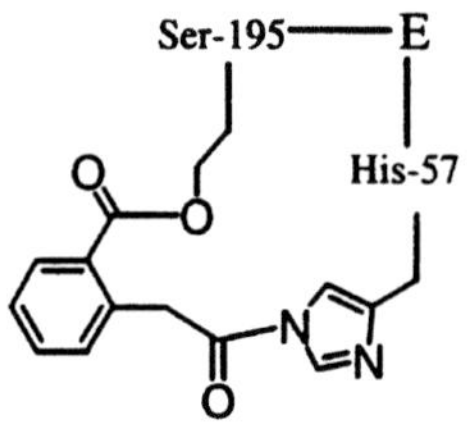
7.4	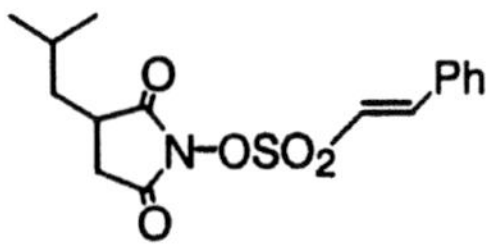	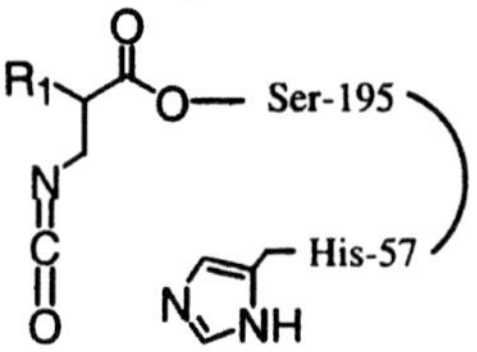	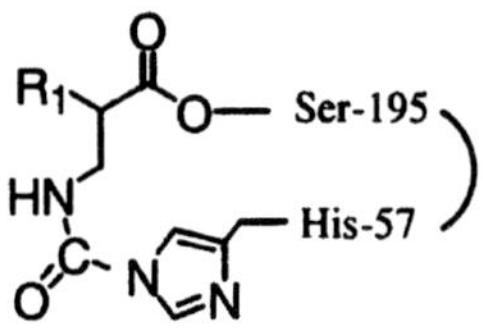
7.5	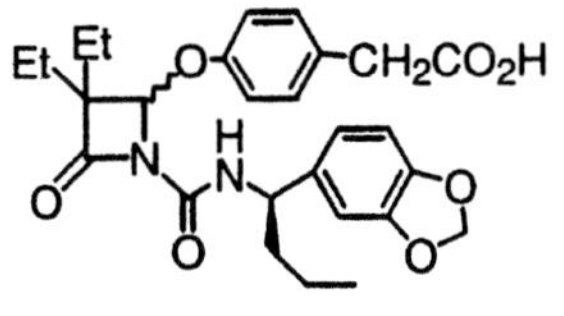	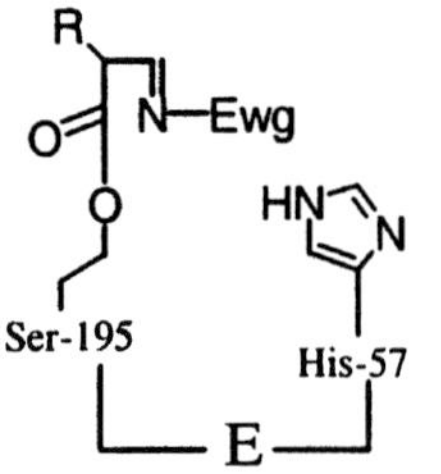	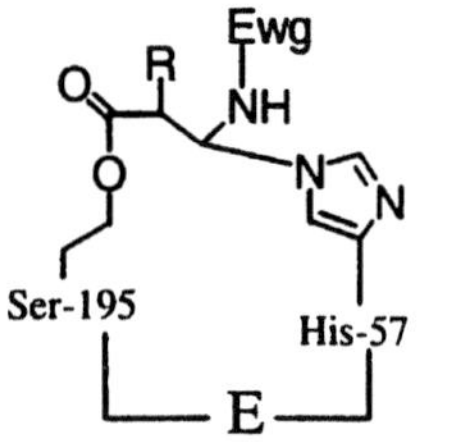

Note: Ewg = Electron withdrawing group.

is not strictly the case. The inhibitor is bound to the enzyme via two labile linkages: an ester to Ser-195 and an amide to His-57. Consequently, enzyme activity may be slowly regained upon standing, and more rapidly upon treatment with a strong nucleophile such as hydroxylamine. Introduction of a second chlorine atom into the 4-position of

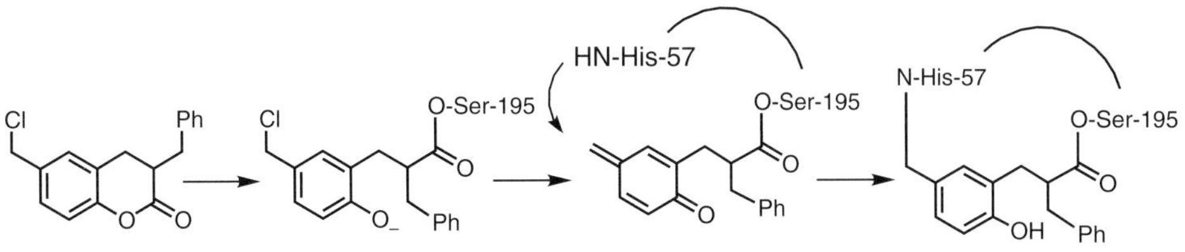

Figure 7.6 Mechanism of inhibition of serine proteases by halomethyl dihydrocoumarins.

3-chloroisocoumarin improves activity by increasing the electrophilicity of the carbonyl group. Replacement of the 3-chloro substituent with an alkoxy group also affords potent inhibitors, although these compounds are incapable of forming a doubly bound intermediate since opening of the lactone ring forms an unreactive ester rather than an acid chloride. However, introduction of a 7-amino substituent into the 4-chloro-3-alkyoxyisocoumarins affords inhibitors that irreversibly inactivated HNE. The proposed mechanism involves the intermediacy of a reactive acyl quinone imine methide which is irreversibly alkylated by His-57, similar to the halomethyl dihydrocoumarins.

A series of inhibitors that demonstrates the powerful masking effect that enzyme-activated inhibitors are capable of are the *N*-sulfonyloxy succinimides (**7.4**, Martyn *et al* 1999). Reaction with Ser-195 forms an acyl-enzyme with concomitant opening of the succinimide ring. The liberated *N*-sulfonyloxy carboxamide undergoes a spontaneous Lössen rearrangement to generate a reactive isocyanate tethered within the catalytic site. Irreversible inhibition results from trapping of the isocyanate by His-57. Isocyanates themselves are poor inhibitors of serine proteases as a consequence of rapid aqueous hydrolysis, and significant inhibition is only observed when the isocyanate is present in very large excess. It is the generation of the isocyanate within the catalytic site that affords it sufficient life-time to alkylate a catalytic-site residue.

The class of enzyme-activated inhibitors that have demonstrated the most potential for development as drugs are the β-lactams pioneered by Merck (**7.5**, Doherty *et al* 1986). The β-lactams are unique in that they are the only class of HNE inhibitors to have been used clinically for disease modification, albeit not as yet for chronic diseases such as those associated with elastase. The mechanism of inhibition of HNE by β-lactams is analogous to that of other heterocyclic enzyme-activated inhibitors. Initial attack of Ser-195 on the lactam carbonyl affords an enzyme-inhibitor tetrahedral adduct. Collapse of the tetrahedral intermediate to generate an acyl-enzyme results in ring opening and elimination of the leaving group in the 4-position forming an enzyme tethered reactive imine. Alkylation of the imine results in irreversible inhibition.

A feature that distinguishes β-lactam inhibitors from other enzyme-activated inhibitors is the breadth of the structure-activity relationships investigated in an attempt to design inhibitors with adequate stability (Hagmann *et al* 1993). In general, enzyme-activated inhibitors suffer from poor aqueous stability as a result of their inherent reactivity. The early cephalosporin esters were hydrolytically unstable and rapidly hydrolyzed in pH 8 buffer to the corresponding inactive C-2 acids. This short half-life is in contrast to that of known cephalosporin antibiotics. Improved stability is afforded by C-2 amides. However, the amides still lacked systemic activity in *in vivo* models as a result of poor stability in blood. This difficulty was overcome by the development of the monocyclic azetidinones which inhibit HNE in a manner similar to the cephalosporins. Members of this class were not only potent *in vitro* inhibitors of HNE, but have also been demonstrated to be orally active in animal models and have proven to be leading candidates for clinical treatment of elastase associated diseases.

7.5.3 Acylating agents

As with enzyme-activated inhibitors, acylating agents form a covalent acyl-enzyme with the active-site Ser-195. However, they lack a second reactive functionality to interact with a second active-site residue, and, therefore, they do not irreversibly inactive the enzyme. Enzyme activity is regained following hydrolysis of the acyl-enzyme. However, acylating agents are not "reversible" inhibitors, in that generally the parent inhibitor is not regenerated but rather it is turned over as a substrate (Figure 7.5b). Indeed, this type of inhibitor is also referred to as a pseudo-substrate. In some cases, however, deacylation can occur via reformation of the heterocyclic ring regenerating the original substrate. Even in such examples, turnover as a substrate is a significant, competing reaction and these inhibitors are thus not truly reversible inhibitors. The relative activity of acylating agents can be evaluated by comparison of a pseudo K_i, K_i^*, which is equal to the rate constant for deacylation (k_{off}) divided by the rate constant for acylation (k_{on}). However, this is not a true K_i, because the inhibitor is not in equilibrium with the inactivated inhibitor–enzyme complex since it is turned over as a substrate. K_i^* is a valuable measure of the relative potency of compounds operating by the same mechanism. Figure 7.7 lists representative examples of acylating agents that have been developed.

The need to balance the reactivity of the parent heterocycle with the stability of the acyl-enzyme has plagued the development of heterocyclic acylating agents: those inhibitors with high reactivity towards the enzyme tend to suffer from limiting hydrolytic instability. Tactics to balance these competing requirements of heterocyclic acylating agents have been most thoroughly applied to benzoxazinone inhibitors of HNE (Krantz *et al* 1990). Simple 2-substituted benzoxazinones such as Figure 7.7b were only weak inhibitors of HNE, in part due to a rapid K_{off}. It was hypothesized that addition of a substituent to the 5-position (e.g. Figure 7.7c) would decrease the rate of deacylation since the acyl-enzyme would be a 2,6-disubstituted benzoate in which both faces of the carbonyl group are hindered towards nucleophilic attack by virtue of the fact that the carbonyl would lie orthogonal to the plane of the ring. However, the rate of acylation was not expected to decrease since the benzoxazinone ring is planar and the 5-substituent should have little effect on nucleophilic attack at the carbonyl group. This tactic has proven to be quite effective, affording as much as a 100-fold increase in activity.

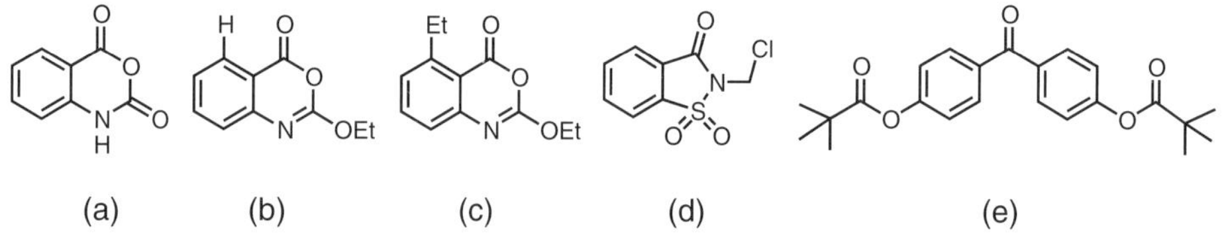

Figure 7.7 Examples of acylating agent inhibitors of HNE: (a) isatoic anhydride; (b) unsubstituted benzoxazinone; (c) 5-substituted benzoxazinone; (d) benzisothiazolinone; (e) aromatic pivalate.

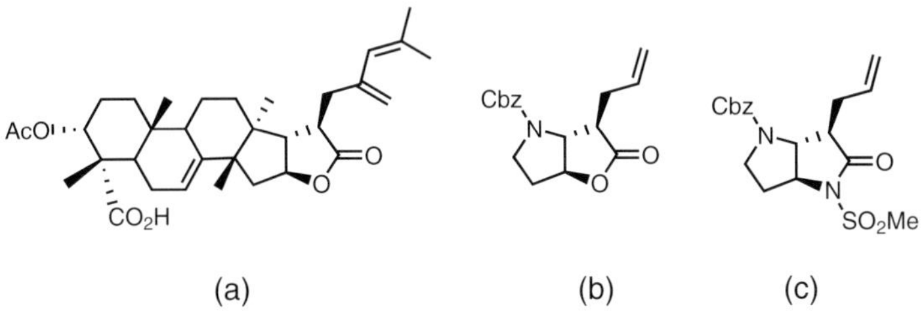

Figure 7.8 Novel 5,5-*trans* lactone and lactam inhibitors of HNE.

Saccharin-based inhibitors of HNE (e.g. Figure 7.7d) have been extensively explored by Groutas *et al* (1996 and 1998). These inhibitors form an acyl-enzyme via attack of Ser-195 on the carbonyl group resulting in ring opening and formation of a sulfonamide. Groutas hypothesized that simple saccharins could be converted into enzyme-activated inhibitors by attachment of a leaving group to the nitrogen atom (Groutas *et al* 1992). Following acyl-enzyme formation, this leaving group would be alkylated by His-57 to form an irreversible enzyme-inhibitor adduct. This strategy proved very successful, and several types of saccharine-based enzyme-activated inhibitors of HNE have been reported (Hlasta *et al* 1996; Martyn *et al* 1999).

The acylating agents that have achieved the most clinical success are the aromatic pivalates (e.g. Figure 7.7e, Kirschenheuter *et al* 1993; Tomizawa *et al* 1999). The structure activity studies are consistent with the formation of a pivaloyl ester bond with Ser-195 and binding of the *tert*-butyl group in the S_1 pocket. An inhibitor from this class of compounds developed by Ono has reached phase III clinical trials in Japan.

Recently, GlaxoWellcome has described a novel class of lactone- and lactam-based inhibitors that are the first new pharmacophores to be developed for HNE in the last 5 years (Macdonald *et al* 1998; Patent Review 1999). High-throughput screening identified the triterpene (Figure 7.8a) as a potent inhibitor of thrombin and a weak inhibitor of HNE. An X-ray crystal structure of the triterpene bound to thrombin revealed that inhibition resulted from ring opening of the lactone and formation of an acyl-enzyme adduct. Focusing on the rare 5,5-*trans* lactone motif, a series of potent low-molecular-weight inhibitors of HNE were developed (e.g. Figure 7.8b; *trans* refers to stereochemistry at the ring fusion). Consistent with the mechanism of inhibition, lactones were significantly more potent than the corresponding lactams. Excellent potency against HNE was obtained in the hydrolytically more stable lactam series by attaching an electron withdrawing group to the lactam nitrogen atom (Figure 7.8c).

7.6 REVERSIBLE INHIBITORS OF HNE

There are two types of reversible inhibitors: those that form a covalent adduct with the enzyme and those that derive all of their binding energies from non-covalent interactions. Elastase-like serine proteases differ from the other major classes of serine proteases (i.e. trypsin-like and chymotrypsin-like proteases) in that they possess a relatively small S_1 binding pocket that only accepts small, hydrophobic P_1 residues such as alanine and valine. Consequently, there has been little success in the development of small, synthetic, reversible inhibitors of HNE that do not form a covalent linkage to the enzyme. While many such inhibitors have been designed and have contributed greatly to our understanding of enzyme-ligand interactions, (Edwards and Bernstein 1994) they have lacked sufficient *in vitro* potency to be considered seriously for clinical development. Indeed, no potent ($K_i < 10^{-6}$ M), competitive, low-molecular-weight reversible inhibitor of HNE has been reported which does not form a covalent adduct with the enzyme. In contrast, the S_1 pocket of trypsin-like enzymes is deep and contains an aspartate residue which can form a salt-bridge with a basic P_1 residue such as arginine. The binding energy that this ionic interaction imparts has allowed successful development of potent, non-covalent inhibitors of trypsin-like enzymes, most notably thrombin (Sanderson and Naylor-Olsen 1998a) and mast-cell tryptase (Rice *et al* 1998).

7.6.1 Electrophilic carbonyl derivatives

With only one exception, potent, reversible inhibitors of HNE have all been ketone- or aldehyde-based compounds (Table 7.2). These inhibitors owe their inhibition to the formation of a transition-state-like hemiketal adduct between the inhibitor carbonyl carbon atom and the hydroxyl group of Ser-195 (Figure 7.9a). The degree of inhibition tends to be correlated with the electrophilicity of the carbonyl group. Simple methyl ketones (**7.6**; Table 7.2) are very weak inhibitors of HNE, and probably do not even form a covalent adduct. In contrast, very electrophilic ketones, such as trifluoromethyl ketones (TFMKs, **7.8**) are extremely potent inhibitors. For this reason, these compounds are also known as electrophilic ketone inhibitors. It has been hypothesized that the high potency of the TFMKs results from the ability of the trifluoromethyl group to lower the pK_a of the hemiketal hydroxyl group such that it is ionized in the active site (Brady *et al* 1989). This anionic oxygen atom can thereby take full advantage of the binding opportunities available in the oxyanion hole.

All reversible inhibitors of HNE are peptide- or peptidomimetic-based compounds. A large number of peptidyl electrophilic ketones have been investigated, including aldehydes (**7.7**; Hassall *et al* 1985), trifluoromethyl ketones (**7.8**; Veale *et al* 1997), pentafluoroethyl ketones (**7.9**; Cregge *et al* 1998), difluoromethylene ketones (**7.10**, **7.11**, **7.12**; Stein *et al* 1989; Govardhan and Abeles 1990), α-ketoheterocycles (**7.13**; Edwards *et al* 1995a/b), α-ketoesters (**7.14**; Peet *et al* 1990; Burkhart *et al* 1998), α-ketoamides (**7.15**, Stein *et al* 1989), and α-diketones (**7.16**; Mehdi *et al* 1990; Stein *et al* 1989, 1990). All of these compounds afford potent inhibitors of HNE. The two most intensively studied classes of inhibitors have been the fluorinated ketones and α-ketoheterocycles. To circumvent the problems generally associated with oral absorption of peptides, the group at Marion-Merrell Dow converted the ketone carbonyl group of pentafluoroethyl ketones to an enol acetate, a tactic that successfully improved oral activity in animals (Burkhart *et al* 1995). Researchers at Zeneca pursued a different tactic, and developed a pyridone-based dipeptide mimetic to replace the P_2–P_3 amino acid residues in peptidyl TFMKs (Figure 7.10a; Andisik *et al* 1997). This dipeptide mimetic was designed using the crystal structure of TFMKs bound to PPE. With a phenyl substituent in the 6-position of the pyridone ring (Figure 7.10b), these non-peptidic TFMKs are potent inhibitors of HNE. The pyridone motif has evolved into a general dipeptide isostere and has found wide spread use in a number of electrophilic ketones, including inhibitors of thrombin (Tamura *et al* 1997a;

Table 7.2 Covalent, reversible inhibitors of HNE

Compound number	*Inhibitor*	K_i (nM) *HNE*
7.6	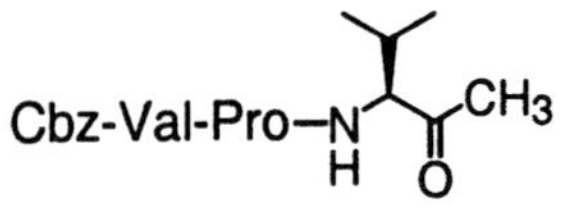	8000
7.7	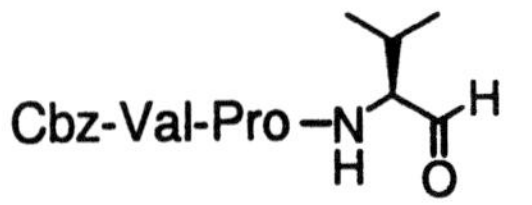	41
7.8	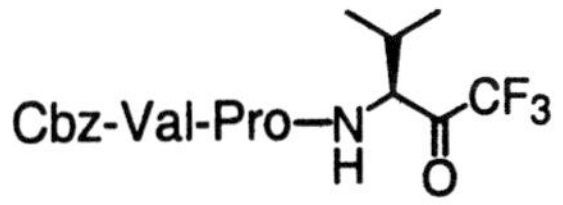	0.8
7.9	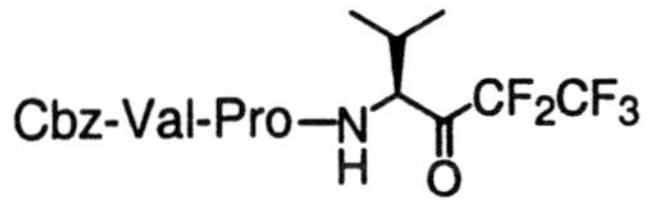	3.0
7.10	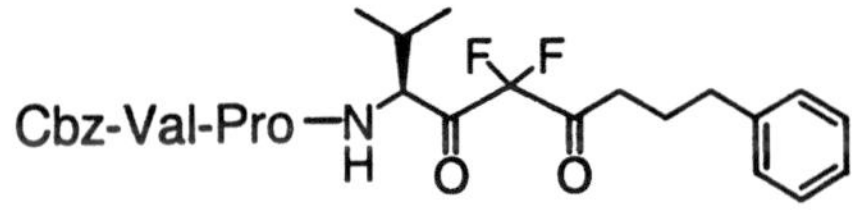	0.23
7.11	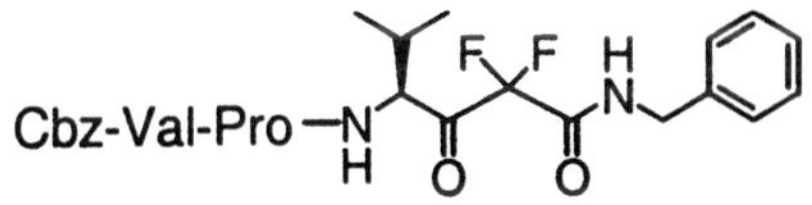	0.4
7.12	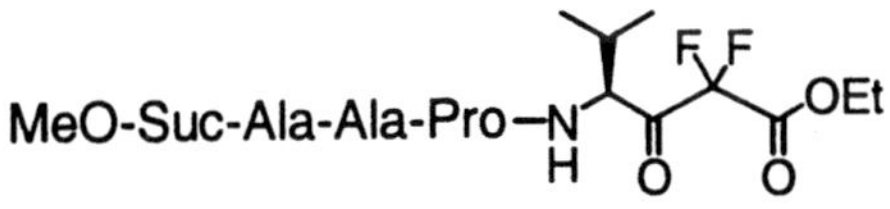	4300
7.13	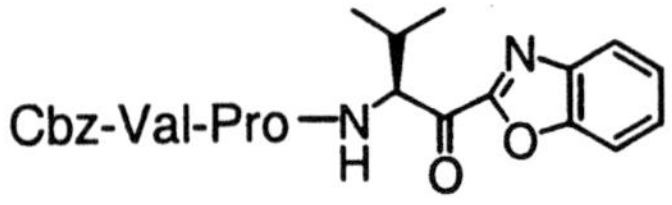	3.0
7.14	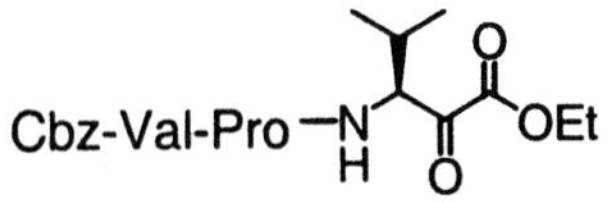	0.6
7.15	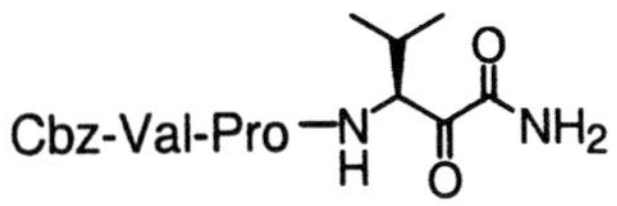	1.8
7.16	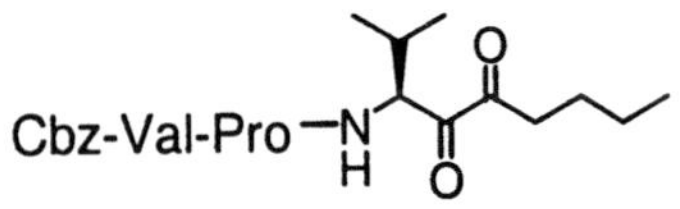	1.6
7.17		0.57

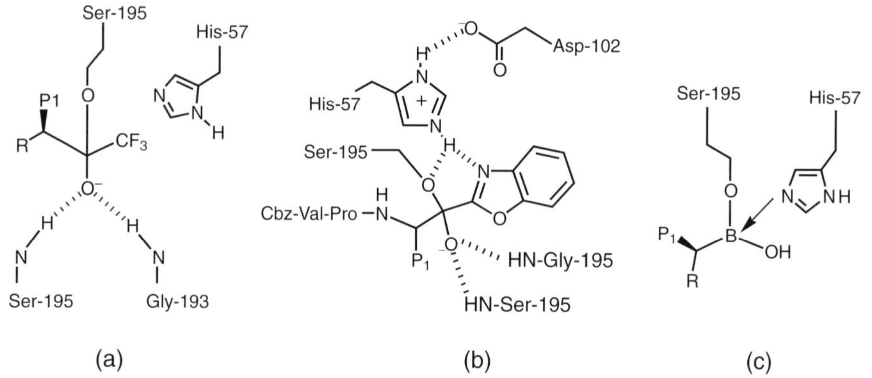

Figure 7.9 Mechanism of inhibition of reversible HNE inhibitors: (a) trifluoromethyl ketones; (b) α-ketoheterocycles; (c) boronic acids.

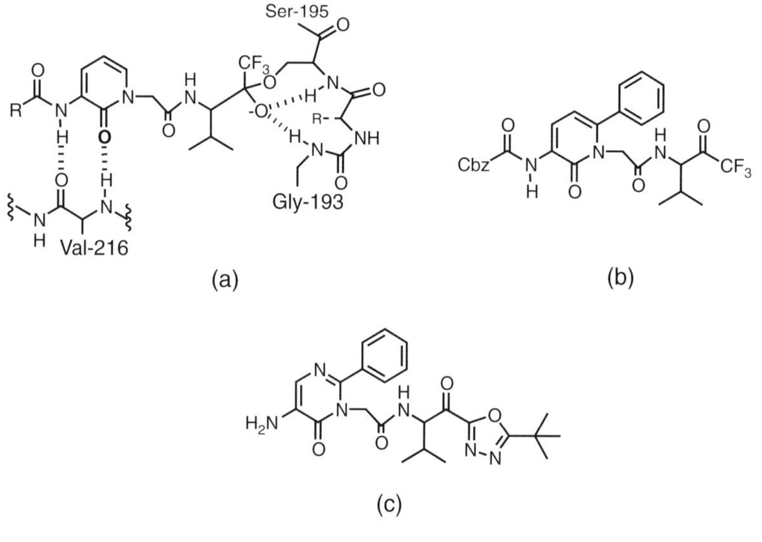

Figure 7.10 Non-peptidic, pyridone-based electrophilic ketone inhibitors of HNE: (a) binding interactions between pyridone inhibitor and HNE; (b,c) 6-phenyl pyridone-based HNE inhibitors.

Sanderson *et al* 1998b) and interleukin converting enzyme (Dolle *et al* 1997; Golec *et al* 1997; Semple *et al* 1998).

The design of the peptidyl α-ketoheterocycles was based on the concept that certain heterocycles would be sufficiently electron withdrawing to activate a ketone carbonyl towards nucleophilic addition by Ser-195, and that an appropriately positioned hydrogen bond acceptor could participate in a hydrogen-bonding interaction with the protonated imidazole ring of His-57 (Figure 7.9b). Both kinetic evidence and an X-ray crystal structure of a peptidyl α-ketobenzoxazole bound to PPE confirm this binding mode. The peptidyl α-ketoheterocycles were the first covalent, reversible inhibitors of HNE that were shown experimentally to form binding interactions with two catalytic residues of serine proteases (Edwards *et al* 1992; Odagaki *et al* 2001). The versatility of this class of inhibitors has been demonstrated by the fact that they have been developed as inhibitors of a wide variety of proteases including thrombin (Costanzo *et al* 1996), factor IIa, factor Xa and Plasmin (Tamura *et al* 1997b), prolyl endopeptidase (Tsutsumi *et al* 1994), human cytomegalovirus protease (Ogilvie *et al* 1997), human neutrophil proteinase 3 (Wieczorek *et al* 1999), chymase (Akahoshi *et al* 2001) and the cysteine proteases calpain (Tao *et al* 1996), interleukin converting enzyme (Batchelor *et al* 1997) and cathepsin K (Yamashita *et al* 1999). Recently, an α-ketooxadiazole has been combined with the pyridone motiff to afford orally active, non-peptidic inhibitors of elastase (Figure 7.5c, Ohmoto *et al* 2000; Ohmoto *et al* 2001).

7.6.2 Boronic acids

The only reversible inhibitors affording potent inhibition of HNE that are not electrophilic ketones are the peptidyl boronic acids (Kettner and Shenvi 1984). Those boronic acids that contain a P_1 group complementary to the S_1 binding pocket form a transition-state-like reversible, tetrahedral adduct with the active-site Ser-195. On the other hand, certain boronic acids undergo elimination of a water molecule and formation of a dative complex with the active-site His-57 (Figure 7.9c). These compounds are very tight binding and afford potent inhibitors of HNE (**7.17**). The *in vitro* potency of boronic esters initially appears to be less than that of the corresponding acids. However, the esters are hydrolytically unstable, and after pre-incubation in assay buffer their activity is the same as the boronic acids. Thus a boronic ester may serve as a prodrug and improve oral bioavailability of this class of HNE inhibitors.

7.7 REVERSIBLE VERSUS IRREVERSIBLE HNE INHIBITORS

For an enzyme inhibitor to succeed as a therapeutic agent, it must possess adequate *in vitro* potency, have acceptable bioavailability via the desired route of administration, possess hydrolytic and metabolic stability, be selective for the target enzyme, and be devoid of toxic effects. Any compound that satisfies these criteria, irrespective of its mechanism of inhibition, can be developed as a drug. However, certain properties of a class of inhibitor can make it more or less attractive as a starting point for a research program. It is our belief that, *a priori*, reversible inhibitors, rather than irreversible inhibitors, offer the best chance of success.

An irreversibly inhibited enzyme may elicit an immunogenic response. When an enzyme is processed, it is degraded to peptide fragments that are presented to the body's immune system. For an irreversibly inhibited enzyme, one of these fragments will contain the inhibitor, and this inhibitor-bound fragment may be recognized as foreign and initiate an immunological response. This is not a possibility with reversible inhibitors, even tight-binding inhibitors, since they are released from the enzyme once it is degraded.

However, the most significant challenge in developing irreversible inhibitors lies in finding the appropriate balance between inhibitor reactivity, enzyme selectivity and stability of the enzyme-inhibitor adduct. While it is desirable for the inhibitor to be reactive so as to facilitate rapid acylation of Ser-195, increased reactivity usually goes hand-in-hand with decreased hydrolytic stability and decreased enzyme selectivity. In addition, for heterocyclic acylating agents, factors which increase the acylation rate of the inhibitor tend to increase the deacylation rate and thus decrease the stability of the acyl-enzyme. It is against these conflicting constraints that the designers of heterocyclic inhibitors have to operate.

Perhaps the most compelling argument against the development of irreversible inhibitors is that even if several orders of magnitude in enzyme selectivity can be demonstrated *in vitro*, chronic dosing may still result in significant inhibition of an undesired enzyme *in vivo*. This is due to the fact that irreversible inhibition is cumulative. Although an enzyme may be only slowly inhibited, if it is subjected to sufficient concentrations of the inhibitor for a sufficient period of time, it will eventually be completely inhibited. This effect does not occur with reversible inhibitors. Thus with irreversible inhibitors, toxic effects due to lack of enzyme selectivity are more difficult to predict and will not become apparent until *in vivo* studies are conducted late in a drug development program.

It is our belief, therefore, that the best starting place for the rational design of clinically useful HNE inhibitors is reversible, electrophilic carbonyl derivatives such as perfluoro ketones and α-ketoheterocycles. These compounds are generally potent, hydrolytically stable, and selective, not suffering from the problems associated with irreversible inhibition following chronic dosing discussed above. The only potential generic problem that has been associated with electrophilic ketones is a propensity for *in vivo* reduction of the carbonyl group. This effect will vary with the reduction potential of the carbonyl, which in most cases will parallel the electrophilicity of the carbonyl carbon atom.

However, one class of electrophilic ketones, the α-ketoheterocycles, affords the opportunity to fine tune the carbonyl electrophilicity via substitution on the heterocyclic ring. Structure–activity studies have demonstrated that substituted peptidyl α-ketobenzoxazoles possessing a range of electron withdrawing and electron donating ring substituents vary little in their *in vitro* potency (Edwards *et al* 1995b). While increasing the electron donation by the ring substituent undoubtedly decreases the electrophilicity of the carbonyl group, and should therefore decrease potency, the increased electron donation also increases the hydrogen bond accepting potential of the ring nitrogen atom (Figure 7.9b). These two properties therefore balance each other and *in vitro* potency is largely unaffected by ring substitution. In addition, a variety of heterocycles with varying electronic properties can be used to activate the ketone carbonyl while still affording potent inhibitors (Edwards *et al* 1995a). Therefore, it should be possible to control any propensity for *in vivo* carbonyl reduction of α-ketoheterocycles by appropriate selection of the ring substituents. Furthermore, the wide variety of heterocycles that have been incorporated into α-ketoheterocyclic inhibitors should also allow modulation of the physiochemical properties and afford compounds with suitable drug-like properties. Thus, it is argued that α-ketoheterocycles offer an ideal starting point for the *de novo*, rational design of serine and cysteine protease inhibitors.

7.8 CLINICAL STATUS OF HNE INHIBITORS

Despite the vast amount of effort that has been applied to the development of elastase inhibitors, no small molecule HNE inhibitors have reached the market. The only marketed elastase inhibitor is Prolastin, a purified form of α_1PI. This natural, proteinaceous elastase inhibitor is used to treat emphysema in individuals genetically deficient in α_1PI. This surprising situation is the result of pharmaceutical health economics, not due to lack of safety or efficacy of small molecule inhibitors of HNE. Long after many companies initiated research programs into the discovery of HNE inhibitors, it became apparent that demonstration of clinical efficacy against emphysema would require expensive, protracted clinical trials since emphysema progresses slowly, with the measures of clinical efficacy being a decreased worsening of lung function in drug-treated patients relative to controls. This fact is compounded by the lack of surrogate biochemical markers or endpoints, which might allow demonstration of decreased lung destruction earlier than actual clinical efficacy. This is often termed "biochemical efficacy". Furthermore, the relative contribution of HNE in emphysema has recently been questioned. There is strong evidence that other proteases such as metallo macrophage elastase (MME) may also play a critical role in the development and progression of emphysema (Pardo and Selman 1999).

As a result, many pharmaceutical companies are pursuing a strategy of initially developing HNE inhibitors to treat acute diseases requiring shorter clinical development. If these inhibitors can be demonstrated to be efficacious in such acute indications, then the probability of an HNE inhibitor successfully treating emphysema will be increased, and thereby balance the financial risks associated with a large and lengthy clinical program. Over the past decade, HNE has been implicated in a number of acute indications including cystic fibrosis, chronic bronchitis, ischemia reperfusion injury and adult (or acute) respiratory distress syndrome (ARDS), and a number of small molecule HNE inhibitors are currently in various stages of clinical development to treat these diseases (Figure 7.11; Anderson and Shinagawa 1999; Adis 2001; IMSworld 2001; Pharmaprojects 2001).

The most advanced is **ONO-5046 (silvistat)**, a pivalate ester acylating agent (Tomizawa *et al* 1999). This compound is in the pre-registration stage in Japan for the treatment of ARDS and septic shock. A second acylating agent, **CE-1037 (MDL-201404**, Kirschenheuter *et al* 1993), has been developed by Cortech and is reported to be in Phase II

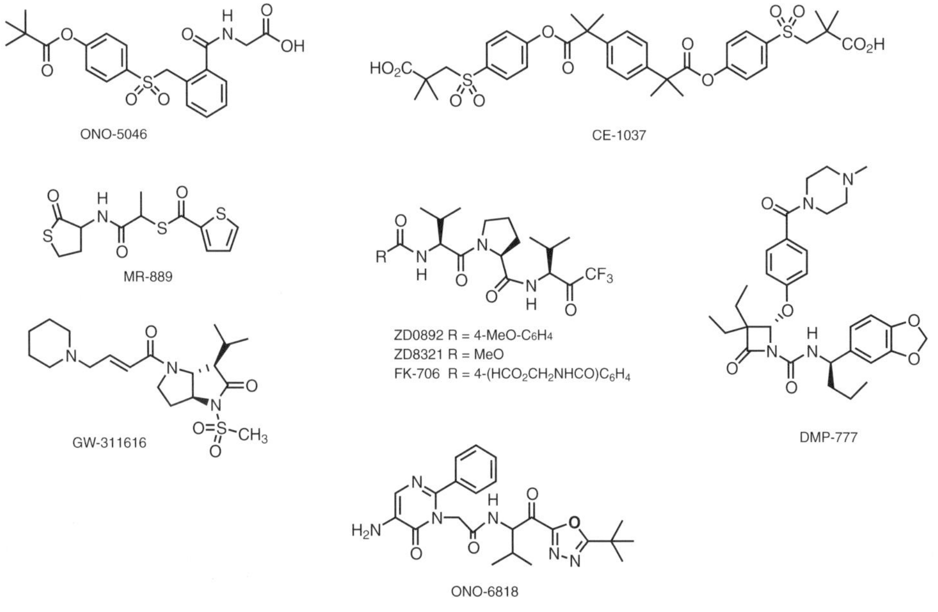

Figure 7.11 Small molecule inhibitors of HNE reported to be in clinical trials.

studies for ARDS and cystic fibrosis. **MR-889 (medesteine**, Luisetti *et al* 1996) is a weak inhibitor of HNE ($K_i = 1.4\,\mu M$) being developed by Medea Research that has shown interesting activity in human trials for chronic pulmonary obstructive disease (COPD) and is currently in Phase III trials in Italy for emphysema. AstraZeneca has two compounds in development. **ZD8321** is in Phase II trials for ARDS and sepsis-induced systemic inflammatory response syndrome, while **ZD0892** is in Phase I trials for COPD and peripheral vascular disease. Both of these compounds are peptidyl trifluoromethyl ketones (Veale *et al* 1997). A third peptidyl trifluoromethyl ketone related to AstraZeneca's TFMKs, **FK706** (Shinguh *et al* 1997), is being developed by Fujisawa for various respiratory disorders and is reported to be in phase II studies. **DMP-777 (L-694,458)** is a β-lactam in Phase II trials for cystic fibrosis, juvenile rheumatoic arthritis and emphysema. This compound was discovered by Merck (Vincent *et al* 1997), and was obtained by DuPont Pharmaceuticals after the dissolution of the joint venture Dupont-Merck Pharmaceuticals. GlaxoWelcome has progressed their new series of 5,5-*trans* lactam-based HNE inhibitors to the clinic, and one of this class of compounds, **GW-311616**, is in Phase I trials for chronic obstructive pulmonary disease. Ono's orally active, pyridone-based α-ketoheterocycle **ONO-6818** is reported to be in Phase I trials for COPD and emphysema. With the number of human neutrophil elastase inhibitors currently progressing through the clinic, it is hoped that therapeutic agents to treat the diseases associated with HNE will finally reach the market place in the near future.

REFERENCES

Aboussouan, L.S. and Stoller, J.K. (1999) New developments in alpha 1-antitrypsin deficiency. *Seminars in Respiratory and Critical Care Medicine*, **20**, 301–310.

Adis R&D Insight. Chester (UK): Adis International Ltd. [Cited April 2001].

Akahoshi, F., Ashimori, A., Sakashita, H., Yoshimura, T., Imada, T., Nakajima, M. *et al* (2001) Synthesis, structure–activity relationships, and pharmacokinetic profiles of nonpeptidic α-ketoheterocycles as novel inhibitors of human chymase. *Journal of Medicinal Chemistry*, **44**, 1286–1296.

Anderson, G.P. and Shinagawa, K. (1999) Neutrophil elastase inhibitors as treatments for emphysema and chronic bronchitis. *Current Opinion in Anti-inflammatory & Immunomodulatory Investigational Drugs*, **1**, 29–38.

Andisik, D., Bernstein, P., Brown, F., Bryant, C., Damewood, J., Edwards, P.D. *et al* (1997) Computer-aided design of novel inhibitors of human leukocyte elastase. *Pharmacochemistry Library*, **28**, 499–509.

Angliker, H., Wikström, P., Rauber, P., Stone, S. and Shaw, E. (1988) Synthesis and properties of peptidyl derivatives of arginylfluoromethanes. *Biochemical Journal*, **256**, 481–486.

An-Zhi, W., Mayr, I. and Bode, W. (1988) The refined 2.3 Å crystal structure of human leukocyte elastase in a complex with a valine chloromethyl ketone inhibitor. *FEBS Letters*, **234**, 367–373.

Batchelor, M.J., Bebbington, D., Bernis, G.W., Fridman, W.H., Gillespie, R.J. *et al* (1997) Inhibitors of interleukin-1β converting enzyme. European Patent Appl. WO 9722619.

Bechét, J.-J., Dupaix, A. and Blagoeva, I. (1977) Inactivation of α-chymotrypsin by new bifunctional reagents: halomethylated derivatives of dihydrocoumarines. *Biochimie*, **59**, 231–239.

Bernstein, P.R., Edwards, P.D. and Williams, J.C. (1994) Inhibitors of human leukocyte elastase. G.P. Ellis and D.K. Luscombe (Eds) *Progress in Medicinal Chemistry*, **31**, 59–120. Amsterdam: Elsevier.

Bieth, J.G. (1986) Elastases: Catalytic and biological properties. In *Regulation of Matrix Accumulation*, edited by R. P. Mecham, pp. 217–320. New York: Academic Press.

Bode, W., An-Zhi, W., Huber, R., Meyer, E. Jr., Travis, J. and Neumann, S. (1986) X-ray crystal structure of the complex of human leukocyte elastase (PMN elastase) and the third domain of the turkey ovomucoid inhibitor. *EMBO Journal*, **8**, 2453–2458.

Brady, K., Liang, T.-C. and Abeles, R.H. (1989) pH Dependence of the inhibition of chymotrypsin by a peptidyl trifluoromethyl ketone. *Biochemistry*, **28**, 9066–9070.

Burkhart, J.P., Koehl, J.R., Mehdi, S., Durham,V., Janusz, M.J., Huber, E.W. *et al* (1995) Inhibition of human neutrophil elastase. 3. An orally active enol acetate prodrug. *Journal of Medicinal Chemistry*, **38**, 223–233.

Burkhart, J.P., Mehdi, S., Koehl, J.R., Angelastro, M.R., Bey, P. and Peet, N.P. (1998) Preparation of α-ketoester enol acetates as potential prodrugs for human neutrophil elastase inhibitors. *Bioorganic & Medicinal Chemistry Letters*, **8**, 63–64.

Copp, L.J., Krantz, A. and Spencer, R.W. (1987) Kinetics and mechanism of human leukocyte elastase inactivation and by ynenol lactones. *Biochemistry*, **26**, 169–178.

Costanzo, M.J., Maryanoff, B.E., Hecker, L.R., Schott, M.R., Yabut, S.C., Zhang, H.-C. *et al* (1996) Potent thrombin inhibitors that probe the S_1' subsite: tripeptide transition state analogues based on a heterocycle-activated carbonyl group. *Journal of Medicinal Chemistry*, **39**, 3039–3043.

Cregge, R.J., Durham, S.L., Farr, R.A., Gallion, S.L., Hare, C.M., Hoffman, R.V. *et al* (1998) Inhibition of human neutrophil elastase. 4. Design, synthesis, X-ray crystallographic analysis, and structure–activity relationships for a series of P_2-modified, orally active pentafluoroethyl ketones. *Journal of Medicinal Chemistry*, **41**, 2461–2480.

Doherty, J.B., Ashe, B.M., Argenbright, L.W., Barker, P.L., Bonney, R.J., Chandler, G.O. *et al* (1986) Cephalosporin antibiotics can be modified to inhibit human leukocyte elastase. *Nature*, **322**, 192–194.

Dolle, R.E., Prasad, C.V.C., Prouty, C.P., Salvino, J.M., Awad, M.M.A., Schmidt, S.J. *et al* (1997) Pyridazinodiazepines as a high-affinity, P_2–P_3 peptidomimetic class of interleukin-1 beta-converting enzyme inhibitor. *Journal of Medicinal Chemistry*, **40**, 1941–1946.

Edwards, P.D., Meyer, E.F. Jr., Vijayalakshmi, J., Tuthill, P.A., Andisik, D.A., Gomes, B. *et al* (1992) Design, synthesis, and kinetic evaluation of a unique class of elastase inhibitors, the peptidyl α-ketobenzoxazoles, and the X-ray crystal structure of the covalent complex between porcine pancreatic elastase and Ac-Ala-Pro-Val-2-Benzoxazole. *Journal of the American Chemical Society*, **114**, 1854–1863.

Edwards, P.D. and Bernstein, P.R. (1994) Synthetic inhibitors of elastase. *Medicinal Research Reviews*, **14**, 127–194.

Edwards, P.D., Wolanin, D.J., Andisik, D.W. and Davis, M.W. (1995a) Peptidyl α-ketoheterocyclic inhibitors of human neutrophil elastase. 2. Effect of varying the heterocyclic ring on *in vitro* potency. *Journal of Medicinal Chemistry*, **38**, 76–85.

Edwards, P.D., Zottola, M.A., Davis, M., Williams, J. and Tuthill, P.A. (1995b) Peptidyl α-keto-heterocyclic inhibitors of human neutrophil elastase. 3. *In vitro* and *in vivo* potency of a series of peptidyl α-ketobenzoxazoles. *Journal of Medicinal Chemistry*, **38**, 3972–3982.

Edwards, P.D. and Veale, C.A. (1997) Inhibitors of human neutrophil elastase. *Expert Opinion in Therapeutic Patents*, **7**, 17–28.

Eriksson, S. (1965) Studies in α_1-antitrypsin deficiency. *Acta Medica Scandinavica*, **177** (Suppl 432), 1–85.

Farley, D., Faller, B. and Nick, H. (1997) Therapeutic protein inhibitors of elastase. *Drugs and the Pharmaceutical Sciences*, **84**, 305–334.

Gold, A.M. and Farney, D.E. (1964) Sulfonyl fluorides as inhibitors of esterases. II. Formation and reactions of phenylmethanesulfonyl α-chymotrypsin. *Biochemistry*, **3**, 783–791.

Golec, J.M.C., Mullican, M.D., Murcko, M.A., Wilson, K.P., Kay, D.P., Jones, S.D. *et al* (1997) Structure-based design of non-peptidic pyridone aldehydes as inhibitors of interleukin-1-beta converting enzyme. *Bioorganic & Medicinal Chemistry Letters*, **7**, 2181–2186.

Govardhan, C.P. and Abeles, R.H. (1990) Structure-activity studies of fluoroketone inhibitors of α-lytic protease and human leukocyte elastase. *Archives of Biochemistry and Biophysics*, **280**, 137–146.

Groutas, W.C., Brubaker, M.J., Venkataraman, R., Epp, J.B., Houser-Archield, N., Chong, L.S. *et al* (1992) Potential mechanism-based inhibitors of proteolytic enzymes. *Bioorganic & Medicinal Chemistry Letters*, **2**, 175–180.

Groutas, W.C., Epp, J.B., Venkataraman, R., Kuang, R., Truong, T.M., McClenahan, J.J. *et al* (1996) Design, synthesis, and *in vitro* inhibitory activity toward human leukocyte elastase, cathepsin G, and proteinase 3 of saccharin-derived sulfones and congeners. *Bioorganic & Medicinal Chemistry*, **4**, 1393–1400.

Groutas, W.C., Kuang, R.Z., Ruan, S.M., Epp, J.B., Venkataraman, R. and Truong, T.M. (1998) Potent and specific inhibition of human leukocyte elastase, cathepsin G and proteinase 3 by sulfone derivatives employing the 1,2,5-thiadiazolidin-3-one 1,1 dioxide scaffold. *Bioorganic & Medicinal Chemistry*, **6**, 661–671.

Hagmann, W.K., Kissinger, A.L., Shah, S.K., Finke, P.E., Dorn, C.P., Brause, K.A. *et al* (1993) Orally active β-lactam inhibitors of human leukocyte elastase. 3. Stereospecific synthesis and structure–activity relationships for 3,3-dialkylazetidin-2-ones. *Journal of Medicinal Chemistry*, **36**, 771–777.

Harper, J.W., Cook, R.R., Roberts, C.J., McLaughlin, B.J. and Powers, J.C. (1984) Active site mapping of the serine protease human leukocyte elastase, cathepsin G, porcine pancreatic elastase, rat mast cell proteases I and II, bovine chymotrypsin A_{α}, and *Staphylococcus aureus* protease V-8 using tripeptide thiobenzyl ester substrates. *Biochemistry*, **23**, 2995–3002.

Harper, J.W. and Powers, J.C. (1985) Reaction of serine proteases with substituted 3-alkoxy-4-chloroisocoumarins and 3-alkoxy-7-amino-4-chloroisocoumarins: new reactive mechanism-based inhibitors. *Biochemistry*, **24**, 7200–7213.

Hassall, C.H., Johnson, W.H., Kennedy, A.J. and Roberts, N.A. (1985) A new class of inhibitors of human leukocyte elastase. *FEBS Letters*, **183**, 201–205.

Hlasta, D.J. and Pagani, E.D. (1994) Human leukocyte elastase inhibitors. *Annual Reports in Medicinal Chemistry*, **29**, 195–204.

Hlasta, D.J., Court, J.J., Desai, R.C., Talomie, T.G. and Shen, J. (1996) A comparative SAR and computer modeling study of benzisothiazolone, mechanism-based inhibitors with porcine pancreatic and human leukocyte elastase. *Bioorganic & Medicinal Chemistry Letters*, **6**, 2941–2946.

IMSworld R&D FOCUS. London: IMSworld Publications Ltd. [Cited August 2001].

Imperiali, B. and Abeles, R.H. (1986) Inhibition of serine proteases by peptidyl fluoromethyl ketones. *Biochemistry*, **25**, 3760–3767.

Jansen, E.F., Nutting, M.-D.F. and Balls, A.K. (1949) Mode of inhibition of chymotrypsin by diisopropyl fluorophosphate. *Journal of Biological Chemistry*, **179**, 201–204.

Katzenellenbogen, J.A., Rai, R. and Dai, W. (1992) Enol lactone derivatives as inhibitors of human neutrophil elastase and trypsin-like proteases. *Bioorganic & Medicinal Chemistry Letters*, **2**, 1399–1404.

Kettner, C.A. and Shenvi, A.B. (1984) Inhibition of the serine proteases leukocyte elastase, pancreatic elastase, cathepsin G, and chymotrypsin by peptide boronic acids. *Journal of Biological Chemistry*, **259**, 15106–15114.

Kirschenheuter, G.P., Oleksyszyn, J., Lyle, W., Wieczorek, M., Kloppel, T.M., Simon, S.R. *et al* (1993) Synthesis and characterization of human neutrophil elastase inhibitors derived from aromatic esters of phenylalkanoic acids. *Agents and Actions*, **42** (*Suppl*), 71–82.

Krantz, A., Spencer, R.W., Tam, T.F., Liak, T.J., Copp, L.J., Thomas, E.M. *et al* (1990) Design and synthesis of 4*H*-3,1-benzoxazin-4-ones as potent alternate substrate inhibitors of human leukocyte elastase. *Journal of Medicinal Chemistry*, **33**, 464–479.

Kraut, J. (1977) Serine proteases: structure and mechanism of catalysis. *Annual Review of Biochemistry*, **46**, 331–358.

Lamden, L.A. and Bartlett, P.A. (1983) Aminoalkylphosphonofluoridate derivatives: rapid and potentially selective inactivators of serine peptidases. *Biochemical and Biophysical Research Communications*, **112**, 1085–1090.

Laurell, C.-B. and Eriksson, S. (1963) The electrophoretic α_1-globulin pattern of serum in α_1-antitrypsin deficiency. *Scandinavian Journal of Clinical & Laboratory Investigation*, **15**, 132–140.

Lively, O. and Powers, J.C. (1978) Specificity and reactivity of human granulocyte elastase and cathepsin G, porcine pancreatic elastase, bovine chymotrypsin and trypsin toward inhibition with sulfonyl fluorides. *Biochimica et Biophysica Acta*, **525**, 171–179.

Luisetti, M., Sturani, C., Sella, D., Madonini, E., Galavotti, V., Bruno, G. *et al* (1996) MR889, a neutrophil elastase inhibitor, in patients with chronic obstructive pulmonary disease: a double-blind, randomized, placebo-controlled clinical trial. *European Respiratory Journal*, **9**, 1482–1486.

Macdonald, S.J.F., Belton, D.J., Buckley, D.M., Spooner, J.E., Anson, M.S., Lee, A.H. (1998) Synthesis of trans-5-oxo-hexahydro-pyrrolo [3,2-*b*]pyrroles (pyrrolidine trans-lactams and trans-lactones): new pharmacophores for elastase inhibition. *Journal of Medicinal Chemistry*, **41**, 3919–3922.

Martyn, D.C., Moore, M.J.B. and Abell, A.D. (1999) Succinimide and saccharin-based enzyme-activated inhibitors of serine proteases. *Current Pharmaceutical Design*, **5**, 405–415.

Matthews, B.W., Sigler, P.B., Henderson, R. and Blow, D.M. (1967) Three-dimensional structure of tosyl-α-chymotrypsin. *Nature*, **214**, 652–656.

Mehdi, S., Angelastro, M.R., Burkhart, J.P., Koehl, J.R., Peet, N.P. and Bey, P. (1990) The inhibition of human neutrophil elastase and cathepsin G by peptidyl 1,2-dicarbonyl derivatives. *Biochemical and Biophysical Research Communications*, **166**, 595–600.

Metz, W.A. and Peet, N.P. (1999) Inhibitors of human neutrophil elastase as a potential treatment of inflammatory diseases. *Expert Opinion in Therapeutic Patents*, **9**, 851–868.

Navia, M.A., McKeever, B.M., Springer, J.P., Lin, T.-Y., Williams, H.R., Fluder, E.M. *et al* (1989) Crystallographic study of a β-lactam inhibitor complex with elastase at 1.84 Å resolution. *Proceedings of the National Academy of Sciences of the U.S.A.*, **86**, 7–11.

Odagaki, Y., Ohmoto, K., Matsuoka, S., Hamanaka, N., Nakai, H., Toda, M. *et al* (2001) The crystal structure of the complex on non-peptidic inhibitor of human neutrophil elastase ONO-6818 and porcine pancreatic elastase. *Bioorganic & Medicinal Chemistry*, **9**, 647–651.

Ogilvie, W., Bailey, M., Poupart, M.-A., Abraham, A., Bhavsar, A., Bonneau, P. *et al* (1997) Peptidomimetic inhibitors of the human cytomegalovirus protease. *Journal of Medicinal Chemistry*, **40**, 4113–4135.

Ohlsson, K. and Olsson, I. (1974) The neutral proteases of human granulocytes. Isolation and partial characterization of granulocyte elastases. *European Journal of Biochemistry*, **42**, 519–527.

Ohmoto, K., Yamamoto, T., Horiuchi, T., Imanishi, H., Odagaki, Y., Kawabata, K. *et al* (2000) Design and synthesis of new orally active nonpeptidic inhibitors of human neutrophil elastase. *Journal of Medicinal Chemistry*, **43**, 4927–4929.

Ohmoto, K., Yamamoto, T., Okuma, M., Horiuchi, T., Imanishi, H., Odagaki, Y. *et al* (2001) Development of orally active nonpeptidic inhibitors of human neutrophil elastase. *Journal of Medicinal Chemistry*, **44**, 1268–1285.

Oleksyszyn, J. and Powers, J.C. (1991) Irreversible inhibition of serine proteases by peptide derivatives of (α-aminoalkyl)phosphonate diphenyl esters. *Biochemistry*, **30**, 485–493.

Pardo, A. and Selman, M. (1999) Proteinase–antiproteinase imbalance in the pathogenesis of emphysema: the role of metalloproteinases in lung damage. *Histology and Histopathology*, **14**, 227–233.

Patent Review (1999) A novel structural motif for serine protease inhibitors. *Expert Opinion in Therapeutic Patents*, **9**, 1133–1137.

Peet, N.P., Burkhart, J.P., Angelastro, M.R., Giroux, E.L., Mehdi, S., Bey, P. *et al* (1990) Synthesis of peptidyl fluoromethyl ketones and peptidyl α-ketoesters as inhibitors of porcine pancreatic elastase, human neutrophil elastase, and rat and human neutrophil cathepsin G. *Journal of Medicinal Chemistry*, **33**, 394–407.

Pharmaprojects. Richmond (UK): PJB Publications Ltd. [Cited April 2001].

Powers, J.C., Gupton, B.F., Harley, A.D., Nishino, N. and Whitley, R.J. (1977a) Specificity of porcine pancreatic elastase, human leukocyte elastase and cathepsin G. Inhibition with peptide chloromethyl ketones. *Biochimica et Biophysica Acta*, **485**, 156–166.

Powers, J.C. (1977b) Haloketone inhibitors of proteolytic enzymes. In *Chemistry and Biochemistry of Amino Acids, Peptides and Proteins*, Vol. 4, edited by B. Weinstein, pp. 65–178. New York: Marcel Dekker, Inc.

Powers, J.C. and Harper, J.W. (1986) Inhibitors of serine proteinases. In *Proteinase Inhibitors*, edited by A.J. Barrett and G. Salvesen, pp. 55–152. Amsterdam: Elsevier.

Powers, J.C., Plaskon, R.R. and Kam, C.-M. (1996) Low-molecular-weight-inhibitors of neutrophil elastase. *Lung Biology in Health and Disease*, **88**, 341–370.

Rasnick, D. (1985) Synthesis of peptide fluoromethyl ketones and the inhibition of human cathepsin-B. *Analytical Biochemistry*, **149**, 461–465.

Rice, K.D., Tanaka, R.D., Katz, B.A., Numerof, R.P. and Moore, W.R. (1998) Inhibitors of tryptase for the treatment of mast cell-mediated diseases. *Current Pharmaceutical Design*, **4**, 381–396.

Sanderson, P.E.J. and Naylor-Olsen, A.M. (1998a) Thrombin inhibitor design. *Current Medicinal Chemistry*, **5**, 289–304.

Sanderson, P.E.J., Lyle, T.A., Cutrona, K.J., Dyer, D.L., Dorsey, B.D., McDonough, C.M. *et al* (1998b) Efficacious, orally bioavailable thrombin inhibitors based on 3-aminopyridinone or 3-aminopyrazinone acetamide peptidomimetic templates. *Journal of Medicinal Chemistry*, **41**, 4466–4474.

Schechter, I. and Berger, A. (1967) On the size of the active site of proteases. I. Papain. *Biochemical and Biophysical Research Communications*, **27**, 157–162.

Semple, G., Ashworty, D.M., Batt, A.R., Baxter, A J., Benzies, D.W.M., Elliot, L.H. *et al* (1998) Peptidomimetic aminomethylene ketone inhibitors of interleukin-1 beta-converting enzyme (ICE). *Bioorganic & Medicinal Chemistry Letters*, **8**, 959–964.

Shaw, E., Angliker, H., Rauber, P., Walker, B. and Wikstrom, P. (1986) Peptidyl fluoromethyl ketones as thiol protease inhibitors. *Biomedica Biochimica Acta*, **45**, 1397–1403.

Shinguh, Y., Imai, K., Yamazaki, A., Inamura, N., Shima, I., Wakabayashi, A. *et al* (1997). Biochemical and pharmacological characterization of FK706, a novel elastase inhibitor. *European Journal of Pharmacology*, **337**, 63–71.

Shotton, D.M. and Watson, H.C. (1970) Three-dimensional structure of tosyl-elastase. *Nature*, **225**, 811–816.

Sinha, S., Watorek, W., Karr, S., Giles, J., Bode, W. and Travis, J. (1987) The Primary structure of human neutrophil elastase. *Proceedings of the National Academy of Sciences of the U.S.A.*, **84**, 2228–2232.

Skiles, J.W. and Jeng, A.Y. (1999) Therapeutic promises of leukocyte elastase and macrophage metalloelastase inhibitors for the treatment of pulmonary emphysema. *Expert Opinion in Therapeutic Patents*, **9**, 869–895.

Snider, G.L. (1992) Emphysema: The first two centuries – and beyond. *American Review of Respiratory Disease*, **146**, 1615–1622.

Stein, M.M., Trainor, D.A., Yee, Y.K., Edwards, P.D., Zottola, M.A., Williams, J. *et al* (1989) Synthesis and evaluation of peptidyl α-diketones as inhibitors of human leukocyte elastase. A new class of peptidyl electrophilic carbonyl containing serine protease inhibitor. In *Abstracts of Papers, Eleventh American Peptide Symposium, University of California, San Diego, CA, July 9–14*.

Stein, M.M., Wildonger, R.A., Trainor, D.A., Edwards, P.D., Yee, Y.K., Lewis, J.J. *et al* (1990) *In vitro* and *in vivo* inhibition of human leukocyte elastase (HLE) by two series of electrophilic carbonyl containing peptides. In *Peptides: Chemistry, Structure, and Biology (Proceedings of the Eleventh American Peptide Symposium)*, edited by J.E. River and G.R. Marshall, pp. 369–370. Leiden: ESCOM.

Stein, R.L., Strimpler, A.M., Edwards, P.D., Lewis, J.J., Mauger, R.C., Schwartz, J.A. (1987) Mechanism of slow-binding inhibition of human leukocyte elastase by trifluoromethyl ketones. *Biochemistry*, **26**, 2682–2689.

Takahashi, H., Nukiwa, T., Yoshimura, K., Quick, C.D., States, D.J., Holmes, M.D. *et al* (1988a) The structure of the human neutrophil gene. *Journal of Biological Chemistry*, **263**, 14739–14747.

Takahashi, L.H., Radhakrishnan, R., Rosenfield, R.E. Jr., Meyer, E.F. Jr., Trainor, D.A. and Stein, M. (1988b) X-Ray diffraction analysis of the inhibition of porcine pancreatic elastase by a peptidyl trifluoromethylketone. *Journal of Molecular Biology*, **201**, 423–428.

Tamura, S.Y., Semple, J.E., Reiner, J.E., Goldman, E.A., Brunck, T.K., Lim-Wilby, M.S. *et al* (1997a) Design and synthesis of a novel class of thrombin inhibitors incorporating heterocyclic dipeptide surrogates. *Bioorganic & Medicinal Chemistry Letters*, **7**, 1543–1548.

Tamura, S.Y., Shamblin, R.M., Brunck, T.K. and Ripka, W.C. (1997b) Rational design, synthesis, and serine protease inhibitory activity of novel P_1-argininoyl heterocycles. *Bioorganic & Medicinal Chemistry Letters*, **7**, 1359–1364.

Tao, M., Bihovsky, R. and Kauer, J.C. (1996) Inhibition of calpain by peptidyl heterocycles. *Bioorganic & Medicinal Chemistry Letters*, **6**, 3009–3012.

Thompson, R.C. and Blout, E.R. (1970) Evidence for an extended active center in elastase. *Proceedings of the National Academy of Sciences of the U.S.A.*, **67**, 1734–1740.

Thompson, R.C. and Blout, E.R. (1973) Dependence of the kinetic parameters for elastase-catalyzed amide hydrolysis on the length of peptide substrates. *Biochemistry*, **12**, 57–65.

Tomizawa, N., Ohwada, S., Ohya, T., Takeyoshi, I., Ogawa, T., Kawashima, Y.M. *et al* (1999) The effects of a neutrophil elastase inhibitor (ONO-5046 Na) and neutrophil depletion using a granulotrap (G-1) column on lung reperfusion injury in dogs. *Journal of Heart and Lung Transplantation*, **18**, 637–645.

Tsutsumi, S., Okonogi, T., Shibahara, S., Ohuchi, S., Hatsushiba, E., Patchett, A.A. and Christensen, B.G. (1994) Synthesis and structure–activity relationships of peptidyl α-ketoheterocycles as novel inhibitors of prolyl endopeptidase. *Journal of Medicinal Chemistry*, **37**, 3492–3502.

Veale, C.A., Bernstein, P.R., Bohnert, C.M., Brown, F.J., Bryant, C., Damewood, J.R. *et al* (1997) Orally active trifluoromethyl ketone inhibitors of human leukocyte elastase. *Journal of Medicinal Chemistry*, **40**, 3173–3181.

Vincent, S.H., Painter, S.K., Luffer-Atlas, D., Karanam, B.V., McGowan, E., Cioffe, C. *et al* (1997) Orally active inhibitors of human leukocyte elastase. II. Disposition of L-694,458 in rats and rhesus monkeys. *Drug Metabolism and Disposition*, **25**, 932–939.

Walsh, C. (1982) Suicide substrates: mechanism-based enzyme inactivators. *Tetrahedron*, **38**, 871–909.

Wieczorek, M., Gyorkos, A., Spruce, L.W., Ettinger, A., Ross, S.E., Kroona, H.S. *et al* (1999) Biochemical characterization of α-ketooxadiazole inhibitors of elastases. *Archives of Biochemistry and Biophysics*, **367**, 193–201.

Wright, C.S., Alden, R.A. and Kraut, J. (1969) Structure of subtilisin BPN′ at 2.5 Å resolution. *Nature*, **221**, 235–242.

Yamashita, D.S., Dong, X., Oh, H.-J., Brook, C.S., Tomaszek, T.A., Szewczuk, L. *et al* (1999) Solid-phase synthesis of a combinatorial array of 1,3-bis(acylamino)-2-butanones, inhibitors of the cysteine proteases cathepsins K and L. *Journal of Combinatorial Chemistry*, **1**, 207–215.

Yoshimura, T., Barker, L.N. and Powers, J.C. (1982) Specificity and reactivity of human leukocyte elastase, porcine pancreatic elastase, human granulocyte cathepsin G, and bovine pancreatic chymotrypsin with arylsulfonyl fluorides. Discovery of a new series of potent and specific irreversible elastase inhibitors. *Journal of Biological Chemistry*, **257**, 5077–5084.

Chapter 8

Thrombin

Jörg Stürzebecher, Jörg Hauptmann and Torsten Steinmetzer

This chapter outlines the discovery and development of synthetic, small molecule inhibitors of the trypsin-like blood clotting proteinase, thrombin, over the last two decades. Thrombin plays a key role in thromboembolic vascular diseases and is, therefore, a major target for drug development. The direct-acting inhibitors, originally derived from tripeptide substrate structures, bind at primary and neighboring binding sites in the active centre of thrombin. Extensive variations of the structural segments (P4-P1') of prototypal inhibitors have led to the development of new lead structures with high potency and selectivity. A number of selected inhibitors are presented illustrating the state and further lines of development in this field. In various animal models it has been demonstrated that active site-directed competitive thrombin inhibitors are effective anticoagulants and antithrombotics and have potential advantages over the indirect-acting thrombin inhibitor heparin. Results of clinical studies of several parenteral thrombin inhibitors show that they are useful alternatives to heparin, with respect to predictable anticoagulation and low bleeding risk, in various cardiovascular disorders.

Design strategies are directed toward conformationally restricted peptidomimetic or non-peptide molecules with high affinity to the target enzyme, high selectivity and properties appropriate for *in vivo* use. Much effort is concentrated at optimization of overall physico-chemical characteristics of the inhibitors (introduction of weakly basic or neutral P1 residues, balanced lipophilicity) in order to improve the pharmacokinetics with the aim to develop orally bioavailable compounds. Such orally active direct thrombin inhibitors would represent a novel class of antithrombotic drugs.

8.1 INTRODUCTION

Thrombin is the proteinase activated last from its zymogen in the pathway leading to the coagulation of blood. The classical view of the blood coagulation process as a cascade-like sequence of proteolytic activation steps has been revised by findings on the complexity of this system. Thrombin is a multifunctional enzyme taking a central position in the coagulation system and having regulatory functions. It not only converts the substrate fibrinogen to fibrin and activates factor XIII, the enzyme responsible for cross-linking of polymerized fibrin monomers, but also accelerates its own generation by activating the protein cofactors, factor V and factor VIII, and by activating blood platelets, which substantially contribute to the generation of thrombin on their surface. The activation of a thrombin receptor (PAR-1, proteinase-activated receptor 1) on various cells is a proteolytic process leading to the appearance of a tethered receptor ligand. In addition, thrombin is able to trigger the release from endothelial cells of several products supporting clot formation. In contrast, thrombin bound to thrombomodulin on the surface of vascular endothelium has an altered substrate specificity, it no longer converts fibrinogen but activates protein C to a proteinase (aPC) which, in turn, inactivates factors Va and VIIIa by proteolytic degradation. The latter negative feed-back process would limit the generation of thrombin. In blood plasma, there are several proteins with a thrombin inhibitory function, such as antithrombin, α_1-antitrypsin, and α_2-macroglobulin. Various mechanisms regulate the formation of thrombin within the circulation by localizing it to sites of vessel injury where it fulfills the physiological role in hemostasis. Multiple receptor-mediated effects of thrombin on cells outside the circulatory system illustrate its role as a multifunctional protease; the physiological/pathological significance of various reactions has still to be clarified. Thrombin as an enzyme target is freely accessible in circulating blood by (small molecule) inhibitors since, first of all, prothrombin activation occurs at sites of injury of the vascular endothelium and/or at (activated) blood cells (platelets, monocytes).

Among the proteinases and peptidases, thrombin emerged relatively early as a target for the development of inhibitors after the anticoagulant and antithrombotic actions of heparin and hirudin had been demonstrated. Today, thrombin serves the role of one of the most widely studied proteinase targets, and the development of various synthetic thrombin inhibitors as drugs has a history longer than that of other proteinase inhibitors, e.g. the HIV-proteinase inhibitors. Biochemical and pharmacological aspects of thrombin inhibitors, either naturally occurring or synthetic ones, as anticoagulant and antithrombotic agents have been studied since the mid-1970s and the strategies for screening and evaluation were developed in the 1980s. Argatroban was only approved for clinical use in 1990 as the first synthetic thrombin inhibitor. Thus, among the proteolytic enzymes, the blood clotting proteinase thrombin is not a "potential" target for inhibitors but rather a well established one. In recent years, a number of comprehensive reviews have dealt with the design, evaluation and development of active site-directed inhibitors of thrombin (Tapparelli *et al* 1993; Balasubramanian 1995; Kimball 1995; Lee 1997; Wiley and Fisher 1997; Menear 1998; Sanderson and Naylor-Olsen 1998; Hauptmann and Stürzebecher 1999; Kimball 1999).

Nowadays, the development of synthetic thrombin inhibitors has reached a new dimension, such that there is no longer a search for whatever a potent thrombin inhibitor but for the thrombin inhibitor with desired and tailored properties. The efforts in the design and development of thrombin inhibitors have been focused on competitive inhibitors with nanomolar K_i values. Irreversible binding, mediated by reactive groups, would bring about limitations with respect to selectivity and *in vivo* applicability (see below). Small molecule, active site-directed reversible inhibitors only would be suited for clinical use as orally active anticoagulants that could complement the therapeutic armamentarium consisting of the parenteral anticoagulants heparin and hirudin and the oral anticoagulants of the vitamin K-antagonist type. For this reason, this chapter will be devoted to this type of thrombin inhibitor only, excluding hirudin and the hirudin-derived bifunctional inhibitors of the hirulog-type, which would not be orally available because of their high molecular weight.

The design strategy for active site-directed thrombin inhibitors was directed, in the first instance, and almost exclusively until recently, to optimal binding to the target, i.e. on affinity and selectivity. There has been much progress, in terms of potency and selectivity, in the recent development of thrombin inhibitors. On the other hand, it is increasingly recognized that the development of a drug from a new lead compound has to end with a congener with pharmacological/toxicological properties compatible with use. The *in vivo* properties, which determine whether a drug candidate will be ultimately developed, are difficult to predict quantitatively, even with a growing knowledge of the various mechanisms involved. Therefore, only a few compounds among the numerous thrombin inhibitors described have reached the early phases of clinical trials. As yet, no actual breakthrough has been reached in the development of a thrombin inhibitor as an orally active antithrombotic drug (Kimball 1995; Rewinkel and Adang 1999). The "first generation" thrombin inhibitors studied in clinical trials had all to be administered parenterally as an intravenous infusion (Clarke *et al* 1991; Andersen *et al* 1996; Bounameaux *et al* 1997).

8.2 CHARACTERISTICS OF TARGET ENZYME AND INHIBITOR BINDING

The trypsin-like serine proteinase thrombin (EC 3.4.21.5., MW 36,600 Da) is a two-chain, arginine-specific endopeptidase and the active site resides in the B chain. Thrombin is formed by specific cleavage of the zymogen prothrombin by the "prothrombinase" complex comprising blood coagulation factors Xa and Va, Ca^{2+}, and phospholipids. Prothrombin, a single-chain glycoprotein (MW 72,500 Da) is synthesized by a vitamin K-dependent process in the liver.

Figure 8.1 shows the X-ray crystal structure of thrombin, in which the irreversible inhibitor PPACK (**8.1**) is bound in the active center (Stubbs and Bode 1993). In contrast to other enzymes of the coagulation pathway, thrombin contains two additional binding regions at a distance of about 20 Å from the active center, named "anionic binding site I and II". The "anionic binding site I" is involved in the binding of natural substrates and of high molecular weight inhibitors such as hirudin and hirulog. The "anionic binding site II" has been demonstrated to be responsible for heparin binding. The physiological substrate, fibrinogen, is bound in the canon-like region of thrombin's surface. High affinity interactions taking place at the "anionic binding site I" ("fibrinogen-recognition exosite") and the aromatic binding site near the active site bring the substrate in close contact to the catalytic apparatus. The P1-Arg is attracted via its guanidino group by Asp189 located in the depth of the "specificity pocket". The catalytic apparatus is formed by the Ser195-His57-Asp102-triad. The "60-insertion loop", in particular Trp60D partially occludes the active site.

Most of the small molecule inhibitors are bound via the "specificity pocket" (S1) and hydrophobic areas near the active site (S2-S4), however, the active center of thrombin is characterized by its ability to accommodate a variety of structures; a circumstance which is of advantage for the design of inhibitors with properties suited for *in vivo* application. Typical structural elements and binding modes of small molecule thrombin inhibitors are exemplified by the "classical" inhibitors PPACK (**8.1**) (Kettner and Shaw 1979), argatroban (**8.2**) (Okamoto *et al* 1981), and NAPAP (**8.3**) (Stürzebecher *et al* 1983). In contrast to the reversible, non-covalently bonded (**8.2**) and (**8.3**), PPACK forms covalent bonds with the active site of thrombin; therefore K_i values cannot be given as kinetic constants.

PPACK and related tripeptides of the D-Phe-Pro-Arg-type bind in an extended conformation (substrate mode-binding) forming an antiparallel β-sheet with Gly216 and Ser214 (Bode *et al* 1992). Figure 8.2 illustrates the interactions of PPACK in the active site region of thrombin. The interaction takes place at three sites: the guanidino group of P1-Arg opposes the carboxylate group of Asp189; the ring of P2-Pro sits in the hydrophobic S2-pocket formed by the side chains of His57, Leu99, Tyr60A, and Trp60D; the D-Phe benzyl ring of PPACK occupies a second hydrophobic pocket, lined by Ile174, Leu99 and

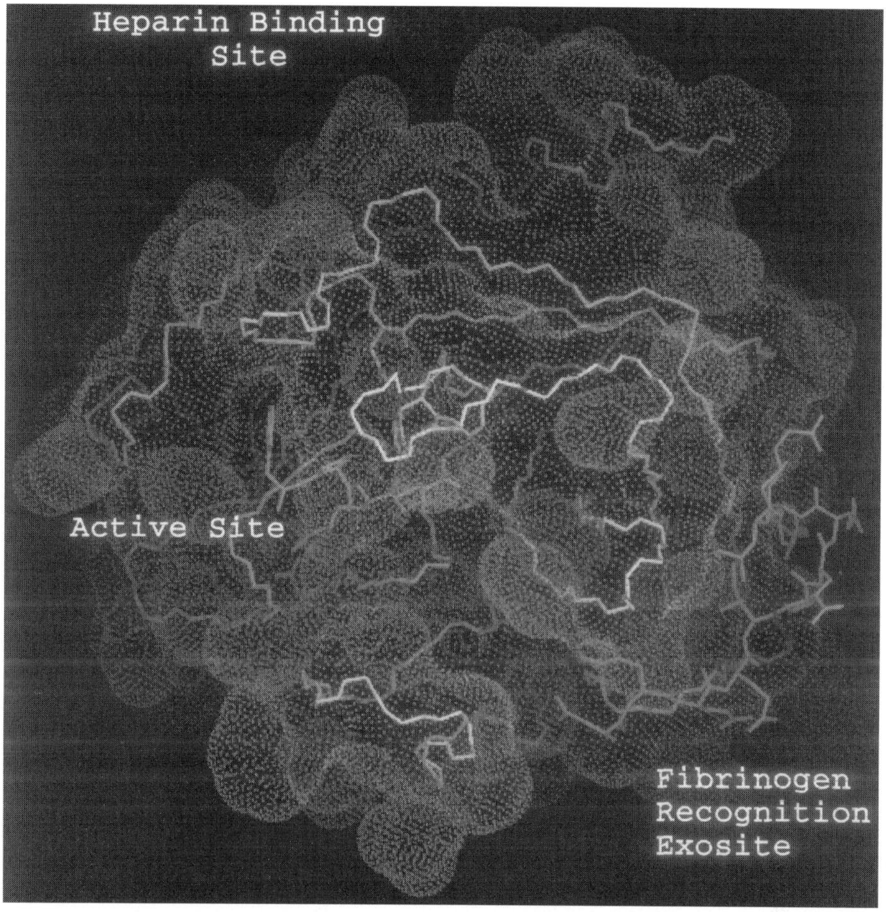

Figure 8.1 Front view of the thrombin molecule (backbone in yellow) in complex with the active site-directed inhibitor PPACK (green) and the C-terminal hirudin tail (residues 55–65 in pink) bound to the "fibrinogen recognition exosite", generated from 1tmu.pdb (Priestle *et al* 1993). Thrombin is diplayed with a Connolly dot surface in blue, red, and yellow for basic, acidic or other residues, resp. The "heparin binding site" is located on the top of the thrombin molecule in this view defined as "standard" orientation (Stubbs and Bode 1993). (*See Color plate 8*)

Trp215. Chloromethyl ketones and transition state analogs like aldehydes, ketones and boroarginines form, additionally, covalent bonds with amino acid residues of the catalytic triad.

Argatroban, NAPAP, and other peptidomimetic and nonpeptidic benzamidine- or arginine-derived inhibitors bind in a more compact, Y-shaped conformation (inhibitor mode-binding); nevertheless, similar interactions occur as with PPACK. The basic amidinophenyl or guanidinoalkyl moiety occupies the S1-pocket and the N-terminal arylsulfonyl secondary group fits into the hydrophobic aryl-binding site (S3-S4). As the main difference, the C-terminal amide moiety (P1′) is orientated backwards and inserts into the S2

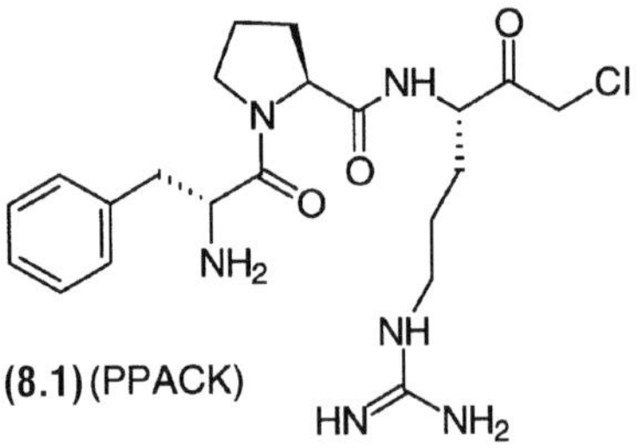
(8.1) (PPACK)

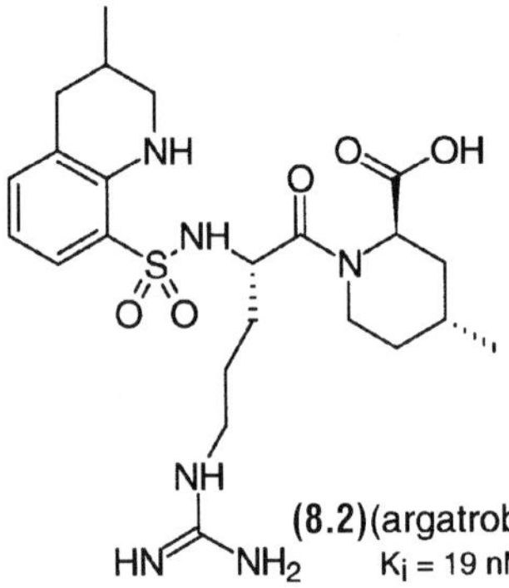
(8.2) (argatroban)
K_i = 19 nM

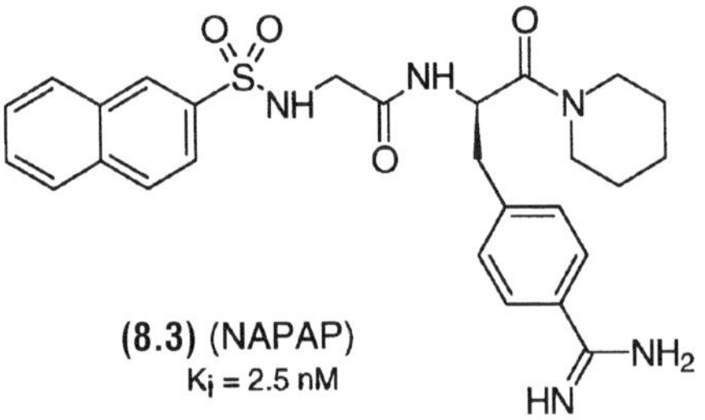
(8.3) (NAPAP)
K_i = 2.5 nM

pocket. Interactions with the catalytic triad do not occur. As examples, argatroban and NAPAP are shown in Figure 8.3.

8.3 DESIGN AND STRUCTURES OF THROMBIN INHIBITORS

The solution of the crystal structure of complexes of thrombin with active site-directed inhibitors (Bode *et al* 1989; Bode *et al* 1990; Banner and Hadvary 1991; Brandstetter *et al* 1992) and the structure–activity relationships derived from previous work drastically accelerated the design and development of potent inhibitors. Many structural analogues of established prototypal compounds were synthesized and evaluated. Various approaches were successful: structure-based molecular design, implemented after solving the structure of the complexes of thrombin with prototypal inhibitors was complemented by screening procedures based on synthesizing derivatives of selected lead structures, whereas random screening yielded compounds different from the "standard" substrate-analog inhibitor structure. Empirical structure optimization, in order to improve overall characteristics of a lead compound, was successful in many instances.

8.3.1 D-Phe-Pro-Arg-derived inhibitors

The D-Phe-Pro-Arg motif has been used and manifoldly modified. Starting with the chloromethyl ketone (**8.1**) (Kettner and Shaw 1979), the aldehyde (**8.4a**) (Bajusz *et al* 1978; Bajusz *et al* 1982) was the first candidate for more extensive studies. The N-terminal

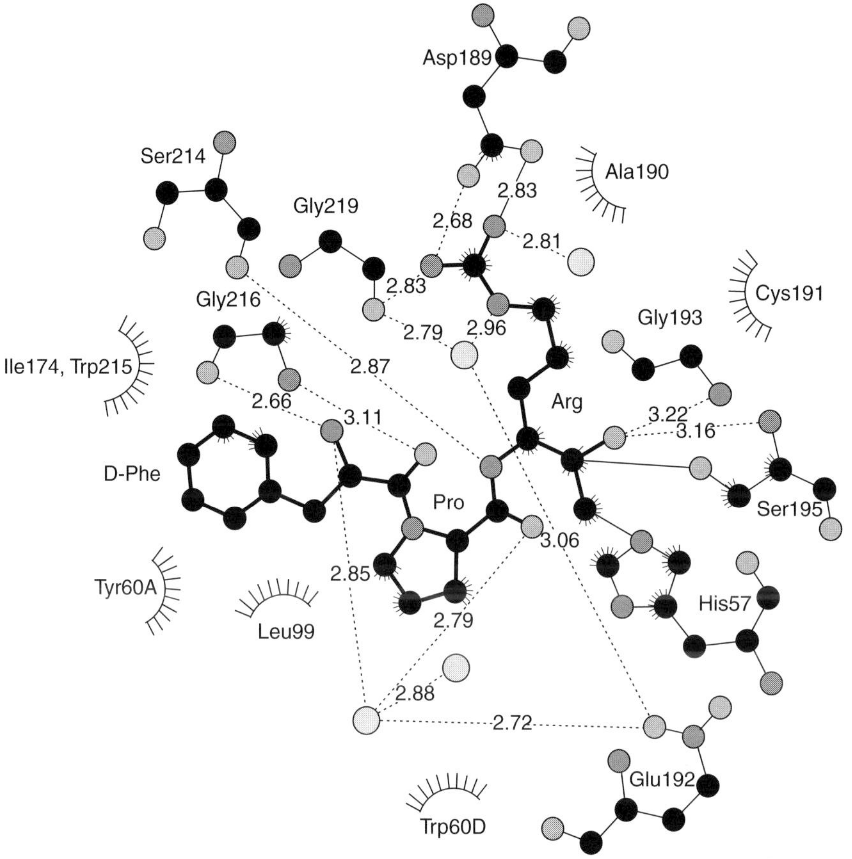

Figure 8.2 Schematic diagram of the PPACK–thrombin complex (1ppb.pdb; Bode *et al* 1989) showing the key interactions, generated by LIGPLOT 4.0 (Wallace *et al* 1995). PPACK is covalently bound to Ser195 and His57. (The distances of the hydrogen bonds are given in Å.) (*See Color plate 9*)

methylated derivative efegatran (GYKI-14766, **8.4b**) (Bajusz *et al* 1990) is less prone to inactivation in solution by cyclization than (**8.4a**).

The development of boroarginine derivatives of D-Phe-Pro-Arg led to DUP 714 (**8.5**) (Kettner and Knabb 1993). SAR studies were performed in order to find compounds with enhanced selectivity and oral absorption. The selectivity was increased by the replacement of proline by a N-cyclopentyl-glycine in S18326 (Rupin *et al* 1997) and with introduction of less basic P1 side chains. In the boronic acid derivative (**8.6a**) (Claeson *et al* 1993), the Arg-side chain is replaced by the neutral methoxypropyl group. With (**8.6a**) it was shown for the first time that thrombin accepts very different residues at the S1 pocket. A closely related analog, the pinacol ester TRI-50b (**8.6b**) is presently under further development (Deadman *et al* 1995). For enhancing oral bioavailability, more hydrophobic residues were introduced at P3 (Quan *et al* 1997). Combination of both a hydrophilic L-amino acid ester

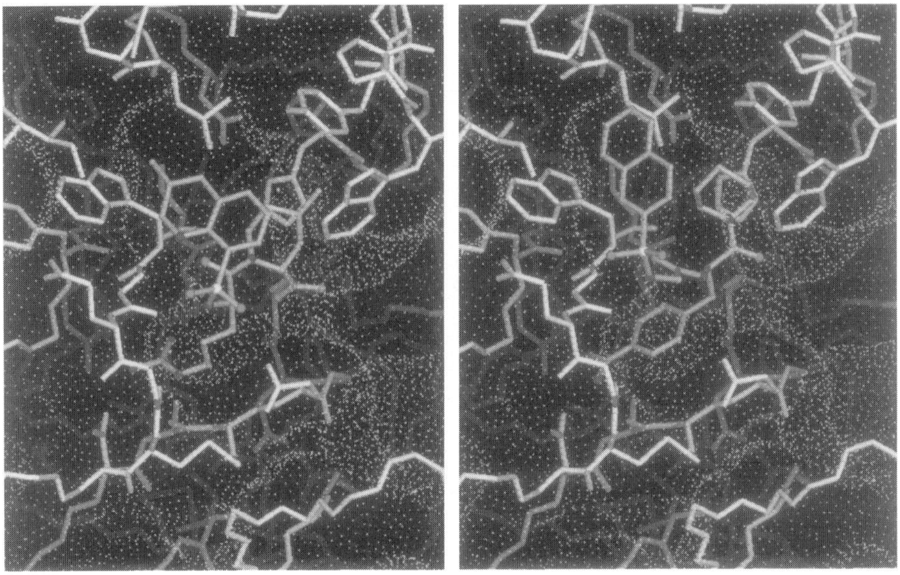

Figure 8.3 Structure of the active site region of the complexes formed between thrombin (yellow) and the inhibitors (orange) argatroban (left, 1etr.pdb) and NAPAP (right, 1ets.pdb) (Brandstetter *et al* 1992). The amino acids of the catalytic triad (Ser195, His57, Asp102) and Asp189 in the S1-pocket are shown in light gray. (*See Color plate 10*)

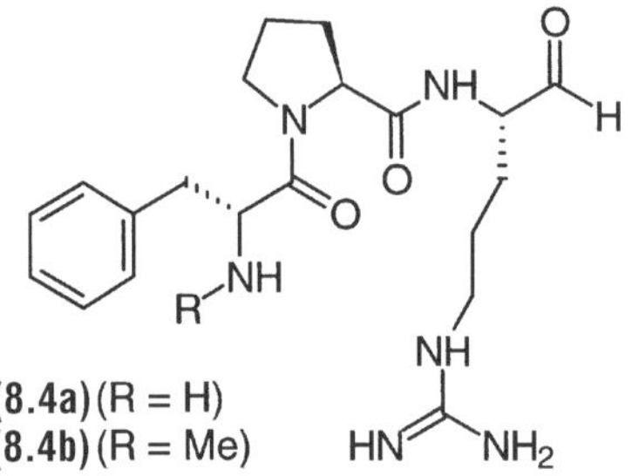

(8.4a) (R = H)
(8.4b) (R = Me)

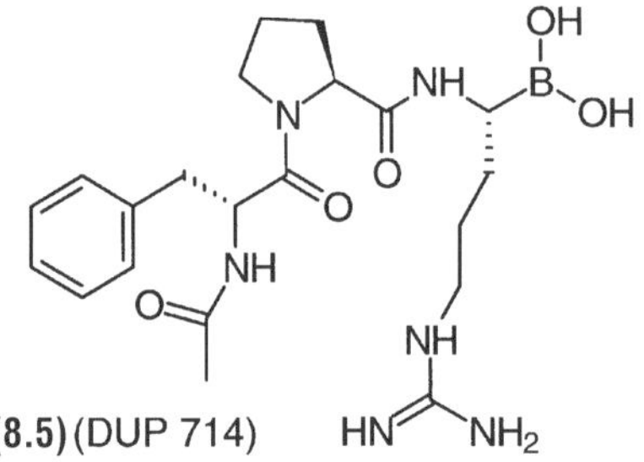

(8.5) (DUP 714)

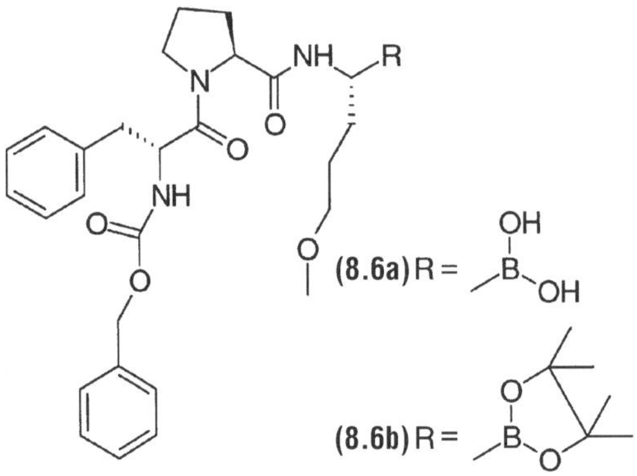

(8.6a) R =

(8.6b) R =

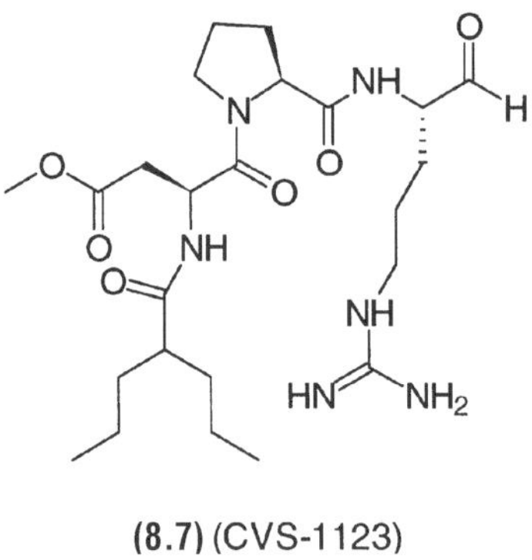

(8.7) (CVS-1123)

at P3 and a hydrophobic N-acyl residue as in the aldehyde CVS-1123 (**8.7**) (Semple *et al* 1996) resulted in increased oral bioavailability.

However, despite the high affinity (nanomolar K_i-values were reported) and some success with regard to selectivity and bioavailability, the transition state-binding mode resulting in slow binding to the enzyme was the reason for relatively low anticoagulant activity. The importance of rapid binding to thrombin for the anticoagulant activity of an inhibitor has been stressed repeatedly (Rupin *et al* 1995; Stone and Tapparelli 1995; Elg *et al* 1997; Nilsson *et al* 1998). *In vitro*, NAPAP was equally effective in assays either starting the reaction with the substrate, following incubation of enzyme and inhibitor, or starting by adding the enzyme, whereas the slow binding DUP 714 had low potency only under the latter condition. *In vivo*, in a study on the relation between inhibition (K_i) or association rate constant (k_{on}) and the antithrombotic effect for various thrombin inhibitors in rats, it was shown that slow-binding inhibitors give steep dose–response curves and a narrow therapeutic range ascribable to the decrease in association time with increasing concentrations (Elg *et al* 1997). In the literature for slow-binding inhibitors very different inhibition constants are given because in many cases the constants were not determined according to the appropriate kinetic conditions without preincubation. Therefore, no K_i values are given here for this type of inhibitors.

A new step in the development was achieved by removing the C-terminal electrophilic part leading to agmatine and noragmatine derivatives without the reactive group. The first example of this widely studied group of competitive, non-covalent inhibitors, D-Phe-Pro-agmatine with a submicromolar K_i has been developed already, in the early 1980s (Bajusz *et al* 1982). Inogatran (**8.8**) (Teger-Nilsson *et al* 1997) is a more potent representative of

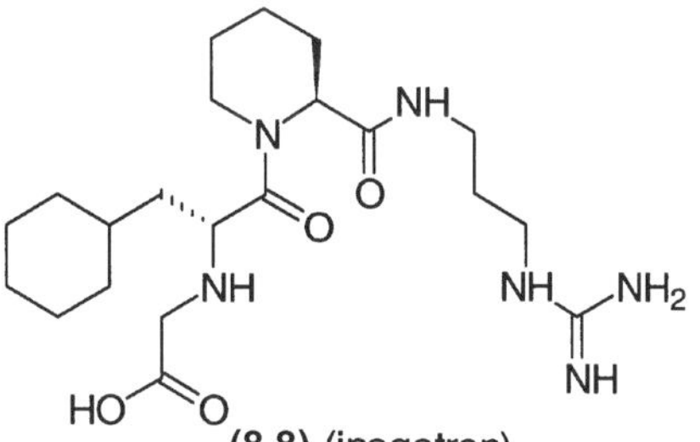

(8.8) (inogatran)
K_i = 15 nM

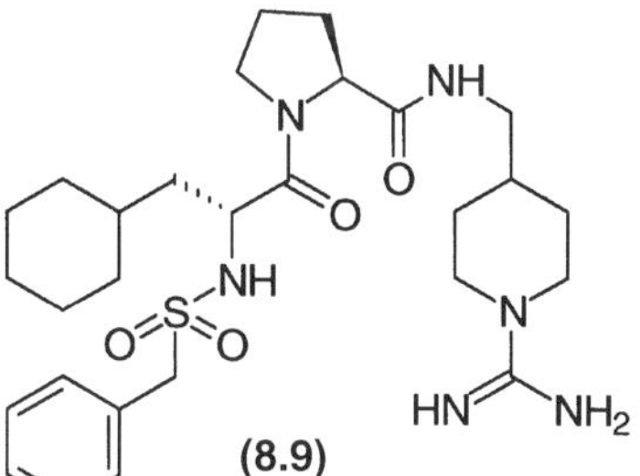

(8.9)
K_i = 0.27 nM

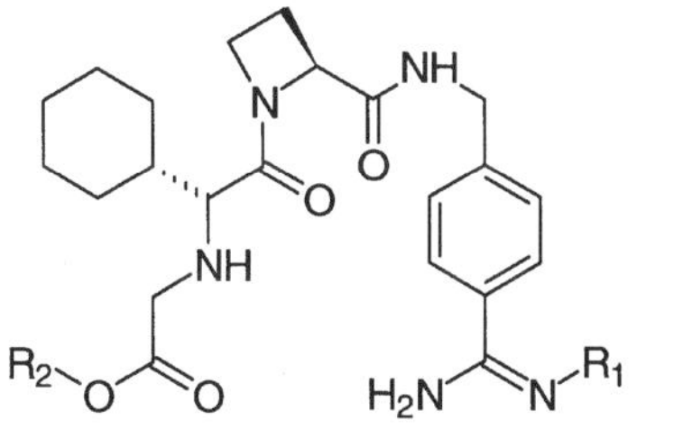

(8.10a) (melagatran) R_1 = H, R_2 = H, K_i = 2.0 nM

(8.10b) (H 376/95) R_1 = OH, R_2 = C_2H_5, K_i = 370 nM

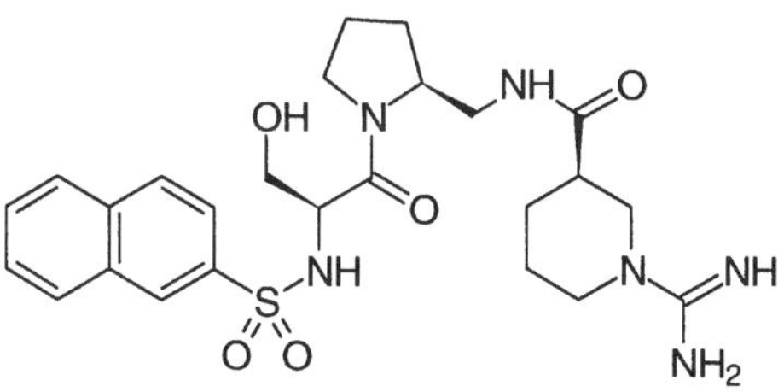

(8.11)
K_i = 3.6 nM

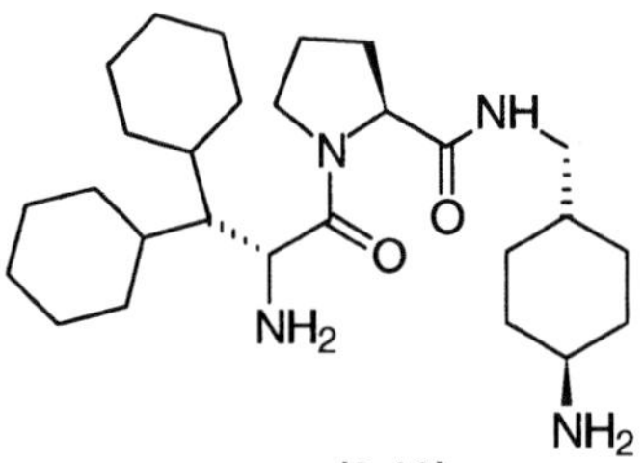

(8.12)
K_i = 0.056 nM

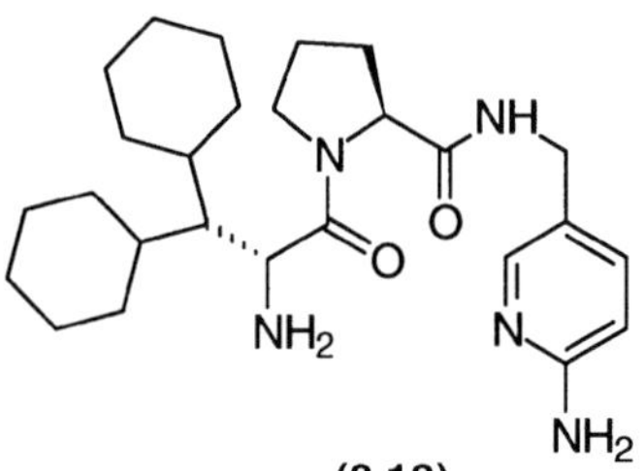

(8.13)
K_i = 0.82 nM

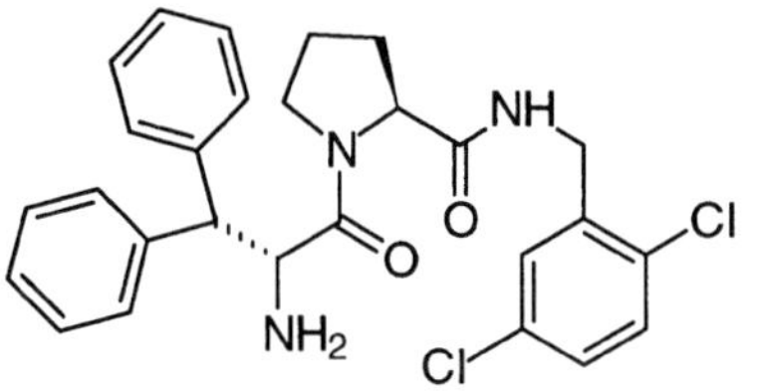

(8.14)
K_i = 3.0 nM

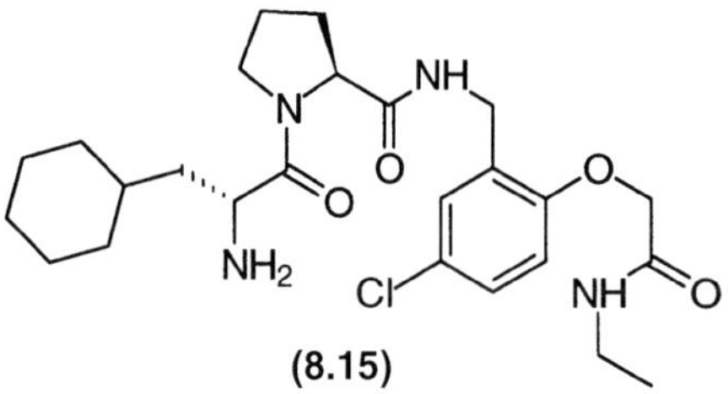

(8.15)
K_i = 0.74 nM

this inhibitor type. Instead of the noragmatine in (**8.8**), N-amidinopiperidine-methyl-amide or 4-amidinobenzylamide were used as P1 elements leading to highly potent inhibitors such as (**8.9**) (Steinmetzer *et al* 1999) and melagatran (**8.10a**) (Elg *et al* 1999). In the melagatran derivative, H 376/95 (**8.10b**), the amidino group and the carboxylic group are modified to the hydroxyamidine and the ethyl ester yielding a prodrug which is converted *in vivo* to the highly active inhibitor (Gustafsson *et al* 2001). In (**8.11**) (Malley *et al* 1996), the hydrophobic P3 D-amino acid was replaced by β-naphthylsulfonylated L-Ser without loss of inhibitory potency. Whereas P2 was only slightly modified, in most cases proline or its homologues were used, many variations were performed with the P3 residue. Highly hydrophobic amino acids such as D-diphenylalanine or D-dicyclohexyl-alanine remarkably increased the potency. However, the extreme lipophilicity resulting from such P3 residues in combination with the lower basicity of P1 in derivatives (**8.12**) (Tucker *et al* 1997a) and (**8.13**) (Feng *et al* 1997) reduced the anticoagulant activity by strong plasma protein binding. In contrast, substitution of P3 (**8.7**, **8.11**) or of the terminal nitrogen (**8.8**, **8.10**) with hydrophilic residues increased both potency and selectivity.

Trans-aminocyclohexylmethylamine-based inhibitors like (**8.12**) were starting points for further development of inhibitors with P1-residues having reduced basicity such as 2-aminopyridine (**8.13**) and imidazole-containing compounds (Wiley *et al* 1999; Lee *et al* 2000). Indeed, tripeptides with uncharged P1 residues even like (**8.14**) (Lumma *et al* 1998) and (**8.15**) (Tucker *et al* 1998) having remarkable affinity were described.

8.3.2 Arginine amides and related inhibitors

The amides of Arg and related amino acids differ in their binding mode within the thrombin active site from the inhibitors of the tripeptide type. In these inhibitors the

C-terminal amide group binds backwards into the S2-pocket of the enzyme in a Y-shaped conformation as described above (see Figure 8.3). The Nα-arylsulfonylated arginine amides were developed by Okamoto *et al* (1981). The leading structure of this type is argatroban (**8.2**), originally named OM-805 or MD-805. With the Arg-derived inhibitors extensive SAR analyses were performed, especially in order to eliminate the four stereogenic centers.

Based on the moderate inhibitory activity of benzamidine toward trypsin-like enzymes, the unnatural amino acids 3- and 4-amidinophenylalanine were introduced into thrombin inhibitors. With a K_i of 2.5 nM for the active D-enantiomer, NAPAP (**8.3**) has been the most potent thrombin inhibitor for quite a time. Variation of the NAPAP structure such as Nα-methylation (Cadroy *et al* 1987), cyclization or exchange of the central Gly does not affect potency and selectivity markedly as was seen with the Asp derivative CRC 220 (**8.16**) (Dickneite *et al* 1995). The Asp in P2 of CRC 220 allows introduction of an additional substituent; derivatives with covalently bound sugar residues or PEG-chains showed delayed elimination from the circulation in experimental animals (Stüber *et al* 1995).

Using L-3-amidinophenylalanine as P1 residue, potent thrombin inhibitors like (**8.17**) were only found when the glycine of the NAPAP-type inhibitors was eliminated (Stürzebecher *et al* 1997). The closely related compound (**8.18**) (Claeson 1994) contains guanidinophenylalanine as a Arg-mimetic.

Relatively early in the development of synthetic thrombin inhibitors, it was assumed that the strongly basic guanidino and amidino groups cause unfavorable *in vivo* characteristics, but the dogma that the basic group is an absolute structural requirement for

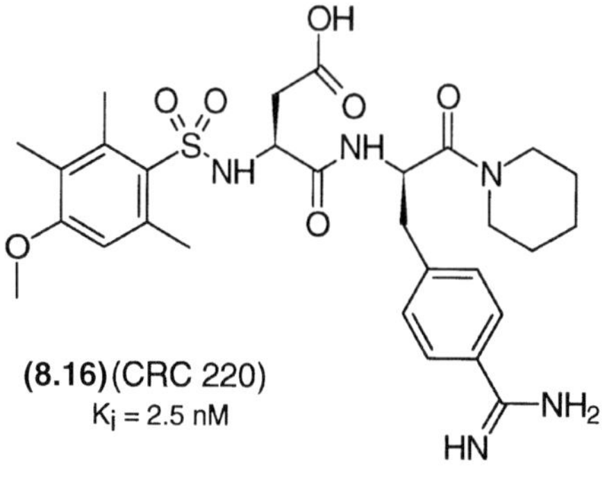

(8.16)(CRC 220)
K_i = 2.5 nM

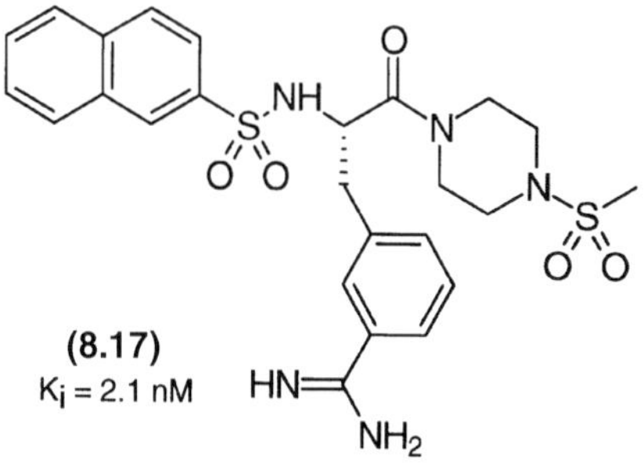

(8.17)
K_i = 2.1 nM

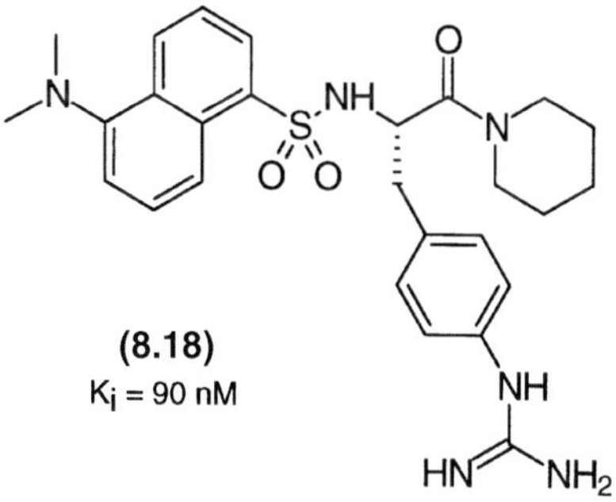

(8.18)
K_i = 90 nM

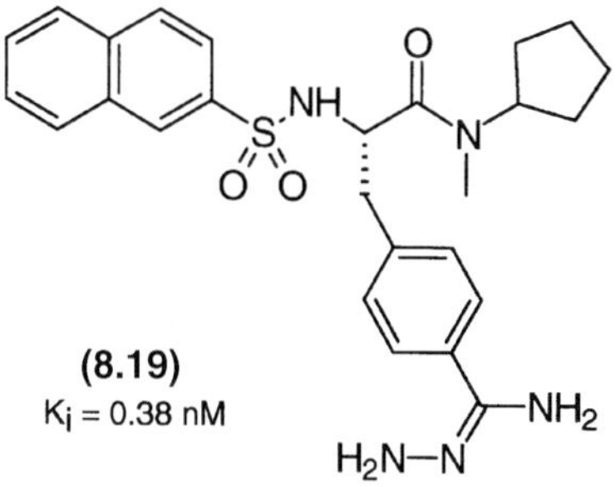

(8.19)
K_i = 0.38 nM

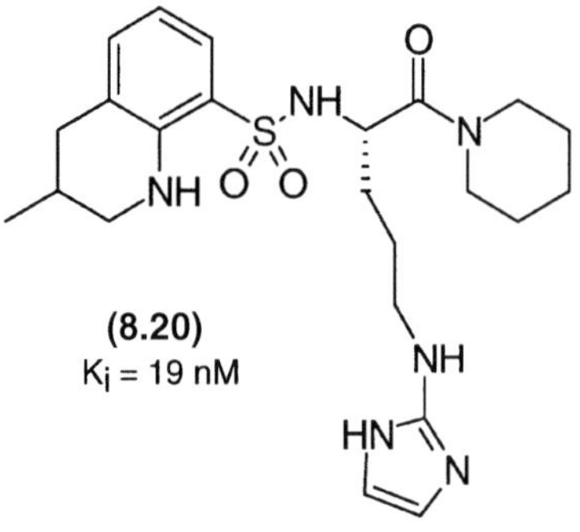

(8.20)
K_i = 19 nM

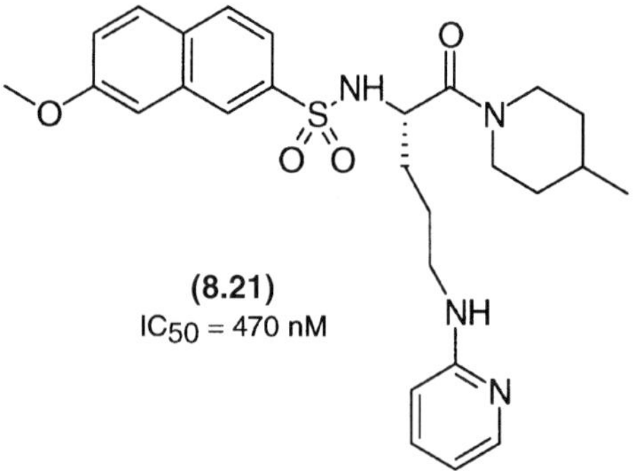

(8.21)
IC_{50} = 470 nM

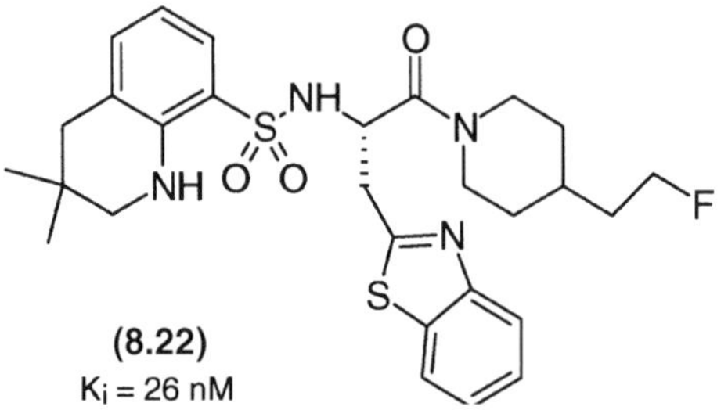

(8.22)
K_i = 26 nM

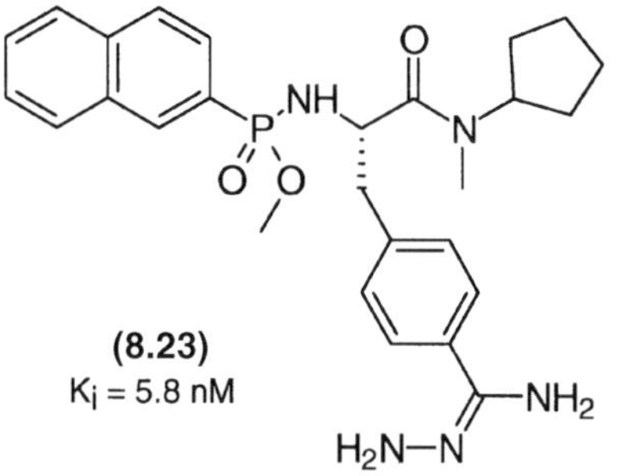

(8.23)
K_i = 5.8 nM

potent thrombin inhibitors was maintained. Meanwhile, numerous approaches to lower the basicity or to eliminate the basic P1 residue have been successful. Some examples are given with tripeptide-type inhibitors; with the amide-type inhibitors similar variations of the basic P1 side chain were performed. The simple modification of the benzamidine into a benzamidrazone moiety (**8.19**) (Oh *et al* 1998) not only reduces the basicity but also enhances the selectivity. Furthermore, like 2-aminoimidazole (**8.20**) and aminopyridine (**8.21**) (Misra *et al* 1994), benzothiazoline (**8.22**) (Ambler *et al* 1999) and several other weakly basic groups were used (e.g. thiopyridine, benzimidazole, aminopyrazine, aminopyrimidine).

In the amide-type inhibitors, the N-terminus needs a protecting group of the sulfonic acid-type because acyl protection reduces the inhibitory potency. However in compound (**8.23**) (Lee and Hwang 1997) the naphthylsulfonyl was successfully replaced by a naphthylphosphonyl residue. The C-terminal amide part is mostly of the piperidine or piperazine type, often substituted D-pipecolic acid is used as in argatroban. However, in the benzamidrazone-based inhibitors acyclic secondary amines were used as in (**8.19**) and (**8.23**).

8.3.3 Inhibitors of other structural types

Despite most of the compounds of this group containing some structural elements of the inhibitors described previously, the peptidic character is essentially abandoned. Napsagatran (Ro 46-6240, (**8.24**) (Hilpert *et al* 1994), based on Nα-2-naphthylsulfonylated Asp, in which both carboxylic groups have an amide structure, is highly potent and selective. Compound (**8.25**) (Engh *et al* 1996; von der Saal *et al* 1997) is a member of a series of inhibitors in which a central 1,3- or 1,3,5-substituted benzene ring fits into the S2 region of thrombin, whereas S1 is occupied by the aminopyridine residue. Both (**8.24**) and (**8.25**) bind to the active site of thrombin like the tripeptides in an extended conformation. Using a central 3-aminopyridinone or 3-aminopyrazinone and a 2-amino-5-aminomethyl-6-methyl-pyridine as S1-binding element, highly potent and orally bioavailable inhibitors like (**8.26**) were developed (Sanderson *et al* 1998a,b).

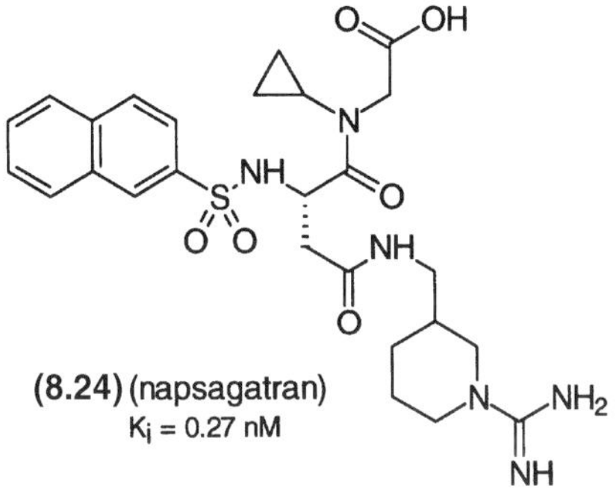

(8.24) (napsagatran)
K_i = 0.27 nM

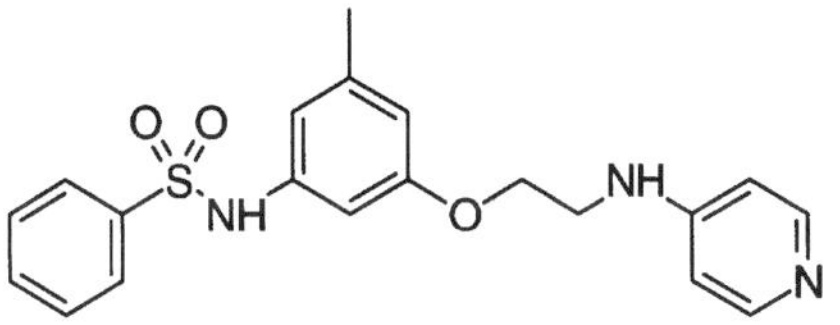

(8.25)
K_i = 23 nM

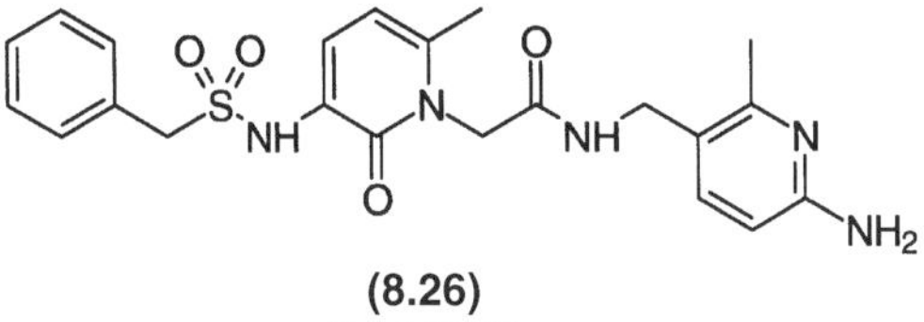

(8.26)
K_i = 0.50 nM

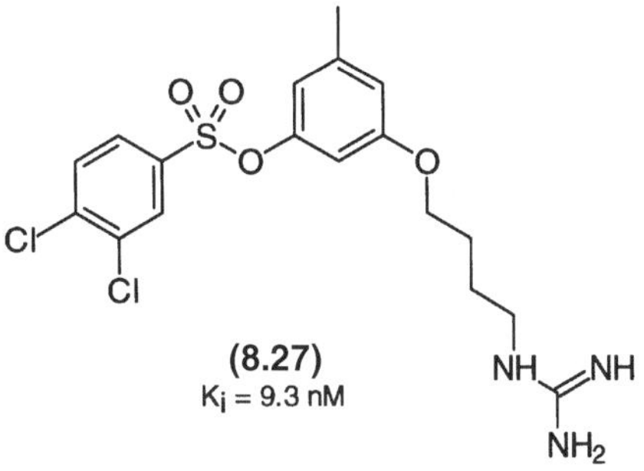

(8.27)
K_i = 9.3 nM

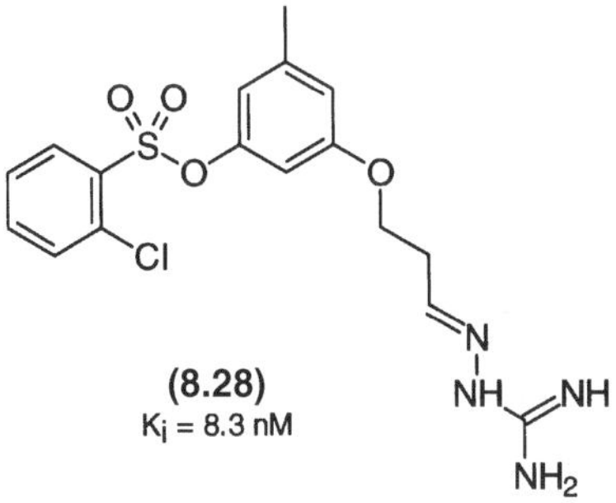

(8.28)
K_i = 8.3 nM

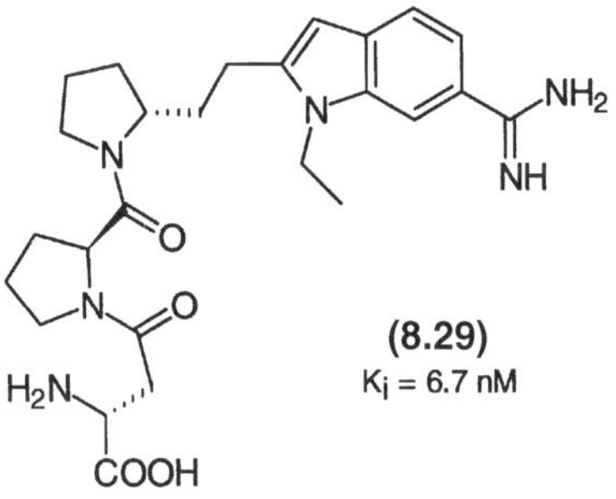

(8.29)
K_i = 6.7 nM

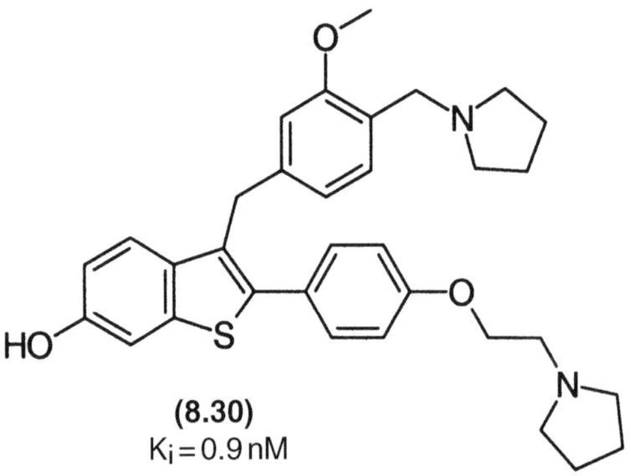

(8.30)
$K_i = 0.9\,nM$

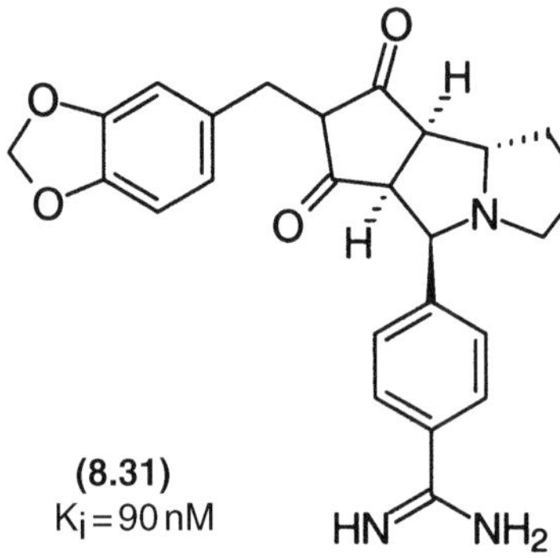

(8.31)
$K_i = 90\,nM$

Closely related inhibitors with guanidino or amidinohydrazone alkyl groups as P1 were described with **(8.27)** and **(8.28)**, the less basic amidinohydrazone derivative showing certain oral bioavailability (Lu *et al* 1998; Soll *et al* 2000). Another basic moiety was introduced with amidinoindole in compounds like **(8.29)** (Cho *et al* 2000). A quite different structure was discovered with the 2,3-disubstituted benzothiophenes as "non-classical" inhibitors of thrombin. From the X-ray crystal structure of one of the most potent compounds **(8.30)** (Zhang *et al* 1999) in complex with thrombin, it was shown that the benzothiophene system binds to the specificity pocket and the hydroxyl interacts with Asp189, whereas one of the side chains interacts with the S2 and S3 regions. No binding interactions were observed with the other substituent. Similarly potent and selective inhibitors were found with further derivatives (Sall *et al* 2000; Takeuchi *et al* 2000). Using 1,3-dipolar cycloaddition, **(8.31)** (Obst *et al* 1995) was the most active compound in a series of heterobi- and heterotricyclic compounds providing a scaffold for combinatorial methods.

8.4 PROPERTIES AND EFFECTS OF THROMBIN INHIBITORS

One of the main criteria for usefulness of a potent thrombin inhibitor is selectivity. The ratio of the affinity of a given inhibitor to thrombin and that to trypsin most commonly serves as a measure of selectivity. It is mandatory that the enzymes of the fibrinolytic system, belonging to the trypsin family, must not be inhibited at therapeutically relevant concentrations (Barabas *et al* 1993; Rupin *et al* 1997; Teger-Nilsson *et al* 1997). PPACK, boroarginines and efegatran lack selectivity with respect to the fibrinolytic system. Moreover, DUP 714 and efegatran were shown to directly inhibit protein Ca (Callas and Fareed 1995).

8.4.1 Anticoagulant and antithrombotic potential

In the evaluation of candidates for further development, an important criterium besides the inhibitory potency and selectivity measured in enzyme inhibition assays is the anti-coagulant potency, measured in plasma clotting assays. The parameters derived from *in vitro* assays are the concentrations of a given thrombin inhibitor effective in prolonging plasma clotting times. They are a measure of the overall *in vitro* performance. The concentration for doubling, compared to controls, the activated partial thromboplastin time (APTT), a clinically used clotting test, is commonly accepted as a measure of the anticoagulant potency of thrombin inhibitors. There is a linear relationship between the K_i and the anticoagulant potency (IC_{200}) over a certain range of K_i values. Table 1 shows for selected inhibitors the inhibition constants and the IC_{200} values. At single-digit nanomolar K_i the APTT values level off, so that the K_i value of hirudin does not translate into a considerably higher anticoagulant activity. The relatively low anticoagulant potency of (**8.12**) is explained by high lipophilicity and plasma protein binding. *Ex vivo*, there is always a correlation between plasma levels and APTT; in this correlation, however, the ratio between K_i and APTT has to be considered.

In contrast to the classical anticoagulant heparin, active site-directed thrombin inhibitors selectively target thrombin, are able to inhibit clot-bound thrombin, do not need a cofactor and are not influenced by platelets factors; in contrast to the vitamin K-antagonists they have an immediate onset of action and affect one enzyme only. The antithrombotic action of active site-directed thrombin inhibitors is fully established. The antithrombotic effect of a given thrombin inhibitor is dependent on the inhibitory potency and the dose, more precisely on the resulting plasma level. Robust models, giving information on the overall *in vivo* performance of a new thrombin inhibitor, are used in the screening of new compounds, such as the rat AV-shunt (Takeuchi *et al* 1999; Zhang *et al* 1999) and the rat ferric chloride arterial thrombosis (Sanderson *et al* 1998a; Tucker *et al* 1998). Antithrombotically effective doses are dependent on the inhibitor potency, the route and mode of administration, the thrombosis model and the species. Thrombin inhibitors are more efficacious in preventing venous thrombosis, compared to arterial thrombosis, in which higher doses are needed. An advantage over heparin is their antithrombotic effect in platelet-dependent thrombosis, which is less sensitive to heparin. Thrombin inhibitors are effective in thrombosis models at lower degrees of anticoagulation, in terms of APTT-prolongation, than heparin and produce less bleeding. They reduce the time to reperfusion in thrombolysis and prevent early reocclusion of vessels.

Table 8.1 Inhibition constant and anticoagulant potency (concentration for doubling of APTT = IC_{200}) of selected thrombin inhibitors in human plasma *in vitro*

Number	*Name*	K_i (nM)	IC_{200} (μM)	*Reference*
8.3	L-NAPAP	1600	>70	Prasa *et al* 1997
8.2	argatroban	19	0.42	Prasa *et al* 1997
8.8	inogatran	15	0.50	Prasa *et al* 1997
8.16	CRC 220	2.5	0.34	Prasa *et al* 1997
8.17		2.1	0.26	Stürzebecher *et al* 1997
8.3	D-NAPAP	2.0	0.25	Prasa *et al* 1997
8.10a	melagatran	2.0	0.59	Gustafsson *et al* 2001
8.30		0.90	0.59	Zhang *et al* 1999
8.15		0.75	0.41	Tucker *et al* 1998
8.26		0.50	0.21	Sanderson *et al* 1998a
8.24	napsagatran	0.27	0.30	Hilpert *et al* 1994
8.12		0.056	1.4	Tucker *et al* 1997a
	hirudin	0.000027	0.092	Prasa *et al* 1997

8.4.2 Thrombin inhibitons and metabolism

The pharmacokinetic profile of choice for oral thrombin inhibitors would be low clearance, terminal half-life appropriate for once or twice-a-day dosing, low liver-extraction ratio (which would give rise to intersubject variability in bioavailability) and high oral bioavailability. Numerous examples show that an unfavorable pharmacokinetic profile may preclude the progression of a newly developed drug candidate to clinical trial. Therefore, it has become increasingly important to study the disposition and metabolism of a novel compound early during its development as a drug (oral). The pharmacokinetics of thrombin inhibitors are closely related to their pharmacodynamics: circulating blood is the primary "effect compartment" for inhibitors of coagulation enzymes, in contrast to the overwhelming majority of drugs which find their targets in tissue/organ compartments. Therefore, the determinants governing the concentrations in plasma are of outstanding importance. The binding to plasma proteins is an item important in characterizing the pharmacokinetics of a given compound. Consequently, the potency in plasma of thrombin inhibitors with a high degree of protein binding is relatively low. For screening of the elimination kinetics it may be sufficient to follow the *ex vivo* "effect" kinetics (anticoagulation) in animals, a procedure not requiring laborious analytical measures.

The pharmacokinetics of most of the substrate-analogous arginine- and benzamidine-derived thrombin inhibitors are characterized by high clearance (short plasma half-life) and low oral bioavailability (for review see Hauptmann and Stürzebecher (1999)). For a number of "first generation" thrombin inhibitors, extensive hepatic extraction governs overall elimination, as is also the case with other peptidomimetic inhibitors of proteinases and peptidases. Hepato-biliary elimination was found for NAPAP, argatroban, efegatran (and similar tripeptides), inogatran and napsagatran in various animal species. Biliary excretion was also demonstrated for napsagatran in humans (Bounameaux *et al* 1997). The active uptake via a multispecific organic anion transporter, Oatp1, of an amidinophenylalanine-type thrombin inhibitor, CRC 220, into isolated rat hepatocytes was described recently (Eckhardt *et al* 1996). Napsagatran showed large interspecies differences for liver and kidney excretion because of involvement of active transport in both organs (Lave *et al* 1999). Hepatic uptake and biliary excretion of thrombin inhibitors may be subject to interactions with other hepato-biliary eliminated drugs resulting in increased plasma levels and lowered biliary clearance (Hauptmann and Stürzebecher 1998).

The low oral bioavailability of most of the "first generation" synthetic thrombin inhibitors has various reasons, one being the low permeation across the intestinal membrane barriers due to insufficient lipophilicity of the compounds bearing strongly basic guanidino or amidino groups and further groups able to form hydrogen bonds. For the majority of xenobiotics, intestinal absorption is accomplished by passive diffusion; usually, hydrophilic molecules cross the mucosal cell layers by the paracellular route, whereas more lipophilic drugs permeate transcellularly. In general, there is an improvement of absorption for more lipophilic molecules. However, increased lipophilicity may, on the one hand, positively influence the pharmacokinetics but, on the other hand it may negatively affect the pharmacodynamics (decrease in anticoagulant activity due to plasma protein binding). Consequently, in the design of thrombin inhibitors the hydrophobicity of a molecule must be balanced (optimization of hydrophobic interactions while maintaining sufficient aqueous solubility) in order not only to be highly active in plasma, but also to provide suitable pharmacokinetic characteristics.

There is a need for models allowing the prediction of oral bioavailability of a candidate drug. The Caco-2 intestinal cell line has been established as an *in vitro* model giving results that correlate with oral bioavailability in humans. The model has been used in several studies on thrombin inhibitors; compared to drugs well absorbed in humans, in most cases their permeation across the cell layer was low. There was a correlation between permeability parameters *in vitro* and the low effect-bioavailability in rats for CRC 220 and related compounds (Walter *et al* 1995). Argatroban analogs with a less basic group showed higher permeability in Caco-2 cells but they had, however, no oral activity in mice (Misra *et al* 1994; Kim *et al* 1996). Compounds with the less basic benzamidine isostere 1-aminoisoquinoline showed better Caco-2 cell permeation than the lead, NA-PAP (Rewinkel *et al* 1999). Compared to melagatran, the less basic melagatran prodrug H 376/95 showed better *in vitro* permeability and higher oral bioavailability in animals and in man (Gustafsson *et al* 2001).

Studies on oral bioavailability in various animal species are far more complex and a number of factors influence the results on a given substance. An important factor is hepatic first-pass metabolism leading to low systemic plasma levels of compounds absorbed readily. Among the thrombin inhibitors described recently, tripeptide derivatives bearing a weakly basic group in P1 showed improved oral bioavailability and prolonged half-life (Tucker *et al* 1997b; Brady *et al* 1998).

Few data have been published on the metabolism of synthetic thrombin inhibitors in animals and humans. Argatroban is metabolized in various species, including man, to several metabolites, one of which is the product of hydroxylation and aromatization of the tetrahydroquinoline ring, having about 30% of the antithrombin activity of the parent compound (Schwarz 1997). Oxidative metabolism of argatroban is unlikely to be an important elimination pathway in man; a cytochrome P450 isoenzyme inhibitor had no influence on the pharmacokinetics of argatroban (Tran *et al* 1999). Efegatran is in part metabolized in rats at the aldehyde group to the corresponding acid and alcohol. In plasma, the epimeric DLD-efegatran is also found (Smith *et al* 1997).

8.5 CLINICAL USE OF THROMBIN INHIBITORS

There are limitations to the established antithrombotic agents with regard to efficacy and safety; heparin and the vitamin K-antagonists have narrow therapeutic windows and variable dose–response relationhips. Small molecule, active site-directed competitive thrombin inhibitors might have potential advantages as antithrombotic drugs (Deutsch *et al* 1993; Fitzgerald 1994; Turpie *et al* 1995; Antman and Braunwald 1996; Verstraete 1997). The thrombin inhibitors used so far in clinical studies, the results of which have been published, are argatroban, efegatran, inogatran, melagatran, H 376/95 and napsagatran. Phase I studies illustrated the importance of the pharmacokinetic profiles of the thrombin inhibitors for dose finding. Argatroban, efegatran, inogatran, and napsagatran were reported with total plasma clearance of 4–7 $ml/min^{-1} kg^{-1}$ in man, whereas the values are considerably higher in animal species used in preclinical studies. The relationships between doses (i.e. the infusion rates), plasma levels and *ex vivo* anticoagulant effects (APTT) were linear. The clearance of melagatran in man is markedly lower (Eriksson *et al* 1999).

Two fields of application can be distinguished. Firstly, the use for anticoagulation over relatively short periods of time in interventional procedures; for instance in percutaneous transluminal coronary angioplasty (PTCA) (Sakamoto *et al* 1995; Suzuki *et al* 1995; Herrman *et al* 1996) or as adjunctive treatment to thrombolysis in patients with myocardial infarction (Vermeer *et al* 1997; Jang *et al* 1999). Secondly, the long-term use in various disease states, such as unstable angina (Gold *et al* 1993; Jackson *et al* 1996; Andersen *et al* 1996; Klootwijk *et al* 1999) and deep venous thrombosis (Bounameaux *et al* 1997; Eriksson *et al* 1999). Since thrombin has emerged as a key player, the therapeutic concept in acute coronary ischemic syndromes is focused on the control of thrombin generation and activity and inhibition of platelet aggregation. However, direct thrombin inhibitors did not bring about apparent long-term benefit in PTCA and, up to now, there is no significant clinical evidence for a superiority of direct thrombin inhibitors over heparin in acute coronary syndromes (Antman 1997; Andersen and Dellborg 1998; Oberhoff and Karsch 1999), despite the unequivocal demonstration of superior antithrombotic effects at lower levels of anticoagulation in a variety of experimental models and the more stable level of anticoagulation reached in patients. The duration of intravenous treatment might have been too short in several studies; orally active thrombin inhibitors would easily allow a prolonged treatment with probably better therapeutic results. The melagatran prodrug, H376/95, was the first thrombin inhibitor to be orally administered in a large clinical trial for prevention of postoperative thrombosis (Gustafsson *et al* 2001). Various other indications, such as prevention of stroke in patients with atrial fibrillation and adjunctive therapy in thrombolysis, are subject to ongoing studies with several thrombin inhibitors.

A special indication of synthetic thrombin inhibitors is given in situations when standard heparin therapy fails or produces even adverse effects. The substitution of heparin by argatroban in patients with heparin-induced thrombocytopenia (HIT) was successful in various clinical settings, i.e. the critical phase of thrombocytopenia associated with thrombosis (Matsuo *et al* 1997; Lewis *et al* 2001), renal stent implant (Lewis *et al* 1997) and haemodialysis (Matsuo *et al* 1992). Argatroban, originally approved in Japan in 1990, was approved for treatment of HIT in the USA in 2000.

8.6 PERSPECTIVES

Safer and more effective oral antithrombotic agents that require less laboratory monitoring are needed for the treatment of thromboembolic disorders. With small molecule direct thrombin inhibitors these criteria should be fulfilled. However, there might be several reasons for the missing breakthrough in the development of oral thrombin inhibitors. It takes place in a field in which other (equally effective?) drugs are available and numerous lead structures have not possessed the clinically required pharmacokinetics/bioavailability, so that further development of several potent inhibitors was terminated. Moreover, a question is whether the missing major advantage over heparin in various clinical trials is due to the characteristics of the inhibitors or to peculiarities of the mode of action of thrombin in the disease processes.

The present design strategies for active site-directed inhibitors of thrombin are no longer focused on optimal binding to the target enzyme only, but also on structural modifications which could modulate the physico-chemical properties and, thus, their fate in the body. Poor enteral absorption and/or extensive hepatic first-pass metabolism are the major obstacles for sufficient oral bioavailability of quite a number of "first generation" thrombin inhibitors. The approach toward balanced lipophilicity of the molecules by structural variations in order to maximize anticoagulant activity and oral bioavailability has led to substantial progress in terms of improved pharmacodynamic and pharmacokinetic profiles of novel thrombin inhibitors. There are highly potent and selective lead compounds providing the basis for further development of "second generation" thrombin inhibitors with high oral bioavailability, appropriate half-life and moderate plasma protein binding. However, despite the powerful new techniques in drug development, e.g. computer-aided molecular design, combinatorial synthesis of large compound libraries and high-throughput screening, there is still a lack of predictability in the discovery of novel compounds with desired properties. The selection of drug candidates for optimization strategies and final development remains to be a difficult task in the drug development process. As with other classes of drugs, various types of oral thrombin inhibitors eventually developed should have the same therapeutic potential and would be distinct only in terms of the dosage regimen. A final answer as to whether the attractive concept of specifically inhibiting the key enzyme in the coagulation cascade is a valid basis for more efficient antithrombotic treatment associated with lower bleeding risk, is expected when the ongoing and further clinical studies on active site-directed small molecule inhibitors of thrombin have been evaluated.

REFERENCES

Ambler, J., Brown, L., Cockcroft, X.L., Grutter, M., Hayler, J., Janus, D. *et al* (1999) Optimisation of the P2 pharmacophore in a series of thrombin inhibitors: ion-dipole interactions with lysine 60G. *Bioorganic & Medicinal Chemistry Letters*, **9**, 1317–1322.

Andersen, K., Dellborg, M., Emanuelsson, H., Grip, L. and Swedberg, K. (1996) Thrombin inhibition with inogatran for unstable angina pectoris: evidence for reactivated ischaemia after cessation of short-term treatment. *Coronary Artery Disease*, **7**, 673–681.

Andersen, K. and Dellborg, M. (1998) Heparin is more effective than inogatran, a low-molecular weight thrombin inhibitor in suppressing ischemia and recurrent angina in unstable coronary disease. Thrombin Inhibition in Myocardial Ischemia (TRIM) Study Group. *American Journal of Cardiology*, **81**, 939–944.

Antman, E.M. and Braunwald, E. (1996) Trials and tribulations of thrombin. *European Heart Journal*, **17**, 971–973.

Antman, E.M. (1997) Another chapter of the antithrombin story has been written. *European Heart Journal*, **18**, 1365–1367.

Bajusz, S., Barabas, E., Tolnay, P., Szell, E. and Bagdy, D. (1978) Inhibition of thrombin and trypsin by tripeptide aldehydes. *International Journal of Peptide and Protein Research*, **12**, 217–221.

Bajusz, S., Szell, E., Barabas, E. and Bagdy, D. (1982) Design and synthesis of peptide inhibitors of blood coagulation. *Folia Haematologica (Leipzig)*, **109**, 16–21.

Bajusz, S., Szell, E., Bagdy, D., Barabas, E., Horvath, G., Dioszegi, M. *et al* (1990) Highly active and selective anticoagulants: D-Phe-Pro-Arg-H, a free tripeptide aldehyde prone to spontaneous inactivation, and its stable N-methyl derivative, D-MePhe-Pro-Arg-H. *Journal of Medicinal Chemistry*, **33**, 1729–1735.

Balasubramanian, B.N. (1995) Advances in the design and development of novel direct and indirect thrombin inhibitors. *Bioorganic & Medicinal Chemistry*, **3**, 999–1156.

Banner, D.W. and Hadvary, P. (1991) Crystallographic analysis at 3.0 Å resolution of the binding to human thrombin of four active site-directed inhibitors. *Journal of Biological Chemistry*, **266**, 20085–20093.

Barabas, E., Szell, E. and Bajusz, S. (1993) Screening for fibrinolysis inhibitory effect of synthetic thrombin inhibitors. *Blood Coagulation & Fibrinolysis*, **4**, 243–248.

Bode, W., Mayr, I., Baumann, U., Huber, R., Stone, S.R. and Hofsteenge, J. (1989) The refined 1.9 Å crystal structure of human α-thrombin: Interaction with D-Phe-Pro-Arg chloromethylketone and significance of the Tyr-Pro-Pro-Trp insertion segment. *EMBO Journal*, **8**, 3467–3475.

Bode, W., Turk, D. and Stürzebecher, J. (1990) Geometry of binding of the benzamidine- and arginine-based inhibitors NAPAP and MQPA to human α-thrombin. X-ray crystallographic determination of the NAPAP-trypsin complex and modeling of NAPAP-thrombin and MQPA-thrombin. *European Journal of Biochemistry*, **193**, 175–182.

Bode, W., Turk, D. and Karshikov, A. (1992) The refined 1.9-Å X-ray crystal structure of D-Phe-Pro-Arg chloromethylketone-inhibited human α-thrombin: Structure analysis, overall structure, electrostatic properties, detailed active-site geometry, and structure-function relationships. *Protein Science*, **1**, 426–471.

Bounameaux, H., Ehringer, H., Hulting, J., Rasche, H., Rapold, H.J. and Zultak, M. (1997) An exploratory trial of two dosages of a novel synthetic thrombin inhibitor (napsagatran, Ro 46–6240) compared with unfractionated heparin for treatment of proximal deep-vein thrombosis. Results of the European Multicenter ADVENT trial. *Thrombosis and Haemostasis*, **78**, 997–1002.

Brady, S.F., Stauffer, K.J., Lumma, W.C., Smith, G.M., Ramjit, H.G., Lewis, S.D. *et al* (1998) Discovery and development of the novel potent orally active thrombin inhibitor N-(9-hydroxy-9-fluorenecarboxy)prolyl trans-4-aminocyclohexylmethyl amide (L-372,460): Coapplication of structure-based design and rapid multiple analogue synthesis on solid support. *Journal of Medicinal Chemistry*, **41**, 401–406.

Brandstetter, H., Turk, D., Hoeffken, H.W., Grosse, D., Stürzebecher, J., Martin, P.D. *et al* (1992) Refined 2.3 Å X-ray crystal structure of bovine thrombin complex formed with the benzamidine and arginine-based thrombin inhibitors NAPAP, 4-TAPAP and MQPA – a starting point for improving antithrombotics. *Journal of Molecular Biology*, **226**, 1085–1099.

Cadroy, Y., Caranobe, C., Bernat, M., Maffrand, J.P., Sieu, P. and Boneu, B. (1987) Antithrombotic and bleeding effects of a new synthetic direct thrombin inhibitor and of standard heparin in rabbits. *Thrombosis and Haemostasis*, **58**, 764–767.

Callas, D.D. and Fareed, J. (1995) Direct inhibition of protein Ca by site directed thrombin Inhibitors: Implications in anticoagulant and thrombolytic therapy. *Thrombosis Research*, **78**, 457–460.

Cho, J.H., Seo, H.S., Yun, C.H., Koo, B., Yoshida, S., Koga, T. *et al* (2000) *In vitro* and *in vivo* studies of AT-1362, a newly synthesized and orally active inhibitor of thrombin. *Thrombosis Research*, **100**, 97–107.

Claeson, G., Philipp, M., Agner, E., Scully, M.F., Metternich, R., Kakkar, V.V. *et al* (1993) Benzyloxycarbonyl-D-Phe-Pro-methoxypropylglycine – a novel inhibitor of thrombin with high selectivity containing a neutral side chain at the P1 position. *Biochemical Journal*, **290**, 309–312.

Claeson, G. (1994) Synthetic peptides and peptidomimetics as substrates and inhibitors of thrombin and other proteases in the blood coagulation system. *Blood Coagulation & Fibrinolysis*, **5**, 411–436.

Clarke, R.J., Mayo, G., FitzGerald, G.A. and Fitzgerald, D.J. (1991) Combined administration of aspirin and a specific thrombin inhibitor in man. *Circulation*, **83**, 1510–1518.

Deadman, J.J., Elgendy, S., Goodwin, C.A., Green, D., Baban, J.A., Patel, G. *et al* (1995) Characterization of a class of peptide boronates with neutral P1 side chains as highly selective inhibitors of thrombin. *Journal of Medicinal Chemistry*, **38**, 1511–1522.

Deutsch, E., Rao, A.K. and Colman, R.W. (1993) Selective thrombin inhibitors: The next generation of anticoagulants. *Journal of the American College of Cardiology*, **22**, 1089–1092.

Dickneite, G., Seiffge, D., Diehl, K.H., Reers, M., Czech, J., Weinmann, E. *et al* (1995) Pharmacological characterization of a new 4-amidinophenylalanine thrombin-inhibitor (CRC 220). *Thrombosis Research*, **77**, 357–368.

Eckhardt, U., Stüber, W., Dickneite, G., Reers, M. and Petzinger, E. (1996) First-pass elimination of a peptidomimetic thrombin inhibitor is due to carrier-mediated uptake by the liver – interaction with bile acid transport systems. *Biochemical Pharmacology*, **52**, 85–96.

Elg, M., Gustafsson, D. and Deinum, J. (1997) The importance of enzyme inhibition kinetics for the effect of thrombin inhibitors in a rat model of arterial thrombosis. *Thrombosis and Haemostasis*, **78**, 1286–1292.

Elg, M., Gustafsson, D. and Carlsson, S. (1999) Antithrombotic effects and bleeding time of thrombin inhibitors and warfarin in the rat. *Thrombosis Research*, **94**, 187–197.

Engh, R.A., Brandstetter, H., Sucher, G., Eichinger, A., Baumann, U., Bode, W. *et al* (1996) Enzyme flexibility, solvent and 'weak' interactions characterize thrombin–ligand interactions: Implications for drug design. *Structure*, **4**, 1353–1362.

Eriksson, H., Eriksson, U.G., Frison, L., Hansson, P.O., Held, P., Holmström, M. *et al* (1999) Pharmacokinetics and pharmacodynamics of melagatran, a novel synthetic LMW thrombin inhibitor, in patients with acute DVT. *Thrombosis and Haemostasis*, **81**, 358–363.

Feng, D.-M., Gardell, S.J., Lewis, S.D., Bock, M.G., Chen, Z., Freidinger, R.M. *et al* (1997) Discovery of a novel, selective, and orally bioavailable class of thrombin inhibitors incorporating amidino-pyridyl moieties at the P1 position. *Journal of Medicinal Chemistry*, **40**, 3726–3733.

Fitzgerald, D. (1994) Specific thrombin inhibitors *in vivo*. *Annals of the New York Academy of Sciences*, **714**, 41–52.

Gold, H.K., Torres, F.W., Garabedian, H.D., Werner, W., Jang, I., Khan, A. *et al* (1993) Evidence for a rebound coagulation phenomenon after cessation of a 4-hour infusion of a specific thrombin inhibitor in patients with unstable angina pectoris. *Journal of the American College of Cardiology*, **21**, 1039–1047.

Gustafsson, D., Nyström, J.E., Carlsson, S., Eriksson, U., Gyzander, E., Elg, M. *et al* (2001) The direct thrombin inhibitor melagatran and its oral prodrug H 376/95: Intestinal absorption properties, biochemical and pharmacodynamic effects. *Thrombosis Research*, **101**, 171–181.

Hauptmann, J. and Stürzebecher, J. (1998) Influence of indocyanine green on plasma disappearance and biliary excretion of a synthetic thrombin inhibitor of the 3-amidino-phenylalanine piperazide-type in rats. *Pharmaceutical Research*, **15**, 751–754.

Hauptmann, J. and Stürzebecher, J. (1999) Synthetic inhibitors of thrombin and factor Xa: from bench to bedside. *Thrombosis Research*, **93**, 203–241.

Herrman, J.P.R., Suryapranata, H., den Heijer, P., Gabriel, L., Kutryk, M.J.B. and Serruys, P.W. (1996) Argatroban during percutaneous transluminal angioplasty; results of a dose verification study. *Journal of Thrombosis & Thrombolysis*, **3**, 367–375.

Hilpert, K., Ackermann, J., Banner, D.W., Gast, A., Gubernator, K., Hadvary, P. *et al* (1994) Design and synthesis of potent and highly selective thrombin inhibitors. *Journal of Medicinal Chemistry*, **37**, 3889–3901.

Jackson, C.V., Satterwhite, J. and Roberts, E. (1996) Preclinical and clinical pharmacology of efegatran (LY294468): a novel antithrombin for the treatment of acute coronary syndromes. *Clinical and Applied Thrombosis/Hemostasis*, **2**, 258–267.

Jang, I.K., Brown, D.F., Giugliano, R.P., Anderson, H.V., Losordo, D., Nicolau, J.C. *et al* (1999) A multicenter, randomized study of argatroban versus heparin as adjunct to tissue plasminogen activator (TPA) in acute myocardial infarction: myocardial infarction with novastan and TPA (MINT) study. *Journal of the American College of Cardiology*, **33**, 1879–1885.

Kettner, C. and Shaw, E. (1979) D-Phe-Pro-ArgCH_2Cl – A selective affinity label for thrombin. *Thrombosis Research*, **14**, 969–973.

Kettner, C. and Knabb, R.M. (1993) Peptide boronic acid inhibitors of thrombin. *Advances in Experimental Medicine and Biology*, **340**, 109–118.

Kim, K.S., Moquin, R.V., Quian, L.G., Morrison, R.A., Seiler, S.M., Roberts, D.G.M. *et al* (1996) Preparation of argatroban analog thrombin inhibitors with reduced basic guanidine moiety, and studies of their permeability and antithrombotic activities. *Medicinal Chemistry Research*, **6**, 377–383.

Kimball, S.D. (1995) Challenges in the development of orally bioavailable thrombin active site inhibitors. *Blood Coagulation & Fibrinolysis*, **6**, 511–519.

Kimball, S.D. (1999) Oral thrombin inhibitors: Challenges and progress. In *Antithrombotics*, edited by A.C.G. Uprichard and K.P. Gallagher, pp. 367–396. Berlin: Springer-Verlag.

Klootwijk, P., Lenderink, T., Meij, S., Boersma, H., Melkert, R., Umans, V.A. *et al* (1999) Anti-coagulant properties, clinical efficacy and safety of efegatran, a direct thrombin inhibitor, in patients with unstable angina. *European Heart Journal*, **20**, 1101–1111.

Lave, T., Portmann, R., Schenker, G., Gianni, A., Guenzi, A., Girometta, M.A. *et al* (1999) Interspecies pharmacokinetic comparisons and allometric scaling of napsagatran, a low molecular weight thrombin inhibitor. *Journal of Pharmacy and Pharmacology*, **51**, 85–91.

Lee, K. (1997) Recent progress in small molecule thrombin inhibitors. *Korean Journal of Medicinal Chemistry*, **7**, 127–144.

Lee, K. and Hwang, S.Y. (1997) Phosphamidate derivatives of 4-amidrazonophenylalanine: A new class of selective thrombin inhibitors. *Korean Journal of Medicinal Chemistry*, **7**, 86–89.

Lee, K., Jung, W.H., Kang, M. and Lee, S.H. (2000) Noncovalent thrombin inhibitors incorporating an imidazolylethynyl P1. *Bioorganic & Medicinal Chemistry Letters*, **10**, 2775–2778

Lewis, B.E., Grassman, E.D., Wrona, L. and Rangel, Y. (1997) Novastan anticoagulation during renal stent implant in a patient with heparin-induced thrombocytopenia. *Blood Coagulation & Fibrinolysis*, **8**, 54–58.

Lewis, B.E., Wallis, D.E., Berkowitz, S.D., Matthai, W.H., Fareed, J., Walenga, J.M. *et al* (2001) Argatroban anticoagulant therapy in patients with heparin-induced thrombocytopenia. *Circulation*, **103**, 1838–1843.

Lu, T.B., Tomczuk, B., Illig, C.R., Bone, R., Murphy, L., Spurlino, J. *et al* (1998) *In vitro* evaluation and crystallographic analysis of a new class of selective, non-amide-based thrombin inhibitors. *Bioorganic & Medicinal Chemistry Letters*, **8**, 1595–1600.

Lumma, W.C., Witherup, K.M., Tucker, T.J., Brady, S.F., Sisko, J.T., Naylor-Olsen, A.M. *et al* (1998) Design of novel, potent, noncovalent inhibitors of thrombin with nonbasic P-1 substructures: rapid structure–activity studies by solid-phase synthesis. *Journal of Medicinal Chemistry*, **41**, 1011–1013.

Malley, M.F., Tabernero, L., Chang, C.Y.Y., Ohringer, S.L., Roberts, D.G.M., Das, J. *et al* (1996) Crystallographic determination of the structures of human α-thrombin complexed with BMS-186282 and BMS-18090. *Protein Science*, **5**, 221–228.

Matsuo, T., Kario, K., Kodama, K. and Okamoto, S. (1992) Clinical applications of the synthetic thrombin inhibitor argatroban (MD-805). *Seminars in Thrombosis and Hemostasis*, **18**, 155–160.

Matsuo, T., Koide, M. and Kario, K. (1997) Application of argatroban, direct thrombin inhibitor, in heparin-intolerant patients requiring extracorporeal circulation. *Artificial Organs*, **21**, 1035–1038.

Menear, K. (1998) Progress towards the discovery of orally active thrombin inhibitors. *Current Medicinal Chemistry*, **5**, 457–468.

Misra, R.N., Kelly, Y.F., Brown, B.R., Roberts, D.G.M., Chong, S. and Seiler, S.M. (1994) Argatroban analogs: Synthesis, thrombin inhibitory activity and cell permeability of aminoheterocyclic guanidine surrogates. *Bioorganic & Medicinal Chemistry Letters*, **4**, 2165–2170.

Nilsson, T., Sjöling-Ericksson, A. and Deinum, J. (1998) The mechanism of binding of low-molecular-weight active site inhibitors to human α-thrombin. *Journal of Enzyme Inhibition*, **13**, 11–29.

Oberhoff, M. and Karsch, K.R. (1999) Synthetic direct thrombin inhibitors in unstable angina – more questions than answers. *European Heart Journal*, **20**, 1058–1060.

Obst, U., Gramlich, V., Diederich, F., Weber, L. and Banner, D.W. (1995) Design of novel, nonpeptidic thrombin inhibitors and structure of a thrombin–inhibitor complex. *Angewandte Chemie-International Edition*, **34**, 1739–1742.

Oh, Y.S., Yun, M., Hwang, S.Y., Hong, S., Shin, Y., Lee, K. *et al* (1998) Discovery of LB30057, a benzamidrazone-based selective oral thrombin inhibitor. *Bioorganic & Medicinal Chemistry Letters*, **8**, 631–634.

Okamoto, S., Hijikata, A., Kikumoto, R., Tonomura, S., Hara, N., Ninomyia, K. *et al* (1981) Potent inhibition of thrombin by the newly synthesized arginine derivative No. 805: the importance of the stereostructure of its hydrophobic carboxamide portion. *Biochemical and Biophysical Research Communications*, **101**, 440–446.

Prasa, D., Svendsen, L. and Stürzebecher, J. (1997) The ability of thrombin inhibitors to reduce the thrombin activity generated in plasma on extrinsic and intrinsic activation. *Thrombosis and Haemostasis*, **77**, 498–503.

Priestle, J.P., Rahuel, J., Rink, H., Tones, M. and Grütter, M.G. (1993) Changes in interactions in complexes of hirudin derivatives and human α-thrombin due to different crystal forms. *Protein Science*, **2**, 1630–1642.

Quan, M.L., Wityak, J., Dominguez, C., Duncia, J.V., Kettner, C.A., Ellis, C.D. *et al* (1997) Biaryl substituted alkylboronate esters as thrombin inhibitors. *Bioorganic & Medicinal Chemistry Letters*, **7**, 1595–1600.

Rewinkel, J.B.M. and Adang, A.E.P. (1999) Strategies and progress towards the ideal orally active thrombin inhibitor. *Current Pharmaceutical Design*, **5**, 1043–1075.

Rewinkel, J.B., Lucas, H., van Galen, P.J., Noach, A.B., van Dinther, T.G., Rood, A.M. *et al* (1999) 1-Aminoisoquinoline as benzamidine isoster in the design and synthesis of orally active thrombin inhibitors. *Bioorganic & Medicinal Chemistry Letters*, **9**, 685–690.

Rupin, A., Mennecier, P., de Nanteuil, G., Laubie, M. and Verbeuren, T.J. (1995) A screening procedure to evaluate the anticoagulant activity and the kinetic behavior of direct thrombin inhibitors. *Thrombosis Research*, **78**, 217–225.

Rupin, A., Mennecier, P., Lila, C., de Nanteuil, G. and Verbeuren, T.J. (1997) Selection of S18326 as a new potent and selective boronic acid direct thrombin inhibitor. *Thrombosis and Haemostasis*, **78**, 1221–1227.

Sakamoto, S., Hirase, T., Suzuki, S., Tsukamoto, T., Miki, T., Yamada, T. *et al* (1995) Inhibitory effect of argatroban on thrombin–antithrombin III complex after percutaneous transluminal angioplasty. *Thrombosis and Haemostasis*, **74**, 801–802.

Sall, D.J., Bailey, D.L., Bastian, J.A., Buben, J.A., Chirgadze, N.Y., Clemens-Smith, A.C. *et al* (2000) Diamino benzo[*b*]thiophene derivatives as a novel class of active site directed thrombin inhibitors. 5. Potency, efficacy, and pharmacokinetic properties of modified C-3 side chain derivatives. *Journal of Medicinal Chemistry*, **43**, 649–663.

Sanderson, P.E.J. and Naylor-Olsen, A.M. (1998) Thrombin inhibitor design. *Current Medicinal Chemistry*, **5**, 289–304.

Sanderson, P.E.J., Cutrona, K.J., Dorsey, B.D., Dyer, D.L., McDonough, C.M., Naylor-Olsen, A.M. *et al* (1998a) L-374,087, an efficacious, orally bioavailable, pyridinone acetamide thrombin inhibitor. *Bioorganic & Medicinal Chemistry Letters*, **8**, 817–822.

Sanderson, P.E.J., Lyle, T.A., Cutrona, K.J., Dyer, D.L., Dorsey, B.D., McDonough, C.M. *et al* (1998b) Efficacous, orally bioavailable thrombin inhibitors based on 3-aminopyridinone or 3-aminopyrazinone acetamide peptidomimetic templates. *Journal of Medicinal Chemistry*, **23**, 4466–4474.

Schwarz, R.P. (1997) The preclinical and clinical pharmacology of Novastan (argatroban). In *New Anticoagulants for the Cardiovascular Patient*, edited by R. Pifarre, pp. 231–249. Philadelphia: Hanley & Belfus.

Semple, J.E., Rowley, D.C., Brunck, T.K., Ha-Uong, T., Minami, N.K., Owens, T.D. *et al* (1996) Design, synthesis and evolution of a novel, selective, and orally bioavailable class of thrombin inhibitors: P1-argininal derivatives incorporating P3-P4 lactam sulfonamide moieties. *Journal of Medicinal Chemistry*, **39**, 4531–4536.

Smith, G.F., Gifford-Moore, D., Craft, T.J., Chirgadze, N., Ruterbories, K.J., Lindstrom, T.D. *et al* (1997) Efegatran: A new cardiovascular anticoagulant. In *New Anticoagulants for the Cardiovascular Patient*, edited by R. Pifarre, pp. 265–300. Philadelphia: Hanley & Belfus.

Soll, R.M., Lu, T., Tomczuk, B., Illig, C.R., Fedde, C., Eisennagel, S. *et al* (2000) Amidinohydrazones as guanidine bioisosteres: application to a new class of potent, selective and orally bioavailable, non-amide-based small-molecule thrombin inhibitors. *Bioorganic & Medicinal Chemistry Letters*, **10**, 1–4.

Steinmetzer, T., Batdorsdhjin, M., Kleinwächter, P., Seyfarth, L., Greiner, G., Reißmann, S. *et al* (1999) New thrombin inhibitors based on D-Cha-Pro-derivatives. *Journal of Enzyme Inhibition*, **14**, 203–216.

Stone, S.R. and Tapparelli, C. (1995) Thrombin inhibitors as antithrombotic agents: The importance of rapid inhibition. *Journal of Enzyme Inhibition*, **9**, 3–16.

Stubbs, M.T. and Bode, W. (1993) A player of many parts: The spotlight falls on thrombin's structure. *Thrombosis Research*, **69**, 1–58.

Stüber, W., Koschinsky, R., Reers, M., Hoffmann, D., Czech, J. and Dickneite, G. (1995) Preparation and evaluation of PEG-bound thrombin inhibitors based on 4-amidinophenylalanine. *Peptide Research*, **8**, 78–85.

Stürzebecher, J., Markwardt, F., Voigt, B., Wagner, G. and Walsmann, P. (1983) Cyclic amides of Nα-arylsulfonaminoacylated 4-amidinophenylalanine – tight binding inhibitors of thrombin. *Thrombosis Research*, **29**, 635–642.

Stürzebecher, J., Prasa, D., Hauptmann, J., Wikström, P. and Vieweg, H. (1997) Synthesis and structure–activity relationships of potent thrombin inhibitors: Piperazides of 3-amidinophenylalanine. *Journal of Medicinal Chemistry*, **40**, 3091–3099.

Suzuki, A., Sakamoto, S., Adachi, K., Mizutani, K., Koide, M., Ohga, N. *et al* (1995) Effect of argatroban on thrombus formation during acute coronary occlusion after balloon angioplasty. *Thrombosis Research*, **77**, 369–373.

Takeuchi, K., Kohn, T.J., Sall, D.J., Denney, M.L., McCowan, J.R., Smith, G.F. *et al* (1999) Dibasic benzo[*b*]thiophene derivatives as a novel class of active site directed thrombin inhibitors: 4. SAR studies on the conformationally restricted C3-side chain of hydroxybenzo[*b*]thiophenes. *Bioorganic & Medicinal Chemistry Letters*, **9**, 759–764.

Takeuchi, K., Kohn, T.J., Harper, R.W., Lin, H.S., Gifford-Moore, D.S., Richett, M.E. *et al* (2000) Diamino benzo[*b*]thiophene derivatives as a novel class of active site directed thrombin inhibitors. Part 6: further focus on the contracted C4′-side chain analogues. *Bioorganic & Medicinal Chemistry Letters*, **10**, 1199–1202.

Tapparelli, C., Metternich, R., Ehrhardt, C. and Cook, N.S. (1993) Synthetic low-molecular weight thrombin inhibitors: Molecular design and pharmacological profile. *Trends in Pharmacological Sciences*, **14**, 366–376.

Teger-Nilsson, A.-C., Bylund, R., Gustafsson, D., Gyzander, E. and Eriksson, U. (1997) *In vitro* effects of inogatran, a selective low molecular weight thrombin inhibitor. *Thrombosis Research*, **85**, 133–145.

Tran, J.Q., Di Cicco, R.A., Sheth, S.B., Tucci, M., Peng, L., Jorkasky, D.K. *et al* (1999) Assessment of the potential pharmacokinetic and pharmacodynamic interactions between erythromycin and argatroban. *Journal of Clinical Pharmacology*, **39**, 513–519.

Tucker, T.J., Lumma, W.C., Lewis, S.D., Gardell, S.J., Lucas, B.J., Baskin, E.P. *et al* (1997a) Potent noncovalent thrombin inhibitors that utilize the unique amino acid D-dicyclohexylalanine in the P3 position. Implications on oral bioavailability and antithrombotic efficacy. *Journal of Medicinal Chemistry*, **40**, 1565–1569.

Tucker, T.J., Lumma, W.C., Lewis, S.D., Gardell, S.J., Lucas, B.J., Sisko, J.T. *et al* (1997b) Synthesis of a series of potent and orally bioavailable thrombin inhibitors that utilize 3,3-disubstituted propionic acid derivatives in the P3 position. *Journal of Medicinal Chemistry*, **40**, 3687–3693.

Tucker, T.J., Brady, S.F., Lumma, W.C., Lewis, S.D., Gardell, S.J., Naylor-Olsen, A.M. *et al* (1998) Design and synthesis of a series of potent and orally bioavailable noncovalent thrombin inhibitors that utilize nonbasic groups in the P1 position. *Journal of Medicinal Chemistry*, **41**, 3210–3219.

Turpie, A.G.G., Weitz, J.I. and Hirsh, J. (1995) Advances in antithrombotic therapy: Novel agents. *Thrombosis and Haemostasis*, **74**, 565–571.

Vermeer, F., Vahanina, A., Fels, P.W., Besse, P., Radzik, D. and Simoons, M.L. (1997) Intravenous argatroban versus heparin as co-medication to alteplase in the treatment of acute myocardial infarction; preliminary results of the ARGAMI pilot study. *Journal of the American College of Cardiology*, **29**, 185–186.

Verstraete, M. (1997) Direct thrombin inhibitors: Appraisal of the antithrombotic/hemorrhagic balance. *Thrombosis and Haemostasis*, **78**, 357–363.

von der Saal, W., Kucznierz, R., Leinert, H. and Engh, R.A. (1997) Derivatives of 4-amino-pyridine as selective thrombin inhibitors. *Bioorganic & Medicinal Chemistry Letters*, **7**, 1283–1288.

Wallace, A.C., Laskowski, R.A. and Thornton, J.M. (1995) LIGPLOT: A program to generate schematic diagrams of protein–ligand interactions. *Protein Engineering*, **8**, 127–134.

Walter, E., Kissel, T., Reers, M., Dickneite, G., Hoffmann, D. and Stüber, W. (1995) Transepithelial transport properties of peptidomimetic thrombin inhibitors in monolayer of human intestinal cell line (Caco-2) and their correlation to *in vivo* data. *Pharmaceutical Research*, **12**, 360–365.

Wiley, M.R. and Fisher, M.J. (1997) Small molecule direct thrombin inhibitors. *Expert Opinion in Therapeutic Patents*, **7**, 1265–1282.

Wiley, M.R., Weir, L.C., Briggs, S.L., Chirgadze, N.Y., Clawson, D., Gifford-Morre, D.S. *et al.* (1999) The design of potent, selective, non-covalent, peptide thrombin inhibitors utilizing imidazole as a S1 binding element. *Bioorganic & Medicinal Chemistry Letters*, **9**, 2767–2772.

Zhang, M., Bailey, D.L., Bastian, J.A., Briggs, S.L., Chirgadze, N.Y., Clawson, D.K. *et al* (1999) Dibasic benzo[*b*]thiophene derivatives as a novel class of active site directed thrombin inhibitors: 2. Sidechain optimization and demonstration of *in vivo* efficacy. *Bioorganic & Medicinal Chemistry Letters*, **9**, 775–780.

Chapter 9

Inhibitors of Factor VIIa, Factor IXa, and Factor Xa as anticoagulants

Robert A. Lazarus and Daniel Kirchhofer

Factor VIIa, Factor IXa, and Factor Xa are three of the key serine proteases in the coagulation cascade. Inhibition of any of these enzymes may prevent the formation of fibrin clots and thus be useful in the management of thrombotic disease. A wide array of strategic approaches to inhibiting the function of these enzymes have been pursued. These approaches involve antagonists that inhibit catalysis directly at the active site, either reversibly or irreversibly, as well as those that impair function by binding to exosites that may interfere with substrate, membrane, or cofactor binding. Antagonists include active site inhibited enzymes, mutagenized enzymes and cofactors, antibodies, naturally-occurring protein inhibitors, peptides, and small molecule active site inhibitors. The characteristics and rationale of each enzyme as a target as well as the properties of various inhibitors are discussed.

9.1 INTRODUCTION

The enzymes, cofactors, and inhibitors involved in the coagulation process are essential in maintaining normal hemostasis. However, pathological thrombosis resulting from uncontrolled intravascular activation of coagulation can lead to serious clinical conditions. These include deep vein thrombosis, pulmonary embolism, myocardial infarction, stroke and disseminated intravascular coagulation during sepsis. Therefore, it is not surprising that the initiation and regulation of coagulation is extremely complex in order to maintain the exquisite balance between hemostasis and thrombosis.

The classical coagulation cascade comprises an intrinsic pathway and an extrinsic pathway (Davie, Fujikawa and Kisiel 1991). Since deficiencies of several proteins in the initiation of the intrinsic pathway do not manifest in bleeding disorders, a revised model for initiating coagulation involving only the extrinsic pathway is now favored (Figure 9.1) (Nemerson 1988; Broze Jr. 1992; Davie 1995; Rapaport and Rao 1995; Mann 1999). The extrinsic pathway is triggered by exposure of zymogen Factor VII (FVII) to its membrane bound cofactor, tissue factor (TF), to form the TF•FVII complex. The TF•FVII complex is converted to the enzymatically active TF•FVIIa complex by FXa or autocatalytically by TF•FVIIa (Figure 9.1). The TF•FVIIa complex can then activate Factor IX to Factor IXa (FIXa), Factor X to Factor Xa (FXa), and Factor VII to FVIIa. FX can also be activated to FXa by the complex of FIXa with its cofactor FVIIIa (Xase complex) and FIX can be activated to FIXa by FXIa via the intrinsic pathway. Thus FIXa and FXa represent points of convergence for the intrinsic and extrinsic pathways. FXa in complex with its cofactor FVa (prothrombinase complex) activates prothrombin to thrombin which cleaves

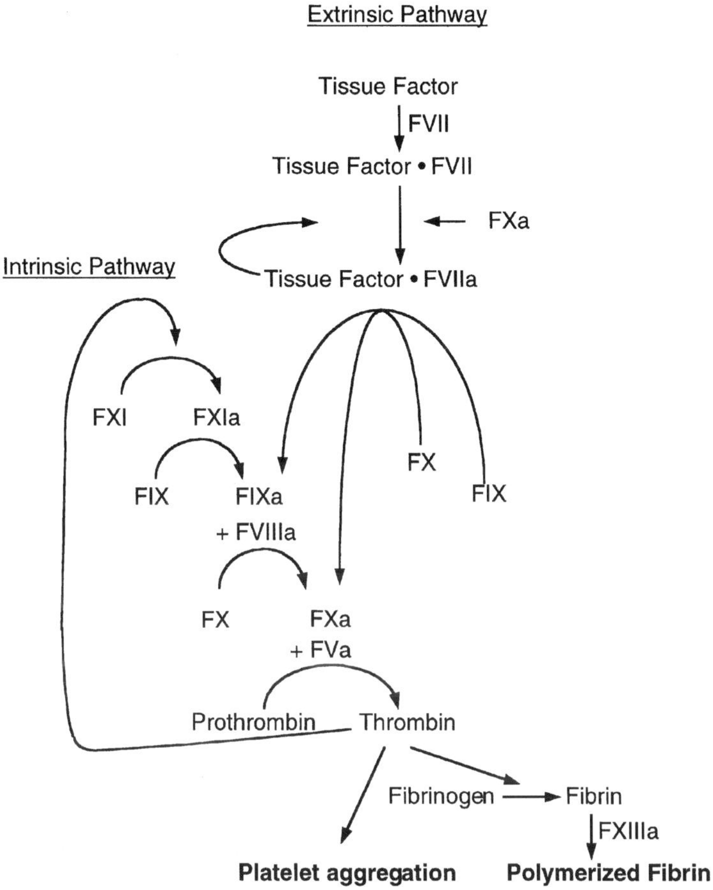

Figure 9.1 Scheme of the coagulation cascade. The depicted role of the TF•FVIIa complex as primary initiator of the coagulation reactions reflects the currently held view of blood coagulation (Nemerson 1988; Broze Jr. 1992; Davie 1995; Rapaport and Rao 1995; Mann 1999). Coagulation is initiated upon tissue injury when TF becomes exposed to blood and the active TF•FVIIa complex is formed. The extrinsic pathway (right) and intrinsic pathway (left) merge at the levels of FIXa and FXa. The intrinsic pathway is activated by the thrombin-mediated conversion of FXI to FXIa. The scheme depicts the two major roles of thrombin in the context of hemostasis and thrombosis, i.e. the cleavage of fibrinogen to fibrin leading to fibrin clot formation and the activation of platelets leading to platelet aggregation.

fibrinogen to fibrin, ultimately resulting in the formation of a fibrin clot. Thrombin also serves to further amplify coagulation by activation of cofactors such as FV and FVIII and zymogens such as Factor XI in the intrinsic pathway. Moreover, thrombin activates platelets leading to platelet aggregation, which is necessary for the formation of a hemostatic plug.

Current anticoagulants in clinical use include heparin, low molecular weight heparin (LMWH), and orally active coumarins such as warfarin (Hirsh 1991; Hirsh, Ginsberg and

Marder 1994; Weitz 1997). Heparins enhance the antithrombin III (ATIII)-mediated inhibition of thrombin and FXa, whereas coumarins impair the function of the vitamin K-dependent proteins including both procoagulants (thrombin, FXa, FIXa, and FVIIa) and anticoagulants (activated protein C and protein S). Vitamin K is an essential cofactor for the post-translational modification of glutamic residues to γ-carboxyglutamate residues, which are critical for the Ca^{2+}-dependent interactions of these proteins with phospholipid membranes. Coumarins inhibit the interconversion of vitamin K and vitamin K epoxide and thus limit the extent of γ-carboxyglutamate formation on these enzymes. Although both unfractionated heparins and coumarins are of great clinical value, they require careful dosing and frequent monitoring. This drawback in the clinical use of unfractionated heparins has been largely eliminated by the introduction of LMWH (Weitz 1997; Clagett *et al* 1998; Bates and Hirsh 1999). The nonselective mode of inhibition by these anticoagulants may account for their therapeutic limitations in maintaining the balance between thrombosis and hemostasis. An ideal anticoagulant would inhibit thrombosis without affecting hemostasis, have acceptable bioavailability and pharmacokinetics, and have a broad therapeutic window with minimal bleeding or other side effects (Sixma and de Groot 1992).

FVIIa, FIXa, and FXa are vitamin K-dependent glycosylated serine proteases derived from their zymogen precursors. Each of these enzymes is ~50 kDa and shares a similar molecular architecture, being composed of an amino-terminal γ-carboxyglutamic acid-rich (Gla) domain, two epidermal growth factor (EGF)-like domains, and a serine protease

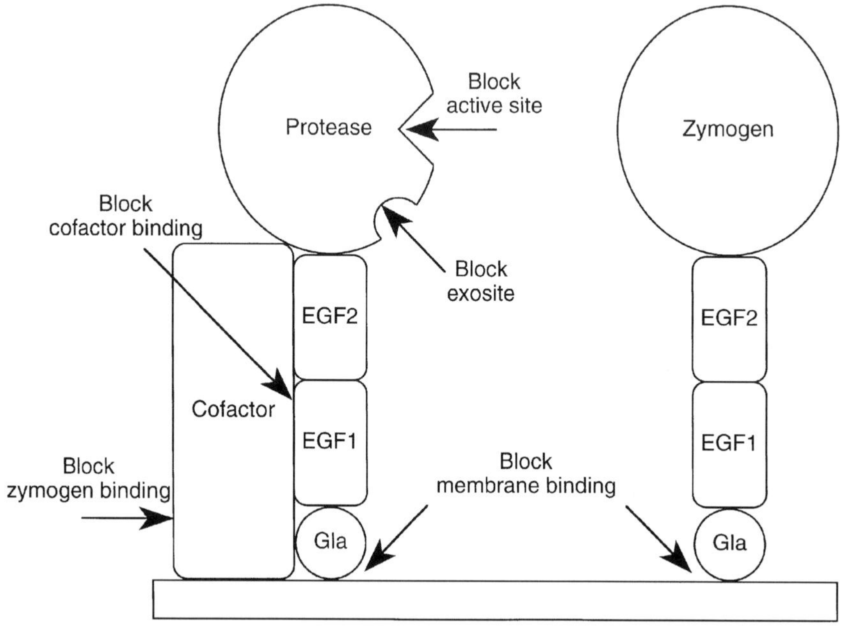

Figure 9.2 Approaches to inhibit the function of FVIIa, FIXa, and FXa. Potential points of intervention to interfere with enzymatic function are depicted. The cofactors for FVIIa, FIXa, and FXa are TF, FVIIIa, and FVa, respectively. The Gla, EGF1, EGF2 and protease (or zymogen) domains are described in the text; the membrane is on the bottom.

domain (Figure 9.2) (Ichinose and Davie 1994). The Gla domains are responsible for orienting these enzymes on the surface of negatively charged phospholipid membranes. The EGF domains are primarily involved in cofactor recognition. The protease domains are structurally homologous to the chymotrypsin family of serine proteases, containing the essential catalytic triad of Ser 195, Asp 102, and His 57 residues (the chymotrypsinogen numbering system is used throughout).

FVIIa, FIXa, and FXa all require the presence of cofactors as well as phospholipid membranes for optimal activity. Although these enzymes have been reasonably well studied from a structural and functional perspective, there are still many unresolved questions. The elaborate nature of these procoagulant complexes suggest a wide variety of strategic approaches to impair their function (Figure 9.2). Recent progress has been made in developing antithrombotic strategies and discovering potent and specific inhibitors of FVIIa, FIXa, and FXa (Leblond and Winocour 1999). The sources and inhibition modes of these inhibitors are diverse and comprise active site inhibited enzymes, mutagenized enzymes and cofactors, antibodies, naturally-occurring protein inhibitors, peptides, and small molecule active site inhibitors. These inhibitors will facilitate our biochemical and biological understanding of the coagulation pathway and aid in the development of potential therapeutic agents for thrombotic diseases.

9.2 FACTOR VIIa

9.2.1 Background and rationale

Zymogen FVII circulates in blood at ~10 nM as a single chain 50 kDa glycoprotein; ca. 1% circulates as the activated form FVIIa (Morrissey *et al* 1993). The proteolytic conversion of zymogen to active enzyme involves cleavage at Arg 152-Ile 153 (Figure 9.3); a number of enzymes can catalyze this conversion *in vitro* including FXa (Radcliffe and Nemerson 1975; Broze Jr. and Majerus 1980), thrombin (Radcliffe and Nemerson 1975), FXIIa (Seligsohn *et al* 1979), FIXa (Seligsohn *et al* 1979), FVIIa (Pedersen *et al* 1989; Nakagaki *et al* 1991) and the TF•FVIIa complex (Neuenschwander, Fiore and Morrissey 1993). FXa has been suggested as the most relevant TF•FVIIa activator *in vivo* (Rao *et al* 1986; Rao and Rapaport 1988; Rapaport and Rao 1995). FVIIa itself is not catalytically competent towards its physiological substrates FIX and FX until it forms a Ca^{2+}-dependent complex with TF on the surface of membrane phospholipids. TF is a 263-residue membrane glycoprotein having a 219-residue extracellular domain that, unlike other cofactors in the coagulation cascade, does not undergo proteolysis to become functional.

Structural and functional studies on TF and FVIIa have added to a molecular understanding of coagulation (Kirchhofer and Banner 1997; Higashi and Iwanaga 1998; Ruf and Dickinson 1998); structures of human TF (Muller *et al* 1994; Harlos *et al* 1994), human FVIIa (Kemball-Cook *et al* 1999; Pike *et al* 1999; Dennis *et al* 2000), and the TF•FVIIa complex (Banner *et al* 1996; Zhang, St. Charles and Tulinsky 1999) have been determined by X-ray crystallography. TF has structural homology to the cytokine receptor family whereas FVIIa is structurally related to the trypsin-like family of serine proteases. Despite the requirement of TF for optimal enzymatic activity, there is little change in the structure of TF or FVIIa in the presence or absence of its partner. The 2.0 Å crystal structure of the TF•FVIIa complex reveals the extended conformation of FVIIa and describes the intricate

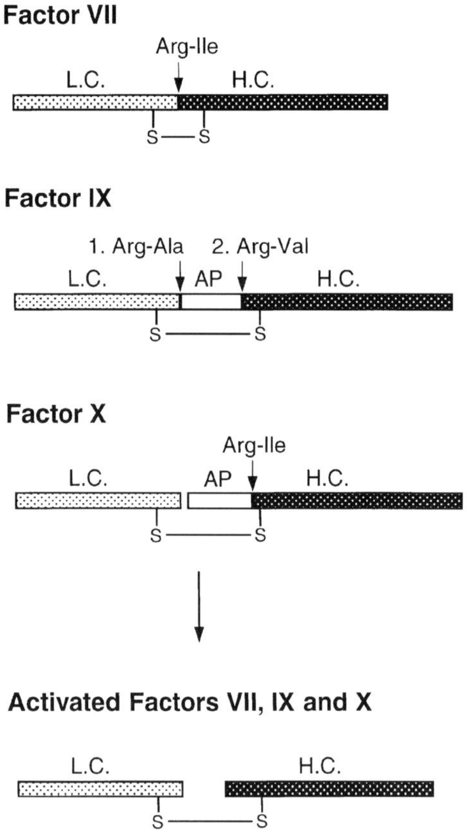

Figure 9.3 Schematic representation of coagulation factors VII, IX, and X. The active enzymes are composed of a light chain (L.C.) and a heavy chain (H.C.), which is also termed the catalytic or protease domain. The L.C. contains the Gla and two EGF domains. FVII is activated to its active enzyme by a single cleavage after Arg 152. For FIX and FX, the zymogen to active enzyme conversion is associated with the liberation of an activation peptide (AP), a 35 residue and 52 residue peptide for FIX and FX, respectively. FIX activation requires two sequential cleavage steps (Arg 145 and Arg 180) whereas FX, which circulates as a two chain molecule, only requires one cleavage (Arg 52).

association between the two fibronectin type III domains of TF and the light chain of FVIIa (Figure 9.4a) (Banner *et al* 1996). Critical contacts with the FVIIa heavy chain (catalytic domain) are thought to induce conformational changes at the FVIIa active site as suggested by the enhancement of FVIIa amidolytic activity by TF (Butenas, Ribarik and Mann 1993; Neuenschwander, Branam and Morrissey 1993). In addition, a TF region that interacts with substrates FX and FIX has been elucidated (Roy *et al* 1991; Ruf *et al* 1992a,b; Huang *et al* 1996; Kirchhofer *et al* 2000a). It is located in the carboxy terminal domain proximal to the membrane, which includes the positively charged residues Lys 165 and Lys 166 (Figure 9.4a), and comprises an area (ca. 1100 Å^2) which is about half the size of the TF•FVIIa contact zone. Therefore, the cofactor function of

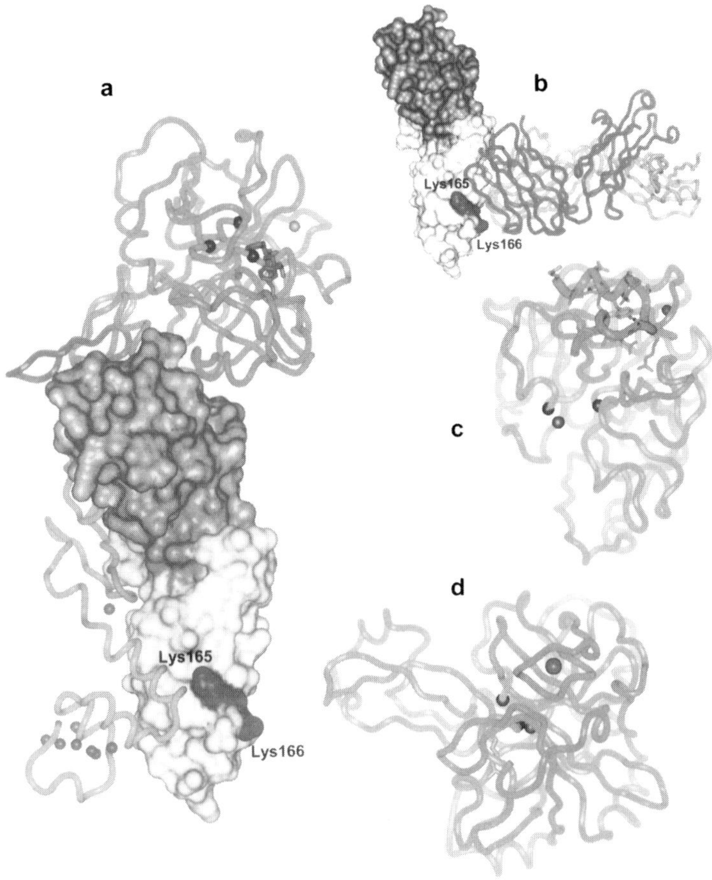

Figure 9.4 Crystal structures of the tissue Factor•Factor VIIa (TF•FVIIa) complex and bound inhibitors. (**a**) The TF•FVIIa complex (Banner *et al* 1996). TF (gray) is shown in a solvent accessible representation with the N- and C-terminal fibronectin type III modules in dark and light gray, respectively. FVIIa light chain comprises the N-terminal γ-carboxyglutamic acid (Gla) domain (yellow) followed by the EGF-1 (orange) and the EGF-2 (salmon) domains. The catalytic serine protease domain (heavy chain) is shown in blue with the D-Phe-L-Phe-L-Arg chloromethyl ketone peptide (red) irreversibly bound in the active site. The positions of the catalytic triad residues His 57, Asp 102 and Ser 195 are indicated (black spheres). The calcium ions are shown in red. The substrate contact region in the C-terminal domain of TF is proximal to the phospholipid membrane and includes residues Lys 165 and Lys 166 (red). (**b**) F(ab) of anti-tissue factor antibody TF8–5G9 bound to tissue factor (Huang *et al* 1998). TF is represented as in Figure 9.4a with Lys 165 and Lys 166 in red. The epitope of the antibody F(ab) overlaps with the substrate interaction region of TF. The antibody heavy chain, comprising the variable and constant domains, is shown in dark purple and the corresponding domains of the light chain in light purple. (**c**) E-76 peptide bound to the FVIIa catalytic domain (Dennis *et al* 2000). The peptide (green) binds to an exosite proximal to the active site and the calcium (red)-binding site of FVIIa (blue). (**d**) BPTI variant 5L15 bound to FVIIa catalytic domain (Zhang, St. Charles and Tulinsky 1999). The inhibitor (green) binds in the active site of FVIIa (blue). Also depicted is the P1 residue (Arg) side chain of the inhibitor, which inserts into the S1 recognition pocket of the enzyme. (*See Color plate 11*)

TF involves several distinct functions, i.e. interaction with substrates, binding and immobilization of FVIIa to the cell surface, correct spatial orientation of FVIIa and FVIIa active site with respect to substrates and induction of conformational changes at/around the active site.

The molecular details of how TF brings about these conformational changes on FVIIa have been difficult to unravel. It was recognized that the conformational state of free FVIIa (i.e. not bound to TF) had some similarities to FVII zymogen (Higashi, Matsumoto and Iwanaga 1996; Dickinson *et al* 1998). Therefore, if we were able to learn more about FVII zymogen we might understand how TF induces the conformational changes on FVIIa. Recently, Eigenbrot *et al* (2001) successfully elucidated the three-dimensional structure of FVII zymogen. The structure differs from that of TF-bound FVIIa in many ways, particularly in the arrangement of four flexible loops around the active site (the so-called 'activation domain') and in the position of two anti-parallel β-strands (residues 134–140 and 153–162; chymotrypsin numbering scheme). These two β-strands on FVIIa connect the TF contact site with the FVIIa active site region and may constitute the elusive molecular linkage that allows TF to effect the conformational state of FVIIa. Therefore, it is conceivable that when TF binds to FVIIa it re-positions these β-strands, which will transmit these changes to the 'activation domain' and the active site. Needless to say, these speculations require further experimental validation.

The homology of TF with cytokine receptors (Bazan 1990), particularly with the interferon-γ receptor (Walter *et al* 1995), has created the notion that TF has functions beyond hemostasis, such as transmembrane signalling. Specifically, the cytoplasmic domain of TF, which contains several serine phosphorylation sites (Mody and Carson 1997; Zioncheck, Roy and Vehar 1992), appears to play a role in some biological processes (reviewed by Camerer *et al* 1996; Prydz *et al* 1999; Ruf and Mueller 1999), for example tumor metastasis (Bromberg *et al* 1995; Mueller and Ruf 1998), cell adhesion and cell migration (Ott *et al* 1998; Ruf and Mueller 1999). The cytoplasmic domain of TF is however dispensable for other processes, such as embryogenesis (Parry and Mackman 2000), signalling (Peterson *et al* 2000) and coagulation. With the exception of a few studies (Abe *et al* 1999), signalling by TF was found to be dependent on catalytically active FVIIa. The TF•FVIIa complex was shown to induce mobilization of intracellular Ca^{2+} stores and phosphorylation of p44/p42 MAPK. None of these reactions seem to involve any of the presently known members of the PAR (protease activated receptor) family (Peterson *et al* 2000). Most studies used supraphysiological concentrations of FVIIa and, therefore, it is not clear whether direct TF•FVIIa signalling is of physiological importance. Another pathway is indirect TF•FVIIa signalling, which involves FXa and thrombin as the ultimate signalling molecules. Transfection experiments using Xenopus oocytes indicated that FXa, generated by TF•FVIIa-mediated conversion of zymogen FX, elicited intracellular signals via interaction with PAR2 (Camerer *et al* 2000). Therefore, by virtue of generating the important signalling molecules FXa and thrombin, the TF•FVIIa complex may initiate PAR-dependent (Coughlin 1999) signalling pathways and thus may link the coagulation reactions with inflammatory processes.

It is important to consider possible effects on some of these biological activities when developing novel anticoagulants which aim at blocking TF•FVIIa function. What makes the TF•VIIa complex a promising antithrombotic target is that TF expression is specifically upregulated in a number of cardiovascular diseases, such as atherosclerosis. For example, atherosclerotic plaques are rich in TF protein (Wilcox *et al* 1989), the cellular expression of which is regulated by the CD40/CD40 ligand pathway (Mach *et al* 1997; Libby *et al* 1999). TF in human plaques is competent to elicit thrombus formation when exposed to flowing blood *in vitro* (Toschi *et al* 1997; Badimon *et al* 1999) and may precipitate myocardial infarction *in vivo*. Furthermore, in unstable angina and venous thrombosis, circulating monocytes are found to express TF (Miller *et al* 1981; Blakowski *et al* 1986; Leatham *et al* 1995) which may initiate coagulation and contribute to the progression of disease. Therefore, therapeutic intervention with the function of TF•FVIIa complex may ameliorate various coagulation-mediated disease processes.

In the following sections the different types of FVIIa inhibitors will be discussed – the endogenous inhibitors tissue factor pathway inhibitor (TFPI) and the serpin antithrombin III (ATIII), active site inhibited FVIIa, TF mutants, naturally-occurring inhibitors from various hematophagous organisms, and small molecule active site inhibitors. Some of these areas have been addressed in previous reviews (Johnson and Hung 1998; Gallagher *et al* 1999).

Inactivated FVIIa (FVIIai), where dansyl-L-Glu-L-Gly-L-Arg chloromethyl ketone or D-Phe-L-Phe-L-Arg chloromethyl ketone have been used to alkylate the active site, competes with FVIIa for binding to TF; the structure of FVIIai is depicted in Figure 9.5a. Since it binds to membranes and TF essentially in the same manner as FVIIa, it serves as a type of competitive inhibitor with regard to FVIIa binding to TF. However, using a similar type of active site inhibited FVIIa (DIP-FVIIa), enzyme kinetic analysis showed that FVIIai acts as a purely noncompetitive inhibitor of FX activation by forming a non-productive ternary TF•FVIIai•FX complex (Nemerson and Gentry 1986). The affinity of TF for FVIIai is 5-fold higher than for FVIIa itself (Sørensen *et al* 1997). FVIIai has shown promising antithrombotic activity in rabbit and baboon models of thrombosis and restenosis (Jang *et al* 1995; Harker *et al* 1996; Golino *et al* 1998; Holst *et al* 1998) and is currently being tested in clinical trials with patients undergoing angioplasty.

An analogous approach was taken with a mutagenized version of the TF cofactor. Recognizing that the two carboxy-terminal TF residues Lys 165 and Lys 166 (Figure 9.4a) are critical for substrate interaction with the TF•FVIIa complex (Roy *et al* 1991; Ruf *et al* 1992a), a soluble TF mutant containing alanines at these positions (TFAA) was constructed for use as a specific anticoagulant (Kelley *et al* 1997). TFAA binds FVIIa and the resulting complex is unable to activate FX. Unlike FVIIai which reduces only V_{max}, full length versions of TFAA were shown to affect both V_{max} and K_m in FX activation assays (Huang *et al* 1996; Dittmar *et al* 1997). In blood, TFAA may form circulating TFAA•FVII complexes and, thus, additionally interfere with FXa-dependent zymogen FVII activation as suggested by *in vitro* studies with full length and truncated TFAA mutants (Dittmar *et al* 1997; Kirchhofer *et al* 2001). TFAA selectively inhibits the TF-dependent coagulation pathway and demonstrates antithrombotic activity with reduced bleeding compared to heparin in arterial thrombosis models in rabbit (Kelley *et al* 1997) and guinea pig (Himber *et al* 2001).

9.2.2 Antibodies

Monoclonal antibodies in humanized or chimeric forms are successfully used to treat a variety of diseases (Vaswani and Hamilton 1998; Vaughan *et al* 1998). Therefore, the use of antibodies that specifically recognize TF at the site of thrombogenesis may be a promising strategy for treating thrombotic disorders. Because the cofactor function of TF involves

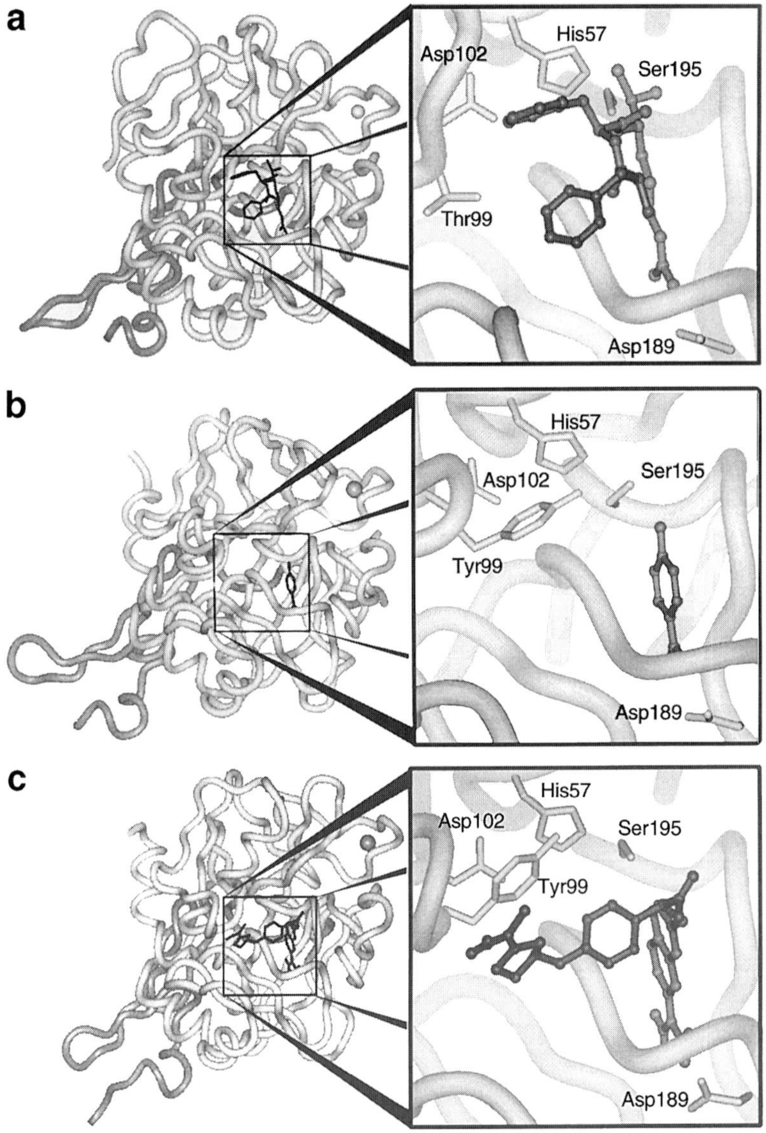

Figure 9.5 Small active site inhibitors bound to Factors VIIa, IXa, and Xa. The figures show the EGF-2 and catalytic domains of the enzymes. The residues of the catalytic triad His 57, Asp 102 and Ser 195 are indicated. Also shown is Asp 189 at the 'bottom' of the S1 pocket and residue 99, which is important for the formation of the S2 pocket. (**a**) Human FVIIa with irreversibly bound D-Phe-L-Phe-L-Arg chloromethyl ketone peptide (Banner *et al* 1996). (**b**) Human FIXa with the reversible inhibitor *p*-aminobenzamidine bound in the S1 pocket (Hopfner *et al* 1999). Note that the side chain of residue 99 (Tyr) is oriented differently as compared to FVIIa or FXa. This unique conformation prevents access of P2 residues to the S2 pocket. (**c**) Human FXa with reversible and selective inhibitor DX-9065a bound in the active site (Brandstetter *et al* 1996). A part of the inhibitor makes specific molecular contacts with the S4 pocket, thereby rendering it selective for FXa.

distinct surface-exposed regions on the TF molecule, the antibody epitope on TF may be a key factor for antithrombotic efficacy *in vivo*. Two fundamentally different types of anti-TF antibodies have been studied in experimental thrombosis models. One type appears to interfere with TF association with FVIIa (Taylor Jr. *et al* 1991; Pawashe *et al* 1994; Golino *et al* 1996; Ragni *et al* 1996; Himber *et al* 1997; Kirchhofer *et al* 2000b) and another type, exemplified by anti-TF antibodies TF8–5G9 (Ruf and Edgington 1991; Fiore *et al* 1992; Huang *et al* 1998) and D3 (Kirchhofer *et al* 2000b), interferes with substrate docking without affecting TF binding to FVIIa. A 3.0 Å resolution crystal structure of the TF8–5G9 Fab complexed to TF provides structural support for the latter type of interaction (Huang *et al* 1998) (Figure 9.4b). Both types of antibodies are effective anticoagulants. Although a side-by-side comparison of these two different types of antibodies has not been done, it appears that based on the epitope location and specific inhibitory mechanism, antibodies like TF8–5G9 and D3 are superior anticoagulants (Ruf and Edgington 1991; Fiore *et al* 1992; Huang *et al* 1998; Kirchhofer *et al* 2000b). Therefore, the murine D3 antibody was 'humanized', i.e. the complementarity determining region (CDR) was grafted onto a human antibody framework (Presta *et al* 2001). Additional mutagenesis in the CDR and the framework region gave a final humanized D3 antibody that had a 100-fold higher binding affinity than its original murine counterpart (K_D 0.1 nM) and displayed very potent anticoagulant activities (Presta *et al* 2001). This humanized antibody was produced in different forms, such as Fab, $F(ab')_2$ and full length variants, which may be useful for short and long term treatment of thrombotic disorders without eliciting immune responses.

A different antibody approach has been used to target FVII/FVIIa instead of TF. The antibody 12D10 binds to the protease domain of FVII/FVIIa in the presence or absence of TF with high affinity and shows mixed-type inhibition for FX activation (Dickinson *et al* 1998). A F(ab) of 12D10 reduced circulating FVII/FVIIa levels and efficiently prevented the procoagulant response in a chimpanzee model of endotoxemia (Biemond *et al* 1995).

9.2.3 Naturally-occurring protein inhibitors

The endogenous inhibition of TF•FVIIa is regulated by TFPI and ATIII. TFPI is a protein of 276 residues containing three tandem Kunitz domains and is a slow tight-binding inhibitor of FXa. The TFPI•FXa complex then forms a quaternary complex with TF•FVIIa (Broze Jr. 1992). The first Kunitz domain binds to TF•FVIIa and the second Kunitz domain binds to FXa; the role of the third domain is unknown. Based on *in vitro* properties of TFPI, it is thought to regulate the tissue factor induced (extrinsic) coagulation pathway by a feedback mechanism (Broze Jr. 1992). A highly homologous protein termed TFPI-2 has also been characterized (Petersen *et al* 1996b). Recombinant TFPI has been studied in various rabbit and canine thrombosis models (Haskel *et al* 1991; Asada *et al* 1998). TFPI is being evaluated in Phase III clinical trials for sepsis. A truncated version containing only the first and second Kunitz domains has also demonstrated antithrombotic effects (Holst *et al* 1994).

ATIII is a 58 kDa glycoprotein belonging to the serpin family of protease inhibitors (Bode and Huber 1992; Olson and Bjork 1994). Although its major role is thought to involve the inhibition of both thrombin and FXa (*vide infra*), it also inhibits FVIIa in the presence of TF (Lawson *et al* 1993; Rao *et al* 1995). Inhibition by ATIII is irreversible since it forms a covalent complex at the active site of the enzyme.

Potent exogenous inhibitors of TF•FVIIa have recently been discovered and isolated from hookworms (Stanssens *et al* 1996). Nematode anticoagulant peptide NAPc2, originally isolated from *Ancylostoma caninum*, is an 84-residue protein that has been expressed recombinantly in *Pichia pastoris*. Like TFPI, NAPc2 inhibits TF•FVIIa by first binding to FXa; the apparent K_i is 8.4 pM. However, on a molecular level their inhibition modes are quite distinct since they share no sequence similarity and NAPc2 binds at an exosite on FXa and not at the active site (Stanssens *et al* 1996; Bergum *et al* 2001). Recombinant NAPc2 has been evaluated in rat and pig thrombosis models (Vlasuk *et al* 1997). NAPc2 showed promising results in preventing venous thrombosis in patients undergoing knee replacement surgery (Lee *et al* 2000).

9.2.4 Peptide inhibitors

Recently, a new class of FVIIa inhibitor has been described by selecting peptides from phage-displayed libraries (Dennis *et al* 2000). This approach resulted in a well structured 18-residue peptide termed E-76, which has a K_d of 8.5 nM for FVIIa or TF•FVIIa. E-76 inhibited activation of FX as well as amidolytic activity with an IC_{50} of 1 nM. E-76 was also very selective, inhibiting only FVIIa enzymatic activity and prolonging only TF-dependent clotting. Structural studies of E-76 complexed with FVIIa revealed that it did not bind to the active site, but rather to an exosite on the protease domain of FVIIa (Figure 9.4c). This was consistent with its noncompetitive mode of inhibition of both FX activation and amidolytic activity.

The mechanism of inhibition has been described as an allosteric 'switch' involving one of the activation loops (the 140s loop), which is part of the canonical activation domain of the trypsin-like serine proteases (Huber and Bode 1978; Wang, Bode and Huber 1985). There is a large change in this loop conformation induced by E-76 binding at the exosite, which may inhibit FX activation by both steric and allosteric components. E-76 may alter the manner in which substrate FX binds and prevent the scissile bond from obtaining the optimal geometry for turnover. In the FVIIa active site, E-76 may also allosterically disrupt the hydrogen bond conformation at the 'oxyanion hole', which is important for stabilization of the transition state of the substrate during catalysis. Using the same phage-display approach another class of peptide inhibitors, exemplified by the 15mer A-183, was discovered (Dennis *et al* 2001). Mutagenesis experiments and structural analysis of A-183 bound to zymogen FVII indicated that the peptide is bound to an exosite on FVIIa, which is distinct from both the active site and the E-peptide binding site (Eigenbrot *et al* 2001; Roberge *et al* 2001). A-183 inhibited the activation of FX by the TF•FVIIa complex with an IC_{50} of 1 nM. These approaches provide an opportunity to develop novel anticoagulants based upon entirely new sites, distinct from the active site and may represent a new paradigm for developing inhibitors of serine proteases.

9.2.5 Active site inhibitors

Kunitz domains are stable proteins containing about 58 residues and three conserved disulfide bonds. They are slow, tight-binding, reversible inhibitors of serine proteases that bind to the active site and inhibit according to the standard mechanism (Laskowski Jr. and Kato 1980; Bode and Huber 1992). The first Kunitz domain of TFPI itself inhibits TF•FVIIa with a K_i of 250 nM (Petersen *et al* 1996a). It has also been linked to the carboxy terminus of the light chain of FX to form a fusion protein that potently inhibits TF•FVIIa (Girard *et al* 1990).

Kunitz domain inhibitors of TF•FVIIa have been selected from Alzheimer's amyloid β-protein precursor inhibitor (APPI) and bovine pancreatic trypsin inhibitor (BPTI) Kunitz domain libraries displayed on phage (Dennis and Lazarus 1994a; Dennis and Lazarus 1994b; Stassen *et al* 1995a). An overall consensus inhibitor, designated TF7I-C, differed from APPI at 4 key residues and inhibited TF•FVIIa with an apparent K_i of 1.9 nM, however it also potently inhibited FXIa and plasma kallikrein, and to a lesser extent plasmin (Dennis and Lazarus 1994a). Competitive selection approaches resulted in inhibitors that were much more selective (Dennis and Lazarus 1994b). For example IV-49C had an apparent K_i of 2.8 nM, but was much less inhibitory towards FXIa, plasma kallikrein, and plasmin. The TF7I-C Kunitz domain has been utilized to construct a bifunctional inhibitor by covalently linking it to the TF mutant TFAA. This bivalent fusion protein inhibited FX activation by TF•FVIIa ca. 200-fold more potently than either of the single components (Lee *et al* 1997).

A BPTI variant called 5L15 also selected from phage-displayed libraries contains 8 mutations and inhibits TF•FVIIa with a K_i of 0.4 nM (Stassen *et al* 1995a). The structure of 5L15 complexed to TF•FVIIa has been solved at 2.1 Å resolution, providing valuable insight into molecular interactions at the active site (Figure 9.4d) (Zhang, St. Charles and Tulinsky 1999). This variant has also been studied in a platelet-rich venous thrombosis model in hamsters (Stassen *et al* 1995b).

The crystal structures of FVIIa (Kemball-Cook *et al* 1999; Pike *et al* 1999; Dennis *et al* 2000) and in complex with TF (Banner *et al* 1996; Zhang, St. Charles and Tulinsky 1999) provide a structural basis to rationally design and develop small molecule active site inhibitors. Since the overall conformation of the S1 recognition pocket as well as the position of the catalytic triad residues is very similar in FVIIa, FIXa, and FXa (Figure 9.5), the generation of a selective FVIIa inhibitor appears to be a formidable challenge. However, there are subtle differences in the adjacent S2 (defined by residues 57 and 99) and S4 (defined by residue 99 and 170 loop) region. The FXa inhibitor, DX-9065a, which selectively binds to the FXa active site, demonstrates that differences in the S2–S4 region can be exploited to design enzyme-specific, reversible inhibitors (*vide infra*). As of yet, little has been published on small molecule active site inhibitors of FVIIa. Combinatorial chemistry has been used to synthesize lead compounds with K_i values in the micromolar range (Roussel *et al* 1999). Recent patent applications have disclosed compounds that inhibit TF•FVIIa in the nanomolar range (Grobke *et al* 1999; Senokuchi and Ogawa 1999; Ackermann *et al* 2000). An advantage of synthetic small molecule inhibitors over protein-based inhibitors is their potential use as oral anticoagulants. However, the generation of molecules with sufficient specificity will be a critical and challenging issue since inhibition of other known or unknown serine proteases may result in unwanted side effects. This may be of particular importance for prolonged treatment periods, such as oral anticoagulation for the prevention of venous thromboembolism.

9.3 FACTOR IXa

9.3.1 Background and rationale

Zymogen FIX circulates in blood at ~90 nM as a 57 kDa single chain protein and is activated either by TF•FVIIa in the extrinsic pathway (Østerud and Rapaport 1977) or

FXIa in the intrinsic pathway (Figure 9.1) (Fujikawa *et al* 1974; Di Scipio *et al* 1978; Østerud *et al* 1978). The proteolytic conversion of zymogen to active enzyme involves two cleavage steps, the first at the Arg 145-Ala 146 and the second at the Arg 180-Val 181 amide bond, resulting in the release of a 35 residue activation peptide (Figure 9.3). The intermediate form, FIXα, is catalytically inactive. The presence of small amounts of FXa significantly amplify the generation of FIXa, mainly by catalyzing the first proteolytic step (Lawson and Mann 1991). Once formed, FIXa assembles with its cofactor FVIIIa into the intrinsic Xase complex on phospholipid surfaces. The membrane surface of activated platelets may be the primary site for Xase assembly, particularly during arterial thrombosis, since platelets accumulate in large numbers at the site of thrombus formation and provide an optimal phospholipid surface for Xase function (Ahmad and Walsh 1994). FIX and FIXa bind to specific high affinity binding sites on platelets and endothelial cells. This interaction is mainly mediated by the FIX-Gla domain (Ryan *et al* 1989; Cheung *et al* 1992; Rawala-Sheikh *et al* 1992; Toomey *et al* 1992; Ahmad *et al* 1994). Therefore, approaches that target the FIX-Gla domain have yielded highly selective and potent inhibitors of FIXa activity. In the following sections five different types of FIXa inhibitors will be discussed – antibodies and a snake venom protein which target the Gla domain, the insect protein nitrophorin-2 or prolixin-S, active site inhibited FIXa, and active site inhibitors.

The first studies to validate FIXa as an antithrombotic target were carried out with active site inhibited FIX (FIXai), where L-Glu-L-Gly-L-Arg chloromethyl ketone has been used to alkylate the active site. FIXai recapitulates all protein–protein and protein–phospholipid interactions of FIXa but is catalytically inactive. FIXai competes with FIX/FIXa for binding to high affinity platelet binding sites (Ahmad *et al* 1989) and, in this way, may bring about its inhibitory effect. FIXai showed remarkable inhibition of thrombus formation in an electric current-induced arterial thrombosis model in dogs without compromising normal hemostasis (Benedict *et al* 1991). Additional studies in canine cardiopulmonary bypass models (Spanier *et al* 1998), rabbit thrombosis models (Wong *et al* 1997) and *in vitro* thrombosis models (Kirchhofer *et al* 1995) further corroborated the strong antithrombotic activity of FIXai.

FIX and FVIII deficiency states (Hemophilia B and A, respectively) are associated with a sometimes severe bleeding diathesis, emphasizing the critical function of the intrinsic Xase complex in normal hemostasis. Therefore, the observed strong antithrombotic activity of FIXai and other FIXa inhibitors discussed below is not surprising. However, what was unexpected was the relative absence of effects on normal hemostasis (Benedict *et al* 1991; Spanier *et al* 1998; Feuerstein *et al* 1999a; Refino *et al* 1999). This may be partially due to the short exposure time of the tested inhibitors. It remains to be seen whether specific inhibition of FIXa is efficacious and safe under clinical conditions.

9.3.2 Antibodies

Two different anti-FIX antibodies, 10C12 and BC2, have been described recently and evaluated in experimental thrombosis models. Both specifically bind to the FIX-Gla domain but not to other Gla-containing coagulation factors (Feuerstein *et al* 1999a; Refino *et al* 1999). BC2 is a murine anti-human antibody (Feuerstein *et al* 1999a), whereas 10C12 is a fully human $F(ab')_2$ generated by phage display technology (Suggett *et al* 2000). 10C12 was shown to interfere with all known FIX/FIXa-dependent coagulation processes and potently inhibited platelet-dependent coagulation *in vitro* (Refino *et al* 1999; Suggett *et al* 2000). Both antibodies displayed excellent antithrombotic activity in animal thrombosis models (Feuerstein *et al* 1999a,b; Refino *et al* 1999). These activities were not associated with any significant interference with normal hemostasis, as judged from cuticle bleeding time and surgical blood loss experiments (Feuerstein *et al* 1999a; Refino *et al* 1999).

9.3.3 Naturally-occurring protein inhibitors

Snake venoms are known to contain proteins that affect coagulation. Proteins which recognize the FIX Gla domain include Factor IX-binding protein (FIX-bp) and Factor IX and X-binding protein (FIX/X-bp), heterodimeric proteins from *Trimeresurus flavoviridis* (Atoda *et al* 1995). The crystal structure of FIX-bp shows two covalently linked (Cys75-Cys79) globular subunits A and B each coordinating one calcium atom (Mizuno *et al* 1999). The concave surface formed between the two subunits is the putative binding site for the FIX-Gla domain. The structure is almost identical with that of FIX/X-bp as expected from the high level of amino acid sequence homology (Mizuno *et al* 1997). When administered to rabbits as an i.v. bolus, FIX-bp potently attenuated thrombus formation under arterial as well as venous flow conditions (De Guzman *et al* 1997; Pater *et al* 1997) .

More than 30 years ago, it was reported that extracts from the salivary gland of the blood-sucking insect *Rhodnius prolixus* inhibited the intrinsic coagulation pathway (Hellmann and Hawkins 1965). Recently, the active component, nitrophorin-2 (NP-2) or prolixin-S, was isolated and purified (Champagne *et al* 1995; Ribeiro *et al* 1995; Sun *et al* 1996). NP-2 is a 19.7 kDa protein, which specifically interferes with the function of the intrinsic Xase complex (Zhang *et al* 1998). Kinetic analysis revealed a rather complex inhibitory mechanism in that NP-2 decreases the V_{max} as well as K_m values of the FIXa/FVIIIa-catalyzed activation of FX. NP-2 binds to both FIX and FIXa and inhibits the conversion of FIX to FIXa by both FXIa and TF•FVIIa (Isawa *et al* 2000).

9.3.4 Active site inhibitors

Protease nexin-2/amyloid β-protein precursor as well as its isolated Kunitz domain APPI (*vide supra*) are potent inhibitors of FIXa, having K_i values in the presence of phospholipid vesicles of 1.9 nM and 2.4 nM, respectively (Schmaier *et al* 1993, 1995). These proteins also potently inhibit FXIa, thereby interfering with the intrinsic coagulation pathway (Smith *et al* 1990).

Although small molecule active site inhibitors have been successfully developed for thrombin and FXa, no such molecules have been reported for FIXa. This may be partly due to the difficulties in obtaining a suitable synthetic substrate for high-throughput screening. The reason for this shortcoming lies in the unique conformational state of the FIXa active site as deduced from FIXa crystal structures with bound inhibitors (Brandstetter *et al* 1995; Hopfner *et al* 1999). The structure of human FIXa with *p*-aminobenzamidine in the active site shows that the S2 pocket is not properly formed (Figure 9.5b); in fact, the side chain of residue Tyr 99 blocks the access of the substrate P2 residue (Hopfner *et al* 1999). Unlike FIXa, the sidechain of residue 99 in FVIIa and FXa (Thr and Tyr, respectively) is oriented differently such that the S2 pocket is properly formed (Figure 9.5). This may explain why, in contrast to FVIIa and FXa, FIXa is virtually an inactive enzyme towards small peptide substrates that primarily occupy the S1-S3 region (Brandstetter *et al* 1995; Hopfner *et al* 1999). Recent evidence has shown that certain alcohols, such as ethyleneglycol, can dramatically enhance catalysis towards chromogenic substrates (Neuenschwander *et al* 1997; Stürzebecher *et al* 1997). This property may be used for evaluating active site inhibitors, yet comes with the caveat that alcohol-induced conformational changes at the active site may not reflect a physiologically relevant state of FIXa. Structural comparison of FIXa with other related serine proteases will aid in structure-based design efforts (Bode *et al* 1997).

9.4 FACTOR Xa

9.4.1 Background and rationale

Zymogen FX circulates in blood at ~170 nM as a 58.8 kDa two chain disulfide-linked protein and can be activated to FXa either by TF•FVIIa or FIXa•FVIIIa (Ichinose and Davie 1994). The proteolytic conversion of zymogen to active enzyme involves cleavage at Arg 52-Ile 53 of the heavy chain, resulting in the release of a 52 residue activation peptide (Figure 9.3). FXa assembles with its cofactor FVa and Ca^{2+} into the prothrombinase complex on membrane phospholipid surfaces. As with the Xase complex, an important site for prothrombinase complex formation is on the membrane surface of activated platelets (Ahmad and Walsh 1994). Since the FX Gla domain mediates these interactions, inhibition strategies similar to those for FIX can be used (*vide supra*).

In the following sections different types of FXa inhibitors will be discussed. These include the endogenous inhibitors TFPI and the serpin ATIII, heparins, active site inhibited FXa, a snake venom protein which binds to the Gla domain, naturally-occurring inhibitors from various hematophagous organisms, and small molecule active site inhibitors (Al-Obeidi and Ostrem 1998; Scarborough 1998; Kunitada *et al* 1999).

The primary endogenous inhibitors of FXa include TFPI and the serpin ATIII, the characteristics of which have been discussed (*vide supra*). The inhibition of both thrombin and FXa by ATIII is potentiated by heparin and low molecular weight heparin, both of which are used clinically (Hirsh 1991; Weitz 1997; Clagett *et al* 1998). The anticoagulant activity of heparins is due to a unique pentasaccharide sequence which binds to ATIII and accelerates the inhibition; the low molecular weight heparins tend to inhibit FXa more than thrombin. The advantages of low molecular weight heparin over heparin include improved bioavailability, prolonged half-life, more predictable clearance and antithrombotic response (Weitz 1997; Clagett *et al* 1998; Bates and Hirsh 1999). These properties permit ease of administration by subcutaneous injection with less laboratory monitoring. Another approach to heparin therapy has relied on drug delivery technology to develop orally active versions of heparin and low molecular weight heparin (Baughman *et al* 1998). An orally active heparin may circumvent the use of coumarins, whose drawbacks include bleeding risk, extensive monitoring, drug interactions, and dose adjustments. Orally active heparin is currently in Phase III clinical trials.

Inactivated FX (FXai) is a competitive inhibitor of FXa in the prothrombinase complex. Two different versions of FXai have been studied where the active site residues Ser 195 and Asp 102 have been mutated as well as a version where L-Glu-L-Gly-L-Arg-chloromethyl ketone has been used to alkylate the active site. Both versions have shown antithrombotic activity in canine and rabbit thrombosis models, validating FXa as a target for antithrombotic therapy (Benedict *et al* 1993; Hollenbach *et al* 1994; Wong *et al* 1997).

9.4.2 Naturally-occurring protein inhibitors

Naturally-occurring protein inhibitors of FXa have been isolated and characterized from various hematophagous organisms including leeches, ticks, hookworms, vampire bats, and mosquitos. Snake venoms also contain proteins that can interfere with FX activity and coagulation.

Antistasin is a potent disulfide-rich 119-residue protein inhibitor of FXa found in the salivary glands of the Mexican leech *Haementeria officinalis* (Tuszynski *et al* 1987). It is a reversible slow tight-binding inhibitor with a K_i of ca. 0.5 nM (Dunwiddie *et al* 1989). The canonical mechanism of inhibition involves a two step process where the Arg 34-Val 35 bond of antistasin is slowly cleaved (Bode and Huber 1992). Antistasin contains two homologous domains; the crystal structure has been solved to 1.9 Å resolution (Lapatto *et al* 1997). Cyclic peptides derived from antistasin inhibit FXa with a K_i of 35 nM (Ohta *et al* 1994) and carboxy terminal peptides can inhibit FXa, however only when it is in the prothrombinase complex (Mao *et al* 1998). Isoforms of antistasin termed ghilantens have also been found in the leech *H. ghilianii* (Brankamp *et al* 1990).

Tick anticoagulant peptide (TAP) is a 60-residue protein derived from the salivary glands of the tick *Ornithodorus moubata* that also reversibly and potently inhibits FXa (Waxman *et al* 1990); K_i values from 0.2 to 0.6 nM have been reported (Jordan *et al* 1990, 1992; Mao *et al* 1995). A protein with 46% identity to TAP isolated from the salivary glands of *Ornithodoros savignyi* ticks is also a competitive and slow tight-binding factor Xa inhibitor with a K_i of 0.8 nM (Joubert *et al* 1998). The structure of TAP has been determined by NMR (Lim-Wilby *et al* 1995) and by crystallography in complex with FXa (Wei *et al* 1998). Both mutational and structural studies have implicated multiple binding sites between TAP and FXa (Dunwiddie *et al* 1992; Mao *et al* 1995; Wei *et al* 1998). The structure of the complex reveals that the amino terminal Tyr of TAP binds in the S1 pocket of the active site and the carboxy terminal helix binds to a secondary binding site (Wei *et al* 1998). TAP has been evaluated in various animal thrombosis models (Schaffer *et al* 1991; Vlasuk *et al* 1991; Jang *et al* 1995; Lefkovits *et al* 1996).

The repertoire of nematode anticoagulant peptides also include proteins that potently inhibit FXa (Stanssens *et al* 1996). NAPc5 and NAPc6 are homologous proteins of 77 and 75 residues that inhibit FXa by binding at the active site with apparent K_i values of 0.04 nM and 1 nM, respectively. They are related to the *Ascaris* family of protease inhibitors (Laskowski Jr. and Kato 1980; Bode and Huber 1992). NAPc5 has demonstrated antithrombotic efficacy in porcine and canine models of arterial and venous thrombosis (Vlasuk *et al* 1995; Rebello *et al* 1997).

Draculin is a glycoprotein isolated from vampire bat (*Desmodus rotundus*) saliva and potently inhibits FXa. Unlike other FXa inhibitors, draculin is a noncompetitive, tight-binding inhibitor of FXa, with a K_i of 14.8 nM (Fernandez *et al* 1999). The inhibition mechanism is thought to involve a two step irreversible process.

The salivary glands of the female yellow fever mosquito *Aedes aegypti* contains a 54 kDa protein termed AFXa with inhibitory activity against FXa (Stark and James 1998). Based on the cDNA translation product, AFXa is related to the serpin family of protease inhibitors. Characterization of AFXa shows reversible, noncompetitive, and noncovalent inhibition of FXa and has been suggested to involve binding at an exosite (Stark and James 1995).

Ecotin is a 142-residue serine protease inhibitor found in the periplasm of *Escherichia coli* and has been characterized as a potent anticoagulant and reversible tight-binding inhibitor of human FXa, having a K_i of 54 pM (Seymour *et al* 1994). Ecotin also potently inhibits trypsin, elastase, Factor XIIa, and plasma kallikrein (Ulmer *et al* 1995). The structure of ecotin has been determined in complex with trypsin and revealed an ecotin dimer interacting with two protease domains at both the active site and an exosite (McGrath *et al* 1994, 1995).

Lefaxin is a 30 kDa protein isolated from the leech *Haementeria depressa* (Faria *et al* 1999). It inhibits FXa activity with a K_i of 4 nM. Lefaxin shares no homology with antistasin or ghilanten, the other leech-derived FXa inhibitors. However, significant homology in the amino terminus is found with prolixin S, the intrinsic Xase inhibitor from the tick *R. prolixus* (*vide supra*).

Yagin is a disulfide-rich 133-residue protein isolated from the saliva of the medicinal leech *Hirudo medicinalis*. It has been evaluated as an adjunct to recombinant tissue-type plasminogen activator in a rabbit thrombosis model of thrombolysis (Kornowski *et al* 1996).

In addition to the Factor IX and X-binding protein (FIX/X-bp) from *T. flavoviridis* described above, homologous proteins with binding characteristics that are more selective towards FX termed Factor X-binding proteins (FX-bp) from *Deinagkistrodon acutus* have been characterized (Atoda *et al* 1998). Changes in amino acid sequence were found on the concave surface thought to be important for binding to FX. FX-bp binds to solid-phase FX and FIX with EC_{50} values of 0.4 nM and 3 nM, respectively. This interaction is Ca^{2+}-dependent and is thought to be mediated primarily by binding to residues 1-44 of the FX Gla domain.

9.4.3 Active site inhibitors

The second Kunitz domain of TFPI itself inhibits FXa with a K_i of 90 nM (Petersen *et al* 1996a) and has been crystallized in complex with FXa (Burgering *et al* 1997). Random and specific mutagenesis of the Kunitz domain BPTI has resulted in mutants with high affinity towards FXa (Stassen *et al* 1995a). For example mutants 4C2 and 7L22, containing 6 and 9 mutations, respectively, inhibited FXa with K_i values of 2.8 and 0.5 nM, respectively; 4C2 had an antithrombotic effect in a platelet-dependent thrombosis model in hamsters (Stassen *et al* 1995b).

In recent years, there has been a great deal of effort directed toward developing small molecule active site inhibitors of FXa. The advantage of this type of inhibitor is its potential use as an orally active anticoagulant with improved pharmacological and clinical effects compared to coumarins, which are currently in clinical use. Prior to investigating FXa inhibitors, a vast effort was applied toward small molecule thrombin inhibitors. Both the chemistry and biology of the thrombin active site effort has provided a significant foundation to develop FXa inhibitors. Small molecule inhibitor strategies have included peptidomimetics, combinatorial chemistry, structure-based design efforts, and traditional medicinal chemistry. Structure based design efforts are facilitated by the structure of FXa and comparison with related serine proteases (Stubbs II 1996; Bode *et al* 1997). The X-ray structures of human Gla-domainless FXa by itself (Padmanabhan *et al* 1993) and complexed with the synthetic inhibitor DX-9065a (Figure 9.5c) (Brandstetter *et al* 1996) have been solved and refined at 2.2 Å and 3.0 Å resolution, respectively.

A discussion of the development of small molecule active site inhibitors of FXa is beyond the scope of this chapter. There are many excellent recent reviews regarding these types of inhibitors (Al-Obeidi and Ostrem 1998; Scarborough 1998; Hauptmann and Stürzebecher 1999; Kunitada *et al* 1999; Sanderson 1999). Examples of some of the more well studied small molecule active site inhibitors of FXa include DX-9065a from Daiichi (Ki of 41 nM), YM-60828 from Yamanouchi (Ki of 1.3 nM), ZK-807191 from Berlex (Ki of 0.1 nM), RPR120844 from Aventis (Ki of 7 nM), Sel-2711 from Selectide (Ki of 3 nM), as well as transition state analogs from Cor Therapeutics, peptidomimetics based on CVS 2371 from Corvas, bis-phenylamidines from Dupont Merck, and 1,2-dibenzamidobenzene-derived amidines and nonamidines from Lilly.

An interesting and somewhat counter-intuitive phenomenon has been observed with a small molecule active site inhibitor of FXa, which likely has implications for FVIIa, FIXa, and perhaps other serine proteases which assemble on membranes and have macromolecular substrates (Krishnaswamy and Betz 1997). While it might be initially expected that an inhibitor that binds at the active site would be competitive with substrate, inhibition was in fact noncompetitive, reducing V_{max}, but not K_m. This can be explained by the realization that most of the binding energy responsible for the K_m effects with macromolecular zymogen substrates is derived from interactions outside of the active site. Thus, substrate interactions with cofactor, the Gla and EGF domains, and membranes are important for determining the mode of inhibition. The specific mode of inhibition – e.g. competitive *vs.* noncompetitive – does not appear to be critical with respect to practical implications for developing protease inhibitors as anticoagulants.

ACKNOWLEDGMENTS

We would like to acknowledge Charles Eigenbrot for many helpful discussions and for preparing Figures 9.4 and 9.5. We also acknowledge Ken Refino, Bob Kelley, and Rick Artis for critical comments on the manuscript.

REFERENCES

Abe, K., Shoji, M., Chen, J., Bierhaus, A., Danave, I., Micko, C. *et al* (1999) Regulation of vascular endothelial growth factor production and angiogenesis by the cytoplasmic domain of tissue factor. *Proceedings of the National Academy of Sciences of the U.S.A.*, **96**, 8663–8668.

Ackermann, J., Alig, J., Chucholowski, A., Groebke, K., Hilpert, K., Kuehne, H. *et al* (2000) Preparation of phenylglycine derivatives as pharmaceuticals. WO 0035858 A1.

Ahmad, S.S., Rawala-Sheikh, R. and Walsh, P.N. (1989) Platelet receptor occupancy with factor IXa promotes factor X activation. *Journal of Biological Chemistry*, **264**, 20012–20016.

Ahmad, S.S., Rawala-Sheikh, R., Cheung, W.-F., Jameson, B.A., Stafford, D.W. and Walsh, P.N. (1994) High-affinity, specific factor IXa binding to platelets is mediated in part by residues 3–11. *Biochemistry*, **33**, 12048–12055.

Ahmad, S.S. and Walsh, P.N. (1994) Platelet membrane-mediated coagulation protease complex assembly. *Trends in Cardiovascular Medicine*, **4**, 271–278.

Al-Obeidi, F. and Ostrem, J.A. (1998) Factor Xa inhibitors by classical and combinatorial chemistry. *Drug Discovery Today*, **3**, 223–231.

Asada, Y., Hara, S., Tsuneyoshi, A., Hatakeyama, K., Kisanuki, A., Marutsuka, K. *et al* (1998) Fibrin-rich and platelet-rich thrombus formation on neointima: Recombinant tissue factor pathway inhibitor prevents fibrin formation and neointimal development following repeated balloon injury of rabbit aorta. *Thrombosis & Haemostasis*, **80**, 506–511.

Atoda, H., Ishikawa, M., Yoshihara, E., Sekiya, F. and Morita, T. (1995) Blood coagulation factor IX-binding protein from the venom of *Trimeresurus flavoviridis*: Purification and characterization. *Journal of Biochemistry*, **118**, 965–973.

Atoda, H., Ishikawa, M., Mizuno, H. and Morita, T. (1998) Coagulation factor X-binding protein from *Deinagkistrodon acutus* venom is a Gla domain-binding protein. *Biochemistry*, **37**, 17361–17370.

Badimon, J.J., Lettino, M., Toschi, V., Fuster, V., Berrozpe, M., Chesebro, J.H. *et al* (1999) Local inhibition of tissue factor reduces the thrombogenicity of disrupted human atherosclerotic plaques. *Circulation*, **99**, 1780–1787.

Banner, D.W., D'Arcy, A., Chéne, C., Winkler, F.K., Guha, A., Konigsberg, W.H. *et al* (1996) The crystal structure of the complex of blood coagulation factor VIIa with soluble tissue factor. *Nature*, **380**, 41–46.

Bates, S.M. and Hirsh, J. (1999) Treatment of venous thromboembolism. *Thrombosis & Haemostasis*, **82**, 870–877.

Baughman, R.A., Kapoor, S.C., Agarwal, R.K., Kisicki, J., Catella-Lawson, F. and FitzGerald, G.A. (1998) Oral delivery of anticoagulant doses of heparin. A randomized, double-blind, controlled study in humans. *Circulation*, **98**, 1610–1615.

Bazan, J.F. (1990) Structural design and molecular evolution of a cytokine receptor superfamily. *Proceedings of the National Academy of Sciences of the U.S.A.*, **87**, 6934–6938.

Benedict, C.R., Ryan, J., Wolitzky, B., Ramos, R., Gerlach, M., Tijburg, P. *et al* (1991) Active site-blocked factor IXa prevents intravascular thrombus formation in the coronary vasculature without inhibiting extravascular coagulation in a canine thrombosis model. *Journal of Clinical Investigation*, **88**, 1760–1765.

Benedict, C.R., Ryan, J., Todd, J., Kuwabara, K., Tijburg, P., Cartwright Jr. J. *et al* (1993) Active site-blocked factor Xa prevents thrombus formation in the coronary vasculature in parallel with inhibition of extravascular coagulation in a canine thrombosis model. *Blood*, **81**, 2059–2066.

Bergum, P.W., Cruikshank, A., Maki, S.L., Kelly, C.R., Ruf, W. and Vlasuk, G.P. (2001) Role of zymogen and activated factor X as scaffolds for the inhibition of the blood coagulation factor VIIa-tissue factor complex by recombinant nematode anticogulant protein c2. *Journal of Biological Chemistry*, **276**, 10063–10071.

Biemond, B.J., Levi, M., ten Cate, H., Soule, H.R., Morris, L.D., Foster, D.L. *et al* (1995) Complete inhibition of endotoxin-induced coagulation activation in chimpanzees with a monoclonal Fab fragment against factor VII/VIIa. *Thrombosis & Haemostasis*, **73**, 223–230.

Blakowski, S.A., Zacharski, L.R. and Beck, J.R. (1986) Postoperative elevation of human peripheral blood monocyte tissue factor coagulant activity. *Journal of Laboratory & Clinical Medicine*, **108**, 117–120.

Bode, W. and Huber, R. (1992) Natural protein proteinase inhibitors and their interactions with proteinases. *European Journal of Biochemistry*, **204**, 433–451.

Bode, W., Brandstetter, H., Mather, T. and Stubbs, M.T. (1997) Comparative analysis of haemostatic proteinases: Structural aspects of thrombin, factor Xa, factor IXa and protein C. *Thrombosis & Haemostasis*, **78**, 501–511.

Brandstetter, H., Bauer, M., Huber, R., Lollar, P. and Bode, W. (1995) X-ray structure of clotting factor IXa: Active site and module structure related to Xase activity and hemophilia B. *Proceedings of the National Academy of Sciences of the U.S.A.*, **92**, 9796–9800.

Brandstetter, H., Kuhne, A., Bode, W., Huber, R., von der Saal, W., Wirthensohn, K. *et al* (1996) X-ray structure of active site-inhibited clotting factor Xa. Implications for drug design and substrate recognition. *Journal of Biological Chemistry*, **271**, 29988–29992.

Brankamp, R.G., Blankenship, D.T., Sunkara, P.S. and Cardin, A.D. (1990) Ghilantens: Anti-coagulant-antimetastatic proteins from the South American leech, *Haementeria ghilianii*. *Journal of Laboratory & Clinical Medicine*, **115**, 89–97.

Bromberg, M.E., Konigsberg, W.H., Madison, J.F., Pawashe, A. and Garen, A. (1995) Tissue factor promotes melanoma metastasis by a pathway independent of blood coagulation. *Proceedings of the National Academy of Sciences of the U.S.A.*, **92**, 8205–8209.

Broze Jr. G.J. and Majerus, P.W. (1980) Purification and properties of human coagulation factor VII. *Journal of Biological Chemistry*, **255**, 1242–1247.

Broze Jr. G.J. (1992) Tissue factor pathway inhibitor and the revised hypothesis of blood coagulation. *Trends in Cardiovascular Medicine*, **2**, 72–77.

Burgering, M.J.M., Orbons, L.P.M., van der Doelen, A., Mulders, J., Theunissen, H.J.M., Grootenhuis, P.D.J. *et al* (1997) The second Kunitz domain of human tissue factor pathway inhibitor: Cloning, structure determination and interaction with factor Xa. *Journal of Molecular Biology*, **269**, 395–407.

Butenas, S., Ribarik, N. and Mann, K.G. (1993) Synthetic substrates for human factor VIIa and factor VIIa-tissue factor. *Biochemistry*, **32**, 6531–6538.

Camerer, E., Kolstø, A.-B. and Prydz, H. (1996) Cell biology of tissue factor, the principal initiator of blood coagulation. *Thrombosis Research*, **81**, 1–41.

Camerer, E., Huang, W. and Coughlin, S.R. (2000) Tissue factor- and factor X-dependent activation of protease-activated receptor 2 by factor VIIa. *Proceedings of the National Academy of Sciences of the U.S.A.*, **97**, 5255–5260.

Champagne, D.E., Nussenzveig, R.H. and Ribeiro, J.M.C. (1995) Purification, partial characterization, and cloning of nitric oxide-carrying heme proteins (nitrophorins) from salivary glands of the blood-sucking insect *Rhodnius prolixus*. *Journal of Biological Chemistry*, **270**, 8691–8695.

Cheung, W.-F., Hamaguchi, N., Smith, K.J. and Stafford, D.W. (1992) The binding of human factor IX to endothelial cells is mediated by residues 3–11. *Journal of Biological Chemistry*, **267**, 20529–20531.

Clagett, C.P., Anderson Jr. F.A., Geerts, W., Heit, J.A., Knudson, M., Lieberman, J.R. *et al* (1998) Prevention of venous thromboembolism. *Chest*, **114**, 531S–560S.

Coughlin, S.R. (1999) How the protease thrombin talks to cells. *Proceedings of the National Academy of Sciences of the U.S.A.*, **96**, 11023–11027.

Davie, E.W., Fujikawa, K. and Kisiel, W. (1991) The coagulation cascade: Initiation, maintenance, and regulation. *Biochemistry*, **30**, 10363–10370.

Davie, E.W. (1995) Biochemical and molecular aspects of the coagulation cascade. *Thrombosis & Haemostasis*, **74**, 1–6.

De Guzman, L., Refino, C.J., Steinmetz, H., Bullens, S., Lipari, T., Smyth, R. *et al* (1997) Inhibition of tissue factor or intrinsic Xase are effective antithrombotic strategies in a new model of venous thrombosis. *Thrombosis & Haemostasis*, **Suppl.**, A292.

Dennis, M.S. and Lazarus, R.A. (1994a) Kunitz domain inhibitors of tissue factor•factor VIIa I. Potent inhibitors selected from libraries by phage display. *Journal of Biological Chemistry*, **269**, 22129–22136.

Dennis, M.S. and Lazarus, R.A. (1994b) Kunitz domain inhibitors of tissue factor•factor VIIa II. Potent and specific inhibitors by competitive phage selection. *Journal of Biological Chemistry*, **269**, 22137–22144.

Dennis, M.S., Eigenbrot, C., Skelton, N.J., Ultsch, M.H., Santell, L., Dwyer, M.A. and Lazarus, R.A. (2000) Peptide exosite inhibitors of factor VIIa as anticoagulants. *Nature*, **404**, 465–470.

Dennis, M.S., Roberge, M., Quan, C. and Lazarus, R.A. (2001) Selection and characterization of a new class of peptide exosite inhibitors of coagulation factor VIIa. *Biochemistry*, **40**, 9513–9521.

Di Scipio, R.G., Kurachi, K. and Davie, E.W. (1978) Activation of human factor IX (Christmas factor). *Journal of Clinical Investigation*, **61**, 1526–1538.

Dickinson, C.D., Shobe, J. and Ruf, W. (1998) Influence of cofactor binding and active site occupancy on the conformation of the macromolecular substrate exosite of factor VIIa. *Journal of Molecular Biology*, **277**, 959–971.

Dittmar, S., Ruf, W. and Edgington, T.S. (1997) Influence of mutations in tissue factor on the fine specificity of macromolecular substrate activation. *Biochemical Journal*, **321**, 787–793.

Dunwiddie, C., Thornberry, N.A., Bull, H.G., Sardana, M., Friedman, P.A., Jacobs, J.W. *et al* (1989) Antistasin, a leech-derived inhibitor of factor Xa. *Journal of Biological Chemistry*, **264**, 16694–16699.

Dunwiddie, C.T., Neeper, M.P., Nutt, E.M., Waxman, L., Smith, D.E., Hofmann, K.J. *et al* (1992) Site-directed analysis of the functional domains in the factor Xa inhibitor tick anticoagulant peptide: Identification of two distinct regions that constitute the enzyme recognition sites. *Biochemistry*, **31**, 12126–12131.

Eigenbrot, C., Kirchhofer, D., Dennis, M.S., Santell, L., Lazarus, R.A., Stamos, J. and Ultsch, M.H. (2001) The factor VII zymogen structure reveals reregistration of β-strands during activation. *Structure*, **9** 627–636.

Faria, F., Kelen, E.M.A., Sampaio, C.A.M., Bon, C., Duval, N. and Chudzinski-Tavassi, A.M. (1999) A new factor Xa inhibitor (lefaxin) from the *Haementeria depressa* leech. *Thrombosis & Haemostasis*, **82**, 1469–1473.

Fernandez, A.Z., Tablante, A., Beguín, S., Hemker, H.C. and Apitz-Castro, R. (1999) Draculin, the anticoagulant factor in vampire bat saliva, is a tight-binding, noncompetitive inhibitor of activated factor X. *Biochimica et Biophysica Acta*, **1434**, 135–142.

Feuerstein, G.Z., Patel, A., Toomey, J.R., Bugelski, P., Nichols, A.J., Church, W.R. *et al* (1999a) Antithrombotic efficacy of a novel murine antihuman factor IX antibody in rats. *Arteriosclerosis Thrombosis & Vascular Biology*, **19**, 2554–2562.

Feuerstein, G.Z., Toomey, J.R., Valocik, R., Koster, P., Patel, A. and Blackburn, M.N. (1999b) An inhibitory anti-factor IX antibody effectively reduces thrombus formation in a rat model of venous thrombosis. *Thrombosis & Haemostasis*, **82**, 1443–1445.

Fiore, M.M., Neuenschwander, P.F. and Morrissey, J.H. (1992) An unusual antibody that blocks tissue factor/factor VIIa function by inhibiting cleavage only of macromolecular substrates. *Blood*, **80**, 3127–3134.

Fujikawa, K., Legaz, M.E., Kato, H. and Davie, E.W. (1974) The mechanism of activation of bovine factor IX (Christmas factor) by bovine factor XIa (activated plasma thromboplastin antecendent). *Biochemistry*, **13**, 4508–4516.

Gallagher, K.P., Mertz, T.E., Chi, L., Rubin, J.R. and Uprichard, A.C.G. (1999) Inhibitors of tissue factor/factor VIIa. In *Antithrombotics*, edited by A.C.G. Uprichard and K.P. Gallagher, pp. 421–445. Berlin: Springer-Verlag.

Girard, T.J., MacPhail, L.A., Likert, K.M., Novotny, W.F., Miletich, J.P. and Broze Jr., G.J. (1990) Inhibition of factor VIIa-tissue factor coagulation activity by a hybrid protein. *Science*, **248**, 1421–1424.

Golino, P., Ragni, M., Cirillo, P., Avvedimento, V.E., Feliciello, A., Esposito, N. *et al* (1996) Effects of tissue factor induced by oxygen free radicals on coronary flow during reperfusion. *Nature Medicine*, **2**, 35–40.

Golino, P., Ragni, M., Cirillo, P., D'Andrea, D., Scognamiglio, A., Ravera, A. *et al* (1998) Antithrombotic effects of recombinant human, active site-blocked factor VIIa in a rabbit model of recurrent arterial thrombosis. *Circulation Research*, **82**, 39–46.

Grobke, K., Ji, Y.-H, Wallbaum, S. and Weber, L. (1999) Preparation of N-(4-amidinophenyl)phenylglycineamides as factor VIIa/tissue factor inhibitors. EP 921116 A1.

Harker, L.A., Hanson, S.R., Wilcox, J.N. and Kelly, A.B. (1996) Antithrombotic and antilesion benefits without hemorrhagic risks by inhibiting tissue factor pathway. *Haemostasis*, **26** (suppl. 1), 76–82.

Harlos, K., Martin, D.M., O'Brien, D.P., Jones, E.Y., Stuart, D.I., Polikarpov, I. *et al* (1994) Crystal structure of the extracellular region of human tissue factor. *Nature*, **370**, 662–666.

Haskel, E.J., Torr, S.R., Day, K.C., Palmier, M.O., Wun, T.-C., Sobel, B.E. *et al* (1991) Prevention of arterial reocclusion after thrombolysis with recombinant lipoprotein-associated coagulation inhibitor. *Circulation*, **84**, 821–827.

Hauptmann, J. and Stürzebecher, J. (1999) Synthetic inhibitors of thrombin and factor Xa: From bench to bedside. *Thrombosis Research*, **93**, 203–241.

Hellmann, K. and Hawkins, R.I. (1965) Prolixin-S and prolixin-G; Two anticoagulants from *Rhodnius prolixus* STÅL. *Nature*, **207**, 265–267.

Higashi, S., Matsumoto, N. and Iwanaga, S. (1996) Molecular mechanism of tissue factor-mediated acceleration of factor VIIa activity. *Journal of Biological Chemistry*, **271**, 26569–26574.

Higashi, S. and Iwanaga, S. (1998) Molecular interaction between factor VII and tissue factor. *International Journal of Hematology*, **67**, 229–241.

Himber, J., Refino, C.J., Burcklen, L., Roux, S. and Kirchhofer, D. (2001) Inhibition of arterial thrombosis by a soluble tissue factor mutant and active site-blocked factors IXa and Xa in the guinea pig. *Thrombosis & Haemostasis*, **85**, 475–481.

Himber, J., Kirchhofer, D., Riederer, M., Tschopp, T.B., Steiner, B. and Roux, S.P. (1997) Dissociation of antithrombotic effect and bleeding time prolongation in rabbits by inhibiting tissue factor function. *Thrombosis & Haemostasis*, **78**, 1142–1149.

Hirsh, J. (1991) Heparin. *New England Journal of Medicine*, **324**, 1565–1574.

Hirsh, J., Ginsberg, J.S. and Marder, V.J. (1994) Anticoagulant therapy with coumarin agents. In *Hemostasis and thrombosis: Basic principles and clinical practice, 3rd edn*, edited by R.W. Colman, J. Hirsh, V.J. Marder and E.W. Salzman, pp. 1567–1583. Philadelphia: J. B. Lippincott Company.

Hollenbach, S., Sinha, U., Lin, P.-H., Needham, K., Frey, L., Hancock, T. *et al* (1994) A comparative study of prothrombinase and thrombin inhibitors in a novel rabbit model of non-occlusive deep vein thrombosis. *Thrombosis & Haemostasis*, **71**, 357–362.

Holst, J., Lindblad, B., Bergqvist, D., Nordfang, O., Østergaard, P.B., Petersen, J.G.L. *et al* (1994) Antithrombotic effect of recombinant truncated tissue factor pathway inhibitor ($TFPI_{1-161}$) in experimental venous thrombosis – a comparison with low molecular weight heparin. *Thrombosis & Haemostasis*, **71**, 214–219.

Holst, J., Kristensen, A.T., Kristensen, H.I., Ezban, M. and Hedner, U. (1998) Local application of recombinant active-site inhibited human clotting factor VIIa reduces thrombus weight and improves patency in a rabbit venous thrombosis model. *European Journal of Vascular & Endovascular Surgery*, **15**, 515–520.

Hopfner, K.-P., Lang, A., Karcher, A., Sichler, K., Kopetzki, E., Brandstetter, H. *et al* (1999) Coagulation factor IXa: The relaxed conformation of Tyr99 blocks substrate binding. *Structure*, **7**, 989–996.

Huang, M., Syed, R., Stura, E.A., Stone, M.J., Stefanko, R.S., Ruf, W. *et al* (1998) The mechanism of an inhibitory antibody on TF-initiated blood coagulation revealed by the crystal structures of human tissue factor, Fab 5G9 and TF•5G9 complex. *Journal of Molecular Biology*, **275**, 873–894.

Huang, Q., Neuenschwander, P.F., Rezaie, A.R. and Morrissey, J.H. (1996) Substrate recognition by tissue factor–factor VIIa. Evidence for interaction of residues Lys^{165} and Lys^{166} of tissue factor with the 4-carboxyglutamate-rich domain of factor X. *Journal of Biological Chemistry*, **271**, 21752–21757.

Huber, R. and Bode, W. (1978) Structural basis of the activation and action of trypsin. *Accounts of Chemical Research*, **11**, 114–122.

Ichinose, A. and Davie, E.W. (1994) The blood coagulation factors: Their cDNAs, genes, and expression. In *Hemostasis and Thrombosis: Basic Principles and Clinical Practice, 3rd edn*, edited by R.W. Colman, J. Hirsch, V.J. Marder and E.W. Salzman, pp. 19–54. Philadelphia: J. B. Lippincott Co.

Isawa, H., Yuda, M., Yoneda, K. and Chinzei, Y. (2000) The insect salivary protein, prolixin-S, inhibits factor IXa generation and Xase complex formation in the blood coagulation pathway. *Journal of Biological Chemistry*, **275**, 6636–6641.

Jang, Y., Guzman, L.A., Lincoff, A.M., Gottsauner-Wolf, M., Forudi, F., Hart, C.E. *et al* (1995) Influence of blockade at specific levels of the coagulation cascade on restenosis in a rabbit atherosclerotic femoral artery injury model. *Circulation*, **92**, 3041–3050.

Johnson, K. and Hung, D. (1998) Novel anticoagulants based on inhibition of the factor VIIa/tissue factor pathway. *Coronary Artery Disease*, **9**, 83–87.

Jordan, S.P., Waxman, L., Smith, D.E. and Vlasuk, G.P. (1990) Tick anticoagulant peptide: Kinetic analysis of the recombinant inhibitor with blood coagulation factor X_a. *Biochemistry*, **29**, 11095–11100.

Jordan, S.P., Mao, S.S., Lewis, S.D. and Shafer, J.A. (1992) Reaction pathway for inhibition of blood coagulation factor Xa by tick anticoagulant peptide. *Biochemistry*, **31**, 5374–5380.

Joubert, A.M., Louw, A.I., Joubert, F. and Neitz, A.W. (1998) Cloning, nucleotide sequence and expression of the gene encoding factor Xa inhibitor from the salivary glands of the tick, *Ornithodoros savignyi*. *Journal of Experimental & Applied Acarology*, **22**, 603–619.

Kelley, R.F., Refino, C.J., O'Connell, M.P., Modi, N., Sehl, P., Lowe, D. *et al* (1997) A soluble tissue factor mutant is a selective anticoagulant and antithrombotic agent. *Blood*, **89**, 3219–3227.

Kemball-Cook, G., Johnson, D.J.D., Tuddenham, E.G.D. and Harlos, K. (1999) Crystal structure of active site-inhibited human coagulation factor VIIa (des-Gla). *Journal of Structural Biology*, **127**, 213–223.

Kirchhofer, D., Tschopp, T.B. and Baumgartner, H.R. (1995) Active site-blocked factors VIIa and IXa differentially inhibit fibrin formation in a human *ex vivo* thrombosis model. *Arteriosclerosis Thrombosis & Vascular Biology*, **15**, 1098–1106.

Kirchhofer, D. and Banner, D.W. (1997) Molecular and structural advances in tissue factor-dependent coagulation. *Trends in Cardiovascular Medicine*, **7**, 316–324.

Kirchhofer, D., Lipari, M.T., Moran, P., Eigenbrot, C. and Kelley, R.F. (2000a) The tissue factor region that interacts with substrates factor IX and factor X. *Biochemistry*, **39**, 7380–7387.

Kirchhofer, D., Moran, P., Chiang, N., Kim, J., Riederer, M.A., Eigenbrot, C. *et al* (2000b) Epitope location on tissue factor determines the anticoagulant potency of monoclonal anti-tissue factor antibodies. *Thrombosis & Haemostasis*, **84**, 1072–1081.

Kirchhofer, D., Eigenbrot, C., Lipari, M.T., Moran, P., Peek, M. and Kelley R.F. (2001) The tissue factor region that interacts with factor Xa in the activation of factor VII. *Biochemistry*, **40**, 675–682.

Kornowski, R., Eldor, A., Werber, M.M., Ezov, N., Zwang, E., Nimrod, A. *et al* (1996) Enhancement of recombinant tissue-type plasminogen activator thrombolysis with a selective factor Xa inhibitor derived from the leech *Hirudo medicinalis*: Comparison with heparin and hirudin in a rabbit thrombosis model. *Coronary Artery Disease*, **7**, 903–909.

Krishnaswamy, S. and Betz, A. (1997) Exosites determine macromolecular substrate recognition by prothrombinase. *Biochemistry*, **36**, 12080–12086.

Kunitada, S., Nagahara, T. and Hara, T. (1999) Inhibitors of factor Xa. In *Antithrombotics*, edited by A.C.G. Uprichard and K.P. Gallagher, pp. 397–420. Berlin: Springer-Verlag.

Lapatto, R., Krengel, U., Schreuder, H.A., Arkema, A., de Boer, B., Kalk, K.H. *et al* (1997) X-ray structure of antistasin at 1.9 Å resolution and its modelled complex with blood coagulation factor Xa. *EMBO Journal*, **16**, 5151–5161.

Laskowski Jr. M. and Kato, I. (1980) Protein inhibitors of proteinases. *Annual Review of Biochemistry*, **49**, 593–626.

Lawson, J.H. and Mann, K.G. (1991) Cooperative activation of human factor IX by the human extrinsic pathway of blood coagulation. *Journal of Biological Chemistry*, **266**, 11317–11327.

Lawson, J.H., Butenas, S., Ribarik, N. and Mann, K.G. (1993) Complex-dependent inhibition of factor VIIa by antithrombin III and heparin. *Journal of Biological Chemistry*, **268**, 767–770.

Leatham, E.W., Bath, P.M.W., Tooze, J.A. and Camm, A.J. (1995) Increased monocyte tissue factor expression in coronary disease. *British Heart Journal*, **73**, 10–13.

Leblond, L. and Winocour, P.D. (1999) The coagulation pathway and antithrombotic strategies. In *Antithrombotics*, edited by A.C.G. Uprichard and K.P. Gallagher, pp. 1–39. Berlin: Springer-Verlag.

Lee, G.F., Lazarus, R.A. and Kelley, R.F. (1997) Potent bifunctional anticoagulants: Kunitz domain-tissue factor fusion proteins. *Biochemistry*, **36**, 5607–5611.

Lee, A., Agnelli, G., Büller, H., Ginsberg, J., Heit, J., Rote, W. *et al* (2000) A dose–response study of the factor VIIa/tissue factor inhibitor rNAPc2 in the prevention of postoperative venous thrombosis in patients undergoing total knee arthroplasty. *Blood* **96**, 491a (abstract).

Lefkovits, J., Malycky, J.L., Rao, J.S., Hart, C.E., Plow, E.F., Topol, E.J. *et al* (1996) Selective inhibition of factor Xa is more efficient than factor VIIa-tissue factor complex blockade at facilitating coronary thrombolysis in the canine model. *Journal of the American College of Cardiology*, **28**, 1858–1865.

Libby, P., Mach, F., Schonbeck, U., Bourcier, T. and Aikawa, M. (1999) Regulation of the thrombotic potential of atheroma. *Thrombosis & Haemostasis*, **82**, 736–741.

Lim-Wilby, M.S.L., Hallenga, K., de Maeyer, M., Lasters, I., Vlasuk, G.P. and Brunck, T.K. (1995) NMR structure determination of tick anticoagulant peptide (TAP). *Protein Science*, **4**, 178–186.

Mach, F., Schonbeck, U., Bonnefoy, J.-Y., Pober, J.S. and Libby, P. (1997) Activation of monocyte/macrophage functions related to acute atheroma complication by ligation of CD40. *Circulation*, **96**, 396–399.

Mann, K.G. (1999) Biochemistry and physiology of blood coagulation. *Thrombosis & Haemostasis*, **82**, 165–174.

Mao, S.-S., Huang, J., Welebob, C., Neeper, M.P., Garsky, V.M. and Shafer, J.A. (1995) Identification and characterization of variants of tick anticoagulant peptide with increased inhibitory potency toward human factor Xa. *Biochemistry*, **34**, 5098–5103.

Mao, S.-S., Przysiecki, C.T., Krueger, J.A., Cooper, C.M., Lewis, S.D., Joyce, J. *et al* (1998) Selective inhibition of factor Xa in the prothrombinase complex by the carboxyl-terminal domain of antistasin. *Journal of Biological Chemistry*, **273**, 30086–30091.

McGrath, M.E., Erpel, T., Bystroff, C. and Fletterick, R.J. (1994) Macromolecular chelation as an improved mechanism of protease inhibition: Structure of the ecotin–trypsin complex. *EMBO Journal*, **13**, 1502–1507.

McGrath, M.E., Gillmor, S.A. and Fletterick, R.J. (1995) Ecotin: Lessons on survival in a protease-filled world. *Protein Science*, **4**, 141–148.

Miller, C.L., Graziano, C., Lim, R.C. and Chin, M. (1981) Generation of tissue factor by patient monocytes: Correlation to thromboembolic complications. *Thrombosis & Haemostasis*, **46**, 489–495.

Mizuno, H., Fujimoto, Z., Koizumi, M., Kano, H., Atoda, H. and Morita, T. (1997) Structure of coagulation factors IX/X-binding protein, a heterodimer of C-type lectin domains. *Nature Structural Biology*, **4**, 438–441.

Mizuno, H., Fujimoto, Z., Koizumi, M., Kano, H., Atoda, H. and Morita, T. (1999) Crystal structure of coagulation factor IX-binding protein from habu snake venom at 2.6 Å: Implication of central loop swapping based on deletion in the linker region. *Journal of Molecular Biology*, **289**, 103–112.

Mody, R.S. and Carson, S.D. (1997) Tissue factor cytoplasmic domain peptide is multiply phosphorylated *in vitro*. *Biochemistry*, **36**, 7869–7875.

Morrissey, J.H., Macik, B.G., Neuenschwander, P.F. and Comp, P.C. (1993) Quantitation of activated factor VII levels in plasma using a tissue factor mutant selectively deficient in promoting factor VII activation. *Blood*, **81**, 734–744.

Mueller, B.M. and Ruf, W. (1998) Requirement for binding of catalytically active factor VIIa in tissue factor-dependent experimental metastasis. *Journal of Clinical Investigation*, **101**, 1372–1378.

Muller, Y.A., Ultsch, M.H., Kelley, R.F. and de Vos, A.M. (1994) Structure of the extracellular domain of human tissue factor: Location of the factor VIIa binding site. *Biochemistry*, **33**, 10864–10870.

Nakagaki, T., Foster, D.C., Berkner, K.L. and Kisiel, W. (1991) Initiation of the extrinsic pathway of blood coagulation: evidence for the tissue factor dependent autoactivation of human coagulation factor VII. *Biochemistry*, **30**, 10819–10824.

Nemerson, Y. and Gentry, R. (1986) An ordered addition, essential activation model of the tissue factor pathway of coagulation: Evidence for a conformational cage. *Biochemistry*, **25**, 4020–4033.

Nemerson, Y. (1988) Tissue factor and hemostasis. *Blood*, **71**, 1–8.

Neuenschwander, P.F., Branam, D.E. and Morrissey, J.H. (1993) Importance of substrate composition, pH and other variables on tissue factor enhancement of factor VIIa activity. *Thrombosis & Haemostasis*, **70**, 970–977.

Neuenschwander, P.F., Fiore, M.M. and Morrissey, J.H. (1993) Factor VII autoactivation proceeds via interaction of distinct protease–cofactor and zymogen–cofactor complexes. *Journal of Biological Chemistry*, **268**, 21489–21492.

Neuenschwander, P.F., McCollough, J., McCallum, C.D. and Johnson, A.E. (1997) A conformational change in the active site of blood coagulation factor IXa is associated with an increase in activity upon treatment with ethylene glycol. *Thrombosis & Haemostasis*, **Suppl.**, A428.

Ohta, N., Brush, M. and Jacobs, J.W. (1994) Interaction of antistasin-related peptides with factor Xa: Identification of a core inhibitory sequence. *Thrombosis & Haemostasis*, **72**, 825–830.

Olson, S.T. and Bjork, I. (1994) Regulation of thrombin activity by antithrombin and heparin. *Seminars in Thrombosis & Haemostasis*, **20**, 373–409.

Østerud, B. and Rapaport, S.I. (1977) Activation of factor IX by the reaction product of tissue factor and factor VII: Additional pathway for initiating blood coagulation. *Proceedings of the National Academy of Sciences of the U.S.A.*, **74**, 5260–5264.

Østerud, B., Bouma, B.N. and Griffin, J.H. (1978) Human blood coagulation factor IX. Purification, properties, and mechanism of activation by activated factor XI. *Journal of Biological Chemistry*, **253**, 5946–5951.

Ott, I., Fischer, E.G., Miyagi, Y., Mueller, B.M. and Ruf, W. (1998) A role of tissue factor in cell adhesion and migration mediated by interaction with actin-binding protein 280. *Journal of Cell Biology*, **140**, 1241–1253.

Padmanabhan, K., Padmanabhan, K.P., Tulinsky, A., Park, C.H., Bode, W., Huber, R. *et al* (1993) Structure of human des(1–45) factor Xa at 2.2 Å resolution. *Journal of Molecular Biology*, **232**, 947–966.

Parry, G.C.N. and Mackman, N. (2000) Mouse embryogenesis requires the tissue factor extracellular domain but not the cytoplasmic domain. *Journal of Clinical Investigation*, **105**, 1547–1554.

Pater, C., Refino, C.J., Nagel, M., Wu, D., Hass, P., Eaton, D. *et al* (1997) Factor IX binding protein inhibits thrombosis in a rabbit deep medial injury model. *Thrombosis & Haemostasis*, **Suppl.**, A578.

Pawashe, A.B., Golino, P., Ambrosio, G., Migliaccio, F., Ragni, M., Pascucci, I. *et al* (1994) A monoclonal antibody against rabbit tissue factor inhibits thrombus formation in stenotic injured rabbit carotid arteries. *Circulation Research*, **74**, 56–63.

Pedersen, A.H., Lund-Hansen, T., Bisgaard-Frantzen, H., Olsen, F. and Petersen, L.C. (1989) Autoactivation of human recombinant coagulation factor VII. *Biochemistry*, **28**, 9331–9336.

Petersen, L.C., Bjørn, S.E., Olsen, O.H., Nordfang, O., Norris, F. and Norris, K. (1996a) Inhibitory properties of separate recombinant Kunitz-type-protease-inhibitor domains from tissue-factor-pathway inhibitor. *European Journal of Biochemistry*, **235**, 310–316.

Petersen, L.C., Sprecher, C.A., Foster, D.C., Blumberg, H., Hamamoto, T. and Kisiel, W. (1996b) Inhibitory properties of a novel human Kunitz-type protease inhibitor homologous to tissue factor pathway inhibitor. *Biochemistry*, **35**, 266–272.

Petersen, L.C., Freskård, P.-O. and Ezban, M. (2000) Tissue factor-dependent factor VIIa signalling. *Trends in Cardiovascular Medicine*, **10**, 47–52.

Pike, A.C.W., Brzozowski, A.M., Roberts, S.M., Olsen, O.H. and Persson, E. (1999) Structure of human factor VIIa and its implications for the triggering of blood coagulation. *Proceedings of the National Academy of Sciences of the U.S.A.*, **96**, 8925–8930.

Presta, L., Sims, P., Meng, Y.G., Moran, P., Bullens, S. Bunting S. *et al* (2001) Generation of a humanized, high affinity anti-tissue factor antibody for use as a novel antithrombotic therapeutic. *Thrombosis & Haemostosis*, **85**, 379–389.

Prydz, H., Camerer, E., Røttingen, J.-A., Wiiger, M.T. and Gjernes, E. (1999) Cellular consequences of the initiation of blood coagulation. *Thrombosis & Haemostasis*, **82**, 183–192.

Radcliffe, R. and Nemerson, Y. (1975) Activation and control of factor VII by activated factor X and thrombin. *Journal of Biological Chemistry*, **250**, 388–395.

Ragni, M., Cirillo, P., Pascucci, I., Scognamiglio, A., D'Andrea, D., Eramo, N. *et al* (1996) Monoclonal antibody against tissue factor shortens tissue plasminogen activator lysis time and prevents reocclusion in a rabbit model of carotid artery thrombosis. *Circulation*, **93**, 1913–1918.

Rao, L.V.M., Rapaport, S.I. and Bajaj, S.P. (1986) Activation of human factor VII in the initiation of tissue factor-dependent coagulation. *Blood*, **68**, 685–691.

Rao, L.V.M. and Rapaport, S.I. (1988) Activation of factor VII bound to tissue factor: A key early step in the tissue factor pathway of coagulation. *Proceedings of the National Academy of Sciences of the U.S.A.*, **85**, 6687–6691.

Rao, L.V.M., Nordfang, O., Hoang, A.D. and Pendurthi, U.R. (1995) Mechanism of antithrombin III inhibition of factor VIIa/tissue factor activity on cell surfaces. Comparison with tissue factor pathway inhibitor/factor Xa-induced inhibition of factor VIIa/tissue factor. *Blood*, **85**, 121–129.

Rapaport, S.I. and Rao, L.V.M. (1995) The tissue factor pathway: How it has become a 'Prima Ballerina'. *Thrombosis & Haemostasis*, **74**, 7–17.

Rawala-Sheikh, R., Ahmad, S.S., Monroe, D.M., Roberts, H.R. and Walsh, P.N. (1992) Role of γ-carboxyglutamic acid residues in the binding of factor IXa to platelets and in factor-X activation. *Blood*, **79**, 398–405.

Rebello, S.S., Blank, H.S., Rote, W.E., Vlasuk, G.P. and Lucchesi, B.R. (1997) Antithrombotic efficacy of a recombinant nematode anticoagulant peptide (rNAP5) in canine models of thrombosis after single subcutaneous administration. *Journal of Pharmacology & Experimental Therapeutics*, **283**, 91–99.

Refino, C.J., Himber, J., Burcklen, L., Moran, P., Peek, M., Suggett, S. *et al* (1999) A human antibody that binds to the γ-carboxyglutamic acid domain of factor IX is a potent antithrombotic *in vivo*. *Thrombosis & Haemostasis*, **82**, 1188–1195.

Ribeiro, J.M.C., Schneider, M. and Guimaraes, J.A. (1995) Purification and characterization of prolixin S (nitrophorin 2), the salivary anticoagulant of the blood-sucking bug *Rhodnius prolixus*. *Biochemical Journal*, **308**, 243–249.

Roberge, M., Santell, L., Dennis, M.S., Eigenbrot, C., Dwyer, M.A. and Lazarus, R.A. (2001) A novel exosite on coagulation factor VIIa and its molecular interactions with a new class of peptide inhibitors. *Biochemistry*, **40**, 9522–9531.

Roussel, P., Bradley, M., Kane, P., Bailey, C., Arnold, R. and Cross, A. (1999) Inhibition of the tissue factor/factor VIIa complex – lead optimization using combinatorial chemistry. *Tetrahedron*, **55**, 6219–6230.

Roy, S., Hass, P.E., Bourell, J.H., Henzel, W.J. and Vehar, G.A. (1991) Lysine residues 165 and 166 are essential for the cofactor function of tissue factor. *Journal of Biological Chemistry*, **266**, 22063–22066.

Ruf, W. and Edgington, T.S. (1991) An anti-tissue factor monoclonal antibody which inhibits TF•VIIa complex is a potent anticoagulant in plasma. *Thrombosis & Haemostasis*, **66**, 529–533.

Ruf, W., Miles, D.J., Rehemtulla, A. and Edgington, T.S. (1992a) Cofactor residues lysine 165 and 166 are critical for protein substrate recognition by the tissue factor–factor VIIa protease complex. *Journal of Biological Chemistry*, **267**, 6375–6381.

Ruf, W., Miles, D.J., Rehemtulla, A. and Edgington, T.S. (1992b) Tissue factor residues 157–167 are required for efficient proteolytic activation of factor X and factor VII. *Journal of Biological Chemistry*, **267**, 22206–22210.

Ruf, W. and Dickinson, C.D. (1998) Allosteric regulation of the cofactor-dependent serine protease coagulation factor VIIa. *Trends in Cardiovascular Medicine*, **8**, 350–356.

Ruf, W. and Mueller, B.M. (1999) Tissue factor signalling. *Thrombosis & Haemostasis*, **82**, 175–182.

Ryan, J., Wolitzky, B., Heimer, E., Lambrose, T., Felix, A., Tam, J.P. *et al* (1989) Structural determinants of the factor IX molecule mediating interaction with the endothelial cell binding site are distinct from those involved in phospholipid binding. *Journal of Biological Chemistry*, **264**, 20283–20287.

Sanderson, P.E.J. (1999) Small, noncovalent serine protease inhibitors. *Medicinal Research Reviews*, **19**, 179–197.

Scarborough, R.M. (1998) Coagulation factor Xa: The prothrombinase complex as an emerging therapeutic target for small molecule inhibitors. *Journal of Enzyme Inhibition*, **14**, 15–25.

Schaffer, L.W., Davidson, J.T., Vlasuk, G.P. and Siegl, P.K.S. (1991) Antithrombotic efficacy of recombinant tick anticoagulant peptide. A potent inhibitor of coagulation factor Xa in a primate model of arterial thrombosis. *Circulation*, **84**, 1741–1748.

Schmaier, A.H., Dahl, L.D., Rozemuller, A.J.M., Roos, R.A.C., Wagner, S.L., Chung, R. *et al* (1993) Protease nexin-2/amyloid β protein precursor. A tight-binding inhibitor of coagulation factor IXa. *Journal of Clinical Investigation*, **92**, 2540–2545.

Schmaier, A.H., Dahl, L.D., Hasan, A.A.K., Cines, D.B., Bauer, K.A. and Van Nostrand, W.E. (1995) Factor IXa inhibition by protease nexin-2/amyloid β-protein precursor on phospholipid vesicles and cell membranes. *Biochemistry*, **34**, 1171–1178.

Seligsohn, U., Østerud, B., Brown, S.F., Griffin, J.H. and Rapaport, S.I. (1979) Activation of human factor VII in plasma and in purified systems. *Journal of Clinical Investigation*, **64**, 1056–1065.

Senokuchi, K. and Ogawa, K. (1999) Preparation of amidinophenylcarbamoylbiphenyl derivatives and heterocyclic analogs thereof as inhibitors of blood coagulation factor VIIa. WO 9941231 A1.

Seymour, J.L., Lindquist, R.N., Dennis, M.S., Moffat, B., Yansura, D., Reilly, D. *et al* (1994) Ecotin is a potent anticoagulant and reversible tight-binding inhibitor of factor Xa. *Biochemistry*, **33**, 3949–3958.

Sixma, J.J. and de Groot, P.G. (1992) The ideal anti-thrombotic drug. *Thrombosis Research*, **68**, 507–512.

Smith, R.P., Higuchi, D.A. and Broze Jr. G.J. (1990) Platelet coagulation factor XIa-inhibitor, a form of Alzheimer amyloid precursor protein. *Science*, **248**, 1126–1128.

Sørensen, B.B., Persson, E., Freskgård, P.-O., Kjalke, M., Ezban, M., Williams, T. *et al* (1997) Incorporation of an active site inhibitor in factor VIIa alters the affinity for tissue factor. *Journal of Biological Chemistry*, **272**, 11863–11868.

Spanier, T.B., Oz, M.C., Minanov, O.P., Simantov, R., Kisiel, W., Stern, D.M. *et al* (1998) Heparinless cardiopulmonary bypass with active-site blocked factor IXa: A preliminary study in the dog. *Journal of Thoracic & Cardiovascular Surgery*, **115**, 1179–1188.

Stanssens, P., Bergum, P.W., Gansemans, Y., Jespers, L., Laroche, Y., Huang, S. *et al* (1996) Anticoagulant repertoire of the hookworm *Ancylostoma caninum*. *Proceedings of the National Academy of Sciences of the U.S.A.*, **93**, 2149–2154.

Stark, K.R. and James, A.A. (1995) A factor Xa-directed anticoagulant from the salivary glands of the yellow fever mosquito *Aedes aegypti*. *Experimental Parasitology*, **81**, 321–331.

Stark, K.R. and James, A.A. (1998) Isolation and characterization of the gene encoding a novel factor Xa-directed anticoagulant from the yellow fever mosquito, *Aedes aegypti*. *Journal of Biological Chemistry*, **273**, 20802–20809.

Stassen, J.M., Lambeir, A.-M., Matthyssens, G., Ripka, W.C., Nyström, A., Sixma, J.J. *et al* (1995a) Characterisation of a novel series of aprotinin-derived anticoagulants. I. *In vitro* and pharmacological properties. *Thrombosis & Haemostasis*, **74**, 646–654.

Stassen, J.M., Lambeir, A.-M., Vreys, I., Deckmyn, H., Matthyssens, G., Nyström, A. *et al* (1995b) Characterisation of a novel series of aprotinin-derived anticoagulants. II. Comparative antithrombotic effects on primary thrombus formation *in vivo*. *Thrombosis & Haemostasis*, **74**, 655–659.

Stubbs II, M.T. (1996) Structural aspects of factor Xa inhibition. *Current Pharmaceutical Design*, **2**, 543–552.

Stürzebecher, J., Kopetzki, E., Bode, W. and Hopfner, K.-P. (1997) Dramatic enhancement of the catalytic activity of coagulation factor IXa by alcohols. *FEBS Letters*, **412**, 295–300.

Suggett, S., Kirchhofer, D., Hass, P., Lipari, T., Moran, P., Nagel, M. *et al* (2000) Use of phage display for the generation of human antibodies that neutralize factor IXa function. *Blood Coagulation & Fibrinolysis*, **11**, 27–42.

Sun, J., Yamaguchi, M., Yuda, M., Miura, K., Takeya, H., Hirai, M. *et al* (1996) Purification, characterization and cDNA cloning of a novel anticoagulant of the intrinsic pathway, (Prolixin-S), from salivary glands of the blood sucking bug, *Rhodnius prolixus*. *Thrombosis & Haemostasis*, **75**, 573–577.

Taylor Jr. F.B., Chang, A., Ruf, W., Morrissey, J.H., Hinshaw, L., Catlett, R. *et al* (1991) Lethal *E. coli* septic shock is prevented by blocking tissue factor with monoclonal antibody. *Circulatory Shock*, **33**, 127–134.

Toomey, J.R., Smith, K.J., Roberts, H.R. and Stafford, D.W. (1992) The endothelial cell binding determinant of human factor IX resides in the γ-carboxyglutamic acid domain. *Biochemistry*, **31**, 1806–1808.

Toschi, V., Gallo, R., Lettino, M., Fallon, J.T., Gertz, S.D., Fernandez-Ortiz, A. *et al* (1997) Tissue factor modulates the thrombogenicity of human atherosclerotic plaques. *Circulation*, **95**, 594–599.

Tuszynski, G.P., Gasic, T.B. and Gasic, G.J. (1987) Isolation and characterization of antistasin: An inhibitor of metastasis and coagulation. *Journal of Biological Chemistry*, **262**, 9718–9723.

Ulmer, J.S., Lindquist, R.N., Dennis, M.S. and Lazarus, R.A. (1995) Ecotin is a potent inhibitor of the contact system proteases factor XIIa and plasma kallikrein. *FEBS Letters*, **365**, 159–163.

Vaswani, S.K. and Hamilton, R.G. (1998) Humanized antibodies as potential therapeutic drugs. *Annals of Allergy, Asthma and Immunology*, **81**, 105–119.

Vaughan, T.J., Osbourn, J.K. and Tempest, P.R. (1998) Human antibodies by design. *Nature Biotechnology*, **16**, 535–539.

Vlasuk, G.P., Ramjit, D., Fujita, T., Dunwiddie, C.T., Nutt, E.M., Smith, D.E. *et al* (1991) Comparison of the *in vivo* anticoagulant properties of standard heparin and the highly selective factor Xa inhibitors antistasin and tick anticoagulant peptide (TAP) in a rabbit model of venous thrombosis. *Thrombosis & Haemostasis*, **65**, 257–262.

Vlasuk, G.P., Dempsey, E.M., Oldeschulte, G.L., Bernardino, V.T., Richard, B.M. and Rote, W.E. (1995) Evaluation of a novel small protein inhibitor of blood coagulation factor Xa (rNAP5) in animal models of thrombosis. *Circulation*, **92** (Suppl. 1), I-685 (abstract).

Vlasuk, G.P., Bergum, P.W., Rote, W.E. and Ruf, W. (1997) Mechanistic and pharmacological evaluation of NAPc2, a novel small protein inhibitor of the factor VIIa/tissue factor complex. *Thrombosis & Haemostasis*, **Suppl.**, A688.

Walter, M.R., Windsor, W.T., Nagabhushan, T.L., Lundell, D.J., Lunn, C.A., Zauodny, P.J. *et al* (1995) Crystal structure of a complex between interferon-γ and its soluble high-affinity receptor. *Nature*, **376**, 230–235.

Wang, D., Bode, W. and Huber, R. (1985) Bovine chymotrypsinogen A. X-ray crystal structure analysis and refinement of a new crystal form at 1.8 Å resolution. *Journal of Molecular Biology*, **185**, 595–624.

Waxman, L., Smith, D.E., Arcuri, K.E. and Vlasuk, G.P. (1990) Tick anticoagulant peptide (TAP) is a novel inhibitor of blood coagulation factor Xa. *Science*, **248**, 593–596.

Wei, A., Alexander, R.S., Duke, J., Ross, H., Rosenfeld, S.A. and Chang, C.H. (1998) Unexpected binding mode of tick anticoagulant peptide complexed to bovine factor Xa. *Journal of Molecular Biology*, **283**, 147–154.

Weitz, J.I. (1997) Low-molecular-weight heparins. *New England Journal of Medicine*, **337**, 688–698.

Wilcox, J.N., Smith, K.M., Schwartz, S.M. and Gordon, D. (1989) Localization of tissue factor in the normal vessel wall and in the atherosclerotic plaque. *Proceedings of the National Academy of Sciences of the U.S.A.*, **86**, 2839–2843.

Wong, A.G., Gunn, A.C., Ku, P., Hollenbach, S.J. and Sinha, U. (1997) Relative efficacy of active site-blocked factors IXa, Xa in models of rabbit venous and arterio-venous thrombosis. *Thrombosis & Haemostasis*, **77**, 1143–1147.

Zhang, E., St. Charles, R. and Tulinsky, A. (1999) Structure of extracellular tissue factor complexed with Factor VIIa inhibited with a BPTI mutant. *Journal of Molecular Biology*, **285**, 2089–2104.

Zhang, Y., Ribeiro, J.M.C., Guimarães, J.A. and Walsh, P.N. (1998) Nitrophorin-2: A novel mixed-type reversible specific inhibitor of the intrinsic factor-X activating complex. *Biochemistry*, **37**, 10681–10690.

Zioncheck, T.F., Roy, S. and Vehar, G.A. (1992) The cytoplasmic domain of tissue factor is phosphorylated by a protein kinase C-dependent mechanism. *Journal of Biological Chemistry*, **267**, 3561–3564.

Chapter 10

The urokinase-type plasminogen activator (uPA) system: a new target for tumor therapy

Markus Bürgle, Stefan Sperl, Jörg Stürzebecher, Achim Krüger, Wolfgang Schmalix, Horst Kessler, Luis Moroder, Viktor Magdolen, Olaf G. Wilhelm and Manfred Schmitt

Metastasis of solid malignant tumors is initiated by detachment of tumor cells from the primary tumor, invasion of tumor cells into the surrounding extracellular matrix and into blood/lymphatic vessels followed by extravasation from the vessels and re-implantation at distant loci, accompanied by neovascularization. Tumor cell migration and invasion into the extracellular matrix is facilitated by a number of proteases including urokinase-type plasminogen activator (uPA) and plasmin (produced by the action of uPA on plasminogen). A new target for tumor therapy to affect tumor invasion and metastasis is uPA and its receptor (uPA-R) and a variety of agents have been developed to interfere with the expression, enzymic activity, or receptor binding activity of the components of the uPA/uPA-R system.

10.1 INTRODUCTION

Metastasis of solid malignant tumors (Figure 10.1) is initiated by detachment of tumor cells from the primary tumor, invasion of tumor cells into the surrounding extracellular matrix and into blood/lymphatic vessels followed by extravasation from the vessels and re-implantation at distant loci, accompanied by neovascularization (Folkman 1995). Tumor cell migration and invasion into the surrounding extracellular matrix is facilitated by a variety of proteolytic enzymes: matrix metalloproteinases (MMPs), including collagenases, gelatinases and stromelysins, cysteine proteases including cathepsins B and L, the aspartyl protease cathepsin D, and serine proteases such as plasmin and the urokinase-type plasminogen activator (uPA) (Andreasen *et al* 1997, 2000; Reuning *et al* 1998; Dano *et al* 1999; Schmitt *et al* 2000).

uPA (E.C. 3.4.21.73) converts the enzymatically inactive pro-enzyme plasminogen (synthesized in the liver) into plasmin (E.C. 3.4.21.7), a serine protease with a broad activity spectrum toward various extracellular matrix components such as fibrin, fibronectin, laminin, and collagen IV. Plasmin, in turn, activates the pro-enzyme form of uPA, pro-uPA, and some pro-enzyme forms of MMPs (Dano *et al* 1999; Schmitt *et al* 1992, 2000). uPA is expressed by a variety of normal cells (e.g. fibroblasts, phagocytic cells, trophoblast cells) but also by tumor cells in solid malignant tumors and binds to a specific high affinity cell surface receptor (uPA-R, CD87), both in its zymogen form (pro-uPA) and its enzymatically active two chain form (HMW-uPA; high-molecular-weight form of uPA)

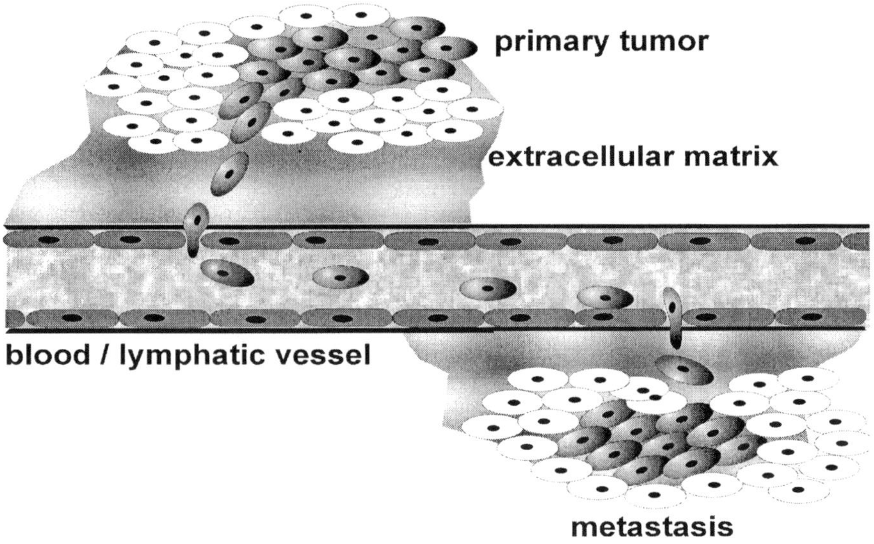

Figure 10.1 Tumor cell invasion and metastasis in solid malignant tumors.

(Fazioli and Blasi 1994; Behrendt *et al* 1995; Blasi 1999). Upon binding to uPA-R, the enzymatically active uPA is focused to the cell surface resulting in a higher state of uPA activity and several fold enhanced rate of conversion of cell surface-associated plasminogen to plasmin (Ellis *et al* 1999). The proteolytic activity of the HMW-uPA is controlled by its inhibitors PAI-1 and PAI-2 (Blasi 1997).

uPA consists of three different protein domains (Figure 10.2): (1) the N-terminally located growth-factor-like domain (GFD), harboring the binding site for uPA-R; (2) the kringle domain which is structurally related to kringle domains of thrombin, tPA (tissue-type plasminogen activator), and some other proteins; (3) the serine protease domain with its catalytic center encompassing amino acids His^{204}, Asp^{255} and Ser^{356}. pro-uPA is activated by different proteases such as plasmin, kallikrein, tryptase, and cathepsins B and L into enzymatically active HMW-uPA (Schmitt *et al* 1992, 2000). The uPA sequence covering domains GFD and kringle is also termed ATF (amino-terminal fragment, uPA_{1-135}).

uPA-R (CD87), the cellular receptor for uPA (Figure 10.3), is a cysteine-rich glycoprotein, inserted into the outer leaflet of the lipid bilayer of the plasma membrane via a glycosylphosphatidylinositol (GPI) anchor (Ploug *et al* 1991). uPA-R comprises three homologous, structurally related protein domains. The main ligand-binding site is located in domain I of uPA-R which most probably functions in concert with domain II and III as uPA-R domain I devoid of domains II and III displays a 100- to 1000-fold lower binding affinity towards uPA than the intact three-domain uPA-R (Mondino *et al* 1999).

In addition to converting plasminogen to plasmin, uPA, via binding to uPA-R, is also implicated in other tumor biological important processes (Figure 10.4) such as signal transduction, cell proliferation, chemotaxis, cell migration, and angiogenesis (Dear and Medcalf 1998; Reuning *et al* 1998). Among other tasks, uPA-R regulates and promotes cell adhesion by interacting with the extracellular matrix protein vitronectin and $\alpha\beta$-type integrins (Chapman 1997).

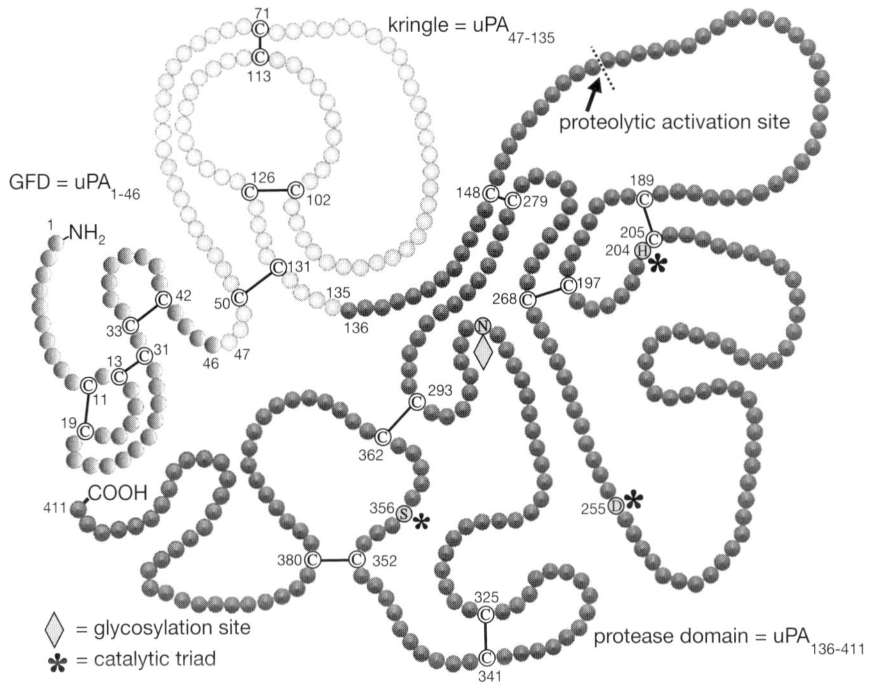

Figure 10.2 The human urokinase-type plasminogen activator (uPA) molecule.

Compared with normal tissues, elevated antigen levels of uPA, uPA-R, and/or PAI-1 but decreased levels of PAI-2 are found in tumor tissues of breast cancer patients. This finding is of considerable clinical importance as a statistically independent prognostic impact has also been attributed to these proteolytic factors in different types of malignancy (Duffy 1996; Schmitt *et al* 1997; Foekens *et al* 2000). Due to the strong correlation of uPA, uPA-R, and/or PAI-1 antigen values, determined in primary cancer tissue, and disease recurrence (metastasis), these proteolytic factors have been chosen as a new target for tumor therapy to affect tumor invasion and metastasis (Fazioli and Blasi 1994; Schmitt *et al* 2000). A variety of substances have been developed and designed to interfere with the expression, enzymatic activity, or receptor binding activity of the components of the uPA/uPA-R-system (Figure 10.5). Some of the reagents (Schmitt *et al* 2000) inhibiting uPA/uPA-R-dependent tumor cell invasion and metastasis are:

- synthetic low-molecular-weight inhibitors directed to the enzymatic site of uPA;
- uPA-derived peptides;
- monoclonal and polyclonal antibodies directed to uPA or uPA-R, either blocking uPA/uPA-R-interaction or lowering uPA enzymatic activity;
- recombinant, soluble form of uPA-R interfering with the binding of uPA to tumor cell surface located uPA-R;
- antisense oligonucleotides or RNA directed against uPA or uPA-R expression.

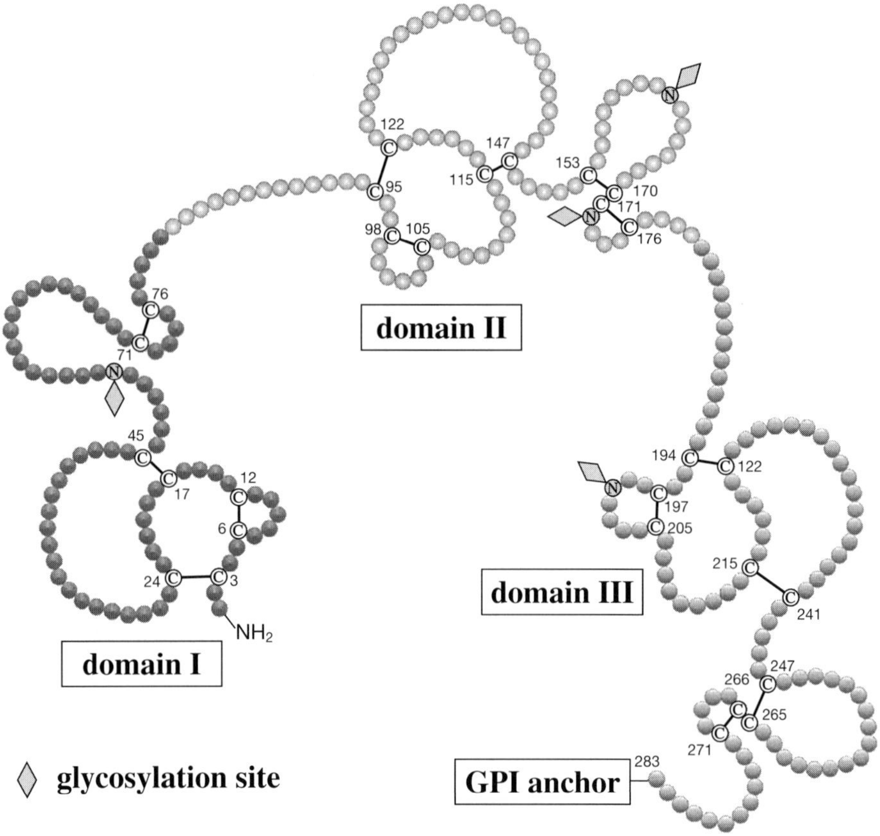

Figure 10.3 The urokinase receptor (uPA-R; CD87) molecule.

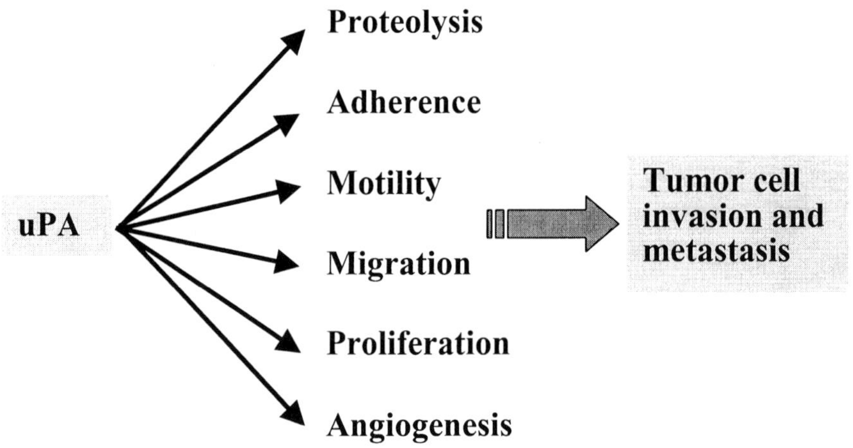

Figure 10.4 Multifunctional potential of urokinase (uPA) to induce various tumor cell responses leading to tumor cell invasion and metastasis.

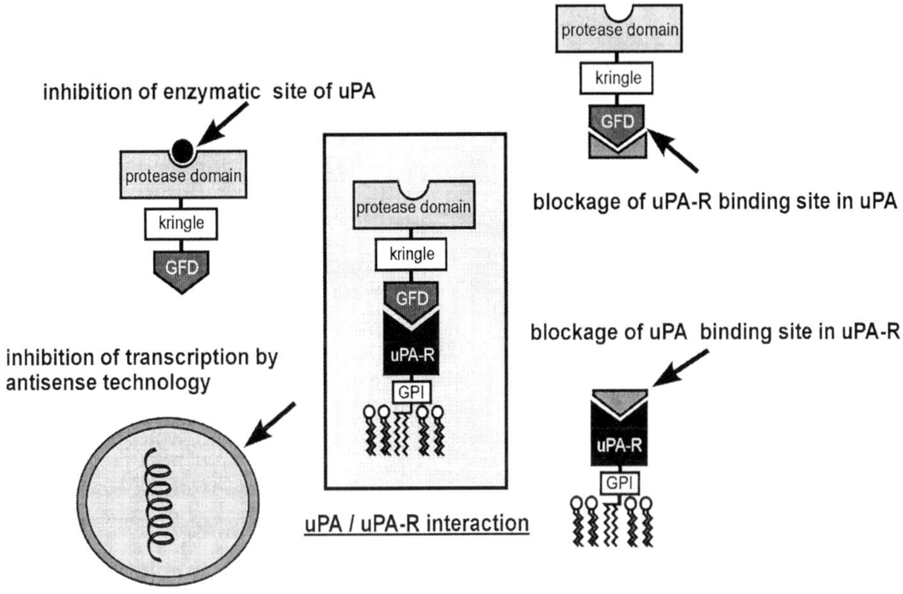

Figure 10.5 Strategies to interfere with the uPA/uPA-R system at the protein and mRNA level.

10.2 INHIBITION OF THE PROTEOLYTIC ACTIVITY OF uPA BY SYNTHETIC INHIBITORS

Besides the naturally occurring uPA-inhibitors PAI-1, PAI-2, protease nexin, and protein C inhibitor, potent synthetic low-molecular-weight inhibitors directed to the proteolytic activity of uPA (Renatus *et al* 1998; Rosenberg 2000) have been developed (Table 10.1). *In vivo* studies employing some of the synthetic uPA inhibitors described in the literature (Table 10.1) have stimulated the development of even more potent and highly selective synthetic uPA inhibitors. One common structural feature of synthetic uPA inhibitors is the presence of an aromatic moiety conjugated with an amidino- or guanidino group, which is expected to interact with Asp^{350} in the arginine-specific S1 pocket of uPA.

The first screening for uPA inhibitors based on benzamidine derivatives was published in 1978 (Stürzebecher and Markwardt 1978). Benzamidine (**10.1**) and substituted benzamidines such as 4-amino-benzamidine (**10.2**) exert rather low anti-uPA activity (Figure 10.6, Table 10.2). Moderately potent inhibitors of uPA were identified among bis-benzamidines containing a cycloalkanone linking bridge. However, such compounds with the same potency inhibit other trypsin-like proteases. Compound (**10.3**), for example, is more likely to inhibit tPA than uPA (Table 10.2). Higher selectivity toward uPA than tPA was obtained with naphthamidine derivatives (Stürzebecher and Markwardt 1978). These compounds inhibit uPA with K_i values in the micromolar range whereas the activity of tPA is not affected. The selectivity within the trypsin-like enzymes is low, however (**10.4**).

Most recently, the 2-naphthamidine lead was successfully improved by Abbott Laboratories (WO-09905096). Introduction of an N-aryl-substituted 6-carboxamide residue increases

Table 10.1 Inhibitors of the enzymatic activity of uPA. References are listed where the effects of uPA inhibitors on tumor invasion and metastasis have been reported. Furthermore, references are listed where the design/synthesis of such inhibitors has been described

Inhibitor of uPA enzymatic activity	*Reference*
PAI-2	Baker *et al* (1990) *Cancer Res.*, **50**, 4676
PAI-1/A-chain cholera toxin	Jankun (1992) *Cancer Res.*, **52**, 5829
GFD-PAI-2	Ballance *et al* (1992) *Eur. J. Biochem.*, **207**, 177
PAI-2	Laug *et al* (1993) *Cancer Res.*, **53**, 6051
dexamethasone	Reeder *et al* (1993) *Terat. Carc. Mutag.*, **13**, 75
substituted benzo[*b*]thiophene-2-carboxamidine (B428 and B623)	Towle *et al* (1993) *Cancer Res.*, **53**, 2553
phenylacetate	Samid *et al* (1993) *J. Clin. Invest.*, **91**, 2288
suramin	Takeuchi *et al* (1993) *Am. J. Gastroenterology*, **88**, 1928
PAI-2	Mueller *et al* (1995) *PNAS*, **92**, 205
PAI-2	Evans and Lin (1995) *Am. Surg.*, **61**, 692
N-(4-hydroxylphenyl)-retinamide	Kim *et al* (1995) *Anticancer Res.*, **15**, 1429
benzo[*b*]thiophene derivative (B428)	Rabbani *et al* (1995) *Int. J. Cancer*, **63**, 840
estramustine, taxol	Santibanez *et al* (1995) *Cell Biol. Funct.*, **13**, 217
p-aminobenzamidine	Billstrom *et al* (1995) *Int. J. Cancer*, **61**, 542
pyroglutamyl-Leu-Arg-CHO	Kawada and Umezawa (1995) *BBRC*, **209**, 25
PAI-1	Soff *et al* (1996) *J. Clin. Invest.*, **96**, 2593
substituted benzo[*b*]thiophene-2-carboxamidine (B428 and B623)	Alonso *et al* (1996) *Breast Cancer Res. Treat.*, **40**, 209
unesterified long chain fatty acids	Higazi *et al* (1996) *Biochemistry*, **35**, 6884
amiloride, p-aminobenzamidine	Jankun *et al* (1997) *Cancer Res.*, **57**, 559
tamoxifen + B428	Xing *et al* (1997) *Cancer Res.*, **57**, 3585
(−)-epigallocatechin gallate	Jankun *et al* (1997) *Nature*, **387**, 561
(−)-*epigallocatechingallate*	Nakachi *et al* (1998) *Jpn. J. Cancer Res.*, **89**, 254
[N,N-dimethylcarbamoylmethyl 4-(4-guanidinobenzoyloxy)-phenylacetate] methanesulfate (FOY-305)ethyl N-allyl-N-[(E)-2-methyl-3- [4-(4-amidino -phenoxycarbonyl)phenyl] propenoyl] amino acetate methanesulfonate (ONO-3403)	Ikeda *et al* (1998) *Anticancer Res.*, **18**, 4259
substituted benzo[*b*]thiophene-2-carboxamidine(B428)	Alonso *et al* (1998) *Anticancer Res.*, **18**, 4499
ecotin M84R + D70R	Yang and Craik (1998) *J. Mol. Biol.*, **279**, 1001
substituted benzo[*b*]thiophene-2-carboxamidine (B428), amiloride	Evans and Sloan-Stakleff (1998) *Invasion Metast.*, **18**, 252
adenovirus encoding either PAI-1 or PAI-2	Praus *et al* (1999) *Gene Ther.*, **6**, 227
substituted benzo[*b*]thiophene-2-carbox amidine (B428 and B623), (-)-epigalloca techin gallate, benzamidine, amiloride	Swiercz *et al* (1999) *Oncol. Rep.*, **6**, 523
ecotin	Takeuchi *et al* (1999) *PNAS*, **96**, 11054
amidinophenylalanine derivative (WX-UK1)	Stürzebecher *et al* (1999) *Bioorg. Med. Chem. Letters*, **9**, 3147
peptide aldehyde uPA inhibitors	Tamura *et al* (2000) *Bioorg. Med. Chem. Lett.*, **10**, 983
(4-aminomethyl)phenylguanidine derivative (WX-293T)	Sperl *et al* (2000) *PNAS*, **97**, 5113
2-naphthamidine, substituted naphthamidines	Nienaber *et al* (2000) *Structure Fold. Des.*, **8**, 553

substituted benzo[*b*]thiophene-2-carboxamidine (B428)	Katz *et al* (2000) *Chem. Biol.*, **7**, 299
substituted benzo[*b*]thiophene-2-carboxamidine (B428), amiloride	Evans and Sloan-Stackleff (2000) *Am. Surg.*, **66**, 460
substituted benzo[*b*]thiophene-2-carboxamidine (B428), amiloride	Nienaber *et al* (2000). *J. Biol. Chem.*, **275**, 7248

inhibition of uPA up to 100-fold (**10.5**). A further increase in inhibitory potency was achieved by substitution at the 8-position, mainly with heterocycles. For the most potent 6,8-bis-substituted 2-naphthamidine (**10.6**) an IC_{50}-value of 0.68 nM was reported. The importance of an 8-substitution for the binding of 2-naphthamidines to uPA was demonstrated by several derivatives lacking a residue in the 6-position (Nienaber *et al* 2000a). The 8-methylcarbamyl derivative (**10.7**) inhibits uPA with a K_i of 0.04 μM. The 8-substituted 2-naphthamidines inhibit uPA selectively (Table 10.2).

In the early 1980s, Geratz and colleagues published results of a screening of several derivatives of bicyclic aromatic amidines (Geratz *et al* 1981). The most potent among

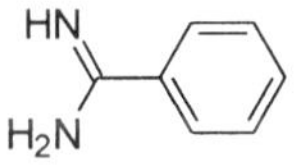

10.1

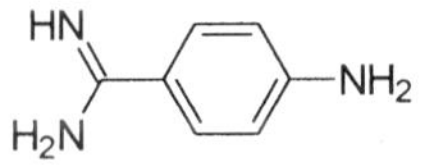

10.2

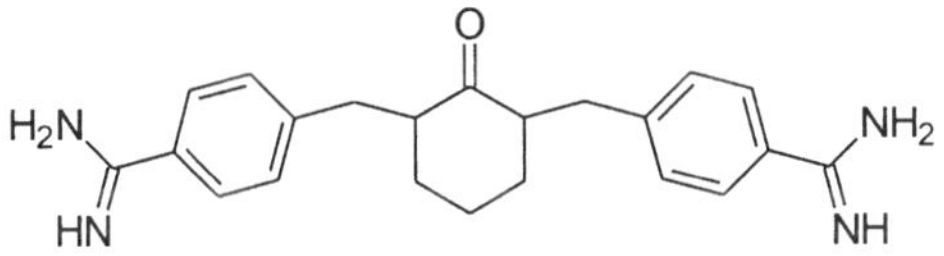

10.3

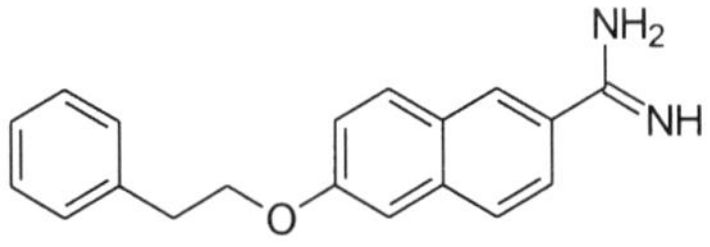

10.4

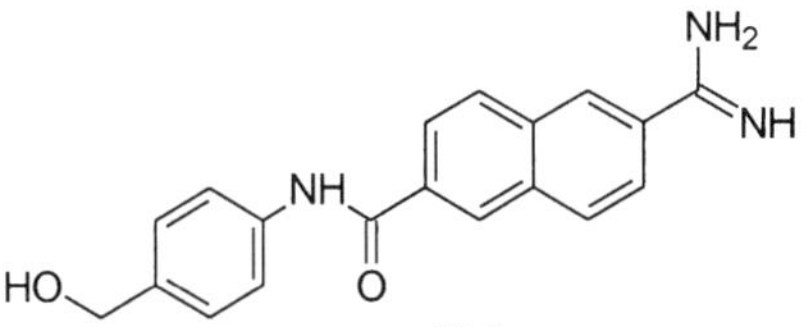

10.5

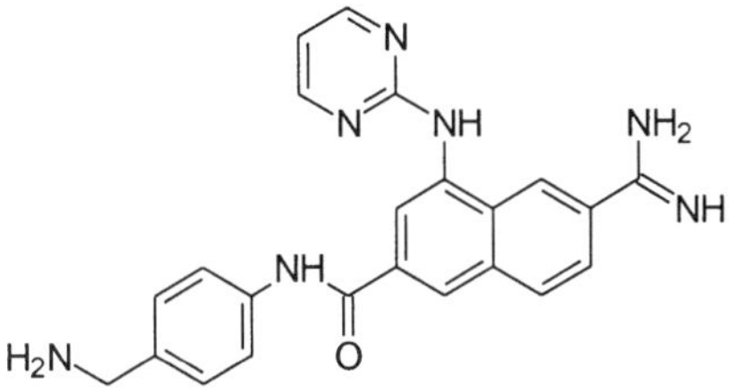

10.6

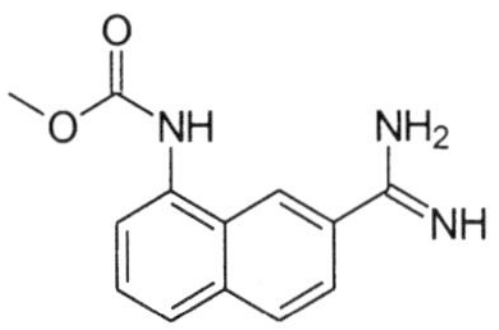

10.7

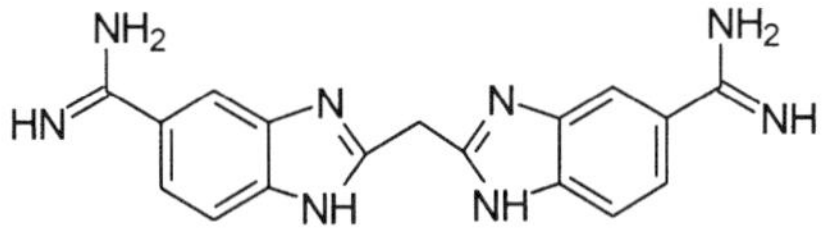

10.8

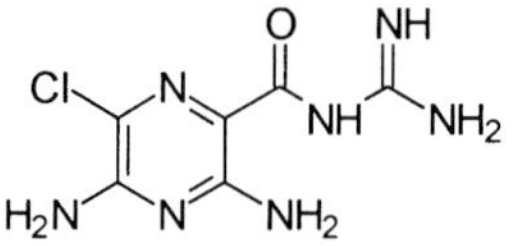

10.9

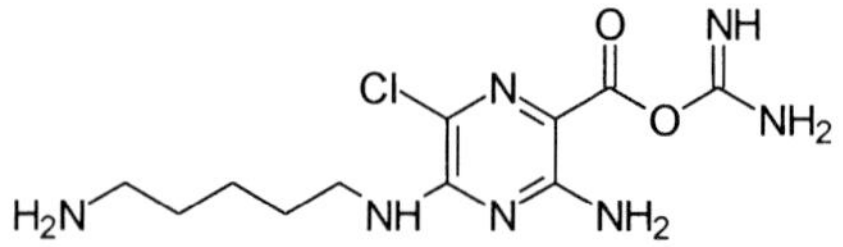

10.10

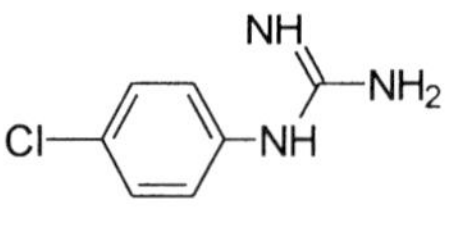

10.11

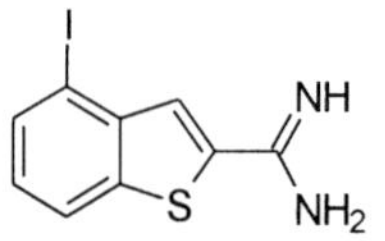

10.12

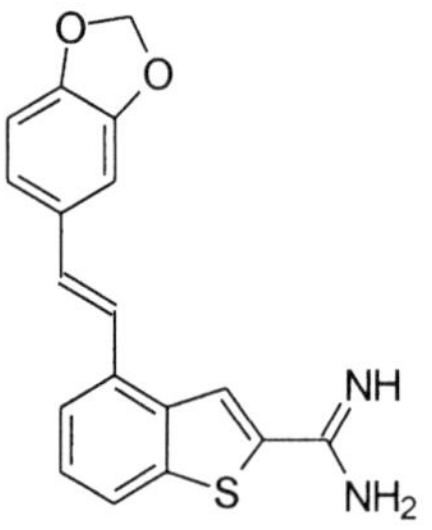

10.13

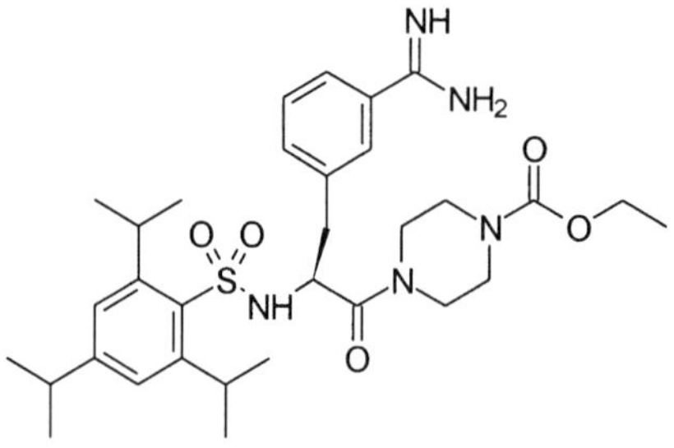

10.14

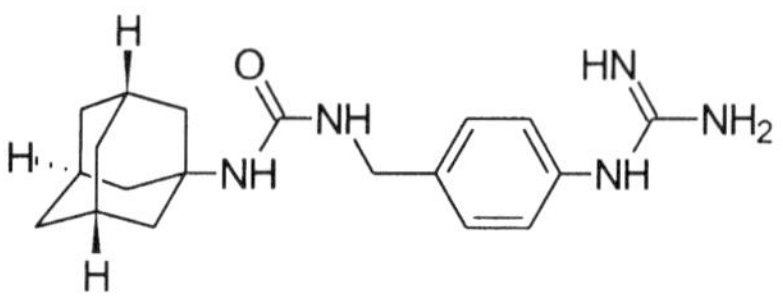

10.15

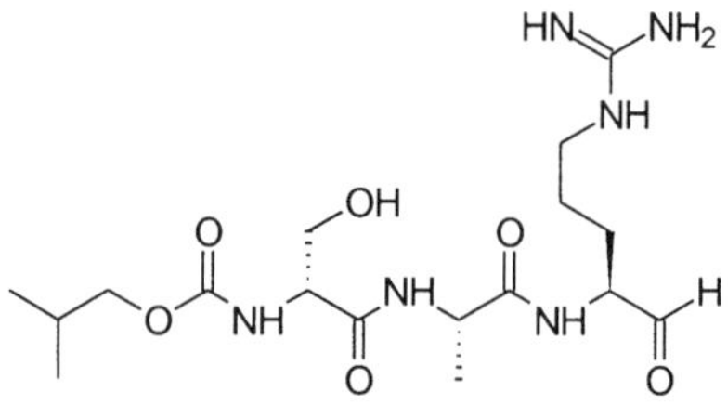

10.16

Figure 10.6 Structures of some synthetic inhibitors.

these inhibitors was (**10.8**). Usually this class of compounds only yields low inhibition of uPA. In 1987, the diuretic drug amiloride (**10.9**), an inhibitor of transepithelial Na^+ transport, was found to efficiently inhibit uPA. Very little structural variation of this molecule is known, leading to improved inhibitor capacity (**10.10**). Monosubstituted phenylguanidines were also tested (Yang *et al* 1990). The phenylguanidines (**10.11**) are selective inhibitors of uPA, besides their ability to inhibit trypsin.

Substituted benzothiophene compounds, the first highly potent inhibitors of uPA, were described by Towle *et al* in 1993. The 4-substituted benzo[*b*]thiophene-2-carboxamidines B428 and B623 (**10.12** and **10.13**) competitively inhibit uPA with K_i-values of 0.53 and

Table 10.2 Inhibition (K_i, μM) of uPA and related serine proteases by synthetic inhibitors (nd = not determined)

Compound	*uPA*	*tPA*	*Plasmin*	*Trypsin*	*Thrombin*	*Reference*
10.1	81	>1000	350	39	890	Stürzebecher *et al* (1978) *Pharmazie*, **33**, 599
10.2	17	>1000	15	12	77	Stürzebecher *et al* (1978) *Pharmazie*, **33**, 599
10.3	2.3	63	50	0.80	3.1	Stürzebecher *et al* (1978) *Pharmazie*, **33**, 599
10.4	2.2	>1000	7.9	1.4	6.4	Stürzebecher *et al* (1978) *Pharmazie*, **33**, 599
10.7	0.04	40	1.8	0.3	5.2	Nienaber *et al* (2000) *Structure Fold. Des.*, **8**, 553
10.8	2.3	nd	2.6	0.017	4.1	Geratz *et al* (1981) *Thromb. Res.*, **24**, 73 Tidwell *et al* (1983) *J. Med. Chem.* **28**, 294
10.9	7.0	>1000	>1000	32	>1000	Vassalli *et al* (1997) *FEBS Letters*, **214**, 187
10.10	2.5	nd	nd	nd	nd	Pato *et al* (1999) *J. Bioac. Comp. Polymers.*, **14**, 99
10.11	6.1	>1000	>1000	120	>1000	Yang *et al* (1990) *J. Med. Chem.*, **33**, 2956
10.12	0.53	nd	nd	4.9	850	Towle *et al* (1993) *Cancer Res.*, **53**, 2553
10.13	0.16	nd	nd	2.8	>250	Towle *et al* (1993) *Cancer Res.* **53**, 2553
10.14	0.41	4.9	0.39	0.037	0.49	Stürzebecher *et al* (1999) *Bioorg. Med. Letters*, **9**, 3147
10.15	2.4	>1000	>1000	46	600	Sperl *et al* (2000) Proc. Natl. Acad. Sci. USA, **97**, 5113

0.16 μM, respectively, and thus represent a potent class of inhibitors of uPA. Another successful approach, in order to design new types of more selective uPA inhibitors, was based on derivatizing the arginine mimetics 3- and 4-amidinophenylalanine. Several derivatives of this type inhibit thrombin and trypsin with K_i values in the nanomolar range but inhibition of uPA is weak. Introduction of N^{α}-triisopropyl-phenylsulfonyl into this molecule resulted in a highly potent uPA inhibitor (**10.14**, WX-UK1) (Stürzebecher *et al* 1999). With $K_i = 0.41$ μM, WX-UK1 is one of the most potent uPA inhibitors described so far. Nevertheless, WX-UK1 exhibits no selectivity toward uPA when compared to other trypsin-like proteases. Most recently, new types of low molecular mass and highly selective uPA inhibitors were generated, consisting of hydrophobic derivatives of 4-aminomethyl-phenylguanidine. The highly selective lead structure (**10.15**, WX-293) inhibits uPA with $K_i = 2.4$ μM (Sperl *et al* 2000).

Irreversible inhibition of uPA occurs with chloromethyl ketones by alkylation of the active site amino acid His^{204} of uPA. Among several compounds tested, chloromethyl ketones with a Glu-Gly-Arg sequence exerted the highest inhibitory activity.

Furthermore, acylation of Ser^{356} of uPA was performed by p-nitrophenyl-p′-guanidinobenzoate, coumarin derivatives, and cyclopeptides. Due to their highly reactive alkylating or acylating function, all of these compounds have in common to react with related enzymes besides uPA. Such irreversible inhibitors are not considered for use as antimetastatic drugs *in vivo*. Recently, the synthesis and biological activity of peptidyl aldehyde uPA inhibitors, based on the sequence (D)-Ser-Ala-Arg, was described. The lead compound of this series, iBoc-(D)-Ser-Ala-Arg-H (**10.16**), selectively inhibits uPA by a slow and tight binding mechanism with an IC_{50}-value in the 10^{-8} molar range. Carbonate prodrugs were prepared and tested as potential drug delivery systems (Tamura *et al* 2000).

The X-ray crystal structure of uPA complexed with the irreversible inhibitor H- Glu-Gly-Arg chloromethyl ketone was published in 1995 by Spraggon *et al*. Recently, the X-ray structures of uPA complexed with reversible inhibitors (e.g. benzamidine, amiloride, WX-293) have been published (Zeslawska *et al* 2000). These structures will be useful for the rational design of more potent and selective uPA inhibitors. WX-293 exerts an unconventional binding mode, involving both the S1 and S1′ pockets of uPA. Despite the crossing of the active site, interactions with the catalytic residues Ser^{356} and His^{204} were not observed (Sperl *et al* 2000). Meanwhile, the X-ray structures of uPA complexed with the benzo[*b*]thiophene derivative B-428 and several 8-substituted 2-naphthamidines have also been solved (Nienaber *et al* 2000a,b).

Studies of the inhibiting effect of the substituted benzo[*b*]thiophene-2-carboxamidines B428 and B623 on tumor growth and metastasis have been published for several types of tumor. B428 and B623 inhibit uPA-mediated processes such as proteolytic degradation of the extracellular matrix, tumor cell adhesion and migration as well as tumor cell invasion. B428 and B632 inhibited local tumor invasion in a murine mammary tumor model but did not reduce spontaneous metastasis or tumor-induced neo-vascularization (Alonso *et al* 1998).

In contrast, in a syngeneic rat prostate cancer model overexpressing uPA, a remarkable decrease in primary tumor growth and metastasis was observed by treatment with B428 (Rabbani *et al* 1995). Moreover, the combination of B428 with the antiestrogen tamoxifen led to a significant reduction in primary tumor volume and metastasis in a syngeneic rat breast cancer model (Xing *et al* 1997).

10.3 PEPTIDIC SUBSTANCES INTERFERING WITH uPA/uPA RECEPTOR INTERACTION

Effective inhibition of the uPA/uPA-R interaction by preventing binding of uPA to cellular uPA-R has been reported (Table 10.3). The first evidence that synthetic uPA-derived peptides are an effective means to block binding of naturally occurring uPA (or fractions thereof) to cellular uPA-R was provided by Appella *et al* in 1987. These authors located the minimal uPA-R-binding epitope to uPA_{10-32} of the uPA_{1-411} full length molecule (Figure 10.2) having synthesized a series of uPA-based peptides of different lengths. A linear peptide spanning amino acids 12–32 of uPA in which Cys^{19} was substituted by Ala was also synthesized, which efficiently competed with the amino-terminal fraction (ATF) of uPA for binding to cell surface uPA-R. Further experiments by Bürgle *et al* (1997) demonstrated that uPA_{19-31} competed with uPA/uPA-R-interaction, whereas peptides uPA_{18-30}, uPA_{20-32} or uPA_{20-30} were not effective. Magdolen *et al* (1996)

Table 10.3 Reagents interfering with the interaction of uPA with uPA-R

uPA/uPA-R interaction directed agent	*Reference*
uPA peptides	Appella *et al* (1987) *J. Biol. Chem.*, **262**, 4437
antibody to uPA	Hearing *et al* (1988) *Cancer Res.*, **48**, 1270
uPA peptides	Schlechte *et al* (1989) *Cancer Res.*, **49**, 6064
ATF	Kirchheimer *et al* (1989) *PNAS*, **86**, 5424
inactive uPA	Cohen *et al* (1991) *Blood*, **78**, 479
ATF, uPA peptides	Rabbani *et al* (1992) *J. Biol. Chem.*, **267**, 14151
suramin, Evans blue, trypan blue	Behrendt *et al* (1993) *J. Biol. Chem.*, **268**, 5985
uPA peptides, ATF	Kobayashi *et al* (1993) *Br. J. Cancer*, **67**, 537
inactive mutant uPA	Crowley *et al* (1993) *PNAS*, **90**, 5021
antibody to uPA-R	Mohanam *et al* (1993) *Cancer Res.*, **53**, 4143
uPA	Howell *et al* (1994) *Blood Coag. Fibrin.*, **5**, 445
ATF-albumin conjugate	Lu *et al* (1994) *FEBS Lett.*, **356**, 56
phage display uPA analogues	Goodson *et al* (1994) *PNAS*, **91**, 7129
uPA peptide	Kobayashi *et al* (1994) *Int. J. Cancer*, **57**, 727
recombinant uPA-R	Wilhelm *et al* (1994) *FEBS Lett.*, **337**, 131
antibody to uPA	Kobayashi *et al* (1994) *Thromb. Haemost.*, **71**, 474
antibody to uPA	Jarrad *et al* (1995) *Invasion Metastasis*, **15**, 34
mutant uPA, uPA peptides	Magdolen *et al* (1996) *Eur. J. Biochem.*, **237**, 743
ATF	Luparello and del Rosso (1996), *Eur. J. Cancer*, **32A**, 702
GFD-IgG conjugate	Min *et al* (1996) *Cancer Res.*, **56**, 2428
cyclic uPA peptide	Bürgle *et al* (1997) *Biol. Chem.*, **378**, 231
mutant uPA	Evans *et al* (1997) *Cancer Res.*, **57**, 3594
ATF-saporin conjugate	Fabbrini *et al* (1997) *FASEB J.*, **11**, 1169
mutant uPA	Quax *et al* (1998) *Arterioscler. Thromb. Vasc. Biol.* **18**, 693
uPA decapeptide antagonist	Ploug *et al* (1998) *Biochemistry* **37**, 3612
ATF-IgG conjugate	Ignar *et al* (1998), *Clin. Exp.* **16**, 9
ATF-UTI conjugate	Kobayashi *et al* (1998) *Eur. J. Biochem.* **253**, 817
ATF-albumin/adenovirus	Li *et al* (1998) *Gene Ther.*, **5**, 1105
uPA, ATF, cyclic uPA peptide	Fischer *et al* (1998) *FEBS Lett.*, **438**, 101
small peptidic uPA antagonists	Tressler *et al* (1999) *APMIS*, **107**, 168
antibodies to uPA-R	List *et al* (1999) *Immunol. Meth.*, **222**, 125
ATF-endotoxin conjugate	Rajagopal and Kreitman (2000) *J. Biol. Chem.*, **275**, 7566
uPA peptide/adenovirus	Drapkin *et al* (2000) *J. Clin. Invest.*, **105**, 589
peptide A6 derived from connecting peptide of uPA	Guo *et al* (2000) *FASEB J.*, **14**, 1400
peptide A6 derived from connecting peptide of uPA	Mishima *et al* (2000) *PNAS*, **97**, 8484
recombinant uPA-R	Krüger *et al* (2000) *Cancer Gene Ther.*, **7**, 292
cystatin-uPA peptide chimerae	Muehlenweg *et al* (2000) *J. Biol. Chem.*, **275**, 33562
cyclic uPA peptide	Magdolen *et al* (2001) *Biol. Chem.*, **382**, 1197

constructed several linear peptides spanning uPA_{14-32} within the ATF domain of uPA in which the naturally occurring amino acids were replaced one at a time by alanine (Ala scan) to identify those amino acids important for binding of uPA to uPA-R. The exchange of Cys^{19}, Lys^{23}, Tyr^{24}, Phe^{25}, Ile^{28}, Trp^{30}, and Cys^{31}, respectively, by Ala resulted in uPA-peptides with strongly impaired uPA-R binding capacities, whereas alteration of the other

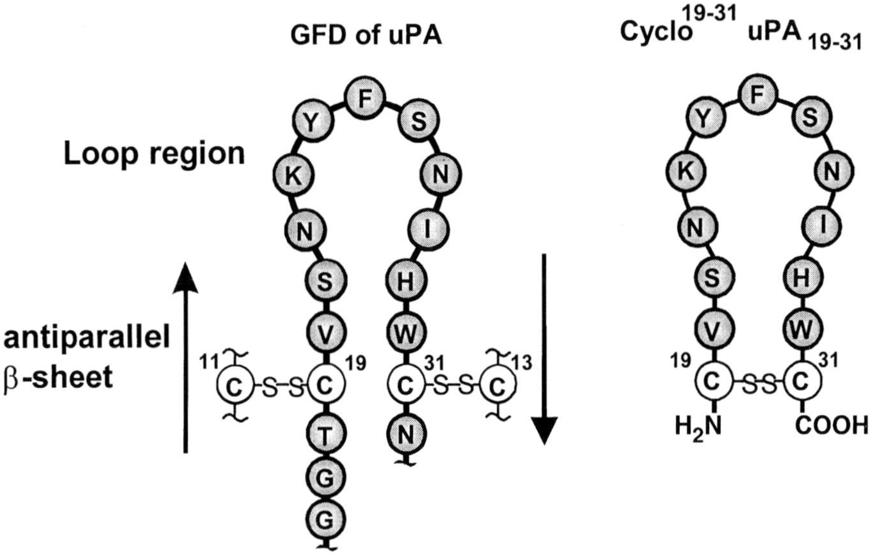

Figure 10.7 In uPA, Cys[19] and Cys[31] are not connected by a disulfide bridge (5.2 Å), but in close proximity (6.1 Å). This may explain why cyclo[19–31] $uPA_{19–31}$ reacts with uPA-R on tumor cells.

amino acids had no or little effect on uPA-R binding. The minimal uPA-R-binding region of uPA was located to $uPA_{19–31}$ spanning Cys^{19} to Cys^{31}.

The peptide region between Thr^{18} and Asn^{32} of uPA consists of a flexible, seven-residue Ω loop (Asn^{22} to Ile^{28}) which by means of a double stranded, antiparallel β-sheet (between Thr^{18} to Ser^{21} and His^{29} to Asn^{32}) is forced into a ring-like structure (Figure 10.7). In uPA, Cys^{19} and Cys^{31}, although in close proximity (6.1 Å), form disulfide bonds with distinct cysteines (Cys^{11}/Cys^{19} and Cys^{13}/Cys^{31}, respectively). Knowing about this close proximity of Cys^{19} and Cys^{31} in the native molecule, we tested whether synthetic cyclic Cys^{19}/Cys^{31} uPA-peptides display any uPA-R binding activities. In fact, not only the linear peptide representing the sequence 19–31 of uPA ($uPA_{19–31}$) but also its disulfide-bridged cyclic form $cyclo^{19–31}$ -$uPA_{19–31}$ efficiently competed with uPA for binding to cell surface-associated uPA-R (Bürgle *et al* 1997; Schmitt *et al* 2000). $Cyclo^{19–31}$-$uPA_{19–31}$ was selected as the lead compound to further improve its binding efficiency to uPA-R. For this purpose, each amino acid was systematically substituted by the corresponding D-amino acid which led to the identification of the more potent uPA analog, $cyclo^{19–31}$ [D-Cys^{19}]-$uPA_{19–31}$, which displays an increased uPA-R-binding affinity.

uPA-R-binding peptides that do not show any obvious similarity to the uPA-R-binding sequence of uPA have been identified by Goodson *et al* (1994) using a bacteriophage peptide display technique. All of the reactive peptides contain two relatively short conserved motifs: LWXXXAr (Ar = Y, W, F or H) and XFXXYLW. None of these sequences is present in uPA, demonstrating that disulfide bond formation (such as in the GFD of uPA) is not a prerequisite for high affinity binding to uPA-R, with clone 20 (AEPMPHSLNFSQYLWYT) being the most potent peptide. For a more precise mapping of the minimal binding region, Tressler *et al* (1999) investigated which amino acids in clone 20 are necessary for binding to human uPA-R, both by truncation and Ala scan of clone 20. Three amino acids present in the parent 17-mer peptide are critical for binding to uPA-R: Phe^{10}, Leu^{14}, and Trp^{15}. The truncation of clone 20 led to several short peptides still containing the C-terminal part of the parent peptide. In their hands, the C-terminal decapeptide LNFSQYLWYT derived from clone 20 is the minimal sequence to retain binding activity in the nM range.

Along the same lines, Ploug *et al* (1998a,b) modified clone 20 by synthesizing a number of peptides with sequential two amino acid deletion, starting from either the amino-terminus or the carboxyl-terminus. As a result, it was found that substantial truncation is allowed at the amino-terminus but not at the carboxyl-terminus. The minimal binding peptide antagonist SLNFSQYLWS derived by this approach binds to uPA-R with an affinity close in order to that of the parent 17-mer antagonist. To determine the contribution of the individual amino acids in SLNFSQYLWS to mediate binding to uPA-R, this peptide was also subjected to a systematic Ala scan. Sequential substitution of Leu^2, Phe^4, Leu^8, or Trp^9 by alanine abolished the antagonistic property of the decapeptide toward uPA-R. Replacement of the other amino acids by alanine was without effect or caused only a moderate reduction in efficacy.

The intellectual property disclosed on inhibitors of the uPA/uPA-R interaction, including modulation of uPA-R/integrin interaction, has been reviewed by Rosenberg (2000). In general, attachment of cells to extracellular matrix components such as fibrin, fibronectin, vitronectin, laminin, collagen, etc. is largely determined by cellular adhesion receptors of the integrin superfamily. To examine the physical association between uPA-R and integrins, and to identify potential peptide inhibitors of uPA-R/integrin interaction, a bacteriophage peptide display library was screened for uPA-R-binding phages to mediate extracellular matrix adhesion and regulate integrin function. A 17-mer peptide (AESTYHHLSLGYMYTLN-NH_2; clone 25) which binds to uPA-R and disrupts the cell membrane-associated uPA-R/caveolin/integrin complex was picked-up by this technique. Simon *et al* (2000) investigated the association of the β_2-integrin Mac-1 (CD11b/CD18) with uPA-R. A critical non-I-domain binding site for uPA-R on CD11b (M25; residues 424–440) was identified by homology with the phage display peptide clone 25 known to bind uPA-R.

10.4 SOLUBLE uPA-R, ANTIBODIES TO uPA/uPA-R, AND ANTISENSE OLIGONUCLEOTIDES AS ADDITIONAL TOOLS TO INTERFERE WITH uPA/uPA-R EXPRESSION OR REACTIVITY

Like other GPI-anchored proteins, uPA-R is present in a cell-associated and a soluble form, both *in vitro* and *in vivo*. Soluble uPA-R (suPA-R) is found in the plasma of healthy human individuals but, moreover, increased levels are detected in the blood and/or ascitic fluid of patients with cancers of the breast, ovary, lung, or colon (Sier *et al* 1998; Stephens *et al* 1999). Recent data have demonstrated that elevated suPA-R levels in blood are related to a poor prognosis for the patient. The functional *in vivo* significance of uPA-R shedding, which is due to the action of cellular GPI-specific phospholipase D and/or proteases, is not known yet (Wilhelm *et al* 1999).

Wilhelm *et al* (1994) designed a soluble recombinant form of uPA-R (suPA-R), lacking the GPI anchor. suPA-R (Figure 10.8) acts as an efficient scavenger of uPA leading to inhibition of tumor cell proliferation and invasion. suPA-R blocks binding of uPA to tumor cell surface-associated uPA-R, suggesting that suPA-R is a candidate protein for treatment of cancer. suPA-R effectively inhibits tumor cell growth and lung colonization in nude mice overexpressing breast or ovarian cancer cells challenged with suPA-R thereby leading to reduction in the plasminogen activation-related proteolytic activity of the cancer cells and metastasis (Krüger *et al* 2000; Lutz *et al* 2001).

Antibodies to uPA or uPA-R have also been employed to affect tumor cell spread. Ossowski and Reich (1983) generated rabbit polyclonal antibodies, which inhibited the catalytic activity of human uPA and tumor cell spread in a chicken tumor model. Since then, a panel of polyclonal and monoclonal antibodies (mAb) has been generated, aimed at interfering with the catalytic activity of uPA or the uPA/uPA-R-interaction or otherwise being used in ELISA or immunohistochemistry (Schmitt *et al* 2000). In general, uPA-directed antibodies, which block binding of uPA to uPA-R, react with a target epitope located in or close to the amino-terminal region of uPA (which harbors the uPA-R-binding site). Likewise, uPA-R-directed mAbs interfering with the uPA/PAR interaction react with domain I of uPA-R. By Ala scan four amino acids in domain I of uPA-R have been identified (Gardsvoll *et al* 1999) which are essential for the interaction of uPA with uPA-R: Arg^{53}, Leu^{55}, Tyr^{57}, and Leu^{66} (Figure 10.3). It has also been reported that mAb IIIF10 directed to $uPA\text{-}R_{52\text{–}60}$ and mAb R3 to a region centered around Leu^{61}/Leu^{62} of uPA-R efficiently block binding of uPA to uPA-R (Luther *et al* 1997; Gardsvoll *et al* 1999).

uPA- and uPA-R-expression facilitates tumor spread and therefore its inhibition by antisense technology is an alternative way to suppress plasminogen activation activity of

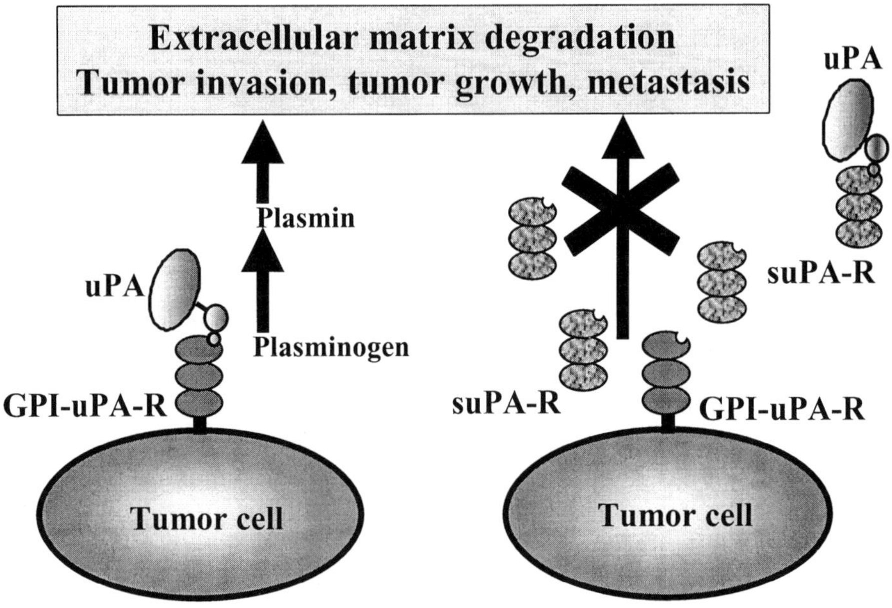

Figure 10.8 Soluble uPA-R (suPA-R) acts as a scavenger for uPA to prevent binding of uPA to cellular uPA-R.

tumor cells. In ovarian cancer cells, Wilhelm *et al* (1995) explored whether uPA-protein expression is impaired by neutralizing uPA-expression using liposome-mediated phosphorothioate antisense oligodeoxynucleotide technology. A significant reduction in invasion of OV-MZ-6 ovarian cancer cells was observed. Reuning *et al* (1995, 1999) examined the role of Rel-related proteins in uPA synthesis by human ovarian cancer cells by inhibiting their expression, also using the antisense oligodeoxynucleotide technology. An even earlier interference with uPA production by tumor cells was achieved by suppression of the uPA synthesis by blocking the expression of transcription factors of the Rel family using antisense probes, accompanied by a significant reduction of the invasive and proteolytic capacity of cancer cells. Quattrone *et al* (1995) abolished the invasive properties of transformed cells switching off uPA-R gene expression by also employing the anti-messenger oligodeoxynucleotide strategy. In esophageal cancer, antisense oligonucleotide inhibition of either uPA or uPA-R expression resulted in a marked reduction in invasiveness of tumor cells which normally coexpress both uPA and uPA-R (Morrissey *et al* 1999). Mohan *et al* (1999) reported that down-regulating uPA-R levels by antisense strategy using an adenoviral construct (Ad-uPA-R) inhibited glioma invasion. Injection of Ad-uPA-R into previously established U87-MG tumors in nude mice resulted in regression of the tumor supporting the therapeutic potential of targeting the uPA/uPA-R system for the treatment of gliomas and other types of cancer.

ACKNOWLEDGEMENTS

This work was supported by grants GR280/4–5 (Klinische Forschergruppe) and SFB 469 (A4) received by the Deutsche Forschungsgemeinschaft (DFG), BIOREGIO M, by the Bundesministerium für Bildung und Forschung, and Wilex A6, Munich, Germany.

REFERENCES

Alonso, D.F., Tejera, A.M., Farias, E.F., Joffe, E.B.D. and Gomez, D.E. (1998) Inhibition of mammary tumour cell adhesion, migration, and invasion by the selective synthetic urokinase inhibitor B428. *Anticancer Research*, **18**, 4499–4504

Andreasen, P.A., Egelund, R. and Petersen, H.H. (2000) The plasminogen activation system in tumour growth, invasion, and metastasis. *Cellular and Molecular Life Sciences*, **57**, 25–40.

Andreasen, P.A., Kjoller, L., Christensen, L. and Duffy, M.J. (1997) The urokinase-type plasminogen activator system in cancer metastasis: a review. *International Journal of Cancer*, **72**, 1–22.

Appella, E., Robinson, E.A., Ullrich, S.J., Stopelli, M.P., Corti, A., Cassani, G. *et al* (1987) The receptor-binding sequence of urokinase. A biological function for the growth-factor module of proteases. *Journal of Biological Chemistry*, **262**, 4437–4440.

Behrendt, N., Ronne, E. and Dano, K. (1995) The structure and function of the urokinase receptor, a membrane protein governing plasminogen activation on the cell surface. *Biological Chemistry Hoppe-Seyler*, **376**, 269–279.

Blasi, F. (1999) Proteolysis, cell adhesion, chemotaxis, and invasiveness are regulated by the u-PA-u-PAR-PAI-1 system. *Thrombosis & Haemostasis*, **82**, 298–304.

Blasi, F. (1997) uPA, uPAR, PAI-1:key intersection of proteolytic, adhesive and chemotactic highways? *Immunology Today*, **18**, 415–417.

Bürgle, M., Koppitz, M., Riemer, C., Kessler, H., Konig, B., Weidle, U.H. *et al* (1997) Inhibition of the interaction of urokinase-type plasminogen activator (uPA) with its receptor (uPAR) by synthetic peptides. *Journal of Biological Chemistry*, **378**, 231–237.

Chapman, H.A. (1997) Plasminogen activators, integrins, and the coordinated regulation of cell adhesion and migration. *Current Opinions in Cell Biology*, **9**, 714–724.

Dano, K., Romer, J., Nielsen, B.S., Bjorn, S., Pyke, C., Rygaard, J. *et al* (1999) Cancer invasion and tissue remodeling – cooperation of protease systems and cell types. *APMIS*, **107**, 120–127.

Dear, A.E. and Medcalf, R.L. (1998) The urokinase-type-plasminogen-activator receptor (CD87) is a pleiotropic molecule. *European Journal of Biochemistry*, **252**, 185–193.

Duffy, M.J. (1996) Proteases as prognostic markers. *Clinical Cancer Research*, **2**, 613–618.

Ellis, V., Whawell, S.A., Werner, F. and Deadman, J.J. (1999) Assembly of urokinase receptor-mediated plasminogen activation complexes involves direct, non-active-site interactions between urokinase and plasminogen. *Biochemistry*, **38**, 65165–65169.

Fazioli, F. and Blasi, F. (1994) Urokinase-type plasminogen activator and its receptor: new targets for anti-metastatic therapy? *Trends in Pharmacological Science*, **15**, 25–29.

Foekens, J.A., Peters, H.A., Look, M.P., Portengen, H., Schmitt, M., Kramer, M.D. *et al* (2000) The urokinase system of plasminogen activation and prognosis in 2780 breast cancer patients. *Cancer Research*, **60**, 636–643.

Folkman, J. (1995) Angiogenesis in cancer, vascular, rheumatoid and other disease. *Nature Medicine*, **1**, 27–31.

Gardsvoll, H., Dano, K. and Ploug, M. (1999) Mapping part of the functional epitope for ligand binding on the receptor for urokinase-type plasminogen activator by site-directed mutagenesis. *Journal of Biological Chemistry*, **274**, 37995–38003.

Geratz, J.D., Shaver, S.R. and Tidwell, R.R. (1981) Inhibitory effect of amidino-substituted heterocyclic compounds on the amidase activity of plasmin and high and low molecular weight urokinase and on urokinase-induced plasminogen activation. *Thrombosis Research*, **24**, 73–83.

Goodson, R.J., Doyle, M.V., Kaufman, S.E. and Rosenberg, S. (1994) High-affinity urokinase receptor antagonists identified with bacteriophage peptide display. *Proceedings of the National Academy of Sciences of the U.S.A.*, **91**, 7129–7133.

Krüger, A., Soeltl, R., Lutz, V., Wilhelm, O.G., Magdolen, V., Rojo, E.E. *et al* (2000) Reduction of breast carcinoma tumour growth and lung colonization by overexpression of the soluble urokinase-type plasminogen activator receptor (CD87) *Cancer Gene Therapy*, **7**, 292–299.

Luther, T., Magdolen, V., Albrecht, S., Lopens, A., Kasper, M., Riemer, C., Vessler, H., Müller, M. and Schmitt, M. (1977) Epitope-mapped monoclonal antibodies as tools for functional and morphological analyses of the human urokinase receptor in tumor tissue. *American Journal of Pathology*, **150**, 1231–1244.

Lutz, V., Reuning, U., Krüger, A., Luther, T., Pildner von Steinburg, S., Graeff, H. *et al* (2001) High level synthesis of recombinant soluble urokinase receptor (CD87) by ovarian cancer cells reduces intraperitoneal tumor growth and spread in nude mice. *Journal of Biological Chemistry*, **382**, 789–798.

Magdolen, V., Rettenberger, P., Koppitz, M., Goretzki, L., Kessler, H., Weidle, U.H. *et al* (1996) Systematic mutational analysis of the receptor-binding region of the human urokinase-type plasminogen activator. *European Journal of Biochemistry*, **237**, 743–751.

Mohan, P.M., Lakka, S.S., Mohanam, S., Kin, Y., Sawaya, R., Kyritsis, A.P. *et al* (1999) Down-regulation of the urokinase-type plasminogen activator receptor through inhibition of translation by antisense oligonucleotide suppresses invasion of human glioblastoma cells. *Clinical & Experimental Metastasis*, **17**, 617–621.

Mondino, A., Resnati, M. and Blasi, F. (1999) Structure and function of the urokinase receptor. *Thrombosis & Haemostasis*, **82**, 19–22.

Morrissey, D., O'Connell, J., Lynch, D., O'Sullivan, G.C., Shanahan, F. and Collins, J.K. (1999) Invasion by esophageal cancer cells: functional contribution of the urokinase plasminogen activation system, and inhibition by antisense oligonucleotides to urokinase or urokinase receptor. *Clinical & Experimental Metastasis*, **17**, 77–85.

Nienaber, V.L., Davidson, D., Edalji, R., Giranda, V.L., Klinghofer, V., Henkin, J. *et al* (2000a) Structure-directed discovery of potent non-peptidic inhibitors of human urokinase that access a novel binding subsite. *Structure with Folding & Design*, **8**, 553–563.

Nienaber, V.L., Wang, J., Davidson, D. and Henkin, J. (2000b) Re-engineering of human urokinase provides a system for structure-based drug design at high resolution and reveals a novel structural subsite. *Journal of Biological Chemistry*, **275**, 7239–7248.

Ossowski, L. and Reich, E. (1983) Antibodies to plasminogen activator inhibit human tumour metastasis. *Cell*, **35**, 611–619.

Ploug, M., Ostergaard, S., Hansen, L.B., Holm, A. and Dano, K. (1998a) Photoaffinity labeling of the human receptor for urokinase-type plasminogen activator using a decapeptide antagonist. Evidence for a composite ligand-binding site and a short interdomain separation. *Biochemistry*, **37**, 3612–3622.

Ploug, M., (1998b) Identification of specific sites involved in ligand binding by photoaffinity labeling of the receptor for the urokinase-type plasminogen activator. Residues located at equivalent positions in uPAR domains I and III participate in the assembly of a composite ligand-binding site. *Biochemistry*, **37**, 16494–16505.

Ploug, M., Ronne, E., Behrendt, N., Jensen, A.L., Blasi, F. and Dano, K. (1991) Cellular receptor for urokinase plasminogen activator: Carboxyl-terminal processing and membrane anchoring by glycosyl-phosphatidylinositol. *Journal of Biological Chemistry*, **266**, 1926–1933.

Quattrone, A., Fibbi, G., Anichini, E., Pucci, M., Zamperini, A., Capaccioli, S. *et al* (1995) Reversion of the invasive phenotype of transformed human fibroblasts by anti-messenger oligonucleotide inhibition of urokinase receptor gene expression. *Cancer Research*, **55**, 90–95.

Rabbani, S.A., Harakidas, P., Davidson, D.J., Henkin, J. and Mazar, A.P. (1995) Prevention of prostate-cancer metastasis *in vivo* by a novel synthetic inhibitor of urokinase-type plasminogen activator (uPA). *International Journal of Cancer*, **63**, 840–845.

Renatus, M., Bode, W., Huber, R., Stürzebecher, J. and Stubbs, M.T. (1998) Structural and functional analyses of benzamidine-based inhibitors in complex with trypsin: implications for the inhibition of factor Xa, tPA, and urokinase. *Journal of Medicinal Chemistry*, **41**, 545–556.

Reuning, U., Guerrini, L., Nishiguchi, T., Page, S., Seibold, H., Magdolen, V. *et al* (1999) Rel transcription factors contribute to elevated urokinase expression in human ovarian carcinoma cells. *European Journal of Biochemistry*, **259**, 143–148.

Reuning, U., Magdolen, V., Wilhelm, O., Fischer, K., Lutz, V., Graeff, H. *et al* (1998) Multifunctional potential of the plasminogen activation system in tumor invasion and metastasis. *International Journal of Oncology*, **13**, 893–906.

Reuning, U., Wilhelm, O., Nishiguchi, T., Guerrini, L., Blasi, F., Graeff, H. *et al* (1995) Inhibition of NF-κB-REL A expression by antisense-oligodeoxynucleotides suppresses synthesis of urokinase-type plasminogen activator (uPA) but not its inhibitor PAI-1. *Nucleic Acids Research*, **23**, 3887–3893.

Rosenberg, S. (2000) Modulators of the urokinase-type plasminogen activation system for cancer. *Expert Opinion on Therapeutic Patents*, 1843–1853.

Schmitt, M., Harbeck, N., Thomssen, C., Wilhelm, O., Magdolen, V., Reuning, U. *et al* (1997) Clinical impact of the plasminogen activation system in tumour invasion and metastasis. Prognostic relevance and target for therapy. *Thrombosis & Haemostasis*, **78**, 285–296.

Schmitt, M., Jänicke, F. and Graeff, H. (1992) Tumor-associated proteases. *Fibrinolysis*, **6** (Suppl. 4), 3–26.

Schmitt, M., Wilhelm, O.G., Reuning, U., Krüger, A., Harbeck, N., Lengyel, E. *et al* (2000) The urokinase plasminogen activator system as a novel target for tumour therapy. *Fibrinolysis & Proteolysis*, **14**, 114–132.

Sier, C.F., Stephens, R., Bizik, J., Mariani, A., Bassan, M., Pedersen, N. *et al* (1998) The level of urokinase-type plasminogen activator receptor is increased in serum of ovarian cancer patients. *Cancer Research*, **58**, 1843–1849.

Simon, D.I., Wei, Y., Zhang, L., Rao, N.K., Xu, H., Chen, Z. *et al* (2000) Identification of a urokinase receptor-integrin interaction site. Promiscuous regulator of integrin function. *Journal of Biological Chemistry*, **275**, 10228–10234.

Sperl, S., Jacob, U., Arroyo de Prada, N., Stürzebecher, J., Wilhelm, O.G., Bode, W. *et al* (2000) (4-aminomethyl) phenylguanidine derivatives as non-peptidic highly selective inhibitors of human urokinase. X-ray crystal structure of a uPA/inhibitor complex at 1.7 Å resolution. *Proceedings of the National Academy of Sciences of the U.S.A.*, **97**, 5113–5118.

Spraggon, G., Phillips, C., Nowak, U.K., Ponting, C.P., Saunders, D., Dobson, C.M. *et al* (1995) The crystal structure of the catalytic domain of human urokinase-type plasminogen activator. *Structure*, **3**, 681–691.

Stephens, R.W., Nielsen, H.J. and Christensen, I.J., Thorlacius-Ussing, O., Sorensen, S., Dano, K. *et al* (1999) Plasma urokinase receptor levels in patients with colorectal cancer: relationship to prognosis. *Journal of the National Cancer Institute*, **91**, 869–874.

Stürzebecher, J. and Markwardt, F. (1978) Synthetic inhibitors of serine proteinases. 17. The effect of benzamidine derivatives on the activity of urokinase and the reduction of fibrinolysis. *Pharmazie*, **33**, 599–602.

Stürzebecher, J., Vieweg, H., Steinmetzer, T., Schweinitz, A., Stubbs, M.T., Renatus, M. *et al* (1999) 3-Amidinophenylalanine-based inhibitors of urokinase. *Bioorganic & Medicinal Chemistry Letters*, **9**, 3147–3152.

Tamura, S.Y., Weinhouse, M.I., Roberts, C.A., Goldman, E.A., Masukawa, K., Anderson, S.M. *et al* (2000) Synthesis and biological activity of peptidyl aldehyde urokinase inhibitors. *Bioorganic & Medicinal Chemistry Letters*, **10**, 983–987.

Towle, M.J., Lee, A., Maduakor, E.C., Schwartz, C.E., Bridges, A.J. and Littlefield, B.A. (1993) Inhibition of urokinase by 4-substituted benzo[*b*]thiophene-2-carboxamidines: an important new class of selective synthetic urokinase inhibitor. *Cancer Research*, **53**, 2553–2559.

Tressler, R.J., Pitot, P.A., Stratton, J.R., Forrest, L.D., Zhuo, S., Drummond, R.J. *et al* (1999) Urokinase receptor antagonists: discovery and application to *in vivo* models of tumour growth. *APMIS*, **107**, 168–173.

Wilhelm, O., Schmitt, M., Höhl, S., Senekowitsch, R. and Graeff, H. (1995) Antisense inhibition of urokinase reduces spread of human ovarian cancer in mice. *Clinical & Experimental Metastasis*, **13**, 296–302.

Wilhelm, O., Weidle, U., Höhl, S., Rettenberger, P., Schmitt, M. and Graeff, H. (1994) Recombinant soluble urokinase receptor as a scavenger for urokinase-type plasminogen activator (uPA). Inhibition of proliferation and invasion of human ovarian cancer cells. *FEBS Letters*, **337**, 131–134.

Wilhelm, O.G., Wilhelm, S., Escott, G.M., Lutz, V., Magdolen, V., Schmitt, M. *et al* (1999) Cellular glycosylphosphatidylinositol-specific phospholipase D regulates urokinase receptor shedding and cell surface expression. *Journal of Cellular Physiology*, **180**, 225–235.

Xing, R.H., Mazar, A., Henkin, J. and Rabbani, S.A. (1997) Prevention of breast cancer growth, invasion, and metastasis by antiestrogen tamoxifen alone or in combination with urokinase inhibitor B-428. *Cancer Research*, **57**, 3585–3593.

Yang, H., Henkin, J., Kim, K.H. and Greer, J. (1990) Selective inhibition of urokinase by substituted phenylguanidines: quantitative structure relationship analyses. *Journal of Medicinal Chemistry*, **33**, 2956–2961.

Zeslawska, E., Schweinitz, A., Karcher, A., Sondermann, P., Sperl, S., Stürzebecher, J. *et al* (2000) Crystals of the urokinase type plasminogen activator variant betac-uPA in complex with small molecule inhibitors open the way towards structure-based drug design. *Journal of Molecular Biology*, **301**, 465–475.

Chapter 11

Proteinases involved in amyloid β-peptide (Aβ) production and clearance

Pari Malherbe, Gerda Huber and Fiona Grueninger

Amyloid-β-protein (Aβ) formation and deposition in brain parenchyma is a key event in the pathogenesis of Alzheimer's disease (AD). The processing of β-amyloid precursor protein (βAPP) to form Aβ requires two sequential cleavages by β- and γ-secretases. The suppression of Aβ levels through direct inhibition of the β- or γ-secretase offers a potential effective therapeutic approach to the treatment of AD. Huge efforts have been made to identify and characterize β- and γ-secretases. An aspartyl protease termed BACE-1 (beta-site APP cleaving enzyme) has been recently reported to be the β-secretase. So far, the nature of γ-secretase is not absolutely clear, because conflicting evidence exists as to whether one or more proteases are involved and, in the latter case, as to whether these enzymes are of the same mechanistic class. However, a number of convincing reports recently suggest that presenilin 1 itself might be a γ-secretase. An alternative, biologically important cleavage of βAPP occurs within the Aβ domain by α-secretase and leads to release of βAPPsα. Because Aβ formation cannot occur in this pathway, augmentation of the α-secretase activity could also be beneficial in AD. Amongst the several α-secretase candidates described to date, the recently isolated ADAM 10, a disintegrin metalloprotease, shows many properties expected of an α-secretase and might indeed be the enzyme responsible for constitutive and regulated secretion of βAPPsα. Finally, proteases are also involved in Aβ degradation and turnover and some progress has been made recently towards identifying these enzymes.

11.1 INTRODUCTION

The characteristic pathological features of AD (Selkoe 2001) are extracellular accumulation of Aβ together with an increased incidence of neurofibrillary tangles. Aβ is a 39–43 amino-acid peptide which is generated by proteolysis from a membrane-anchored glycoprotein, βAPP. βAPP is processed in the cell through two principal metabolic pathways (Selkoe 1994; Mills and Reiner 1999). In the secretory processing by α-secretase, the soluble βAPPsα ectodomain is released by cleavage within Aβ-domain. In the "β-secretase" pathway, two proteolytic cleavages, one by β-secretase at the N-terminus, the other by γ-secretase at the C-terminus of Aβ, generate $A\beta_{1-40/42}$. The β-secretase cleavage, which is a rate limiting step, occurs sequence-specifically between Met and Asp and precedes the γ-secretase cleavage in the Aβ production cascade (Figure 11.1). Several compelling lines of evidence indicate that the βAPP metabolism, Aβ production and Aβ deposition play a pivotal role in the progression of AD (Selkoe 1999). Therefore, the inhibition of processes leading to the generation of Aβ or enhance its clearance should halt the excessive

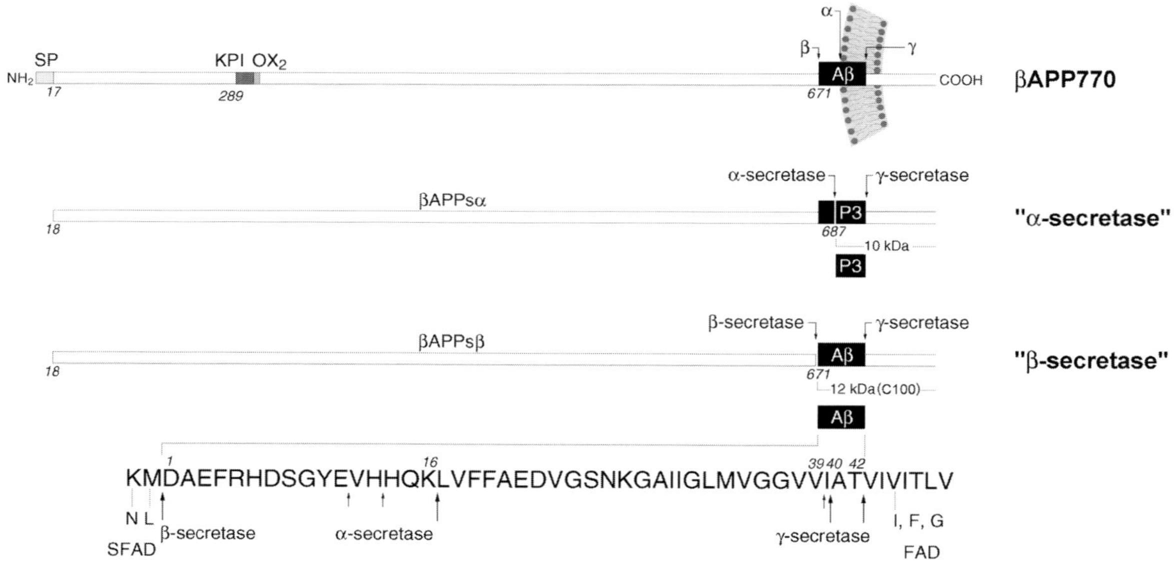

Figure 11.1 Schematic diagram of different secretase cleavage sites within βAPP. In the major constitutive "α-secretase" pathway, the cleavage by α-secretase releases the βAPPsα into the extracellular space and precludes the generation of amyloidogenic Aβ. In the "β-secretase" pathway, the cleavages by β- and γ-secretases result in Aβ generation. Familial AD (FAD, V717I, F, G) and Swedish FAD (SFAD, K670N, M671L) are indicated.

production and accumulation of Aβ. The β- and γ-secretases are in this respect extremely interesting therapeutic targets for AD. The aim of the present report is to review potential secretase candidates which are proposed to be involved in βAPP metabolism.

11.2 α-SECRETASE

In the non-amyloidogenic "α-secretase" pathway, βAPP can undergo endoproteolytic cleavage within the Aβ sequence at the cell surface by a putative α-secretase. This secretase has a relaxed substrate sequence specificity, cleaving βAPP at a defined distance from the cell membrane, with its major cleavage site occurring between Lys-16 and Leu-17 of Aβ (Sisodia 1992). The resulting soluble βAPPsα (105–125 kDa), which has neurotrophic and neuroprotective activities (Qiu *et al* 1995), constitutes the major form of βAPP found in cerebrospinal fluid (CSF) and brain homogenates (Palmert *et al* 1989). The α-secretase cleavage and secretion of βAPP ectodomain are regulated processes that can be stimulated by a number of agents such as phorbol esters, elevated iron levels, calcium ionophores, interleukin-1, cholinergic agonists, estrogen, cholinesterase inhibitors, etc. (Buxbaum *et al* 1992; *Rossner et al* 1998). It has been shown in cell cultures that the activation of protein kinase C (PKC) caused 80–95% of βAPP to undergo α-secretase cleavage, consequently reducing the level of Aβ (Buxbaum *et al* 1993). Therefore, the profile of an α-secretase would predict that the relevant enzyme must be an ubiquitous membrane-bound protease with high sensitivity to agents activating PKC and to Ca^{2+} ionophores. An increase in "α-secretase" pathway activity should preclude the production of amyloidogenic Aβ. Thus, any therapeutic strategy aiming at enhancing α-secretase activity could prove beneficial in AD treatment. Considerable progress has been made over the last two years in identifying potential α-secretase candidates.

11.2.1 Caveolae microdomains as mediated α-site-processing of βAPP

Caveolae are plasma membrane microdomains where multiple signalling molecules, such as G-protein-mediated signalling cascades, are concentrated. A major protein component of caveolae is caveolin-1, which has been shown to co-localize and is physically associated with βAPP. The report by Ikezu *et al* (1998) indicates that caveolae and caveolin protein might play a role in the α-secretase-mediated cleavage of βAPP *in vivo*. Overexpression of recombinant caveolin-1 in intact cells greatly increases the secretion of the βAPPsα, while the disruption of endogenous caveolin-1 expression by antisense oligonucleotides abolishes this βAPP cleavage/release. At its C-terminal cytoplasmic tail, βAPP has a predicted caveolin-binding motif (20-amino acids) which promotes an interaction with caveolin-1. This allows the sequestering of βAPP within caveolae microdomains where the α-secretase cleavage occurs. It is interesting to note that cholesterol, which has been shown to down-regulate the secretion of βAPPsα (Bodovitz and Klein 1996), binds directly to caveolin-1.

11.2.2 GPI-linked aspartyl proteases

The report by Komano *et al* (1998) demonstrates that two structurally and functionally related glycosyl-phosphatidylinositol (GPI)-linked yeast aspartyl proteases, Mkc7p and

Yap3p, are involved in α-secretase cleavage of βAPP expressed in yeast. Both enzymes, localized at the cell surface, cleaved yeast expressed βAPP at the correct α-secretase cleavage site. Deletion of both enzymes in yeast abolished this α-site cleavage event. However, mammalian homologues have not yet been reported.

11.2.3 ADAM family of zinc metalloproteases

Many membrane-anchored proteins such as growth factors, growth factor receptors, ecto-enzymes and cell adhesion molecules similar to βAPP, undergo a process referred to as "ectodomain shedding" in which the cleavage by a membrane-associated protease results in the release of the extracellular domain into the extracellular space (Arribas *et al* 1997). The members of the ADAM family (a disintegrin and metalloprotease-family) of proteases are apparently responsible for many of these ectodomain cleavage activities (Black and White 1998). The characteristic features of ADAM proteases are an autoinhibitory domain that needs to be cleaved for activity, a metalloprotease domain, a disintegrin domain, a cysteine-rich domain and a transmembrane domain. Since the regulated secretion of βAPP was strongly inhibited by hydroxamic acid-based compounds that are effective inhibitors of ADAM family proteases, several ADAM proteases have been examined for their role as α-secretase.

11.2.3.1 ACE-secretase

Parvathy *et al* (1998a) have observed that hydroxamic acid-based zinc metalloprotease inhibitors, batimastat, marimastat and BB2116 strongly block the release of βAPPsα from both SH-SY5Y and IMR-32 Neuroblastoma cell lines. The inhibition pattern of βAPPsα release from IMR-32 cells by these inhibitors was identical to that of ACE (angiotensin converting enzyme) release from ACE-transfected IMR-32 cells. ACE-secretase has some similarities to α-secretase. It is an integral membrane protease and cleaves ACE at the cell surface between a basic and a hydrophobic residue (Arg-Leu). ACE-secretase, like α-secretase, is regulated by agents activating PKC. It seems therefore, that although ACE-secretase and α-secretase are not identical, they both belong to a closely related zinc metalloprotease family (Parvathy *et al* 1998b).

11.2.3.2 TACE (=ADAM 17)

Membrane-bound tumor necrosis factor-α (TNF-α), like βAPP, undergoes ectodomain shedding by TACE (tumor necrosis factor-α converting enzyme) which is a member of the ADAM family (also called ADAM 17). In their report, Buxbaum *et al* (1998) indicate that TACE might be responsible for the major regulated α-secretase cleavage of βAPP in cell culture. They have studied the basal and regulated secretions of βAPPsα in primary embryonic fibroblasts derived from control mice or TACE-knockout mice. They observed that only the regulated secretion, but not basal secretion, was stimulated by activation of PKC through phorbol 12-myristate 13-acetate (PMA) and/or by inhibition of protein phosphatases 1 and 2A through okadaic acid in the cells derived from control mice. These PMA and/or okadaic acid effects were completely blocked by Immunex compound 3 (IC3), whereas PMA or okadiac acid had no effects on the basal and regulated secretion in the cells derived from TACE-knockout mice. In addition, recombinant TACE was able to cleave a synthetic peptide spanning the α-site between Lys and Leu. In CHO cells stably expressing hβAPP$_{751}$, PMA or okadaic acid have also caused an increase in βAPPsα secretion together with a concomitant decrease in the secretion of Aβ. This result suggests the presence of two classes of α-secretases, a constitutive and a regulated one, and that the features of TACE are consistent with a regulated α-secretase.

11.2.3.3 ADAM 10

Recently, Lammich *et al* (1999) showed convincing evidence that a disintegrin metalloprotease classified as ADAM 10, a type I membrane protein, is involved in constitutive and regulated α-secretase cleavage of βAPP. The basal and PMA-regulated secretion of βAPPsα was increased in HEK-293 cells overexpressing ADAM 10. These effects were blocked by 70% by the hydroxamic acid-based inhibitor BB-3103. The protease inhibitor profile of ADAM 10 was identical to that of an integral membrane protease reported earlier (Roberts *et al* 1994). The cellular localization of ADAM 10 in these transfected HEK-293 cells showed that the active form of the enzyme was present in the plasma membrane, although much of the enzyme is found as proenzyme in the Golgi. ADAM 10 cleaved an Aβ peptide at the correct α-secretase site. In HEK-293 cells overexpressing a dominant negative ADAM 10 mutant (point mutation E384A in the zinc-binding site), the basal and regulated α-secretase cleavages of βAPP were significantly reduced. All these experimental results indicate that ADAM 10 could be a relevant α-secretase.

11.3 β-SECRETASE

In recent years many different proteases have been proposed as potential β-secretases. These included metalloproteases MP78 and MP100 (Huber *et al* 1999a,b; Thompson *et al* 1997), the aspartyl proteases cathepsin D and E (Chevallier *et al* 1997; Thompson *et al* 1997; Gruninger-Leitch *et al* 2000), serine proteases such as CASP and Zyme (Little *et al* 1997; Meckelein *et al* 1998) and the cysteine proteases cathepsin B and bleomycin hydrolase (Marks *et al* 1995; Montoya *et al* 1998; Malherbe *et al* 2000). None of these proteases possessed all of the functional and physical criteria expected for a true β-secretase, which included tissue specificity (should be expressed in brain), subcellular localization (should co-localize to the same compartments as βAPP and should preferably be membrane-bound), substrate specificity (should cleave both wild-type βAPP and βAPP$_{SFAD}$ at the Leu/Met-Asp site but should exhibit a strong preference for βAPP$_{SFAD}$) and optimum pH (should be acidic since Aβ is generated in an acidic compartment). At the end of 1999, the intensive search for β-secretase culminated in the simultaneous announcement by five different research groups of the discovery of a membrane-bound aspartic protease which finally seemed to posess all of these prerequisite properties.

11.3.1 Membrane-anchored aspartic protease of the pepsin family

11.3.1.1 BACE-1 (memapsin-2)

Several groups using different approaches (e.g. expression cloning, homology search, EST database mining, biochemical techniques) reported the identification of the same enzyme as β-secretase (Hussain *et al* 1999; Sinha *et al* 1999; Vasar *et al* 1999; Yan *et al* 1999; Lin *et al* 2000). This enzyme BACE-1 (beta-site APP cleaving enzyme), also known as memapsin-2 or Asp-2, is a type I transmembrane protein with 501 amino acids. It is synthesized as preproenzyme before undergoing sequential proteolytic processing to remove the signal peptide and the pro-peptide. The protease responsible for propeptide removal seems to be furin, a member of the proprotein convertase family. Furin removes 24 residues from the N-terminus of pro-BACE-1 to produce mature BACE-1 within the Golgi and trans-Golgi network (Benjannet *et al* 2001). BACE-1 has a large pepsin-like domain which is extended at the C-terminus by a transmembrane region of 17 residues and a short cytoplasmic tail of 24 residues. The two classical active site aspartyl motifs occur at residues 93–96 (DTGS) and residues 289–292 (DSGT). The protein sequence of BACE-1 displays about 30% identity with other members of the pepsin family. However, the presence of membrane anchor places BACE-1 in a new class of aspartic proteases. BACE-1 is also unique amongst the mammalian aspartyl proteases in that its enzymatic activity is not inhibited by pepstatin. The crystal structure of BACE-1 complexed with an eight-residue transition-state inhibitor has been determined at 1.9 Å resolution (Hong *et al* 2000). BACE-1 has been mapped to chromosomal location 11q23.2–11q23.3 (Saunders *et al* 1999). In humans, the BACE-1 transcript is detected with the highest level in pancreas and a moderately high level in the brain as compared to other tissues. In the brain, the majority of transcripts were found in neurons and, at subcellular level, were co-localized with βAPP in Golgi (Vassar *et al* 1999; Cai *et al* 2001). Interestingly, *in situ* hybridization revealed the co-expression of βAPP with BACE-1 and ADAM 10 in human cortical neurons (Marcinkiewicz and Seidah 2000). Luo *et al* (2001) have reported the generation of BACE-1 knockout mice. They have shown that these β-secretase deficient mice have a completely normal phenotype, but are unable to form Aβ in their neurons.

11.3.1.2 BACE-2 (memapsin-1)

By homology searching of public EST databases, Saunders *et al* (1999) identified a homolog of BACE-1 which is termed BACE-2 or Asp-1. It is mapped to chromosomal location 21q22.2–21q22.3, the obligate Down syndrome region. BACE-2 has 52% protein sequence identity to BACE-1 and is able to cleave APP at the β-secretase site (Farzan *et al* 2000). In human tissues, the BACE-2 transcript is found mainly in the peripheral organs, with the highest level of expression in heart and pancreas. The level of BACE-2 mRNA is extremely low or undetectable in human brain (Bennett *et al* 2000) and for this reason is unlikely to function as a β-secretase *in vivo*.

11.4 γ-SECRETASE

11.4.1 Properties of γ-secretase

The substrate for γ-secretase is the C-terminal domain of βAPP, C100 (12 kDa), generated by β-secretase cleavage of the full-length βAPP. That β-secretase cleavage of βAPP is a prerequisite for γ-secretase cleavage was demonstrated by Paganetti *et al* (1996) who showed that Aβ cannot be generated from βAPP mutants truncated at the γ-secretase cleavage site. The simple reason for this is that γ-truncated βAPP does not insert into the cell membrane. These data also imply that γ-secretase is a membrane-bound protease. γ-Secretase cleaves C100 in the transmembrane region to generate $A\beta_{40}$ or $A\beta_{42}$. A number of publications have shown that γ-secretase has a fairly broad sequence specificity. Lichtenthaler *et al* (1999), for instance, performed phenylalanine scanning of residues in the C-terminal portion of the transmembrane domain and found that all mutants tested were still processed to Aβ with approximately similar amounts of Aβ produced in all cases. Intramembrane cleavage is energetically highly unfavorable and, therefore, an unusual event. Similar cleavages are known to occur in only two other transmembrane proteins (Notch and SREBP, see section below). It has been speculated, therefore, that the γ-cleavage may not actually occur intramembranously even though both γ-secretase cleavage sites are well within the predicted membrane-spanning region of C100. Murphy *et al* (1999) proposed a "cut-expose-cut" model in which β-secretase cleavage causes a movement of the transmembrane segment in the cytoplasmic direction, perhaps due to conformational rearrangement. This would then expose the cryptic γ-secretase cleavage sites.

11.4.2 Subcellular location

The subcellular location of γ-secretase can only be defined once the enzyme (or enzymes) itself has been unequivocally identified. In principle, however, its approximate location in the cell can be inferred from the production site of Aβ. A large number of publications have led to the conclusion that Aβ production occurs at two distinct locations in the cell (Koo and Squazzo 1994; Higaki *et al* 1995; Cook *et al* 1997; Peraus *et al* 1997). A portion is produced somewhere on the endosomal-lysosomal pathway from βAPP reinternalized from the cell surface. Presumably processing occurs at the level of the early/recycling endosomes since Aβ is rapidly released into the medium (Koo and Squazzo 1994). Aβ is also generated along the secretory pathway, within the ER and/or in the Golgi/TGN. A recent publication has suggested that reinternalized βAPP might be delivered to the TGN after trafficking through an endocytic recycling compartment, raising the possibility that all Aβ is ultimately produced in the secretory pathway (Greenfield *et al* 1999). The existence of two routes of production for Aβ is suggestive of the involvement of two different enzymes operating in different subcellular compartments. There are two other complicating issues, however: (1) $A\beta_{40}$ and $A\beta_{42}$ are produced at different locations in the cell; $A\beta_{40}$ seems to be produced primarily by the endocytic pathway but also late in the secretory pathway while $A\beta_{42}$ appears to be produced earlier in the secretory pathway, probably in the ER (Cook *et al* 1997); and (2) it is unclear if $A\beta_{40}$ and $A\beta_{42}$ are generated by specific γ-secretases (see below). Thus, it is possible that a specific $A\beta_{42}$ γ-secretase resides in the ER and that a $A\beta_{40}$ specific γ-secretase is present in Golgi/TGN and within the endosomal-lysosomal pathway.

11.4.3 Specific $A\beta_{40}$ and $A\beta_{42}$ γ-secretases?

The major Aβ species in both conditioned cell culture media and human CSF is $A\beta_{40}$ (50–70%). Small amounts of $A\beta_{42}$ (5–20%) are also found, along with minor amounts of other N- or C-terminally truncated Aβ species. A number of publications have addressed the issue of whether discrete γ-secretases are responsible for the production of $A\beta_{40}$ and $A\beta_{42}$ or whether the two species are produced by a single secretase with loose sequence specificity (see also section below). Tischer and Cordell (1996) proposed that membrane thickness might have an influence on where γ-secretase cleaves βAPP. Membrane thickness varies between the ER, where it is relatively thin, and the plasma membrane, where it is relatively thick. Presumably the same γ-secretase could produce $A\beta_{40}$ or $A\beta_{42}$ depending on where exactly it encounters βAPP along the secretory pathway. Murphy *et al* (1999) agreed with this basic tenet but showed that two different types of protease activity are involved in γ-processing of βAPP, namely a pepstatin-sensitive (aspartic protease) activity and a pepstatin-insensitive activity. The pepstatin-sensitive protease cleaves preferentially but not exclusively at $A\beta_{40}$. A recent study, using a newly designed difluoro ketone peptidomimetic (Wolfe *et al* 1999a) inhibitor of γ-secretase, also came to the conclusion that an aspartyl protease might be responsible for γ-secretase activity although in this case a secretase with loose sequence specificity was proposed (Wolfe *et al* 1999a). In stark contrast, Figuereido-Pereira *et al* (1999) claimed that two distinct and specific enzymes are indeed involved and that these are a cysteine protease ($A\beta_{40}$-secretase) and possibly a serine protease ($A\beta_{42}$-secretase). Several groups also showed that the peptide aldehyde protease inhibitor N-acetyl-leucyl-leucyl-norleucinal differentially affected production of $A\beta_{40}$ and $A\beta_{42}$ (Higaki *et al* 1995; Klafki *et al* 1996). However, in none of the above-mentioned publications was it conclusively shown that the inhibitors in question bind directly to γ-secretase and so the differential effects observed might have alternative explanations (as suggested, for instance, by Zhang *et al* (1999) for N-acetyl-leucyl-leucyl-norleucinal).

11.4.4 Non-specific generation of Aβ

Tjernberg *et al* (1997) showed that C100, the βAPP fragment generated by β-secretase, can be efficiently processed to Aβ using proteinase K. They performed these experiments using recombinant C100 *in situ* in cell membranes or using purified C100. The only prerequisite for Aβ production from purified C100 was that the C100 had to be in an oligomeric state, otherwise it was completely degraded by proteinase K. They then showed that C100 expressed in Sf9 cells forms large oligomers, probably polymerizing via its Aβ domain. The data are consistent with the observation that Aβ acquires protease resistance following polymerization (Nordstedt *et al* 1994). The results of Tjernberg *et al* (1997) suggest that the C-terminus of Aβ can be generated by non-specific proteases, acting on a polymerized substrate, rather than a specific γ-secretase. This situation could conceivably prevail in the endosomal-lysosomal pathway of Aβ production where lysosomal cathepsins could degrade C100 relatively non-specifically. It would also explain the C-terminal heterogeneity observed for Aβ without having to invoke the existence of different γ-secretases.

11.4.5 Candidate enzymes

11.4.5.1 SREB2 protease

The sterol regulatory element-binding protein (SREBP) is initially synthesized as an ER-bound precursor. Following a two-step cleavage by the site 1 and site 2 proteases, the cytosolic domain of SREBP is released and targeted to the nucleus where it transcriptionally regulates genes containing the sterol regulatory element. Brown and Goldstein (1997) pointed out the striking similarities between the site 2 protease (S2P) and γ-secretase. First, their respective substrates, SREBP and βAPP (both membrane-bound proteins), are cleaved within the transmembrane region, a highly unusual event. Second, both S2P and γ-secretase catalyze the second step of a two-step cleavage reaction. Third, processing of SREBP is regulated by a multi-transmembrane domain protein (SCAP), in analogy to the presenilin regulation of βAPP processing (see below). However, subsequent work showed that normal βAPP processing occurs in cells which are deficient in S2P activity and expression and, therefore, it is unlikely that S2P itself is γ-secretase (Ross *et al* 1998).

11.4.5.2 PS1

Presenilin 1 (PS-1) knockout mice show defective βAPP processing. These mice accumulate β-stubs in the ER membrane, suggesting that presenilin itself is the γ-secretase or a component of γ-secretase (De Strooper *et al* 1998). Alternatively, presenilin may be required for transport of βAPP to the γ-secretase for processing. However, a number of publications lend support to the hypothesis that PS-1 has indeed a γ-secretase activity (De Strooper *et al* 1999). These studies describe the effect of presenilin deficiency/modulation on the Notch signaling pathway in *Drosophila* and *C. elegans*. The membrane-bound receptor protein, Notch, is proteolytically processed in a manner analogous to βAPP. The protease which releases the cytoplasmic domain of Notch is, in many respects, similar to γ-secretase; for example, in both Notch and βAPP an intramembranous cleavage occurs to release the cytoplasmic domain. Presenilin mutations that influence γ-secretase cleavage of βAPP have parallel effects on Notch cleavage. The data from these publications are highly indicative of presenilin, Notch protease and γ-secretase being one and the same protein. Wolfe *et al* (1999a,b) showed that two intramembrane aspartates are crucial for γ-secretase activity associated with human PS-1 and suggest that these two aspartates interact to form the active site of an aspartic protease. The sequence of PS-1 does not bear any resemblance to the protein sequence of any known mammalian aspartic protease (not even at the level of the short sequence motifs observed in the region around the catalytic aspartates), and the suggestion that γ-secretase might be an intramembrane aspartic protease initially generated considerable scepticism. However, a recent publication by Steiner *et al* (2000) showed that the active site of PS-1 is homologous to a newly described family of bacterial aspartic proteases, namely the type-4 prepilin peptidases (TFPPs). The TFPPs are also polytopic membrane proteins, like the presenilins, and contain two critical aspartates in the transmembrane regions (LaPointe and Taylor 2000). Esler *et al* (2000) demonstrated that aspartyl protease transition-state analogue inhibitors of γ-secretase bind directly to PS-1. Li *et al* (2000) have purified the γ-secretase activity in the detergent solubilized form and have also shown that this γ-secretase activity is associated with PS-1 and that it can be inhibited by pepstatin and by a aspartyl protease transition-state analogue inhibitor of γ-secretase. In summary, a substantial body of evidence now exists to support the hypothesis that PS-1, itself, is the true γ-secretase.

11.5 AMYLOID DEGRADATION AND CLEARANCE

Since amyloid is produced intracellularly and subsequently secreted, both intracellular and extracellular enzymes could be involved in Aβ degradation and clearance.

11.5.1 Extracellular enzymes

Several publications have proposed a role for insulin-degrading enzyme (IDE) in amyloid turnover. IDE is a well-characterized thiol-dependent non-matrix metalloprotease which degrades small peptides such as insulin, glucagon and atrial natriuretic peptide. It was shown to be the main soluble Aβ-degrading enzyme at neutral pH in extracts of human brain and microglial cell lines (Qiu *et al* 1997). Insulin is an efficient competitive inhibitor of Aβ degradation by IDE. Although IDE expression levels do not seem to be significantly different in individuals with Alzheimer's disease compared to healthy controls, IDE activity might be competitively inhibited by higher insulin levels in the former group (Qiu *et al* 1998). Insulin levels tend to rise with age in humans and an association between increased serum insulin levels and decreased cognitive function has been pointed out (Stolk *et al* 1997). Moreover, a recent report by Chesneau *et al* (2000) has shown that the purified recombinant IDE efficiently degrades Aβ. The Ca^{2+}-dependent matrix metalloprotease, MMP-9 (also known as human Type IV collagenase), may also be involved in Aβ clearance (Backstrom *et al* 1992, 1996). In contrast to IDE, which is a glial enzyme, MMP-9 is produced by certain types of neurons. The distribution of MMP-9 immunostaining in the hippocampus showed a good correlation with regions that are most affected in Alzheimer's Disease (Backstrom *et al* 1996). It has also been found in close proximity to plaques. MMP-9 cleaves predominantly in the membrane-spanning region of Aβ, unlike IDE, which appears to cleave in the N-terminal half of Aβ. MMP-9 expression levels are higher than normal in Alzheimer's Disease (Backstrom *et al* 1992) but the enzyme seems to accumulate as a 'latent' (pro) form.

11.5.2 Intracellular enzymes

Aβ is rapidly degraded at acidic pH by a pepstatin-sensitive protease present in soluble brain extracts. This activity is likely to be cathepsin D since purified human cathepsin D cleaves at identical sites in Aβ (McDermott and Gibson 1996). Because cathepsin D is localized in endosomes and lysosomes, where Aβ is generated from reinternalized βAPP, it is possible that the protease can eliminate some Aβ at the site of generation. It is known, however, that cathepsin D levels are upregulated in Alzheimer's Disease, suggesting that Aβ levels should be lower rather than higher if this enzyme significantly affects Aβ metabolism. In addition, as pointed out above in the context of the secretases, cathepsin D knockout mice make normal amounts of Aβ (Saftig *et al* 1996).

11.6 CONCLUSION

Due to their central role in Aβ generation, β- and γ-secretases are of enormous interest to the AD research community and the pharmaceutical industry as drug targets. Development of highly specific inhibitors will only be possible once these enzymes have been better characterized. Recent progress in the identification and characterisation of BACE-1 as β-secretase should facilitate the development of highly selective small-molecule inhibitors for this key enzyme. The potential perils of inhibiting presenilin 1-associated γ-secretase have been pointed out (De Strooper *et al* 1999) but on a more general level it may be undesirable to raise intracellular levels of C100, as two recent publications on C100 transgenic mice have demonstrated (Jin *et al* 1998; Berger-Sweeney *et al* 1999)

ACKNOWLEDGEMENTS

We thank Dr. J. Grayson Richards for critical reading of the review and Mr. Juerg Messer for the artwork.

REFERENCES

Arribas, J., Lopez-Casillas, F. and Massague, J. (1997) Role of the juxtamembrane domains of the transforming growth factor-alpha precursor and the beta-amyloid precursor protein in regulated ectodomain shedding. *Journal of Biological Chemistry* **272**, 17160–17165.

Backstrom, J.R., Miller, C.A. and Tokes, Z.A. (1992) Characterization of neutral proteinases from Alzheimer-affected and control brain specimens: identification of calcium-dependent metalloproteinases from the hippocampus. *Journal of Neurochemistry*, **58**, 983–992.

Backstrom, J.R., Lim, G.P., Cullen, M.J. and Tokes, Z.A. (1996) Matrix metalloproteinase-9 (MMP-9) is synthesized in neurons of the human hippocampus and is capable of degrading the amyloid-beta peptide (1–40). *Journal of Neuroscience*, **16**, 7910–7919.

Benjannet, S., Elagoz, A., Wickham, L., Mamarbachi, M., Munzer, J.S., Basak, A. *et al* (2001) Post-translational processing of beta-secretase (beta-amyloid-converting enzyme) and its ectodomain shedding. *Journal of Biological Chemistry*, **276**, 10879–10887.

Bennett, B.D., Babu-Khan, S., Loeloff, R., Louis, J.C., Curran, E., Citron, M. *et al* (2000) Expression analysis of BACE2 in brain and peripheral tissues. *Journal of Biological Chemistry*, **275**, 20647– 20651.

Berger-Sweeney, J., McPhie, D.L., Arters, J.A., Greenan, J., Oster-Granite, M.L. and Neve, R.L. (1999) Impairments in learning and memory accompanied by neurodegeneration in mice transgenic for the carboxyl-terminus of the amyloid precursor protein. *Brain Research Molecular Brain Research*, **66**, 150–162.

Black, R.A. and White, J.M. (1998) ADAMs: focus on the protease domain. *Current Opinion in Cell Biology*, **10**, 654–659.

Bodovitz, S. and Klein, W.L. (1996) Cholesterol modulates alpha-secretase cleavage of amyloid precursor protein. *Journal of Biological Chemistry*, **271**, 4436–4440.

Brown, M.S. and Goldstein, J.L. (1997) The SREBP pathway: regulation of cholesterol metabolism by proteolysis of a membrane-bound transcription factor. *Cell*, **89**, 331–340.

Buxbaum, J.D., Oishi, M., Chen, H.I., Pinkas-Kramarski, R., Jaffe, E.A., Gandy, S.E. *et al* (1992) Cholinergic agonists and interleukin 1 regulate processing and secretion of the Alzheimer beta/A4 amyloid protein precursor. *Proceedings of the National Academy of Sciences of the U.S.A.*, **89**, 10075–10078.

Buxbaum, J.D., Koo, E.H. and Greengard, P. (1993) Protein phosphorylation inhibits production of Alzheimer amyloid beta/A4 peptide. *Proceedings of the National Academy of Sciences of the U.S.A.*, **90**, 9195–9188.

Buxbaum, J.D., Liu, K.N., Luo, Y., Slack, J.L., Stocking, K.L., Peschon, J.J. *et al* (1998) Evidence that tumor necrosis factor alpha converting enzyme is involved in regulated alpha-secretase cleavage of the Alzheimer amyloid protein precursor. *Journal of Biological Chemistry*, **273**, 27765–27767.

Cai, H., Wang, Y., McCarthy, D., Wen, H., Borchelt, D.R., Price, D.L. *et al* (2001) BACE1 is the major beta-secretase for generation of Abeta peptides by neurons. *Nature Neuroscience*, **4**, 233–234.

Chesneau, V., Vekrellis, K., Rosner, M.R. and Selkoe, D.J. (2000) Purified recombinant insulin-degrading enzyme degrades amyloid beta-protein but does not promote its oligomerization. *Biochemistry Journal*, **351**, 509–516.

Chevallier, N., Vizzavona, J., Marambaud, P., Baur, C.P., Spillantini, M., Fulcrand, P. *et al* (1997) Cathepsin D displays *in vitro* beta-secretase-like specificity. *Brain Research*, **750**, 11–19.

Cook, D.G., Forman, M.S., Sung, J.C., Leight, S., Kolson, D.L., Iwatsubo, T. *et al* (1997) Alzheimer's A beta(1–42) is generated in the endoplasmic reticulum/intermediate compartment of NT2N cells. *Nature Medicine*, **3**, 1021–1023.

De Strooper, B., Saftig, P., Craessaerts, K., Vanderstichele, H., Guhde, G., Annaert, W. *et al* (1998) Deficiency of presenilin-1 inhibits the normal cleavage of amyloid precursor protein [see comments]. *Nature*, **391**, 387–390.

De Strooper, B., Annaert, W., Cupers, P., Saftig, P., Craessaerts, K., Mumm, J.S. *et al* (1999) A presenilin-1-dependent gamma-secretase-like protease mediates release of Notch intracellular domain [see comments]. *Nature*, **398**, 518–522.

Esler, W.P., Kimberly, W.T., Ostaszewski, B.L., Diehl, T.S., Moore, C.L., Tsai, J.Y. *et al* (2000) Transition-state analogue inhibitors of gamma-secretase bind directly to presenilin-1. *Nature Cell Biology*, **2**, 428–434.

Farzan, M., Schnitzler, C.E., Vasilieva, N., Leung, D. and Choe, H. (2000) BACE2, a beta -secretase homolog, cleaves at the beta site and within the amyloid-beta region of the amyloid-beta precursor protein. *Proceedings of the National Academy of Sciences of the U.S.A.*, **97**, 9712–9717.

Figueiredo-Pereira, M.E., Efthimiopoulos, S., Tezapsidis, N., Buku, A., Ghiso, J., Mehta, P. *et al* (1999) Distinct secretases, a cysteine protease and a serine protease, generate the C termini of amyloid beta-proteins Abeta1-40 and Abeta1-42, respectively. *Journal of Neurochemistry*, **72**, 1417–1422.

Greenfield, J.P., Tsai, J., Gouras, G.K., Hai, B., Thinakaran, G., Checler, F. *et al* (1999) Endoplasmic reticulum and trans-Golgi network generate distinct populations of Alzheimer beta-amyloid peptides. *Proceedings of the National Academy of Sciences of the U.S.A.*, **96**, 742–747.

Gruninger-Leitch, F., Berndt, P., Langen, H., Nelboeck, P. and Dobeli, H. (2000) Identification of beta-secretase-like activity using a mass spectrometry-based assay system. *Nature Biotechnology*, **18**, 66–70.

Higaki, J., Quon, D., Zhong, Z. and Cordell, B. (1995) Inhibition of beta-amyloid formation identifies proteolytic precursors and subcellular site of catabolism. *Neuron*, **14**, 651–659.

Hong, L., Koelsch, G., Lin, X., Wu, S., Terzyan, S., Ghosh, A.K. *et al* (2000) Structure of the protease domain of memapsin 2 (beta-secretase) complexed with inhibitor. *Science*, **290**, 150–153.

Huber, A.B., Brosamle, C., Mechler, H. and Huber, G. (1999a) Metalloprotease MP100: a synaptic protease in rat brain. *Brain Research*, **837**, 193–202.

Huber, G., Thompson, A., Gruninger, F., Mechler, H., Hochstrasser, R., Hauri, H.P. *et al* (1999b) cDNA cloning and molecular characterization of human brain metalloprotease MP100: a beta-secretase candidate? *Journal of Neurochemistry*, **72**, 1215–1223.

Hussain, I., Powell, D., Howlett, D.R., Tew, D.G., Meek, T.D., Chapman, C. *et al* (1999) Identification of a novel aspartic protease (Asp 2) as beta-secretase. *Molecular and Cellular Neurosciences*, **14**, 419–427.

Ikezu, T., Trapp, B.D., Song, K.S., Schlegel, A., Lisanti, M.P. and Okamoto, T. (1998) Caveolae, plasma membrane microdomains for alpha-secretase-mediated processing of the amyloid precursor protein. *Journal of Biological Chemistry*, **273**, 10485–10495.

Jin, L.W., Hearn, M.G., Ogburn, C.E., Dang, N., Nochlin, D., Ladiges, W.C. *et al* (1998) Transgenic mice over-expressing the C-99 fragment of betaPP with an alpha-secretase site mutation develop a myopathy similar to human inclusion body myositis. *American Journal of Pathology*, **153**, 1679–1686.

Klafki, H., Abramowski, D., Swoboda, R., Paganetti, P.A. and Staufenbiel, M. (1996) The carboxyl termini of beta-amyloid peptides 1-40 and 1-42 are generated by distinct gamma-secretase activities. *Journal of Biological Chemistry*, **271**, 28655–28659.

Komano, H., Seeger, M., Gandy, S., Wang, G.T., Krafft, G.A. and Fuller, R.S. (1998) Involvement of cell surface glycosyl-phosphatidylinositol-linked aspartyl proteases in alpha-secretase-type cleavage and ectodomain solubilization of human Alzheimer beta-amyloid precursor protein in yeast. *Journal of Biological Chemistry*, **273**, 31648–31651.

Koo, E.H. and Squazzo, S.L. (1994) Evidence that production and release of amyloid beta-protein involves the endocytic pathway. *Journal of Biological Chemistry*, **269**, 17386–17389.

Lammich, S., Kojro, E., Postina, R., Gilbert, S., Pfeiffer, R., Jasionowski, M. *et al* (1999) Constitutive and regulated alpha-secretase cleavage of Alzheimer's amyloid precursor protein by a disintegrin metalloprotease. *Proceedings of the National Academy of Sciences of the U.S.A.*, **96**, 3922–3927.

LaPointe, C.F. and Taylor, R.K. (2000) The type 4 prepilin peptidases comprise a novel family of aspartic acid proteases. *Journal of Biological Chemistry*, **275**, 1502–1510.

Li, Y.M., Lai, M.T., Xu, M., Huang, Q., DiMuzio-Mower, J., Sardana, M.K. (2000) Presenilin 1 is linked with gamma-secretase activity in the detergent solubilized state. *Proceedings of the National Academy of Sciences of the U.S.A.*, **97**, 6138–6143.

Lichtenthaler, S.F., Wang, R., Grimm, H., Uljon, S.N., Masters, C.L. and Beyreuther, K. (1999) Mechanism of the cleavage specificity of Alzheimer's disease gamma-secretase identified by phenylalanine-scanning mutagenesis of the transmembrane domain of the amyloid precursor protein. *Proceedings of the National Academy of Sciences of the U.S.A.*, **96**, 3053–3058.

Lin, X., Koelsch, G., Wu, S., Downs, D., Dashti, A. and Tang, J. (2000) Human aspartic protease memapsin 2 cleaves the beta-secretase site of beta-amyloid precursor protein. *Proceedings of the National Academy of Sciences of the U.S.A.*, **97**, 1456–1460.

Little, S.P., Dixon, E.P., Norris, F., Buckley, W., Becker, G.W., Johnson, M. *et al* (1997) Zyme, a novel and potentially amyloidogenic enzyme cDNA isolated from Alzheimer's disease brain. *Journal of Biological Chemistry*, **272**, 25135–25142.

Luo, Y., Bolon, B., Kahn, S., Bennett, B.D., Babu-Khan, S., Denis, P. *et al* (2001) Mice deficient in BACE1, the Alzheimer's β-secretase, have normal phenotype and abolished β-amyloid generation. *Nature Neuroscience*, **4**, 231–232.

Malherbe, P., Faull, R.L. and Richards, J.G. (2000) Regional and cellular distribution of bleomycin hydrolase mRNA in human brain: comparison between Alzheimer's diseased and control brains. *Neuroscience Letters*, **281**, 37–40.

Marcinkiewicz, M. and Seidah, N.G. (2000) Coordinated expression of beta-amyloid precursor protein and the putative beta-secretase BACE and alpha-secretase ADAM10 in mouse and human brain. *Journal of Neurochemistry*, **75**, 2133–2143.

Marks, N., Berg, M.J., Sapirstein, V.S., Durrie, R., Swistok, J., Makofske, R.C. *et al* (1995) Brain cathepsin B but not metalloendopeptidases degrade rAPP751 with production of amyloidogenic fragments. Comparison with synthetic peptides emulating beta- and gamma-secretase sites. *International Journal of Peptide & Protein Research*, **46**, 306–313.

McDermott, J.R. and Gibson, A.M. (1996) Degradation of Alzheimer's beta-amyloid protein by human cathepsin D. *Neuroreport*, **7**, 2163–2166.

Meckelein, B., Marshall, D.C., Conn, K.J., Pietropaolo, M., Van Nostrand, W. and Abraham, C.R. (1998) Identification of a novel serine protease-like molecule in human brain. *Brain Research Molecular Brain Research*, **55**, 181–197.

Mills, J. and Reiner, P.B. (1999) Regulation of amyloid precursor protein cleavage. *Journal of Neurochemistry*, **72**, 443–460.

Montoya, S.E., Aston, C.E., DeKosky, S.T., Kamboh, M.I., Lazo, J.S. and Ferrell, R.E. (1998) Bleomycin hydrolase is associated with risk of sporadic Alzheimer's disease [letter] [published erratum appears in Nat Genet 1998 August 19(4): 404]. *Nature Genetics*, **18**, 211–212.

Murphy, M.P., Hickman, L.J., Eckman, C.B., Uljon, S.N., Wang, R. and Golde, T.E. (1999) gamma-secretase, evidence for multiple proteolytic activities and influence of membrane positioning of substrate on generation of amyloid beta peptides of varying length. *Journal of Biological Chemistry*, **274**, 11914–11923.

Nordstedt, C., Naslund, J., Tjernberg, L.O., Karlstrom, A.R., Thyberg, J. and Terenius, L. (1994) The Alzheimer A beta peptide develops protease resistance in association with its polymerization into fibrils. *Journal of Biological Chemistry*, **269**, 30773–30776.

Paganetti, P.A., Lis, M., Klafki, H.W. and Staufenbiel, M. (1996) Amyloid precursor protein truncated at any of the gamma-secretase sites is not cleaved to beta-amyloid. *Journal of Neuroscience Research*, **46**, 283–293.

Palmert, M.R., Podlisny, M.B., Witker, D.S., Oltersdorf, T., Younkin, L.H., Selkoe *et al* (1989) The beta-amyloid protein precursor of Alzheimer disease has soluble derivatives found in human brain and cerebrospinal fluid. *Proceedings of the National Academy of Sciences of the U.S.A.*, **86**, 6338–6342.

Parvathy, S., Hussain, I., Karran, E.H., Turner, A.J. and Hooper, N.M. (1998a) Alzheimer's amyloid precursor protein alpha-secretase is inhibited by hydroxamic acid-based zinc metalloprotease inhibitors: similarities to the angiotensin converting enzyme secretase. *Biochemistry*, **37**, 1680–1685.

Parvathy, S., Karran, E.H., Turner, A.J. and Hooper, N.M. (1998b) The secretases that cleave angiotensin converting enzyme and the amyloid precursor protein are distinct from tumour necrosis factor-alpha convertase. *FEBS Letters*, **431**, 63–65.

Peraus, G.C., Masters, C.L. and Beyreuther, K. (1997) Late compartments of amyloid precursor protein transport in SY5Y cells are involved in beta-amyloid secretion. *Journal of Neuroscience*, **17**, 7714–7724.

Qiu, W.Q., Ferreira, A., Miller, C., Koo, E.H. and Selkoe, D.J. (1995) Cell-surface beta-amyloid precursor protein stimulates neurite outgrowth of hippocampal neurons in an isoform-dependent manner. *Journal of Neuroscience*, **15**, 2157–2167.

Qiu, W.Q., Ye, Z., Kholodenko, D., Seubert, P. and Selkoe, D.J. (1997) Degradation of amyloid beta-protein by a metalloprotease secreted by microglia and other neural and non-neural cells. *Journal of Biological Chemistry*, **272**, 6641–6646.

Qiu, W.Q., Walsh, D.M., Ye, Z., Vekrellis, K., Zhang, J., Podlisny, M.B. *et al* (1998) Insulin-degrading enzyme regulates extracellular levels of amyloid beta-protein by degradation. *Journal of Biological Chemistry*, **273**, 32730–32738.

Roberts, S.B., Ripellino, J.A., Ingalls, K.M., Robakis, N.K. and Felsenstein, K.M. (1994) Non-amyloidogenic cleavage of the beta-amyloid precursor protein by an integral membrane metallo-endopeptidase. *Journal of Biological Chemistry*, **269**, 3111–3116.

Ross, S.L., Martin, F., Simonet, L., Jacobsen, F., Deshpande, R., Vassar, R. *et al* (1998) Amyloid precursor protein processing in sterol regulatory element-binding protein site 2 protease-deficient Chinese hamster ovary cells. *Journal of Biological Chemistry*, **273**, 15309–15312.

Rossner, S., Ueberham, U., Schliebs, R., Perez-Polo, J.R. and Bigl, V. (1998) The regulation of amyloid precursor protein metabolism by cholinergic mechanisms and neurotrophin receptor signaling. *Progress in Neurobiology*, **56**, 541–569.

Saftig, P., Peters, C., von Figura, K., Craessaerts, K., Van Leuven, F. and De Strooper, B. (1996) Amyloidogenic processing of human amyloid precursor protein in hippocampal neurons devoid of cathepsin D. *Journal of Biological Chemistry*, **271**, 27241–27244.

Saunders, A.J., Kim, T. and Tanzi, R.E. (1999) BACE maps to chromosome 11 and a BACE homolog, BACE2, reside in the obligate Down Syndrome region of chromosome 21. *Science*, **286**, 1255–1255a.

Selkoe, D.J. (1994) Cell biology of the amyloid beta-protein precursor and the mechanism of Alzheimer's disease. *Annual Review of Cell Biology*, **10**, 373–403.

Selkoe, D.J. (1999) Translating cell biology into therapeutic advances in Alzheimer's disease [In Process Citation]. *Nature*, **399**, A23–31.

Selkoe, D.J. (2001) Alzheimer's disease: genes, proteins, and therapy. *Physiological Reviews*, **81**, 741–766.

Sinha, S., Anderson, J.P., Barbour, R., Basi, G.S., Caccavello, R., Davis, D. *et al* (1999) Purification and cloning of amyloid precursor protein beta-secretase from human brain. *Nature*, **402**, 537–540.

Sisodia, S.S. (1992) Beta-amyloid precursor protein cleavage by a membrane-bound protease. *Proceedings of the National Academy of Sciences of the U.S.A.*, **89**, 6075–6079.

Steiner, H., Kostka, M., Romig, H., Basset, G., Pesold, B., Hardy, J. *et al* (2000) Glycine 384 is required for presenilin-1 function and is conserved in bacterial polytopic aspartyl proteases. *Nature Cell Biology*, **2**, 848–851.

Stolk, R.P., Breteler, M.M., Ott, A., Pols, H.A., Lamberts, S.W., Grobbee, D.E. *et al* (1997) Insulin and cognitive function in an elderly population. The Rotterdam Study. *Diabetes Care*, **20**, 792–795.

Thompson, A., Grueninger-Leitch, F., Huber, G. and Malherbe, P. (1997) Expression and characterization of human beta-secretase candidates metalloendopeptidase MP78 and cathepsin D in beta APP-overexpressing cells. *Brain Research Molecular Brain Research*, **48**, 206–214.

Tischer, E. and Cordell, B. (1996) Beta-amyloid precursor protein. Location of transmembrane domain and specificity of gamma-secretase cleavage. *Journal of Biological Chemistry*, **271**, 21914–21919.

Tjernberg, L.O., Naslund, J., Thyberg, J., Gandy, S.E., Terenius, L. and Nordstedt, C. (1997) Generation of Alzheimer amyloid beta peptide through nonspecific proteolysis. *Journal of Biological Chemistry*, **272**, 1870–1875.

Vassar, R., Bennett, B.D., Babu-Khan, S., Kahn, S., Mendiaz, E.A., Denis, P. *et al* (1999) Beta-secretase cleavage of Alzheimer's amyloid precursor protein by the transmembrane aspartic protease BACE. *Science*, **286**, 735–741.

Wolfe, M.S., Xia, W., Moore, C.L., Leatherwood, D.D., Ostaszewski, B., Rahmati, T. *et al* (1999a) Peptidomimetic probes and molecular modeling suggest that Alzheimer's gamma-secretase is an intramembrane-cleaving aspartyl protease. *Biochemistry*, **38**, 4720–4727.

Wolfe, M.S., Xia, W., Ostaszewski, B.L., Diehl, T.S., Kimberly, W.T. and Selkoe, D.J. (1999b) Two transmembrane aspartates in presenilin-1 required for presenilin endoproteolysis and gamma-secretase activity [see comments]. *Nature*, **398**, 513–517.

Yan, R., Bienkowski, M.J., Shuck, M.E., Miao, H., Tory, M.C., Pauley, A.M. *et al* (1999) Membrane-anchored aspartyl protease with Alzheimer's disease beta-secretase activity. *Nature*, **402**, 533–537.

Zhang, L., Song, L. and Parker, E.M. (1999) Calpain inhibitor I increases beta-amyloid peptide production by inhibiting the degradation of the substrate of gamma-secretase. Evidence that substrate availability limits beta-amyloid peptide production. *Journal of Biological Chemistry*, **274**, 8966–8972.

Chapter 12

Herpes virus and cytomegalovirus proteinase

Richard L. Jarvest and Christine E. Dabrowski

The eight human herpesviruses cause a variety of different diseases and are characterized by their ability to enter a latent state in the human host, reactivating at a later date to cause recurrent disease. In 1991, a virally encoded serine protease was identified in herpes simplex virus type 1. Characterization of the protease identified a number of features that made it attractive as a target for potential new antiviral agents: homologues of the protease occur in all the herpesviruses; the protein does not have sequence homology to any other proteases; and the proteolytic activity is essential for generation of infectious viral particles. Crystal structures of the protease from several herpesviruses have been solved and reveal a novel catalytic mechanism but an open, shallow active site. Potent mechanism based inhibitors and peptidomimetic inhibitors have been designed and their mode of inhibition characterized. Inhibitors have also been derived from high-throughput screening, including natural product inhibitors. In cell-based mechanistic assays, some compounds have demonstrated inhibition of proteolytic processing in viral infected cells. Few compounds have demonstrated effective antiviral activity in cell culture. Inhibitor potency, stability, cell penetration, and cytotoxicity are parameters that need to be addressed in order to assess the real therapeutic utility of herpes protease inhibitors.

12.1 INTRODUCTION

12.1.1 Herpesvirus biology

The herpesviruses are enveloped, linear double-strand DNA viruses with a distinctive morphology, containing an icosadeltahedral capsid surrounded by an amorphous tegument. The viral genomes range in size from 125 to 230 kilobases, encoding approximately 70 to more than 200 gene products. Much of the general biology of the herpesviruses has been established through experimentation with the well-characterized herpes simplex virus type 1 (HSV-1). During lytic infection the virus attaches to cellular receptors through glycoproteins located on the viral envelope, followed by fusion with the cell membrane and transport of the viral capsid and tegument to the nucleus of the cell. The viral DNA is then extruded into the nucleus at the nuclear pore, circularizes, and a regulated pattern of gene expression progresses resulting in replication of the viral DNA and the synthesis of late viral proteins. Newly synthesized viral genomes are packaged into pre-formed viral capsids which then egress, with lysis of the infected cell and resultant acute primary disease. Following the primary infection, herpesviruses are characterized by the ability to become latent in the human host with very limited gene expression and with little obvious consequence to the host cell, reactivating in response to a multitude of stimuli to again generate infectious virus which may result in recurrent disease (Roizman and Sears 1993).

Eight human herpesviruses have been identified to date, which are divided into three sub-families according to the biology of the viruses, including DNA and protein homology (Moore *et al* 1996). The alphaherpesviruses, HSV-1, HSV-2 and varicella-zoster virus (VZV), cause widespread disease in both the immunocompetent and immunocompromised populations. HSV-1 and HSV-2 are the primary causative agents of mucosal-labial (cold sores) and genital herpes, respectively. Approximately 70% of the population of industrialized countries are infected by HSV-1, with approximately 20% infected by HSV-2. Prenatal infection with HSV-2 may result in neonatal herpes, with significant morbidity and mortality in new-borns (Whitley and Gnann 1993). VZV infection is also widespread in the human population, causing chickenpox following primary infection in children and shingles (herpes zoster) upon reactivation in adults. A common complication following resolution of zoster lesions is post-herpetic neuralgia, a condition associated with persistent pain in the affected areas which can be debilitating in some individuals (McCrary *et al* 1999). The betaherpesviruses include the prototype, human cytomegalovirus (CMV), as well as human herpesviruses 6 and 7 (HHV-6, HHV-7). Primary infection with CMV occurs in approximately 40–100% of the population and is usually asymptomatic, although mild symptoms of mononucleosis may develop. Viral reactivation is typically not a problem in the immunocompetent population, but significant disease is associated with viral infection and reactivation in immunocompromised and immunosuppressed individuals. In particular, CMV retinitis has been identified in the AIDS population, and CMV disease may cause serious complications in transplant recipients, for example, pneumonitis in bone marrow transplants and hepatitis in liver transplants (Rawlinson 1999). Congenital disease is associated with recurrent or, more commonly, primary CMV infection of pregnant women, with approximately 1% of new-borns born shedding virus. Approximately 10% of these infected babies are born with Cytomegalic Inclusion Disease (CID) and other severe neurological manifestations, resulting in significant morbidity and mortality (Alford *et al* 1990). In addition, congenital CMV infection causes morbidity in approximately 15% of babies asymptomatic at birth, and in particular is a leading cause of progressive hearing loss in babies and young children (Fowler *et al* 1997). HHV-6 is the primary cause of roseola (exanthem subitum), infecting greater than 95% of babies, and HHV-7 may also cause roseola in a minority of cases in young children. Reactivation of HHV-6 has been associated with complications in transplant recipients, and has also been identified as a potential factor in multiple sclerosis (Campadelli-Fiume *et al* 1999). The gammaherpesviruses include Epstein-Barr virus (EBV) and the newly identified human herpesvirus 8 (HHV-8). Primary EBV infection is the leading cause of infectious mononucleosis in adolescence and early adulthood. EBV is also associated with Burkitt's lymphoma, nasopharyngeal carcinoma and Hodgkin's disease, as well as other proliferative diseases (Liebowitz and Kieff 1993). HHV-8 is associated with Kaposi's sarcoma in the AIDS and solid organ transplant populations, and has been postulated to be a significant factor in the development of multiple myeloma (Rettig *et al* 1997; Greenblatt 1998).

12.1.2 Antiviral therapy

The alphaherpesviruses are safely and effectively treated with a number of antiviral drugs, available in oral, topical and intravenous formulations. Penciclovir and its prodrug, famciclovir, as well as acyclovir and its valine ester prodrug, valacyclovir, are used to treat alphaherpesvirus-associated disease including herpes labialis (cold sores), genital herpes and herpes zoster, as well as to suppress viral shedding in recurrent genital herpes. These antiviral drugs are nucleoside analogues which inhibit viral DNA replication by targeting the viral DNA polymerase, following phosphorylation by the viral thymidine kinase and by cellular kinases. The nucleoside analogues significantly reduce symptoms such as lesion duration and severity and acute and chronic pain. They have also been shown to significantly inhibit symptomatic and asymptomatic viral shedding following primary infection and reactivation from latency. However, treatment with these antivirals does not result in the elimination of symptoms, and in particular post-herpetic neuralgia associated with herpes zoster continues to be a significant problem, particularly in the aging population (Wutzler 1997). While resistance to acyclovir has not been shown to be a serious issue in immunocompetent populations to date, antiviral compounds directed against a different target would provide an alternative treatment and may also result in increased efficacy against these viruses (Christophers *et al* 1998).

The antiviral compounds ganciclovir, foscarnet and cidofovir have been approved for treatment of CMV related disease (Walmsley and Tseng 1999). Fomivirsen has also recently been approved for treatment of retinitis by intravitreal injection, and is the first approved antisense therapy for any indication (Perry and Balfour 1999). Ganciclovir, a nucleoside analogue, is used to prevent and to treat CMV retinitis in AIDS patients, as well as to prevent CMV disease in bone marrow and solid organ transplant recipients. While oral, intravenous and intravitreal implant formulations are available, intravenous ganciclovir is the current therapy of choice. Valganciclovir, the prodrug of ganciclovir, has been shown to have significantly increased bioavailability and is currently in Phase III clinical trials. Intravenous foscarnet, a pyrophosphate analogue, is used as monotherapy and in combination with ganciclovir to treat CMV retinitis in the AIDS population. Intravenous cidofovir, a nucleotide analog, has also recently been approved to treat CMV retinitis. However, toxicity issues limit treatment with all three of these systemic antivirals. Ganciclovir treatment commonly results in bone marrow suppression, while foscarnet causes renal impairment and serum electrolyte imbalance. Cidofovir has been shown to cause nephrotoxicity, and may also result in neutropenia (Walmsley and Tseng 1999). Because of the significant toxicity associated with these therapies, none of these antivirals has been approved to treat congenital CMV although ganciclovir has been tested in clinical trials (Whitley *et al* 1997). These three systemic antivirals ultimately target the viral DNA polymerase to inhibit viral DNA replication, and both resistance and cross-resistance are significant problems in treated populations (Vogel *et al* 1997). Low-level resistance to ganciclovir typically first involves the viral protein kinase, UL97, which is required to phosphorylate ganciclovir, while high-level ganciclovir resistance also includes mutations within the viral DNA polymerase (Smith *et al* 1997). Thus, new antivirals with increased safety, bioavailability, and efficacy, with a different mechanism of action, are needed for treatment of CMV-related disease.

12.1.3 Role of the serine protease

The viral serine protease was identified in HSV-1 in 1991, following the characterization of a complex transcriptional unit encoding the protease within the UL26 gene and the protease substrate, the scaffold protein precursor, within UL26.5. These genes are co-terminal at the 3′-end, with initiation of UL26.5 within the UL26 coding region. The proteins are translated in-frame; therefore, the scaffold protein is identical to the

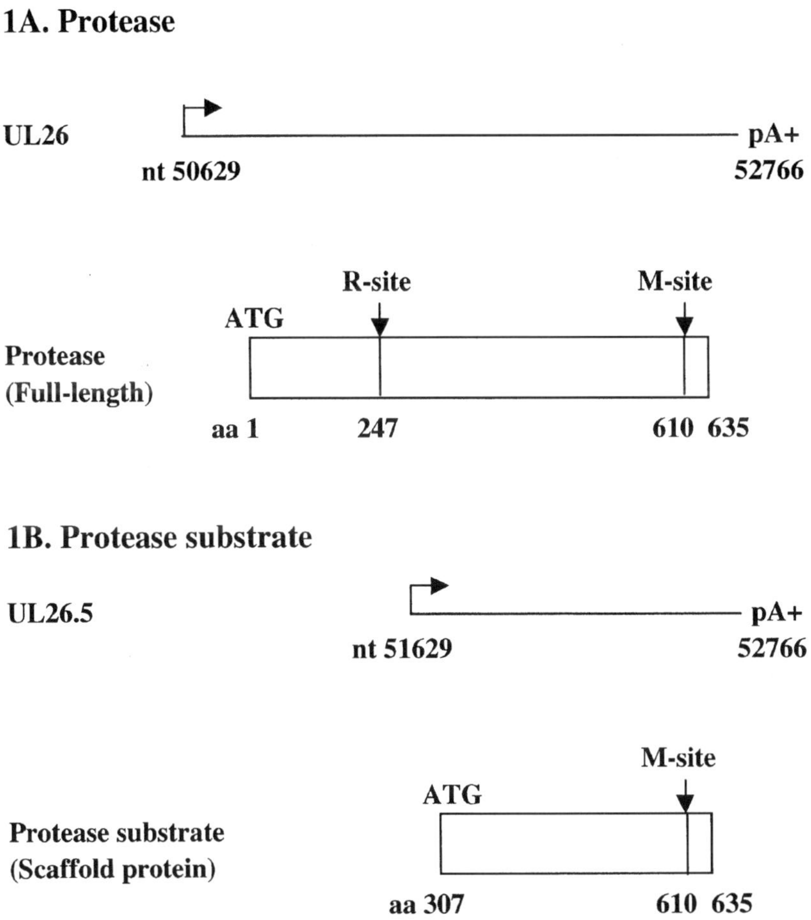

Figure 12.1 HSV-1 protease and substrate. 1A. Protease. (Top) The HSV-1 protease is encoded within the UL26 transcript (Liu and Roizman 1991a,b), with nucleotide (nt) numbering according to McGeoch *et al* (1988). The 5′ end of the transcript is indicated by an arrow, while the 3′ end is shown at the pA+ site. (Bottom) The structure of the protease is shown, including amino acid (aa) numbers for the amino-terminus (aa 1), carboxy-terminus (aa 635), and the protease cleavage sites (R: aa247; M: aa 610). 1B. Protease substrate. (Top) The protease substrate, the scaffold protein, is encoded within the UL26.5 transcript (Liu and Roizman 1991a,b) with 5′ and 3′ ends shown as described above. (Bottom) The structure of the protease substrate is shown as described above, with aa numbers given in relation to protease.

C-terminal half of the protease (Liu and Roizman 1991a,b; see Figure 12.1). Evidence supporting the identification of the UL26-encoded gene product as necessary and sufficient for protease activity was based on a number of *in vitro* assays, demonstrating in particular the requirement for the unique N-terminal half of UL26 for protease activity (Deckman *et al* 1992; Liu and Roizman 1992). The *in vitro* data supported results from previous studies of a temperature-sensitive HSV-1 mutant, 17*ts*VP1201, which was shown to be defective in the processing of specific nucleocapsid proteins at a late stage in the viral life-cycle, and which failed to encapsidate newly synthesized viral DNA at the nonpermissive temperature (Preston *et al* 1983). The mutations in *ts*1201 were later identified within the catalytic domain of the UL26 encoded gene product, and were confirmed to be critical for protease activity (Gao *et al* 1994). Homologues of the HSV-1 UL26 and UL26.5 genes have been identified in all herpesviruses characterized to date although many of the transcriptional units are more complex, potentially encoding additional as yet uncharacterized proteins.

The basic structure of the protease and substrate is conserved in all of the herpes proteases (Gibson and Hall 1997; Holwerda 1997; Figure 12.1). Thus, the catalytic domain of the protease is encoded in the N-terminus, which is bounded by a protease cleavage site, the R (release)-site. A second cleavage site, the M (maturation)-site, is found at the C-terminus of the protease; the R- and M-sites flank the minor scaffold protein. The M-site is also contained within the protease substrate, the scaffold protein precursor (Holwerda 1997). Proteolytic cleavage results in the generation of the (major) scaffold protein and a C-terminal tail, which has been shown to be essential for the generation of sealed viral capsids (Matusick-Kumar *et al* 1995). The C-terminal tail is also generated following M-site cleavage in the protease. Protease activity is present in both the full-length protein and in the catalytic domain following cleavage at the R-site. CMV has a third cleavage site, the I (internal)-site, within the protease catalytic domain. This cleavage occurs both *in vitro* and at late times *in vivo*, although the significance of this cleavage event is unclear. Protease activity is retained in the I-site cleaved heterodimer, but not in either isolated subunit (Holwerda 1997).

The work of numerous laboratories has resulted in a model of capsid assembly in herpesviruses, whereby the major capsid protein is assembled around a core comprised of the scaffold protein and full-length protease. This results in the formation of an immature, or pre-B capsid (procapsid) with a "large-core" phenotype, which is a transient structure not generally observed following infection with wild-type virus. The protease subsequently cleaves itself and the scaffold protein, resulting in a conformational change and the more angular mature B capsid, which has a "small-core" phenotype (Newcomb *et al* 1996; Trus *et al* 1996). Newly synthesized viral DNA is packaged into the mature B capsid with release of the scaffold proteins. The catalytic domain is retained in the viral capsid although its function is unknown. Mutation of the protease active site results in a lethal phenotype in the mutant virus, with the primary defect at the capsid maturation step (Robertson *et al* 1996). Capsids containing an inactive protease remain as large-core pre-B capsids which are typically intranuclear.

Thus, the herpesvirus serine protease is an attractive antiviral target, with enzymatic activity which has been shown to be essential for the generation of infectious virus. This target differs from the currently approved alphaherpesvirus and betaherpesvirus antivirals, all of which ultimately target the viral DNA polymerase resulting in the inhibition of viral DNA synthesis within the nucleus of the infected cell. However, protease inhibitors may face an additional hurdle in that the viral protease has been shown to be active within the newly formed viral capsid (Robertson *et al* 1996). Thus, compounds may be required to cross the nuclear membrane and access the viral capsid for efficacy. Further experimentation with protease inhibitors will be required to compare the effect of antivirals directed against the different aspects of the viral life-cycle.

12.2 STRUCTURE AND MECHANISM

A number of studies have described the determination of key residues by mutagenesis, the identification of minimal peptide substrates, and the development of various assay formats. These have been summarized in earlier reviews (Flynn *et al* 1997; Holwerda 1997; Dabrowski *et al* 2000).

The three dimensional structure of the cytomegalovirus protease (EC 3.4.21.97) was solved by four groups independently (Chen *et al* 1996; Qiu *et al* 1996; Shieh *et al* 1996; Tong *et al* 1996). Subsequently the structures of the VZV protease (Qiu *et al* 1997) and the HSV-1 protease adduct formed from diisopropyl fluorophosphate (Hoog *et al* 1997) were also solved. Whilst it had already been apparent that the herpes proteases had no sequence homology with other protease families, the crystal structures revealed that they also have a unique protein fold. Confirming biochemical studies, the structures showed the herpes proteases to be dimeric. Within the family there are significant differences in the geometry of the dimer interface and low conservation of residues at the interface.

Perhaps most significantly for the development of inhibitors, the herpes protease structures had a number of characteristic features of the active site. Firstly they have a novel catalytic triad comprised of Ser-132, His-63 and His-157; the use of histidine as the third residue of the triad may be a major contributor to the low catalytic efficiency of the herpes proteases. Secondly the active site is very shallow in geometry, with implications for the design of small molecule inhibitors. Thirdly the active site contains other functionalised amino acid residues, such as Cys-161 which might be exploited in inhibitor action.

Binding of peptidomimetic inhibitors to the cytomegalovirus protease results in a conformational change that can be monitored by changes in the tryptophan fluorescence (Bonneau *et al* 1997). Most recently, a crystal structure of this enzyme complexed to a peptidomimetic inhibitor was solved (Tong *et al* 1998). Large-scale conformational changes are observed, including loop movements and increased ordering. The bound inhibitor effectively spans the P_4 to P_1' positions and is in an extended conformation, an observation confirmed by NMR studies (LaPlante *et al* 1998). The binding is similar to that of peptidomimetic inhibitors of the classic serine protease families represented by chymotrypsin and subtilisin. In this respect at least, the herpes proteases demonstrate convergent evolution with the major serine protease families.

12.3 INHIBITORS

Generic inhibitors of serine proteases have only weak activity against the herpes proteases. The low catalytic efficiency, shallow active site, and large substrate requirement of this protease family have resulted in the design and discovery of highly potent inhibitors being very challenging. This is reflected in few reports of really potent (<50 nM) inhibitors, despite much interest in this target within the pharmaceutical industry.

Some of the earliest reports of inhibition of herpes proteases essentially described thiol-reactive reagents that formed disulfide linkages with the cysteines around the active site of CMV protease (Baum *et al* 1996a,b) or other thiol-reactive agents (Flynn *et al* 1996). The majority of inhibitors have been based on the design of mechanism-based strategies or peptidomimetic transition state analogs. Other classes of inhibitor have been identified by high-throughput screening of synthetic compound and natural product collections. The dimer interface has been postulated to be a potential target for inhibition but no inhibitors with this mechanism have been described.

12.3.1 Mechanism-based inhibitors

Mechanism-based inhibitors of the herpes proteases rely on acylation of the active site serine to afford a relatively stable acyl-enzyme adduct. In some cases a "second hit" mechanism occurs where reaction with one of the active site cysteines also occurs.

12.3.1.1 Benzoxazinones

The first report of mechanism-based inhibitors of a herpes protease described benzoxazinone inhibitors of the HSV-1 protease which had potency down to 1 μM (Jarvest *et al* 1996). Structure-activity relationships were noted and it was shown that a 1:1 enzyme–inhibitor complex was formed. Benzoxazinones have also been shown to inhibit the CMV protease with affinity down to 0.24 μM (**12.1**) and activity in cell culture is described (see below) (Abood *et al* 1997). An issue for benzoxazinones and other mechanism-based inhibitors is that of stability in aqueous solution and in plasma; the results of stability monitoring are described in both publications.

Oxazinone inhibitors

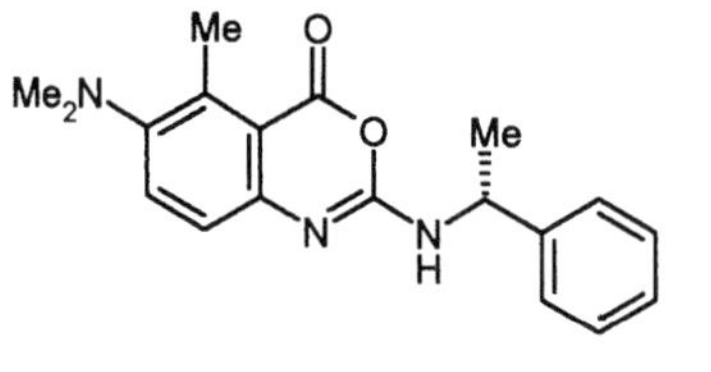

(12.1)

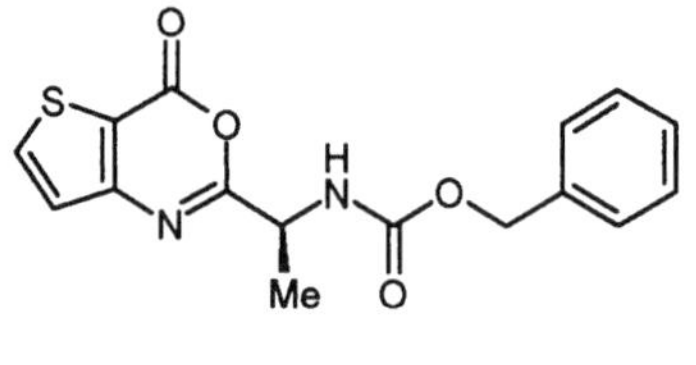

(12.2)

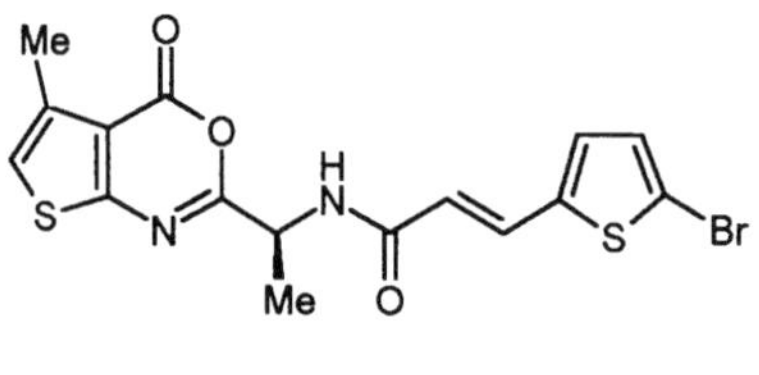

(12.3)

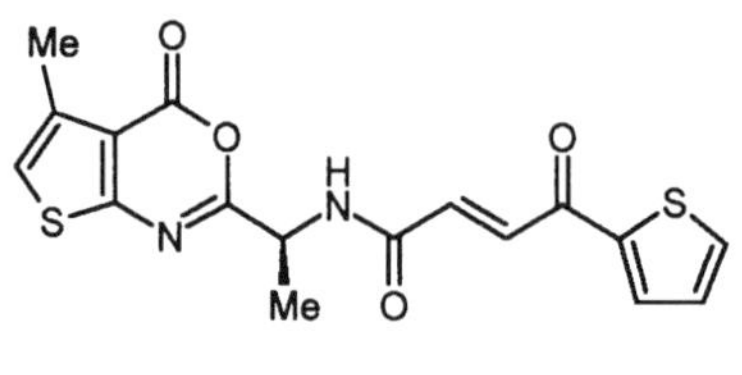

(12.4)

12.3.1.2 Thienoxazinones

Further development of the oxazinone inhibitors has led to thieno-fused inhibitors. Thieno[3,2-*d*]oxazinones (**12.2**) inhibit HSV-1, HSV-2 and CMV proteases with IC_{50} values down to sub-micromolar (Jarvest *et al* 1997). These compounds demonstrate enhanced aqueous stability and selectivity with respect to chymotrypsin family proteases. Thieno[2,3-*d*]oxazinone isomers are also potent inhibitors of HSV-2, VZV and CMV proteases. The mechanism of inhibition has been characterized in detail (Dabrowski *et al* 1999). The presence of the ring methyl substituent *peri* to the carbonyl group greatly enhances stability in serum and allows inhibition of protease processing in cell culture to be measured. SAR studies in this series (**12.3**) afforded the most potent reported inhibitors of α-herpesvirus proteases with IC_{50} values for VZV protease down to <10 nM, the limit determined by the enzyme concentration in the assay (Jarvest *et al* 1999). Determination of the acyl-enzyme deacylation rates showed that deacylation half-lives were up to 25 hours.

12.3.1.3 Dual mechanism oxazinones

A series of enedione derivatives of thieno[2,3-d]oxazinones derivatives have been reported as potent and selective CMV protease inhibitors (Pinto *et al* 1999). These compounds act by not only acylating the catalytic serine but also by alkylating cysteine-161 *via* a Michael addition reaction. Structures such as (**12.4**), IC_{50} 30 nM, are the most potent CMV inhibitors described to date, but they were not developed further due to unacceptable cytotoxicity.

12.3.1.4 Oxazolones and imidazolones

In contrast to benzoxazinones, which were known as a class to be mechanism-based inhibitors of other serine protease, these inhibitors, discovered by directed mechanism-based screening, represent novel types of mechanism-based serine protease inhibitors (Pinto *et al* 1996). Oxazolones (**12.5**) had submicromolar IC_{50} values against HSV-2 and CMV proteases and were shown to acylate the active site serine. Imidazolone (**12.6**) was a selective CMV protease inhibitor and mass measurements and model studies suggested that loss of sulfonyloxy occurred with a novel rearrangement to afford a new stable acyl enzyme adduct. Further exploitation of these classes was limited by relatively poor aqueous stability.

Oxazolones and imidazolones

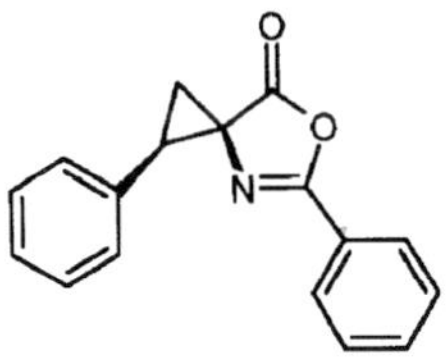

(12.5)

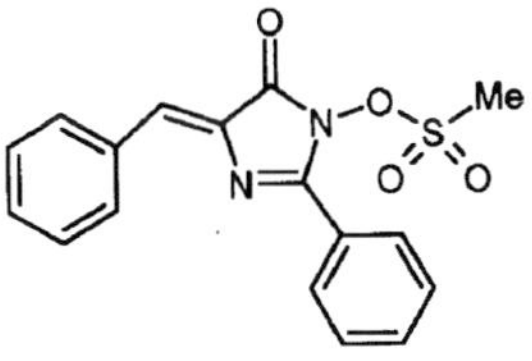

(12.6)

Lactam inhibitors

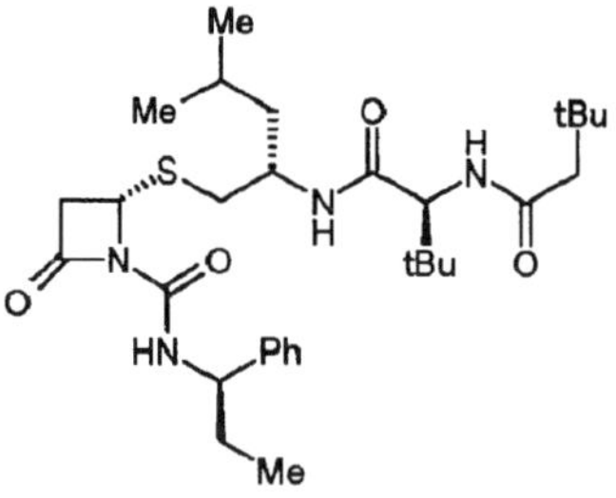

(12.7)

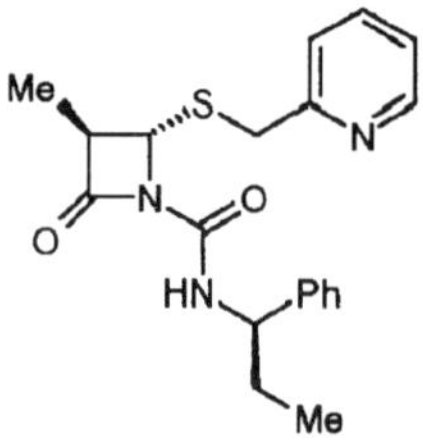

(12.8)

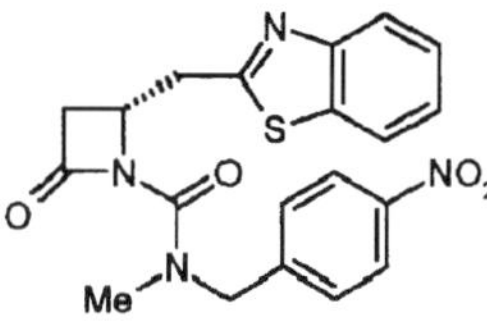

(12.9)

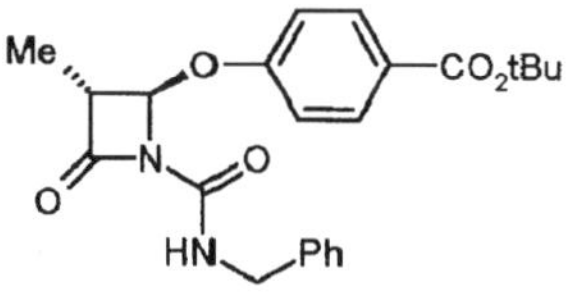

(12.10)

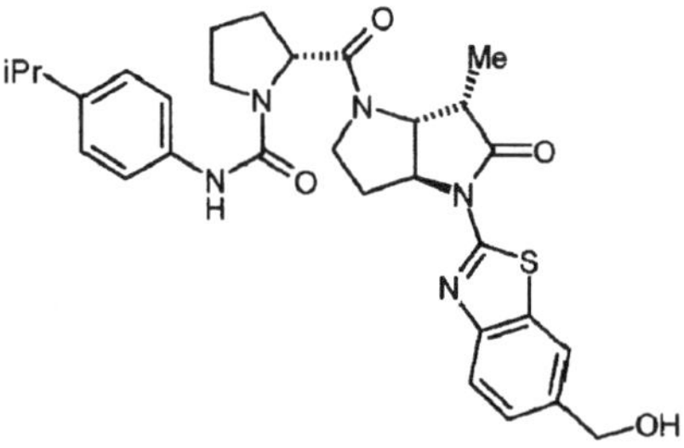

(12.11)

12.3.1.5 Lactams

Two groups have reported monocyclic β-lactam inhibitors of cytomegalovirus protease. A series of substrate based inhibitors incorporating a β-lactam was optimized at structure (**12.7**), which was a potent CMV protease inhibitor with IC_{50} 70 nM (Déziel and Malenfant 1998). Attempts to reduce the peptidic character and molecular weight with a 4-carba or 4-thio substituent resulted in inhibitors with potency down to 1 μM and a range of aqueous stability (Déziel and Malenfant 1998; Yoakim *et al* 1998b). It is suggested that the formation of an acyl-enzyme adduct from these compounds is reversible, and that a "double-hit" mechanism is not involved. Introducing a methyl group as in (**12.8**) afforded increased potency (IC_{50} 350 nM) and stability, but a loss of selectivity against elastase (Déziel and Malenfant 1998). Heterocyclic derivatives such as (**12.9**) had IC_{50} values against CMV protease down to about 1 μM but showed some activity on cell culture (Ogilvie *et al* 1999). A study of 4-oxa β-lactams identified (**12.10**) as the best inhibitor with a modest IC_{50} of 17 μM and, intriguingly, the opposite C3-stereochemistry to (**12.8**) (Borthwick *et al* 1998a); mechanistic studies confirm that acylation of the active site serine occurs (Haley *et al* 1998).

A patent application describes bicyclic γ-lactams as herpesvirus protease inhibitors (Borthwick *et al* 1998b). These compounds have CMV protease IC_{50} values down to 200 nM. (**12.11**) is described as having an IC_{50} of 310 nM, K_i of 42 nM and antiviral activity against CMV with antiviral activity in cell culture.

Peptidomimetics

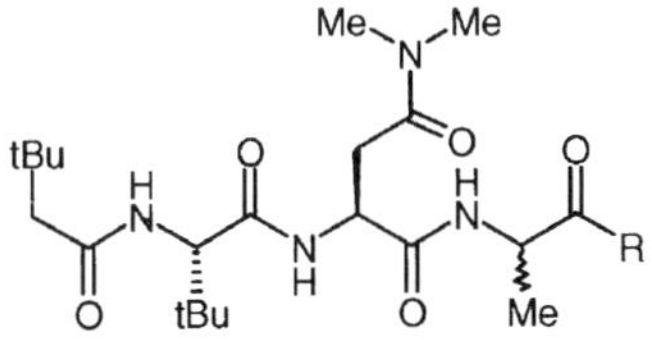

(12.12) R = CF_2CF_3

(12.13) R =

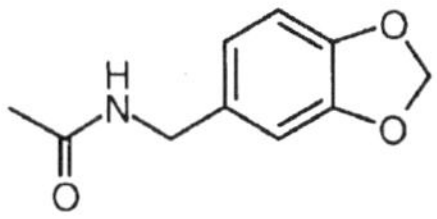

(12.14) R =

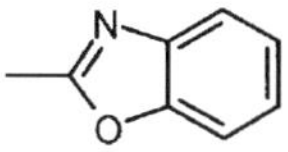

12.3.2 Peptidomimetic inhibitors

Peptidomimetic inhibitors of CMV protease bind in an extended conformation (Tong *et al* 1998). Optimization of the aminoacyl residues led to the acyl tripeptide inhibitors (**12.12–12.14**) (Ogilvie *et al* 1997). Utilization of various activated carbonyl transition state analog motifs afforded potencies down to 100 nM for the fluoroalkyl ketone (**12.12**) and α-ketoamide (**12.13**), and down to 600 nM for the heterocyclic ketone (**12.14**). Whilst the fluoroketone was selective with respect to other serine proteases the heterocyclic ketones inhibited elastase at similar concentrations. The peptidomimetic inhibitors did not show significant antiviral activity in cell culture, probably due to problems of cell penetration and/or degradation.

12.3.3 Other inhibitors

The most active of the remaining inhibitors of the herpes proteases are the benzothiopyran-4-ones (Dhanak *et al* 1998). The lead compound was discovered by high-throughput screening and optimization studies afforded (**12.15**), with an IC_{50} of 60 nM.

The trifluoromethyl ketone inhibitor (**12.16**) inhibits CMV protease with an IC_{50} value of 5.3 μM but its mechanism of action has not been determined (Flynn *et al* 1997). The compound is reported to exhibit weak antiviral activity.

The cobalt complex salcomine (**12.17**) and some analogs inhibited CMV protease with IC_{50} values down to about 1 μM and good selectivity with respect to chymotrypsin family proteases (Watanabe *et al* 1998). Antiviral activity was also reported.

Six natural product inhibitors of the herpes proteases have been described but all are of modest potency. The fungal metabolite Sch 65676 (**12.19**) (Chu *et al* 1996) and the Streptomycete natural product bripiodionen (**12.19**) (Shu *et al* 1997) inhibited CMV protease with IC_{50} values of 24 μM and 30 μM respectively. Two naphthacenequinone

Other inhibitors

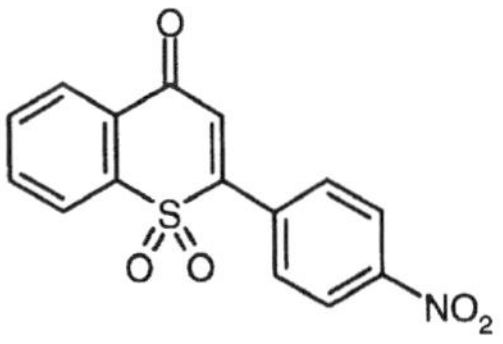

(12.15)

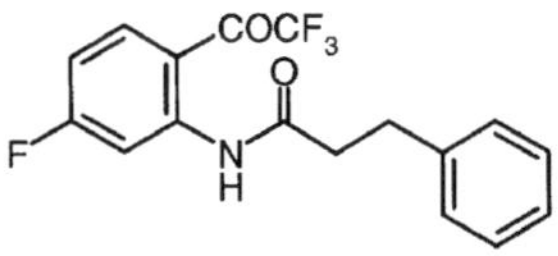

(12.16)

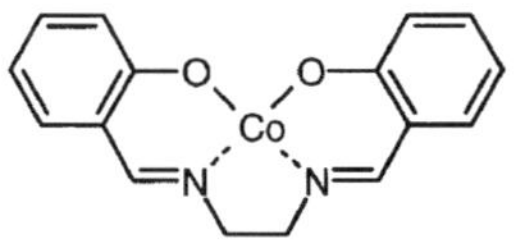

(12.17)

Natural products

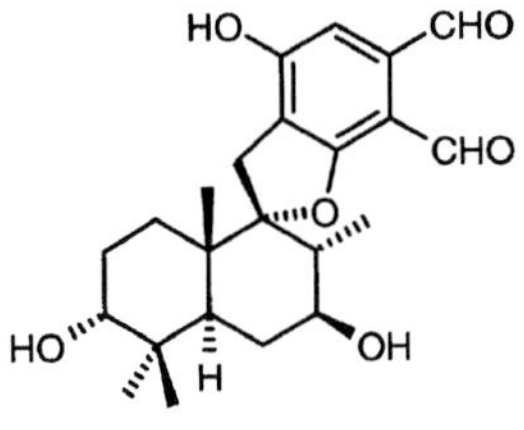

(12.18)

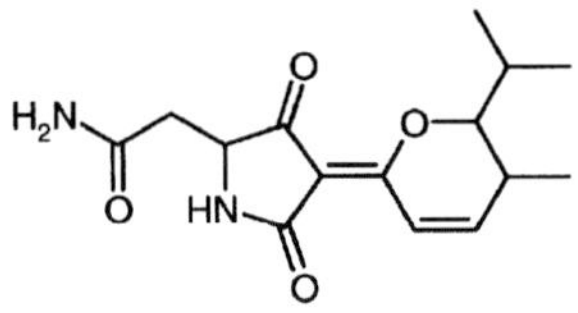

(12.19)

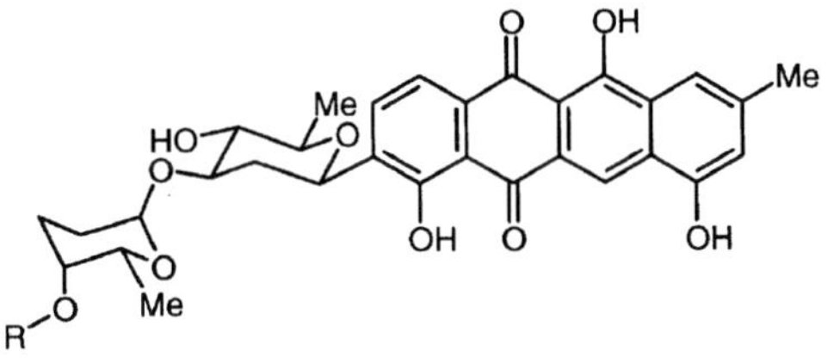

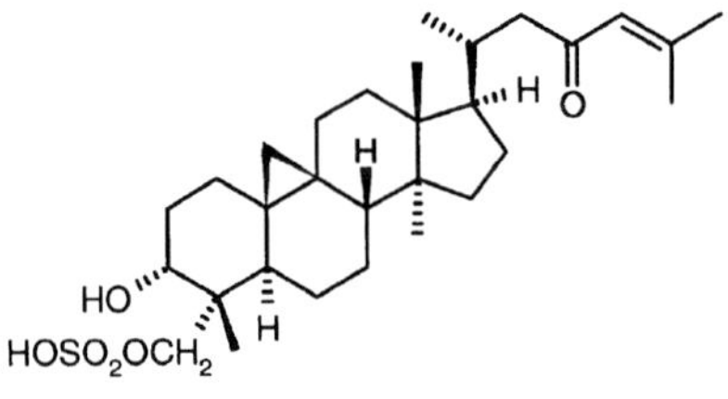

(12.22)

(12.20) R = 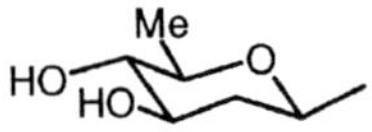 ; **(12.21)** R = H

glycosides, quanolirones I and II (**12.20**) and (**12.21**) were also isolated from *Streptomyces* and shown to have weak activity against the CMV protease, with IC_{50} values of 14 and 35 μM respectively (Qian-Cutrone *et al* 1998). A novel cycloartanol sulfate metabolite (**12.22**) isolated from a green alga was found to inhibit both CMV and VZV proteases with IC_{50} values in the range 4–7 μM (Patil *et al* 1997).

12.4 CELL-BASED ASSAYS

Assays designed to assess the antiviral activity of compounds have been established for the human herpesviruses, with well-characterized assays having been developed for testing against HSV-1, HSV-2, VZV and CMV. The most commonly utilized assays are the plaque reduction assay (PRA) and the yield reduction assay (YRA), both of which measure the inhibition of infectious virus in the presence of increasing concentrations of compound. In broad terms, the PRA measures infectious virus generated following multiple rounds of viral replication, whereas the YRA measures infectious virus generated following a single round of replication (Boyd *et al* 1987). The alphaherpesviruses HSV-1 and HSV-2 have short life-cycles of approximately 18–20 hours. Human CMV, in contrast, has an extended life-cycle of approximately 5 days, although infectious virus may be identified as early as day 3 post-infection. In addition to antiviral assays, compounds may be tested in assays designed to confirm the mechanism of action, which assess the ability of compounds to inhibit protease-directed cleavage of a protein substrate in the context of the infected cell. For both the antiviral and mechanistic cell-based assays the analysis of compound cytotoxicity in the uninfected cell is required for data interpretation. Cytotoxicity is measured by various methods, including for example colorimetric assays based on the reduction of tetrazolium salts (MTT, XTT assays) (Mosmann 1983; Weislow *et al* 1989). Surprisingly, although potent inhibitors of the HSV and CMV proteases have been identified *in vitro*, relatively few have shown activity in cell-based mechanistic and antiviral assays.

12.4.1 HSV-2

A series of 5-methylthieno[2,3-*d*]oxazinone based inhibitors were identified which demonstrated potent submicromolar activity against the HSV-2, VZV and CMV protease *in vitro*, utilizing quenched fluorescence assays. These compounds were tested for mechanistic inhibition in HSV-2 infected Vero cells (African green monkey kidney; ATCC CCL-81) by pulse-chase assay, to follow the conversion of full-length to proteolytically-cleaved product within a defined period of time. Using the pulse-chase assay, thieno[2,3-*d*]oxazinones such as (**12.3**), with submicromolar activity against the purified protease in *in vitro* assays, demonstrated inhibition of viral protease processing in the context of HSV-2-infected cells, with good separation from cytotoxic effects (Jarvest *et al* 1999). Compounds were also tested for antiviral activity by YRA in which Vero cells were infected with HSV-2 in the presence of increasing concentrations of compound, and the amount of infectious virus was quantitated by plaque assay on Vero cells. However, little separation between cytotoxicity and antiviral activity was observed in these assays (unpublished data). Thus, while these compounds demonstrated protease-specific inhibition following HSV-2 infection, no reduction in infectious virus could be documented.

12.4.2 CMV

Cell-based activities of CMV protease inhibitors have been identified for a number of different structural classes. β-lactams and β-lactam derivatives were shown to exhibit potent inhibition of protease cleavage of a M-site based peptide substrate by *in vitro* quenched fluorescence assays (Yoakim *et al* 1998a). One of the β-lactams was shown to have weak antiviral activity against CMV by PRA (EC_{50} 78 μM), with some separation from cytotoxicity ($TC_{50} > 219$ μM). Additional β-lactam derivatives have also been described with weak antiviral activity (Déziel and Malenfant 1998; Yoakim *et al* 1998b). Recently, β-lactams such as (**12.9**) were described which demonstrated antiviral activity down to 11 μM, with some separation from cytotoxicity ($TC_{50} > 25$ μM), as well as inhibition of protease processing in cell-based transfection assays (Ogilvie *et al* 1999). Additionally, potent bicyclic γ-lactam inhibitors of CMV protease, such as (**12.11**), have been described in a patent application, with the best antiviral activity by plaque reduction assay demonstrating an EC_{50} of 2.4 μM, although cytotoxicity was not measured in the same cell line (Borthwick *et al* 1998b). Overall, however, the β-lactams and derivatives were limited in cell-based assays by cytotoxicity and instability in culture medium.

Salcomine, (**12.17**), and its derivatives were identified as potent inhibitors of CMV protease in *in vitro* assays (Watanabe *et al* 1998), using HPLC to identify cleavage products from a M-site based substrate. When tested for antiviral activity salcomine and a derivative, RD3–0174, were shown by PRA to effectively inhibit the generation of infectious virus against a laboratory strain and clinical strains of CMV, with EC_{50}s of approximately 2–3 μM. The antiviral activity of these compounds was separated from cytotoxicity, with selective indices (TC_{50}/EC_{50}) of 13 and 15 for salcomine and RD3-0174, respectively, against the laboratory CMV strain AD169.

Benzoxazinones such as (**12.1**), showing submicromolar or micromolar inhibition *in vitro*, have been identified as having weak antiviral activity in CMV infected cells (Abood *et al* 1997). Antiviral activity was determined against a recombinant human CMV strain expressing the β-galactosidase gene in a rapid (2 day) 96-well assay, with cytotoxicity assessed by comparison of cell number in the absence of compound. Moderate to weak antiviral activity was identified, 23–63 μM, in the absence of apparent cytotoxicity ($TC_{50} > 100$ μM). Additional compounds were identified which were limited by cytotoxic effects at concentrations below 100 μM.

In summary, many compounds in multiple structural classes have been identified which are potent inhibitors of the HSV or CMV proteases in *in vitro* assays, while few have proven to be effective in cell-based antiviral or mechanistic assays. The problems associated with these compounds appear to be primarily due to compound cytotoxicity as well as instability under cell culture conditions. Further progress in this area will be dependent on addressing these issues, as well as the potential need for increased potency against these protease targets.

12.5 PROSPECTS

The herpesvirus encoded protease is an essential enzyme and a potentially attractive target for new antiviral agents. However, factors such as the shallow active site geometry and extended substrate interaction area have raised considerable challenges to the discovery of small molecule inhibitors with drug-like properties. Although a number of inhibitor classes have now been reported with enzyme potency at the sub-micromolar level, very few reports of really potent inhibitors ($IC_{50} < 50$ nM) have appeared. In addition, potent ($EC_{50} < 5$ μM) antiviral activity which can be separated from compound cytotoxic effects in cell culture has yet to be described. Increased stability in biological milieu, improved cell penetration, and/or reduced cytotoxicity are properties that need to be achieved in new inhibitors. As these parameters are addressed, the real therapeutic potential of herpes protease inhibitors can be more fully assessed.

NOTE

The most recent reports of inhibitors of the herpes proteases have further developed the established inhibitor classes and in one case applied a new inhibition concept to this protease family.

Naphthoquinone inactivators of CMV protease have been described which irreversibly alkylate residue Cys202 (Ertl *et al* 1999). However, protease inhibition correlates with Michael acceptor reactivity and the compounds react readily with simple thiols such as glutathione, making them of little further interest.

Further studies in the oxazinone series resulted in highly potent hydroxylamine-containing CMV protease inhibitors with IC_{50} values down to 14 nM (Smith *et al* 1999). Elucidation of the mechanism of the hydroxylamine analogs indicated standard acylation of Ser132 when the oxazinone moiety was present, but in the absence of the oxazinone, inhibition occurred by covalent addition to Cys138.

The initial structure activity relationships of the γ-lactam inhibitor series has been reported (Borthwick *et al* 2000a). These compounds are covalent, reversible CMV protease inhibitors, consistent with Ser132 acylation. The examples described did not show selectivity with respect to other serine protease such as elastase but distinctive structure activity relationships for CMV protease suggested that selectivity is attainable. A second patent application extends the bicyclic γ-lactam series and describes apparently interesting whole cell antiherpesvirus activity for some examples (Borthwick *et al* 2000b).

A new class of natural product inhibitors has been reported, but as with those previously described the potency is relatively modest, the best compound having an IC_{50} value against CMV protease of 11 μM (Guo *et al* 2000).

A new approach to herpes protease inhibition is the application of the zinc ion mediated inhibitor binding concept to CMV protease (Dhanak *et al* 2000). Although inhibition was shown to be increased in the presence of zinc ions, the enhancement was fairly modest and the best inhibitors from this approach only achieved IC_{50} values of about 5 μM.

REFERENCES

Abood, N.A., Schretzman, L.A., Flynn, D.L., Houseman, K.A., Wittwer, A.J., Dilworth, V.M. *et al* (1997) Inhibition of human cytomegalovirus protease by benzoxazinones and evidence of antiviral activity in cell culture. *Bioorganic & Medicinal Chemistry Letters*, **7**, 2105–2108.

Alford, C.A., Stagno, S., Pass, R.F. and Britt, W.J. (1990) Congenital and perinatal cytomegalovirus infections. *Reviews of Infectious Diseases*, **12** (Suppl. 7), S745–S753.

Baum, E.Z., Siegel, M.M., Bebernitz, G.A., Hulmes, J.D., Sridharan, L., Sun, L. *et al* (1996a) Inhibition of human cytomegalovirus UL80 protease by specific intramolecular disulfide bond formation. *Biochemistry*, **35**, 5838–5846.

Baum, E.Z., Ding, W.-D., Siegel, M.M., Hulmes, J., Bebernitz, G.A., Sridharan, L. *et al* (1996b) Flavins inhibit human cytomegalovirus UL80 protease via disulfide bond formation. *Biochemistry*, **35**, 5847–5855.

Bonneau, P.R., Grand-Maître, C., Greenwood, D.J., Lagacé, L., LaPlante, S.R., Massariol, M.-J. *et al* (1997) Evidence of a conformational change in the human cytomegalovirus protease upon binding of peptidyl-activated carbonyl inhibitors. *Biochemistry*, **36**, 12644–12652.

Borthwick, A.D., Weingarten, H.M., Haley, T.M., Tomaszewski, M., Wang, W., Hu, Z. *et al* (1998a) Design and synthesis of monocyclic β-lactams as mechanism-based inhibitors of human cytomegalovirus protease. *Bioorganic & Medicinal Chemistry Letters*, **8**, 365–370.

Borthwick, A.D., Davies, D.E., Exall, A.M., Jackson, D.L., Mason, A.M., Pennell, A.M.K. *et al* (1998b) WO9843975, (Glaxo Group Ltd.).

Borthwick, A.D., Angier, S.J., Crame, A.J., Exall, A.M., Haley, T.M., Hart, G.J. *et al.* (2000a) Design and synthesis of pyrrolidine-5,5-*trans*-lactams (5-oxo-hexahydro-pyrrolo[3,2-*b*]pyrroles) as novel mechanism-based inhibitors of human cytomegalovirus protease. 1. The α-Methyl-*trans*-lactam template. *Journal of Medicinal Chemistry*, **43**, 4452–4464.

Borthwick, A.D., Davies, D.E., Exall, A.M., Leahy, J.H., Rahim, G.S., Shah, P. *et al.* (2000b) WO200018770, (Glaxo Group Ltd.).

Boyd, M.R., Bacon, T.H., Sutton, D. and Cole, M. (1987) Antiherpesvirus activity of 9-(4-hydroxy-3-hydroxy methylbut-1-yl)guanine (BRL 39123) in cell culture. *Antimicrobial Agents & Chemotherapy*, **31**, 1238–1242.

Campadelli-Fiume, G., Mirandola, P. and Menotti, L. (1999) Human herpesvirus 6: an emerging pathogen. *Emerging Infectious Diseases*, **5**, 353–366.

Chen, P., Tsuge, H., Almassy, R.J., Gribskov, C.L., Katoh, S., Vanderpool, D.L. *et al* (1996) Structure of the human cytomegalovirus protease catalytic domain reveals a novel serine protease fold and catalytic triad. *Cell*, **86**, 835–843.

Christophers, J., Clayton, J., Craske, J., Ward, R., Collins, P., Trowbridge, M. *et al.* (1998) Survey of resistance of herpes simplex virus to acyclovir in northwest England. *Antimicrobial Agents & Chemotherapy*, **42**, 868–872.

Chu, M., Mierzwa, R., Truumees, I., King, A., Patel, M., Pichardo, J. *et al* (1996) Sch 65676: A novel fungal metabolite with the inhibitory activity against the cytomegalovirus protease. *Tetrahedron Letters*, **37**, 3943–3946.

Dabrowski, C.E., Ashman, S.M., Fernandez, A.V., Gorczyca, M., Lavery, P., Parratt, M.J. *et al* (1999) Inhibition of herpesvirus proteases by novel thieno[2,3-*d*]oxazinones: demonstration of inhibition of virus protein processing in cell culture. *Journal of Biological Chemistry*, (submitted).

Dabrowski, C.E., Qiu, X. and Abdel-Meguid, S.S. (2000) The Human Herpesvirus Proteases. In *Proteases as Targets for Therapy*. Berlin: Springer-Verlag (in press).

Deckman, I.C., Hagen, M. and McCann III, P.J. (1992) Herpes simplex virus type 1 protease expressed in *Escherichia coli* exhibits autoprocessing and specific cleavage of the ICP35 assembly protein. *Journal of Virology*, **66**, 7362–7367.

Déziel, R. and Malenfant, E. (1998) Inhibition of human cytomegalovirus protease N_0 with monocyclic γ-lactams. *Bioorganic & Medicinal Chemistry Letters*, **8**, 1437–1442.

Dhanak, D., Keenan, R.M., Burton, G., Kaura, A., Darcy, M.G., Shah, D.H. *et al* (1998) Benzothiopyran-4-one based reversible inhibitors of the human cytomegalovirus (HCMV) protease. *Bioorganic & Medicinal Chemistry Letters*, **8**, 3677–3682.

Dhanak, D., Burton, G., Christmann, L.T., Darcy, M.G., Elrod, K.C., Kaura, A. *et al.* (2000) Metal mediated protease inhibition: design and synthesis of inhibitors of the human cytomegalovirus (hCMV) protease. *Bioorganic & Medicinal Chemistry Letters*, **10**, 2279–2282.

Ertl, P., Cooper, D., Allen, G., and Slater, M.J. (1999) 2-Chloro-3-substituted-1,4-naphthoquinone inactivators of human cytomegalovirus protease. *Bioorganic & Medicinal Chemistry Letters*, **9**, 2863–2866.

Flynn, D.L., Becker, D.P., Hippenmeyer, P., Hockerman, S., Houseman, K., Moormann, A. *et al* (1996) Selective inhibitors of herpesvirus serine protease assemblin. Abstract 241, 211th American Chemical Society National Meeting, New Orleans.

Flynn, D.L., Abood, N.A. and Holwerda, B.C. (1997) Recent advances in antiviral research: identification of inhibitors of the herpesvirus proteases. *Current Opinion in Chemical Biology*, **1**, 190–196.

Fowler, K.B., McCollister, F.P., Dahle, A.J., Boppana, S., Britt, W.J. and Pass, R.F. (1997) Progressive and fluctuating sensorineural hearing loss in children with asymptomatic congenital cytomegalovirus infection. *Journal of Pediatrics*, **130**, 624–630.

Gao, M., Matusick-Kumar, L., Hurlburt, W., DiTusa, S.F., Newcomb, W.W., Brown, J.C. *et al* (1994) The protease of herpes simplex virus type 1 is essential for functional capsid formation and viral growth. *Journal of Virology*, **68**, 3702–3712.

Gibson, W. and Hall, M.R.T. (1997) Assemblin, an essential herpesvirus proteinase. *Drug Design and Discovery*, **15**, 39–47.

Greenblatt, R.M. (1998) Kaposi's sarcoma and human herpesvirus-8. *Infectious Disease Clinics of North America*, **12**, 63–82.

Guo, B., Dai, J.-R., Ng, S., Huang, Y., Leong, C., Ong, W. *et al.* (2000) Cytonic acids A and B: novel tridepside inhibitors of hCMV protease from the endophytic fungus *Cytonaema* species. *Journal of Natural Products*, **63**, 602–604.

Haley, T.M., Angier, S.J., Borthwick, A.D., Montgomery, D.S., Purvis, I.J., Smart, D.H. *et al* (1998) Investigation of the covalent modification of the catalytic triad of human cytomegalovirus protease by pseudo-reversible beta-lactam inhibitors and a peptide chloromethylketone. *Journal of Mass Spectrometry*, **33**, 1246–1255.

Holwerda, B.C. (1997) Herpesvirus proteases: targets for novel antiviral drugs. *Antiviral Research*, **35**, 1–21.

Hoog, S.S., Smith, W.W., Qiu, X., Janson, C.A., Hellmig, B., McQueney, M.S. *et al* (1997) Active site cavity of herpesvirus proteases revealed by the crystal structure of herpes simplex virus protease/inhibitor complex. *Biochemistry*, **36**, 14023–14029.

Jarvest, R.L., Parratt, M.J., Debouck, C.M., Gorniak J.G., Jennings, L.J., Serafinowska, H.T. *et al* (1996) Inhibition of HSV-1 protease by benzoxazinones. *Bioorganic & Medicinal Chemistry Letters*, **6**, 2463–2466.

Jarvest, R.L., Connor, S.C., Gorniak, J.G., Jennings, L.J., Serafinowska, H.T. and West, A. (1997) Potent selective thienoxazinone inhibitors of herpes proteases. *Bioorganic & Medicinal Chemistry Letters*, **7**, 1733–1738.

Jarvest, R.L., Pinto, I.L., Ashman, S.M., Dabrowski, C.E., Fernandez, A.V., Jennings, L.J. *et al* (1999) Inhibition of herpes proteases and antiviral activity of 2-substituted thieno[2,3-*d*]oxazinones. *Bioorganic & Medicinal Chemistry Letters*, **9**, 443–448.

LaPlante, S.R., Aubry, N., Bonneau, P.R., Cameron, D.R., Lagace, L., Massariol, M.J. *et al* (1998) Human cytomegalovirus protease complexes its substrate recognition sequences in an extended peptide conformation. *Biochemistry*, **37**, 9793–9801.

Liebowitz, D. and Kieff, E. (1993) Epstein-Barr virus. In *The Human Herpesviruses*, edited by B. Roizman, R.J. Whitley and C. Lopez, pp. 107–172. New York: Raven Press, Ltd.

Liu, F. and Roizman, B. (1991a) The herpes simplex virus 1 gene encoding a protease also contains within its coding domain the gene encoding the more abundant substrate. *Journal of Virology*, **65**, 5149–5156.

Liu, F. and Roizman, B. (1991b) The promoter, transcriptional unit, and coding sequence of herpes simplex virus 1 family 35 proteins are contained within and in frame with the UL26 open reading frame. *Journal of Virology*, **65**, 206–212.

Liu, F. and Roizman, B. (1992) Differentiation of multiple domains in the herpes simplex virus 1 protease encoded by the UL26 gene. *Proceedings of the National Academy of Sciences of the U.S.A.*, **89**, 2076–2080.

Matusick-Kumar, L., Newcomb, W.W., Brown, J.C., McCann III, P.J., Hurlburt, W., Weinheimer, S.P. *et al* (1995) The C-terminal 25 amino acids of the protease and its substrate ICP35 of herpes simplex virus type 1 are involved in the formation of sealed capsids. *Journal of Virology*, **69**, 4347–4356.

McCrary, M.L., Severson, J. and Tyring, S.K. (1999) Varicella zoster virus. *Journal of the American Academy of Dermatology*, **41**, 1–14.

McGeoch, D.J., Dalrymple, M.A., Davison, A.J., Dolan, A., Frame, M.C., McNab, D. *et al* (1988) The complete DNA sequence of the long unique region in the genome of herpes simplex virus type 1. *Journal of General Virology*, **69**, 1531–1574.

Moore, P.S., Gao, S.-J., Dominquez, G., Cesarman, E., Lungu, O., Knowles, D.M. *et al* (1996) Primary characterization of a herpesvirus agent associated with Kaposi's sarcoma. *Journal of Virology*, **70**, 549–558.

Mosmann, T. (1983) Rapid colorimetric assay for cellular growth and survival application to proliferation and cytotoxicity assays. *Journal of Immunological Methods*, **65**, 55–63.

Newcomb, W.W., Homa, F.L., Thomsen, D.R., Booy, F.P., Trus, B.L., Steven, A.C. *et al* (1996) Assembly of the herpes simplex virus capsid: characterization of intermediates observed during cell-free capsid formation. *Journal of Molecular Biology*, **263**, 432–446.

Ogilvie, W., Bailey, M., Poupart, M.-A., Abraham, A., Bhavsar, A., Bonneau, P. *et al* (1997) Peptidomimetic inhibitors of the human cytomegalovirus protease. *Journal of Medicinal Chemistry*, **40**, 4113–4135.

Ogilvie, W.W., Yoakin, C., Dô, F., Haché, B., Lagacé, L., Naud, J. *et al* (1999) Synthesis and antiviral activity of monobactams inhibiting the human cytomegalovirus protease. *Bioorganic & Medicinal Chemistry*, **7**, 1521–1531.

Patil, A.D., Freyer, A.J., Killmer, L., Breen, A. and Johnson, R.K. (1997) A new cycloartanol sulfate from the green alga *Tuemoya* sp.: an inhibitor of VZV protease. *Natural Product Letters*, **9**, 209–215.

Perry, C.M. and Balfour, J.A.B. (1999) Fomivirsen. *Drugs*, **57**, 375–380.

Pinto, I.L., West, A., Debouck, C.M., DiLella, A.G., Gorniak J.G., O'Donnell, K.C. *et al* (1996) Novel, selective mechanism-based inhibitors of the herpes proteases. *Bioorganic & Medicinal Chemistry Letters*, **6**, 2467–2472.

Pinto, I.L., Jarvest, R.L., Clarke, B., Dabrowski, C.E., Fenwick, A., Gorczyca, M.M. *et al* (1999) Inhibition of human cytomegalovirus protease by enedione derivatives of thieno[2,3-*d*]oxazinones through a novel dual acylation/alkylation mechanism. *Bioorganic & Medicinal Chemistry Letters*, **9**, 449–452.

Preston, V.G., Coates, J.A.V. and Rixon, F.J. (1983) Identification and characterization of a herpes simplex virus gene product required for encapsidation of virus DNA. *Journal of Virology*, **45**, 1056–1064.

Qian-Cutrone, J., Kolb, J.M., McBrien, K., Huang, S., Gustavson, D., Lowe, S.E. and Manly, S.P. (1998) Quanolirones I and II, two new human cytomegalovirus protease inhibitors produced by *Streptomyces* sp. WC76535. *Journal of Natural Products*, **61**, 1379–1382.

Qiu, X., Culp, J.S., DiLella, A.G., Hellmig, B., Hoog, S.S., Janson, C.A. *et al* (1996) Unique fold and active site in cytomegalovirus protease. *Nature*, **383**, 275–279.

Qiu, X., Janson, C.A., Culp, J.S., Richardson, S.B., Debouck, C., Smith, W.W. *et al* (1997) Crystal structure of varicella-zoster virus protease. *Proceedings of the National Academy of Sciences of the U.S.A.*, **94**, 2874–2879.

Rawlinson, W.D. (1999) Diagnosis of human cytomegalovirus infection and disease. *Pathology*, **31**, 109–115.

Rettig, M.B., Ma, H.J., Vescio, R.A., Pold, M., Schiller, G., Belson, D. *et al* (1997) Kaposi's sarcoma-associated herpesvirus infection of bone marrow dendritic cells from multiple myeloma patients. *Science*, **276**, 1851–1854.

Robertson, B.J., McCann, III, P.J., Matusick-Kumar, L., Newcomb, W.W., Brown, J.C., Colonno, R.J. *et al* (1996) Separate functional domains of the herpes simplex virus type 1 protease: evidence for cleavage inside capsids. *Journal of Virology*, **70**, 4317–4328.

Roizman, B. and Sears, A.E. (1993) Herpes simplex viruses and their replication. In *The Human Herpesviruses*, edited by B. Roizman, R.J. Whitley and C. Lopez, pp. 11–68. New York: Raven Press, Ltd.

Shieh, H.-J., Kurumbail, R.G., Stevens, A.M., Stegeman, R.A., Sturman, E.J., Pak, J.Y. *et al* (1996) Three-dimensional structure of human cytomegalovirus protease. *Nature*, **383**, 279–282.

Shu, Y.-Z., Ye, Q., Kolb, J.M., Huang, S., Veitch, J.A., Lowe, S.E. and Manly, S.P. (1997) Bripiodionen, a new inhibitor of human cytomegalovirus protease from *Streptomyces* sp. WC96599. *Journal of Natural Products*, **60**, 529–532.

Smith, I.L., Cherrington, J.M., Jiles, R.E., Fuller, M.D., Freeman, W.R. and Spector, S.A. (1997) High-level resistance of cytomegalovirus to ganciclovir is associated with alterations in both the UL97 and DNA polymerase genes. *Journal of Infectious Diseases*, **176**, 69–77.

Smith, D.G., Gribble, A.D., Haigh, D., Ife, R.J., Lavery, P., Skett P. *et al.* (1999) The inhibition of human cytomegalovirus (hCMV) protease by hydroxylamine derivatives. *Bioorganic & Medicinal Chemistry Letters*, **9**, 3137–3142.

Tong, L., Qian, C., Massariol, M.-J., Bonneau, P.R., Cordingley, M.G. and Lagacé, L. (1996) A new serine-protease fold revealed by the crystal structure of human cytomegalovirus protease. *Nature*, **383**, 272–275.

Tong, L., Qian, C., Massariol, M.-J., Déziel, R., Yoakim, C. and Lagacé, L. (1998) Conserved mode of peptidomimetic inhibition and substrate recognition of human cytomegalovirus protease. *Nature Structural Biology*, **5**, 819–826.

Trus, B.L., Booy, F.P., Newcomb, W.W., Brown, J.C., Homa, F.L., Thomsen, D.R. and Steven, A.C. (1996) The herpes simplex virus procapsid – structure, conformational changes upon maturation, and roles of the triplex proteins VP19C and VP23 in assembly. *Journal of Molecular Biology*, **263**, 447–462.

Vogel, J.-U., Scholz, M. and Cinatl, Jr. J. (1997) Treatment of cytomegalovirus diseases. *Intervirology*, **40**, 357–367.

Walmsley, S. and Tseng, A. (1999) Comparative tolerability of therapies for cytomegalovirus retinitis. *Drug Safety*, **21**, 203–224.

Watanabe, S., Konno, K., Shigeta, S. and Yokata, T. (1998) Inhibition of human cytomegalovirus proteinase by salcomine derivatives. *Antiviral Chemistry & Chemotherapy*, **9**, 269–274.

Weislow, O.S., Kiser, R., Fine, D.L., Bader, J., Shoemaker, R.H. and Boyd, M.R. (1989) New soluble-formazan assay for HIV-1 cytopathic effects: application to high-flux screening of synthetic and natural products for AIDS-antiviral activity. *Journal of the National Cancer Institute*, **81**, 577–586.

Whitley, R.J. and Gnann, Jr. J.W. (1993) The epidemiology and clinical manifestations of herpes simplex virus infections. In *The Human Herpesviruses*, edited by B. Roizman, R.J. Whitley and C. Lopez, pp. 69–105. New York: Raven Press, Ltd.

Whitley, R.J., Cloud, G., Gruber, W., Storch, G.A., Demmler, G.J., Jacobs, R.F. *et al* (1997) Ganciclovir treatment of symptomatic congenital cytomegalovirus infection: results of a Phase II study. *Journal of Infectious Diseases*, **175**, 1080–1086.

Wutzler, P. (1997) Antiviral therapy of herpes simplex and varicella-zoster virus infections. *Intervirology*, **40**, 343–356.

Yoakim, C., Ogilvie, W.W., Cameron, D.R., Chabot, C., Grand-Maitre, C., Guse, I. *et al* (1998a) Potent β-lactam inhibitors of human cytomegalovirus protease. *Antiviral Chemistry & Chemotherapy*, **9**, 379–387.

Yoakim, C., Ogilvie, W.W., Cameron, D.R., Chabot, C., Guse, I., Haché, B. *et al* (1998b) β-Lactam derivatives as inhibitors of human cytomegalovirus protease. *Journal of Medicinal Chemistry*, **41**, 2882–2891.

Chapter 13

Human rhinovirus 3C proteinase inhibitors

Peter S. Dragovich and Stephen E. Webber

Human rhinoviruses (HRVs) are a major cause of the common cold. For most people, contracting a rhinoviral infection merely brings about annoying discomfort, but for many young children and individuals suffering from obstructive respiratory diseases serious complications can arise. Ethical treatments for this viral malady have been sought for many years, but to date no clinically proven antirhinoviral agents exist. Recently, significant attention has been paid to the human rhinovirus 3C proteinase (3CP), a cysteine-containing enzyme essential for the life cycle of the virus. Advancements in cloning and expression techniques have made it possible to study this critical proteinase in more detail, while crystallographic data has provided valuable information with regard to structure and mechanism. The design, discovery, and development of HRV 3CP inhibitors, some of which have the potential to function as novel antirhinoviral agents, are comprehensively reviewed in this chapter.

13.1 INTRODUCTION†

13.1.1 Background

The human rhinoviruses (HRVs) are small, non-enveloped, RNA viruses belonging to the picornavirus family and are the single most significant cause of the common cold (Couch 1996; Rueckert 1996). These viruses, of which there are more than 100 known serotypes (Hamparian *et al* 1987), typically infect the upper respiratory tract in humans and specifically target nasal epithelial cells. In spite of considerable efforts by a number of research groups (Couch 1996; Turner *et al* 1999 and references therein), to date no effective antirhinoviral therapy has been approved for clinical use. In addition, the large number of rhinovirus serotypes makes development of a vaccine unlikely. Recently, however, the identification of chemical entities which inhibit a critical viral proteinase has provided new possibilities for obtaining novel antirhinoviral agents. This critical viral enzyme, the human rhinovirus 3C proteinase, is described below along with a summary of efforts to develop inhibitors of its proteolytic activity. Due to space limitations, this review will focus only on the HRV 3C proteinase and not the related viral enzymes of hepatitis A and poliovirus.

†The following chemical abbreviations are utilized throughout: Ac = acetyl, Boc = *tert*-butyloxycarbonyl, Cbz = benzyloxycarbonyl, Met(SO_2) = methionine sulfone, HPhe = homophenylalanine. In addition, all amino acids are of L (natural)-configuration.

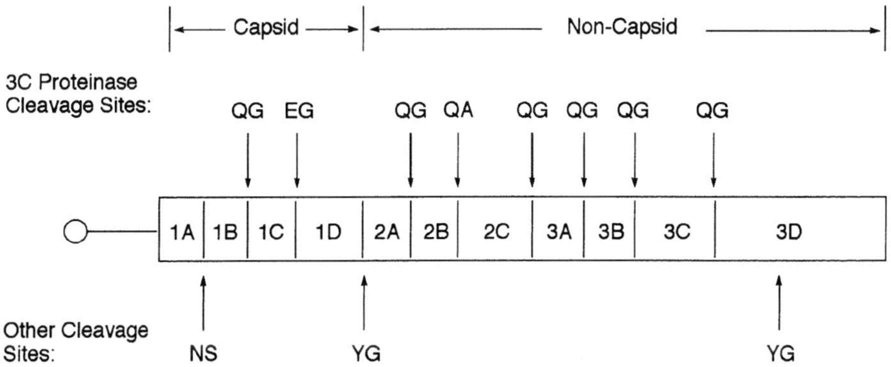

Figure 13.1 Gene organization and proteolytic processing in rhinovirus. Eight of the eleven proteolytic processing sites are cleaved by 3C proteinase, principally at glutamine/glycine (QG) junctions. The virally encoded 2A proteinase processes the 1D/2A site and also cleaves within the 3D protein precursor to form mature 3D RNA polymerase. The 1A/1B cleavage occurs through an intramolecular event. Reproduced with permission from Kati *et al* (1999). Copyright 1999 Academic Press.

13.1.2 3C proteinase and the HRV replication cycle

All HRV serotypes require attachment to a cell surface receptor to initiate infection, and most bind to the intracellular adhesion molecule 1 (ICAM-1) (Couch 1996). The remainder, with the possible exception of serotype 87, interact with a less-well-characterized 120-kDa cell surface protein. Upon gaining entry to a host cell and subsequent uncoating, the positive-strand viral RNA genome is directly translated by the host cell into a large polyprotein. This polyprotein undergoes several co- and post-translational cleavage events that are effected by three virus-specific proteinase activities to produce the structural and enzymatic proteins required for subsequent viral replication (Palmenberg 1987; Kräusslich and Wimmer 1988; Porter 1993). Specifically, the 2A proteinase intramolecularly cleaves the 1D/2A junction to separate the structural and non-structural protein precursors (1A through 1D and 2A through 3D, respectively) (Figure 13.1). This enzyme also effects an intermolecular peptide cleavage event within the 3D protein. Along with the 2A processing described above, eight additional intermolecular viral polyprotein cleavages are performed by the 3C proteinase (3CP) or its 3C–3D precursor (Figure 13.1). Thus, 3CP is of critical importance for the completion of the rhinovirus infection cycle, and this importance led to the extensive biochemical characterization of the enzyme.

13.1.3 Biochemistry of HRV 3C proteinase

The cloning and expression of HRV-14 3CP in *Escherichia coli* was accomplished by several research groups and provided sufficient protein quantities for subsequent biological studies (Cordingley *et al* 1989; Knott *et al* 1989; Leong *et al* 1992; Birch *et al* 1995). More recently, similar productions of HRV-1A 3CP (Aschauer *et al* 1991), HRV-1B 3CP (Kati *et al* 1999) and HRV-2, 16, and 89 3CPs (Webber *et al* 1996) were also reported. Analysis of the known 3CP cleavage sites within the viral polypeptide indicated that the enzyme primarily cleaves between glutamine and glycine peptidyl residues (Figure 13.1). Accordingly, the ability of HRV-14 3CP to catalyze the cleavage of several synthetic peptide substrates containing a Gln-Gly sequence was extensively studied (Cordingley *et al* 1989, 1990; Long *et al* 1989; Orr *et al* 1989). Some of these experiments characterized the 3CP enzyme as a cysteine proteinase based on its inactivation by several known cysteine proteinase inhibitors (Orr *et al* 1989). In addition, site-directed mutagenesis of the suspected active site cysteine residue to serine afforded inactive enzyme (Orr *et al* 1989; Cheah *et al* 1990). Similar experiments indicated that several other 3CP residues participated in substrate catalysis as well (Cheah *et al* 1990). An RNA-binding activity, independent of the above proteolytic function, was also associated with 3CP (Leong *et al* 1993), but is not believed to be relevant to the discussion of 3CP inhibitors provided below. A deamidated isoform of the enzyme was recently described as well (Cox *et al* 1999) and characterization of the 3C–3D proteinase precursor was also reported (Davis *et al* 1997). Importantly, sequence analysis of 3CPs from several rhinovirus serotypes (Lee *et al* 1994, and references therein) indicated a minimal homology with other known cysteine proteinases and prevalent mammalian enzymes. These studies suggested that 3CP belongs to a new class of structurally unique cysteine proteinases and this suggestion was confirmed by subsequent crystallographic analysis.

13.1.4 Crystallographic studies of HRV 3C proteinase

A 2.3 Å X-ray crystal structure of HRV-14 3CP was reported by Matthews and coworkers (1994) and revealed that the enzyme was indeed a structurally novel cysteine proteinase. In agreement with the above mutagenesis studies, the catalytically important residues Cys-146, His-40, and Glu-71 (HRV-14 numbering) were observed to form a linked cluster with an overall geometry similar to that of the Ser-His-Asp triad found in members of the trypsin-like serine proteinase family. Other portions of the 3CP structure, including several double β-barrel polypeptide folding motifs, were also noted to resemble known trypsin-like serine proteinases. A model of a putative octapeptide 3CP substrate bound to the proteinase was also constructed and proved to be useful in the design and evaluation of many of the 3CP inhibitors described below.

13.2 HRV 3C PROTEINASE INHIBITORS AS ANTIRHINOVIRAL AGENTS

13.2.1 Background

Previous approaches toward the identification of antirhinoviral therapeutics include the use of interferon (Couch 1996, and references therein), the disruption of virus-host cell interactions with a soluble form of the ICAM-1 receptor (Turner *et al* 1999, and references therein), the examination of capsid-binding antipicornaviral compounds which inhibit cellular attachment and/or viral uncoating (McKinlay *et al* 1992; Oren *et al* 1996; Fromtling and Castañer 1997, and references therein), and the use of other miscellaneous agents (Carrasco 1994; Hamdouchi *et al* 1999, and references therein). To date, however, none of these approaches has afforded a marketed drug for the treatment of rhinovirus infections. Due to the importance of 3CP in the viral replication cycle and the suspected conservation of its active site residues among all rhinovirus serotypes (Gorbalenya *et al* 1986; Lee *et al* 1994, and references therein; Matthews *et al* 1999), inhibitors of this enzyme were also sought which might function as broad-spectrum antirhinoviral agents (Wang 1998a, 1999).

Several biological assays were utilized to evaluate the various chemical entities described below as both 3CP inhibitors and antirhinoviral agents. Descriptions of most of these assays are provided in the literature reports of the inhibitors themselves, although several have been independently disclosed (Hopkins *et al* 1991; Heinz *et al* 1996; Wang *et al* 1997). In addition, certain anti-3CP compounds were evaluated against the 3C–3D precursor and shown to posses similar inhibitory properties (Davis *et al* 1997).

13.2.2 Reversible HRV 3C proteinase inhibitors

Reversible inhibitors of serine and cysteine proteinases can generally be divided into two categories: those which form covalent adducts with the target enzyme and those which do not (covalent and noncovalent inhibitors, respectively). Molecules belonging to the former group typically combine a molecular recognition element (binding moiety, binding element) with an electrophile that can form the desired covalent adduct with an active site enzyme residue. Examples of many reversible covalent cysteine proteinase inhibitors along with general descriptions of their inhibition mechanisms have been reviewed in the literature (Rich 1986; Rich 1990; Shaw 1990; Rasnick 1996; Otto and Schirmeister 1997). Most of the reversible 3CP inhibitors reported to date are believed to form covalent 3CP-inhibitor adducts, and several classes of such molecules are detailed below.

13.2.2.1 Peptide aldehydes

Molecules which contain peptidyl binding elements and aldehyde electrophiles exemplify the covalent proteinase inhibitor design strategy. The first reports of peptide aldehyde 3CP inhibitors which appeared in the literature examined the commercially available compounds leupeptin (**13.1**) and chymostatin (**13.2**). Both molecules inhibited the proteolytic activity of 3CP derived from HRV serotype 14 and these experiments helped characterize the enzyme as a cysteine proteinase (Orr *et al* 1989).

More recently, a hexapeptide previously determined to be an effective 3CP substrate was employed as a starting point for peptide aldehyde inhibitor design. Replacement of the scissile amide linkage in the substrate with an aldehyde moiety and protection of the N-terminus afforded a tetrapeptide (**13.3**) which displayed moderately potent 3CP inhibition properties (Table 13.1) (Kaldor *et al* 1995). The molecule was also active in an *in vitro* assay which examined its ability to inhibit rhinoviral RNA translation and displayed weak antirhinoviral activity in cell culture ($ED_{50} = 0.4\,mM$; Shepherd *et al* 1996). In addition to compound (**13.3**), several other peptide aldehydes of varying size and structure were reported, but none exhibited greater 3CP inhibition properties than (**13.3**) (Table 13.1, compounds **13.5–13.8**).

Due to the reactive nature of the aldehyde electrophile, the presence of the glutamine side chain in molecules such as (**13.3**) may attenuate 3CP inhibition activity through intramolecular hemiaminal formation (structure **13.4**). In fact, the designers of (**13.3**) reported that the ^{1}H NMR spectrum of the compound did not exhibit the expected aldehyde resonance (Kaldor *et al* 1995). In addition, purposeful preparation of the hemiaminal

Table 13.1 Peptide aldehyde 3CP inhibitors. Reprinted in part from: Kaldor *et al* (1995). Copyright 1995, with permission from Elsevier Science

Compd.	*Structure*	*3CP IC_{50} (μM)*[a]	*Translation IC_{50} (μM)*[a, b]
13.3	Boc-Val-Leu-Phe-Gln-CHO	0.6	10
13.5	Boc-Val-Leu-Phe-Gly-CHO	10	180
13.6	H_2N-Leu-Phe-Gln-CHO	1.0	toxic
13.7	Ac-Leu-Phe-Gln-CHO	2.7	170
13.8	Ac-Gly-Phe-Gln-CHO	16	ND

[a]Inhibition of HRV-14 3CP; [b]Assay described in Heinz *et al* (1996); ND = not determined.

corresponding to inhibitor (**13.3**) (structure **13.4**) afforded material that was identical to the aldehyde constructed earlier by a synthetic route that did not involve hemiaminal intermediates. These results suggested that glutamine-mediated intramolecular hemiaminal formation could indeed occur in molecules such as (**13.3**), and prompted efforts to identify suitable glutamine isosteres for incorporation into other 3CP inhibitors.

Early attempts at such glutamine isostere identification involved incorporation of an *N,N*-dimethylglutamine-derived aldehyde into the 3CP inhibitor design (Malcolm *et al* 1995). Thus, the tetrapeptide aldehyde (**13.9**) displayed measurable 3CP inhibition activity when tested against enzyme derived from HRV serotype-14. Although such activity

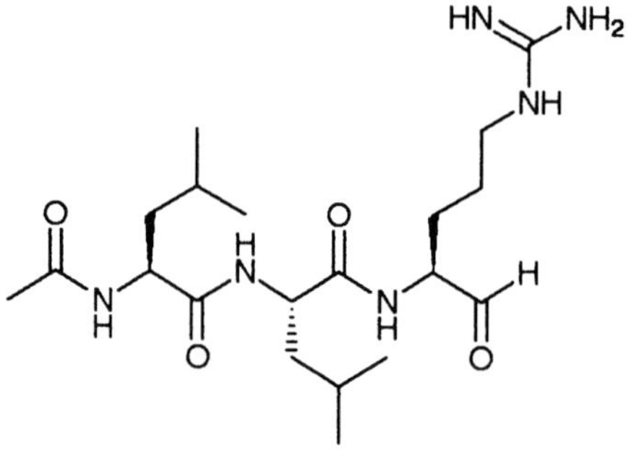

13.1
85% inhib. at 10 μg/mL
(HRV-14 3CP)

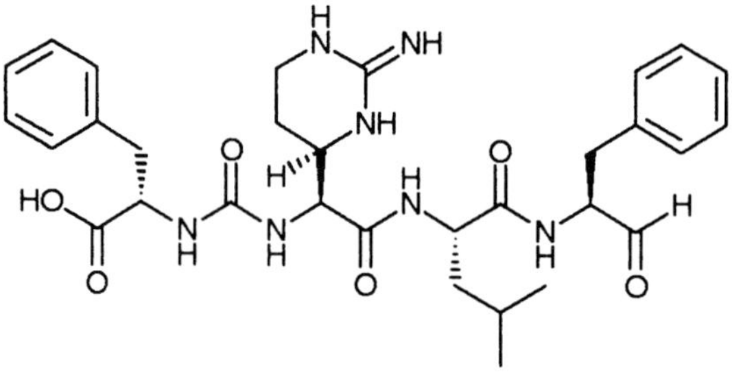

13.2
30% inhib. at 10 μg/mL
(HRV-14 3CP)

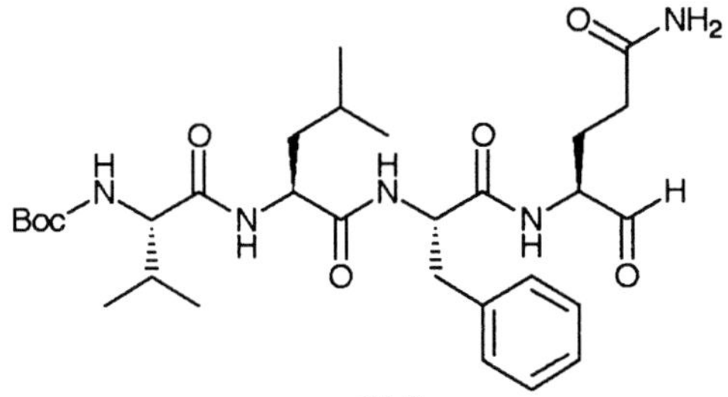

13.3

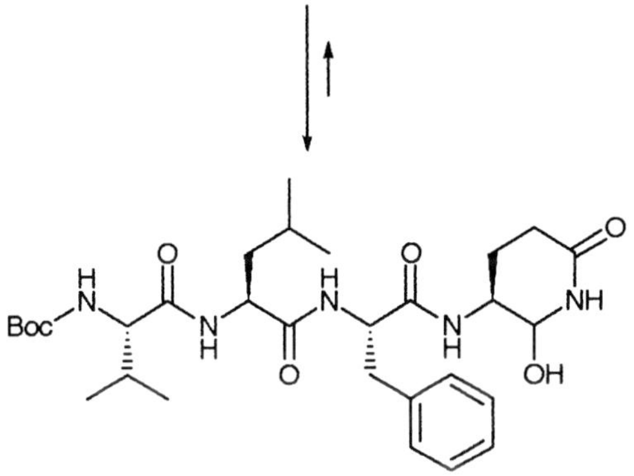

13.4

was not quite as potent as that exhibited by the related glutamine-containing molecule (**13.3**) described above, it was nevertheless noteworthy since a suspected hydrogen bond inhibitor-enzyme interaction would be forfeited by the *N,N*-dimethylglutamine-containing compound. Even more striking was the observation that a tetrapeptide aldehyde related to (**13.9**) (compound **13.10**) displayed 3CP inhibition activity comparable to the glutamine-containing molecule (**13.3**). The antirhinoviral activities of (**13.9**) and (**13.10**) were not reported, but the above studies indicated that relatively potent 3CP inhibitors could indeed be obtained by the incorporation of appropriate glutamine replacements into peptide aldehyde inhibitors.

A second approach to glutamine isostere-containing peptide aldehyde inhibitor design combined a methionine sulfone aldehyde moiety with an acylated Leu-Leu dipeptide binding element (compound **13.11**, Table 13.2) (Shepherd *et al* 1996). This molecule exhibited

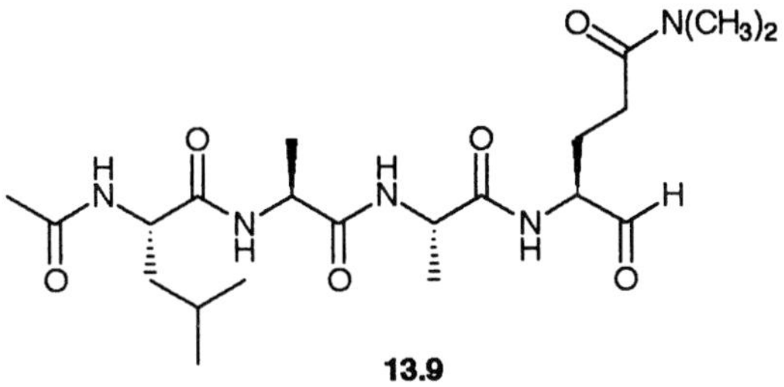

13.9
$K_i = 2.7\ \mu M$
(HRV-14 3CP)

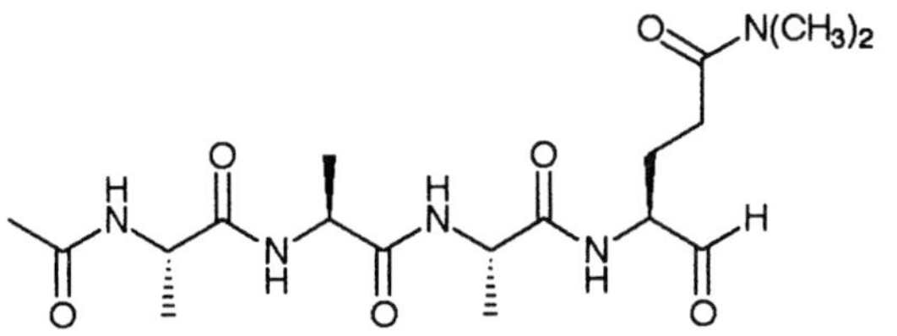

13.10
$K_i = 0.11\ \mu M$
(HRV-14 3CP)

potent, reversible inhibition of HRV-14 3CP and displayed apparent antirhinoviral activity when tested in cell culture. However, the designers of (**13.11**) sought to more closely mimic known 3CP substrates and therefore prepared the phenylalanine-containing dipeptide aldehyde inhibitor (**13.12**). This compound also exhibited potent 3CP inhibition properties along with moderate antiviral activity without observed cytotoxicity (Table 13.2). Several other dipeptide aldehydes incorporating the methionine sulfone moiety were also reported but none of these entities displayed greater 3CP inhibition activity or antirhinoviral properties than exhibited by compound (**13.12**) (compounds **13.13–13.15**, Table 13.2).

The most recent example of peptide aldehyde 3CP inhibitors described in the literature combined a substrate-inspired Leu-Phe dipeptide binding element with a variety of aldehyde-containing moieties (Webber *et al* 1998). One such tripeptidyl inhibitor that contains a glutamine-derived aldehyde displayed moderate inhibition of HRV-14 3CP (compound (**13.16**), Table 13.3). The molecule also exhibited weak antirhinoviral activity in cell culture, but this property could not be entirely dissociated from the observed cytotoxicity of the compound. In addition, ^{1}H NMR experiments indicated that the molecule predominantly exists as the hemiaminal (**13.17**) in direct analogy with studies of the related tetrapeptide aldehyde inhibitor (**13.3**) described above (Kaldor *et al* 1995). In order to circumvent intramolecular hemiaminal formation in tripeptide aldehydes related to (**13.16**), a number of glutamine replacements were incorporated into the Leu-Phe-containing tripeptide inhibitor design.

Inversion of the glutamine amide moiety present in the lead tripeptide aldehyde (**13.16**) afforded an *N*-acetyl-containing molecule which displayed greatly improved HRV-14 3CP inhibition properties and antiviral activity (**13.18**, Table 13.3). The compound also potently inhibited 3CPs obtained from several other HRV serotypes and

Table 13.2 Methionine sulfone-containing peptide aldehyde 3CP inhibitors. Reprinted in part from Shepherd *et al* (1996). Copyright 1996, with permission from Elsevier Science

Compd.	*Structure*	*3CP K_i (μM)*[a]	*IC_{50}(μM)*[b]	*TC_{50}(μM)*[c]
13.11	Ac-Leu-Leu-Met(SO_2)-CHO	0.49	81.9	>224
13.12	Cbz-Phe-Met(SO_2)-CHO	0.47	3.4	>224
13.13	Boc-Phe-Met(SO_2)-CHO	60[d]	71.3	147.6
13.14	Cbz-Leu-Met(SO_2)-CHO	ND	39.3	>224
13.15	Cbz-HPhe-Met(SO_2)-CHO	ND	81.2	161.5

[a]Inhibition of HRV-14 3CP; [b]Antiviral activity against HRV-14; [c]Cytotoxicity; [d]IC_{50} value; ND = not determined.

Table 13.3 Tripeptide aldehyde 3CP inhibitors containing glutamine replacements. Reproduced in part from Webber *et al* (1998). Copyright 1998 American Chemical Society

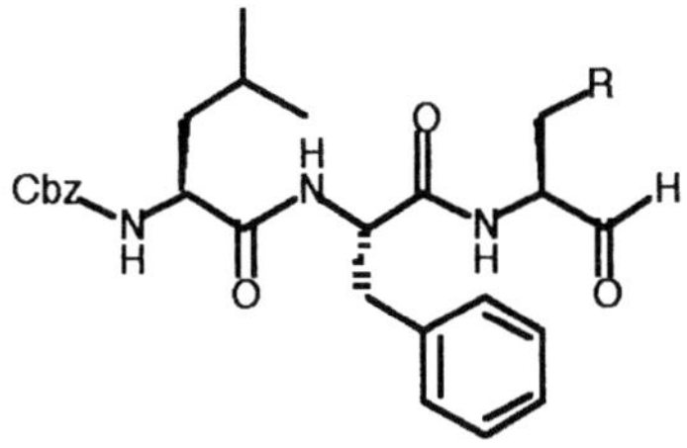

Compd.	*R*	*Serotype*	*3CP* $K_i(\mu M)^a$	$EC_{50}(\mu M)^b$	$TC_{50}(\mu M)^c$
13.16	CH_2CONH_2	14	3.6	66	398
13.18	$NHCOCH_3$	14	0.006	2.4	316
		2	0.16	3.1	>100
		16	0.069	2.1	>100
		89	0.036	2.2	>100
13.19	$CH_2CON(CH_3)_2$	14	0.005	1.3	63
13.20	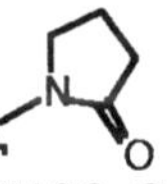	14	0.052	4.0	>100
13.21	$NHCO_2C(CH_3)_3$	14	0.066	2.0	56.2
13.22	$CH_2CONHCPh_3$	14	0.040	>15.8	15.8
13.23	NHCOPh	14	0.012	1.0	51.2
13.24	$CH_2SOCH_3^d$	14	0.005	4.5	>100
13.25	CH_2CN	14	0.19	8.0	31

[a]Inhibition of 3CP from indicated HRV serotype; [b]Antiviral activity against indicated HRV serotype; [c]Cytotoxicity; [d]1:1 Mixture of diastereomers.

exhibited moderate antiviral activity when tested against these serotypes in cell culture (Table 13.3). Incorporation of the *N*,*N*-dimethylglutamine moiety employed above by Malcolm and coworkers into the tripeptidyl inhibitor series also afforded an extremely potent, reversible anti-3CP agent which displayed moderate levels of antiviral activity in cell culture (compound **13.19**). The difference in activities between (**13.19**) and the Malcolm tetrapeptide (**13.10**) illustrates the importance of proper substrate mimicry for obtaining potent C-terminal aldehyde-containing peptidyl 3CP inhibitors. In addition to (**13.18**) and (**13.19**), many other compounds incorporating glutamine replacements and/or modifications were disclosed (**13.20–13.25**) and indicated a wide tolerance for structural variation. Several X-ray crystal structures of tripeptide aldehydes complexed with HRV-2 3CP were also reported (Webber *et al* 1998; Matthews *et al* 1999). These structures revealed a covalent hemithioacetal linkage between the 3CP active site cysteine residue and the inhibitors and identified numerous protein-ligand interactions.

13.2.2.2 Peptidyl ketones

In addition to the peptide aldehydes described above, several other classes of reversible 3CP inhibitors have been described in the literature. First among these was a series of tetrapeptidyl methyl ketones (Kettner and Korant 1987). One such compound from this series

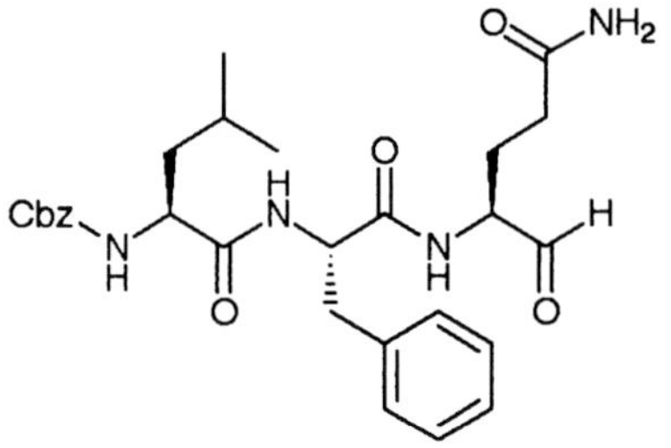

13.16

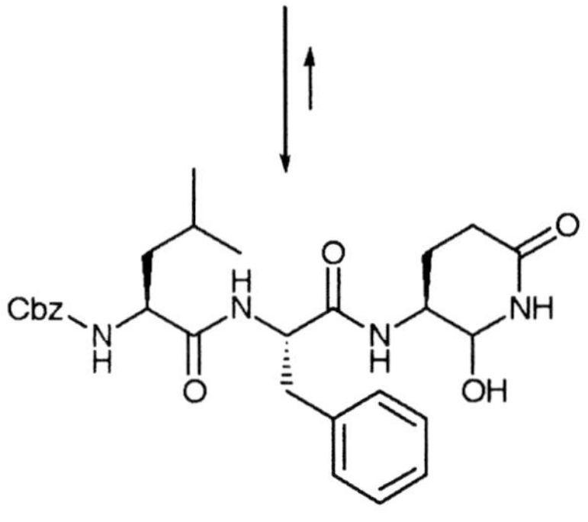

13.17

(**13.26**) was active in a viral cleavage assay against HRV-1A, although specific inhibition of 3CP was not rigorously confirmed. The molecule also displayed antirhinoviral activity that was clearly distinguishable from cytotoxicity against HRV-1A in a plaque reduction assay.

More recently, molecules belonging to an unrelated series of substrate-derived tripeptides bearing C-terminal ketone moieties were also shown to inhibit 3CP (Dragovich *et al* 2000). One such example is compound (**13.27**) which displayed good levels of anti-HRV-14 3CP inhibition and antirhinoviral activity when tested against the same serotype in cell culture. An optimized ketone-containing molecule (**13.28**) exhibited more potent HRV-14 3CP inhibition and antiviral properties and was also active against several other rhinovirus serotypes. Both (**13.27**) and (**13.28**) were nontoxic (CC_{50}) when tested to relatively high concentrations.

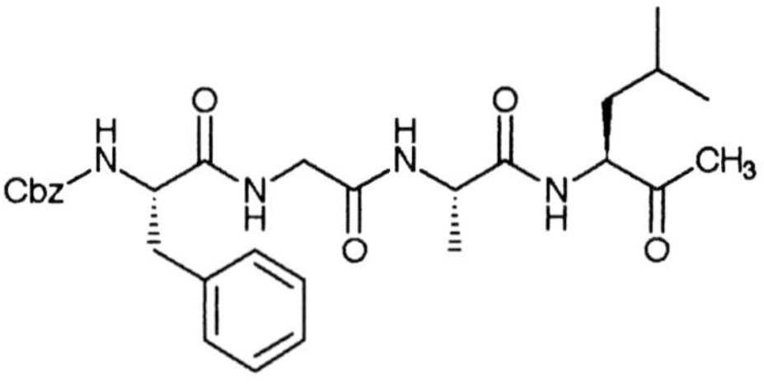

13.26
"active" at 100 µg/mL (viral cleavage assay)
90% plaque reduction at 50 µg/mL
nontoxic to 1000 µg/mL
(HRV-1A)

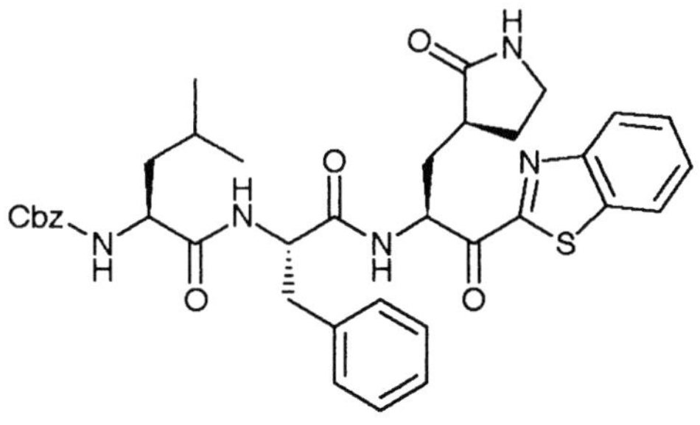

13.27
$K_i = 0.065\ \mu M$
$EC_{50} = 3.2\ \mu M$
$CC_{50} = >320\ \mu M$
(HRV-14)

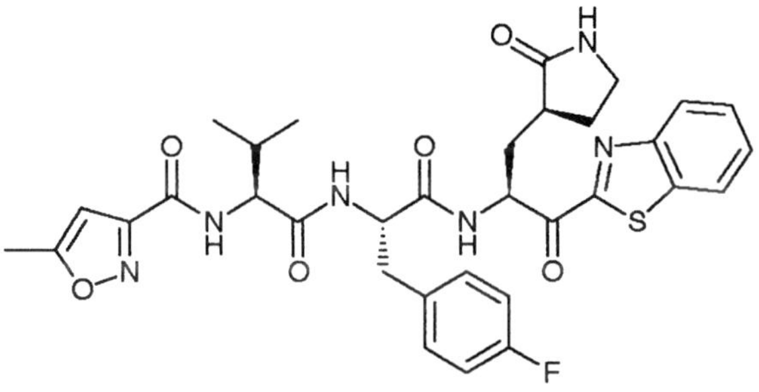

13.28
$K_i = 0.0045\ \mu M$
$EC_{50} = 0.34\ \mu M$
$CC_{50} = 250\ \mu M$
(HRV-14)

13.2.2.3 Isatins

The reversible 3CP inhibitors described above are all derived in part from known peptide substrates of the viral proteinase. However, several reports have also appeared in the literature detailing non-peptidic, reversible 3CP inhibitors. The first of these disclosed the use of isatins to inhibit 3CP proteolytic function (Webber *et al* 1996), and an initially prepared compound displayed fairly potent inhibition of 3CPs derived from several HRV serotypes (**13.29**, Table 13.4). However, the molecule was inactivated by the addition of dithiothreitol to the enzyme assay suggesting that the observed 3CP inhibition resulted from the inherent reactive nature of the isatin moiety towards thiols. Therefore, a number of other *N*-methyl isatins containing various C-5 substituents were also examined and all displayed significantly less potent 3CP inhibition properties than the molecule incorporating the C-5 carboxamide functionality (compounds **13.30–13.33**, Table 13.4). These results indicated that effective recognition of the 3CP active site by isatins such as (**13.29**) contributed significantly to the observed inhibition properties.

In order to further improve the ability of the isatins to recognize the 3CP active site, the *N*-methyl moiety contained in (**13.29**) was replaced with larger substituents intended to

Table 13.4 Isatin 3CP inhibitors. Reproduced in part from Webber *et al* (1996). Copyright 1996 American Chemical Society

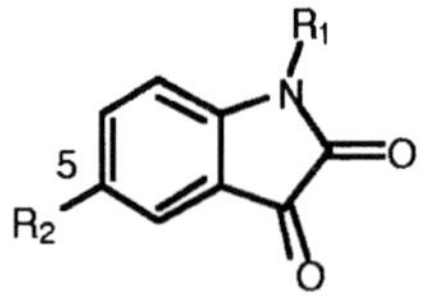

Compd.	R_1	R_2	Serotype	3CP K_i(μM)[a]	EC_{50}(μM)[b]	TC_{50}(μM)[c]
13.29	CH_3	$CONH_2$	14	0.051	66	270
			2	0.077	ND	ND
			16	0.040	ND	ND
			89	0.035	ND	ND
13.30	CH_3	H	14	>100	ND	ND
13.31	CH_3	NO_2	14	9.7	ND	ND
13.32	CH_3	CO_2CH_3	14	30.0	56	>100
13.33	CH_3	CN	14	51.0	ND	ND
13.34	CH_2-2-benzo [*b*]thiophene	$CONH_2$	14	0.002	>5.6	5.6
			2	0.005	ND	ND
			16	0.012	ND	ND
			89	0.001	ND	ND
13.35	$(CH_2)_3Ph$	$CONH_2$	14	0.027	>7.1	7.1
13.36	CH_2-β-naphthyl	$CONH_2$	14	0.004	>10	10
13.37	CH_2(3,5-di-OH C_6H_3)	$CONH_2$	14	0.004	>100	>100

[a]Inhibition of 3CP from indicated HRV serotype; [b]Antiviral activity against indicated HRV serotype; [c]Cytotoxicity; ND = not determined.

mimic the phenylalanine side chain present in known 3CP substrates. Many of the isatins containing these larger moieties displayed significantly improved anti-3CP properties compared to the *N*-methyl analog (compounds **13.34–13.37**, Table 13.4). An X-ray crystal structure of (**13.34**) complexed with HRV-2 3CP was determined and showed that a covalent hemithioketal adduct was formed between the 3CP active site cysteine residue and the isatin ketone carbonyl moiety (Webber *et al* 1996; Matthews *et al* 1999). In addition, several representative isatins were shown not to significantly inhibit a variety of other serine and cysteine proteinases, indicating a high level of selectivity for 3CP. Unfortunately, all of the reported isatins displayed only weak antirhinoviral activity that could not be dissociated from the observed toxicities of the compounds (Table 13.4). The origin of the poor antiviral activity displayed by these particular 3CP inhibitors was not identified.

13.2.2.4 Homophthalimides

The homophthalimides (**13.38**) and (**13.39**) (Table 13.5) were identified from a blind screening program by Jungheim and coworkers (1997) as moderate inhibitors of HRV-14 3CP. These non-peptidic molecules were considered suitable for further development and a variety of additional homophthalimides were prepared and evaluated as 3CP inhibitors (Table 13.5). Several of these compounds (**13.40–13.43**) also displayed measurable 3CP

Table 13.5 Homophthalimide 3CP inhibitors. Reprinted in part from Jungheim *et al* (1997). Copyright 1997, with permission from Elsevier Science

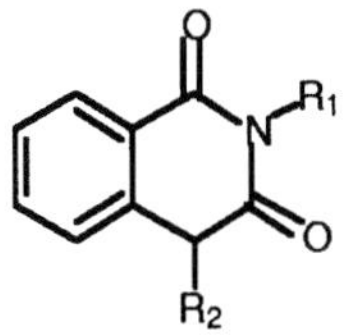

Compd.	R_1	R_2	*3CP % Inhib.*[a]	*3CP* $IC_{50}(\mu M)$[b]	*HRV* IC_{50} (μM)[c]	TC_{50} (μM)[d]
13.38	CH_3	H	14	ND	ND	ND
13.39	CH_3	$CH_2COPh(p\text{-}F)$	72	41.1	>10	>100
13.40	$(CH_2)_2CO_2Et$	CH_2COPh	71	55.3	6.0	>100
13.41	$(CH_2)_2SO_2Me$	CH_2COPh	100	22.1	8.5	71
13.42	$(CH_2)_3SMe$	CH_2COPh	35	131	>10	63
13.43	$(CH_2)_3SO_2Me$	CH_2COPh	70	25	ND	ND

[a]Inhibition of HRV-14 3CP at compound concentration of 25μg/mL (assay described in Wang *et al* (1997)); [b]Inhibition of HRV-14 3CP (assay described in Wang *et al* (1997)); [c]Antiviral activity against HRV-14; [d]Cytotoxicity; ND = not determined.

inhibition properties, and one exhibited non-toxicity-related antirhinoviral activity (compound **13.40**, Table 5). However, it was also noted that the antiviral potency of **13.40** exceeded its measured 3CP inhibitory activity. This discrepancy may be due to simultaneous inhibition of the HRV 2A proteinase by such homophthalimide-derived compounds (Wang *et al* 1998b). The interaction of inhibitors (**13.41**) and (**13.43**) with HRV-14 3CP was studied by mass spectroscopy and was suggested to occur via the active site cysteine residue. A computational model constructed using a NMR structure of HRV-14 3CP also supported similar homophthalimide-3CP interactions. However, whether these inhibitors form reversible, covalent 3CP adducts was not rigorously determined and their precise mechanism of 3CP inhibition remains unknown.

13.2.3 Irreversible HRV 3C proteinase inhibitors

Cysteine proteinase inhibitors, which irreversibly inhibit their enzyme targets, are well known in the literature and their various mechanisms of action have been extensively reviewed (Rich 1986; Rich 1990; Shaw 1990; Rasnik 1996; Otto and Schirmeister 1997). These compounds resemble the reversible, covalent inhibitors discussed above (e.g. peptide aldehydes) in that they typically combine an enzyme recognition element (binding moiety, binding element) with an electrophile that can form a covalent adduct with an active site enzyme residue. However, they differ from the reversible inhibitors in that they permanently modify the enzymes which they inhibit. Several examples of irreversible 3CP inhibitors have appeared in the literature, and these compounds are detailed below. It should be noted that true irreversible 3CP inhibition by several of the following molecules was not rigorously confirmed through biological testing. These compounds are included in this section based on their structural resemblance to other known irreversible cysteine proteinase inhibitors.

13.2.3.1 α-Halomethylcarbonyl compounds

Early literature reports of α-halomethylcarbonyl 3CP inhibitors (Orr *et al* 1989) described the activity of two commercially available compounds TLCK (**13.44**) and TPCK (**13.45**) that were known to non-specifically inactivate a variety of cysteine proteinases (Rich 1986). Both molecules displayed measurable inhibition of 3CP derived from HRV serotype-14 as assessed by a peptide cleavage assay. However, the precise nature of this inhibition was not rigorously determined.

Later publications described a series of tetrapeptide chloromethylketone-containing 3CP inhibitors and one such molecule (compound **13.46**, Table 13.6) was indicated by Kettner and Korant (1987) to exhibit non-toxicity-related antirhinoviral activity when tested in a plaque reduction assay against HRV serotype 1A (90% plaque reduction at 1 μg/mL; toxicity noted at 15 μg/mL). However, the identical compound was later reported by Kati and coworkers (1999) to display no detectable antiviral activity against HRV-1A, although weak irreversible inhibition of HRV-1B 3CP was noted (Table 13.6). The discrepancy in the reported antiviral activities of (**13.46**) may result from different cell lines chosen by the Kettner and Kati research groups for their antiviral assays (human HeLa-O and MRC-5, respectively). In addition to compound (**13.46**), Kati and coworkers (1999) also reported several other chloromethylketone-containing 3CP inhibitors (compounds **13.47–13.50**, Table 13.6). All of these compounds exhibited more potent 3CP inhibition properties than the initially studied molecule (**13.46**), although none were observed to be active antirhinoviral agents.

Along with the chloromethylketones depicted above, several α-bromomethyl-carbonyl-containing 3CP inhibitors were also described by Kati and coworkers (1999) (see also: Sham *et al* 1995). These compounds incorporate an azaglutamine (AzaGln) moiety to avoid intramolecular cyclization analogous to that observed with related glutamine-containing peptide aldehydes discussed above. Many of these molecules exhibited relatively potent irreversible inhibition of HRV-1B 3CP and several were reported to be active antirhinoviral agents when tested against the same HRV serotype in cell culture (compounds **13.51–13.55**, Table 13.7). In addition, all compounds in Table 13.7 were shown not to significantly inhibit the serine proteinases chymotrypsin and elastase and the cysteine proteinase cathepsin B. These results suggested that the azaglutamine-containing molecules described in Table 13.7 may be fairly specific inhibitors of HRV 3CP.

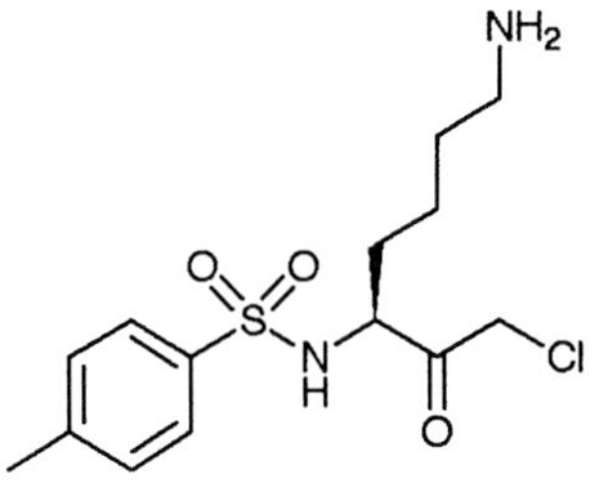

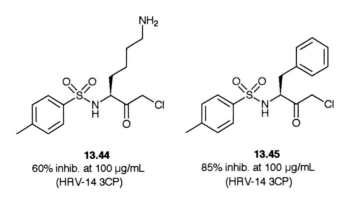

13.44
60% inhib. at 100 μg/mL
(HRV-14 3CP)

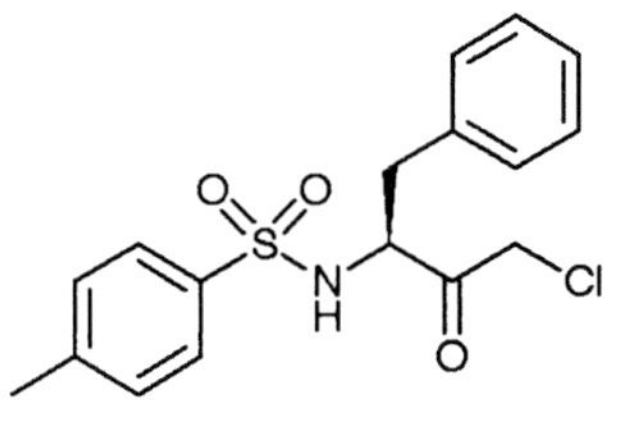

13.45
85% inhib. at 100 μg/mL
(HRV-14 3CP)

Table 13.6 Peptidyl chloromethylketone 3CP inhibitors. Reproduced with permission from Kati *et al* (1999). Copyright 1999 Academic Press

Compd.	*Structure*	$k_{inact}/K_{inact}(M^{-1}s^{-1})^a$
13.46	Cbz-Phe-Gly-Leu-Leu-CH_2Cl	100
13.47	Cbz-Ala-Ile-Leu-Leu-CH_2Cl	1 170
13.48	Boc-Ala-Ile-Leu-Leu-CH_2Cl	1 150
13.49	Boc-Ala-Ile-Leu-Phe-CH_2Cl	420
13.50	Cbz-Ala-Ile-Leu-Phe-CH_2Cl	800

[a]Inhibition of HRV-1B 3CP.

Table 13.7 Peptidyl α-bromomethylcarbonyl 3CP inhibitors. Reproduced in part with permission from Kati *et al* (1999). Copyright 1999 Academic Press

Compd.	*Structure*	k_{inact}/K_{inact} $(M^{-1}s^{-1})^a$	ID_{50} $(\mu g/mL)^b$	*MTC* $(\mu g/mL)^c$	*Virus rating*[d]
13.51	Boc-Phe-AzaGln-CH_2Br	310	0.5	1.5	1.0
13.52	Cbz-Phe-AzaGln-CH_2Br	4 800	0.9	5.1	1.3
13.53	Boc-Ile-Phe-AzaGln-CH_2Br	1 540	2.3	5.8	1.0
13.54	Boc-Ala-Ile-Phe-AzaGln-CH_2Br	23 400	2.5	20.7	2.9
13.55	Boc-Ile-Thr-Thr-AzaGln-CH_2Br	14 500	11.1	64	1.1

[a]Inhibition of HRV-1B 3CP; [b]Antiviral activity against HRV-1B; [c]Cytotoxicity; [d]See Kati *et al* (1999) for definition. A compound with VR score ≥ 1.0 is considered to have true antiviral activity. AzaGln = -NH-N($CH_2CH_2CONH_2$)CO-.

13.2.3.2 Peptidyl Michael acceptors

The first report of Michael acceptor-containing 3CP inhibitors described several substrate-derived tetrapeptidyl compounds which incorporate C-terminal vinylogous esters (molecules **13.56–13.62**, Table 13.8) (Kong *et al* 1998). Many of these compounds rapidly inactivated the HRV-14-derived 3C enzyme and exhibited potent antirhinoviral activity when tested against the same HRV serotype in cell culture. However, a molecule incorporating a C-terminal vinylogous acid (**13.63**, Table 13.8) was a much weaker 3CP inhibitor than compounds containing the corresponding methyl and ethyl esters (**13.56** and **13.57**, respectively). In addition, replacement of the glutamine side chain of one of the most active inhibitors with a methionine sulfone moiety also drastically reduced 3CP inhibitory potency (compound **13.61**). This result was in contrast to that noted during the study of peptide aldehyde 3CP inhibitors described above in which incorporation of a methionine sulfone moiety afforded relatively potent anti-3CP agents. A similar reduction in 3CP inhibitor potency was observed for a dipeptide-derived molecule (**13.62**). Biochemical and mass spectral studies conducted with several of the above inhibitors supported the formation of an irreversible covalent enzyme-inhibitor complex.

Shortly after the initial disclosure of the Michael acceptor-containing compounds described above, a series of reports appeared by Dragovich and coworkers detailing their efforts to optimize a similar series of 3CP inhibitors. The first of these (Dragovich *et al*

Table 13.8 Peptide Michael acceptor 3CP inhibitors. Reproduced in part with permission from Kong *et al* (1998). Copyright 1998 American Chemical Society

Compd.	*Structure*[a]	*3CP* IC_{50}(μM)[b]	*HRV* IC_{50}(μM)[c]
13.56	Boc-Val-Leu-Phe-Gln-CH=CHCO$_2$Me	0.25	0.74
13.57	Boc-Val-Leu-Phe-Gln-CH=CHCO$_2$Et	0.13	0.41
13.58	Cbz-Val-Leu-Phe-Gln-CH=CHCO$_2$Me	0.17	0.93
13.59	Boc-Glu(tBu)-Val-Leu-Phe-Gln-CH=CHCO$_2$Me	0.2	4.5
13.60	H$_2$N-Glu-Val-Leu-Phe-Gln-CH=CHCO$_2$Me	0.49	>10
13.61	Boc-Val-Leu-Phe-Met(SO$_2$)-CH=CHCO$_2$Me	13.6	>10
13.62	Cbz-Phe-Gln-CH=CHCO$_2$Me	9.5	3.2
13.63	Boc-Val-Leu-Phe-Gln-CH=CHCO$_2$H	17.7	>10

[a]All olefins are of *E* (trans) geometry; [b]Inhibition of HRV-14 3CP; [c]Antiviral activity against HRV-14; Compounds (**13.56**) and (**13.57**) were slightly cytotoxic at 10 μg/mL; All others did not exhibit significant cytotoxicity when tested to 10 μg/mL.

Table 13.9 Tripeptide Michael acceptor 3CP inhibitors. Reproduced in part from Dragovich *et al* (1998a). Copyright 1998 American Chemical Society

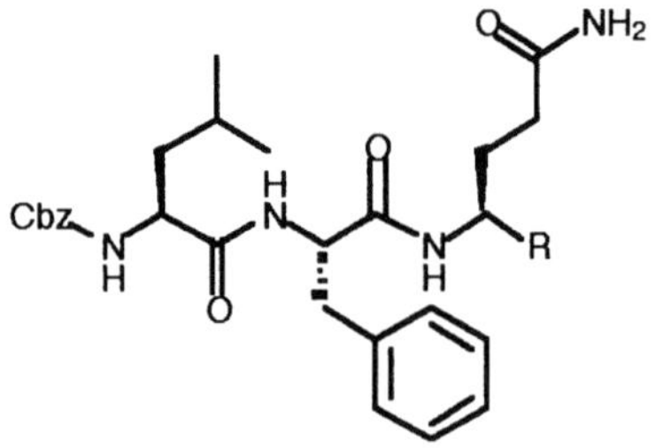

Compd.	*R*[a]	*Serotype*	$k_{obs}/[I](M^{-1}s^{-1})$[b]	EC_{50}(μM)[c]	CC_{50}(μM)[d]
13.64	CH=CHCO$_2$Me	14	20 000	1.3	>320
13.65	CH=CHCO$_2$Et	14	25 000	0.54	>320
		16	6 500	2.3	>320
		2	2 000	1.6	>320
13.66	CH=CHCO$_2$H	14	30	>100	>100
13.67	CH=CHCONHEt	14	350	>320	>320
13.68	CH=CHSO$_2$Me	14	160	>320	>320
13.69	CH=CHCOMe	14	54 000	2.0	60
		16	8 900	ND	ND
		2	2 700	ND	ND
13.70	CH=CHCOPh	14	500 000	4.0	16
13.71	CH$_2$CH$_2$CO$_2$Et	14	20 μM(K_i)	>100	>100

[a]All olefins are of *E* (trans) geometry; [b]Inhibition of 3CP from indicated HRV serotype; [c]Antiviral activity against indicated HRV serotype; [d]Cytotoxicity; ND = not determined.

1998a) reported extensive variation of a Michael acceptor moiety that was incorporated into a substrate-derived tripeptidyl molecule (Table 13.9). As was noted by Kong and coworkers above, molecules containing C-terminal vinylogous esters (**13.64** and **13.65**) displayed potent HRV-14 3CP inhibition and antirhinoviral properties while a compound incorporating a corresponding vinylogous acid did not (**13.66**). In addition, introduction

Table 13.10 Peptide Michael acceptor 3CP inhibitors. Reproduced from Dragovich *et al* (1998b). Copyright 1998 American Chemical Society

Compd.	*Structure*[a]	$k_{obs}/[I](M^{-1}s^{-1})^b$	$EC_{50}(\mu M)^c$	$CC_{50}(\mu M)^d$
13.65	Cbz-Leu-Phe-Gln-CH=CHCO$_2$Et	25 000	0.54	>320
13.72	Cbz-Phe-Gln-CH=CHCO$_2$Et	400	5.6	>100
13.73	Cbz-Gln-CH=CHCO$_2$Et	4.5	>100	>320

[a]All olefins are of *E* (trans) geometry; [b]Inhibition of HRV-14 3CP; [c]Antiviral activity against HRV-14; [d]Cytotoxicity.

of C-terminal vinylogous amides and sulfones into the inhibitor design also afforded poor 3CP inhibitors and weak antiviral agents (compounds **13.67** and **13.68**, respectively). Several ketone-containing Michael acceptors were also examined and these proved to be highly active 3CP inhibitors (compounds **13.69** and **13.70**). However, all the reported ketone-containing molecules were inactivated by treatment with dithiothreitol prior to testing in the 3CP inhibition assay, suggesting that the compounds could react readily with non-enzymatic thiols. Such inhibitors also did not typically display antirhinoviral activity in cell culture that was significantly distinguishable from cytotoxicity. A compound which lacked a Michael acceptor moiety was tested as well and was shown to be a weak, reversible 3CP inhibitor and poor antirhinoviral agent (**13.71**). An X-ray crystal structure of inhibitor (**13.65**) complexed with HRV-2 3CP was also described (Dragovich *et al* 1998a; Matthews *et al* 1999) which confirmed the expected existence of a covalent enzyme–inhibitor linkage.

A subsequent publication by Dragovich and coworkers (1998b) detailed modification of the peptidyl binding element contained within inhibitor (**13.65**). As observed by Kong and coworkers above, truncation of the tripeptide structure present in the lead compound resulted in significant loss of HRV-14 3CP inhibitory properties (compounds **13.72** and **13.73**, respectively, Table 13.10). An extensive structure activity study of tripeptidyl Michael acceptor-containing 3CP inhibitors was also reported in which each amino acid residue and the *N*-terminal functionality was varied while the others remained constant. Briefly, many modifications to the glutamine side chain, including those shown above to afford potent aldehyde-containing 3CP inhibitors, resulted in drastic loss of 3CP inhibition activity (compounds **13.74–13.76**, Table 13.11). Truncation of the phenylalanine side chain contained in (**13.65**) also reduced anti-3CP activity (compound **13.77**), but small substituents at the 4-position of the phenylalanine aryl ring were tolerated (compounds **13.78** and **13.79**, Table 13.11). Virtually any amino acid could be incorporated in place of the Leu residue of (**13.65**) without loss of 3CP inhibitory properties (compounds **13.80–13.82**, Table 13.11) in agreement with the crystallographic observation that this residue's side chain did not significantly interact with the 3CP enzyme. Many substituents were also tolerated at the *N*-terminus of the molecules with *S*-alkyl thiocarbamates or a 5-methylisoxazole-3-carboxamide affording the most potent inhibitors (compounds **13.83** and **13.84**, Table 13.11). The latter substituent was identified in a separate study, which employed solid phase synthesis techniques to rapidly prepare a large number of *N*-terminal amide-containing tripeptidyl 3CP inhibitors (Dragovich *et al* 1999a).

More recently, Dragovich and coworkers (1999b) reported an extension of the above inhibitor series with the introduction of a ketomethylene isostere into the inhibitor

Table 13.11 Tripeptide and related Michael acceptor 3CP inhibitors. Reproduced in part from Dragovich *et al* (1998b, 1999a,b,c,d).

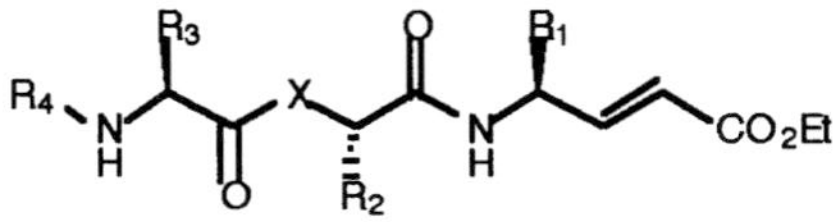

Compd.	*X*	R_1	R_2	R_3	R_4	$k_{obs}/[I](M^{-1}s^{-1})$[a]	$EC_{50}(\mu M)$[b,c]
13.65	NH	$CH_2CH_2CONH_2$	CH_2Ph	$CH_2CH(CH_3)_2$	CO_2CH_2Ph	25 000	0.54
13.74	NH	$CH_2CH_2CON(CH_3)_2$	CH_2Ph	$CH_2CH(CH_3)_2$	CO_2CH_2Ph	60	4.0
13.75	NH	$CH_2CH_2SO_2CH_3$	CH_2Ph	$CH_2CH(CH_3)_2$	CO_2CH_2Ph	60	>100
13.76	NH	$CH_2NHCOCH_3$	CH_2Ph	$CH_2CH(CH_3)_2$	CO_2CH_2Ph	800	2.2
13.77	NH	$CH_2CH_2CONH_2$	CH_3	$CH_2CH(CH_3)_2$	CO_2CH_2Ph	1 300	20
13.78	NH	$CH_2CH_2CONH_2$	$CH_2Ph(4\text{-}F)$	$CH_2CH(CH_3)_2$	CO_2CH_2Ph	46 000	1.8
13.79	NH	$CH_2CH_2CONH_2$	$CH_2Ph(4\text{-}OCH_3)$	$CH_2CH(CH_3)_2$	CO_2CH_2Ph	29 000	1.7
13.80	NH	$CH_2CH_2CONH_2$	CH_2Ph	$CH(CH_3)_2$	CO_2CH_2Ph	62 500	0.38
13.81	NH	$CH_2CH_2CONH_2$	CH_2Ph	$(CH_2)_4NH_2$	CO_2CH_2Ph	18 600	205
13.82	NH	$CH_2CH_2CONH_2$	CH_2Ph	CH_2CO_2H	CO_2CH_2Ph	35 000	2.4
13.83	NH	$CH_2CH_2CONH_2$	CH_2Ph	$CH_2CH(CH_3)_2$	$C(O)SCH_2Ph$	280 000	0.27
13.84	NH	$CH_2CH_2CONH_2$	CH_2Ph	$CH_2CH(CH_3)_2$	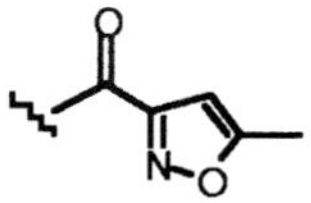	260 000	0.25
13.85	CH_2	$CH_2CH_2CONH_2$	CH_2Ph	$CH_2CH(CH_3)_2$	CO_2CH_2Ph	17 400	0.36
13.86	$N(CH_3)$	$CH_2CH_2CONH_2$	CH_2Ph	$CH_2CH(CH_3)_2$	CO_2CH_2Ph	5 300	1.0
13.87	NH	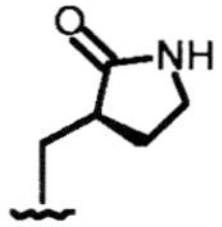	CH_2Ph	$CH_2CH(CH_3)_2$	CO_2CH_2Ph	257 000	0.10
13.88	CH_2	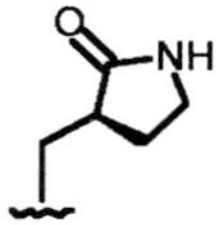	$CH_2Ph(4\text{-}F)$	$CH(CH_3)_2$	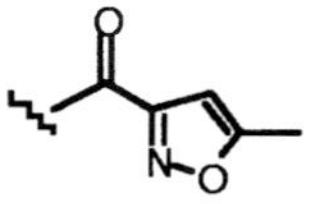	1 090 000	0.005

[a]Inhibition of HRV-14 3CP; [b]Antiviral activity against HRV-14; [c]Compounds were non-cytotoxic to 100 μM.

design. This modification resulted in somewhat reduced HRV-14 3CP inhibitory properties but improved the antiviral activities of ketomethylene-containing compounds relative to the corresponding peptides (compare compounds **13.65** and **13.85**, Table 13.11). The same peptidyl linkage was also substituted with an *N*-methyl amide moiety to afford a different series of 3CP inhibitors and antirhinoviral agents (e.g. compound **13.86**, Table 13.11) (Dragovich *et al* 1999c). Additional improvement to the inhibitor series was realized by the incorporation of lactam moieties as glutamine replacements (e.g., compound **13.87**, Table 13.11) (Dragovich *et al* 1999d). Such modification significantly improved both the anti-3CP and the antiviral properties of compounds to which it was applied. Introduction of both the ketomethylene isostere and lactam modifications into a single compound afforded a highly active, irreversible 3CP inhibitor (compound **13.88**). This molecule Ruprintrivir (AG7088) displayed extremely potent antiviral properties without observed cytotoxicity when tested against 48 rhinovirus serotypes in cell culture (mean $EC_{50} = 0.023\mu M$; mean $EC_{90} = 0.082\mu M$; Patick *et al* 1999), and also reduced the levels of inflammatory cytokines produced by bronchial epithelial cells infected with several HRV serotypes (Zalman *et al* 2000). The compound is currently being evaluated for the treatment of human rhinovirus infections in phase II clinical trials, and an X-ray crystal structure of its covalent complex with HRV-2 3CP was recently described (Matthews *et al* 1999).

13.2.4 Miscellaneous HRV 3C proteinase inhibitors

13.2.4.1 Natural products

Some of the earliest 3CP inhibitors reported in the literature are natural products of microbial origin. For example, thysanone (**13.89**), a novel lactol-containing naphthoquinone, was isolated from fermentation extracts of *Thysanophora penicilloides* (MF 5636, Merck Culture collection) and exhibited moderate activity when tested against HRV 3CP (serotype unknown) (Singh *et al* 1991). A total synthesis of the natural product was recently reported which identified its absolute stereochemistry as (1*S*,3*R*) (Donner and Gill 1999). The nature of the observed 3CP inhibition (reversible or irreversible) as well as the antirhinoviral activity of thysanone have not been disclosed.

Related screening efforts also identified citrinin hydrate (**13.90**) and radicinin (**13.91**) as HRV 3CP inhibitors (Kadam *et al* 1994). The former molecule was isolated from a *Penicillium* sp. strain (AB 2089ZZD-62; NRRL 22560) while the latter was obtained from the mycelium of a *Curvularia* sp. strain (AB 2090A-11, NRRL 22559). Both (**13.90**) and (**13.91**) displayed relatively weak 3CP inhibition activity and, as with thysanone (**13.89**) above, the nature of this inhibition as well as the antirhinoviral activity of the compounds were not described.

A third natural product 3CP inhibitor discovered by microbial screening efforts is A-108835 (**13.92**) (Brill *et al* 1996). This novel triperpene sulfate was isolated from a stationary fermentation extract of *Fusarium compactum* (AB 2194I-103, NRRL 25020) and exhibited 3CP inhibitory activity when tested against enzyme derived from HRV serotype 1B.

13.2.4.2 Other HRV 3C proteinase inhibitors

A spiro indoline beta-lactam was also synthesized and evaluated as a 3CP inhibitor (Skiles and McNeil 1990). This molecule was prepared in order to determine if

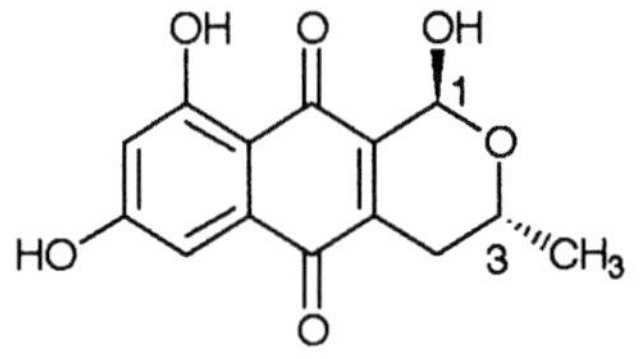

13.89
IC_{50} = 13 µg/mL
(Serotype unknown)

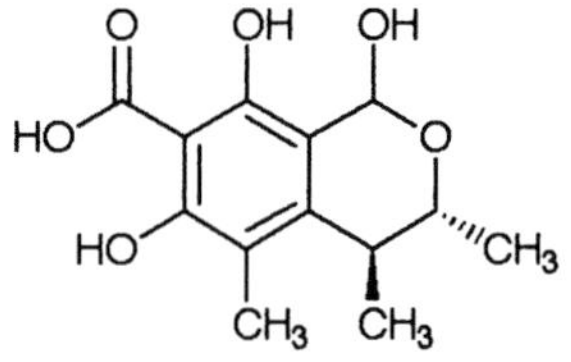

13.90
IC_{50} = 1 mM
(HRV-14 3CP)

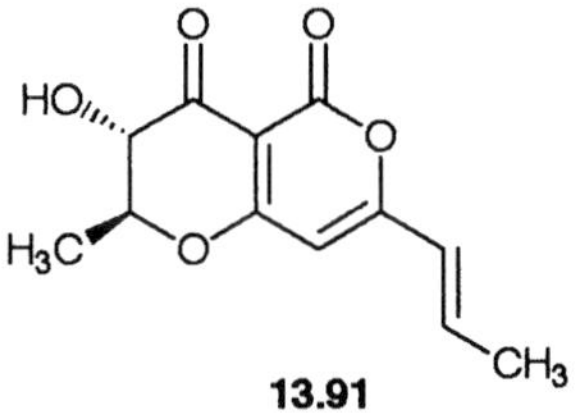

13.91
IC_{50} = 0.7 mM
(HRV-14 3CP)

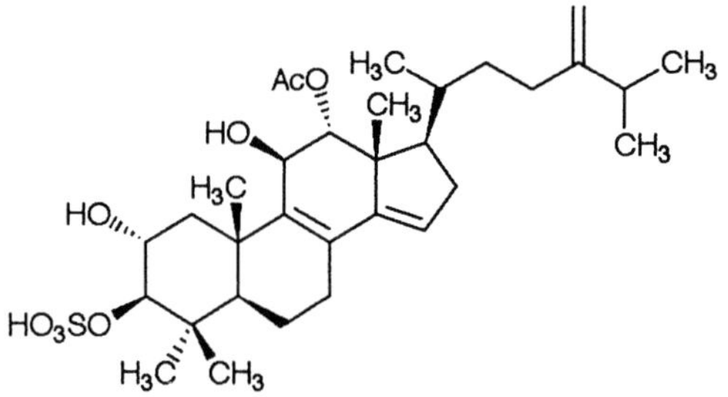

13.92
IC_{50} = 48 µg/mL
(HRV-1B 3CP)

beta-lactam-containing compounds, which are known to inhibit serine proteinases, could also affect cysteine-dependent enzymes such as 3CP. This compound was found to inhibit HRV 3CP (serotype unknown), but the nature of this 3CP inhibition (reversible or irreversible) was not disclosed. The compound also inhibited human leukocyte elastase (HLE) and Cathepsin G with IC_{50} values of 0.4 µg/mL and 4.0 µg/mL, respectively.

13.3 SUMMARY

Numerous challenges are present in the identification of an efficacious antirhinoviral therapy. Ideally, any such treatment must be extremely safe, readily available, and easy to administer. The pursuit of HRV 3CP as a target for antirhinoviral therapeutic development has provided several new chemical entities which display antirhinoviral activity in cell culture. Most of these inhibitors contain some type of electrophilic functionality that is capable of forming a reversible or irreversible covalent bond with the catalytic cysteine residue of the 3C enzyme. Many also incorporate a tri- or tetrapeptide (or peptidomimetic) binding element which imparts good 3CP recognition properties. To date, 3CP inhibitors which contain a Michael acceptor moiety display the most potent *in vitro* antirhinoviral activity against a range of tested serotypes. Many of these compounds also exhibit minimal cytotoxicity in cell culture, and one (compound **13.88**) has advanced to the stage of clinical testing in man. Although the clinical effectiveness of HRV 3CP inhibitors remains to be proven, the research described in this chapter has clearly provided new opportunities for the treatment of this prevalent human pathogen.

NOTES

Pleconaril, an HRV capsid-binding agent, was examined in a large phase III clinical trial and was reported to shorten the duration of naturally acquired colds (Viropharma press release, 2001). An NDA for this antipicornaviral agent was submitted to the FDA for approval in August, 2001 (Viropharma press release, 2001). In addition, several azodicarboxamides and azapeptides were independently reported as active, irreversible 3CP inhibitors (Hill, R.D. and Vederas, J.C. (1999) *J. Med. Chem.*, **64**, 9538–9546 and Venkatraman, S. *et al* (1999) *Biorg. Med. Chem. Lett.*, **9**, 577–580) as were a number of benzamide Michael acceptor-containing molecules (Reich, S.H. *et al* (2000) *J. Med. Chem.*, **43**, 1670–1683). Independently, several nitrosothiols were described which reversibly inhibited HRV 3CP (Xian, M. *et al* (2000) *Bioorg. Med. Chem. Lett.*, **10**, 2097–2100). Finally, several monocyclic and bicyclic 2-pyridone-containing 3CP inhibitors were reported and one member of the former group was shown to be orally bioavailable in the dog (Dragovich, P.S. *et al J. Med. Chem.* submitted and Dragovich, P.S. *et al Bioorg. Med. Chem. Lett.* submitted, respectively).

ACKNOWLEDGEMENTS

We would like to thank all of our collaborators on the Agouron human rhinovirus 3C proteinase project for their valuable scientific contributions over the years. We are also indebted to Thomas Prins and Drs. David Matthews and Stephen Worland for their critical reading of this work.

REFERENCES

Aschauer, B., Werner, G., McCray, J., Rosenwirth, B. and Bachmayer, H. (1991) Biologically active protease 3C of human rhinovirus 1A is expressed from a cloned cDNA segment in *Escherichia coli. Virology*, **184**, 587–594.

Birch, G.M., Black, T., Malcolm, S.K., Lai, M.T., Zimmerman, R.E. and Jaskunas, S.R. (1995) Purification of recombinant human rhinovirus 14 3C protease expressed in *Escherichia coli. Protein Expression and Purification*, **6**, 609–618.

Brill, G.M., Kati, W.M., Montgomery, D., Karwowski, J.P., Humphrey, P.E., Jackson, M. *et al* (1996) Novel triterpene sulfates from *Fusarium compactum* using a rhinovirus 3C protease inhibitor screen. *Journal of Antibiotics*, **49**, 541–546.

Carrasco, L. (1994) Picornavirus inhibitors. *Pharmacology and Therapeutics*, **64**, 215–290.

Cheah, K.-C., Leong, L.E.-C. and Porter, A.G. (1990) Site-directed mutagenesis suggests close functional relationship between human rhinovirus 3C protease and cellular trypsin-like serine proteases. *Journal of Biological Chemistry*, **265**, 7180–7187.

Cordingley, M.G., Register, R.B., Callahan, P.L., Garsky, V.M. and Colonno, R.J. (1989) Cleavage of small peptides *in vitro* by human rhinovirus 14 3C protease expressed in *Escherichia coli. Journal of Virology*, **63**, 5037–5045.

Cordingley, M.G., Callahan, P.L., Sardana, V.V., Garsky, V.M. and Colonno, R.J. (1990) Substrate requirements of human rhinovirus 3C protease for peptide cleavage *in vitro. Journal of Biological Chemistry*, **265**, 9062–9065.

Couch, R.B. (1996) Rhinoviruses. In *Fields Virology, 3rd Ed.*, edited by B.N. Fields, D.M. Knipe, P.M. Howley, R.M. Chanock, J.L. Melnick, T.P. Monath *et al*, pp. 713–734. Philadelphia: Lippincott-Raven.

Cox, G.A., Johnson, R.B., Cook, J.A., Wakulchick, M., Johnson, M.G., Villarreal, E.C. *et al* (1999) Identification and characterization of human rhinovirus-14 3C protease deamidation isoform. *Journal of Biological Chemistry*, **274**, 13211–13216.

Davis, G.J., Wang, Q.M., Cox, G.A., Johnson, R.B., Wakulchick, M., Dotson, C.A. *et al* (1997) Expression and purification of recombinant rhinovirus 14 3CD proteinase and its comparison to the 3C proteinase. *Archives of Biochemistry and Biophysics*, **346**, 125–130.

Donner, C.D. and Gill, M. (1999) Synthesis and absolute stereochemistry of thysanone. *Tetrahedron Letters*, **40**, 3921–3924.

Dragovich, P.S., Webber, S.E., Babine, R.E., Fuhrman, S.A., Patick, A.K., Matthews, D.A. *et al* (1998a) Structure-based design, synthesis, and biological evaluation of irreversible human rhinovirus 3C protease inhibitors. 1. Michael acceptor structure–activity studies. *Journal of Medicinal Chemistry*, **41**, 2806–2818.

Dragovich, P.S., Webber, S.E., Babine, R.E., Fuhrman, S.A., Patick, A.K., Matthews, D.A. *et al* (1998b) Structure-based design, synthesis, and biological evaluation of irreversible human rhinovirus 3C protease inhibitors. 2. Peptide structure–activity studies. *Journal of Medicinal Chemistry*, **41**, 2819–2834.

Dragovich, P.S., Zhou, R., Skalitzky, D.J., Fuhrman, S.A., Patick, A.K., Ford, C.E. *et al* (1999a) Solid-phase synthesis of irreversible human rhinovirus 3C protease inhibitors. Part 1. Optimization of tripeptides incorporating N-terminal amides. *Bioorganic & Medicinal Chemistry*, **7**, 589–598.

Dragovich, P.S., Prins, T.J., Zhou, R., Fuhrman, S.A., Patick, A.K., Matthews, D.A. *et al* (1999b) Structure-based design, synthesis, and biological evaluation of irreversible human rhinovirus 3C protease inhibitors. 3. Structure–activity studies of ketomethylene-containing peptidomimetics. *Journal of Medicinal Chemistry*, **42**, 1203–1212.

Dragovich, P.S., Webber, S.E., Prins, T.J., Zhou, R., Marakovits, J.T., Tikhe, J.G. *et al* (1999c) Structure-based design of irreversible, tripeptidyl human rhinovirus 3C protease inhibitors containing *N*-methyl amino acids. *Bioorganic & Medicinal Chemistry Letters*, **9**, 2189–2194.

Dragovich, P.S., Prins, T.J., Zhou, R., Webber, S.E., Marakovits, J.T., Fuhrman, S.A. *et al* (1999d) Structure-based design, synthesis, and biological evaluation of irreversible human rhinovirus 3C protease inhibitors. 4. Incorporation of P_1 lactam moieties as L-glutamine replacements. *Journal of Medicinal Chemistry*, **42**, 1213–1224.

Dragovich, P.S., Zhou, R., Webber, S.E., Prins, T.J., Kwok, A.K., Okano, K. *et al* (2000) Structure-based design of ketone-containing, tripeptidyl human rhinovirus 3C protease inhibitors. *Bioorganic & Medicinal Chemistry Letters*, **10**, 45–48.

Fromtling, R.A. and Castañer, J. (1997) VP-63843. *Drugs of the Future*, **22**, 40–44.

Gorbalenya, A.E., Blinov, V.M. and Donchenko, A.P. (1986) Poliovirus-encoded proteinase 3C: a possible evolutionary link between cellular serine and cysteine proteinase families. *FEBS Letters*, **194**, 253–259.

Hamdouchi, C., Ezquerra, J., Vega, J.A., Vaquero, J.J., Alvarez-Builla, J. and Heinz, B.A. (1999) Short synthesis and anti-rhinoviral activity of imidazo[1,2-*a*]pyridines: the effect of acyl groups at 3-position. *Bioorganic & Medicinal Chemistry*, **9**, 1391–1394.

Hamparian, V.V., Colonno, R.J., Cooney, M.K., Dick, E.C., Gwaltney, J.M., Jr. Hughes, J.H. *et al* (1987) A collaborative report: rhinoviruses-extension of the numbering system from 89 to 100. *Virology*, **159**, 191–192.

Heinz, B.A., Tang, J., Labus, J.M., Chadwell, F.W., Kaldor, S.W. and Hammond, M. (1996) Simple in vitro translation assay to analyze inhibitors of rhinovirus proteases. *Antimicrobial Agents and Chemotherapy*, **40**, 267–270.

Hopkins, J.L., Betageri, R., Cohen, K.A., Emmanuel, M.J., Joseph, C.R., Bax, P.M. *et al* (1991) Rhinovirus 3C protease catalyzes efficient cleavage of a fluorescein-labeled peptide affording a rapid and robust assay. *Journal of Biochemical and Biophysical Methods*, **23**, 107–113.

Jungheim, L.N., Cohen, J.D., Johnson, R.B., Villarreal, E.C., Wakulchik, M., Loncharich, R.J. *et al* (1997) Inhibitors of human rhinovirus 3C protease by homophthalimides. *Bioorganic & Medicinal Chemistry Letters*, **7**, 1589–1594.

Kadam, S., Poddig, J., Humphrey, P., Karwowsky, J., Jackson, M., Tennett, S. *et al* (1994) Citrinin hydrate and radicinin: human rhinovirus 3C-protease inhibitors discovered in a target-directed microbial screen. *Journal of Antibiotics*, **47**, 836–839.

Kaldor, S.W., Hammond, M., Dressman, B.A., Labus, J.M., Chadwell, F.W., Kline, A.D. *et al* (1995) Glutamine-derived aldehydes for the inhibition of human rhinovirus 3C protease. *Bioorganic & Medicinal Chemistry Letters*, **5**, 2021–2026.

Kati, W.M., Sham, H.L., McCall, J.O., Montgomery, D.A., Wang, G.T., Rosenbrook, W. *et al* (1999) Inhibition of 3C protease from human rhinovirus strain 1B by peptidyl bromomethylketonehydrazides. *Archives of Biochemistry and Biophysics*, **362**, 363–375.

Kettner, C.A. and Korant, B.D. (1987) Method for preparing specific inhibitors of virus-specified proteases. United States Patent No. 4,644,055.

Knott, J.A., Orr, D.C., Montgomery, D.S., Sullivan, C.A. and Weston, A. (1989) The expression and purification of human rhinovirus protease 3C. *European Journal of Biochemistry*, **182**, 547–555.

Kong, J.-S., Venkatraman, S., Furness, K., Nimkar, S., Shepherd, T.A., Wang, Q.M. *et al* (1998) Synthesis and evaluation of peptidyl Michael acceptors that inactivate human rhinovirus 3C protease and inhibit virus replication. *Journal of Medicinal Chemistry*, **41**, 2579–2587.

Kräusslich, H.-G. and Wimmer, E. (1988) Viral proteinases. *Annual Review of Biochemistry*, **57**, 701–754.

Lee, W.-M., Wang, W. and Rueckert, R.R. (1994) Complete sequence of the RNA genome of human rhinovirus 16, a clinically useful common cold virus belonging to the ICAM-1 receptor group. *Virus Genes*, **9**, 177–181.

Leong, L.E.-C., Walker, P.A. and Porter, A.G. (1992) Efficient expression and purification of a protease from the common cold virus, human rhinovirus type 14. *Journal of Crystal Growth*, **122**, 246–252.

Leong, L.E.-C., Walker, P.A., Porter, A.G. (1993) Human rhinovirus-14 protease 3C (3Cpro) binds specifically to the 5′-noncoding region of the viral RNA. *Journal of Biological Chemistry*, **268**, 25735–25739.

Long, A.C., Orr, D.C., Cameron, J.M., Dunn, B.M. and Kay, J. (1989) A consensus sequence for substrate hydrolysis by rhinovirus 3C proteinase. *FEBS Letters*, **258**, 75–78.

McKinlay, M.A., Pevear, D.C. and Rossmann, M.G. (1992) Treatment of the picornavirus common cold by inhibitors of viral uncoating and attachment. *Annual Reviews of Microbiology*, **46**, 635–654.

Malcolm, B.A., Lowe, C., Shechosky, S., McKay, R.T., Yang, C.C., Shah, V.J. *et al* (1995) Peptide aldehyde inhibitors of hepatitis A virus 3C proteinase. *Biochemistry*, **34**, 8172–8179.

Matthews, D.A., Smith, W.W., Ferre, R.A., Condon, B., Budahazi, G., Sisson, W. *et al* (1994) Structure of human rhinovirus 3C protease reveals a trypsin-like polypeptide fold, RNA binding site, and means for cleaving precursor polyprotein. *Cell*, **77**, 761–771.

Matthews, D.A., Dragovich, P.S., Webber, S.E., Fuhrman, S.A., Patick, A.K., Zalman, L.S. *et al* (1999) Structure-assisted design of mechanism based irreversible inhibitors of human rhinovirus 3C protease with potent antiviral activity against multiple rhinovirus serotypes. *Proceedings of the National Academy of Sciences, U.S.A.*, **96**, 11000–11007.

Oren, D.A., Zhang, A., Nesvadba, H., Rosenwirth, B. and Arnold, E. (1996) Synthesis and activity of piperazine-containing antirhinoviral agents and crystal structure of SDZ 880-061 bound to human rhinovirus 14. *Journal of Molecular Biology*, **259**, 120–134.

Orr, D.C., Long, A.C., Kay, J., Dunn, B.M. and Cameron, J.M. (1989) Hydrolysis of a series of synthetic peptide substrates by the human rhinovirus 14 3C proteinase, cloned and expressed in *Escherichia coli*. *Journal of General Virology*, **70**, 2931–2942.

Otto, H.-H. and Schirmeister, T. (1997) Cysteine proteases and their inhibitors. *Chemical Reviews*, **97**, 133–171.

Palmenberg, A.C. (1987) Picornaviral processing: some new ideas. *Journal of Cellular Biochemistry*, **33**, 191–198.

Patick, A.K., Binford, S.L., Brothers, M.A., Jackson, R.L., Ford, C.E., Diem, M.D. *et al* (1999) Antiviral activity of AG7088, a potent inhibitor of human rhinovirus 3C protease. *Antimicrobial Agents and Chemotherapy*, **43**, 2444–2450.

Porter, A.G. (1993) Picornavirus nonstructural proteins: emerging roles in virus replication and inhibition of host cell functions. *Journal of Virology*, **67**, 6917–6921.

Rasnick, D. (1996) Small synthetic inhibitors of cysteine proteases. *Perspectives in Drug Discovery and Design*, **6**, 47–63.

Rich, D.H. (1986) Inhibitors of cysteine proteases. In *Proteinase Inhibitors*, edited by A.J. Barrett and G. Salvesen, pp. 153–178. Amsterdam: Elsevier.

Rich, D.H. (1990) Peptidase inhibitors. In *Comprehensive Medicinal Chemistry*, edited by P.G. Sammes and J.B. Taylor, pp. 391–441. Oxford: Pergamon.

Rueckert, R.R. (1996) *Picornaviridae*: the viruses and their replication. In *Fields Virology, 3rd Ed.*, edited by B.N. Fields, D.M. Knipe, P.M. Howley, R.M. Chanock, J.L. Melnick, T.P. Monath *et al*, pp. 609–654. Philadelphia: Lippincott-Raven.

Sham, H.L., Rosenbrook, W., Kati, W., Betebenner, D.A., Wideburg, N.E., Saldivar, A. *et al* (1995) Potent inhibitor of the human rhinovirus (HRV) 3C protease containing a backbone modified glutamine. *Journal of the Chemical Society, Perkin Transactions 1*, 1081–1082.

Shaw, E. (1990) Cysteinyl proteinases and their selective inactivation. *Advances in Enzymology*, **63**, 271–347.

Shepherd, T.A., Cox, G.A., McKinney, E., Tang, J., Wakulchik, M., Zimmerman, R.E. *et al* (1996) Small peptidic aldehyde inhibitors of human rhinovirus 3C protease. *Bioorganic & Medicinal Chemistry Letters*, **6**, 2893–2896.

Singh, S.B., Cordingley, M.G., Ball, R.G., Smith, J.L., Dombrowski, A.W. and Goetz, M.A. (1991) Structure and stereochemistry of thysanone: a novel human rhinovirus 3C-protease inhibitor from *Thysanophora penicilloides*. *Tetrahedron Letters*, **32**, 5279–5282.

Skiles, J.W. and McNeil, D. (1990) Spiro indolinone beta-lactams, inhibitors of poliovirus and rhinovirus 3C-proteinases. *Tetrahedron Letters*, **31**, 7277–7280.

Turner, R.B., Wecker, M.T., Pohl, G., Witek, T.J., McNally, E., St. George, R. *et al* (1999) Efficacy of Tremacamra, a soluble intercellular adhesion molecule 1, for experimental rhinovirus infection. *Journal of the American Medical Association*, **281**, 1797–1804.

Wang, Q.M., Johnson, R.B., Cox, G.A., Villarreal, E.C. and Loncharich, R.J. (1997) A continuous colorimetric assay for rhinovirus-14 3C protease using peptide *p*-nitroanilides as substrates. *Analytical Biochemistry*, **252**, 238–245.

Wang, Q.M. (1998a) Human rhinovirus 3C protease inhibitors: recent developments. *Expert Opinion on Therapeutic Patents*, **8**, 1151–1156.

Wang, Q.M., Johnson, R.B., Jungheim, L.N., Cohen, J.D. and Villarreal, E.C. (1998b) Dual inhibition of human rhinovirus 2A and 3C proteases by homophthalimides. *Antimicrobial Agents and Chemotherapy*, **42**, 916–920.

Wang, Q.M. (1999) Protease inhibitors as potential antiviral agents for the treatment of picornaviral infections. *Progress in Drug Research*, **52**, 197–219.

Webber, S.E., Tikhe, J., Worland, S.T., Fuhrman, S.A., Hendrickson, T.F., Matthews, D.A. *et al* (1996) Design, synthesis, and evaluation of nonpeptidic inhibitors of human rhinovirus 3C protease. *Journal of Medicinal Chemistry*, **39**, 5072–5082.

Webber, S.E., Okano, K., Little, T.L., Reich, S.H., Xin, Y., Fuhrman, S.H. *et al* (1998) Tripeptide aldehyde inhibitors of human rhinovirus 3C protease: design, synthesis, biological evaluation, and cocrystal structure solution of P_1 glutamine isosteric replacements. *Journal of Medicinal Chemistry*, **41**, 2786–2805.

Zalman, L.S., Brothers, M.A., Dragovich, P.S., Zhou, R., Prins, T.J., Worland, S.T. *et al* (2000) Inhibition of human rhinovirus induced cytokine production by AG7088, a human rhinovirus 3C protease inhibitor. *Antimicrobial Agents and Chemotherapy*, **44**, 1236–1241.

Chapter 14

Aminopeptidases

Allen Taylor and Jason Warner

Many physiological functions of aminopeptidases (APs) have been identified. These include essential roles in protein maturation, degradation of nonhormonal and hormonal peptides and, possibly, determination of protein stability. Other recent reports indicate functions for aminopeptidases including trimming of antigenic peptides for presentation on MHC 1. Recently it was shown that immunization of sheep with LAP protects against fascioliasis. In cats, aminopeptidase N acts as a common receptor for coronaviruses in group 1. Many disease states are associated with impaired proteolytic function. Several APs also catalyze reactions in addition to peptide hydrolysis. The enzymes have long been used for diagnosis of various physiological states and disease conditions and recently they have found laboratory and industrial uses as well. Administration of bestatin and other inhibitors, which presumably have their effect by inhibition of APs, has also been used to alter physiological status and disease progress.

14.1 INTRODUCTION

Aminopeptidases (AP) catalyze the hydrolysis of amino acid residues from the amino terminus of peptide substrates. These enzymes generally have broad specificity, occur in several forms in many tissues or cells, on cell surfaces, in soluble cytoplasmic or secreted forms (Ledeme *et al* 1983; Taylor *et al* 1984; McDonald 1986; Chang and Smith 1989; Stirling *et al* 1989; Watt and Yip 1989; Ahmad and Ward 1990; Aoyagi 1996; Bradshaw and Arvin 1996; Chang 1996; Fitzpatrick and Orning 1996; Taylor *et al* 1996; Van Wart 1996; Walling and Gu 1996), and are widely distributed throughout the plant and animal kingdoms. Over 150 APs have been purified, many of which have been cloned and characterized. In some cells they constitute a substantial proportion of enzyme protein (Taylor *et al* 1982a,b, 1983; Watt and Yip 1989). The last compendium of this topic was in 1996 (Taylor 1996a). This book is now available from the author. Since then there have been over 750 publications regarding aminopeptidases. It is impossible for this brief review to cover all the articles. For more information I refer readers to the previous reviews and to a useful website, *www.merops.co.uk*.

14.2 NOMENCLATURE

As reviewed in depth earlier (Taylor 1996b), a unique nomenclature system for APs remains elusive. Instead, classification of AP has been done primarily according to the interest of the investigator. The following criteria have been used to classify aminopeptidases. (1) Number of amino acids cleaved from the N-terminus of substrates. (2) With respect to the relative efficiency with which NH_2-terminal residues are removed from peptides or peptide analogs. (3) Location. Some peptidases are secreted, but most are not. There are cytosolic and microsomal enzymes, integral-membrane-bound or membrane-associated enzymes. (4) Inhibitor susceptibility. (5) Metal ion content. (6) Conditions of maximal activity (i.e. pH). (7) Size. (8) Thermostability. (9) Number of functions. While for most aminopeptidases only one function has been elucidated, there are now several APs for which two functions have been demonstrated (Chang 1996; Fitzpatrick and Orning 1996; Walling and Gu 1996).

These nomenclature or classification systems serve to identify the protease with respect to a topic of interest, and they are not mutually exclusive. Thus, it is not surprising that in recent years several enzymes, which were previously thought to be distinct, were shown to be identical (Taylor *et al* 1996). For example, hog kidney leucine aminopeptidase (LAP) (which was described as not able to hydrolyze prolyl bonds) and hog intestinal prolyl AP are indistinguishable (Matsushima *et al* 1991). The same pertains to rat kidney and brain prolyl AP and hog kidney LAP (Taylor *et al* 1981; Sanderink *et al* 1988; Turzynski and Mentlein 1990; Gibson *et al* 1991). The *Escherichia coli xerB* gene product, *E. coli* AP I, now called AP A, and *Salmonella typhimurium* aminopeptidase appear to be the same (Stirling *et al* 1989), as are APs N and M (McDonald 1986).

14.3 BIOLOGICAL, PHARMACOLOGICAL, LABORATORY AND INDUSTRIAL USES OF AMINOPEPTIDASES AND THEIR INHIBITORS

Many physiological functions of APs have been identified. These include essential roles in protein maturation (Moerschell *et al* 1990), degradation of nonhormonal (Botbol and Scornik 1991) and hormonal peptides, and, possibly, determination of protein stability (Figure 14.1) (Bachmair *et al* 1986). Other recent reports indicate functions for aminopeptidases including trimming of antigenic peptides for presentation on MHC I (Beninga *et al* 1998; Mo *et al* 1999; York *et al* 1999). Recently it was shown that immunization of sheep with LAP protects against fascioliasis (Piacenza *et al* 1999). In cats, aminopeptidase N acts as a common receptor for coronaviruses in group I (Tresnan and Holmes 1998). Many disease states are associated with impaired proteolytic function (Umezawa *et al* 1976; Nishizawa *et al* 1977; Taylor and Davies 1987; Ayoyagi 1990; Dice 1993; Taylor *et al* 1993a; Zuo *et al* 1995). Several APs also catalyze reactions in addition to peptide hydrolysis (described below).

Some aminopeptidase activities, such as methionine aminopeptidases, are essential. Others, while not essential, affect cells in profound ways. In all cells, protein synthesis is initiated at an AUG codon specifying methionine in the cytosol of eukaryotes, or formylmethionine in prokaryotes, mitochondria and chloroplasts. Protein maturational events, including NH_2-terminal modifications of nascent peptides, are by far the most common processing events, occurring on nearly all proteins (Moerschell *et al* 1990). In this process where formylmethionine is used to initiate protein synthesis, the formyl group is usually removed cotranslationally by a deformylase, leaving methionine bearing a free NH_2 group (Takeda and Webster 1968; Hausman *et al* 1972; Ball and Kaesberg 1973;

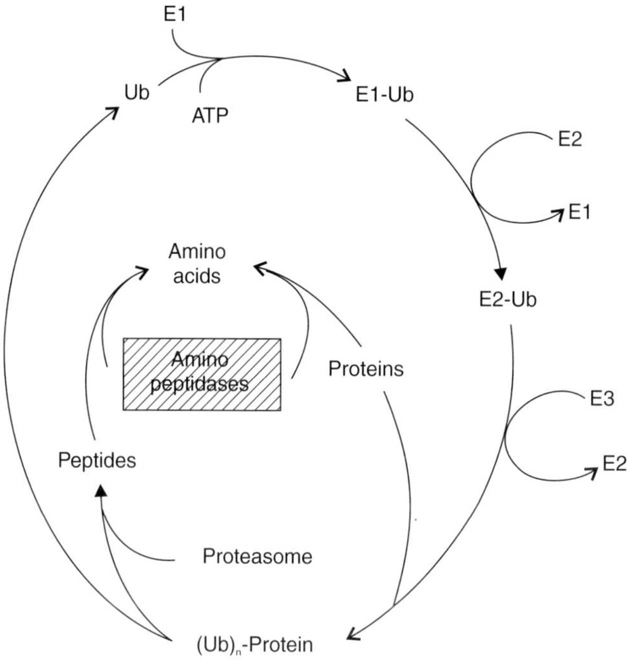

Figure 14.1 Role of aminopeptidases in determining rates of degradation of proteins. According to the "N-end rule," the rate at which a protein is degraded is determined (at least in part) by its amino-terminal amino acid residue. Since aminopeptidases remove amino-terminal amino acids, thus revealing new amino-terminal amino acid residues, the aminopeptidases may play a significant role in determining rates of protein degradation. In order to dramatize their role, the aminopeptidases are shown in a central position in this schematic representation of the ubiquitin-dependent proteolytic pathway. E1, E2, and E3 are enzymes which activate and sequentially transfer ubiquitin to protein substrates.

Bachmair *et al* 1986; Wilcox *et al* 1987; Moerschell *et al* 1990; Tsunasawa 1995). In both eukaryotes and prokaryotes the NH_2-terminal methionine may be removed by a methionine aminopeptidase(s) (MAP) (Bachmair *et al* 1986; Ben-Bassat *et al* 1987; Miller *et al* 1987; Wilcox *et al* 1987; Moerschell *et al* 1990; Bradshaw *et al* 1998; Kitamura *et al* 1999). Removal of the NH_2-terminal methionine is required in order to reveal functionally important NH_2-terminal residues and/or to allow NH_2-terminal modification (such as myristoylation) which is required for physiological function. It is not surprising that deletion of methionine AP (*pepM* gene product) is lethal in *S. typhimurium* (Miller *et al* 1989). Consistent with essentiality of some methionine APs is the observation that *pepM* mutants of *E. coli* or *S. typhimurium* could not be obtained (Miller *et al* 1987, 1989). Met AP also appears to play a role in p53 mediated cell-cycle inhibition by the anti-angiogenic agent TNP-470 (Zhang *et al* 2000) and are required for EGF-induced cell-cycle control (Takahashi *et al* 1989).

Since, except in disease states or upon aging, protein fragments rarely accumulate (Dice 1993; Taylor *et al* 1993a), a "housekeeping" role for APs is indicated in continuous protein turnover and/or regulation of protein levels and in selective elimination of obsolete or defective proteins including degradation via the ubiquitin-dependent proteolytic pathway (Taylor 1996a). In some cases the N-end amino acid is involved in determining the rate of proteolysis of substrates by this pathway (Johnson *et al* 1995). A central role for APs in defining the stability of proteins can be envisioned for those proteins which are degraded by such an "N-end ubiquitin-dependent degradation pathway" (Figure 14.1). A corollary is that retention of Met or other stabilizing residues may protect short-lived proteins from premature protein degradation (Bachmair *et al* 1986; Chang *et al* 1992).

Proteolytic capabilities also allow cells to adapt to changing environmental conditions. Supply of amino acids and energy during starvation and/or differentiation and degradation of transported exogenous peptides to amino acids for nutrition also represent functions of aminopeptidases (Botbol and Scornik 1991). This is confirmed by the time-related reduction in viability of some strains mutated for several AP coding genes (Reeve *et al* 1984). In *E. coli*, AP-A (*pepA*, the *xerB* gene product) is absolutely required for ColE1 stabilization of unstable plasmid multimers, which occurs via site-specific recombination (at the cer locus) into monomer form (Stirling *et al* 1989). This is also identical with the carP gene which is involved in pyrimidine-specific regulation of the upstream P1 promoter of *E. coli* (Charlier *et al* 1995).

PepA and the arginine repressor (ArgR) serve as accessory proteins, ensuring that recombination is exclusively intramolecular. In contrast, *pepA* homologs from other species have no known DNA binding activity and are not implicated in transcriptional regulation or site-specific recombination (Strater *et al* 1999a). New phosphinic peptides are potent inhibitors of these enzymes (Georgiadis *et al* 2000) and may have uses in resolving physiologic function or be of pharmaceutical interest.

In contrast with the essentiality of the APs noted above, in *Saccharomyces* deletion of the *MAP1* gene is associated with retarded growth but is not lethal (Chang *et al* 1992). This suggests that alternative NH_2-terminal processing pathway(s) exist for cleaving methionine from nascent polypeptide chains in eucaryotic cells (Chang and Smith 1989; Cueva *et al* 1989). Redundancy of AP activities in procaryotic and eucaryotic cells may provide yet another, albeit teleological, support for the important cellular functions for APs (Chang and Smith 1989; Cueva *et al* 1989).

*BLH*1 codes for a yeast thiol aminopeptidase which is homologous to mammalian bleomycin hydrolase (Enenkel and Wolf 1993). Whereas deletion of the *BLH*1 gene is not lethal under normal growth conditions, *BLH*1 mutants show hypersensitivity to bleomycin. This indicates that bleomycin hydrolase is able to inactivate bleomycin *in vivo* and to protect cells from bleomycin-induced toxicity (Enenkel and Wolf 1993).

To the extent to which physiological functions which are affected by bestatin involve aminopeptidases, it would appear that APs are involved in the delayed-type hypersensitivity (Umezawa *et al* 1976), murine tumor growth rate and enhanced antitumor activity of antibiotics (Ishizuka *et al* 1980), DNA metabolism in spleen and thymus T-cells (Muller *et al* 1979), and stimulated polysome assembly (Muller *et al* 1981). Roles for AP in antibiotic activation and transport have been documented (Gonzales and Robert-Baudouy 1996). Three-fold higher levels of LAP have been found in HIV patients as compared to uninfected individuals and bestatin reduced infection in cell-infection assays (Pulido-Cejudo *et al* 1997).

Leukotriene hydrolase activity of an aminopeptidase suggests roles for APs in inflammation (Minami *et al* 1990; Orning *et al* 1991; Izumi *et al* 1993; Fitzpatrick and Orning 1996). Also suggestive of relations between AP activity and inflammation are observations that AP P (aminoacylprolyl-peptide hydrolase EC 3.4.11.9) may be involved in hydrolyzing bradykinin (Simmons and Orawski 1992; Orawski and Simmons 1995; Kitamura *et al* 1999; Kim *et al* 2000).

Recent data indicate that bestatin-inhibitable aminopeptidases are involved in conversion of procollagenase to collagenase with additional roles in growth and differentiation (Yoneda *et al* 1992; Riemann *et al* 1999). In related research, it was demonstrated that AP N/CD13 plays a role in degradation and invasion of extracellular matrix since A375M melanoma cells transfected with full length cDNA of AP N/CD13 show enhanced degradation of type IV collagen (Fujii *et al* 1995). This association between AP activity and tissue invasion is corroborated by data which show that monoclonal antibodies to AP N/CD13 inhibit invasion of metastatic renal cells into Matrigel-coated filters and degradation of type IV collagen, and that this activity is also inhibited by bestatin (Saiki *et al* 1993). Aminopeptidase N from pig contains sequences which can act as cellular virus binding receptors (Delmas *et al* 1994). An extension of this information indicates that bestatin, which inhibits several aminopeptidases (Taylor 1996a; Taylor *et al* 1996) exhibits direct antileukemic effects against human leukemic cells through induction of apoptosis (Sekine *et al* 1999). Drugs aimed at MetAps also have anti-cancer potential l (Lowther and Matthews 2000). Inhibition of aminopeptidase N also has growth modulatory effects in Karpas-299 cells, an effect elicited via p24/ERK Map Kinase (Lendeckel *et al* 1998).

Aminopeptidases are also involved in regulation of blood pressure (Reaux *et al* 1999). In the blood clotting cascade, aminopeptidase A appears to liberate angiotensin from angiotensin II (Chauvel *et al* 1994). Aminopeptidases N and A participate in inactivation of angiotensin III (Song and Healy 1999). This information has been exploited for the design of drugs to regulate rates of blood clotting. Aminopeptidase N is also involved in angiogenesis and can serve as a target for delivering drugs into tumors and for inhibiting angiogenesis (Pasqualini *et al* 2000). A new potent aminopeptidase inhibitor is a phosphinylpropanoyl iodotyrosine derivative (Noble *et al* 2000). Similarly, fumagillin, an inhibitor of MetAP-2 shows antiangiogenic activity (Sin *et al* 1997).

Aminopeptidases also participate in metabolism of secreted regulatory molecules including hormones and neurotransmitters (Taylor and Dixon 1978; Malfroy *et al* 1989; Ahmad and Ward 1990; Nyberg *et al* 1990; Gibson *et al* 1991; Squire *et al* 1991). This includes partial degradation of enkephalin by cerebral pericytic AP N at the blood–brain interface (Kunz *et al* 1994), and by murine macrophages. Another function of APs appears to be in modulation of cell–cell interactions (Watt and Yip 1989).

Evidence for dual functions or activities of aminopeptidases has been mounting. In some cases, mutational analysis has been cleverly exploited to separate functions in the APs. For example, mutation of the *pepA* gene which inactivates the hydrolytic activity does not eliminate the role of AP A in the recombination process (McCulloch *et al* 1994). Second functions for APs, which are separable from AP activity, are also reported for leukotriene A4 hydrolase which has arginyl aminopeptidase activity (Izumi *et al* 1993) and *Ochrobactrum anthropi* D-aminopeptidase (also see Fitzpatrick and Orning 1996; Taylor *et al* 1996). *Saccharomyces cerevisiae* leucyl aminopeptidase transforms leukotriene A_4 to an eicosatetraenoic acid and unlike the mammalian enzyme, both activities appear to be due to a similar, if not identical, active site (Kull *et al* 1999). Dual function may also be implied by structural studies of methionine AP from *S. cerevisiae*. Whereas a form of MAP missing the Zn^{2+} fingers is as active as wild type, the truncated form is significantly less active in rescuing the slow growth phenotype of the *map* mutant than wild type MAP (Zuo *et al*

1995; Bradshaw and Arvin 1996; Chang 1996; Taylor *et al* 1996). Thus, initiation factor-associated proteins may include methionine aminopeptidases. Another example of a second function for aminopeptidases is found in aminopeptidase N which serves as a coronavirus receptor (Delmas *et al* 1992; Yeager *et al* 1992).

Another recent area of research regarding the aminopeptidases is in trafficking in response to insulin (Waters *et al* 1997; Kandror 1999; Garza and Birnbaum 2000). To this end, endothelin-1 stimulated the translocation of insulin-responsive aminopeptidase and GLUT4 (glucose transporter) to the plasma membrane with associated stimulation of glucose uptake (Wu-Wong *et al* 1999).

In contrast with the catabolic roles for aminopeptidases, a biosynthetic or hydrolytic role in peptidoglycan has been suggested (Gonzales and Robert-Baudouy 1996). Further advances regarding physiological functions of APs should be possible, since the availability of new fluorogenic substrates for aminopeptidases makes it possible to detect their activity *in vivo* (Ganesh *et al* 1995).

Aminopeptidases have frequently been used to sequentially remove amino-terminal amino acids from proteins; i.e. AP M was used for studies of structure-biological function of peptides that bind the thrombin receptor (Godin *et al* 1994). Pyroglutamate AP has found use in determining sequence and content of pyroglutamate in proteins and for deprotection prior to sequencing of proteins (Klebert *et al* 1993; Kim and Kim 1995). A recent use of aminopeptidases is in removal of polyhistidine tags from recombinant proteins (Pedersen *et al* 1999). In the dairy industry the terminal degradation of peptides derived from casein is accomplished with APs (Gonzales and Robert-Baudouy 1996).

Aminopeptidases have also found use in peptide synthesis. Prolyldipeptidyl AP from *Lactococcus lactis* (*pepX*) was used as a catalyst in the kinetically controlled synthesis of peptide bonds involving proline (Yoshpe-Besanco *et al* 1994). Aminopeptidase A was used for reversible protection of the αNH_2 group of amino acids (Yoshpe-Besancon *et al* 1994).

14.4 REGULATION OF EXPRESSION

There is a bourgeoning in reports regarding effects of temperature (Moser *et al* 1970; Stoll *et al* 1972, 1973; Lazdunski *et al* 1975b; Roncari *et al* 1976; Miller *et al* 1991), oxygen limitation (Lazdunski *et al* 1975a; Murgier *et al* 1976; Gharbi *et al* 1985; Strauch *et al* 1985; Foglino and Lazdunski 1987; Brown *et al* 1993) and other stresses, endogenous inhibitors and drugs on expression and regulation of expression of aminopeptidases. Much of that work was reviewed previously (Ledeme *et al* 1983; Taylor *et al* 1983, 1996; Eisenhauer *et al* 1988; Watt and Yip 1989; Harris *et al* 1992; Wallner *et al* 1993; Aoyagi 1996; Taylor 1996b).

Several experiments are consistent with regulation of aminopeptidase expression at the transcriptional level. There is a marked enhancement of LAP activity (Eisenhauer *et al* 1988); and LAP mRNA (Wallner *et al* 1993) upon removal of serum from culture media in which *in vitro* aged and/or transforming lens epithelial cells were grown. Concerted induction of LAP mRNA and LAP protein by interferon gamma was also noted in human ACHN renal carcinoma, A549 lung carcinoma, HS153 fibroblasts, and A375 melanoma (Harris *et al* 1992). Induction of LAP mRNA is a secondary response to interferon, blocked by inhibition of protein synthesis.

Genetic elements involved in expression of both transcripts for membrane-bound AP N/CD13 have been identified, and it appears that physically distinct promoters have evolved to regulate expression of this enzyme in different tissues (Shapiro *et al* 1991). Many of the bacterial aminopeptidases display promotor consensus sequences characteristic of genes transcribed by an RNA polymerase associated with $\sigma 70$ (Gonzales and Robert-Baudouy 1996). An AT motif binding factor 1-A negatively regulates transcription of intestinal ApN (Kataoka *et al* 2000). Several genes encoding APs are part of operons.

Keller recently demonstrated that a member of the mammalian zinc-dependent membrane APs is vesicle protein 165, the cellular distribution of which is at least in part insulin regulated (Keller *et al* 1995). A proposed function as a receptor for aminopeptidase N has been confirmed by molecular genetic techniques in M. *sexta* (Knight *et al* 1995).

The yeast *yscI* gene product, APE1, is the vacuolar glycoprotein AP (Cueva *et al* 1989), which appears to facilitate amino acid uptake by hydrolyzing peptides prior to absorption. APE1 synthesis is subject to carbon catabolite levels i.e. APE1 is repressed in media containing more than 1% glucose. But as cells reach stationary phase, the increase in APE1 activity may indicate release from carbon catabolite repression. Other examples of induction by amino acid limitation or catabolite repression (Carter and Miller 1984; Ludewig *et al* 1987; Conlin *et al* 1994), as well as during different phases of cell cycle, (Mayo *et al* 1991; Yan *et al* 1992; Arora and Lee 1994; Vesanto *et al* 1994) have been noted. Studies regarding the APE1 promoter are in progress (Bordallo *et al* 1995). It is curious that the enzyme isolated from stationary cells has 4 amino acids removed from the N-terminus (Chang and Smith 1989). Activity is enhanced several fold when ammonia rather than peptone is used as the sole source of nitrogen (Frey and Rohm 1978). Expression of *yscI* is dependent upon levels of *yscA* and *pep4* gene product (Jones *et al* 1982).

LAP (Taylor *et al* 1983; Eisenhauer *et al* 1988) and some plant AP (Couton *et al* 1991) levels are also enhanced during development and growth. In *Lactobacillus* one open reading frame is co-transcribed with the *pepC* gene at the exponential phase of growth, whereas, at the stationary growth phase, transcripts from the *pepC* promoter were below the detection limit and the ORF2 was expressed by its own promoter (Vesanto *et al* 1994).

Regulation at the post transcriptional level is reported for *Aeromonas proteolytica* aminopeptidase (Rawlings and Barrett 1995). The enzyme is synthesized as a 43 kDa precursor. Maintenance of the organism at elevated temperatures results in double cleavages and frees a mature and active 32 kDa enzyme which is active at 70 °C. The precursor is active but is inactivated at 70 °C. Similar regulation might be anticipated for lens and kidney LAP since it is also synthesized as a precursor protein (Wallner *et al* 1993).

14.5 COMPOSITION AND STRUCTURE

Lens leucine aminopeptidase is the aminopeptidase for which structural, kinetic, and mechanistic information is most developed. Thus, this enzyme is described as a prototype for other aminopeptidases which share similar structure and mechanism of action. Crystallographic, electron micrographic, NMR, and photoaffinity labeling and modeling studies indicate that lens LAP protomers are bilobal, and that inhibitors and substrates are bound in an active site which is found in the larger lobe of each protomer (Figures 14.2a and 14.3). Zn^{2+} is involved in substrate liganding and presumably in catalysis of hydrolysis. Homologies are extensive between mammalian lens LAP and APs in organisms as diverse as *E. coli* and plants, particularly in catalytically important residues, or in residues involved in metal ion binding. For additional details regarding composition of some other APs, readers are referred to previous reviews and the world wide web.

Like several of the other APs, beef lens (bl) LAP is synthesized as a larger precursor of 514 amino acids (Wallner *et al* 1993) . This is reduced to an oligomer containing 487 amino acid residues and 2 (or possibly 3) Zn^{2+} (Wallner *et al* 1993; Strater and Lipscomb 1995a). Other aminopeptidases also show prosequences which include amino terminal extensions (Chang and Smith 1989; Cueva *et al* 1989; Watt and Yip 1989; Guenet *et al* 1992). Propeptides of the proenzymes may have intramolecular chaperone functions with roles in assuring correct folding of the enzyme (Nirasawa *et al* 1999). The distribution of protein in the blLAP protomer is two-thirds and one-third between the larger and smaller lobes, respectively. (Jurnak *et al* 1977; Taylor *et al* 1979; 1993a,b, 1996; Taylor 1993a,b). Trypsinization of the enzyme or reaction with hydroxylamine result in unique cleavages (Cuypers *et al* 1982). Hexamers of identical protomers appear as two concentric triangles, the smaller being offset from the larger "less dense" triangle by 60° (Figure 14.2b,c).

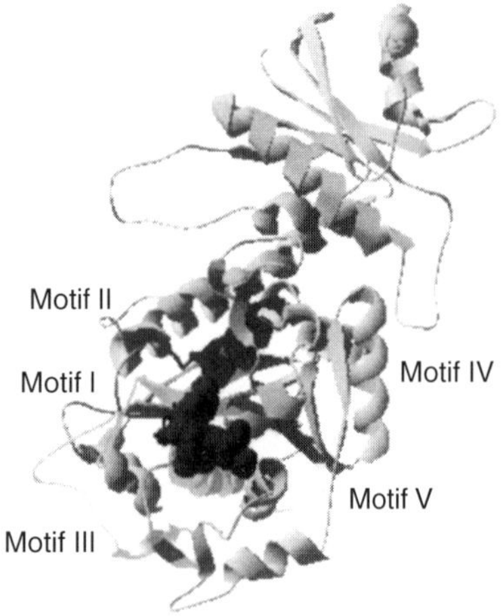

(a)

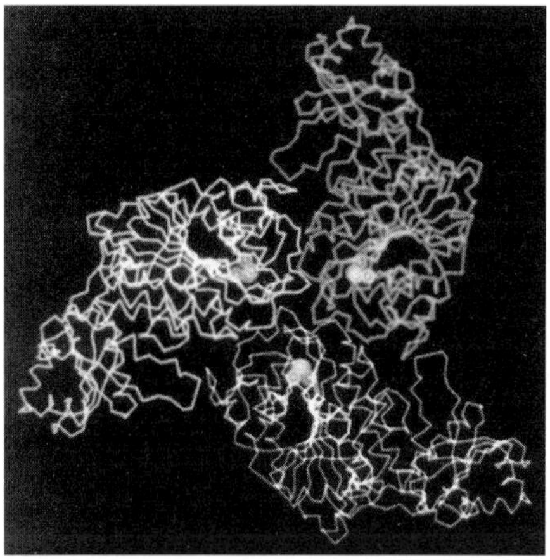

(b)

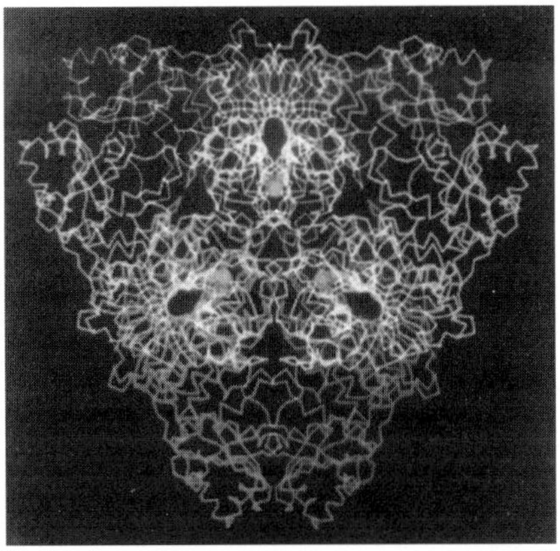

(c)

Figure 14.2 (a) Schematic ribbon drawing of the LAP monomer. Blue spheres show the amastatin (or substrate) binding site. The five motifs which are common to many aminopeptidases are shown in color and are labeled I–V. The zinc ions are behind the amastatin and are shown in pink. (b) and (c) Tracing of the carbon backbone of the blLAP trimer (b) and hexamer (c) as viewed down the threefold axis. (See *Color plate 12*)

A search of the data bases indicates that lens LAP shares 5 motifs which comprise portions of the active site with other aminopeptidases (Figure 14.2a). Motif I, is a portion of a sheet and loop which is shared by 11 of 30 (numbers indicate similarity within 59 022 Swiss-Prot registries/and similarities within 228 865GenPep entries) other proteins most of which are listed as aminopeptidases from the plant and animal kingdoms. Motif II is a loop and helix which is shared by 9 of 17 other proteins. Motif III includes a long loop and part of a sheet which is found in 8 of 18 proteins. Motif IV includes a long helix which makes close contact with the substrate and is found in 11 of 28 proteins. Motif V is a shorter element comprised of helix and sheet which is found in 9 of 26 proteins.

There is no significant structural change induced in the enzyme upon binding of LeuP (Strater and Lipscomb 1995a). The same pertains to binding of other inhibitors including bestatin, [(2*S*,3*R*)-3-amino-2-hydroxy-4-phenylbutanoyl]-L-leucine, and amastatin, [(2*S*,3*R*)-3-amino-2-hydroxy-5-methylhexanoyl]-L-valyl-L-valyl-aspartic acid (Nishizawa *et al* 1977), as well. The shape of blLAP protomers and their arrangement within the highly homologous hog kidney (hk) LAP hexamer (Taylor *et al* 1984) is indistinguishable from that described for the beef enzyme.

14.6 INHIBITOR BINDING SITE AND PROPOSED MECHANISM OF ACTION

The discovery of a relatively tight-binding transition state inhibitor of LAP, bestatin ($K_i^* = 1.3 \times 10^{-9}$ M, $Ki = 1.1 \times 10^{-7}$ M), provided new opportunities to characterize the active site (Figures 14.2a and 14.3) (Nishizawa *et al* 1977; Taylor *et al* 1992, 1993b). Kinetic and binding studies indicated that bestatin, a transition state analog, is a slow,

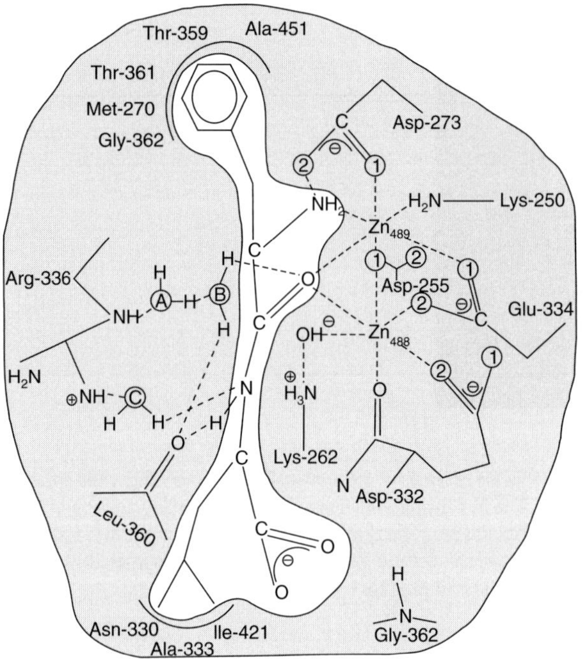
Thr-359
Ala-451
Thr-361
Met-270
Gly-362
Asp-273
Lys-250
Arg-336
Asp-255
Glu-334
Lys-262
Asp-332
Leu-360
Asn-330
Ala-333
Ile-421
Gly-362

(a)

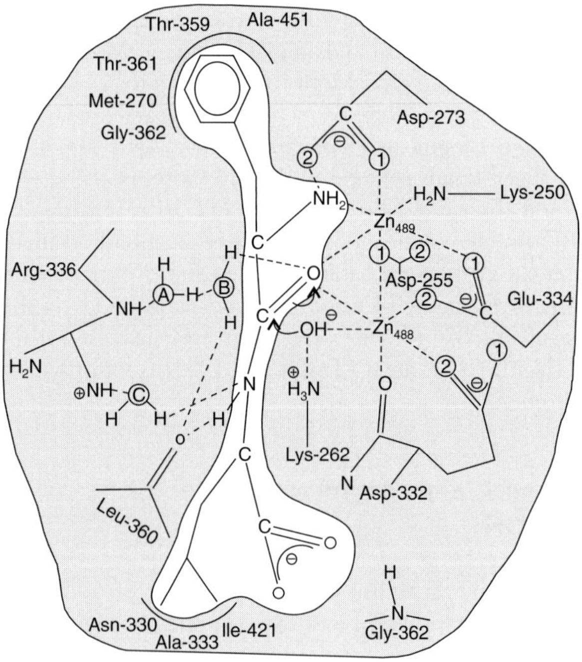
Thr-359
Ala-451
Thr-361
Met-270
Gly-362
Asp-273
Lys-250
Arg-336
Asp-255
Glu-334
Lys-262
Asp-332
Leu-360
Asn-330
Ala-333
Ile-421
Gly-362

(b)

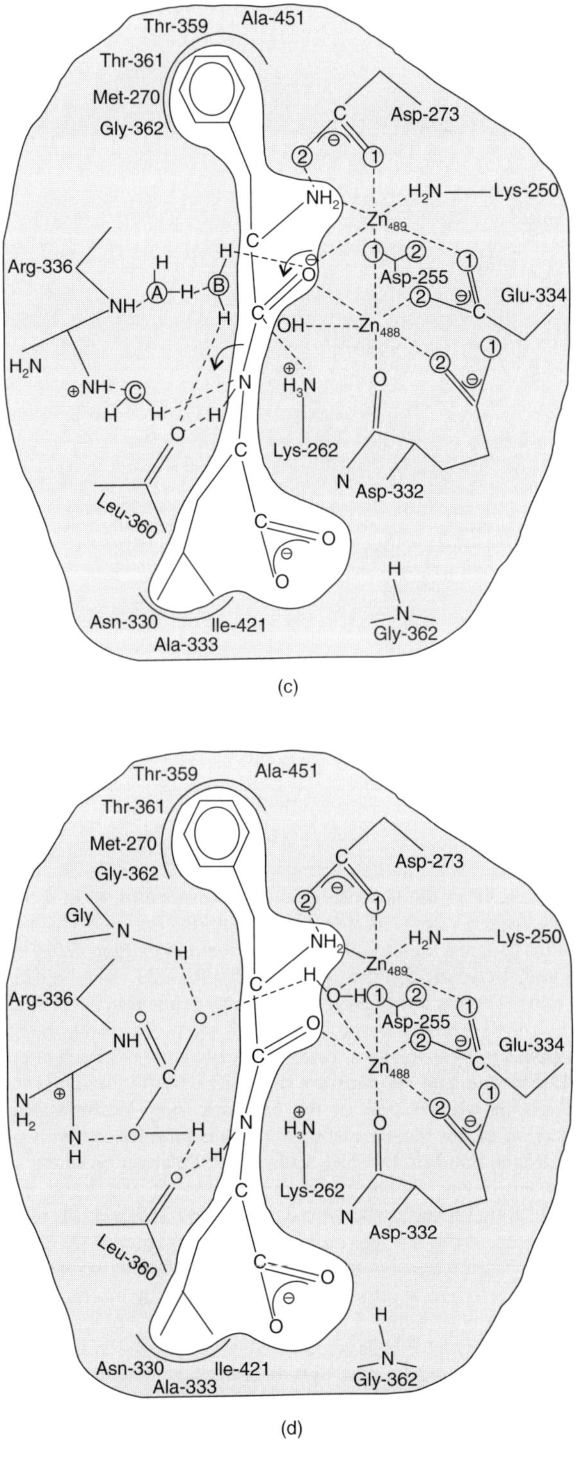

Figure 14.3 (Continued)

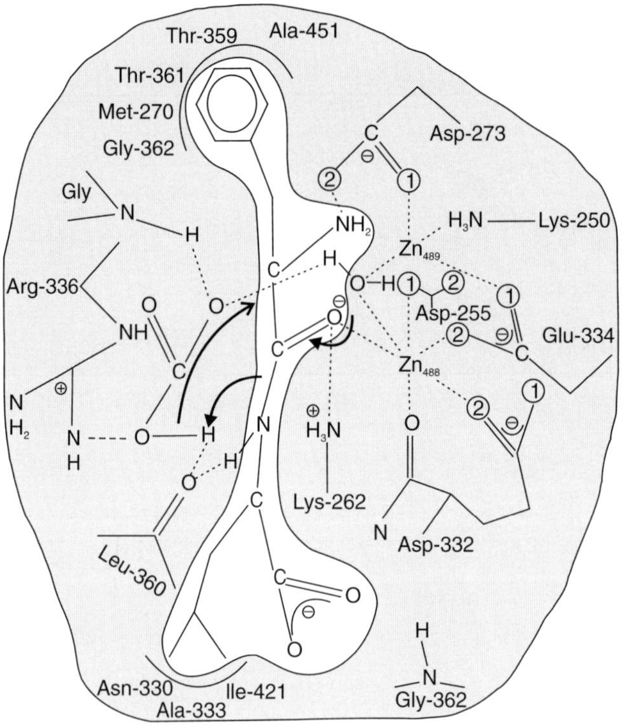

(e)

Figure 14.3 Binding of inhibitors and substrates to blLAP and proposed mechanism of hydrolysis of a model substrate, Phe-Leu Proposed mode of binding of the model substrate, Phe-Leu to blLAP in the absence (a–c), or presence (d, e) of a bicarbonate ion. By analogy to the observed binding of bestatin, the side chain of the putative substrate Phe-Leu, are found within pockets formed by Met-270, Thr-359, Gly-362, Ala-451, and Met-454 (S_1), and by Ala-330, Ala-333, and Ile-421 (S_1'), respectively (see text) The carbonyl oxygen assumes a position similar to the position occupied by the hydroxyl in bestatin. Zn488 is the more readily exchanged ion. Both Zn ions appear to be involved in catalysis. Enzyme residues are in outline, and α-carbons of LAP amino acid residues are black. The blLAP Phe-Leu model is rotated slightly to permit visualization of the aromatic ring. In the absence of bicarbonate, polarization of the scissile carbonyl involves interactions with one or both Zn^{2+} and a hydrogen bond to the ϵNH_2 of Lys-262. Polarization of the scissile C–N bond and enhanced electrophilicity of the carbon of this bond involves interactions of the P_1'NH with the carbonyl oxygen of Leu-360 and the electrons of this amide with a proton from a water molecule. (b) Formation of the transition state involves nucleophilic attack (arrows) at the scissile carbon by stabilized OH^- or H_2O (see text). The tetrahedral intermediate is stabilized by interaction of the negatively charged oxygen with either Zn^{2+}488 and the ϵNH_2 of Lys-262 or by both ions and water. (c) Hydrolysis is accomplished when the transition state collapses (arrows). (d) In the model that uses bicarbonate, that ion coordinates the bridging water (dashed lines), and it acts as a general base. This generates a nucleophilic OH^-. (e) The transition state is stabilized by coordination of the negatively charged oxygen with the zinc ions and with Lys-262. Collapse of the transition state is facilitated by proton transfer (arrows) via the bicarbonate.

tight binding inhibitor of LAP. The reaction for binding of bestatin to LAP is described by the following scheme 14.1:

$$E + S \rightleftarrows ES \rightarrow E + P$$

$+$ Kon

k_3 k_5

$$I \rightleftarrows EI \rightleftarrows EI^*$$

k_4 k_6

k_{off}

Scheme 14.1 Kinetics of binding of bestatin to LAP.

where K_i and K_i^*, respectively, are 1.1×10^{-7} M and 1.3×10^{-9} M. These data indicate that bestatin and LAP are bound approximately 84-fold more tightly in the final EI* complex than in the initial collision complex. The apparent rate constant is $k = 3.4 \times 10^{-4}\ sec^{-1}$. The rate constants for the formation and deformation of the final complex from the initial collision complex are $k_5 = 1.5 \times 10^{-2}\ sec^{-1}$ and $k_6 = 2 \times 10^{-4}\ sec^{-1}$, respectively. Finally, the rate constant for the dissociation of the collision complex, $k_4 = 1\ sec^{-1}$. This is corroborated by a value for k_{off} of $8.29(\pm 1.4) \times 10^{-5}\ sec^{-1}$. These data indicate that slow achievement of steady state involves slow deformation of the EI*. Thus, the slow binding of bestatin involves rapid formation of the initial collision complex (EI), slow transformation of the EI to the tight complex EI*, and even slower deformation of that complex (Taylor *et al* 1993b). Binding constants for bestatin to other aminopeptidases are probably within the range of these numerical values after taking into consideration differences in conditions used for the experiments. (Taylor *et al* 1992, 1993b, 1996), and it is plausible that APs share modes of bestatin binding which are even more similar than presently indicated, despite their diverse apparent specificity.

Direct binding measurements, kinetic determinations, structural and affiniity labeling studies indicate that six equivalents of bestatin are bound per LAP hexamer (Peltier and Taylor 1986; Burley *et al* 1992; Taylor *et al* 1992, 1993b, 1996). The same pertains for amastatin (Kim and Lipscomb 1993a,) and leucinephosphonic acid (Strater and Lipscomb 1995a). A ratio of 1:1 bestatin/monomer was also obtained in *Aeromonas* AP (Wilkes and Prescott 1985) and yeast AP I (Rohm 1984). This information is in contrast with data which indicate that there is 80% inhibition of hkLAP when only 1 bestatin is bound per hexamer (Wilkes and Prescott 1985).

Kinetic, NMR and structural studies also identified the active site and allowed proposals for a mechanism of action (Smith EL 1960; Delange RJ 1971; Carpenter and Vahl 1973; Hanson and Frohne 1976; Thompson and Carpenter 1976; Taylor *et al* 1981; 1982b; 1992, 1993a,b, 1996; Burley *et al* 1992; Kim and Lipscomb 1993a,b).

Among the features that distinguish the "statin"- containing moieties (i.e. bestatin and amastatin) is a tetrahedral carbon atom positioned between the scissile carbonyl and the carbon which bears the α-NH_2 group (Figure 14.4). This tetrahedral carbon has a hydroxyl

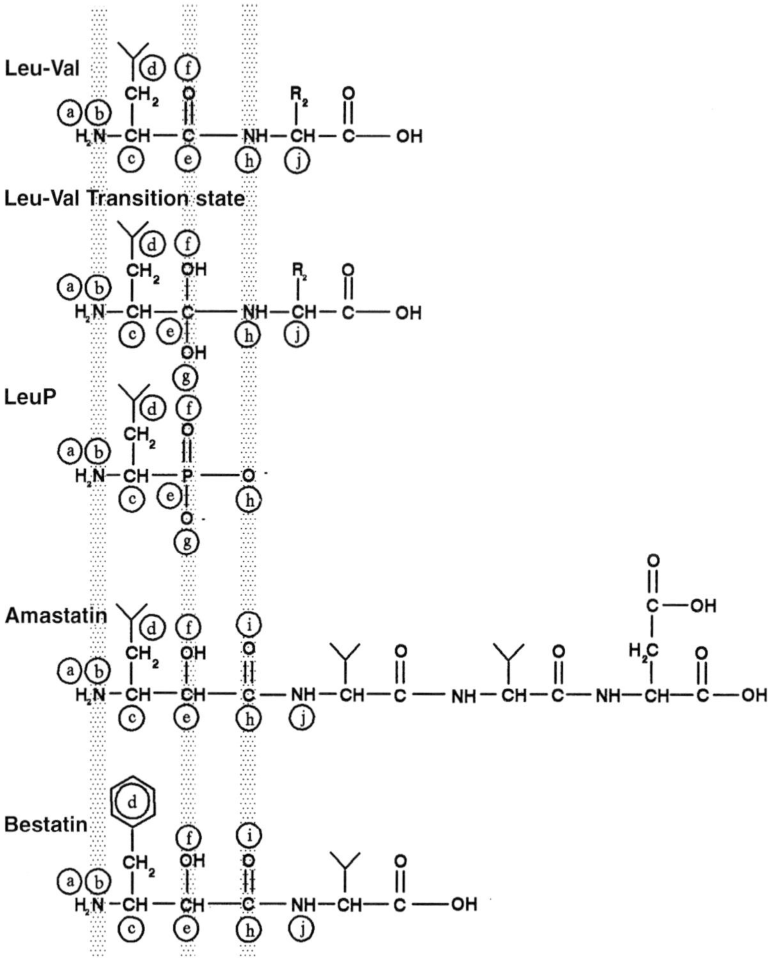

Figure 14.4 Structures of a theoretical substrate Leu-Val, the putative Leu-Val transition state, leucinephosphonic acid, amastatin, and bestatin with positions of comparable functional groups indicated. The substrate Phe-Leu has the same functional groups as shown in bestatin other than the tetrahedral carbon labeled (e). However, the stereochemistry is reversed.

group which mimics the presumptive tetrahedral transition state which is formed after nucleophilic addition of OH^-. The importance of this tetrahedral carbon is indicated by the K_i($K_i \approx 10^{-8}$ M) value of such inhibitors (Taylor *et al* 1996), which is $\approx 10^{-5}$ the K_i of peptide substrates (Taylor *et al* 1981). Bestatin, epibestatin, and epiamastatin, which have their C2 in the *R*-configuration instead of the *S*-configuration, bind to hkLAP $> 10^2$ less tightly than do bestatin and amastatin (Rich *et al* 1984). Thus, stereochemistry at C2 is of importance in determining avidity of binding of these inhibitors.

The carbonyl also retains importance as indicated by an increase in K_i of 5×10^4 when the carbonyl group is changed to a methylene (Harbeson and Rich 1988). This is consistent with interpretation that this dramatic difference in binding is due to loss of the carbonyl interaction with Zn^{2+} (see below). However, it should be noted that in bestatin and amastatin co-crystal structures, the P_1 carbonyl oxygen atom is not liganded to Zn^{2+}488.

A variety of other transition state analogs have been shown to bind to LAP. Amastatin has the α-NH_2-terminal and a Leu side chain as part of the "statin" component, as well as an extended peptide chain (Figure 14.4). It bounds less tightly to blLAP but more tightly to other APs (Rich *et al* 1984). L-Leucinephosphonic acid mimics the putative *gem*-diolate of a putative transition state which might be found during hydrolysis of peptides and, as expected, it is bound (0.23 μM to hkLAP) more tightly than peptides but less tightly that some "statins."

The availability of a relatively tightly-bound LAP inhibitor such as bestatin and the atomic coordinates obtained from the single crystal structure determination of this transition state analog inhibitor (Ricci *et al* 1982) provided further details with respect to identification of the active site. The bestatin binding site and presumably the active site are within the carboxyl third of the subunit data (Taylor *et al* 1982, 1992, 1993b; Burley *et al* 1992). Enzyme residues which are involved in substrate and inhibitor binding have been described and are shown in Figure 14.3 (Burley *et al* 1992; Taylor *et al* 1996; Strater *et al* 1999a,b). The data suggests that the binding pockets are larger than most of the peptide side chains used. This would appear to be consistent with the ability of LAP to hydrolyze aminoacyl naphthyl amides and large polypeptides (Taylor *et al* 1982). The apparent hydrophobic nature of this binding pocket should be more thoroughly studied. The backbone of the inhibitor is stabilized in the enzyme by hydrogen bonds involving Lys-262, Asp-273, and Leu-360.

Many, but not all, APs, have been identified as zinc or cobalt metalloenzymes (Carpenter and Vahl 1973; Taylor *et al* 1996). Native lens LAP protomers bind 2 zinc ions, each with different avidity. BlLAP is active only when both of these metal-ion binding sites are occupied. Stoichiometric and kinetic data indicate that Mn^{2+}, Mg^{2+}, and Co^{2+} can readily be exchanged for one zinc ion (Carpenter and Vahl 1973; Allen *et al* 1983). In various publications the Zn^{2+} which can be replaced by Mg^{2+} is called the more readily exchangeable ion. By analogy it would appear that Mn^{2+} binds to this site as well. That substitution at both metal ion sites in blLAP affects both K_m and k_{cat}(Allen *et al* 1983) suggested that both metal ions have a (possibly direct) role in binding and catalysis (Taylor *et al* 1982), a suggestion that is gaining support (Strater *et al* 1999a,b).

Based on crystallographic studies (Burley *et al* 1992; Strater and Lipscomb 1995a,b), Zn^{2+}488 is shown liganded by residues Asp-255, Asp-332 (one oxygen from a carboxylate and an oxygen from the backbone peptide bond), as well as by Glu-334 (Figure 14.3). The essential requirement for Glu-334 is indicated by the observation that mutation of Glu-334 to Ala in a homologous *pepA*-encoded AP was associated with inactivation (McCulloch *et al* 1994). Other ligands of the metal ions in native enzyme include a water molecule. In the blLAP substrate complex, and in the putative transition state, the nucleophilic hydroxide bound at the tetrahedral carbon and the negatively charged oxygen derived from the scissile carbonyl also appear to be liganded to metal ions.

Zn^{2+}489 is shown liganded by Asp-255, Lys-250, Asp-273, and Glu-334 (Burley *et al* 1992; Taylor *et al* 1992; Kim and Lipscomb 1993a; Taylor 1993a,b; Strater and Lipscomb 1995a,b). It is to Zn^{2+}489 that the α-NH_2 group in inhibitors (Figure 14.3), and presumably in substrates, is coordinated. A bridging water also appears to be coordinated to this ion in the native enzyme. In blLAP-inhibitor complexes this position would be occupied by one of the hydroxyls which is anticipated in the tetrahedral intermediate.

An unexpected observation was that of a third metal ion binding site in blLAP (Strater and Lipscomb 1995a). While unremarkable in a structural context, this is surprising since no prior compositional analyses indicated more than 2 Zn^{2+} per protomer (Carpenter and Vahl 1973; Thompson and Carpenter 1976; Allen *et al* 1983). This metal is coordinated by Leu-170, Met-171, Thr-173, Arg-271, a water molecule, and possibly Met-274, and is 12 Å away from Zn^{2+}489. Any relation of this ion to the active site ions would presumably be via coordination to Arg-271 which is also coordinated to Asp-273. The latter is coordinated to Zn^{2+}489. It appears likely that the third metal ion, if it exists, is involved in stabilization of part of the interface between the NH_2-terminal and the catalytic domain in the protomer (Strater and Lipscomb 1995a).

Ideally, a complete description of the mechanism should explain pH-optima for hydrolysis of esters, peptides, peptide analogs, large polypeptides, and why peptides are hydrolyzed at rapid rates despite considerable variability in their inhibition constants or K_m (Taylor *et al* 1981). It should also elucidate the roles of both ions in K_m and k_{cat}, why some inhibitors bind in the usual fashion whereas others bind in slow and/or tight fashion, and why some substrates also act as activators of the hydrolysis of other substrates (Taylor *et al* 1981). The proposed mechanism of action should also explain how or why inhibitors with D-αNH_2-residues bind tightly to the enzyme; however, to be substrates for hydrolysis, an unblocked L-αNH_2 amino residue is required (Smith and Spackman 1955; Hanson and Frohne 1976; McDonald JK 1986). Substantial progress toward a mechanism of action which explains many of these phenomena has been achieved (Figure 14.3).

The scissile carbonyl oxygen of substrates is coordinated to the more readily exchanged Zn^{2+} (Taylor *et al* 1982a,b; Kim and Lipscomb 1993a; Taylor 1993a,b; Strater and Lipscomb 1995a,b; Strater *et al* 1999b) (Figure 14.3a and d). Recent data suggest that polarization of the carbonyl oxygen of the scissile peptide bond may also involve interaction with one or both Zn^{2+} ions and a hydrogen bond to the ϵNH_2 of Lys-262 (Taylor *et al* 1982a,b, 1993a,b; Burley *et al* 1992; Taylor 1993a,b; Strater and Lipscomb 1995a). Polarization of the scissile C-N bond and enhanced electrophilicity of the C of this bond involves interaction of the P_1'NH with the carbonyl oxygen of Leu-360 and, possibly, the electrons of this amide with a proton from a water molecule (Figure 14.3b and e). This water is coordinated to and probably activated by Arg-336 in the bicarbonate-free mechanism. By analogy with PepA, it is also possible that a bicarbonate is coordinated to Arg-336 (Strater *et al* 1999b). This bicarbonate might be a general base which participates in activating a water which bridges the Zn ions and shuttles a proton to the leaving group (Figure 14.3d and e).

Hydrolysis involves nucleophilic attack at the carbon of the scissile carbonyl (Figure 14.3b and e). The absence of an enzyme-bound nucleophile is consistent with an inability to inactivate or label the active site of blLAP using a variety of affinity labels, which required attack by an enzyme-bound nucleophile for covalent attachment (Taylor *et al* 1981, 1982b, 1992). Accordingly, general base catalysis was suggested for the mechanism of hydrolysis of peptides (Nishizawa *et al* 1977). The most plausible nucleophile would appear to be H_2O or stabilized OH^-. The nucleophilic OH^- is presumably, (a) either generated by simultaneous dissociation of H_2O and stabilization of the incipient hydroxide ion by either both Zn^{2+} ions and a water, or by Zn^{2+}488 and ϵNH_2 of Lys-262 (Figure 14.3b) (Strater and Lipscomb 1995a) or (b) the bridging water (Figure 14.3e). This origin of the OH^- complements prior proposals which implicated an unidentified enzyme-bound base for generation of the OH^- (Taylor 1993b). The tetrahedral intermediate formed upon hydroxide attack is stabilized by interaction of the negatively charged O with either Zn^{2+}488 and the ϵNH_2 of Lys-262 or by both Zn^{2+} ions and a water molecule (Figure 14.3b and d). This interaction may explain in part the effect of both ions on hydrolytic rates. The added OH^- remains stabilized by the interactions which aided its formation. Stabilization and departure of the leaving groups could be accomplished by donation of a proton to the new NH_2 from a water or the bicarbonate (Figure 14.3c or e). Upon hydrolysis, an increase in pK of the α-NH_2 to >9 would result in protonation and release from the active site. The proposed function of the bicarbonate ion in LAP and *PepA* is analogous to the role of glutamate side chains in the active sites of other zinc enzymes (Chen *et al* 1997). It is also plausible that the new leaving amine is protonated by Lys-262. For further considerations of these mechanisms see Strater *et al* (1999a,b).

These mechanistic discussions may also have a bearing on several other observations. A two-step binding process was proposed (Kim and Lipscomb 1994) in order to rationalize the two K_is indicated for the two steps of formation of the tightly bound LAP-bestatin complex (Taylor *et al* 1993b). Initially a complex is formed between the P_1 hydroxyl and Zn^{2+}488 (Kim and Lipscomb 1994), subsequently this complex dissociates and a new complex is formed in which the hydroxyl and the α-NH_2 of bestatin is bound to Zn^{2+}489 (Burley *et al* 1992; Kim and Lipscomb 1993b, 1994). It is curious that a delay in blLAP-catalyzed hydrolysis is also observed using leucylamide as substrate (Taylor, unpublished). The two-step binding mechanism is also consistent with the observation of two presteady state intermediates in dimetal ion-containing hkLAP and with only one intermediate when hkLAP had one ion. Enhancement of blLAP activity by bicarbonate was not reported (Strater *et al* 1999b). Thus, it is not likely that equilibration of assay solutions with airborne CO_2 would result in the observed binding or activations noted when leucyl amide was used as a substrate or when other molecules proved to be activators (see below). For arginine AP the different binding kinetics and absence of metal content indicate other binding mechanisms exist (Harbeson and Rich 1988).

Space limitations prohibit describing the structure of MetAP in great detail. It should be noted that the structure of MetAP-2 complexed with fumagellin has been solved (Liu *et al* 1998). This solution offers many opportunities for drug design. Readers are referred to Bradshaw (1998) for details of the mechanism.

14.7 HOMOLOGIES

Homologies between aminopeptidases were reviewed in 1995 (Taylor *et al* 1996). However, since then a sufficient body of information has become available so as to make it more profitable to look up such information using bioinformatic tools such as the world wide web sites Merops or Swiss Prot. Such searches indicate and confirm several unanticipated results including the identities noted in the Introduction. Only the most salient similarities are noted here.

Residues which appear to be involved in zinc binding or catalysis in blLAP, residues (except Met-454) used in inhibitor and presumably substrate-binding, organized secondary structure, and resides in the loop regions comprising the active site are conserved between blLAP and *E. coli pepA* (Stirling *et al* 1989) and the *xerB* gene product shows (Song *et al* 1994; Taylor *et al* 1996) identity with *Arabidopsis thaliana* (Bartling and Weiler 1992). Thus, there is homology of APs from plants to mammals.

The data presented here suggest that blLAP, bkLAP, human lens and liver LAPs, hlLAP, hkLAP, hog intestine LAP, prolyl-AP, *E. coli* AP A, AP I and the *S. typhimurium pepA* gene product are part of a family of zinc APs which utilize the zinc-binding (and probably much of the substrate binding) amino acid constellations described for beef LAP (Wallner *et al* 1993). Structural and kinetic data suggest that *Aeromonas* LAP and the detergent-resistant alkaline exoprotease of *Vibrio alginolyticus* (Guenet *et al* 1992) may also be part of this group (Rich *et al* 1984; Wilkes and Prescott 1985). A new tightly bound mercaptoacyl-leucyl-*p*-nitroanilide provides further opportunities to study inhibitor binding in aminopeptidases (Huntington *et al* 1999). These peptidases can be distinguished from another recently identified superfamily of zinc proteases which appear to use Glu in catalysis, 2 His and Glu to bind zinc, and Arg in substrate binding (Watt and Yip 1989). These include rat kidney zinc protease (rKZP), AP N, thermolysin *B.T.* (*Bacillus thermoproteolyticus*), thermolysin *B.S.* (*B. stearothermophilus*), protease *B.A.* (*B. amyloliquefaciens*), protease *Serratia*, rat enkephalinase, carboxypeptidases A and B and, possibly APs M and N, some collagenases, angiotensin-converting enzyme, human AP M, and leukotriene A4 hydrolase (Malfroy *et al* 1989).

There are two classes of methionine aminopeptidases, each of which has two subgroups which are distinguished by the presence or absence of N-terminal domains (Bradshaw *et al* 1998). The yeast enzyme consists of two functional domains: a unique NH_2-terminal domain containing two motifs resembling zinc fingers, which may allow the protein to interact with ribosomes and function co-translationally (also see references; Ben-Bassat *et al* 1987; Arfin and Bradshaw 1988), and a catalytic COOH-terminal domain resembling other procaryotic methionine APs. Most of the similarity between these enzymes is localized to the COOH region. These enzymes share little homology with blLAP (Ben-Bassat *et al* 1987) and, like AP A, some are not inhibited by bestatin but there are similarities with respect to mechanisms of action.

In *E. coli* methionine AP, the 2 Co^{2+} are separated by 2.9 Å (Roderick and Matthews 1993). The ions are liganded by the side chains of Asp-97, Asp-108, Glu-204, Glu-235, and His-171 with approximately octahedral coordination (Roderick and Matthews 1993). In terms of both the novel backbone fold and the constitution of the active site, *E. coli* MAP appears to represent another class of proteolytic enzymes.

Streptococcus thermophilus CNR 302 contains at least 3 aminopeptidases. One gene codes for a 445 amino acid residue, 50.4 kDa protein (total mass of 300 kDa; thus it is a hexamer) which is 70% identical to PepC from *Lactoccocus lactis cremoris* and 38% identity to eukaryotic bleomycin hydrolase (Chapot-Chartier *et al* 1994). It also contains regions of strong similarity to cysteine proteinases and appears to be a thiol AP. The *Lactobacillus helveticus pepC* gene product is encoded in two open reading frames, is 51.4 kDa, and shares 48 and 98% sequence identity with the *PepC* proteins from *Lactococcus lactis* and *L. helviticus* CNRZ32, respectively (Vesanto *et al* 1994).

Despite many common features between methionine AP, *Aeromonas* AP and leucine AP there are sufficient differences to conclude that these enzymes probably did not evolve from a common ancestral protein, and they are representative of different classes of enzymes.

In contrast with the metallopeptidases, and prolyl APs which are homologous to them, a critical active site serine is indicated in the D-amino acid AP of *Ochrobactrum anthropi* and other prolyl APs (Asano *et al* 1989, 1992; Gonzales and Robert-Baudouy 1996).

14.8 REMAINING MYSTERIES AND CONCLUSIONS

The enzymes have long been used for diagnosis of various physiological states and disease conditions, and recently they have found laboratory and industrial uses as well. Administration of bestatin and other inhibitors, which presumably have their effect by inhibition of APs, has also been used to alter physiological status and disease progress. Aminopeptidases play critical roles in protein maturation, antigen presentation, regulation of hormone levels, selective or homeostatic protein turnover, and plasmid stabilization. The availability of further genetic, structural, and kinetic data, as well as substrates which can be used to monitor AP activity in whole cell systems, should aid in elucidating more physiological functions and mechanisms of action of these enzymes. It would appear that activity assays which employ physiologically relevant dipeptides or tripeptides as substrates may be more informative for nomenclature and characterization purposes than assays which utilize synthetic peptide analogs.

Some questions remain. Complete rationalization of the slow and tight binding of the "statin" type or other inhibitors is also not available and specific enzyme–inhibitor interactions or substrate–inhibitor interactions remain to be established. It would be interesting to know if "slow" events for the inhibitors and substrate involve similar interactions in the enzyme, i.e. whether there is a two step binding process involving formation of initial and tight complexes for each (Taylor *et al* 1993). Several compounds, some of which are substrates, were anticipated to be inhibitors of LAP. Surprisingly, several of these compounds are activators of blLAP or related enzymes. These include Leu- and Ala-*p*-aminobenzenesulfonates (which are also substrates), orthanilic and aminobenzenesulfonic acid (Taylor *et al* 1981), and an amastatin analog [(3*S*,4*S*)-Statin-Val-Val-Asp] (Rich *et al* 1984). The observed activation was interpreted as evidence for a second, perhaps adjacent, binding site, but this remains to be established (Woolford *et al* 1986). Structural information, perhaps with dynamic aspects, is needed to pursue these questions and to elucidate differences in structure between enzymes specific for d-, as opposed to l-amino acids, and the questions noted above. The availability of structural, kinetic, and genetic data will foster further appreciation of roles of APs in physiology.

ACKNOWLEDGMENTS

This project has been funded in part with Federal funds from the United States Department of Agriculture under agreement number 58-1950-9-001-3K06-0-1. The help of Esther Epstein, Tom Nowell and Dr. Fu Shang in preparation of the manuscript is greatly appreciated.

REFERENCES

Ahmad, S. and Ward, P.E. (1990) Role of aminopeptidase activity in the regulation of the pressor activity of circulating angiotensins. *Journal of Pharmacology and Experimental Therapeutics*, **252**, 643–650.

Allen, M.P., Yamada, A.H. and Carpenter, F.H. (1983) Kinetic parameters of metal-substituted leucine aminopeptidase from bovine lens. *Biochemistry*, **22**, 3778–3783.

Aoyagi, T. (1996) Physiological roles of ectoenzymes indicated by the use of aminopeptidase inhibitors. In *Aminopeptidases*, edited by A. Taylor, pp. 155–172. Austin, TX: R.G. Landes Company.
Arfin, S.M. and Bradshaw, R.A. (1988) Cotranslational processing and protein turnover in eukaryotic cells. *Biochemistry*, **27**, 7979–7984.
Arora, G. and Lee, B.H. (1994) Purification and characterization of an aminopeptidase from *Lactobacillus casei* subsp. *rhamnosus* S93. *Biotechnology & Applied Biochemistry*, **19**, 179–192.
Asano, Y., Nakazawa, A., Kato, Y. and Kondo, K. (1989) Properties of a novel D-stereospecific aminopeptidase from *Ochrobactrum anthropi*. *Journal of Biological Chemistry*, **264**, 14233–14239.
Asano, Y., Kato, Y., Yamada, A. and Kondo, K. (1992) Structural similarity of D-aminopeptidase to carboxypeptidase DD and beta-lactamases. *Biochemistry*, **31**, 2316–2328.
Ayoyagi, T. (1990) Small molecular protease inhibitors and their biological effects. In *Biochemistry of Peptide Antibiotics*, edited by H. Kleinkauf and H. Dohren, pp. 311–363. Berlin, NY: Walter de Greyter.
Bachmair, A.B.A., Finley, D. and Varshavsky, A. (1986) *In vivo* half-life of a protein is a function of its amino-terminal residue. *Science*, **234**, 179–186.
Ball, L.A. and Kaesberg, P. (1973) Cleavage of the N-terminal formylmethionine residue from a bacteriophage coat protein *in vitro*. *Journal of Molecular Biology*, **79**, 531–537.
Bartling, D. and Weiler, E.W. (1992) Leucine aminopeptidase from *Arabidopsis thaliana*. Molecular evidence for a phylogenetically conserved enzyme of protein turnover in higher plants. *European Journal of Biochemistry*, **205**, 425–431.
Ben-Bassat, A., Bauer, K., Chang, S.Y., Myambo, K., Boosman, A. and Chang, S. (1987) Processing of the initiation methionine from proteins: properties of the *Escherichia coli* methionine aminopeptidase and its gene structure. *Journal of Bacteriology*, **169**, 751–757.
Beninga, J., Rock, K.L. and Goldberg, A.L. (1998) Interferon-gamma can stimulate post-proteasomal trimming of the N terminus of an antigenic peptide by inducing leucine aminopeptidase. *Journal of Biological Chemistry*, **273**, 18734–18742.
Bordallo, J., Cueva, R. and Suarez-Rendueles, P. (1995) Transcriptional regulation of the yeast vacuolar aminopeptidase yscI encoding gene (APE1) by carbon sources [published erratum appears in FEBS Letters 1995 Aug 7; **369**(2–3), 353]. *FEBS Letters*, **364**, 13–16.
Botbol, V. and Scornik, O.A. (1991) Measurement of instant rates of protein degradation in the livers of intact mice by the accumulation of bestatin-induced peptides. *Journal of Biological Chemistry*, **266**, 2151–2157.
Bradshaw, R.A. and Arvin, S.M. (1996) Methionine aminopeptidase: structure and function. In *Aminopeptidases*, edited by A. Taylor, pp. 91–106. Austin, TX: R.G. Landes Company.
Bradshaw, R.A., Brickey, W.W. and Walker, K.W. (1998) N-terminal processing: the methionine aminopeptidase and N alpha-acetyl transferase families. *Trends in Biochemical Sciences*, **23**, 263–267.
Brown, S.B., Krause, D. and Ellem, K.A. (1993) Low fluences of ultraviolet irradiation stimulate HeLa cell surface aminopeptidase and candidate "TGF alpha ase" activity. *Journal of Cell Biochemistry*, **51**, 102–115.
Burley, S.K., David, P.R., Sweet, R.M., Taylor, A. and Lipscomb, W.N. (1992) Structure determination and refinement of bovine lens leucine aminopeptidase and its complex with bestatin. *Journal of Molecular Biology*, **224**, 113–140.
Carpenter, F.H. and Vahl, J.M. (1973) Leucine aminopeptidase (Bovine lens). Mechanism of activation by Mg^{2+} and Mn^{2+} of the zinc metalloenzyme, amino acid composition, and sulfhydryl content. *Journal of Biological Chemistry*, **248**, 294–304.
Carter, T.H. and Miller, C.G. (1984) Aspartate-specific peptidases in *Salmonella typhimurium*: mutants deficient in peptidase E. *Journal of Bacteriology*, **159**, 453–459.
Chang, Y.H. and Smith, J.A. (1989) Molecular cloning and sequencing of genomic DNA encoding aminopeptidase I from *Saccharomyces cerevisiae*. *Journal of Biological Chemistry*, **264**, 6979–6983.
Chang, Y.H., Teichert, U. and Smith, J.A. (1992) Molecular cloning, sequencing, deletion, and overexpression of a methionine aminopeptidase gene from *Saccharomyces cerevisiae*. *Journal of Biological Chemistry*, **267**, 8007–8011.

Chang, Y.-H. (1996) Genetic and biochemical analysis of yeast aminopeptidases. In *Aminopeptidases*, edited by A. Taylor, pp. 107–127. Austin, TX: R.G. Landes Company.

Chapot-Chartier, M.P., Rul, F., Nardi, M. and Gripon, J.C. (1994) Gene cloning and characterization of PepC, a cysteine aminopeptidase from *Streptococcus thermophilus*, with sequence similarity to the eucaryotic bleomycin hydrolase. *European Journal of Biochemistry*, **224**, 497–506.

Charlier, D., Hassanzadeh, G. and Kholti, A. (1995) carP, involved in pyrimidine regulation of the *Escherichia coli* carbamoylphosphate synthetase operon encodes a sequence-specific DNA-binding protein identical to XerB and PepA, also required for resolution of Co1EI multimers. *Journal of Molecular Biology*, **250**, 392–406.

Chauvel, E.N., Coric, P., Llorens-Cortes, C., Wilk, S., Roques, B.P. and Fournie-Zaluski, M.C. (1994) Investigation of the active site of aminopeptidase A using a series of new thiol-containing inhibitors. *Journal of Medicinal Chemistry*, **37**, 1339–1346.

Chen, G., Edwards, T., D'Souza V,M. and Holz, R.C. (1997) Mechanistic studies on the aminopeptidase from *Aeromonas proteolytica*: a two-metal ion mechanism for peptide hydrolysis. *Biochemistry*, **36**, 4278–4286.

Conlin, C.A., Hakensson, K., Liljas, A. and Miller, C.G. (1994) Cloning and nucleotide sequence of the cyclic AMP receptor protein-regulated *Salmonella typhimurium* pepE gene and crystallization of its product, an alpha-aspartyl dipeptidase. *Journal of Bacteriology*, **176**, 166–172.

Couton, J.M., Sarath, G. and Wagner, F.W. (1991) Purification and characterization of a soybean cotyledon aminopeptidase. *Plant Science*, **75**, 9–17.

Cueva, R., Garcia-Alvarez, N. and Suarez-Rendueles, P. (1989) Yeast vacuolar aminopeptidase yscI. Isolation and regulation of the APE1 (LAP4) structural gene. *FEBS Letters*, **259**, 125–129.

Cuypers, H.T., van Loon-Klaassen, L.A., Egberts, W.T., de Jong, W.W. and Bloemendal, H. (1982) The primary structure of leucine aminopeptidase from bovine eye lens. *Journal of Biological Chemistry*, **257**, 7077–7085.

Delange RJ, S.E. (1971) Leucine aminopeptidase and other N-terminal exopeptidases. In *The Enzymes*, edited by Boyer P.D., pp. 81–103. New York: Academic Press.

Delmas, B., Gelfi, J.,L'Haridon, R., Vogel, L.K., Sjostrom, H., Noren, O. *et al* (1992) Aminopeptidase N is a major receptor for the entero-pathogenic coronavirus TGEV. *Nature*, **357**, 417–420.

Delmas, B., Gelfi, J., Kut, E., Sjostrom, H., Noren, O. and Laude, H. (1994) Determinants essential for the transmissible gastroenteritis virus-receptor interaction reside within a domain of aminopeptidase-N that is distinct from the enzymatic site. *Journal of Virology*, **68**, 5216–5224.

Dice, J.F. (1993) Cellular and molecular mechanisms of aging. *Physiological Reviews*, **73**, 149–159.

Eisenhauer, D.A., Berger, J.J., Peltier, C.Z. and Taylor, A. (1988) Protease activities in cultured beef lens epithelial cells peak and then decline upon progressive passage. *Experimental Eye Research*, **46**, 579–590.

Enenkel, C. and Wolf, D.H. (1993) BLH1 codes for a yeast thiol aminopeptidase, the equivalent of mammalian bleomycin hydrolase. *Journal of Biological Chemistry*, **268**, 7036–7043.

Fitzpatrick, F.A. and Orning, L. (1996) Alternate functions for aminopeptidases: hydrolysis of Leukotriene A_4. In *Aminopeptidases*, edited by A. Taylor, pp. 129–154. Austin, TX: R.G. Landes Company.

Foglino, M. and Lazdunski, A. (1987) Deletion analysis of the promoter region of the *Escherichia coli* pepN gene, a gene subject *in vivo* to multiple global controls. *Molecular & General Genetics*, **210**, 523–527.

Frey, J. and Rohm, K.H. (1978) Subcellular localization and levels of aminopeptidases and dipeptidase in *Saccharomyces cerevisiae*. *Biochimica et Biophysica Acta*, **527**, 31–41.

Fujii, H., Nakajima, M., Saiki, I., Yoneda, J., Azuma, I. and Tsuruo, T. (1995) Human melanoma invasion and metastasis enhancement by high expression of aminopeptidase N/CD13. *Clinical & Experimental Metastasis*, **13**, 337–344.

Ganesh, S., Klingel, S., Kahle, H. and Valet, G. (1995) Flow cytometric determination of aminopeptidase activities in viable cells using fluorogenic rhodamine 110 substrates. *Cytometry*, **20**, 334–340.

Georgiadis, D., Vazeux, G., Llorens-Cortes, C., Yiotakis, A. and Dive, V. (2000) Potent and selective inhibition of zinc aminopeptidase A (EC 3.4.11.7, APA) by glutamyl aminophosphinic peptides: importance of glutamyl aminophosphinic residue in the P1 position. *Biochemistry*, **39**, 1152–1155.

Garza, L.A. and Birnbaum, M.J. (2000) Insulin-responsive aminopeptidase trafficking in 3T3-L1 adipocytes. *Journal of Biological Chemistry*, **275**, 2560–2567.

Gharbi, S., Belaich, A., Murgier, M. and Lazdunski, A. (1985) Multiple controls exerted on *in vivo* expression of the pepN gene in *Escherichia coli*: studies with pepN-lacZ operon and protein fusion strains. *Journal of Bacteriology*, **163**, 1191–1195.

Gibson, A.M., Biggins, J.A., Lauffart, B., Mantle, D. and McDermott, J.R. (1991) Human brain leucyl aminopeptidase: Isolation, characterization and specificity against some neuropeptides. *Neuropeptides*, **19**, 163–168.

Godin, D., Marceau, F., Beaule, C., Rioux, F. and Drapeau, G. (1994) Aminopeptidase modulation of the pharmacological responses to synthetic thrombin receptor agonists. *European Journal of Pharmacology*, **253**, 225–230.

Gonzales, T. and Robert-Baudouy, J. (1996) Bacterial aminopeptidases: properties and functions. *FEMS Microbiology Reviews*, **18**, 319–344.

Guenet, C., Lepage, P. and Harris, B.A. (1992) Isolation of the leucine aminopeptidase gene from *Aeromonas proteolytica*. Evidence for an enzyme precursor. *Journal of Biological Chemistry*, **267**, 8390–8395.

Hanson, H. and Frohne, M. (1976) Crystalline leucine aminopeptidase from lens (alpha-aminoacyl-peptide hydrolase; EC 3.4.11.1). *Methods in Enzymology*, **45**, 504–520.

Harbeson, S.L. and Rich, D.H. (1988) Inhibition of arginine aminopeptidase by bestatin and arphamenine analogues. Evidence for a new mode of binding to aminopeptidases. *Biochemistry*, **27**, 7301–7310.

Harris, C.A., Hunte, B., Krauss, M.R., Taylor, A. and Epstein, L.B. (1992) Induction of leucine aminopeptidase by interferon-gamma. Identification by protein microsequencing after purification by preparative two-dimensional gel electrophoresis. *Journal of Biological Chemistry*, **267**, 6865–6869.

Hausman, M.S., Snyderman, R. and Mergenhagen, S.E. (1972) Humoral mediators of chemotaxis of mononuclear leukocytes. *Journal of Infectious Diseases*, **125**, 595–602.

Huntington, K.M., Bienvenue, D.L., Wei, Y., Bennett, B., Holz, R.C. and Pei, D. (1999) Slow-binding inhibition of the aminopeptidase from *Aeromonas proteolytica* by peptide thiols: synthesis and spectroscopic characterization. *Biochemistry*, **38**, 15587–15596.

Ishizuka, M., Masuda, T., Kanbayashi, N., Fukasawa, S., Takeuchi, T., Aoyagi, T. *et al* (1980) Effect of bestatin on mouse immune system and experimental murinetumors. *Journal of Antibiotics*, **33**, 642–652.

Izumi, T., Minami, M., Ohishi, N., Bito, H. and Shimizu, T. (1993) Site-directed mutagenesis of leukotriene A_4 hydrolase distinction of leukotriene A_4 hydrolase and aminopeptidase activities. *Journal of Lipid Mediators*, **6**, 53–58.

Johnson, E.S., Ma, P.C.M., Ota, I.M. and Varshavsky (1995) A proteolytic pathway that recognizes ubiquitin as a degradation signal. *Journal of Biological Chemistry*, **270**, 17442–17456.

Jones, E.W., Zubenko, G.S. and Parker, R.R. (1982) PEP4 gene function is required for expression of several vacuolar hydrolases in *Saccharomyces cerevisiae*. *Genetics*, **102**, 665–677.

Jurnak, F., Rich, A., van Loon-Klaassen, A., Bloemendal, H., Taylor, A. and Carpenter, F.H. (1977) Preliminary X-ray study of leucine aminopeptidase (bovine lens), and oligomeric metalloenzyme. *Journal of Molecular Biology*, **112**, 149–153.

Kandror, K.V. (1999) Insulin regulation of protein traffic in rat adipose cells. *Journal of Biological Chemistry*, **274**, 25210–25217.

Kataoka, H., Joh, T., Miura, Y., Tamaoki, T., Senoo, K., Ohara, H. *et al* (2000) AT motif binding factor 1-A (ATBF1-A) negatively regulates transcription of the aminopeptidase N gene in the crypt-villus axis of small intestine. *Biochemical & Biophysical Research Communications*, **267**, 91–95.

Keller, S.R., Scott, H.M., Mastick, C.C., Aebersold, R. and Lienhard, G.E. (1995) Cloning and characterization of a novel insulin-regulated membrane aminopeptidase from Glut4 vesicles [published erratum appears in *Journal of Biological Chemistry*, 1995, **270**(50),30236]. *Journal of Biological Chemistry*, **270**, 23612–23618.

Kim, H. and Lipscomb, W.N. (1993a) Differentiation and identification of the two catalytic metal binding sites in bovine lens leucine aminopeptidase by x-ray crystallography. *Proceedings of the National Academy of Sciences of the U.S.A.*, **90**, 5006–5010.

Kim, H. and Lipscomb, W.N. (1993b) X-ray crystallographic determination of the structure of bovine lens leucine aminopeptidase complexed with amastatin: formulation of a catalytic mechanism featuring a gem-diolate transition state. *Biochemistry*, **32**, 8465–8478.

Kim, H. and Lipscomb, W.N. (1994) Structure and mechanism of bovine lens leucine aminopeptidase. *Advances in Enzymology & Related Areas of Molecular Biology*, **68**, 153–213.

Kim, J. and Kim, K. (1995) The use of FAB mass spectrometry and pyroglutamate aminopeptidase digestion for the structure determination of pyroglutamate in modified peptides. *Biochemistry & Molecular Biology International*, **35**, 803–811.

Kim, K.S., Kumar, S., Simmons, W.H. and Brown, N.J. (2000) Inhibition of aminopeptidase P potentiates wheal response to bradykinin in angiotensin-converting enzyme inhibitor-treated humans. *Journal of Pharmacology and Experimental Therapeutics*, **292**, 295–298.

Kitamura, S., Carbini, L.A., Simmons, W.H. and Scicli, A.G. (1999) Effects of aminopeptidase P inhibition on kinin-mediated vasodepressor responses. *American Journal of Physiology*, **276**, H1664–H1671.

Klebert, S., Kratzin, H.D., Zimmermann, B., Vaesen, M., Frosch, M., Weisgerber, C. *et al* (1993) Primary structure of the murine monoclonal IgG2a antibody mAb735 against alpha (2–8) polysialic acid. 2. Amino acid sequence of the heavy (H-) chain Fd' region. *Biological Chemistry Hoppe Seyler*, **374**, 993–1000.

Knight, P.J., Knowles, B.H. and Ellar, D.J. (1995) Molecular cloning of an insect aminopeptidase N that serves as a receptor for *Bacillus thuringiensis* CryIA(c) toxin. *Journal of Biological Chemistry*, **270**, 17765–17770.

Kull, F., Ohlson, E. and Haeggstrom, J.Z. (1999) Cloning and characterization of a bifunctional leukotriene A(4) hydrolase from *Saccharomyces cerevisiae*. *Journal of Biological Chemistry*, **274**, 34683–34690.

Kunz, J., Krause, D., Kremer, M. and Dermietzel, R. (1994) The 140-kDa protein of blood-brain barrier-associated pericytes is identical to aminopeptidase N. *Journal of Neurochemistry*, **62**, 2375–2386.

Lazdunski, A., Pellissier, C. and Lazdunski, C. (1975a) Regulation of *Escherichia coli* K10 aminoendopeptidase synthesis. Effects of mutations involved in the regulation of alkaline phosphatase. *European Journal of Biochemistry*, **60**, 357–362.

Lazdunski, C., Busuttil, J. and Lazdunski, A. (1975b) Purification and properties of a periplasmic aminoendopeptidase from *Escherichia coli*. *European Journal of Biochemistry*, **60**, 363–369.

Ledeme, N., Vincent-Fiquet, O., Hennon, G. and Plaquet, R. (1983) Human liver L-leucine aminopeptidase: evidence for two forms compared to pig liver enzyme. *Biochimie*, **65**, 397–404.

Lendeckel, U., Kahne, T., Arndt, M., Frank, K. and Ansorge, S. (1998) Inhibition of alanyl aminopeptidase induces MAP-kinase p42/ERK2 in the human T cell line KARPAS-299. *Biochemical & Biophysical Research Communications*, **252**, 5–9.

Liu, S., Widom, J., Kemp, C.W., Crews, C.M. and Clardy, J. (1998) Structure of human methionine aminopeptidase-2 complexed with fumagillin. *Science*, **282**, 1324–1327.

Lowther, W.T. and Matthews, B.W. (2000) Structure and function of the methionine aminopeptidases. *Biochimica et Biophysica Acta*, **1477**, 157–167.

Ludewig, M., Fricke, B. and Aurich, H. (1987) Leucine aminopeptidase in intracytoplasmic membranes of *Acinetobacter calcoaceticus*. *Journal of Basic Microbiology*, **27**, 557–563.

Malfroy, B., Kado-Fong, H., Gros, C., Giros, B., Schwartz, J.C. and Hellmiss, R. (1989) Molecular cloning and amino acid sequence of rat kidney aminopeptidase M: a member of a super family of zinc-metallohydrolases. *Biochemical & Biophysical Research Communications*, **161**, 236–241.

Matsushima, M., Takahashi, T., Ichinose, M., Miki, K., Kurokawa, K. and Takahashi, K. (1991) Structural and immunological evidence for the identity of prolyl aminopeptidase with leucyl aminopeptidase. *Biochemical and Biophysical Research Communications*, **178**, 1459–1464.

Mayo, B., Kok, J., Venema, K., Bockelmann, W., Teuber, M., Reinke, H. *et al* (1991) Molecular cloning and sequence analysis of the X-prolyl dipeptidyl aminopeptidase gene from *Lactococcus lactis* subsp. *cremoris*. *Applied & Environmental Microbiology*, **57**, 38–44.

McCulloch, R., Burke, M.E. and Sherratt, D.J. (1994) Peptidase activity of *Escherichia coli* aminopeptidase A is not required for its role in Xer site-specific recombination. *Molecular Microbiology*, **12**, 241–251.

McDonald JK, B.A. (1986) *Mammalian Proteases*. New York, Academic Press.

Miller, C.G., Strauch, K.L., Kukral, A.M., Miller, J.L., Wingfield, P.T., Mazzei, G.J. *et al* (1987) N-terminal methionine-specific peptidase in *Salmonella typhimurium*. *Proceedings of the National Academy of Sciences of the U.S.A.*, **84**, 2718–2722.

Miller, C.G., Kukral, A.M., Miller, J.L. and Movva, N.R. (1989) pepM is an essential gene in *Salmonella typhimurium*. *Journal of Bacteriology*, **171**, 5215–5217.

Miller, C.G., Miller, J.L. and Bagga, D.A. (1991) Cloning and nucleotide sequence of the anaerobically regulated pepT gene of Salmonella typhimurium. *Journal of Bacteriology*, **173**, 3554–3558.

Minami, M., Ohishi, N., Mutoh, H., Izumi, T., Bito, H., Wada, H. *et al* (1990) Leukotriene A4 hydrolase is a zinc-containing aminopeptidase. *Biochemical & Biophysical Research Communications*, **173**, 620–626.

Mo, X.Y., Cascio, P., Lemerise, K., Goldberg, A.L. and Rock, K. (1999) Distinct proteolytic processes generate the C and N termini of MHC class I-binding peptides. *Journal of Immunology*, **163**, 5851–5859.

Moerschell, R.P., Hosokawa, Y., Tsunasawa, S. and Sherman, F. (1990) The specificities of yeast methionine aminopeptidase and acetylation of amino-terminal methionine *in vivo*. Processing of altered iso-1-cytochromes c created by oligonucleotide transformation. *Journal of Biological Chemistry*, **265**, 19638–19643.

Moser, P., Roncari, G. and Zuber, H. (1970) Thermophilic aminopeptidases from *Bac. stearothermophilus*. II. Aminopeptidase I (AP I): physico-chemical properties; thermostability and activation; formation of the apoenzyme and subunits. *International Journal of Protein Research*, **2**, 191–207.

Muller, W.E.G., Zahn, R.K. and Arendes, J. (1979) Activation of DNA metabolism in T-cells by bestatin. *Biochemical Pharmacology*, **28**, 3131–3137.

Muller, W.E., Zahn, R.K., Maidhof, A., Schroder, H.C. and Umezawa, H. (1981) Bestatin, a stimulator of polysome assembly in T cell lymphoma (L 5178y). *Biochemical Pharmacology*, **30**, 3375–3377.

Murgier, M., Pelissier, C., Lazdunski, A. and Lazdunski, C. (1976) Existence, localization and regulation of the biosynthesis of aminoendopeptidase in gram-negative bacteria. *European Journal of Biochemistry*, **65**, 517–520.

Nirasawa, S., Nakajima, Y., Zhang, Z.Z., Yoshida, M. and Hayashi, K. (1999) Intramolecular chaperone and inhibitor activities of a propeptide from a bacterial zinc aminopeptidase. *Biochemical Journal*, **341**, 25–31.

Nishizawa, R., Saino, T., Takita, T., Suda, H. and Aoyagi, T. (1977) Synthesis and structure–activity relationships of bestatin analogues, inhibitors of aminopeptidase B. *Journal of Medicinal Chemistry*, **20**, 510–515.

Noble, F., Luciani, N., Da Nascimento, S., Lai-Kuen, R., Bischoff, L., Chen, H. *et al* (2000) Binding properties of a highly potent and selective iodinated aminopeptidase N inhibitor appropriate for radioautography. *FEBS Letters*, **467**, 81–86.

Nyberg, F., Thornwall, M. and Hetta, J. (1990) Aminopeptidase in human CSF which degrades delta-sleep inducing peptide (DSIP). *Biochemical & Biophysical Research Communications*, **167**, 1256–1262.

Orawski, A.T. and Simmons, W.H. (1995) Purification and properties of membrane-bound aminopeptidase P from rat lung. *Biochemistry*, **34**, 11227–11236.

Orning, L., Krivi, G., Bild, G., Gierse, J., Aykent, S. and Fitzpatrick, F.A. (1991) Inhibition of leukotriene A4 hydrolase/aminopeptidase by captopril. *Journal of Biological Chemistry*, **266**, 16507–16511.

Pasqualini, R., Koivunen, E., Kain, R., Lahdenranta, J., Sakamoto, M., Stryhn, A. *et al* (2000) Aminopeptidase N is a receptor for tumor-homing peptides and a target for inhibiting angiogenesis. *Cancer Research*, **60**, 722–727.

Pedersen, J., Lauritzen, C., Madsen, M.T. and Weis Dahl, S. (1999) Removal of N-terminal polyhistidine tags from recombinant proteins using engineered aminopeptidases. *Protein Expression & Purification*, **15**, 389–400.

Peltier, C.Z. and Taylor, A. (1986) Kinetic paramaters for the slow binding of bestatin to beef lens leucine aminopeptidase (bLAP) (abstr.). *Federal Proceedings*, **45**, 856.

Piacenza, L., Acosta, D., Basmadjian, I., Dalton, J.P. and Carmona, C. (1999) Vaccination with cathepsin L proteinases and with leucine aminopeptidase induces high levels of protection against fascioliasis in sheep. *Infection & Immunity*, **67**, 1954–1961.

Pulido-Cejudo, G., Conway, B., Proulx, P., Brown, R. and Izaguirre, C.A. (1997) Bestatin-mediated inhibition of leucine aminopeptidase may hinder HIV infection. *Antiviral Research*, **36**, 167–177.

Rawlings, N.D. and Barrett, A.J. (1995) Evolutionary families of metallopeptidases. *Methods in Enzymology*, **248**, 183–228.

Reaux, A., Fournie-Zaluski, M.C., David, C., Zini, S., Roques, B.P., Corvol, P. *et al* (1999) Aminopeptidase A inhibitors as potential central antihypertensive agents. *Proceedings of the National Academy of Sciences of the U.S.A.*, **9**, 13415–13420.

Reeve, C.A., Bockman, A.T. and Matin, A. (1984) Role of protein degradation in the survival of carbon-starved *E. coli* and *Salmonella typhimurium*. *Journal of Bacteriology*, **157**, 758–763.

Ricci, J., Bousvaros, A. and Taylor, A. (1982) The crystal and molecular structure of bestatin. Interaction of this transition state analog with leucineaminopeptidase. *Journal of Organic Chemistry*, **46**, 3063–3065.

Rich, D.H., Moon, B.J. and Harbeson, S. (1984) Inhibition of aminopeptidases by amastatin and bestatin derivatives. Effect of inhibitor structure on slow-binding processes. *Journal of Medicinal Chemistry*, **27**, 417–422.

Riemann, D., Kehlen, A. and Langner, J. (1999) CD13-not just a marker in leukemia typing. *Immunology Today*, **20**, 83–88.

Roderick, S.L. and Matthews, B.W. (1993) Structure of the cobalt-dependent methionine aminopeptidase from *Escherichia coli*: a new type of proteolytic enzyme. *Biochemistry*, **32**, 3907–3912.

Rohm, K.H. (1984) Interaction of the (2S,3S)-isomer of bestatin with yeast aminopeptidase I. Kinetic and binding studies. *Hoppe Seylers Zeitschrift fur Physiologische Chemie*, **365**, 1235–1246.

Roncari, G., Stoll, E. and Zuber, H. (1976) Thermophilic aminopeptidase I. *Methods in Enzymology*, **45**, 522–530.

Saiki, I., Fujii, H., Yoneda, J., Abe, F., Nakajima, M., Tsuruo, T. *et al* (1993) Role of aminopeptidase N (CD13) in tumor-cell invasion and extracellular matrix degradation. *International Journal of Cancer*, **54**, 137–143.

Sanderink, G.J., Artur, Y. and Siest, G. (1988) Human aminopeptidases: A review of the literature. *Journal of Clinical Chemistry & Clinical Biochemistry*, **26**, 795–807.

Sekine, K., Fujii, H. and Abe, F. (1999) Induction of apoptosis by bestatin (ubenimex) in human leukemic cell lines. *Leukemia*, **13**, 729–734.

Shapiro, L.H., Ashmun, R.A., Roberts, W.M. and Look, A.T. (1991) Separate promoters control transcription of the human aminopeptidase N gene in myeloid and intestinal epithelial cells. *Journal of Biological Chemistry*, **266**, 11999–12007.

Simmons, W.H. and Orawski, A.T. (1992) Membrane-bound aminopeptidase P from bovine lung. Its purification, properties, and degradation of bradykinin. *Journal of Biological Chemistry*, **267**, 4897–4903.

Sin, N., Meng, L., Wang, M.Q., Wen, J.J., Bornmann, W.G. and Crews, C.M. (1997) The anti-angiogenic agent fumagillin covalently binds and inhibits the methionine aminopeptidase, MetAP-2. *Proceedings of the National Academy of Sciences of the U.S.A.*, **94**, 6099–6103.

Smith, E.L. and Spackman, D.H. (1955) Leucine aminopeptidase V. activation, specificity, and mechanism of action. *Journal of Biological Chemistry*, **212**, 271–299.

Smith, E.L., Hill, R.L. (1960) Leucine aminopeptidase. In *The Enzymes*, edited by Boyer, P.D., 2nd ed, vol 4 (part A), pp. 37–63. New York; Academic Press.

Song, L., Ye, M., Troyanovskaya, M., Wilk, E., Wilk, S. and Healy, D.P. (1994) Rat kidney glutamyl aminopeptidase (aminopeptidase A): molecular identity and cellular localization. *American Journal of Physiology*, **267**, F546–F557.

Song, L. and Healy, D.P. (1999) Kidney aminopeptidase A and hypertension, part II: effects of angiotensin II. *Hypertension*, **33**, 746–752.

Squire, C.R., Talebian, M., Menon, J.G., Dekruyff, S., Lee, T.D., Shively, J.E. *et al* (1991) Leucine aminopeptidase-like activity in Aplysia hemolymph rapidly degrades biologically active alpha-bag cell peptide fragments. *Journal of Biological Chemistry*, **266**, 22355–22363.

Stirling, C.J., Colloms, S.D., Collins, J.F., Szatmari, G. and Sherratt, D.J. (1989) xerB, an *Escherichia coli* gene required for plasmid ColE1 site-specific recombination, is identical to pepA, encoding aminopeptidase A, a protein with substantial similarity to bovine lens leucine aminopeptidase. *Embo Journal*, **8**, 1623–1627.

Stoll, E., Hermodson, M.A., Ericsson, L.H. and Zuber, H. (1972) Subunit structure of the thermophilic aminopeptidase I from *Bacillus stearothermophilus*. *Biochemistry*, **11**, 4731–4735.

Stoll, E., Ericsson, L.H. and Zuber, H. (1973) The function of the two subunits of thermophilic aminopeptidase I. *Proceedings of the National Academy of Sciences of the U.S.A.*, **70**, 3781–3784.

Strater, N. and Lipscomb, W.N. (1995a) Transition state analogue L-leucinephosphonic acid bound to bovine lens leucine aminopeptidase: X-ray structure at 1.65 Å resolution in a new crystal form. *Biochemistry*, **34**, 9200–9210.

Strater, N. and Lipscomb, W.N. (1995b) Two-metal ion mechanism of bovine lens leucine aminopeptidase: active site solvent structure and binding mode of L-leucinal, a gem-diolate transition state analogue, by X-ray crystallography. *Biochemistry*, **34**, 14792–14800.

Strater, N., Sherratt, D.J. and Colloms, S.D. (1999a) X-ray structure of aminopeptidase A from *Escherichia coli* and a model for the nucleoprotein complex in Xer site-specific recombination. *EMBO Journal*, **18**, 4513–4522.

Strater, N., Sun, L., Kantrowitz, E.R. and Lipscomb, W.N. (1999b) A bicarbonate ion as a general base in the mechanism of peptide hydrolysis by dizinc leucine aminopeptidase. *Proceedings of the National Academy of Sciences of the U.S.A.*, **96**, 11151–11155.

Strauch, K.L., Lenk, J.B., Gamble, B.L. and Miller, C.G. (1985) Oxygen regulation in *Salmonella typhimurium*. *Journal of Bacteriology*, **161**, 673–680.

Takahashi, S., Ohishi, Y., Kato, H., Noguchi, T., Naito, H. and Aoyagi, T. (1989) The effects of bestatin, a microbial aminopeptidase inhibitor, on epidermal growth factor-induced DNA synthesis and cell division in primary cultured hepatocytes of rats. *Experimental Cell Research*, **183**, 399–412.

Takeda, M. and Webster, R.E. (1968) Protein chain initiation and deformylation in *B. subrilis* homogenates. *Proceedings of the National Academy of Sciences of the U.S.A.*, **60**, 1487–1494.

Taylor, W.L. and Dixon, J.E. (1978) Characterization of a pyroglutamate aminopeptidase from rat serum that degrades thyrotropin-releasing hormone. *Journal of Biological Chemistry*, **253**, 6934–6940.

Taylor, A., Carpenter, F.H. and Wlodawer, A. (1979) Leucine aminopeptidase (bovine lens): an electron microscopic study. *Journal of Ultrastructure Research*, **68**, 92–100.

Taylor, A., Tisdell, F.E. and Carpenter, F.H. (1981) Leucine aminopeptidase (bovine lens): synthesis and kinetic properties of ortho-, meta-, and para-substituted leucyl-anilides. *Archives of Biochemistry & Biophysics*, **210**, 90–97.

Taylor, A., Daims, M., Lee, J. and Surgenor, T. (1982a) Identification and quantification of leucine aminopeptidase in aged normal and cataractous human lenses and ability of bovine lens LAP to cleave bovine crystallins. *Current Eye Research*, **2**, 47–56.

Taylor, A., Sawan, S. and James, T. (1982b) On the binding of leucyl-p-sulfonic acid in leucine aminopeptidase. Interaction between this substrate analog and the activation site metal-viewed by NMR. *Journal of Biological Chemistry*, **257**, 11571–11576.

Taylor, A., Brown, M.J., Daims, M.A. and Cohen, J. (1983) Localization of leucine aminopeptidase in normal hog lens by immunofluorescence and activity assays. *Investigative Ophthalmology & Visual Sciences*, **24**, 1172–1180.

Taylor, A., Surgenor, T., Thomson, D.K., Graham, R.J. and Oettgen, H. (1984) Comparison of leucine aminopeptidase from human lens, beef lens and kidney, and hog lens and kidney. *Experimental Eye Research*, **38**, 217–229.

Taylor, A. and Davies, K.J.A. (1987) Protein oxidation and loss of protease activity may lead to cataract formation in the aged lens. *Free Radical Biology & Medicine*, **3**, 371–377.

Taylor, A., Peltier, C.Z., Jahngen, E.G., Jr., Laxman, E., Szewczuk, Z. and Torre, F.J. (1992) Use of azidobestatin as a photoaffinity label to identify the active site peptide of leucine aminopeptidase. *Biochemistry*, **31**, 4141–4150.

Taylor, A., Jacques, P.F. and Dorey, C.K. (1993a) Oxidation and aging: impact on vision. *Journal of Toxicology Industries Health*, **9**, 349–371.

Taylor, A., Peltier, C.Z., Torre, F.J. and Hakamian, N. (1993b) Inhibition of bovine lens leucine aminopeptidase by bestatin: number of binding sites and slow binding of this inhibitor. *Biochemistry*, **32**, 784–790.

Taylor, A., Sanford, D. and Nowell, T. (1996) Structure and function of bovine lens aminopeptidase and comparison with homologous aminopeptidases. In *Aminopeptidases*, edited by A. Taylor, pp. 21–67. Austin, TX: R.G. Landes Company.

Taylor, A. (1996a) *Aminopeptidases*. Auxtin, TX: R.G. Landes Company.

Taylor, A. (1996b) Aminopeptidases, occurrence, regulation and nomenclature. In *Aminopeptidases*, edited by A. Taylor, pp 1–20. Austin, TX: R.G. Landes Company.

Thompson, G.A. and Carpenter, F.H. (1976) Leucine aminopeptidase (bovine lens) The relative binding of cobalt and zinc to leucine aminopeptidase and the effect of cobalt substitution on specific activity. *Journal of Biological Chemistry*, **251**, 1618–1624.

Tresnan, D.B. and Holmes, K.V. (1998) Feline aminopeptidase N is a receptor for all group I coronaviruses. *Advances in Experimental Medicines & Biology*, **440**, 69–75.

Tsunasawa, S. (1995) Amino-terminal processing of nascent proteins: their role and implication on biological function. *Tanpakushitsu Kakusan Koso-Protein, Nucleic Acid, Enzyme*, **40**, 389–398.

Turzynski, A. and Mentlein, R. (1990) Prolyl aminopeptidase from rat brain and kidney. Action on peptides and identification as leucyl aminopeptidase. *European Journal of Biochemistry*, **190**, 509–515.

Umezawa, H., Ishizuka, M., Aoyagi, T. and Takeuchi, T. (1976) Enhancement of delayed-type hypersensitivity by bestatin, an inhibitor of aminopeptidase B and leucine aminopeptidase. *Journal of Antibiotics (Tokyo)*, **29**, 857–859.

Van Wart, H.E. (1996) Metallobiochemistry of aminopeptidases. In *Aminopeptidases*, edited by A. Taylor, pp. 69–90. Austin, TX: R.G. Landes Company.

Vesanto, E., Varmanen, P., Steele, J.L. and Palva, A. (1994) Characterization and expression of the *Lactobacillus helveticus* pepC gene encoding a general aminopeptidase. *European Journal of Biochemistry*, **224**, 991–997.

Walling, L.L. and Gu, Y.-Q. (1996) Plant aminopeptidases: occurrence, function and characterization. In *Aminopeptidases*, edited by A. Taylor, pp. 173–218. Austin, TX: R.G. Landes Company.

Wallner, B.P., Hession, C., Tizard, R., Frey, A.Z., Zuliani, A., Mura, C. *et al* (1993) Isolation of bovine kidney leucine aminopeptidase cDNA: comparison with the lens enzyme and tissue-specific expression of two mRNAs. *Biochemistry*, **32**, 9296–9301.

Waters, S.B., D'Auria, M., Martin, S.S., Nguyen, C., Kozma, L.M. and Luskey, K.L. (1997) The amino terminus of insulin-responsive aminopeptidase causes Glut4 translocation in 3T3-L1 adipocytes. *Journal of Biological Chemistry*, **272**, 23323–23327.

Watt, V.M. and Yip, C.C. (1989) Amino acid sequence deduced from a rat kidney cDNA suggests it encodes the Zn-peptidase aminopeptidase N. *Journal of Biological Chemistry*, **264**, 5480–5487.

Wilcox, C., Hu, J.-S. and Olson, E.N. (1987) Acylation of proteins with myristicacid occurs cotranslationally. *Science*, **238**, 1275–1278.

Wilkes, S.H. and Prescott, J.M. (1985) The slow, tight binding of bestatin and amastatin to aminopeptidases. *Journal of Biological Chemistry*, **260**, 13154–13162.

Woolford, C.A., Daniels, L.B., Park, F.J., Jones, E.W., Van Arsdell, J.N. and Innis, M.A. (1986) The PEP4 gene encodes an aspartyl protease implicated in the post translational regulation of *Saccharomyces cerevisiae* vacuolar hydrolases. *Molecular & Cell Biology*, **6**, 2500–2510.

Wu-Wong, J.R., Berg, C.E., Wang, J., Chiou, W.J. and Fissel, B. (1999) Endothelin stimulates glucose uptake and GLUT4 translocation via activation of endothelin ETA receptor in 3T3-L1 adipocytes. *Journal of Biological Chemistry*, **274**, 8103–8110.

Yan, T.R., Ho, S.C. and Hou, C.L. (1992) Catalytic properties of X-prolyl dipeptidyl aminopeptidase from *Lactococcus lactis* subsp. *cremoris* nTR. *Bioscience Biotechnology & Biochemistry*, **56**, 704–707.

Yeager, C.L., Ashmun, R.A., Williams, R.K., Cardellichio, C.B., Shapiro, L.H., Look, A.T. *et al* (1992) Human aminopeptidase N is a receptor for human coronavirus 229E. *Nature*, **357**, 420–422.

Yoneda, J., Saiki, I., Fujii, H., Abe, F., Kojima, Y. and Azuma, I. (1992) Inhibition of tumor invasion and extracellular matrix degradation by ubenimex (bestatin) *Clinical & Experimental Metastasis*, **10**, 49–59.

York, I.A., Goldberg, A.L., Mo, X.Y. and Rock, K.L. (1999) Proteolysis and class I major histocompatibility complex antigen presentation. *Immunological Reviews*, **172**, 49–66.

Yoshpe-Besancon, I., Gripon, J.C. and Ribadeau-Dumas, B. (1994) Xaa-Pro-dipeptidyl-aminopeptidase from *Lactococcus lactis* catalyses kinetically controlled synthesis of peptide bonds involving proline. *Biotechnology & Applied Biochemistry*, **20**, 131–140.

Zhang, Y., Griffith, E.C., Sage, J., Jacks, T. and Liu, J.O. (2000) Cell cycle inhibition by the anti-angiogenic agent TNP-470 is mediated by p53 and $p21^{WAF1/CIP1}$. *Proceedings of the National Academy of Sciences of the U.S.A.*, **97**, 6427–6432.

Zuo, S., Guo, Q., Ling, C. and Chang, Y.H. (1995) Evidence that two zinc fingers in the methionine aminopeptidase from Saccharomyces cerevisiae are important for normal growth. *Molecular & General Genetics*, **246**, 247–253.

Chapter 15

The hepatitis C virus NS3 serine-type proteinase

Ralf Bartenschlager and Jan-Oliver Koch

The hepatitis C virus (HCV) is a leading cause of acute and chronic liver disease worldwide. HCV possesses a plusstrand RNA genome encoding a ~330 kDa polyprotein that undergoes proteolytic cleavage. Some of these cleavages are mediated by a serine proteinase located in the amino terminal domain of nonstructural protein (NS) 3. Activity of this enzyme is enhanced by the viral cofactor NS4A forming a stable heterodimer with the NS3 domain. Intensive biochemical studies and the determination of the three-dimensional structure of this enzyme revealed a number of peculiar features like the complex requirement for a viral peptide cofactor, a structural zinc ion and an unusual substrate specificity. These results opened several possibilities for the development of NS3-specific inhibitors. Potential strategies are interference with the NS3:4A interaction, interference with binding of the zinc ion and development of peptidomimetic drugs interfering with substrate binding and/or catalysis.

15.1 INTRODUCTION

The hepatitis C virus (HCV) is the major etiologic agent of sporadic and transfusion-associated non-A, non-B hepatitis worldwide. In spite of a mild clinical course and a high proportion of asymptomatic cases, owing to the high rate of chronicity, HCV infections are a severe medical problem. About 50% of all infections lead to chronic liver disease with variable clinical manifestations ranging from an apparently healthy carrier state to chronic active hepatitis, liver cirrhosis and hepatocellular carcinoma (Houghton 1996; Booth 1998).

Blood and blood products as well as associated risk factors like use of intravenous drugs or needle stick accidents are the main routes of transmission whereas infections *in utero* or by sexual contacts appear to be rare events. HCV is distributed worldwide with prevalences of 1% in Western Europe and the US, 2% in Asian countries and a peak up to 15% in Egypt (Booth 1998). It is estimated that ~2% of the total world population is infected with this virus (WHO fact sheet 164, 2000). At present, the only therapy available is treatment with the immune-stimulatory cytokine interferon-α either alone or in combination with ribavirin (Moussalli *et al* 1998). However, less than 50% of treated patients benefit from this therapy, which has also a number of side effects. There is thus an obvious need to develop efficient antiviral drugs.

Since the molecular cloning of the virus genome in 1989 (Choo *et al* 1989), significant progress with respect to the elucidation of the genome structure and the function of individual viral proteins has been made and this led to the identification of several prime targets for causal therapeutics. Among these targets, the most intensively studied is the NS3 proteinase that will be described in detail in this chapter.

15.2 HCV GENOME ORGANISATION AND REPLICATION

HCV is a distinct member of the family *Flaviviridae* to which the animal pathogenic pestiviruses, the flaviviruses and, although this awaits official confirmation, a group of recently cloned viruses designated GBV-A, GBV-B, and GBV-C (the latter also called hepatitis G virus) belong (Murphy *et al* 1995). These viruses have in common a plusstrand RNA genome that in case of HCV has a length of about 9600 nucleotides. It carries a single long open reading frame (ORF) encoding an ~3000 amino acids long polyprotein (for recent reviews see Reed and Rice 2000; Bartenschlager and Lohmann 2000) (Figure 15.1). The ORF is flanked at the 5′ and the 3′ end by non-translated regions (NTRs) essential for replication, and, in the case of the 5′ NTR, also for translation of the viral genome.

The polyprotein is cleaved co- and posttranslationally into at least ten individual products (Hijikata *et al* 1991, 1993b; Bartenschlager *et al* 1993; Grakoui *et al* 1993c; Selby *et al* 1993; Tomei *et al* 1993; Lin *et al* 1994a; Mizushima *et al* 1994) (Fig. 15.1). The first one is the core protein (C), an RNA binding protein forming the major constituent of the nucleocapsid. Envelope protein 1 (E1) and E2 are highly glycosylated type 1 transmembrane proteins forming stable heterodimers. They are embedded into the lipid envelope surrounding the viral nucleocapsid. p7 is a small hydrophobic peptide of unknown function. Most of the nonstructural proteins 2–5B, the term indicates that these proteins are not constituents of the virus particle, are required for multiplication of the viral RNA (Lohmann *et al* 1999). NS2 and the amino terminal domain of NS3 constitute the NS2/3 autoproteinase responsible for cleavage at the NS2/3 junction (Grakoui *et al* 1993a; Hijikata *et al* 1993a; Hirowatari *et al* 1993). NS3 is a bifunctional molecule carrying three different enzymatic activities. The aminoterminal domain forms a serine-type proteinase responsible for processing at 4 different sites and the carboxy terminal domain carries NTPase/helicase activities (Bartenschlager *et al* 1993; Eckart *et al* 1993; Grakoui *et al* 1993b; Hirowatari *et al* 1993; Suzich *et al* 1993; Tomei *et al* 1993; Manabe *et al* 1994; Kim *et al* 1995; Gwack *et al* 1996; Hong *et al* 1996; Preugschat *et al* 1996; Tai *et al* 1996; Morgenstern *et al* 1997). NS4A, a 54 amino acids long peptide, is an essential cofactor of the NS3 proteinase and will be described in detail below. NS4B is a highly hydrophobic protein with unknown function. The role phosphoprotein NS5A plays with respect to RNA replication is not known. However, there is accumulating evidence that at least for some HCV isolates NS5A is involved in the interferon resistance, presumably *via* inhibition of the interferon-induced double stranded RNA-activated protein kinase PKR (Gale *et al* 1997, 1998). NS5B is the RNA-dependent RNA polymerase (Behrens *et al* 1996; Lohmann *et al* 1997).

15.3 PROCESSING OF THE POLYPROTEIN PRECURSOR

Cleavage of the HCV polyprotein is mediated by two classes of enzymes. The first one are host cell signal peptidases cleaving in the lumen of the endoplasmic reticulum after

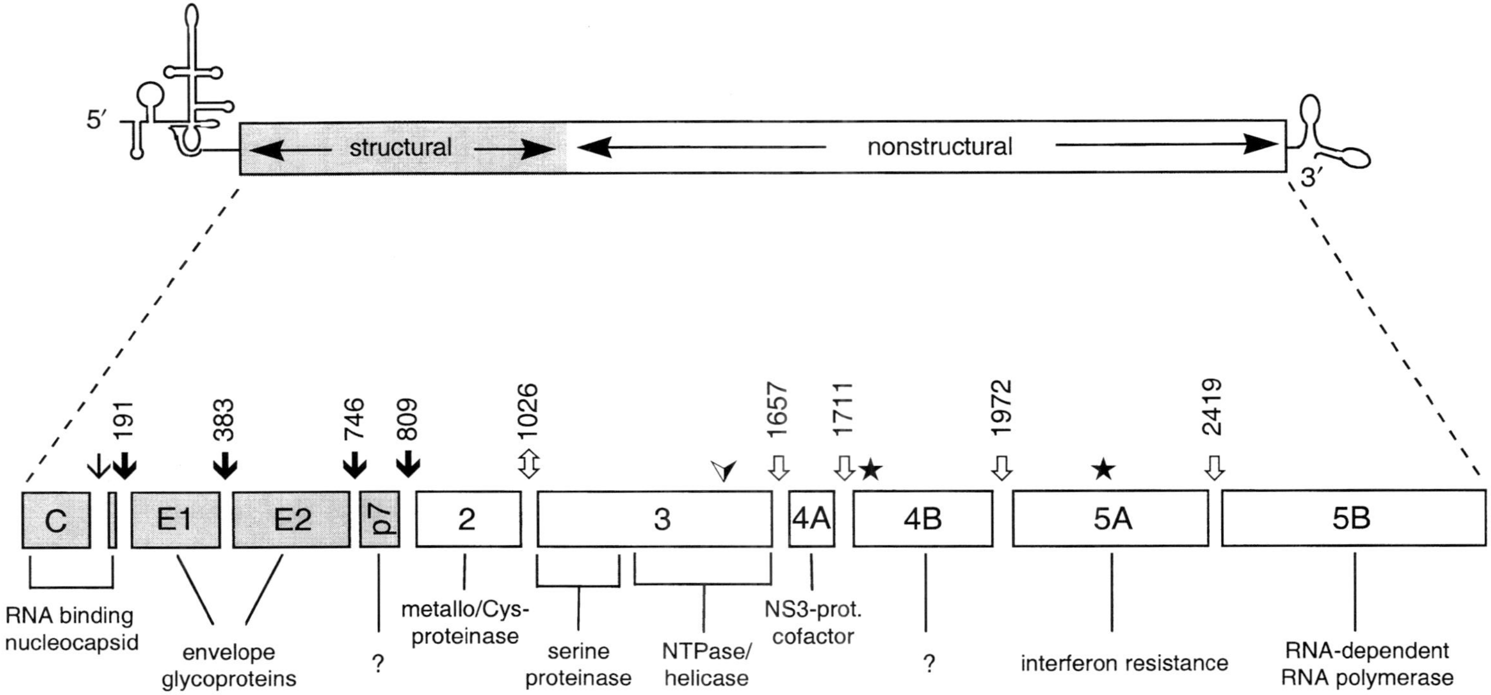

Figure 15.1 HCV genome organization, polyprotein processing and properties of cleavage products. A schematic representation of the HCV genome indicating the positions of the structural and the nonstructural proteins within the polyprotein is given at the top. 5′ and 3′ NTRs are indicated with secondary structures. The polyprotein cleavage products are drawn below with their potential functions indicated underneath. Cleavage sites for host cell signalases (⬇), the NS2-3 proteinase (⇳), the NS3 proteinase (⇩) and an unknown cellular proteinase (↓) are marked. (★) indicates additional NS3-sites and (⋎) the cleavage site within the helicase domain.

internal hydrophobic signal sequences (Figure 15.1). These enzymes are responsible for processing of the C-NS2 region. Cleavage of the remainder is accomplished by two viral proteinases. The first one is the NS2/3 autoproteinase, responsible for processing at the NS2/3 junction. Efficient cleavage at this site requires the 130 carboxy terminal residues of NS2 and the first 180 amino acids of NS3 (Grakoui *et al* 1993a; Hijikata *et al* 1993a; Santolini *et al* 1995). The same NS3 domain carries a serine-type proteinase, responsible for cleavage at the NS3/4A, NS4A/B, NS4B/5A and the NS5A/B sites. In addition, two further cleavage sites for the NS3 proteinase have been described, one located close to the amino terminus of NS4B and the other in the middle of NS5A (Figure 15.1) (Kolykhalov *et al* 1994; Markland *et al* 1997). However, since they are only found under certain experimental conditions their importance for the viral life cycle remains to be clarified. Another cleavage site was found within the helicase domain. Whether this cleavage is mediated by NS3 or a cellular enzyme has been discussed controversially (Gallinari *et al* 1999; Shoji *et al* 1999).

Processing of the NS-region by the NS3 proteinase is accomplished in the following prefered but not obligatory order: NS3/4A → NS5A/B → NS4A/B → NS4B/5A (Bartenschlager *et al* 1994; Lin *et al* 1994b; Tanji *et al* 1994a). The absence of detectable precursors and the first order reaction kinetics suggest that cleavage at the NS3/4A site occurs intramolecularly (in *cis*) (Tomei *et al* 1993; Bartenschlager *et al* 1994; Lin *et al* 1994b; Tanji *et al* 1994a). In contrast, processing at the remaining sites can be mediated intermolecularly (in *trans*) (Tomei *et al* 1993; Bartenschlager *et al* 1994; Failla *et al* 1994; Lin *et al* 1994b; Tanji *et al* 1994a,b). However, the HCV proteins form a stable higher-order complex associated with intracellular membranes (Hijikata *et al* 1993b; Lin *et al* 1997; Koch and Bartenschlager 1999). Therefore, even after liberation of NS3 from the polyprotein, the enzyme and its substrate are in very close vicinity.

15.4 BIOCHEMICAL PROPERTIES OF THE NS3/4A PROTEINASE

Despite significant sequence heterogeneity between different HCV isolates it was possible to accurately predict a classical catalytic triad of a serine-type proteinase in the amino terminal NS3 region. This prediction was confirmed in several studies showing that amino acid substitutions affecting residues His-57, Asp-81 or Ser-139 reduce or completely block processing at the NS3-dependent sites whereas cleavage at the NS2/3 junction is not affected (Bartenschlager *et al* 1993; Eckart *et al* 1993; Grakoui *et al* 1993b; Hijikata *et al* 1993a; Hirowatari *et al* 1993; Tomei *et al* 1993). Thus, although the NS2/3 autoproteinase and the NS3 proteinase overlap, their enzymatic activities can be separated genetically.

15.4.1 The NS4A cofactor

Although the amino terminal NS3 domain possesses intrinsic proteolytic activity, polyprotein cleavage by NS3 is strongly enhanced in the presence of NS4A (Bartenschlager *et al* 1994; Failla *et al* 1994; Lin *et al* 1994b; Tanji *et al* 1995). For example, coexpression of an NS4B-5B substrate with the NS3 proteinase domain in cell culture leads to inefficient cleavage at the NS5A/B site whereas processing between NS4B and NS5A is not detectable. However, upon coexpression of NS4A, the substrate is completely cleaved. Based on this type of analyses it was found that NS4A is essential for processing at the NS3/4A, NS4A/B and NS4B/5A sites and, for most isolates, greatly stimulates cleavage between NS5A and NS5B (although some processing at this site is found even in the absence of the cofactor; Bartenschlager *et al* 1994; Failla *et al* 1994; Lin *et al* 1994b; Tanji *et al* 1995). Subsequent coprecipitation studies have demonstrated the formation of a detergent-stable NS3/4A complex (Hijikata *et al* 1993b; Bartenschlager *et al* 1995b; Failla *et al* 1995; Lin *et al* 1995; Satoh *et al* 1995). Since mutational ablation of the interaction also reduces or blocks processing at the NS3-dependent sites, complex formation is an essential prerequisite for proteinase activation. The interaction domains have been mapped to the ~30 amino terminal residues of NS3 and a 12-residue sequence in the centre of NS4A (Bartenschlager *et al* 1995b; Failla *et al* 1995; Lin *et al* 1995; Satoh *et al* 1995; Tanji *et al* 1995), which can be supplied as a synthetic peptide without loss of activation function (Butkiewicz *et al* 1996; Koch *et al* 1996; Shimizu *et al* 1996; Steinkühler *et al* 1996a; Tomei *et al* 1996).

Depending on the reaction conditions and peptide substrates used the major effect exerted by NS4A is either an increase of catalysis (k_{cat}) or an increase of binding affinity (lower K_m) concomitant with an increase of k_{cat} (Steinkühler *et al* 1996a,b; Shimizu *et al* 1996; Landro *et al* 1997; Bianchi *et al* 1997; Urbani *et al* 1997). Since the presence of NS4A alters the physicochemical requirements for substrate cleavage, it was suggested that NS4A-binding induces conformational changes in the NS3 proteinase domain leading to higher enzymatic activity (Steinkühler *et al* 1996b) (see below).

Two additional effects might contribute *in vivo* to an enhancement of polyprotein cleavage by NS4A. First, stabilization of the NS3 proteinase, which in the absence of NS4A is degraded very rapidly (Tanji *et al* 1995) and second, association of the NS3 domain with intracellular membranes via the complexed NS4A moiety and in this way an increase of the local concentration of proteinase and polyprotein substrate (Hijikata *et al* 1993b; Tanji *et al* 1995; Yan *et al* 1998).

15.4.2 Substrate specificity

Determination of the amino terminal sequence of cleavage products has allowed the identification of amino acid residues crucial for substrate cleavage (Grakoui *et al* 1993b; Pizzi *et al* 1994; Leinbach *et al* 1994). A sequence comparison of the residues flanking the scissile bonds of the NS3-dependent sites identified an acidic amino acid at the P6 position, a P1-Cys in case of the *trans*-sites but a P1-Thr at the NS3/4A *cis*-site and an amino acid with a small side chain (alanine or serine) at the P1′-position (Figure 15.2). Results from site-directed mutagenesis and structure modelling studies have shown that the P1-residue is the major determinant of substrate specificity. In contrast, the P1′-residue is highly tolerant towards substitutions and the acidic P6-residue is not essential (Kolykhalov *et al* 1994; Komoda *et al* 1994; Leinbach *et al* 1994; Tanji *et al* 1994a; Bartenschlager *et al* 1995a). Alanine scanning experiments performed with peptide substrates confirmed the major importance of the P1 residue and showed that the P6, P3 and P4′ residues contribute to substrate recognition by the NS3 proteinase to a lesser extent (Urbani *et al* 1997; Zhang *et al* 1997).

Amino acid substitutions introduced at the various NS3-dependent cleavage sites revealed a differential sensitivity. While the *trans*-sites, in particular the NS5A/B-site were readily inactivated, the NS3/4A site was very tolerant and only amino acid residues with a high potential to influence the structure, like proline or multiple alanine residues, rendered

Cleavage Site	P site	↓	P′ site
	6 5 4 3 2 1		1′2′3′4′
NS3/4A	**D** L E V V **T**		**S** T W V
NS4A/B	**D** E M E E **C**		**A** S H L
NS4B/5A	**D** C S T P **C**		**S** G S W
NS5A/B	**E** D V V C **C**		**S** M S Y
Consensus	**D** . . . . **C** **E** . . . . **T**		**S** **A**

Figure 15.2 Primary sequence around the NS3-cleavage sites. Amino acids are given in single letter code and they refer to the primary sequence of the HCV-BK isolate (Takamizawa *et al* 1991). According to the nomenclature of Schechter and Berger (1968) the amino acid amino terminal of the cleavage site is designated P1, followed by the P2, P3 etc. residues and the amino acid carboxy terminal of the scissile bond is designated P1′, followed by P2′ etc. Amino acids conserved around the scissile bonds are indicated with bold letters.

this site uncleavable. These results most likely reflect the different mechanisms operating at the NS3-dependent sites. Cleavage at the *cis*-site primarily is governed by polyprotein folding bringing the enzyme in close proximity to its substrate and thereby compensating for less favourable P1 residues. In contrast, processing at the *trans*-sites appears to require an intermolecular protein:protein interaction and therefore more stringent conditions.

15.5 THREE-DIMENSIONAL STRUCTURE OF THE NS3/4A PROTEINASE

The recent determination of the three-dimensional structure of the NS3 proteinase with (Kim *et al* 1996; Yan *et al* 1998) or without (Love *et al* 1996) complexed NS4A peptide spanning the central activation domain has explained and extended most of the observations described above. The enzyme adopts a chymotrypsin-like fold and is constituted of two β-barrel domains separated by a deep cleft where the amino acid residues forming the catalytic triad are located (Figure 15.3). In the structures of the NS3/4A complex, the amino terminal domain forms an eight-stranded β-barrel with one strand contributed by NS4A. This β-strand is clamped between the two β-strands A0 and A1 of the NS3 proteinase that are separated by the α0-helix (Figure 15.3). Several side chains emanating from the NS4A β-strand are buried in hydrophobic pockets formed by the NS3 amino terminus and they contribute to the hydrophobic nature of the amino terminal proteinase domain. These results confirm mutation studies showing that NS3/4A interaction and proteinase activation require multiple hydrophobic residues in the central NS4A domain (Bartenschlager *et al* 1995b; Failla *et al* 1995; Butkiewicz *et al* 1996; Shimizu *et al* 1996; Tomei *et al* 1996). Based on this tight association NS4A can be regarded as an integral component of the NS3 proteinase.

The carboxy terminal domain forms a six-stranded β-barrel and ends with a structurally conserved helix (α3; Figure 15.3). In agreement with a homology modelling study

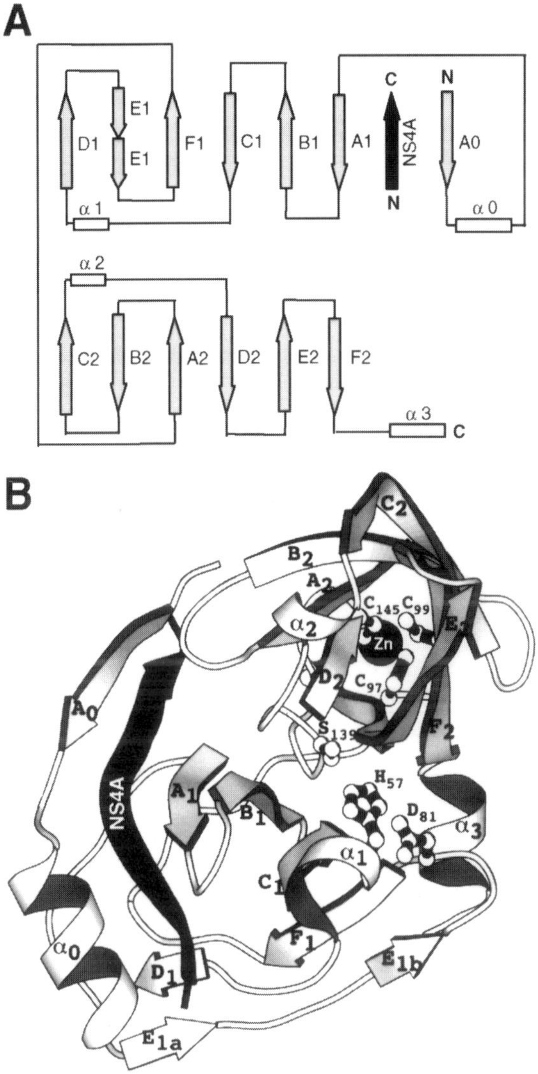

Figure 15.3 Three-dimensional structure of the NS3 proteinase. (A) Secondary structure topology of the NS3 proteinase complexed with a synthetic NS4A cofactor peptide. Arrows indicate β-strands and boxes refer to α-helices. The β-strand contributed by NS4A is shown in black. The α-helices and β-strands are labeled according to the scheme employed for trypsin (Bode *et al* 1976). (B) X-ray crystal structure of the NS3/4A complex as viewed from above the active site. The NS4A cofactor is shown in black. Numbering of α-helices and β-strands is as in A. Side-chains of active site residues His-57, Asp-81 and Ser-139, as well as the three Cys-residues complexing the zinc ion (black sphere) are shown. The side-chain of the fourth zinc-ligand (His-149) is masked by strand β-D2. This figure was kindly provided by R. De Francesco at the IRBM in Rome, Italy and it is printed with his permission. The coordinates of the structure can be found in the Brookhaven Protein Data Bank under the code 1jxp.

(De Francesco *et al* 1996) a tetrahedrally coordinated zinc-ion is found in this domain. It is complexed with three Cys-residues and, through a water-molecule, with one His-residue. The distance of the zinc-ion from the catalytic serine residue (23 Å) suggests that zinc plays a structural role and is not directly involved in catalysis. This assumption is further supported by several observations: (1) Co^{2+} and Cd^{2+} can substitute for Zn^{2+}, (2) removal of Zn^{2+} by denaturation and subsequent refolding only gives soluble protein when zinc is present during renaturation, (3) the amount of soluble NS3 expressed in *Escherichia coli* increases when the cells are cultured in zinc-containing media, (4) mutations affecting the coordinating Cys- or His-residues greatly affect solubility of the protein and, (5) catalytically active zinc is usually coordinated by nitrogen and oxygen atoms, but with thiolates when zinc is required for structural integrity (Berg and Shi 1996; De Francesco *et al* 1996; Stempniak *et al* 1997).

The X-ray crystal structure also sheds light on the substrate specificity of the NS3/4A proteinase. Several loops involved in shaping the substrate binding pocket (SBP) in the case of the structurally related enzymes chymotrypsin and elastase are missing, rendering the NS3-SBP rather flat and solvent exposed. The S_1 subsite is determined primarily by residues Leu-135, Phe-154 and Ala-157 (Pizzi *et al* 1994; Kim *et al* 1996; Love *et al* 1996; Yan *et al* 1998). The phenylalanine is located at the bottom of the S_1 pocket and thereby delineates the length of the P1-side chain that fits into it. Since a sulfhydryl group forms favourable interactions with the aromatic ring of a phenylalanine, this structure provides an explanation why cysteine is the prefered P1-residue. The importance of the residues shaping the S_1 pocket was confirmed by mutation analyses showing that certain substitutions affecting Phe-154 and Ala-157 generate proteinases with altered substrate specificities (Failla *et al* 1996; Koch and Bartenschlager 1997). Apart from the S_1 pocket, the substrate binding channel is rather featureless. All surface loops connecting β-strands, that in other serine proteinases form the S_2, S_3 and S_4 subsites, are very short or absent. The only strong interaction might be between the acidic P6-residue and Arg-161 and Lys-165 in the NS3-domain. Otherwise, the S-binding region involves the loop connecting strands E2 and F2 (Cicero *et al* 1999) and a hydrophobic cavity, formed by NS4A, contributes to binding of P′-residues of the substrate. This lack of involvement of amino acid side-chains in proteinase:substrate interactions is compensated by main-chain interactions over a long distance which explains why P6-P4′ decameric peptide substrates are required for efficient cleavage. Based on these results and the observation that poorly cleavable peptide substrates are still bound efficiently by the proteinase it was suggested that ground state binding of peptidic substrates is mediated by multiple weak interactions over a long distance. The efficiency with which the bound substrate will proceed through the transition state is strongly influenced by the nature of the P1-residue (Urbani *et al* 1998).

15.6 MECHANISM OF PROTEINASE ACTIVATION BY NS4A

A comparison of the proteinase structure crystallized in the absence or presence of the NS4A peptide reveals several differences which shed light on the mechanism of proteinase activation (Yan *et al* 1998; Love *et al* 1998). In the absence of NS4A the ~30 amino terminal NS3 residues are flexible and extended away from the protein while in the presence of the cofactor these residues are folded into helix α0 and β-strand A0, the latter involved in interaction with the cofactor (Figure 15.3). A second alteration affects the geometry of the catalytic triad. In the absence of NS4A, the side-chain of Asp-81, expected to provide charge stabilization to His-57 after deprotonation of Ser-139, is rotated away from His-57 and forms a hydrogen bond with the guanidinium moiety of Arg-155 (Love *et al* 1996). Furthermore, the distance between His-57 and Ser-139 is too large to allow engagement in a bonding interaction. This orientation apparently is less favourable for catalysis explaining the low activity of the free NS3 proteinase. However, this model was recently challenged by the unraveling of the solution structure of the NS3 proteinase by NMR (Barbato *et al* 1999). It was found that even in the absence of the NS4A cofactor, the relative positioning of the catalytic triad corresponds to the one of a fully active proteinase. The discrepancy to the structure determined by X-ray diffraction is probably due to distortions induced by crystal packing forces that interfere with formation of helix $\alpha3$ (Figure 15.3; Barbato *et al* 2000). This helix is important for the correct packing of the strand F1 that positions the catalytic Asp-81. The proper positioning of the catalytic triad in the absence of NS4A would explain why NS3 retains enzymatic activity without cofactor (cleavage at the NS5A/B site; see above). Accordingly, NS4A may play another role in proteinase activation. It was suggested that the primary action of the cofactor is the stabilization of the fold of the S′ subsite of the enzyme (Barbato *et al* 2000). This stabilization may facilitate the exit of the substrate leaving group after cleavage, leading to the observed increase in K_{cat}.

Apart from NS4A, the substrate itself plays an important role for enzyme activation. Activity of the serine proteinase requires a stable network of hydrogen bonds between residues of the catalytic triad. However, owing to the solvent exposure, bonding between His-57 and Asp-81 is not stable. By analyzing the solution structure of the NS3 proteinase complexed with a peptidic inhibitor derived from a P1-P4 substrate (see below), Barbato and coworkers (2000) found that binding of the substrate to the enzyme leads to an expulsion of solvent from the surface of the His-Asp interaction and thereby stabilizes the conformation of the active site. This stabilization of the correct geometry of the catalytic triad by substrate binding is without precedence among chymotrypsin-like proteinases.

15.7 POTENTIAL APPROACHES FOR THE DEVELOPMENT OF NS3/4A-SPECIFIC INHIBITORS

At least two prerequisites must be fullfilled before the NS3/4A proteinase complex can be considered as a target for development of antiviral drugs. First, it must be essential for virus replication and second, appropriate test systems must be available. Since both criteria are fulfilled, the NS3 proteinase is a valid drug target (Bartenschlager 1997; Kolykhalov *et al* 2000). Based on the biochemical properties of the enzyme described above, at least three different strategies for proteinase inhibition can be envisaged.

15.7.1 Removal of zinc

Owing to the importance of zinc for structural integrity of the enzyme and the high conservation of the zinc-binding site, interference with zinc-binding is one possibility for enzyme inhibition. However, the weak inhibitory activity of metal chelators, such as EDTA or 1,10-phenanthroline, and the inability to remove zinc by dialysis against chelators at neutral pH, demonstrate its strong association with the enzyme (De Francesco

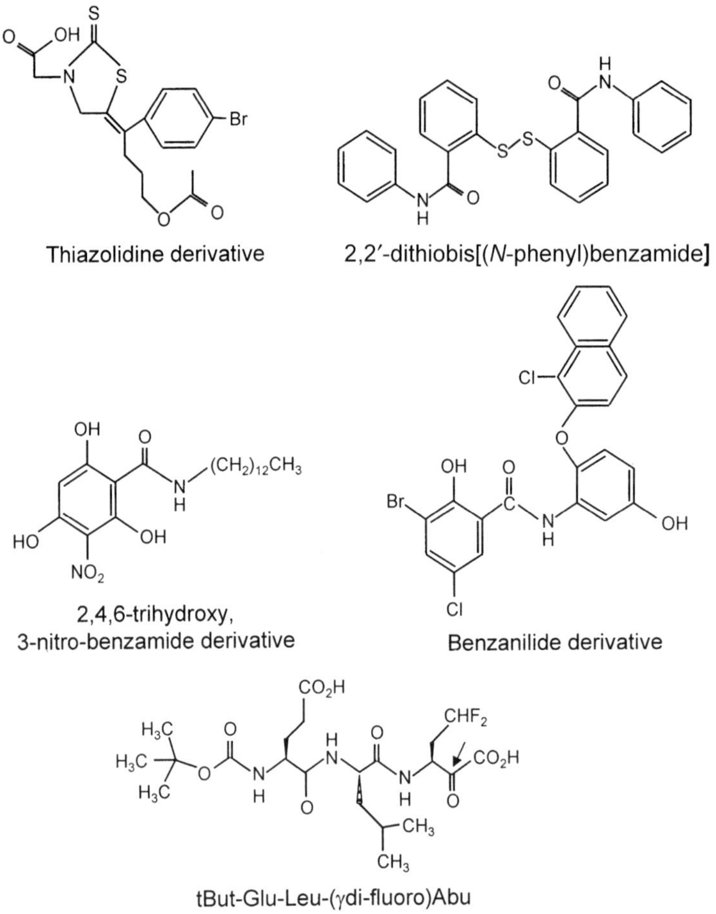

Figure 15.4 Representative structures of the following selected NS3/4A proteinase inhibitors: the zinc-ejecting compound 2,2′-dithiobis[(*N*-phenyl)benzamide] (De Francesco *et al* 1999); a thiazolidine derivative inhibiting the NS3/4A complex with an IC_{50} of 6.4 μg/ml (Sudo *et al* 1997a); a 2,4,6-trihydroxy,3-nitro-benzamide derivative blocking the HCV proteinase with an IC_{50} of 5.8 μM (Sudo *et al* 1997b); a benzanilide derivative inhibiting the enzyme with an IC_{50} of 6.2 μM (Kakiuchi *et al* 1998); the α-ketoacid inhibitor [tBut-Glu-Leu-(di-fluoro)Abu] derived from a P1-P4 peptide (Narjes *et al* 2000). The activated carbonyl moiety which acts as binding group is marked with an arrow.

et al 1996). The binding site itself offers only few features for specific recognition by metal chelators and since zinc is an essential cofactor for many cellular enzymes it will be difficult to develop compounds interfering with zinc binding to the NS3/4A proteinase with sufficient specificity and potency and with low toxicity. Accordingly, zinc-ejecting compounds like 2,2′-dithiobis[(N-phenyl)benzamide] (Figure 15.4) able to interfere with the proteinase activity *in vitro* were shown to be highly cytotoxic (De Francesco *et al* 1999).

15.7.2 Inhibition of the NS3/4A interaction

In principle, every interaction between the proteinase, its cofactor and the substrate could be envisaged as drug target. A first example was reported by Shimizu and coworkers (1996). They described an NS4A peptide carrying a glutamine instead of an arginine at position 28 of the NS4A cofactor that inhibited activation of the proteinase *in vitro*. Based on the model described above, this peptide still binds to the NS3 domain but does not stabilize the S′ subsite of the enzyme leading to a drastic reduction of its catalytic activity. However, this approach is limited by the fact that *in vivo* cleavage at the NS3/4A site is a rapid intramolecular reaction leading to the (cotranslational) formation of a very stable complex. Indeed, ~2400 Å^2 of surface area are buried by the interaction between the NS3 domain and the cofactor and it will be very difficult to displace NS4A once the complex has been formed.

15.7.3 Inhibitors targeted to the substrate binding site

Classically, enzyme inhibitors are identified by high-throughput random screening of large peptide libraries or collections of natural compounds isolated from various sources. With the availability of convenient *in vitro* test formats this approach was also pursued to identify NS3 inhibitors and at least four of these have been reported up to now (Figure 15.4): (1) a phenanthrenequinone compound with an IC_{50} of ~2.5 µg/ml isolated from a *Streptomyces* (Chu *et al* 1997), (2) thiazolidine derivatives with IC_{50}s in the micromolar range (Sudo *et al* 1997a), (3) a 2,4,6-trihydroxy-3-nitro-benzamide derivative with an IC_{50} of ~2 µg/ml (Sudo *et al* 1997b) and, (4) halogenated benzanilides (Kakiuchi *et al* 1998).

The first report on the development of peptide-based inhibitors of the NS3 proteinase is from Landro and coworkers (1997). They described a competitive hexapeptide aldehyde of the sequence Glu-Asp-Val-Val-α-aminobutyricacid-Val-CHO inhibiting the NS3 proteinase in an NS4A-independent way with a K_i of 50 µM. Interestingly, when they extended the peptide to four residues on the P′-side, inhibition became NS4A-dependent and the K_i was lowered to 340 nM. This observation is in agreement with the assumption that NS4A contributes to substrate binding on the P′-side (see above).

One of the most remarkable observations is the extraordinary susceptibility of the NS3/4A proteinase complex to product inhibition. During detailed kinetic analyses it was found that the amino terminal cleavage products of peptides corresponding to the *trans*-sites are potent competitive inhibitors of proteinase activity (Llinas-Brunet *et al* 1998a; Steinkühler *et al* 1998) with the following order of potency: NS4A > NS5A ≫ NS4B(K_is = 0.6 µM, 1.4 µM, 180 µM). Interestingly, peptides corresponding to the *cis*-site do not have an inhibitory effect (Steinkühler *et al* 1998) corroborating that primary sequence around the scissile bond is one determinant for the different mechanisms operating at the *cis*- and the *trans*-sites. Since the K_i-values of the amino terminal cleavage products are up to one order of magnitude lower than the K_m values of the respective substrates, this property was exploited for the design of potent proteinase inhibitors. These studies have shown that efficient binding requires two "anchor residues": an acidic residue in P6/P5 and a "P1 anchor" at the carboxy terminus of the peptide (Ingallinella *et al* 1998; Llinas-Brunet *et al* 1998a; Martin *et al* 1998; Steinkühler *et al* 1998). The latter contributes most to enzyme binding with the negatively charged carboxylate making an ionic interaction with the positively charged Lys-136 proximal to the active site Ser-139 and the P1 side chain interacting with the S1 pocket. Owing to the shape of the S1 subsite, accomodating best a P1-Cys, most efficient inhibition is achieved with a peptide carrying a carboxy terminal cysteine residue (Ingallinella *et al* 1998; Cicero *et al* 1999). By introducing a series of chemical modifications at the P2-P5 positions, a peptide hexamer (Ac-Asp-D-γ-carboxyglutamic acid-Leu-Ile-β-cyclohexylalanine-Cys-OH) corresponding to the P6-P1 position of the NS4A/B cleavage site was developed with an IC_{50} of 1.5 nM. This corresponds to a 400-fold increase in inhibitory potency compared to the parental peptide with a K_i of 600 nM (Ingallinella *et al* 1998). In a similar approach, Llinas-Brunet and coworkers (1998b) developed a hexapeptide N-terminal cleavage product of a dodecamer substrate derived from the NS5A/B cleavage site with an IC_{50} of 640 nM. Structure-activity studies on this hexapeptide led to the identification of competitive inhibitors of the NS3 proteinase with an IC_{50} of 3 nM and a high specificity (e.g. IC_{50} of α-chymotrypsin $>300\,\mu M$; Llinas-Brunet *et al* 2000). Based on the observation that hexapeptide inhibitors bind to the enzyme in an extended conformation from P1-P4 with little involvement of the P5 and P6 side chains (LaPlante *et al* 1999), low-micromolar tetrapeptide inhibitors with high specificity were developed (Llinas-Brunet *et al* 2000). More recently, capped tri-peptide α-ketoacids with difluoro aminobutyric acid derivatives at the P1 position were generated, which are potent slow-binding inhibitors of the NS3 proteinase (Narjes *et al* 2000; Figure 15.4). These compounds bind in an extended backbone conformation and they form an antiparallel β-sheet with a β-strand of the enzyme (Di-Marco *et al* 2000). As expected from the genetic data described above, the P1 residue contributes most to binding energy whereas the side chains of the P2-P4 residues are partially solvent exposed. Unfortunately, even in the presence of those compounds the substrate binding region of the enzyme remains largely featureless and solvent-exposed, underscoring the difficulties in developing efficient small molecule inhibitors.

An alternative approach is the random affinity selection of inhibitors efficiently binding to the proteinase. Dimasi and coworkers (1997) constructed two different libraries for screening of proteinase binding molecules using the phage-display technique. In both cases they introduced random sequences into loops exposed in the context of a stable protein scaffold (either antibody-like proteins or a derivative of the human pancreatic secretory trypsin inhibitor). After affinity selection with the NS3 proteinase as a ligate molecule, they isolated two different inhibitors with marked specificity and potency in the micromolar range. A similar approach, based on the selection of a "camelized" variable domain antibody fragment, was used to develop a competitive inhibitor with a K_i of 150 nM (Martin *et al* 1997). Affinity selection with the NS3 proteinase as a ligate molecule was also used to identify RNA aptamers from a completely random RNA library (Kumar *et al* 1997). Two RNAs were found that inhibited the proteinase in the micromolar range and in addition also the helicase. A similar approach was used by Fukuda and coworkers (2000) to identify RNA aptamers effective in the nanomolar range. Finally, Martin and coworkers (1998) described the development of proteinase inhibitors generated by engineering the active site-binding loop of the general serine proteinase inhibitor eglin c. They obtained an inhibitor with an IC_{50} of 60 nM.

Although these results provide an encouraging starting point for the development of peptidomimetic drugs, owing to the peptide nature of such compounds and their lengths required for efficient inhibition, such compounds are likely to encounter problems with cell penetration, bioavailability and drug stability.

15.8 PROBLEMS ASSOCIATED WITH THE DEVELOPMENT OF HCV-SPECIFIC PROTEINASE INHIBITORS

In spite of the rapid progress made with the development of the first NS3 proteinase inhibitors, there are still several problems that have to be solved. The first one relates to the enormous variability of the virus likely to accelerate the development of therapy-resistant HCV variants. In addition, drugs will be required with broad reactivities effective against all HCV genotypes. The second problem is the lack of a convenient animal model. The only animal infectable with HCV is the chimpanzee, a fact that will render the evaluation of antiviral drugs *in vivo* much more difficult. One alternative is the use of a virus closely related to HCV and replicating in a small animal. The most attractive is GBV-B, a virus replicating in tamarins (*Sanguinus* sp.) and with a genomic organization very closely related to HCV (Muerhoff *et al* 1995). The close evolutionary relationship between these two viruses is emphasized by the observation that the GBV-B proteinase can cleave an HCV peptide substrate, albeit without requirement for an NS4A cofactor (Scarselli *et al* 1997). Until recently, the third and perhaps most serious problem for drug development was the lack of an efficient cell culture system. This limitation has now been overcome by the development of subgenomic HCV RNAs (replicons) amplifying autonomously in a hepatoma cell line (Lohmann *et al* 1999). These RNAs contain all nonstructural proteins required for self-replication and therefore all prime targets for antiviral therapy: the NS3/4A proteinase complex, the NS3 NTPase/helicase and the NS5B RNA-dependent RNA polymerase. Owing to the high replication of these RNAs in transfected cells, allowing the easy detection of viral nucleic acids and proteins, this cell-based system may have an enormous impact for the identification and evaluation of antiviral drugs. Moreover, we and others identified cell culture adaptive mutations that greatly enhance replication of these RNAs (Blight *et al* 2000; Krieger *et al* 2001; Lohmann *et al* 2001). This discovery allowed the construction of more efficiently replicating subgenomic RNAs replicons carrying easily measurable reporter genes and selectable self-replicating full length HCV genomes (Lohmann, Krieger, Pietschmann and Bartenschlager, unpublished).

15.9 CONCLUSION

The cloning of the HCV genome in 1989 has initiated worldwide research efforts that have tremendously increased our knowledge about HCV molecular biology. Apparently led by the enthusiasm resulting from the development of successful inhibitors of the HIV (human immunodeficiency virus) proteinase, intensive efforts have been invested by industrial and academic groups into the biochemical characterization of the HCV NS3/4A proteinase complex, the determination of its three-dimensional structure and into the development of drugs specifically inhibiting this enzyme. Unfortunately, several unexpected problems were encountered such as the complex requirement for a cofactor, the poorly structured substrate binding channel and the length of peptide substrates required for efficient binding. It is obvious that these properties will render the generation of potent NS3-inhibitors more difficult. On the other hand, the enormous progress made in the last few years should keep us optimistic that efficacious anti-HCV therapeutics will not remain fiction but become reality.

REFERENCES

Barbato, G., Cicero, D.O., Nardi, M.C., Steinkühler, C., Cortese, R., De Francesco, R., and Bazzo, R. (1999) The solution structure of the N-terminal protease domain of the hepatitis C virus (HCV) NS3 protein provides new insights into its activation and catalytic mechanism. *Journal of Molecular Biology*, **289**, 370–384.

Barbato, G., Cicero, D.O., Cordier, F., Narjes, F., Gerlach, B., Sambucini, S., *et al* (2000) Inhibitor binding induces active site stabilization of the HCV NS3 protein serine protease domain. *EMBO Journal*, **19**, 1195–1206.

Bartenschlager, R., Ahlborn-Laake, L., Mous, J. and Jacobsen, H. (1993) Nonstructural protein 3 of the hepatitis C virus encodes a serine-type proteinase required for cleavage at the NS3/4 and NS4/5 junctions. *Journal of Virology*, **67**, 3835–3844.

Bartenschlager, R., Ahlborn-Laake, L., Mous, J. and Jacobsen, H. (1994) Kinetic and structural analyses of hepatitis C virus polyprotein processing. *Journal of Virology*, **68**, 5045–5055.

Bartenschlager, R., Ahlborn-Laake, L., Yasargil, K., Mous, J. and Jacobsen, H. (1995a) Substrate determinants for cleavage in *cis* and in *trans* by the hepatitis C virus NS3 proteinase. *Journal of Virology*, **69**, 198–205.

Bartenschlager, R., Lohmann, V., Wilkinson, T. and Koch, J.O. (1995b) Complex formation between the NS3 serine-type proteinase of the hepatitis C virus and NS4A and its importance for polyprotein maturation. *Journal of Virology*, **69**, 7519–7528.

Bartenschlager, R. (1997) Molecular targets in inhibition of hepatitis C virus replication. *Antiviral Chemistry and Chemotherapy*, **8**, 281–301.

Bartenschlager, R. and Lohmann, V. (2000) Replication of hepatitis C virus. *Journal of General Virology*, **81**, 1631–1648.

Behrens, S.E., Tomei, L. and De Francescso, R. (1996) Identification and properties of the RNA-dependent RNA polymerase of hepatitis C virus. *EMBO Journal*, **15**, 12–22.

Berg, J.M. and Shi, Y. (1996) The galvanization of biology: a growing appreciation for the roles of zinc. *Science*, **271**, 1081–1085.

Bianchi, E., Urbani, A., Biasiol, G., Brunetti, M., Pessi, A., De Francesco, R. *et al* (1997) Complex formation between the hepatitis C virus serine protease and a synthetic NS4A cofactor peptide. *Biochemistry*, **36**, 7890–7897.

Blight, K.J., Kolykhalov, A.A. and Rice, C.M. (2000) Efficient initiation of HCV RNA replication in cell culture. *Science*, **290**, 1972–1974.

Bode, W., Fehlhammer, H. and Huber, R. (1976) Crystal structure of bovine trypsinogen at 1-8 A resolution. I. Data collection, application of patterson search techniques and preliminary structural interpretation. *Journal of Molecular Biology*, **106**, 325–335.

Booth, J.L. (1998) Chronic hepatitis C: the virus, its discovery and the natural history of the disease. *Journal of Viral Hepatitis*, **5**, 213–222.

Butkiewicz, N.J., Wendel, M., Zhang, R., Jubin, R., Pichardo, J., Smith, E.B. *et al* (1996) Enhancement of hepatitis C virus NS3 proteinase activity by association with NS4A-specific synthetic peptides: identification of sequence and critical residues of NS4A for the cofactor activity. *Virology*, **225**, 328–338.

Choo, Q.L., Kuo, G., Weiner, A.J., Overby, L.R., Bradley, D.W. and Houghton, M. (1989) Isolation of a cDNA clone derived from a blood-borne non-A, non-B viral hepatitis genome. *Science*, **244**, 359–362.

Chu, M., Mierzwa, R., Truumees, I., King, A., Patel, M., Berrie, R. *et al* (1997) Structure of Sch 68631: A new hepatitis C virus proteinase inhibitor from *Streptomyces* sp. *Tetrahedron Letters*, **38**, 901–901.

Cicero, D.O., Barbato, G., Koch, U., Ingallinella, P., Bianchi, E., Nardi, M.C. *et al* (1999) Structural characterization of the interactions of optimized product inhibitors with the N-terminal proteinase domain of the hepatitis C virus (HCV) NS3 protein by NMR and modelling studies. *Journal of Molecular Biology*, **289**, 385–396.

De Francesco, R., Urbani, A., Nardi, M.C., Tomei, L., Steinkühler, C. and Tramontano, A. (1996) A zinc binding site in viral serine proteinases. *Biochemistry*, **35**, 13282–13287.

De Francesco, R., Pessi, A. and Steinkühler, C. (1999) Mechanisms of hepatitis C virus NS3 proteinase inhibitors. *Journal of Viral Hepatitis*, **6** (Suppl. 1), 23–30.

Dimasi, N., Martin, F., Volpari, C., Brunetti, M., Biasiol, G., Altamura, S. *et al* (1997) Characterization of engineered hepatitis C virus NS3 protease inhibitors affinity selected from human pancreatic secretory trypsin inhibitor and minibody repertoires. *Journal of Virology*, **71**, 7461–7469.

Di Marco, S., Rizzi, M., Volpari, C., Walsh, M.A., Narjes, F., Colarusso, S. *et al* (2000) Inhibition of the hepatitis C virus NS3/4A protease. *Journal of Biological Chemistry*, **275**, 7152–7157.

Eckart, M.R., Selby, M., Masiarz, F., Lee, C., Berger, K., Crawford, K. *et al* (1993) The hepatitis C virus encodes a serine protease involved in processing of the putative nonstructural proteins from the viral polyprotein precursor. *Biochemical and Biophysical Research Commununications*, **192**, 399–406.

Failla, C., Tomei, L. and De Francesco, R. (1994) Both NS3 and NS4A are required for proteolytic processing of hepatitis C virus nonstructural proteins. *Journal of Virology*, **68**, 3753–3760.

Failla, C., Tomei, L. and De Francesco, R. (1995) An amino-terminal domain of the hepatitis C virus NS3 protease is essential for interaction with NS4A. *Journal of Virology*, **69**, 1769–1777.

Failla, C.M., Pizzi, E., De Francesco, R. and Tramontano, A. (1996) Redesigning the substrate specificity of the hepatitis C virus NS3 protease. *Folding and Design*, **1**, 35–42.

Fukuda, K., Vishnuvardhan, D., Sekiya, S., Hwang, J., Kakiuchi, N. Taira, K., *et al* (2000) Isolation and characterization of RNA aptamers specific for the hepatitis C virus nonstructural protein 3 protease. *European Journal of Biochemistry*, **267**, 3685–3694.

Gale, M.J., Korth, M.J., Tang, N.M., Tan, S.L., Hopkins, D.A., Dever, T.E. *et al* (1997) Evidence that hepatitis C virus resistance to interferon is mediated through repression of the PKR protein kinase by the nonstructural 5A protein. *Virology*, **230**, 217–227.

Gale, M.J. Jr., Blakely, S.M., Kwieciszewski, B., Tan, S.L., Dossett, M., Tang, N.M. *et al* (1998) Control of PKR protein kinase by hepatitis C virus nonstructural 5A protein: molecular mechanism of kinase regulation. *Molecular and Cellular Biology*, **18**, 5208–5218.

Gallinari, P., Paolini, C., Brennan, D., Nardi, C., Steinkühler, C. and De Francesco, R. (1999) Modulation of hepatitis C virus NS3 protease and helicase activities through the interaction with NS4A. *Biochemistry*, **38**, 5620–5632.

Grakoui, A., McCourt, D.W., Wychowski, C., Feinstone, S.M. and Rice, C.M. (1993a) A second hepatitis C virus-encoded proteinase. *Proceedings of the National Academy of Sciences of the U.S.A.*, **90**, 10583–10587.

Grakoui, A., McCourt, D.W., Wychowski, C., Feinstone, S.M. and Rice, C.M. (1993b) Characterization of the hepatitis C virus-encoded serine proteinase: determination of proteinase-dependent polyprotein cleavage sites. *Journal of Virology*, **67**, 2832–2843.

Grakoui, A., Wychowski, C., Lin, C., Feinstone, S.M. and Rice, C.M. (1993c) Expression and identification of hepatitis C virus polyprotein cleavage products. *Journal of Virology*, **67**, 1385–1395.

Gwack, Y., Kim, D.W., Han, J.H. and Choe, J. (1996) Characterization of RNA binding activity and RNA helicase activity of the hepatitis C virus NS3 protein. *Biochemical and Biophysical Research Communications*, **225**, 654–659.

Hijikata, M., Kato, N., Ootsuyama, Y., Nakagawa, M. and Shimotohno, K. (1991) Gene mapping of the putative structural region of the hepatitis C virus genome by *in vitro* processing analysis. *Proceedings of the National Academy of Sciences of the U.S.A.*, **88**, 5547–5551.

Hijikata, M., Mizushima, H., Akagi, T., Mori, S., Kakiuchi, N., Kato, N. *et al* (1993a) Two distinct proteinase activities required for the processing of a putative nonstructural precursor protein of hepatitis C virus. *Journal of Virology*, **67**, 4665–4675.

Hijikata, M., Mizushima, H., Tanji, Y., Komoda, Y., Hirowatari, Y., Akagi, T. *et al* (1993b) Proteolytic processing and membrane association of putative nonstructural proteins of hepatitis C virus. *Proceedings of the National Academy of Sciences of the U.S.A.*, **90**, 10773–10777.

Hirowatari, Y., Hijikata, M., Tanji, Y., Nyunoya, H., Mizushima, H., Kimura, K. *et al* (1993) Two proteinase activities in HCV polypeptide expressed in insect cells using baculovirus vector. *Archives of Virology*, **133**, 349–356.

Hong, Z., Ferrari, E., Wright-Minogue, J., Chase, R., Risano, C., Seelig, G. *et al* (1996) Enzymatic characterization of hepatitis C virus NS3/4A complexes expressed in mammalian cells by using the herpes simplex virus amplicon system. *Journal of Virology*, **70**, 4261–4268.

Houghton, M. (1996) Hepatitis C viruses. In *Virology*, edited by B.N. Fields, P.M. Knipe and P.M. Howley, pp. 1035–1058. Philadelphia: Lippincott-Raven.

Ingallinella, P., Altamura, S., Bianchi, E., Taliani, M., Ingenito, R., Cortese, R. *et al* (1998) Potent peptide inhibitors of human hepatitis C virus NS3 protease are obtained by optimizing the cleavage products. *Biochemistry*, **37**, 8906–8914.

Kakiuchi, N., Komoda, Y., Komoda, K., Takeshita, N., Okada, S., Tani, T. *et al* (1998) Non-peptide inhibitors of HCV serine proteinase. *FEBS Letters*, **421**, 217–220.

Kim, D.W., Gwack, Y., Han, J.H. and Choe, J. (1995) C-terminal domain of the hepatitis C virus NS3 protein contains an RNA helicase activity. *Biochemical and Biophysical Research Communications*, **215**, 160–166.

Kim, J.L., Morgenstern, K.A., Lin, C., Fox, T., Dwyer, M.D., Landro, J.A. *et al* (1996) Crystal structure of the hepatitis C virus NS3 protease domain complexed with a synthetic NS4A cofactor peptide *Cell*, **87**, 343–355.

Koch, J.O., Lohmann, V., Herian, U. and Bartenschlager, R. (1996) *In vitro* studies on the activation of the hepatitis C virus NS3 proteinase by the NS4A cofactor. *Virology*, **221**, 54–66.

Koch, J.O. and Bartenschlager, R. (1997) Determinants of substrate specificity in the NS3 serine proteinase of the hepatitis C virus. *Virology*, **237**, 78–88.

Koch, J.O. and Bartenschlager, R. (1999) Modulation of hepatitis C virus NS5A hyperphosphorylation by nonstructural proteins NS3, NS4A, and NS4B. *Journal of Virology*, **73**, 7138–7146.

Kolykhalov, A.A., Agapov, E.V. and Rice, C.M. (1994) Specificity of the hepatitis C virus NS3 serine protease: effects of substitutions at the 3/4A, 4A/4B, 4B/5A, and 5A/5B cleavage sites on polyprotein processing. *Journal of Virology*, **68**, 7525–7533.

Kolykhalov, A.A., Mihalik, K., Feinstone, S.M. and Rice, C.M. (2000) Hepatitis C virus-encoded enzymatic activities and conserved RNA elements in the 3′ nontranslated region are essential for virus replication *in vivo*. *Journal of Virology*, **74**, 2046–2051.

Komoda, Y., Hijikata, M., Sato, S., Asabe, S., Kimura, K. and Shimotohno, K. (1994) Substrate requirements of hepatitis C virus serine proteinase for intermolecular polypeptide cleavage in *Escherichia coli*. *Journal of Virology*, **68**, 7351–7357.

Krieger, N., Lohmann, V. and Bartenschlager, R. (2001) Enhancement of hepatitis C virus RNA replication by cell culture adaptive mutations. *Journal of Virology*, **75**, 4614–4624.

Kumar, P.K., Machida, K., Urvil, P.T., Kakiuchi, N., Vishnuvardhan, D., Shimotohno, K. *et al* (1997) Isolation of RNA aptamers specific to the NS3 protein of hepatitis C virus from a pool of completely random RNA. *Virology*, **237**, 270–282.

Landro, J.A., Raybuck, S.A., Luong, Y.C., O'Malley, E.T., Harbeson, S.L., Morgenstern, K.A. *et al* (1997) Mechanistic role of an NS4A peptide cofactor with the truncated NS3 protease of hepatitis C virus: elucidation of the NS4A stimulatory effect *via* kinetic analysis and inhibitor mapping. *Biochemistry*, **36**, 9340–9348.

LaPlante, S.R., Cameron, D.R., Aubry, N., Lefebvre, S., Kukolj, G., Maurice, R., *et al* (1999) Solution structure of substrate-based ligands when bound to hepatitis C virus NS3 protease domain. *Journal of Biological Chemistry*, **274**, 18618–18624.

Leinbach, S.S., Bhat, R.A., Xia, S.M., Hum, W.T., Stauffer, B., Davis, A.R. *et al* (1994) Substrate specificity of the NS3 serine proteinase of hepatitis C virus as determined by mutagenesis at the NS3/NS4A junction. *Virology*, **204**, 163–169.

Lin, C., Lindenbach, B.D., Pragai, B.M., McCourt, D.W. and Rice, C.M. (1994a) Processing in the hepatitis C virus E2-NS2 region: identification of p7 and two distinct E2-specific products with different C termini. *Journal of Virology*, **68**, 5063–5073.

Lin, C., Pragai, B.M., Grakoui, A., Xu, J. and Rice, C.M. (1994b) Hepatitis C virus NS3 serine proteinase: trans-cleavage requirements and processing kinetics. *Journal of Virology*, **68**, 8147–8157.

Lin, C., Thomson, J.A. and Rice, C.M. (1995) A central region in the hepatitis C virus NS4A protein allows formation of an active NS3-NS4A serine proteinase complex *in vivo* and *in vitro*. *Journal of Virology*, **69**, 4373–4380.

Lin, C., Wu, J.W., Hsiao, K. and Su, M.S. (1997) The hepatitis C virus NS4A protein: interactions with the NS4B and NS5A proteins. *Journal of Virology*, **71**, 6465–6471.

Llinas-Brunet, M., Bailey, M., Deziel, R., Fazal, G., Gorys, V., Goulet, S. *et al* (1998a) Studies on the C-terminal hexapeptide inhibitors of the hepatitis C virus serine protease. *Bioorganic & Medicinal Chemistry Letters*, **8**, 2719–2724.

Llinas-Brunet, M., Bailey, M., Fazal, G., Goulet, S., Halmos, T., Laplante, S. *et al* (1998b) Peptide-based inhibitors of the hepatitis C virus serine protease. *Bioorganic & Medicinal Chemistry Letters*, **8**, 1713–1718.

Llinas-Brunet, M., Bailey, M., Fazal, G., Ghiro, E., Gorys, V., Goulet, S. *et al* (2000) Highly potent and selective peptide-based inhibitors of the hepatitis C virus serine protease: towards smaller inhibitors. *Bioorganic & Medicinal Chemistry Letters*, **10**, 2267–2270.

Lohmann, V., Körner, F., Herian, U. and Bartenschlager, R. (1997) Biochemical properties of hepatitis C virus NS5B RNA-dependent RNA polymerase and identification of amino acid sequence motifs essential for enzymatic activity. *Journal of Virology*, **71**, 8416–8428.

Lohmann, V., Körner, F., Koch, J.O., Herian, U., Theilmann, L. and Bartenschlager, R. (1999) Replication of subgenomic hepatitis C virus RNAs in a hepatoma cell line. *Science*, **285**, 110–113.

Lohmann, V., Krieger, N., Dobierzewska, A., and Bartenschlager, R. (2001) Mutations in hepatitis C virus RNAs conferring cell culture adaptation. *Journal of Virology*, **75**, 1437–1449.

Love, R.A., Parge, H.E., Wickersham, J.A., Hostomsky, Z., Habuka, N., Moomaw, E.W. *et al* (1996) The crystal structure of hepatitis C virus NS3 proteinase reveals a trypsin-like fold and a structural zinc binding site. *Cell*, **87**, 331–342.

Love, R.A., Parge, H.E., Wickersham, J.A., Hostomsky, Z., Habuka, N., Moomaw, E.W. *et al* (1998) The conformation of hepatitis C virus NS3 proteinase with and without NS4A: a structural basis for the activation of the enzyme by its cofactor. *Clinical and Diagnostic Virology*, **10**, 151–156.

Manabe, S., Fuke, I., Tanishita, O., Kaji, C., Gomi, Y., Yoshida, S. *et al* (1994) Production of nonstructural proteins of hepatitis C virus requires a putative viral protease encoded by NS3. *Virology*, **198**, 636–644.

Markland, W., Petrillo, R.A., Fitzgibbon, M., Fox, T., McCarrick, R., McQuaid, T. *et al* (1997) Purification and characterization of the NS3 serine protease domain of hepatitis C virus expressed in *Saccharomyces cerevisiae*. *Journal of General Virology*, **78**, 39–43.

Martin, F., Volpari, C., Steinkühler, C., Dimasi, N., Brunetti, M., Biasiol, G. *et al* (1997) Affinity selection of a camelized V(H) domain antibody inhibitor of hepatitis C virus NS3 protease. *Protein Engineering*, **10**, 607–614.

Martin, F., Dimasi, N., Volpari, C., Perrera, C., Di, M.S., Brunetti, M. *et al* (1998) Design of selective eglin inhibitors of HCV NS3 proteinase. *Biochemistry*, **37**, 11459–11468.

Mizushima, H., Hijikata, M., Asabe, S., Hirota, M., Kimura, K. and Shimotohno, K. (1994) Two hepatitis C virus glycoprotein E2 products with different C termini. *Journal of Virology*, **68**, 6215–6222.

Morgenstern, K.A., Landro, J.A., Hsiao, K., Lin, C., Gu, Y., Su, M.S. *et al* (1997) Polynucleotide modulation of the protease, nucleoside triphosphatase, and helicase activities of a hepatitis C virus NS3-NS4A complex isolated from transfected COS cells. *Journal of Virology*, **71**, 3767–3775.

Moussalli, J., Opolon, P. and Poynard, T. (1998) Management of hepatitis C. *Journal of Viral Hepatitis*, **5**, 73–82.

Muerhoff, A.S., Leary, T.P., Simons, J.N., Pilot, M.T., Dawson, G.J., Erker, J.C. *et al* (1995) Genomic organization of GB viruses A and B: two new members of the Flaviviridae associated with GB agent hepatitis. *Journal of Virology*, **69**, 5621–5630.

Murphy, F.A., Fauquet, C.M., Bishop, D.H.L., Ghabrial, S.A., Jarvis, A.W., Martelli, G.P. *et al* (editors) (1995) Classification and nomenclature of viruses. Sixth Report of the International Committee on Taxonomy of Viruses. pp. 424–426. Vienna/New York: Springer-Verlag.

Narjes, F., Brunetti, M., Colarusso, S., Gerlach, B., Koch, U., Biasiol, G. *et al* (2000) α-ketoacids are potent slow binding inhibitors of the hepatitis C virus NS3 protease. *Biochemistry*, **39**, 1849–1861.

Pizzi, E., Tramontano, A., Tomei, L., La Monica, N., Failla, C., Sardana, M. *et al* (1994) Molecular model of the specificity pocket of the hepatitis C virus protease: implications for substrate recognition. *Proceedings of the National Academy of Sciences of the U.S.A.*, **91**, 888–892.

Preugschat, F., Averett, D.R., Clarke, B.E. and Porter, D.T. (1996) A steady-state and pre-steady-state kinetic analysis of the NTPase activity associated with the hepatitis C virus NS3 helicase domain. *Journal of Biological Chemistry*, **271**, 24449–24457.

Reed, K.E. and Rice, C.M. (2000) Overview of hepatitis C virus genome structure, polyprotein processing, and protein properties. *Current Topics in Microbiology and Immunology*, **242**, 55–84.

Santolini, E., Pacini, L., Fipaldini, C., Migliaccio, G. and La Monica, N. (1995) The NS2 protein of hepatitis C virus is a transmembrane polypeptide. *Journal of Virology*, **69**, 7461–7471.

Satoh, S., Tanji, Y., Hijikata, M., Kimura, K. and Shimotohno, K. (1995) The N-terminal region of hepatitis C virus nonstructural protein 3 (NS3) is essential for stable complex formation with NS4A. *Journal of Virology*, **69**, 4255–4260.

Scarselli, E., Urbani, A., Sbardellati, A., Tomei, L., De Francesco, R. and Traboni, C. (1997) GB virus B and hepatitis C virus NS3 serine proteases share substrate specificity. *Journal of Virology*, **71**, 4985–4989.

Schechter, I. and Berger, A. (1968) On the active site of proteases. 3. Mapping the active site of papain; specific peptide inhibitors of papain. *Biochemical and Biophysical Research Communications*, **32**, 898–902.

Selby, M.J., Choo, Q.L., Berger, K., Kuo, G., Glazer, E., Eckart, M. *et al* (1993) Expression, identification and subcellular localization of the proteins encoded by the hepatitis C viral genome. *Journal of General Virology*, **74**, 1103–1113.

Shimizu, Y., Yamaji, K., Masuho, Y., Yokota, T., Inoue, H., Sudo, K. *et al* (1996) Identification of the sequence on NS4A required for enhanced cleavage of the NS5A/5B site by hepatitis C virus NS3 protease. *Journal of Virology*, **70**, 127–132.

Shoji, I., Suzuki, T., Sato, M., Aizaki, H., Chiba, T., Matsuura, Y. *et al* (1999) Internal processing of hepatitis C virus NS3 protein. *Virology*, **254**, 315–323.

Steinkühler, C., Tomei, L. and De Francesco, R. (1996a) *In vitro* activity of hepatitis C virus protease NS3 purified from recombinant baculovirus-infected Sf9 cells. *Journal of Biological Chemistry*, **271**, 6367–6373.

Steinkühler, C., Urbani, A., Tomei, L., Biasiol, G., Sardana, M., Bianchi, E. *et al* (1996b) Activity of purified hepatitis C virus protease NS3 on peptide substrates. *Journal of Virology*, **70**, 6694–6700.

Steinkühler, C., Biasiol, G., Brunetti, M., Urbani, A., Koch, U., Cortese, R. *et al* (1998) Product inhibition of the hepatitis C virus NS3 protease. *Biochemistry*, **37**, 8899–8905.

Stempniak, M., Hostomska, Z., Nodes, B.R. and Hostomsky, Z. (1997) The NS3 proteinase domain of hepatitis C virus is a zinc-containing enzyme. *Journal of Virology*, **71**, 2881–2886.

Sudo, K., Matsumoto, Y., Matsushima, M., Fujiwara, M., Konno, K., Shimotohno, K. *et al* (1997a) Novel hepatitis C virus protease inhibitors: Thiazolidine derivatives. *Biochemical and Biophysical Research Communications*, **238**, 643–647.

Sudo, K., Matsumoto, Y., Matsushima, M., Konno, K., Shimotohno, K., Shigeta, S. *et al* (1997b) Novel hepatitis C virus protease inhibitors: 2,4,6-trihydroxy,3-nitro- benzamide derivatives. *Antiviral Chemistry and Chemotherapy*, **8**, 541–544.

Suzich, J.A., Tamura, J.K., Palmer, H.F., Warrener, P., Grakoui, A., Rice, C.M. *et al* (1993) Hepatitis C virus NS3 protein polynucleotide-stimulated nucleoside triphosphatase and comparison with the related pestivirus and flavivirus enzymes. *Journal of Virology*, **67**, 6152–6158.

Tai, C.L., Chi, W.K., Chen, D.S. and Hwang, L.H. (1996) The helicase activity associated with hepatitis C virus nonstructural protein 3 (NS3). *Journal of Virology*, **70**, 8477–8484.

Takamizawa, A., Mori, C., Fuke, I., Manabe, S., Murakami, S., Fujita, J. *et al* (1991) Structure and organization of the hepatitis C virus genome isolated from human carriers. *Journal of Virology*, **65**, 1105–1113.

Tanji, Y., Hijikata, M., Hirowatari, Y. and Shimotohno, K. (1994a) Hepatitis C virus polyprotein processing: kinetics and mutagenic analysis of serine proteinase-dependent cleavage. *Journal of Virology*, **68**, 8418–8422.

Tanji, Y., Hijikata, M., Hirowatari, Y. and Shimotohno, K. (1994b) Identification of the domain required for trans-cleavage activity of hepatitis C viral serine proteinase. *Gene*, **145**, 215–219.

Tanji, Y., Hijikata, M., Satoh, S., Kaneko, T. and Shimotohno, K. (1995) Hepatitis C virus-encoded nonstructural protein NS4A has versatile functions in viral protein processing. *Journal of Virology*, **69**, 1575–1581.

Tomei, L., Failla, C., Santolini, E., De Francesco, R. and La Monica, N. (1993) NS3 is a serine protease required for processing of hepatitis C virus polyprotein. *Journal of Virology*, **67**, 4017–4026.

Tomei, L., Failla, C., Vitale, R.L., Bianchi, E. and De Francesco, R. (1996) A central hydrophobic domain of the hepatitis C virus NS4A protein is necessary and sufficient for the activation of the NS3 protease. *Journal of General Virology*, **77**, 1065–1070.

Urbani, A., Bianchi, E., Narjes, F., Tramontano, A., De Francesco, R., Steinkühler, C. *et al* (1997) Substrate specificity of the hepatitis C virus serine protease NS3. *Journal of Biological Chemistry*, **272**, 9204–9209.

Urbani, A., Bazzo, R., Nardi, M.C., Cicero, D.O., De Francesco, R., Steinkühler, C. *et al* (1998) The metal binding site of the hepatitis C virus NS3 protease. A spectroscopic investigation. *Journal of Biological Chemistry*, **273**, 18760–18769.

Yan, Y.W., Li, Y., Munshi, S., Sardana, V., Cole, J.L., Sardana, M. *et al* (1998) Complex of NS3 protease and NS4A peptide of BK strain hepatitis C virus: A 2.2 angstrom resolution structure in a hexagonal crystal form. *Protein Science*, **7**, 837–847.

Zhang, R.M., Durkin, J., Windsor, W.T., McNemar, C., Ramanathan, L. and Le, H.V. (1997) Probing the substrate specificity of hepatitis C virus NS3 serine protease by using synthetic peptides. *Journal of Virology*, **71**, 6208–6213.

Chapter 16

Zinc metallopeptidases

Nigel M. Hooper

The zinc metallopeptidases angiotensin converting enzyme (ACE), neprilysin (NEP) and endothelin converting enzyme (ECE) play key roles in cardiovascular and renal function. ACE is involved in the generation of the vasoconstrictor angiotensin II and inactivates the vasodilator bradykinin. Since the development of the first synthetic, non-peptide inhibitor of ACE, captopril, numerous ACE inhibitors have been developed and used clinically for the treatment of hypertension and congestive heart failure. Inhibition of NEP potentiates the hypotensive effects of atrial natriuretic factor, while inhibition of ECE prevents the formation of the potent vasoconstrictor endothelin. Dual inhibitors of ACE and NEP, and of NEP and ECE have been developed which, through combining inhibition of the two enzymes in a single compound, often have better effects than the individual enzyme inhibitors. The latest developments in this area are triple inhibitors of all three related zinc metallopeptidases.

16.1 INTRODUCTION

One of the most successful applications of zinc metalloprotease inhibitors is in the treatment of hypertension and congestive heart failure. Research in this area has resulted in potent and selective inhibitors of certain zinc metallopeptidases, some of which are now in routine clinical use. Important lessons can be learned from the rational design of these compounds, particularly considering that the 3-dimensional crystal structures of the key enzymes have yet to be determined, which have a significant bearing on the development and application of inhibitors against other zinc metalloproteases. Within the field of blood pressure regulation much attention has been focused on angiotensin converting enzyme (ACE; peptidyl dipeptidase A; EC 3.4.15.1) which plays a key role in the renin-angiotensin system. However, more attention is now being paid to two other zinc metallopeptidases, neprilysin (NEP; endopeptidase-24.11; enkephalinase; EC 3.4.24.11) and endothelin converting enzyme (ECE; EC 3.4.24.71), inhibitors of which also have antihypertensive properties. In this chapter, the properties of these enzymes are briefly reviewed and the development of inhibitors of them, with the focus being on the more recent developments. For more detailed reviews of the initial development of ACE and NEP inhibitors the reader is referred to Thorsett and Wyvratt (1987); Roques and Beaumont (1990); Roques *et al* (1993); Beaumont *et al* (1996).

16.1.1 Basis for use of inhibitors of zinc metallopeptidases

For many years, the renin-angiotensin system has been known to play a key role in the regulation of blood pressure in mammals (Vallotton 1987). Renin, secreted from the juxtoglomerular apparatus of the kidneys in response to a variety of stimuli, acts on the circulating precursor angiotensinogen to generate the decapeptide angiotensin I (Figure 16.1). Angiotensin I has little, if any, effect on blood pressure, but is converted by ACE into the potent vasopressor octapeptide, angiotensin II (Figure 16.1). ACE also inactivates the vasodilator nonapeptide bradykinin (Figure 16.1) which is derived from its precursor kininogen through the action of kallikrein (Erdos *et al* 1999). Thus ACE has a dual role in blood pressure regulation, and its inhibition would be expected to prevent the formation of the hypertensive agent angiotensin II and to potentiate the hypotensive properties of bradykinin, leading to a lowering of the blood pressure.

In the early 1980s, the natriuretic family of peptides was identified and characterized (Ruskoaho 1992; Wilkins and Kenny 1997). One of these, α-human atrial natriuretic peptide (ANP), a 28 amino acid peptide with an intramolecular disulfide loop, is the major circulating form and plays a key role in blood pressure regulation. ANP is secreted in response to stretching of the myocardial wall, such as occurs with volume or pressure overload of the heart. The released circulating peptide then has potent diuretic and natriuretic effects, inhibits the secretion of renin (Figure 16.1) and aldosterone and causes vasodilation which result in a decrease in arterial pressure. ANP is metabolized *in vitro* and *in vivo* by NEP (Kenny and Stephenson 1988). Thus blockade of the metabolism of circulating ANP, through the inhibition of NEP (Figure 16.1), was also investigated as a possible therapeutic approach to the control of hypertension. Previously, inhibitors of this enzyme had been developed as potential analgesics following the observation that NEP is the key enzyme responsible for the metabolism of the endogenous opioid peptides, Leu- and Met-enkephalin (Roques *et al* 1980).

The identification of another peptide family which is involved in blood pressure regulation in the late 1980s opened up the way for a further therapeutic target. Endothelin-1 (ET-1) is a 21 amino acid peptide with two intramolecular disulfide bonds that is derived from a larger precursor, big endothelin-1 (big ET-1), through the action of ECE (Figure 16.1) (Turner and Murphy 1996). Synthetic ET-1 is a powerful constrictor of coronary arteries *in vitro* and a potent hypertensive agent in *vivo* (Yanagisawa *et al* 1988). Thus prevention of the production of ET-1, through inhibition of ECE, would be predicted to lower blood pressure. Three distinct ET genes have been identified which encode three closely related peptides (ET-1, ET-2 and ET-3).

16.1.2 Problems in drug design

As with the development of inhibitors of most proteases, one of the major problems to overcome is the inherent lability of compounds which are essentially substrate analogues, that is peptides or peptide derivatives. In general, the bioavailability of peptide drugs is extremely low due to extensive extra- and intracellular proteolytic metabolism. In the oral route, peptide drugs have to survive the battery of proteases and peptidases present in and lining the gastrointestinal tract (Bai 1993; Hooper 1993). In addition, the drugs have to be efficiently absorbed across the intestinal epithelium. Obviously to be effective *in vivo*,

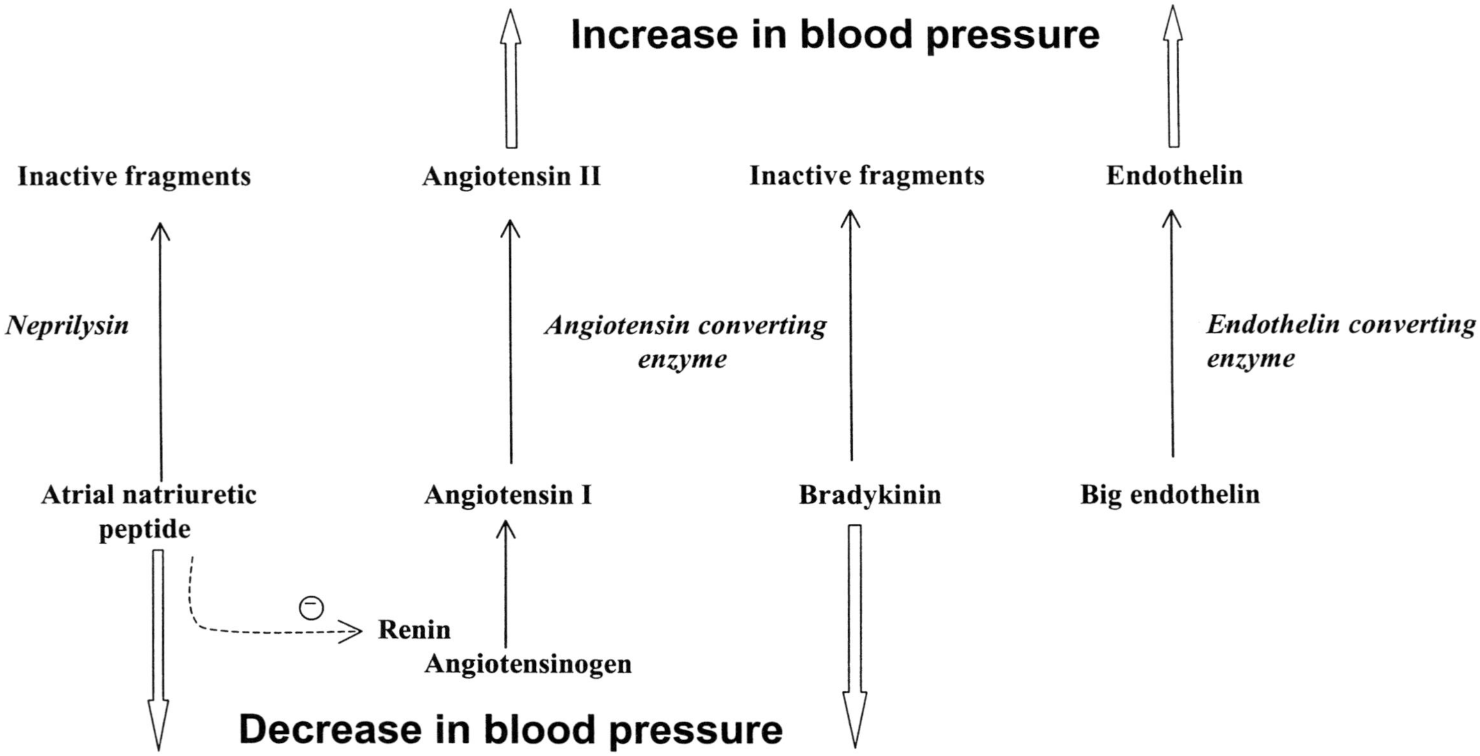

Figure 16.1 The relationship between angiotensin converting enzyme, neprilysin and endothelin converting enzyme in the control of blood pressure. Angiotensin II and endothelin cause an increase in blood pressure, while atrial natriuretic peptide and bradykinin cause a decrease in blood pressure. The formation of the former two peptides is prevented by the inhibition of angiotensin converting enzyme and endothelin converting enzyme, and the breakdown of the latter two is prevented by the inhibition of angiotensin converting enzyme and neprilysin, resulting in a lowering of the blood pressure.

compounds must be relatively potent and selective inhibitors of the target enzyme. These and other problems have arisen and been overcome with the development of inhibitors of the zinc metallopeptidases ACE, NEP and ECE.

16.2 CHARACTERISTICS OF THE PEPTIDASES

16.2.1 Angiotensin converting enzyme

ACE is a type I integral membrane glycoprotein (Hooper *et al* 1987). Two isoforms of the enzyme exist; the somatic isoform of 150–180 kDa which is found throughout the body on endothelial and epithelial cells, and the testicular isoform of 90–110 kDa which is only found in germinal cells in the testis. cDNA cloning and sequencing revealed that the somatic isoform has two HEXXH zinc binding motifs (Soubrier *et al* 1988), and subsequent mutagenesis and expression studies have shown that both catalytic sites are enzymically active (Wei *et al* 1991; Jaspard *et al* 1993). The testicular isoform corresponds to the C-terminal half of the somatic form apart from the presence of a unique 67 residue N-terminus (Ehlers *et al* 1989). Both isoforms are transcribed from the same gene through the use of tissue-specific initiation sites (Howard *et al* 1990). In general ACE has the specificity of a dipeptidyl carboxypeptidase, removing C-terminal dipeptides from its substrates, although it displays endopeptidase action on certain substrates, such as luteinizing hormone-releasing hormone and substance P (Hooper 1991). Although angiotensin I and bradykinin are metabolized by both the N- and C-domain catalytic sites, recently it has been reported that the N-domain catalytic site selectively cleaves the haemoregulatory peptide *N*-acetyl-Ser-Asp-Lys-Pro (Rousseau *et al* 1995; Azizi *et al* 1996; Michaud *et al* 1999).

16.2.2 Neprilysin

NEP is a type II integral membrane glycoprotein of 90 kDa that contains a single HEXXH zinc binding motif. The enzyme was first isolated from the brush border membranes of rabbit kidney (Kerr and Kenny 1974a,b), but is now known to have a widespread tissue distribution, including the vascular wall of the aorta (Gee *et al* 1985). NEP is an endopeptidase, cleaving peptides on the amino side of hydrophobic residues (Kerr and Kenny 1974b). This specificity is similar to that of the bacterial enzyme thermolysin, and the natural compound phosphoramidon inhibits both enzymes with K_i values in the low nanomolar range (Beaumont *et al* 1996). cDNA cloning and subsequent mutagenesis and expression studies have revealed that NEP has a very similar active site organization to thermolysin (Devault *et al* 1988; Malfroy *et al* 1988; Le Moual *et al* 1991). Thus, the availability of the crystal structure of thermolysin has aided the design of NEP inhibitors. NEP hydrolyses a range of bioactive peptides including the enkephalins, substance P and other tachykinins, ANP, and cholecystokinin.

16.2.3 Endothelin converting enzyme

An enzyme with the capacity to convert big ET-1 into ET-1 was first isolated from rat lung (Takahashi *et al* 1993) and bovine aorta (Ohnaka *et al* 1993), and subsequently the cDNA encoding the human enzyme was cloned and sequenced (Schmidt *et al* 1994). ECE-1 is a type II integral membrane protein with a short N-terminal cytoplasmic tail, a transmembrane hydrophobic domain and a large extracellular domain containing the catalytic site and a HEXXH zinc-binding motif (Turner and Tanzawa 1997). ECE-1 shows significant sequence similarity, especially in the C-terminal region, to NEP. In particular important structural, substrate-binding and catalytic residues previously identified in neprilysin are conserved in ECE-1 (Sansom *et al* 1995). Until recently the activity of ECE-1 appeared to be restricted to the big endothelins. However, the enzyme does have activity towards bradykinin, neurotensin and substance P (Hoang and Turner 1997; Johnson *et al* 1999). To date two other endothelin converting enzymes have been isolated (ECE-2 and ECE-3) (Emoto and Yanagisawa 1995; Hasegawa *et al* 1998). Although both ECE-1 and ECE-2 show a similar preference towards big ET-1, ECE-3 is highly specific for big ET-3.

16.3 DESIGN OF INHIBITORS

16.3.1 Angiotensin converting enzyme inhibitors

In the early 1970s, it was reported that the venom of the South American pit viper could potentiate the action of bradykinin (Ferreira *et al* 1970; Ondetti *et al* 1971). The active compounds within the venom were subsequently identified as short proline-rich peptides, of which teprotide (pGlu-Trp-Pro-Arg-Pro-Gln-Ile-Pro-Pro) had the best *in vivo* action, not only in preventing the breakdown of bradykinin, but also in blocking the conversion of angiotensin I to angiotensin II. The observation that intravenous injection of teprotide caused a lowering of blood pressure was the first evidence that inhibiting the renin-angiotensin system would be a valid therapeutic approach to control high blood pressure (Gavras *et al* 1974). However, early on in these studies a significant problem was encountered: teprotide was not effective when administered orally, presumably because it was degraded in the gastrointestinal system by soluble proteases and the battery of membrane-bound peptidases (Hooper 1993).

However, with information derived from structural and mechanistic studies on the best characterized zinc metalloprotease at the time, carboxypeptidase A, Ondetti and Cushman at the Squibb Institute proposed a hypothetical model for the active site of ACE (Cushman and Ondetti 1999) and made a series of synthetic compounds with proline in the P'_2 position (for nomenclature see Schechter and Berger 1967). Ultimately this resulted in the Ala-Pro based dipeptide captopril which contains a sulfydryl group to coordinate to the active site zinc ion (Table 16.1) (Ondetti and Cushman 1982; Thorsett and Wyvratt 1987; Unger *et al* 1990). In 1981, captopril was approved for use in most countries for the treatment of hypertension and congestive heart failure.

The first potent non-sulfydryl-containing ACE inhibitor was enalaprilat (MK422), a substituted *N*-carboxymethyl-dipeptide (Table 16.1) (Patchett *et al* 1980). However, the bioavailability of enalaprilat is very low, and enalapril (MK421) an ethyl-ester of the compound (Table 16.1), was developed as a pro-drug form. Enalapril, although having a very low potency towards ACE, is fairly well absorbed from the gut and subsequently hydrolyzed to the active diacid enalaprilat by esterases in the liver, blood and other tissues (Unger *et al* 1982; Grima *et al* 1991). Numerous other potent dipeptide and tripeptide inhibitors of ACE have since been synthesized with either thiol, carboxylate (e.g. cilazapril

Table 16.1 Inhibitors of angiotensin converting enzyme

Compound	*Structure*	K_i (nM)
Captopril	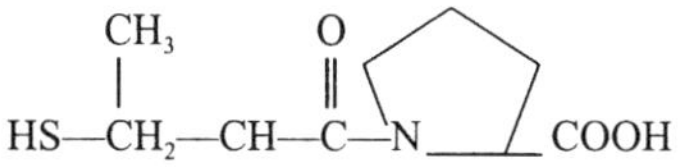	1.7
Enalaprilat	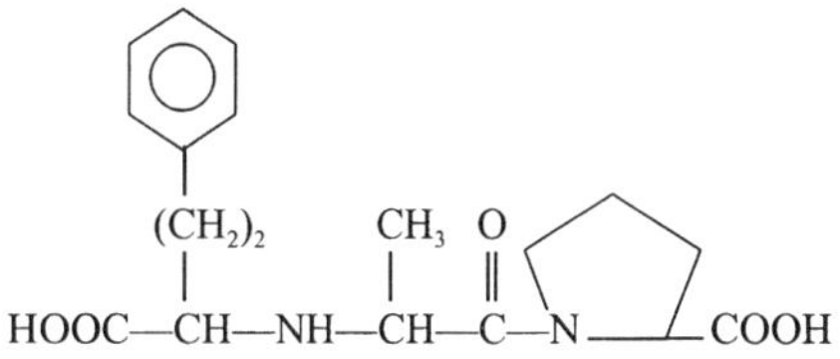	0.15
Enalapril	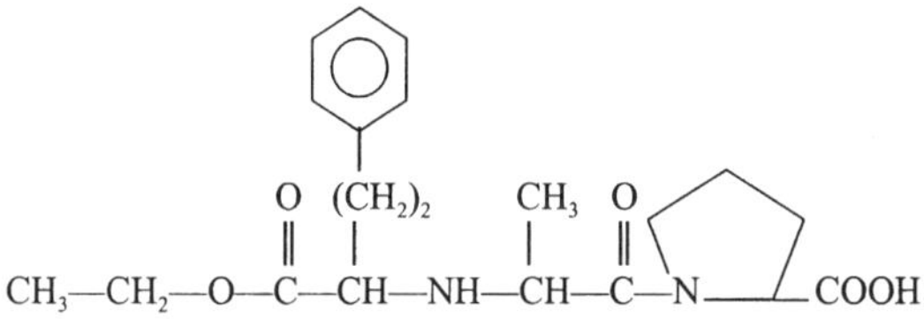	>10,000
Ramipril	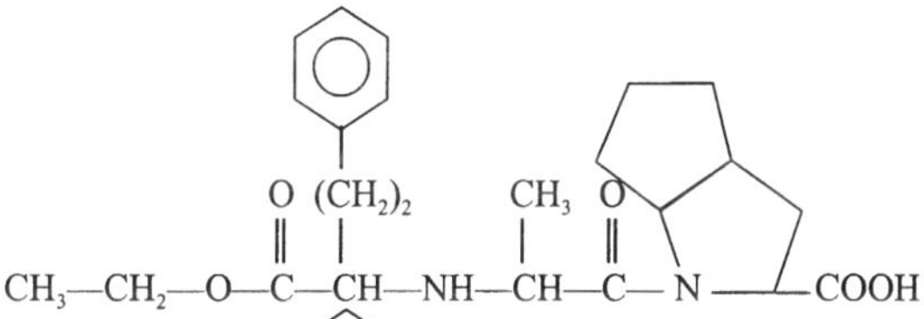	(0.007)*
Cilazapril	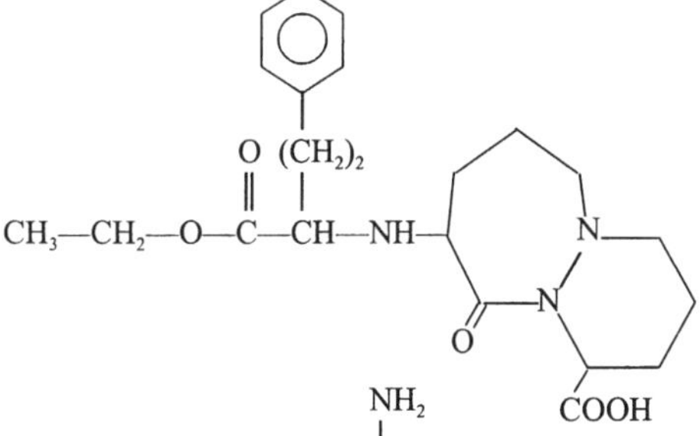	(0.05)*
Ceranapril	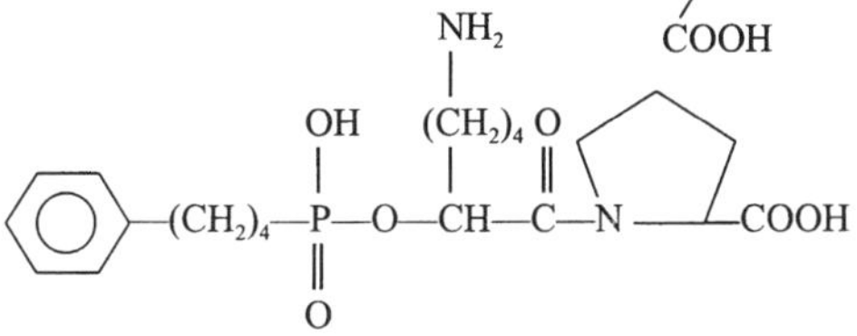	I_{50} 36.0

*Ramipril and cilazapril are prodrugs; the K_i values are for the active components.

and ramipril) or phosphate (e.g. ceranapril) groups as zinc chelators (Table 16.1) (Natoff *et al* 1985; Becker and Scholkens 1987; DeForrest *et al* 1990; Todd and Benfield 1990). Many of these compounds have a small hydrophobic residue in the P'_1 position and a negative charge in the P'_2 position, often on a proline residue whose positioning is important for inhibition of ACE. Several reviews have detailed the development and clinical uses of these ACE inhibitors (Salvetti 1990; Jackson *et al* 1991; Mancia 1991).

Recently, a combinatorial chemistry approach has been used to identify a phosphinic peptide Ac-Asp-*L*-Pheψ(PO_2-CH_2)*L*-Ala-Ala-NH_2 (RXP 407) which has a K_i of 12 nM for the N-domain catalytic site, but with a dissociation constant three orders of magnitude higher for the C-domain catalytic site (Dive *et al* 1999). This compound, which is metabolically stable *in vivo*, may lead to a new generation of ACE inhibitors able to block *in vivo* only a subset of the different functions regulated by ACE.

ACE inhibitors are still amongst the most potent synthetic protease inhibitors with K_i values in the low nanomolar range. Something which is all the more remarkable when one considers that the 3-dimensional structure of the active site of ACE has yet to be elucidated.

16.3.2 Neprilysin inhibitors

The first described synthetic potent NEP inhibitor was thiorphan (Roques *et al* 1980), which is however only about 50-fold more potent in inhibiting NEP than ACE (Table 16.2). However, the two enantiomers (*R* and *S*) of thiorphan have the same inhibitory potency towards NEP, but not for ACE, indicating large differences in the stereochemical requirements for optimal interactions in the active sites of the two enzymes. To try to increase NEP selectivity, various structural modifications of the P'_1 and/or P'_2 moieties of thiorphan were made (Gordon *et al* 1983; Fournie-Zaluski *et al* 1984). However, this generally resulted in the synthesis of highly potent mixed inhibitors of NEP and ACE (see below). Retro-inversing the amide bond of thiorphan, as in retrothiorphan (Table 16.2), was more successful in producing selective NEP inhibitors. One drawback of the thiol

Table 16.2 Inhibitors of neprilysin

Compound	*Structure*	K_i (nM) *NEP*	*ACE*
Thiorphan	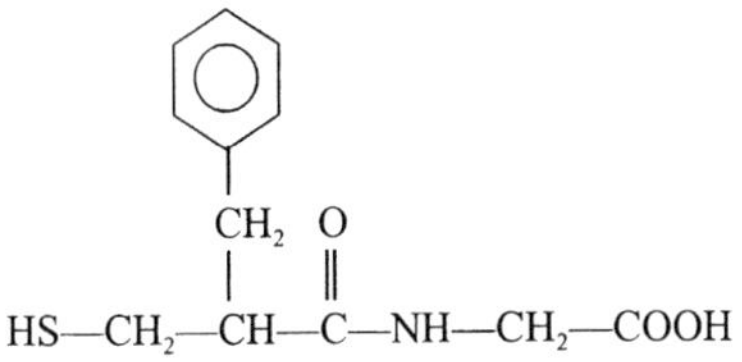	S 4 R 4	S 140 R 860
Retrothiorphan	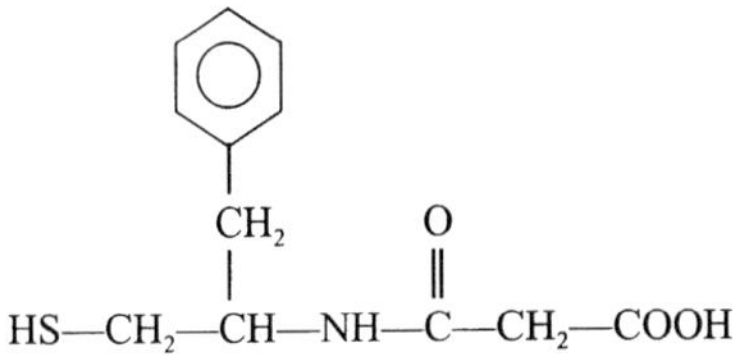	S 210 R 2.3	>10,000
Acetorphan (*R*, *S*) Sinorphan (*S*)	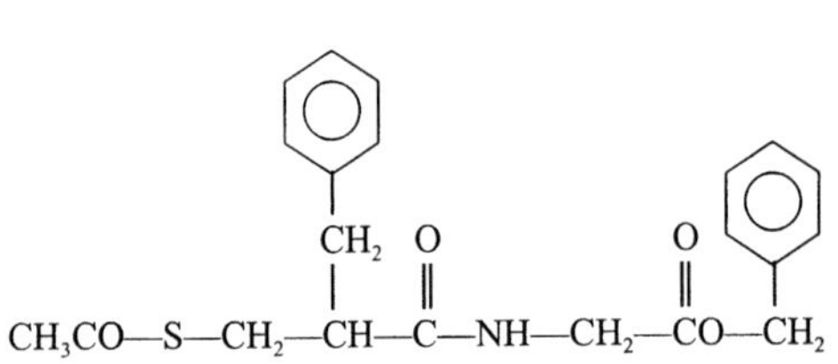	10,000	n.d.

inhibitors is that they are relatively hydrophilic and do not readily cross the gastrointestinal or blood–brain barriers. An improvement in the bioavailability of thiorphan was obtained by protecting its thiol and carboxyl hydrophilic groups. The resulting prodrugs, acetorphan and sinorphan (Table 16.2), are rapidly transformed to thiorphan by esterases in the brain and blood. Numerous other NEP inhibitors have been developed including carboxyl, bidentate and phosphorus containing inhibitors (Beaumont *et al* 1996).

16.3.3 Dual inhibitors of angiotensin converting enzyme and neprilysin

Inhibitors of NEP, such thiorphan, as well as other more potent and selective inhibitors, have significant diuretic and natriuretic effects *in vivo*, primarily through potentiating the actions of ANP (Schwartz *et al* 1990; Beaumont *et al* 1996). The idea then arose to combine the renal effects of ANP with the vasodilation resulting from ACE inhibition, by simultaneously inhibiting NEP and ACE (Roques and Beaumont 1990). Although co-administration of selective inhibitors of each enzyme showed that in principle the idea would work (Seymour *et al* 1991), clear advantages are gained in terms of pharmacokinetics and bioavailability with a single molecule that inhibits both enzymes. Thus dual inhibitors of ACE and NEP were synthesized and their pharmacological properties assessed. Interestingly the first molecule to be tested, ES 34 (SQ 28,133) (Table 16.3), was initially synthesized as part of studies aimed at elucidating the structural requirements for selective inhibition of both enzymes (Gordon *et al* 1983). The two compounds alatrioprilat and glycoprilat (Table 16.3) inhibit both ACE and NEP with similar nanomolar potencies *in vitro*, while *in vivo*, alatriopril and glycopril, the corresponding diester prodrugs exert typical actions of ACE inhibitors and NEP inhibitors, indicating promise for the treatment of various cardiovascular and salt-retention disorders (Gros *et al* 1991).

More recently, rigid constrained inhibitors such as MDL 100 173, a mercapto-derivative of a tricyclic Phe-Leu mimetic which has subnanomolar affinities for both enzymes (Table 16.3), and its thioester prodrug MDL 100 240, have been shown to lower blood pressure *in vivo* (French *et al* 1995). Another compound, RB105, which is non-cyclic and therefore easier and cheaper to synthesize than the cyclic compounds, has a K_i of 1.7 nM for NEP and 4.5 nM for ACE (Table 16.3). A lipophilic prodrug of RB105, mixanpril, in which the thiol is protected by a benzoyl derivative, elicited dose-dependent hypotensive effects in spontaneously hypertensive rats after oral administration, clearly indicating the potential of dual ACE/NEP inhibitors for clinical investigations (Fournie-Zaluski *et al* 1994). Clinical studies are beginning to demonstrate the potential use of dual inhibitors of ACE and NEP (Rousso *et al* 1998, 1999; Venn *et al* 1998; Norton *et al* 1999). For a recent commentary on the potential role of dual ACE/NEP inhibitors in the treatment of heart failure see Hu and Ertl (1999).

16.3.4 Inhibitors of endothelin converting enzyme

A few selective inhibitors of ECE-1 have been reported (Takaishi *et al* 1998; Asai *et al* 1999; Hanessian and Rogel 1999; Russell and Davenport 1999). These include PD069185 (Table 16.4), a non-peptidic trisubstituted quinazoline compound with an IC_{50} of 0.9 μM, which does not inhibit NEP or ECE-2 (Ahn *et al* 1998), and CGS 31447 (Table 16.4) which inhibits ECE-1 with an IC_{50} of 21 nM and which inhibited the mean increase in big ET-1-induced pressor responses in isolated and perfused rat kidneys (Shetty *et al* 1998).

Table 16.3 Dual inhibitors of angiotensin converting enzyme and neprilysin

Compound	*Structure*	K_i (nM) *ACE*	K_i (nM) *NEP*
ES 34 (SQ 28,133)	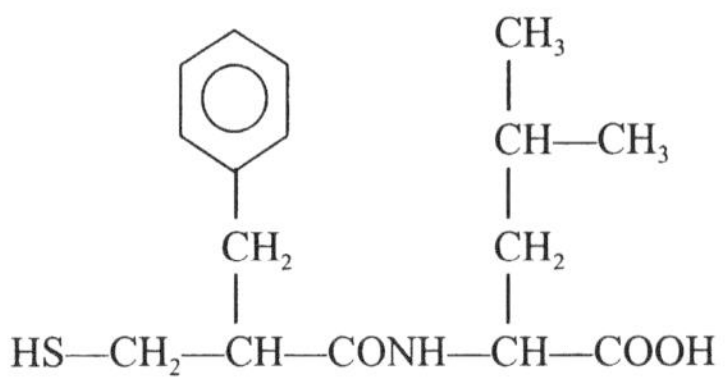	55	4.5
RB105	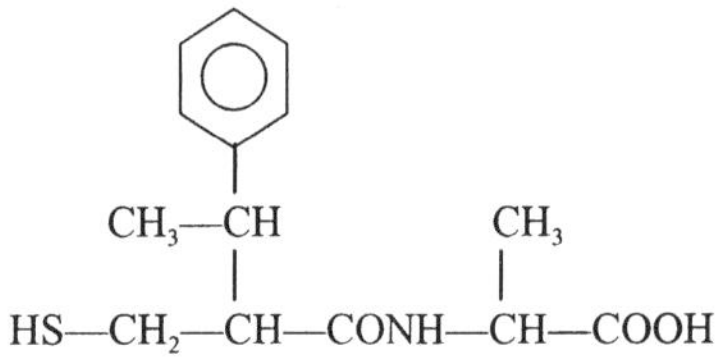	4.5	1.7
MDL 100 173	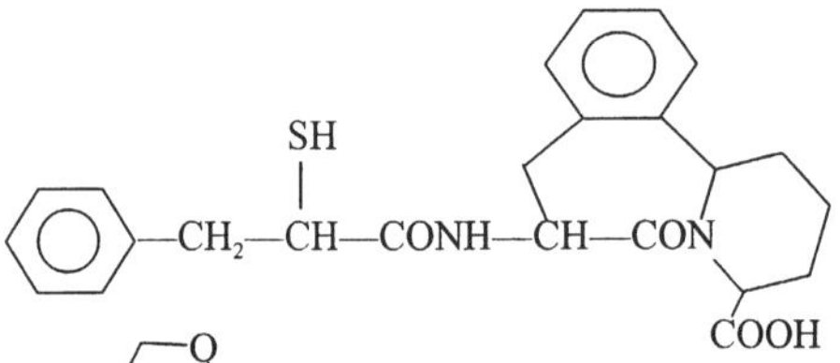	0.11	0.08
Alatrioprilat	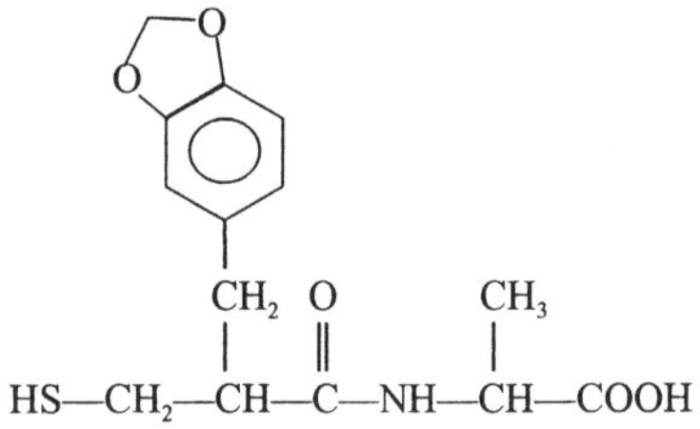	*S*, *S* 9.8 *R*, *S* 215	5.1 13.7
Glycoprilat	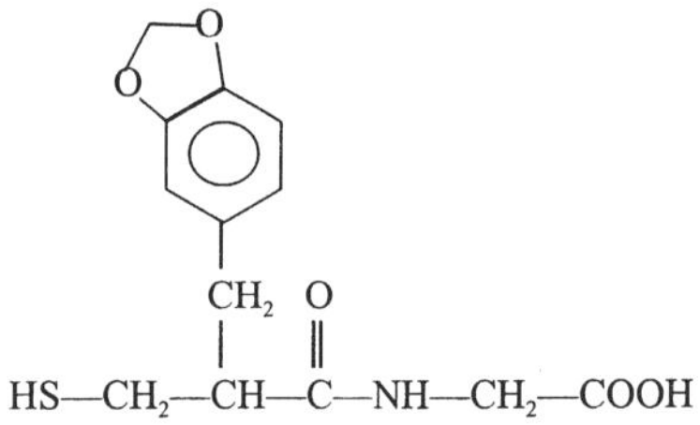	*S* 6.5 *R* 420	5.6 12.0

The potential use of ECE-1 inhibitors in the treatment of hypertension and congestive heart failure has been demonstrated in animal models (Takahashi *et al* 1998; Wada *et al* 1999). With the high degree of similarity between the catalytic sites of ECE-1 and NEP (Sansom *et al* 1998), it is unsurprising that dual ECE/NEP inhibitors have been developed. One of these, CGS 26303 (Table 16.4), was shown on chronic subcutaneous administration to reduce the mean arterial pressure in the spontaneously hypertensive rat more effectively than its NEP selective analogue (De Lombaert *et al* 1994).

Table 16.4 Inhibitors of endothelin converting enzyme

Compound	*Structure*	IC_{50} *ECE*	*NEP*	*ACE*
PD069185	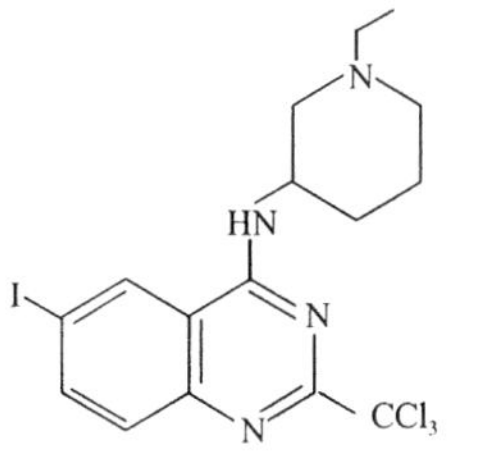	0.9 μM	n.i.	–
CGS31447	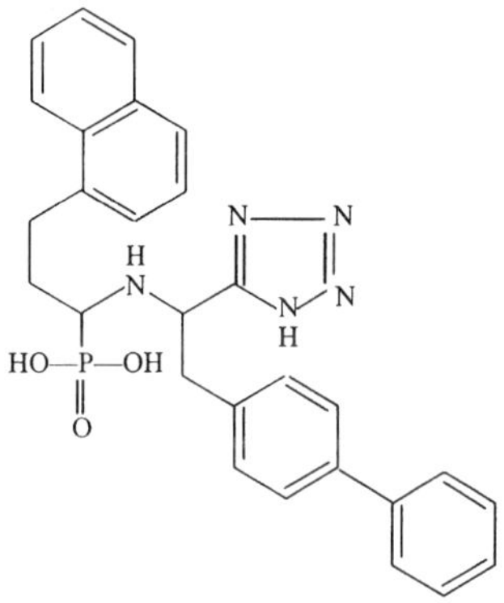	21 nM	–	–
CGS26303	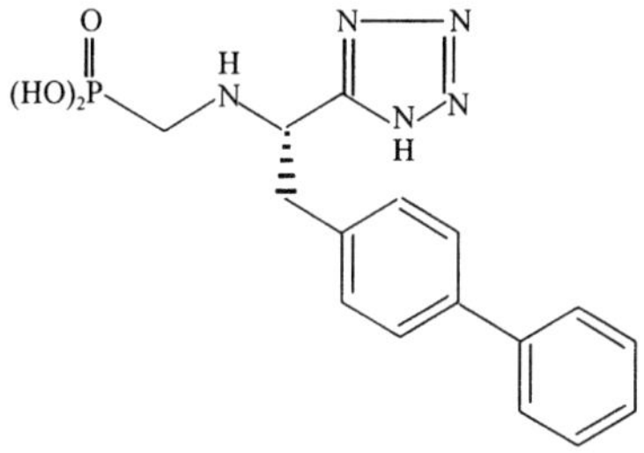	1.1 μM	0.9 nM	–
CGS26582	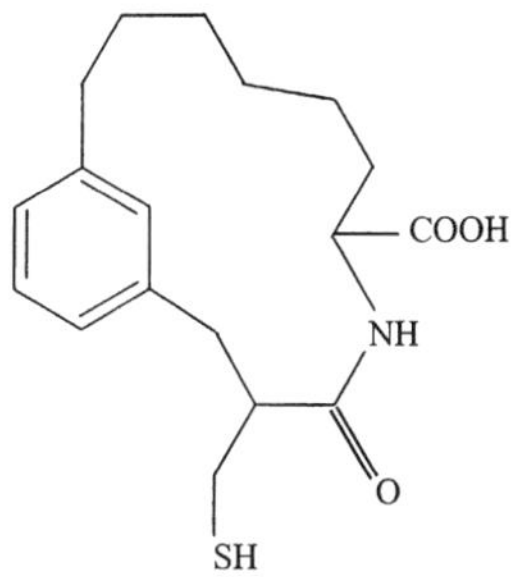	620 nM	4 nM	175 nM

n.i. = no inhibition.

16.3.5 Triple inhibitors of angiotensin converting enzyme, neprilysin and endothelin converting enzyme

With the similarities between the active sites of ACE, NEP and ECE, it is probably not surprising that triple inhibitors have been reported that have nanomolar affinities for all three enzymes, for example the benzofused macrocyclic lactams such as CGS 26582 (Table 16.4) (McKittrick *et al* 1996; Ksander *et al* 1998).

16.4 CLINICAL ASPECTS

The importance of the applications of oral ACE inhibitors in clinical medicine is now self evident, with many thousands of patients being treated with these drugs at the present time for either hypertension, congestive heart failure, postmyocardial infarction or diabetic nephropathy. The NEP inhibitor acetorphan is now on the market as an antidiarrhoeal agent under the registered trademark TIORFAN (Bergmann *et al* 1992) due to the role played by NEP in the metabolism of endogenous enkephalin peptides in the gastrointestinal tract. The therapeutic value of triple inhibitors of ACE/NEP/ECE, of dual inhibitors of ECE/NEP, or indeed of selective inhibitors of ECE in cardiovascular or renal medicine awaits the outcome of more extensive clinical trials.

ACKNOWLEDGMENTS

I gratefully acknowledge the Medical Research Council of Great Britain for the financial support of my studies on angiotensin converting enzyme. I thank Diane Baldwin for preparation of the figures and tables.

REFERENCES

Ahn, K., Sisneros, A.M., Herman, S.B., Pan, S.M., Hupe, D., Lee, C. *et al* (1998) Novel selective quinazoline inhibitors of endothelin converting enzyme-1. *Biochemical and Biophysical Research Communications*, **243**, 184–190.

Asai, Y., Nonaka, N., Suzuki, S., Nishio, M., Takahashi, K., Shima, H. *et al* (1999) TMC-66, a new endothelin converting enzyme inhibitor produced by *Streptomyces* sp. A5008. *Journal of Antibiotics (Tokyo)*, **52**, 607–612.

Azizi, M., Rousseau, A., Ezan, E., Guyene, T.-T., Michelet, S., Grognet, J.-M. *et al* (1996) Acute angiotensin-converting enzyme inhibition increases the plasma level of the natural stem cell regulator *N*-acetyl-seryl-aspartyl-lysyl-proline. *Journal of Clinical Investigation*, **97**, 839–844.

Bai, J.P.F. (1993) Distribution of brush-border membrane peptidases along the rabbit intestine: implication for oral delivery of peptide drugs. *Life Sciences*, **52**, 941–947.

Beaumont, A., Fournie-Zaluski, M.-C. and Roques, B.P. (1996) Neutral endopeptidase-24.11: structure, and design and clinical use of inhibitors. In *Zinc Metalloproteases in Health and Disease*, edited by N.M. Hooper, pp. 105–129. London: Taylor and Francis.

Becker, R.H.A. and Scholkens, B. (1987) Ramipril: review of pharmacology. *American Journal of Cardiology*, **59**, 30–110.

Bergmann, J.F., Chaussade, S., Couturier, D., Baumer, P., Schwartz, J.C. and Lecomte, J.M. (1992) Effects of acetorphan, an antidiarrhoeal enkephalinase inhibitor, on oro-caecal and colonic transit times in healthy volunteers. *Alimentary Pharmacology and Therapeutics*, **6**, 305–313.

Cushman, D.W. and Ondetti, M.A. (1999) Design of angiotensin converting enzyme inhibitors. *Nature Medicine*, **5**, 1110–1112.

DeForrest, J.M., Waldron, T.L., Harvey, C., Scalese, R., Hammerstone, S., Powell, J.R. *et al* (1990) Ceranapril (SQ 29,852) an orally active inhibitor of angiotensin converting enzyme (ACE). *Journal of Cardiovascular Pharmacology*, **16**, 121–127.

De Lombaert, S., Ghai, R.D., Jeng, A.Y., Trapani, A.J. and Webb, R.L. (1994) Pharmacological profile of a non-peptidic dual inhibitor of neutral endopeptidase 24.11 and endothelin-converting enzyme. *Biochemical and Biophysical Research Communications*, **204**, 407–412.

Devault, A., Sales, V., Nault, C., Beaumont, A., Roques, B., Crine, P. *et al* (1988) Exploration of the catalytic site of endopeptidase 24.11 by site-directed mutagenesis. Histidine residues 583 and 587 are essential for catalysis. *Federation of European Biochemical Societies Letters*, **231**, 54–58.

Dive, V., Cotton, J., Yiotakis, A.M.A., Vassiliou, S., Jiracek, J., Vazeux, G. *et al* (1999) RXP 407, a phosphinic peptide, is a potent inhibitor of angiotension I converting enzyme able to differentiate between its two active sites. *Proceedings of the National Academy of Sciences of the U.S.A.*, **96**, 4330–4335.

Ehlers, M.R.W., Fox, E.A., Strydom, D.J. and Riordan, J.F. (1989) Molecular cloning of human testicular angiotensin-converting enzyme: the testis isozyme is indentical to the C-terminal half of endothelial ACE. *Proceedings of the National Academy of Sciences of the U.S.A.*, **86**, 7741–7745.

Emoto, N. and Yanagisawa, M. (1995) Endothelin-converting enzyme-2 is a membrane-bound, phosphoramidon-sensitive metalloprotease with acidic pH optimum. *Journal of Biological Chemistry*, **250**, 15262–15268.

Erdos, E.G., Deddish, P.A. and Marcic, B.M. (1999) Potentiation of bradykinin actions by ACE inhibitors. *Trends in Endocrinology and Metabolism*, **10**, 223–229.

Ferreira, S.H., Bartelt, D.C. and Greene, L.J. (1970) Isolation of bradykinin-potentiating peptides from *Bothrops jararaca* venom. *Biochemistry*, **9**, 2583–2593.

Fournie-Zaluski, M.C., Lucas, E., Waksman, G. and Roques, B.P. (1984) Differences in the structural requirements for selective interaction with neutral metalloendopeptidase (enkephalinase) or angiotensin-converting enzyme. (Molecular investigation by use of new thiol inhibitors). *European Journal of Biochemistry*, **139**, 267–274.

Fournie-Zaluski, M.C., Gonzalez, W., Turcaud, S., Pham, I., Roques, B.P. and Michel, J.B. (1994) Dual inhibition of angiotensin-converting enzyme and neutral endopeptidase by the orally active inhibitor mixanpril: a potential therapeutic approach in hypertension. *Proceedings of the National Academy of Sciences of the U.S.A.*, **91**, 4072–4076.

French, J.F., Anderson, B.A., Downs, R.T. and Dage, R.C. (1995) Dual inhibition of angiotensin converting enzyme and neutral endopeptidase in rats with hypertension. *Journal of Cardiovascular Pharmacology*, **26**, 107–113.

Gavras, H., Brunner, H.R., Laragh, J.H., Sealey, J.E., Gavras, I. and Vukovich, R.A. (1974) An angiotensin-converting enzyme inhibitor to identify and treat vasoconstrictor and volume factors in hypertensive patients. *New England Journal of Medicine*, **291**, 817–821.

Gee, N.S., Bowes, M.A., Buck, P. and Kenny, A.J. (1985) An immunoradiometric assay for endopeptidase-24.11 shows it to be a widely distributed enzyme in pig tissues. *Biochemical Journal*, **228**, 119–126.

Gordon, E.M., Cushman, D.W., Tung, R., Cheung, H.S., Wang, F.L. and Delaney, N.G. (1983) Rat brain enkephalinase: characterisation of the active site using mercapropropanoyl amino acid inhibitors and comparison with angiotensin converting enzyme. *Life Sciences*, **33**, 113–116.

Grima, M., Welsch, C., Michel, B., Barthelmebs, M. and Imbs, J.-L. (1991) *In vitro* tissue potencies of converting enzyme inhibitors. Prodrug activation by kidney esterase. *Hypertension*, **17**, 492–496.

Gros, C., Noel, N., Souque, A., Schwartz, J.-C., Danvy, D., Plaquevent, J.-C. *et al* (1991) Mixed inhibitors of angiotensin converting enzyme (EC 3.4.15.1) and enkephalinase (EC 3.4.24.11):

rational design, properties and potential cardiovascular application of glycopril and alatriopril. *Proceedings of the National Academy of Sciences of the U.S.A.*, **88**, 4210–4214.

Hanessian, S. and Rogel, O. (1999) Synthesis of a phostone glycomimetic of the endothelin converting enzyme inhibitor phosphoramidon. *Bioorganic & Medicinal Chemistry Letters*, **9**, 2441–2446.

Hasegawa, H., Hiki, K., Sawamura, T., Aoyama, T., Okamoto, Y., Miwa, S. *et al* (1998) Purification of a novel endothelin-converting enzyme specific for big endothelin-3. *Federation of European Biochemical Societies Letters*, **428**, 304–308.

Hoang, M.V. and Turner, A.J. (1997) Novel activity of endothelin-converting enzyme: hydrolysis of bradykinin. *Biochemical Journal*, **327**, 23–26.

Hooper, N.M., Keen, J., Pappin, D.J.C. and Turner, A.J. (1987) Pig kidney angiotensin converting enzyme. Purification and characterization of amphipathic and hydrophilic forms of the enzyme establishes C-terminal anchorage to the plasma membrane. *Biochemical Journal*, **247**, 85–93.

Hooper, N.M. (1991) Angiotensin converting enzyme: implications from molecular biology for its physiological functions. *International Journal of Biochemistry*, **23**, 641–647.

Hooper, N.M. (1993). Ectopeptidases. In *Biological Barriers to Protein Delivery*, vol. 5, edited by K.L. Audus and T.J. Raub, pp. 23–50. New York: Plenum Press.

Howard, T.E., Shai, S.Y., Langford, K.G., Martin, B.M. and Bernstein, K.E. (1990) Transcription of testicular angiotensin-converting enzyme (ACE) is initiated within the 12th intron of the somatic ACE gene. *Molecular and Cellular Biology*, **10**, 4294–4302.

Hu, K. and Ertl, G. (1999) Potential role of mixed ACE and neutral endopeptidase inhibitor in the treatment of heart failure. *Cardiovascular Research*, **41**, 503–505.

Jackson, B., Mendelsohn, F.A.O. and Johnston, C.I. (1991) Angiotensin-converting enzyme inhibition: prospects for the future. *Journal of Cardiovascular Pharmacology*, **18 (suppl. 7)**, 54–58.

Jaspard, E., Wei, L. and Alhenc-Gelas, F. (1993) Differences in the properties and enzymatic specificities of the two active sites of angiotensin I-converting enzyme (kininase II). Studies with bradykinin and other natural peptides. *Journal of Biological Chemistry*, **268**, 9496–9503.

Johnson, G.D., Stevenson, T. and Ahn, K. (1999) Hydrolysis of peptide hormones by endothelin-convering enzyme-1. *Journal of Biological Chemistry*, **274**, 4053–4058.

Kenny, A.J. and Stephenson, S.L. (1988) Role of endopeptidase-24.11 in the inactivation of atrial natriuretic peptide. *Federation European Biochemical Societies Letters*, **232**, 1–8.

Kerr, M.A. and Kenny, A.J. (1974a) The molecular weight and properties of a neutral metalloendopeptidase from rabbit kidney brush border. *Biochemical Journal*, **137**, 489–495.

Kerr, M.A. and Kenny, A.J. (1974b) The purification and specificity of a neutral endopeptidase from rabbit kidney brush border. *Biochemical Journal*, **137**, 477–488.

Ksander, G.M., Savage, P., Trapani, A.J., Balwierczak, J.L. and Jeng, A.Y. (1998) Benzofused macrocyclic lactams as triple inhibitors of endothelin-converting enzyme, neutral endopeptidase 24.11, and angiotensin-converting enzyme. *Journal of Cardiovascular Pharmacology*, **31**, S71–S73.

Le Moual, H., Devault, A., Roques, B.P., Crine, P. and Boileau, G. (1991) Identification of glutamic acid 646 as a zinc-co-ordinating residue in endopeptidase-24.11. *Journal of Biological Chemistry*, **266**, 15670–15674.

Malfroy, B., Kuang, W.-J., Seeburg, P.H., Mason, A.J. and Schofield, P.R. (1988) Molecular cloning and amino acid sequence of human enkephalinase (neutral endopeptidase). *Federation European Biochemical Societies Letters*, **229**, 206–210.

Mancia, G. (1991) Angiotensin-converting enzyme inhibitors in the treatment of hypertension. *Journal of Cardiovascular Pharmacology*, 18 (Suppl. 7), S1–S3.

McKittrick, B.A., Stamford, A.W., Weng, X., Ma, K., Chackalamanil, S., Czarniecki, M. *et al* (1996) Design and synthesis of phosphinic acids that triply inhibit endothelin converting enzyme and neutral endopeptidase 24.11. *Bioorganic Chemistry Letters*, **6**, 1629–1634.

Michaud, A., Chauvet, M.-T. and Corvol, P. (1999) N-domain selectivity of angiotensin I-converting enzyme as assessed by structure-function studies of its highly selective substrate, *N*-acetyl-seryl-aspartyl-lysyl-proline. *Biochemical Pharmacology*, **57**, 611–618.

Natoff, I.L., Nixon, J.S., Francis, R.J., Klevans, L.R., Brewster, M., Budd, J. *et al* (1985) Biological properties of the angiotensin-converting enzyme inhibitor cilazapril. *Journal of Cardiovascular Pharmacology*, **7**, 569–580.

Norton, G.R., Woodiwiss, A.J., Hartford, C., Trifunovic, B., Middlemost, S., Lee, A. *et al* (1999) Sustained antihypertensive actions of a dual angiotensin-converting enzyme neutral endopeptidase inhibitor, sampatrilat, in black hypertensive subjects. *American Journal of Hypertension*, **12**, 563–571.

Ohnaka, K., Takayanagi, R., Nishikawa, M., Haji, M. and Nawata, H. (1993) Purification and characterization of a phosphoramidon-sensitive endothelin-converting enzyme in porcine aortic endothelium. *Journal of Biological Chemistry*, **268**, 26759–26766.

Ondetti, M.A., Williams, N.J., Sabo, E.F., Pluscec, L., Weaver, E.R. and Kocy, O. (1971) Angiotensin converting enzyme inhibitors from the venom of *Bothrops jararaca*. Isolation, elucidation of structure, and synthesis. *Biochemistry*, **10**, 4033–4039.

Ondetti, M.A. and Cushman, D.W. (1982) Enzymes of the renin-angiotensin system and their inhibitors. *Annual Review of Biochemistry*, **51**, 283–308.

Patchett, A.A., Harris, E., Tristam, E.W., Wyvratt, M.J., Wu, M.T., Taub, D. *et al* (1980) A new class of angiotensin-converting enzyme inhibitors. *Nature*, **288**, 280–283.

Roques, B.P., Fournie-Zaluski, M.C., Soroca, E., Lecomte, J.M., Malfroy, B., Llorens, C. *et al* (1980) The enkephalinase inhibitor thiorphan shows antinociceptive activity in mice. *Nature*, **288**, 286–288.

Roques, B.P. and Beaumont, A. (1990) Neutral endopeptidase-24.11 inhibitors: from analgesics to antihypertensives. *Trends in Pharmacological Sciences*, **11**, 245–249.

Roques, B.P., Noble, F., Dauge, V., Fournie-Zaluski, M.-C. and Beaumont, A. (1993) Neutral endopeptidase 24.11: structure, inhibition, and experimental and clinical pharmacology. *Pharmacological Reviews*, **45**, 87–146.

Rousseau, A., Michaud, A., Chauvet, M.-T., Lenfant, M. and Corvol, P. (1995) The hemoregulatory peptide *N*-acetyl-Ser-Asp-Lys-Pro is a natural and specific substrate of the N-terminal active site of human angiotensin converting enzyme. *Journal of Biological Chemistry*, **270**, 3656–3661.

Rousso, P., Buclin, T., Nussberger, J., Brunner-Ferber, F., Brunner, H.R. and Biollaz, J. (1998) Effects of MDL 100,240, a dual inhibitor of angiotensin-converting enzyme and neutral endopeptidase on the vasopressor response to exogenous angiotensin I and angiotensin II challenges in healthy volunteers. *Journal of Cardiovascular Pharmacology*, **31**, 408–417.

Rousso, P., Buclin, T., Nussberger, J., Decosterd, L.A., La Roche, S.D., Brunner-Ferber, F. *et al* (1999) Effects of a dual inhibitor of angiotensin converting enzyme and neutral endopeptidase, MDL 100,240, on endocrine and renal functions in healthy volunteers. *Journal of Hypertension*, **17**, 427–437.

Ruskoaho, R. (1992) Atrial natriuretic peptide: synthesis, release and metabolism. *Pharmacological Reviews*, **44**, 479–602.

Russell, F.D. and Davenport, A.P. (1999) Evidence for intracellular endothelin-converting enzyme-2 expression in cultured human vascular endothelial cells. *Circulation Research*, **84**, 891–896.

Salvetti, A. (1990) Newer ACE inhibitors. A look at the future. *Drugs*, **40**, 800–828.

Sansom, C., Hoang, V.M. and Turner, A.J. (1995) Molecular modelling of the active site of endothelin-converting enzyme. *Journal of Cardiovascular Pharmacology*, **26**, S75–S77.

Sansom, C.E., Hoang, M.V. and Turner, A.J. (1998) Molecular modelling and site-directed mutagenesis of the active site of endothelin-converting enzyme. *Protein Engineering*, **11**, 1235–1241.

Schechter, I. and Berger, A. (1967) On the size of the active site in proteases. I. Papain. *Biochemical and Biophysical Research Communications*, **27**, 157–162.

Schmidt, M., Kroger, B., Jacob, E., Seulberger, H., Subkowski, T., Otter, R. *et al* (1994) Molecular characterization of human and bovine endothelin converting enzyme (ECE-1). *Federation European Biochemical Societies Letters*, **356**, 238–243.

Schwartz, J.-C., Gros, C., Lecomte, J.-M. and Bralet, J. (1990) Minireview: Enkephalinase (EC 3.4.24.11) inhibitors: protection of endogenous ANF against inactivation and potential therapeutic applications. *Life Sciences*, **47**, 1279–1297.

Seymour, A.A., Swerdel, J.N. and Abboa-Offei, B. (1991) Antihypertensive activity during inhibition of neutral endopeptidase and angiotensin converting enzyme. *Journal of Cardiovascular Pharmacology*, **17**, 456–465.

Shetty, S.S., Savage, P., DelGrande, D., De Lombaert, S. and Jeng, A.Y. (1998) Characterization of CGS 31447, a potent and nonpeptidic endothelin-converting enzyme inhibitor. *Journal of Cardiovascular Pharmacology*, **31**, S68–S70.

Soubrier, F., Alhenc-Gelas, F., Hubert, C., Allegrini, J., John, M., Tregear, G. *et al* (1988) Two putative active centers in human angiotensin I-converting enzyme revealed by molecular cloning. *Proceedings of the National Academy of Sciences of the U.S.A.*, **85**, 9386–9390.

Takahashi, M., Matsushita, Y., Iijima, Y. and Tanzawa, K. (1993) Purification and characterization of endothelin-converting enzyme from rat lung. *Journal of Biological Chemistry*, **268**, 21394–21398.

Takahashi, T., Kanda, T., Inoue, M., Sumino, H., Kobayashi, I., Iwamoto, A. *et al* (1998) Endothelin converting enzyme inhibitor protects development of right ventricular overload and medial thickening of pulmonary arteries in rats with monocrotaline-induced pulmonary hypertension. *Life Sciences*, **63**, L137–L143.

Takaishi, S., Tuchiya, N., Sato, A., Negishi, T., Takamatsu, Y., Matsushita, Y. *et al* (1998) B-90063, a novel endothelin converting enzyme inhibitor isolated from a new marine bacterium, *Blastobacter* sp. SANK 71894. *Journal of Antibiotics (Tokyo)*, **51**, 805–815.

Thorsett, E.D. and Wyvratt, M.J. (1987). Inhibition of zinc peptidases that hydrolyse neuropeptides. In *Neuropeptides and their Peptidases*, edited by A.J. Turner, pp. 229–292. Chichester: Ellis Horwood.

Todd, P.A. and Benfield, P. (1990) Ramipril. A review of its pharmacological properties and therapeutic efficacy in cardiovascular disorders. *Drugs*, **40**, 110–135.

Turner, A.J. and Murphy, L.J. (1996) Molecular pharmacology of endothelin converting enzymes. *Biochemical Pharmacology*, **51**, 91–102.

Turner, A.J. and Tanzawa, K. (1997) Mammalian membrane metallopeptidases: NEP, ECE, KELL, and PEX. *FASEB Journal*, **11**, 355–364.

Unger, T., Schull, B., Rascher, W., Lang, R.E. and Ganten, D. (1982) Selective activation of the converting enzyme inhibitor MK-421 and comparison of its active diacid form with captopril in different tissues of the rat. *Biochemical Pharmacology*, **19**, 3063–3070.

Unger, T., Gohlke, P. and Gruber, M.-G. (1990) Converting enzyme inhibitors. In *Pharmacology of Anti-hypersensitive Therapeutics*, edited by D. Ganten and P.J. Mulrow, pp. 379–481. Berlin: Springer-Verlag.

Vallotton, M.B. (1987) The renin-angiotensin system. *Trends in Pharmacological Sciences*, **8**, 69–74.

Venn, R.F., Barnard, G., Kaye, B., Macrae, P.V. and Saunders, K.C. (1998) Clinical analysis of sampatrilat, a combined renal endopeptidase and angiotensin-converting enzyme inhibitor II: assay in the plasma and urine of human volunteers by dissociation enhanced lanthanide fluorescence immunoassay (DELFIA). *Journal of Pharmaceutical and Biomedical Analysis*, **16**, 883–892.

Wada, A., Tsutamoto, T., Ohnishi, M., Sawaki, M., Fukai, D., Maeda, Y. *et al* (1999) Effects of a specific endothelin-converting enzyme inhibitor on cardiac, renal, and neurohumoral functions in congestive heart failure: comparison of effects with those of endothelin A receptor antagonism. *Circulation*, **99**, 570–577.

Wei, L., Alhenc-Gelas, F., Corvol, P. and Clauser, E. (1991) The two homologous domains of human angiotensin I converting enzyme are both catalytically active. *Journal of Biological Chemistry*, **266**, 9002–9008.

Wilkins, M.R. and Kenny, A.J. (1997) Natriuretic peptide metabolism: inhibitors of endopeptidase-24.11 as possible therapeutic agents for cardiovascular disease. In *Cell-surface Peptidases in Health and Disease*, edited by A.J. Kenny and C.M. Boustead, pp. 303–322. Oxford: BIOS Scientific Publishers.

Yanagisawa, M., Kurihara, H., Kimura, S., Tomobe, Y., Kobayashi, Y., Mitsui, Y. *et al* (1988) A novel potent, vasoconstrictor peptide produced by vascular endothelial cells. *Nature*, **332**, 411–415.

Chapter 17

HIV aspartate proteinase: resistance to inhibitors

Paul J. Ala and Chong-Hwan Chang

The primary cause of resistance to the currently available HIV protease inhibitors is an accumulation of mutations in the viral protease that reduces the protease's affinity for inhibitors. So far more than 20 substitutions have been observed in the active site, dimer interface, surface loops, and flap of the protein. This high degree of genetic variability has made the protease an elusive drug target. In this chapter, the design of the next generation of HIV protease inhibitors will be discussed in light of the resistance problem.

17.1 INTRODUCTION

The phenomenon of acquired drug resistance has existed ever since natural and synthetic chemical compounds have been used as therapeutic agents. In the case of infectious diseases, resistance often develops as a result of suboptimal therapy, which selects for mutants that have a growth advantage over wild type. Even more worrisome than the development of acquired resistance, however, is the possibility that subsequent transmissions of resistant variants to uninfected individuals may lead to infections that are drug-resistant from the onset. An outbreak of resistant mutants will therefore significantly delay progress in eradicating the disease because a completely new set of drugs will have to be developed.

Mechanisms responsible for drug resistance can be quite different depending on the type of disease. Parasites, for example, become resistant by (1) chemically modifying the target or the drug, (2) developing pathways that bypass the inhibited process, (3) increasing the production of the drug target, or (4) reducing uptake or increasing efflux of the drug by altering the number of transmembrane pumps in the cell membrane. In the case of most viral infections, the principal cause of resistance is the spontaneous occurrence of mutations in the drug target followed by amplification under conditions of drug pressure.

Given the number and diversity of resistance pathways available to most infectious agents, it is very important to develop drugs that exert little or no selective pressure. In this chapter, we describe the current efforts to elucidate the mechanisms of drug resistance of the human immunodeficiency virus (HIV) to protease inhibitors and attempts to incorporate this information into the development of the next generation of inhibitors.

ANTIRETROVIRAL AGENTS APPROVED FOR TREATMENT

In the early 1980s, HIV was identified as the causal agent of the acquired immunodeficiency syndrome (Gallo *et al* 1983; Barre-Sinoussi *et al* 1983), and in 1987 the Food and Drug Administration (FDA) approved the use of the first anti-HIV agent AZT, a nucleoside analog that inhibits reverse transcriptase (RT) (De Clercq 1995). Over the next twelve years, eight additional nucleoside and non-nucleoside RT inhibitors were approved to help prevent the progression of wild-type and mutant infections. By the mid-1990s, a new class of inhibitors that target the viral protease was approved. The protease plays a pivotal role in the maturation step of viral particles by processing the polyprotein gene products of *gag* and *gag-pol* into active structural and replicative proteins (Robins and Plattner 1993). The crystal structure of the 99 amino acid protease reveals that it is structurally distinct from mammalian aspartyl proteases: its active form is a homodimer and both monomers contribute equally to its active site (Figure 17.1) (Navia *et al* 1989; Wlodawer *et al* 1989; Debouck 1992). This information was particularly important because it allowed researchers to design and optimize many potent inhibitors. In just four years (1995–1999), the FDA approved the use of five protease inhibitors: saquinavir

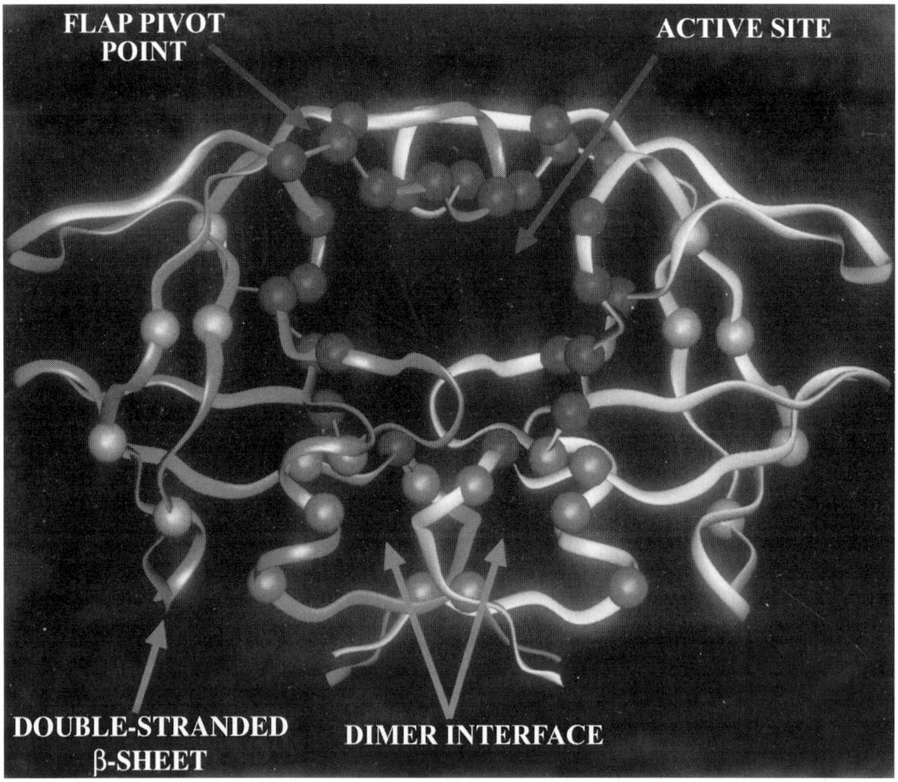

Figure 17.1 Residues associated with HIV protease drug resistance. The color-coded spheres represent point mutations in different regions of the homodimer: red, active site; pink, flap pivot point; blue, dimer interface; and green, double-stranded β-sheet (Val56-Gly78). The C_α atoms of the monomers are drawn as blue and yellow ribbons. (*See Color plate 13*)

Table 17.1 HIV protease inhibitors and their resistance profiles

HIV protease inhibitors		*Key mutations*	*Other mutations*[g]	
FDA APPROVED				
SAQUINAVIR (Hoffmann-La Roche)	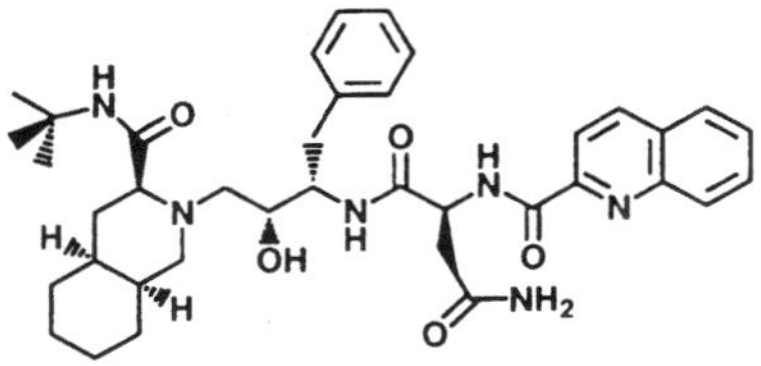	G48V[a] L90M	L10I I54V V82A I84V	
RITONAVIR (Abbott Laboratories)	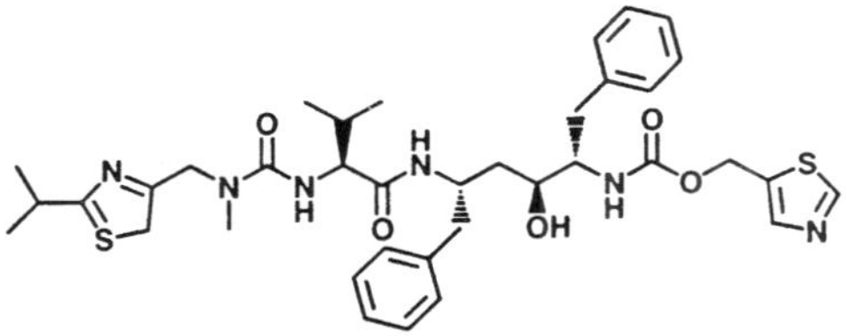	V82F[b] I84V	K20R V32I L33F M36I,L E35D M46I,L	I54V,L L63P A71V,T V82A,S,T I84V L90M
INDINAVIR (Merck)	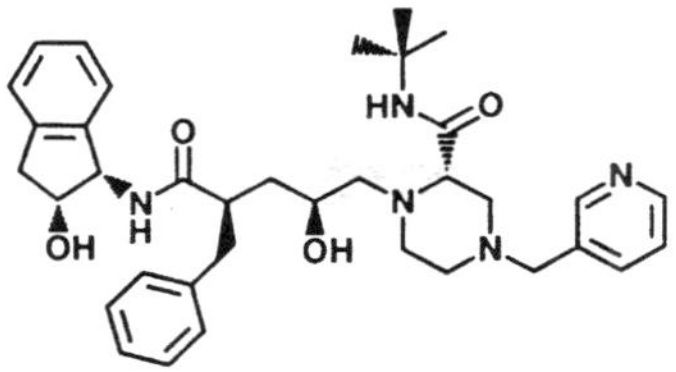	L10R,I,V[c] M46I,L L63P V82T I84V	K20M,R L24I V32I G48V	I54V A71T G73S V82A,F

Table 17.1 continued

HIV protease inhibitors		*Key mutations*	*Other mutations*[g]
NELFINAVIR (Agouron)	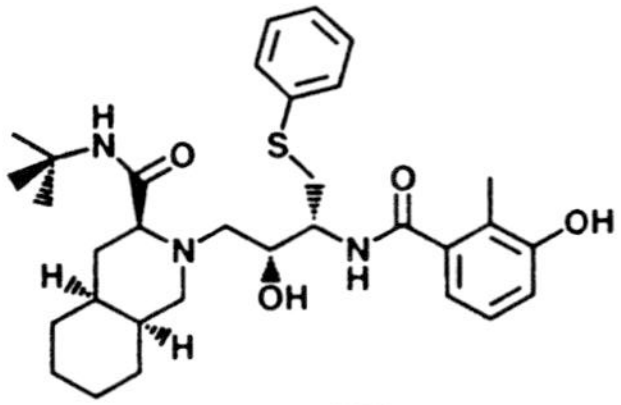	D30N[d]	M36I M46I L63P A71V V77I I84A,V N88D L90M
AMPRENAVIR (Vertex/GlaxoWellcome)	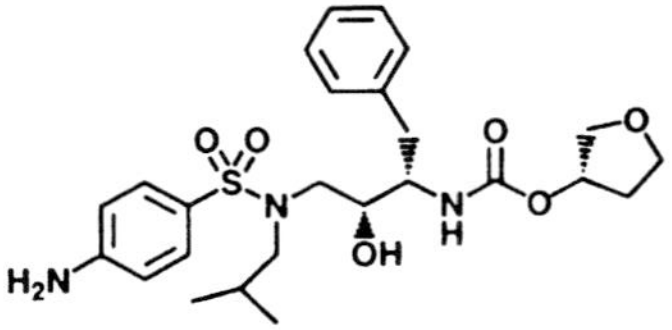	I50V[e] M46I I47V	L10F I84V
EXPERIMENTAL			
DMP-323 (DuPont)	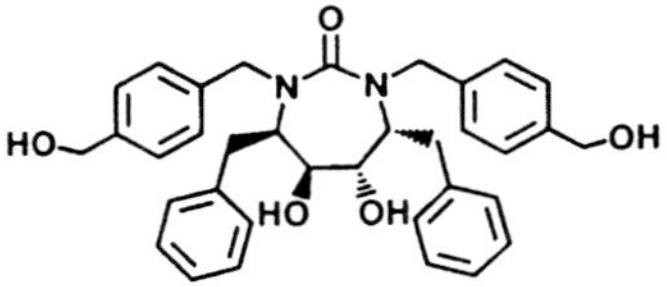	V82F[f] I84V	L10F M45I M46L V82A,I L97V
SD-146 (DuPont)	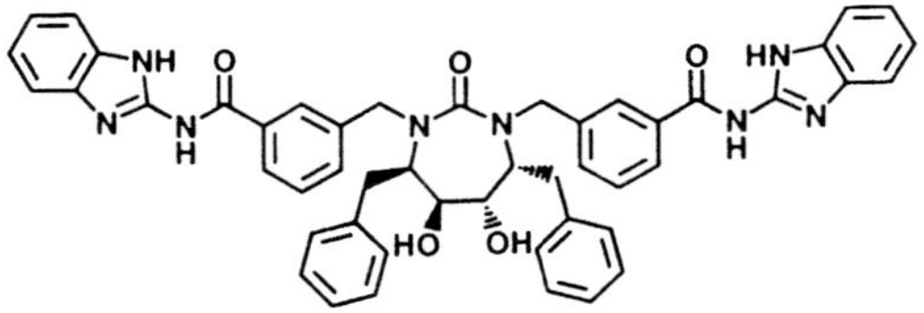		K41I[h] A71T

References: [a]Ives *et al* (1997) and Jacobsen *et al* (1995); [b]Molla *et al* (1996) and Markowitz *et al* (1995); [c]Condra *et al* (1995); [d]Patick *et al* (1996); [e]Partaledis *et al* (1995); [f]King *et al* (1995) and Hodge *et al* (1996); [g]Hammond *et al* (1998); [h]S. Garber, DuPont Pharmaceuticals Company, personal communication.

(Invirase/Fortovase; Hoffmann-LaRoche), ritonavir (Norvir; Abbott Laboratories), indinavir (Crixivan; Merck & Co), nelfinavir (Viracept; Agouron Pharmaceuticals) and amprenavir (Agenerase; Vertex/GlaxoWellcome) (Table 17.1). Altogether, fourteen drugs are now being used to combat HIV. The rapid development of these therapeutic agents has been remarkable and represents one of the most successful structure-based drug design stories of our time (Wlodawer and Erickson 1993; Wlodawer and Vondrasek 1998).

17.3 HIV DRUG RESISTANCE

The daunting ability of the virus to mutate has recently cast doubt on the long-term therapeutic benefit of antiretroviral inhibitors. Even in the presence of the highly anticipated protease inhibitors, resistant variants of HIV often emerged during the first year of monotherapy (Condra *et al* 1995; Molla *et al* 1996; Patick *et al* 1996; Ives *et al* 1997). The ability of the virus to persist is based on its high rate of genetic evolution, which is primarily due to its rapid rate of replication (Williams and Loeb 1992) and the high error rate of RT (Wei *et al* 1995). Unlike human DNA polymerases, RT does not proofread during transcription and makes one error per 10,000 bases copied (or one error per HIV replication cycle). Since errors occur randomly, mutations either leave the virus nonviable, unchanged, or with an altered growth habit. Prior to treatment, patients probably harbor a minor population of mutants in addition to the wild-type strain. In the presence of protease inhibitors, however, variants that encode mutant proteases that have a reduced affinity for inhibitors and retain enough enzymatic activity to process the viral precursors have a growth advantage over wild type.

Efforts to prevent the virus from acquiring resistance mutations have led to the use of combinations of RT and protease inhibitors. The FDA recommended the use of the first RT inhibitor combination dideoxycytidine/zidovudine in 1992 and the first combination of a nucleoside analog with the protease inhibitor saquinavir in 1995. By 1996, multidrug regimens had become the standard protocol, as they were able to maintain viral loads at an undetectable level for two years (Gulick *et al* 1998). Recently, the need to completely suppress viral replication was re-emphasized following the discovery of a latently infected reservoir of HIV in some patients (Chun *et al* 1997; Finzi *et al* 1997; Wong *et al* 1997). The existence of a viral reservoir coupled with the daunting ability of the virus to mutate further highlights the need to elucidate the molecular basis of resistance.

17.3.1 Molecular basis of HIV protease drug resistance

Sequence analyses of HIV protease obtained from drug-resistant strains have revealed over twenty substitutions clustered in the active site, flap pivot point, C-terminal dimer interface, and double-stranded β-sheet (Val56-Gly78) of the protein (Figure 17.1). Approximately half of these mutations are located in the active site where they reduce van der Waals (vdw) contacts, increase steric hindrance, or create unfavorable electrostatic interactions between the protease and inhibitor. All other mutations are called compensatory changes because they are thought to increase the catalytic efficiency of the active-site mutants. Therefore, active-site mutations reduce the protease's affinity for inhibitors whereas compensatory changes help restore enzymatic activity.

17.3.1.1 Active site mutations

The three-dimensional crystal structures of a few mutant proteases have shed some light on the molecular basis of resistance. The first reported structure was the single mutant V82A, which has a four-fold higher K_i value than wild type for the inhibitor A-77003; the loss of affinity was attributed to a reduction in vdw interactions between the protein and inhibitor (Baldwin *et al* 1995). The structure of the quadruple mutant M46I/L63P/V82T/I84V revealed an unfavorable electrostatic interaction and a loss of vdw contacts as a result of the Val82 to Thr and Ile84 to Val substitutions, respectively (Chen *et al* 1995). The effects of the other two mutations were not obvious but the authors speculated that they alter the activity of the protease. In the triple mutant V32I/I47V/V82I, a loss of vdw contacts was observed between Ile82/82′ and the P1/P1′ groups of the inhibitor; the other

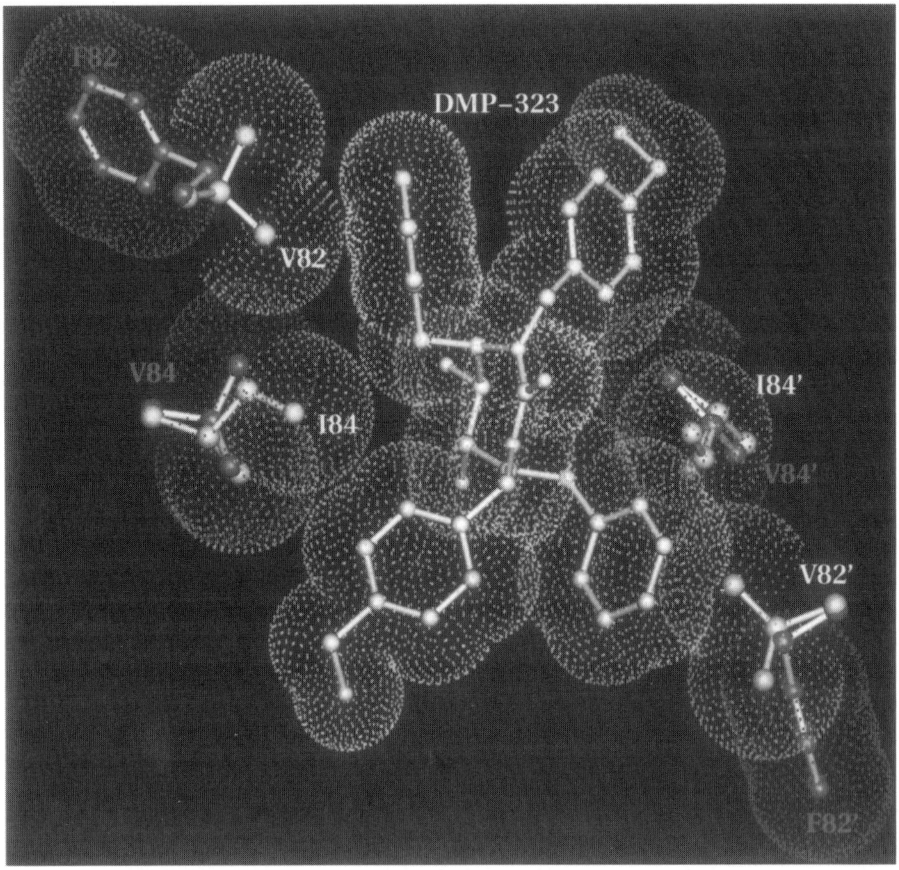

Figure 17.2 Loss of vdw interactions. Conformations of Val82/82′ and Ile84/84′ in the wild type (yellow) and in the double mutant V82F/I84V (blue), when complexed to DMP-323 (white). A drawing of the vdw spheres (dotted spheres) of the atoms clearly shows that the two mutations enlarge the substrate binding pocket and reduce the extent of vdw interactions to the inhibitor; note that Val84 is too small to contact the inhibitor and the side-chain rotamer of Phe82 packs outside of the S1 subsite. (*See Color plate 14*)

two mutations did not cause any significant structural change (Hoog *et al* 1996). In the A71T/V82A mutant, the Val82 to Ala substitution reduces vdw contacts by enlarging the S1 subsite and the Ala71 to Thr change is thought to increase the stability of the protein (Kervinen *et al* 1996). Finally, the DMP-323-selected double mutant V82F/I84V exhibits a 1000-fold reduction in affinity for the cyclic urea (CU) inhibitor because the mutations enlarge the S1 subsite: the side chain of Val84 is smaller than that of Ile and the side chain of Phe82 packs outside of the active site (Figure 17.2) (Ala *et al* 1997).

To better understand the relationship between active-site mutations and resistance, an extensive structure–activity analysis was performed using CU inhibitors. The K_i values for a series of P1-analog range from 2000 to 1 nM, decreasing as the substituent increases in size (Table 17.2). The inhibitor XN-127 is the least potent because it contains an isobutyl group at P1 which is too small to fill the S1 subsite, whereas SC-133 is the most potent inhibitor because it contains a bulky benzodioxan group that interacts extensively with the protease. The size of P1 also appears to influence the probability of selecting for V82F/I84V since the resistance values (the ratio of the double mutant and wild-type K_i values) increase from 2 to 687 for XN-127 and SC-133, respectively (Table 17.2). This suggests that inhibitors with small P1 groups have low affinity and low resistance values because

Table 17.2. K_i values for cyclic urea-based HIV protease inhibitors

P1-substituted CUs				P2-substituted CUs			
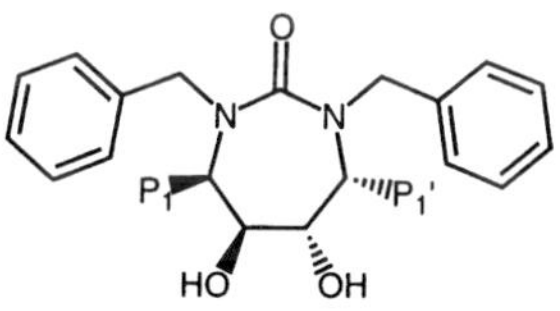				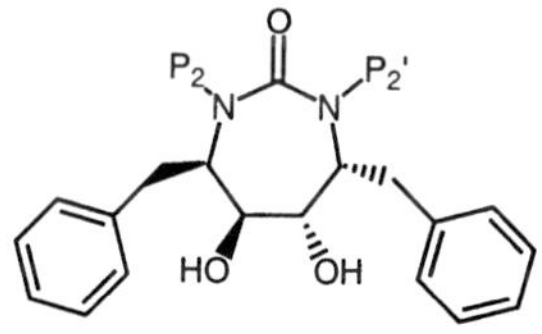			
Cyclic Urea	*P1/P1′*	K_i *(nM)*	$\frac{K_i(82F/84V)}{K_i(WT)}$	*Cyclic Urea*	P2/P2′	K_i (nM)	$\frac{K_i(82F/84V)}{K_i(WT)}$
XN127	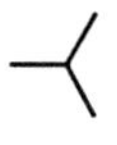	2047	2	XV638	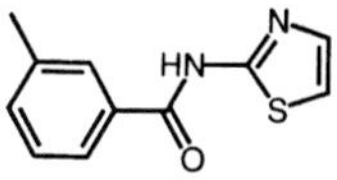	0.1	25
XR808	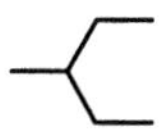	220	3	SD146	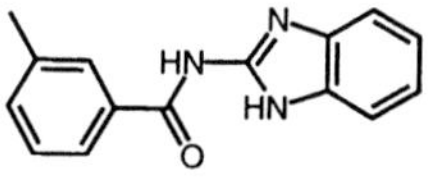	0.1	38
XV076	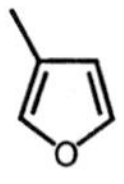	38	92	XN974	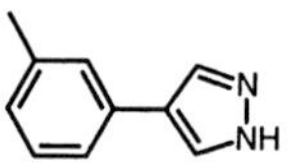	0.1	114
XL075	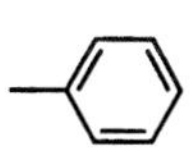	5	146	XZ885	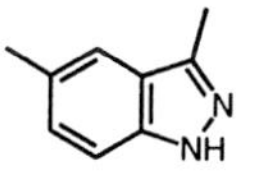	0.1	298
SC120	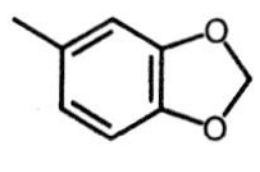	2	656	DMP323	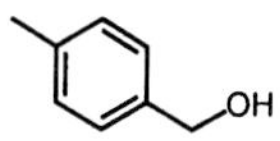	0.8	1016
SC133	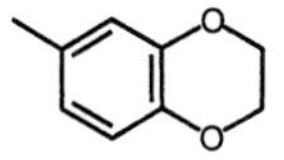	1	687	XK234	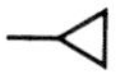	6.0	1645

they do not contact Val82 and Ile84. In contrast, inhibitors that contain large P1 substituents that interact extensively with Val82 and Ile84 have high affinity and high resistance values. These observations illustrate the complexity of the resistance problem in that an increase in potency is often accompanied by an increase in the risk of selecting for resistant mutants.

The P2-S2 compatibility was assessed by measuring the K_i values for a series of P2-analogs of XL-075 (Table 17.2). The values range from 6 to 0.1 nM, decreasing as P2 increases in size. Similar to the observations for the P1-analogs, the K_i decreases as the total number of P2 interactions increases but levels off at 0.1 nM (the detection limit for the assay was 0.01 nM). The resistance values, however, continue to drop to a value of 25 for XV-638. This suggests that the additional P2–S2 interactions observed with the large inhibitors XN-974, SD-146, and XV-638 do not contribute significantly to the wild-type binding affinities but instead play an important role in maintaining potency against the double mutant. It is interesting to note that the P2 substituents do not appear to influence how the P1 groups interact with the S1 subsite, as the same interactions are lost whether the mutants are complexed to DMP-323, XV-638, or SD-146. The high potency of SD-146 and XV-638 against the double mutant is thus, not a result of unique structural perturbations in the protein but rather an increase in the number of P2-S2 vdw contacts and hydrogen bonds.

17.3.1.2 Compensatory mutations

Mutations also occur outside of the protease substrate-binding pocket. These substitutions are thought to produce conformational changes that compensate for the impaired activity of the active-site mutants by preferentially increasing the mutant protease's affinity for substrates over inhibitors or increasing its rate of catalysis (Pazhanisamy *et al* 1996; Schock *et al* 1996). Unfortunately, these mutations do not significantly perturb the structure of the protease and so the cause of these effects remains to be elucidated.

A superposition of the uncomplexed and inhibitor-bound structures of the protease, however, reveals that inhibitor binding induces two large conformational changes: (1) the well documented movement (7 Å) of the flaps, which close over the active site (Miller *et al* 1989), and (2) the rotation of each monomer into the active site (Ala *et al* 1998b; Rose *et al* 1998). The latter change is best viewed when two monomers, one from each state, are superimposed; the relative shift between the other two monomers emphasizes the rigid-body movement that occurs during complex formation (Figure 17.3). Typically, the root mean square deviation of the C_α atoms for the superimposed monomers is only 0.7 Å but the distance between the C_α atoms of Gly17 in the other two monomers is 6 Å. These two conformational changes suggest that binding occurs at least as a two step process, similar to that proposed for peptidomimetics: inhibitors bind to the open state of the protein to form a loose complex which then becomes more compact as the flaps close over the active site (Furfine *et al* 1992). Since the flap is an extension of a large double-stranded β-sheet (Val56-Gly78), its movement might pull on the β-sheet, which in turn pulls the rest of the monomer towards the active site. An important consequence of this rearrangement is the movement of the Pro79-Val82 loop into the active site – creating the S1 subsite. This proposed mechanism is consistent with the assumption that mutations in the flap pivot point, hydrophobic pockets near the C terminus, or along the β-sheet disrupt the coordinated movement that occurs upon complex formation. For example, mutating Lys45, Met46, or Ile54 in the flap pivot point might allow the flap to open and close

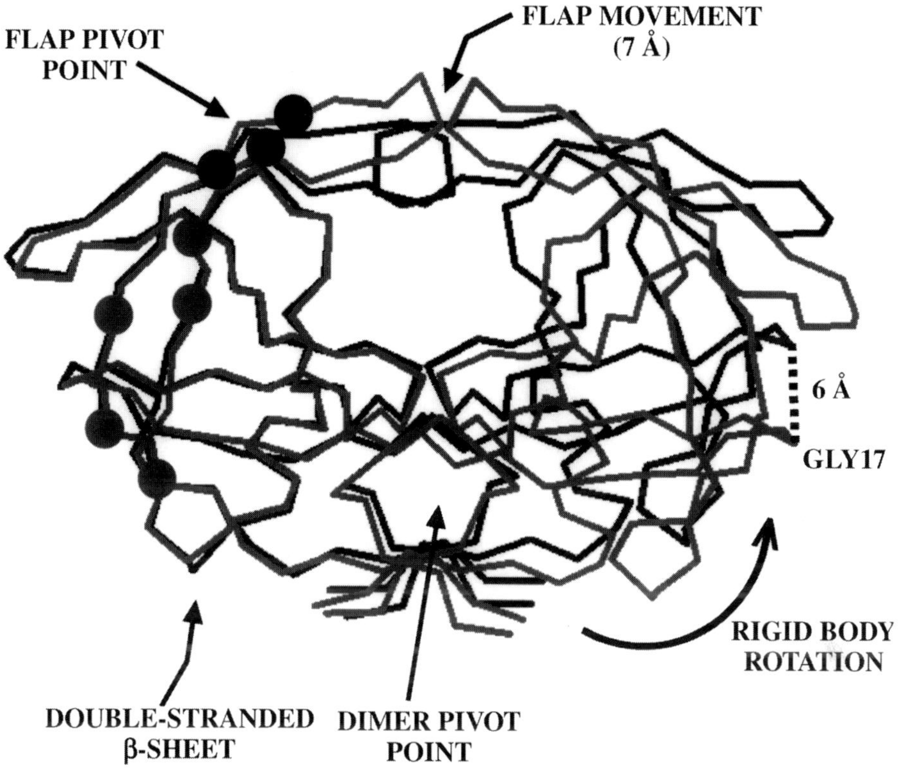

Figure 17.3 Inhibitor-induced conformational changes in HIV protease. Overlay of the C_α atoms of the uncomplexed (blue) and DMP-323-complexed (red) structures of the protease. Note the contraction of the active site in the complexed state as the flaps close and the monomers rotate as two rigid bodies around the C-terminal dimer pivot point. (*See Color plate 15*)

independently of the β-sheet movement; mutations along the β-sheet might allow the latter to slide more freely over the rest of the monomer; and mutating Leu24 and Leu97 – two residues in vdw contact – to valines, might weaken the dimer interface and allow the monomers to move towards the active site independently of the flap and β-sheet movements. The function of compensatory mutations would thus be to allow active-site mutants to more easily discriminate against inhibitors by allowing the flap to close without forming an S1 subsite and *vice versa*.

17.3.1.3 Cleavage-site mutations

Mutations that complement active-site substitutions have been found in the substrate cleavage sites of the viral polyprotein precursors. This is an important finding because these mutations enhance viral replication and represent yet another resistance pathway HIV can exploit to evade active-site directed inhibitors. (Doyon *et al* 1996; Rose *et al* 1996; Zhang *et al* 1997; Carrillo *et al* 1998; Mammano *et al* 1998).

17.4 IMPACT ON DRUG DISCOVERY

17.4.1 Inhibitor design

The surprising ability of the protease to accommodate multiple mutations has clearly changed the way new compounds are being designed. Current attempts to create broadly-active inhibitors include some of the features listed below. The hope is that variants resistant to these newly designed inhibitors will encode mutant proteases that are unable to support viral replication.

17.4.1.1 Increase interaction with catalytic and substrate binding residues

Inhibitors that interact with key residues involved in catalysis or substrate binding are less likely to select for resistant variants, since mutations at these sites will inactivate the protease or significantly reduce its catalytic efficiency. Currently, many potent inhibitors take advantage of this feature by hydrogen bonding to the catalytic aspartates (Asp25 and

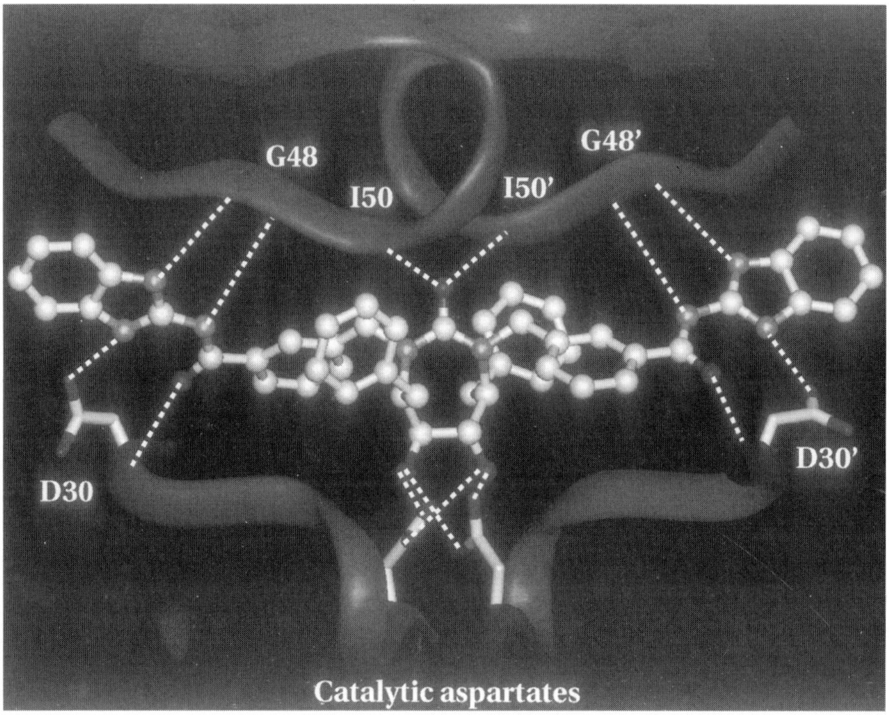

Figure 17.4 Broad specificity of SD-146. This compound is a potent inhibitor of wild type and V82F/I84V because, (1) its diols form an extensive hydrogen bonding network (dashed lines) with the catalytic aspartates, (2) its urea oxygen hydrogen bonds directly to the backbone amides of Ile50 and 50′ of the flaps, (3) its P2 and *P2*′ substituents participate in six hydrogen bonds with the backbone atoms of G48, *G48*′, *D30*′, and *D30*′, and (4) it interacts with a total of 31 residues and participates in 177 vdw contacts. (*See Color plate 16*)

Asp25′) and the amide backbones of Ile50 and 50′ of the flaps via the structural water or, as in the case of CU inhibitors, the urea oxygen (Figure 17.4) (Ala *et al* 1998a).

17.4.1.2 Design asymmetric compounds

Asymmetric inhibitors might be better suited than symmetric ones to reduce drug resistance because a single mutation at the genetic level will result in a pair of mutations related by the two-fold symmetry axis of the homodimer. Consequently, virus will have to mutate additional residues to overcome the inhibitory activity of compounds that interact with different residues in each monomer.

17.4.1.3 Increase flexibility

Flexible substitutions can reorganize in response to a structural change. For example, the P1 substituent of DMP-323 moves deeper into the enlarged S1 subsite of V82F/I84V and picks up additional vdw interactions with the protease (Ala *et al* 1997); the P1′ substituent of CGP-53820 rotates 10° towards the flaps in HIV-2 protease compared to its position in HIV-1 protease (Priestle *et al* 1995); and the flexible P2′ substituent of JE-2147 is thought to be responsible of maintaining potency against several mutants (Yoshimura *et al* 1999). In these cases, inhibitor reorganization is likely responsible for recruiting new interactions and preventing a much larger decrease in binding affinity.

17.4.1.4 Increase interaction with backbone atoms

Inhibitors that interact with backbone atoms might be less sensitive to structural change (Jadhav *et al* 1997). This assumption is based on the fact that these atoms often reorganize only slightly in response to a mutation. Therefore, inhibitors that form hydrogen bonds with the backbone atoms C=O and NH might have a better chance of inhibiting a broad range of mutants (Figure 17.4).

17.4.1.5 Increase total number of interactions

Increasing the number of interactions between the protease and inhibitors appears to minimize the loss of binding affinity caused by resistance mutations. Mutating Val82 and Ile84, for example, does not provide the virus with as much of a growth advantage in the presence of SD-146 as compared to DMP-323 because the number of interactions lost in the SD-146 complex represents only a small percentage of the total. In contrast, a significant percentage (~10%) of the total number of vdw contacts is lost in the DMP-323 mutant complex. Large inhibitors are thus more likely to have an extended therapeutic benefit because their total contact surface area with respect to the protease is not significantly reduced by one or two point mutations. A more complex set of mutations will therefore be needed to overcome their inhibition.

17.4.2 Cross resistance

Until broadly active inhibitors become available, the use of drugs in combination has become the only logical therapeutic alternative. Drug candidates for combination therapy should not have overlapping toxicity profiles and should bind to different targets or at least different subsites of the same target. In the light of this strategy, a chemically diverse pool of inhibitors is needed to minimize cross-resistance. Although the currently available HIV protease drugs are potent (<0.3 nM), asymmetric, flexible inhibitors that interact with the catalytic aspartates and flaps, they all contain a linear scaffold and most of their substituents bind in the same subsites. For example, all of the FDA approved inhibitors are mono-ols; one half of nelfinavir is identical to saquinavir; and many of the inhibitors possess a benzyl group at P1 or P1′. Furthermore, the bound structures of indinavir, VX-478, and saquinavir are very similar, only deviating at P2(P2′) and P3(P3′). These similarities create a situation where it is highly probable that cross-resistant variants will emerge and reduce the effectiveness of any combination of the currently available protease drugs.

17.4.3 Future targets

Given that cross-resistant variants already exist, new classes of HIV drugs must be added to the current multidrug regimens in order to achieve a long-term suppression of viral replication. The three most promising new drug targets are involved in essential steps in the viral life cycle. The first target is HIV integrase, an enzyme that catalyses the integration of viral DNA into the host genome. The availability of its crystal structure (Wlodawer 1999), mutagenesis analysis (Engelman *et al* 1997), and small molecule inhibitors (Neamati *et al* 1999) make it an attractive drug target.

The second target is gp120, an envelope glycoprotein that recognises and binds to cellular receptors (Dimitrov 1997). According to recent reports, viral entry has been prevented by interfering with the interaction between gp120 and the host cell receptor by either co-expressing wild-type and defective gp120 (Lund *et al* 1998), or using antibodies against epitopes of gp120/gp41 (Verrier *et al* 1997). Furthermore, the crystal structure of gp120 has recently been determined (Kwong *et al* 1998; Rizzuto *et al* 1998; Wyatt *et al* 1998) and a small molecule inhibitor, which blocks viral entry and membrane fusion *via* the CXCR4 co-receptor, has been designed (Donzella *et al* 1998).

The third target is gp41, a membrane-spanning glycoprotein that associates with gp120 and mediates the fusion of the viral and host membranes. The availability of its crystal structure (Chan *et al* 1997; Tan *et al* 1997; Weissenhorn *et al* 1997), mutagenesis analysis (Weng and Weiss 1998), and ligands that inhibit gp41-mediated cell membrane fusion (Debnath *et al* 1999; Eckert *et al* 1999; Ferrer *et al* 1999) will hopefully initiate a rapid development of potent inhibitors.

17.5 CONCLUSION

HIV will continue to persist in the presence of potent inhibitors until we raise the genetic barrier to resistance to a point where resistant variants no longer produce active protease. This should be feasible because the protease can accommodate only a limited number of active-site mutations, which are key to high levels of resistance, before losing processing activity. A combination of broadly-active inhibitors should thus completely block viral replication. The effectiveness of this therapy, however, will depend on our ability to develop accurate prediction methods for determining resistance profiles. Until then, additional viral targets will have to be inhibited in order to reduce HIV's ability to evade treatment.

REFERENCES

Ala, P.J., Huston, E.E., Klabe, R.M., McCabe, D.D., Duke, J.L., Rizzo, C.J. *et al* (1997) Molecular basis of HIV-1 protease drug resistance: Structural analysis of mutant proteases complexed with cyclic urea inhibitors. *Biochemistry*, **36**, 1573–1580.

Ala, P.J., DeLoskey, R.J., Huston, E.E., Jadhav, P.K., Lam, P.Y.S., Eyermann, C.J. *et al* (1998a) Molecular recognition of cyclic urea HIV-1 protease inhibitors. *Journal of Biological Chemistry*, **273**, 12325–12331.

Ala, P., Huston, E.E., Klabe, R., Jadhav, P.K., Lam, P. and Chang, C.H. (1998b) Counteracting HIV-1 protease drug resistance: Structural analysis of mutant proteases complexed with XV638 and SD146, cyclic urea amides with broad specificity. *Biochemistry*, **37**, 15042–15049.

Baldwin, E.T., Bhat, T.N., Liu, B., Pattabiraman, N. and Erickson, J.W. (1995) Structural basis of drug resistance for the V82A mutant of HIV-1 proteinase. *Nature Structural Biology*, **2**, 244–249.

Barre-Sinoussi, F., Chermann, J.C., Rey, F., Nugeyre, M.T., Chamaret, S., Gruest, J. *et al* (1983) Isolation of a T-lymphotropic retrovirus from a patient at risk for acquired immune deficiency syndrome (AIDS). *Science*, **220**, 868–871.

Carrillo, A., Stewart, K.D., Sham, H.L., Norbeck, D.W., Kohlbrenner, W.E., Leonard, J.M. *et al* (1998) *In vitro* selection and characterization of human immuno deficiency virus type 1 variants with increased resistance to ABT-378, a novel protease inhibitor. *Journal of Virology*, **72**, 7532–7541.

Chan, D.C., Fass, D., Berger, J.M. and Kim, P.S. (1997) Core structure of gp41 from the HIV envelope glycoprotein. *Cell*, **89**, 263–273.

Chen, Z., Li, Y., Schock, H.B., Hall, D., Chen, E. and Kuo, L.C. (1995) Three-dimensional structure of a mutant HIV-1 protease displaying cross-resistance of all protease inhibitors in clinical trials. *Journal of Biological Chemistry*, **270**, 21433–21436.

Chun, T.W., Stuyver, L., Mizell, S.B., Ehler, L.A., Mican, J.A., Baseler, M. *et al* (1997) Presence of an inducible HIV-1 latent reservoir during highly active antiretroviral therapy. *Proceedings of the National Academy of Sciences of the U.S.A.*, **94**, 13193–13197.

Condra, J.H., Schleif, W.A., Blahy, O.M., Gabryelski, L.J., Graham, D.J., Quintero, J.C. *et al* (1995) *In vivo* emergence of HIV-1 variants resistant to multiple protease inhibitors. *Nature*, **374**, 569–571.

Debnath, A.K., Radigan, L. and Jiang, S. (1999) Structure-based identification of small molecule antiviral compounds targeted to the gp41 core structure of the human immunodeficiency virus type 1. *Journal of Medicinal Chemistry*, **42**, 3203–3209.

Debouck, C. (1992) The HIV-1 protease as a therapeutic target for AIDS. *AIDS Research of Human Retroviruses*, **8**, 153–164.

De Clercq, E. (1995) Toward improved anti-HIV chemotherapy: Therapeutic strategies for intervention with HIV infections. *Journal of Medicinal Chemistry*, **38**, 2491–2517.

Dimitrov, D.S. (1997) How do viruses enter cells? The HIV coreceptors teach us a lesson of complexity. *Cell*, **91**, 721–730.

Donzella, G.A., Schols, D., Lin, S.W., Este, J.A., Nagashima, K.A., Maddon, P.J. *et al* (1998) AMD3100, a small molecule inhibitor of HIV-1 entry via the CXCR4 co-receptor. *Nature Medicine*, **4**, 72–77.

Doyon, L., Croteau, G., Thibeault, D., Poulin, F., Pilote, L. and Lamarre, D. (1996) Second locus involved in human immunodeficiency virus type 1 resistance to protease inhibitors. *Journal of Virology*, **70**, 3763–3769.

Eckert, D.M., Malashkevich, V.N., Hong, L.H., Carr, P.A. and Kim, P.S. (1999) Inhibiting HIV-1 entry: discovery of D-peptide inhibitors that target the gp41 coiled-coil pocket. *Cell*, **99**, 103–115.

Engelman, A., Liu, Y., Chen, H., Farzan, M. and Dyda, F. (1997) Structure-based mutagenesis of the catalytic domain of human immunodeficiency virus type 1 integrase. *Journal of Virology*, **71**, 3507–3514.

Ferrer, M., Kapoor, T.M., Strassmaier, T., Weissenhorn, W., Skehel, J.J., Oprian, D. *et al* (1999) Selection of gp41-mediated HIV-1 cell entry inhibitors from biased combinatorial libraries of non-natural binding elements. *Nature Structural Biology*, **6**, 953–960.

Finzi, D., Hermankova, M., Pierson, T., Carruth, L.M., Buck, C., Chaisson, R.E. *et al* (1997) Identification of a reservoir for HIV-1 in patients on highly active antiretroviral therapy. *Science*, **278**, 1295–1300.

Furfine, E.S., D'Souza, E., Ingold, K.J., Leban, J.J., Spector, T. and Porter, D.J.T. (1992) Two-step binding mechanism for HIV protease inhibitors. *Biochemistry*, **31**, 7886–7891.

Gallo, R.C., Sarin, P.S., Gelmann, E.P., Robert-Guroff, M., Richardson, E., Kalyanaraman, V.S. *et al* (1983) Isolation of human T-cell leukemia virus in acquired immune deficiency syndrome (AIDS). *Science*, **220**, 865–867.

Gulick, R.M., Mellors, J.W., Havlir, D., Eron, J.J., Gonzalez, C., McMahon, D. *et al* (1998) Simultaneous vs sequential initiation of therapy with indinavir, zidovudine, and lamivudine for HIV-1 infection: 100-week follow-up. *Journal of American Medical Association*, **280**, 35–41.

Hammond, J., Larder, B.A., Schinazi, R.F. and Mellors, J.W. (1998) Mutations in retroviral genes associated with drug resistance. *The PRN Notebook online edition* (http://hiv-web.lanl.gov./HTML/reviews/Mellors.html).

Hodge, C.N., Aldrich, P.E., Bacheler, L.T., Chang, C.H., Eyermann, C.J., Garber, S. *et al* (1996) Improved cyclic urea inhibitors of the HIV-1 protease: synthesis, potency, resistance profile, human pharmacokinetics and X-ray crystal structure of DMP 450. *Chemistry & Biology*, **3**, 301–314.

Hoog, S.S., Towler, E.M., Zhao, B., Doyle, M.L., Debouck, C. and Abdel-Meguid, S.S. (1996) Human immunodeficiency virus protease ligand specificity conferred by residues outside of the active site cavity. *Biochemistry*, **35**, 10279–10286.

Ives, K.J., Jacobsen, H., Galpin, S.A., Garaev, M.M., Dorrell, L., Mous, J. *et al* (1997) Emergence of resistant variants of HIV *in vivo* during monotherapy with the proteinase inhibitor saquinavir. *Journal of Antimicrobial Chemotherapy*, **39**, 771–779.

Jacobsen, H., Yasargil, K., Winslow, D.L., Craig, J.C., Kröhn, A., Duncan, I.B. *et al* (1995) Characterization of human immunodeficiency virus type 1 mutants with decreased sensitivity to proteinase inhibitor Ro 31–8959. *Virology*, **206**, 527–534.

Jadhav, P.K., Ala, P., Woerner, F.J., Chang, C.H., Garber, S.S., Anton, E.D. *et al* (1997) Cyclic urea amides: HIV-1 protease inhibitors with low nanomolar potency against both wild type and protease inhibitor resistant mutants of HIV. *Journal of Medicinal Chemistry*, **40**, 181–191.

Kervinen, J., Thanki, N., Zdanov, A., Tino, J., Barrish, J., Lin, P.F. *et al* (1996) Structural analysis of the native and drug-resistant HIV-1 proteinases complexed with an aminodiol inhibitor. *Protein and Peptide Letters*, **3**, 399–406.

King, R.W., Winslow, D.L., Garber, S., Scarnati, H.T., Bachelor, L., Stack, S. *et al* (1995) Identification of a clinical isolate of HIV-1 with an isoleucine at position 82 of the protease which retains susceptibility to protease inhibitors. *Antiviral Research*, **28**, 13–24.

Kwong, P.D., Wyatt, R., Robinson, J., Sweet, R.W., Sodroski, J. and Hendrickson, W.A. (1998) Structure of an HIV gp120 envelope glycoprotein in complex with the CD4 receptor and a neutralizing human antibody. *Nature*, **393**, 648–659.

Lund, O.S., Losman, B., Schonning, K., Bolmstedt, A., Olofsson, S. and Hansen, J.E. (1998) Inhibition of HIV type 1 infectivity by coexpression of a wild-type and a defective glycoprotein 120. *AIDS Research of Human Retroviruses*, **14**, 1445–1450.

Mammano, F., Petit, C. and Clavel, F. (1998) Resistance-associated loss of viral fitness in human immunodeficiency virus type 1: phenotypic analysis of protease and gag coevolution in protease inhibitor-treated patients. *Journal of Virology*, **72**, 7632–7637.

Markowitz, M., Mo, H., Kempf, D.J., Norbeck, D.W., Bhat, T.N., Erickson, J.W. *et al* (1995) Selection and analysis of human immunodeficiency virus type 1 variants with increased resistance to ABT-538, a novel protease inhibitor. *Journal of Virology*, **69**, 701–706.

Miller, M., Schneider, J., Sathyanarayana, B.K., Toth, M.V., Marshall, G.R., Clawson, L. *et al* (1989) Structure of complex of synthetic HIV-1 protease with a substrate-based inhibitor at 2.3 Å resolution. *Science*, **246**, 1149–1152.

Molla, A., Korneyeva, M., Gao, Q., Vasavanonda, S., Schipper, P.J., Mo, H.M. *et al* (1996) Ordered accumulation of mutations in HIV protease confers resistance to ritonavir. *Nature Medicine*, **2**, 760–766.

Navia, M.A., Fitzgerald, P.M., McKeever, B.M., Leu, C.T., Heimbach, J.C., Herber, W.K. *et al* (1989) Three-dimensional structure of aspartyl protease from human immunodeficiency virus HIV-1. *Nature*, **337**, 615–620.

Neamati, N., Turpin, J.A., Winslow, H.E., Christensen, J.L., Williamson, K., Orr, A. *et al* (1999) Thiazolothiazepine inhibitors of HIV-1 integrase. *Journal of Medicinal Chemistry*, **42**, 3334–3341.

Partaledis, J.A., Yamaguchi, K., Tisdale, M., Blair, E.E., Falcione, C., Maschera, B. *et al* (1995) *In vitro* selection and characterization of human immunodeficiency virus type 1 (HIV-1) isolates with reduced sensitivity to hydroxyethylamino sulfonamide inhibitors of HIV-1 aspartyl protease. *Journal of Virology*, **69**, 5228–5235.

Patick, A.K., Mo, H., Markowitz, M., Appelt, K., Wu, B., Musick. L. *et al* (1996) Antiviral and resistance studies of AG1343, an orally bioavailable inhibitor of human immunodeficiency virus protease. *Journal of Antimicrobial Chemotherapy*, **40**, 292–297.

Pazhanisamy, S., Stuver, C.M., Cullinan, A.B., Margolin, N., Rao, B.G. and Livingston, D.J. (1996) Kinetic characterization of human immunodeficiency virus type-1 protease-resistant variants. *Journal of Biological Chemistry*, **271**, 17979–17985.

Priestle, J.P., Fässler, A., Rösel, J., Tintelnot-Blomley, M., Strop, P. and Grütter, M.G. (1995) Comparative analysis of the X-ray structures of HIV-1 and HIV-2 proteases in complex with CGP 53820, a novel pseudosymmetric inhibitor. *Structure*, **3**, 381–389.

Rizzuto, C.D., Wyatt, R., Hernandez-Ramos, N., Sun, Y., Kwong, P.D. and Hendrickson, W.A. (1998) A conserved HIV gp120 glycoprotein structure involved in chemokine receptor binding. *Science*, **280**, 1949–1953.

Robins, T. and Plattner, J. (1993) HIV protease inhibitors: their anti-HIV activity and potential role in treatment. *Journal of Acquired Immune Deficiency Syndromes*, **6**, 162–170.

Rose, R.E., Gong, Y.F., Greytok, J.A., Bechtold, C.M., Terry, B.J., Robinson, B.S. *et al* (1996) Human immunodeficiency virus type 1 viral background plays a major role in development of resistance to protease inhibitors. *Proceedings of the National Academy of Sciences of the U.S.A.*, **93**, 1648–1653.

Rose, R.B., Craik, C.S. and Stroud, R.M. (1998) Domain flexibility in retroviral proteases: structural implications for drug resistant mutations. *Biochemistry*, **37**, 2607–2621.

Schock, H.B., Garsky, V.M. and Kuo, L.C. (1996) Mutational anatomy of an HIV-1 protease variant conferring cross-resistance to protease inhibitors in clinical trials. Compensatory modulations of binding and activity. *Journal of Biological Chemistry*, **271**, 31957–31963.

Tan, K., Liu, J., Wang, J., Shen, S. and Lu, M. (1997) Atomic structure of a thermostable subdomain of HIV-1 gp41. *Proceedings of the National Academy of Sciences of the U.S.A.*, **94**, 12303–12308.

Verrier, F.C., Charneau, P., Altmeyer, R., Laurent, S., Borman, A.M. and Girard, M. (1997) Antibodies to several conformation-dependent epitopes of gp120/gp41 inhibit CCR-5-dependent cell-to-cell fusion mediated by the native envelope glycoprotein of a primary macrophage-tropic HIV-1 isolate. *Proceedings of the National Academy of Sciences of the U.S.A.*, **94**, 9326–9331.

Wei, X., Ghosh, S.K., Taylor, M.E., Johnson, V.A., Emini, E.A., Deutsch, P. *et al* (1995) Viral dynamics in human immunodeficiency virus type 1 infection. *Nature*, **373**, 117–122.

Weissenhorn, W., Dessen, A., Harrison, S.C., Skehel, J.J. and Wiley, D.C. (1997) Atomic structure of the ectodomain from HIV-1 gp41. *Nature*, **387**, 426–430.

Weng, Y. and Weiss, C.D. (1998) Mutational analysis of residues in the coiled-coil domain of human immunodeficiency virus type 1 transmembrane protein gp41. *Journal of Virology*, **72**, 9676–9682.

Williams, K.J. and Loeb, L.A. (1992) Retroviral reverse transcriptases: Error frequencies and mutagenesis. *Current Topics in Microbiology and Immunology*, **176**, 165–180.

Wlodawer, A., Miller, M., Jaskolski, M., Sathyanarayana, B.K., Baldwin, E., Weber, I.T. *et al* (1989) Conserved folding in retroviral proteases: crystal structure of a synthetic HIV-1 protease. *Science*, **245**, 616–621.

Wlodawer, A. and Erickson, J.W. (1993) Structure-based inhibitors of HIV-1 protease. *Annual Reviews Biochemistry*, **62**, 543–585.

Wlodawer, A. and Vondrasek, J. (1998) Inhibitors of HIV-1 protease: a major success of structure-assisted drug design. *Annual Reviews Biophysics & Biomolecular Structure*, **27**, 249–284.

Wlodawer, A. (1999) Crystal structures of catalytic core domains of retroviral integrases and role of divalent cations in enzymatic activity. *Advances in Virus Research*, **52**, 335–350.

Wong, J.K., Hezareh, M., Günthard, H.F., Havlir, D.V., Ignacio, C.C., Spina, C.A. *et al* (1997) Recovery of replication-competent HIV despite prolonged suppression of plasma viremia. *Science*, **278**, 1291–1295.

Wyatt, R., Kwong, P.D., Desjardins, E., Sweet, R.W., Robinson, J., Hendrickson, W.A. *et al* (1998) The antigenic structure of the HIV gp120 envelope glycoprotein. *Nature*, **393**, 705–711.

Yoshimura, K., Kato, R., Yusa, K., Kavlick, M., Maroun, V., Nguyen, H. *et al* (1999) JE-2147: A dipeptide protease inhibitor (PI) that potently inhibits multi-PI-resistant HIV-1. *Proceedings of the National Academy of Sciences of the U.S.A.*, **96**, 8675–8680.

Zhang, Y.M., Imamichi, H., Imamichi, T., Lane, H.C., Falloon, J., Vasudevachari, M.B. *et al* (1997) Drug resistance during indinavir therapy is caused by mutations in the protease gene and in its Gag substrate cleavage sites. *Journal of Virology*, **71**, 6662–6670.

Chapter 18

Proteases of protozoan parasites

Philip J. Rosenthal

Many proteases of medically important protozoan parasites have been identified and characterized. All of these organisms appear to express cysteine proteases, which act either extracellularly or intracellularly in lysosome-like organelles. Protozoan serine proteases are less well characterized, but they appear to play key roles in some organisms. Individual protozoan metalloproteases and aspartic proteases have been well characterized, although there is not yet evidence for the expression of these classes of proteases in most protozoans. Important functions of protozoan proteases likely include the invasion of host cells and tissues, the degradation of components of the host immune response, and the hydrolysis of host proteins for nutritional purposes. Initial results from multiple systems suggest that the inhibition of protozoan proteases may be an important new means of antiparasitic chemotherapy.

18.1 INTRODUCTION

Protozoan parasites are responsible for some of the most important infectious diseases in the world, including malaria, leishmaniasis, trypanosomiasis, amebiasis, and toxoplasmosis. For many of these diseases, which principally affect individuals in underdeveloped countries, new therapies are urgently needed. New therapeutic approaches include the inhibition of parasite proteases, as proteases appear to play important roles in the life cycles of all medically important protozoan parasites. This chapter will review available information on well characterized protozoan proteases and on studies of their inhibition. A recent more detailed and more extensively referenced review on this subject is also available (Rosenthal 1999). Each medically important protozoan parasite will be discussed separately, but in many cases similar enzymes appear to play key roles in a number of organisms. Thus, the potential exists for the development of antiprotozoan protease inhibitors with activity against multiple human parasites.

18.2 *LEISHMANIA*

Leishmania species cause serious diseases in many parts of the developing world. Leishmanial syndromes include visceral leishmaniasis (kala-azar), which is often fatal, cutaneous disease, which causes chronic skin ulcers, and mucosal leishmaniasis, which can cause severe destructive lesions of mucosal membranes. Available treatments for leishmaniasis are limited by drug resistance, significant toxicities, and high cost. Two important classes of leishmanial proteases have been extensively characterized, a surface metalloprotease and intracellular cysteine proteases.

18.2.1 *Leishmania* metalloprotease (leishmanolysin)

Leishmania parasites contain a major surface glycoprotein with protease activity that has been referred to as gp63, promastigote surface protease, and leishmanolysin (Bouvier *et al* 1985, 1995). Leishmanolysin appears to be present in all parasite stages, but it is principally expressed in promastigotes, the parasite stage that develops in the sandfly vector and then circulates in humans after inoculation (Bouvier *et al* 1985). Leishmanolysin is a surface protein in promastigotes, while in intracellular amastigotes the protein appears to be localized to the flagellar pocket (Medina-Acosta *et al* 1989) and lysosomes (Ilg *et al* 1993).

Leishmanolysin has the biochemical properties of a zinc metalloprotease, including incorporation of radiolabelled zinc, inhibition by zinc chelators, and reversal of this inhibition with added zinc (Bouvier *et al* 1989; Chaudhuri *et al* 1989). It is processed from a proform to its active mature form, possibly by an autocatalytic process similar to that in other metalloproteases (Button *et al* 1993). Proteolytic activity was maximal at slightly basic pH for the promastigote surface form of leishmanolysin and at acid pH for intracellular amastigote forms (Ilg *et al* 1993; Bouvier *et al* 1995).

The leishmanolysin sequence predicts a typical zinc metalloprotease (Bouvier *et al* 1989; Chaudhuri *et al* 1989). Studies utilizing site-directed mutagenesis demonstrated that conserved residues at the zinc binding site are required for activity and stability of the protease, that asparagine-glycosylation at three sites contributes to stability, and that Asn-577 acts as the glycosyl phosphatidylinositol addition site and is required for the membrane anchoring of the protease (McGwire and Chang 1996). However, deglycosylation of leishmanolysin did not block appropriate folding, transport, and activity of the protease (Funk *et al* 1994).

The function of leishmanolysin is unknown, but it is believed to act as a *Leishmania* virulence factor (Chang *et al* 1990). Possible functions of leishmanolysin include mediation of binding between *Leishmania* promastigotes and macrophages (Russell and Wilhelm 1986; Liu and Chang 1992), cleavage of C3 to expedite complement receptor-mediated endocytosis of promastigotes (Chaudhuri and Chang 1988; Brittingham *et al* 1995), and cleavage of membranolytic complement components to protect circulating promastigotes. Mutants lacking functional leishmanolysin (Brittingham *et al* 1995) or lacking six of seven genes encoding leishmanolysin (Joshi *et al* 1998) showed increased sensitivity to complement-mediated lysis. Available results suggest that leishmanolysin may allow *Leishmania* to utilize the opsonic properties of complement to mediate entry into macrophages while avoiding the lytic effects of activation of the complete complement pathway. Intracellular amastigote leishmanolysin may support organisms within macrophage lysosomes by cleaving microbicidal factors (Chang *et al* 1990), by degrading proteins to provide nutrients, or by preventing intracellular degradation of the parasites (Chaudhuri *et al* 1989; Seay *et al* 1996). Leishmanolysin-deficient *Leishmania* mutants have markedly decreased intracellular survival, and transfection of wild-type leishmanolysin into these mutants significantly improved survival (McGwire and Chang 1994). The recent determination of the structure of leishmanolysin (Schlagenhauf *et al* 1998) should aid in the evaluation of its function and in the development of specific inhibitors.

18.2.2 *Leishmania* cysteine protease

A number of distinct cysteine protease activities have been identified in *Leishmania* (Robertson and Coombs 1990), mostly within amastigote lysosome-like organelles known as megasomes (Pupkis *et al* 1986). Leishmanial cysteine protease genes can be grouped into a single-copy *lmcpa* gene (Mottram *et al* 1992), an array of 19 *lmcpb* genes (Souza *et al* 1992), and a single-copy *lmcpc* gene (Robertson and Coombs 1994). The *lmcpa* and *lmcpb* genes encode cathepsin L-like proteases. The *lmcpc* gene product is more cathepsin B-like in sequence, but also has cathepsin L-like substrate specificity (Chan *et al* 1999). Most of these proteases are expressed principally in amastigotes (Mottram *et al* 1997). The products of the *lmcpb* genes contain unusual carboxy-terminal extensions (Mottram *et al* 1998). The function of the carboxy terminal extensions, which are also present in related proteases of other trypanasomatid protozoans and in some plant cysteine proteases, are unknown. They may determine life-cycle stage-dependent expression of proteases, as two genes that are uniquely expressed in metacyclic promastigotes have truncated extensions (Mottram *et al* 1997). Processing of *lmcpa* was inhibited by cysteine protease inhibitors, suggesting autohydrolytic enzyme activation (Duboise *et al* 1994).

The specific functions of the leishmanial cysteine proteases and the reason that so many protease genes are expressed are unexplained (Mottram *et al* 1998). Disruption of the *lmcpa* gene did not alter parasite viability, growth rate, or pathogenicity in a murine model (Souza *et al* 1994), and disruption of the entire 19 gene *lmcpb* tandem array did not eliminate the viability of cultured parasites (Mottram *et al* 1996). However, parasites lacking the *lmcpb* array were markedly less virulent than wild-type parasites, apparently due to decreased survival in macrophages. The product of a single *lmcpb* gene re-expressed in the mutant restored macrophage infectivity to wild-type levels. Mutants in which both the *lmcpb* array and the *lmcpa* gene were disrupted had macrophage infectivity similar to that of the *lmcpb* mutant, although they were unable to form lesions in mice (Mottram *et al* 1996). For mutants with a disrupted *lmcpc* gene, infectivity to macrophages was reduced, and skin lesions in mice were smaller than those caused by wild-type parasites (Bart *et al* 1997). In summary, available results suggest that leishmanial cysteine proteases, although not essential for infectivity, are important virulence determinants.

Peptidyl cysteine protease inhibitors blocked the growth of amastigotes in mouse peritoneal macrophages (Coombs and Baxter 1984). Recently, both nonpeptide hydrazides identified via a molecular modeling approach (Selzer *et al* 1997) and pseudopeptide substrate analog (Selzer *et al* 1999) were shown to inhibit leishmanial cysteine proteases, to block parasite growth at concentrations that were nontoxic to cultured mammalian cells, and to ameliorate the pathology associated with experimental infection in mice. The inhibitors appeared to act by interfering with parasite lysosomal function, as inhibitor-treated parasites accumulated undigested debris in lysosomes and in the flagellar pocket (Selzer *et al* 1999). The results suggest that cysteine protease inhibitors have promise as antileishmanial drugs.

18.3 AFRICAN TRYPANOSOMES

Organisms of the *Trypanosoma brucei* complex cause African trypanosomiasis, a severe febrile illness that is commonly fatal. All available therapies for African trypanosomiasis are inadequate due to poor efficacy and significant toxicity.

18.3.1 *T. brucei* cysteine protease (trypanopain)

The *T. brucei* protease trypanopain is a lysosomal papain-family enzyme that has an acid pH optimum and a cathepsin L-like substrate preference (Robertson *et al* 1990; Troeberg *et al* 1996). Genes encoding trypanopain predict typical papain family proteases except for a carboxy-terminal extension that is rich in proline (Mottram *et al* 1989) and is not required for activity (Pamer *et al* 1991). More than 20 copies of the trypanopain gene are arranged in a long tandem array (Mottram *et al* 1989).

The function of trypanopain is not known. Unlike the related trypanosomatids *Leishmania* and *T. cruzi*, *T. brucei* replicates extracellularly and does not invade host cells. Trypanopain may play a role in the development of bloodstream trypanosomes (Pamer *et al* 1989) and it may inhibit parasite opsonization by degrading antibody-bound trypanosome antigens (Russo *et al* 1993). However, trypanopain is probably not active in the bloodstream, as it is inhibited by circulating cystatins (Troeberg *et al* 1996).

Cysteine protease inhibitors may be appropriate new treatments for African trypanosomiasis. In recent studies, a number of peptidyl chloromethyl ketone, diazomethyl ketone, fluoromethyl ketone, and vinyl sulfone trypanopain inhibitors killed cultured bloodstream forms of *T. brucei* at low micromolar concentrations (Troeberg *et al* 1999) and diazomethyl ketones inhibited the progression of murine trypanosomiasis (Scory *et al* 1999). Studies with biotinylated peptidyl inhibitors verified that trypanopain was their likely target (Troeberg *et al* 1999).

18.3.2 Other proteases of African trypanosomes

Another *T. brucei* protease is the serine oligopeptidase, an atypical trypsin-like serine protease that cleaves typical trypsin peptide substrates, but is not inhibited by some trypsin inhibitors, is inhibited by agents that modify cysteine residues, and it is stimulated by reducing agents (Troeberg *et al* 1996). These data suggest that the serine oligopeptidase contains a cysteine residue that must be reduced for optimal activity. The function of the protease is unknown. It may be a target for the existing (albeit inadequate) antitrypanosomal drugs pentamidine and suramin (Morty *et al* 1998). A leishmanial analog of the serine oligopeptidase has also been identified (de Andrade *et al* 1998).

T. brucei has recently been shown to contain genes encoding homolog of leishmanolysin (El-Sayed and Donelson 1997). Typical metalloprotease sequences are conserved between the leishmanial and trypanosomal genes. The function of the newly identified trypanosomal leishmanolysin homolog is unknown, but their expression in obligate extracellular parasites suggests a different role than that proposed for leishmanolysin in leishmanial amastigotes. One possible role is protection against complement-mediated cell lysis.

A cytosolic proteasome was recently purified from bloodstream and insect stages of *T. brucei* (Hua *et al* 1996). As is typical for mammalian proteasomes, the complex demonstrated chymotrypsin-like, trypsin-like, and peptidylglutamylpeptide-hydrolysing activities against synthetic substrates (Hua *et al* 1996). The proteasome inhibitor lactacystin inhibited parasite development, suggesting that the proteasome may be required for trypanosomal cell cycle progression (Mutomba *et al* 1997).

18.4 *TRYPANOSOMA CRUZI*

T. cruzi infection causes American trypanosomiasis (Chagas disease) in much of Central and South America. Chagas disease causes a febrile illness that is followed years later by serious heart and gastrointestinal tract disease. No satisfactory therapies for Chagas disease are available.

18.4.1 Cysteine protease of *T. cruzi* (cruzain)

Cruzain (Eakin *et al* 1992; also known as cruzipain (Cazzulo *et al* 1989)) is a papain-family protease with an acidic pH optimum and a cathepsin L-like substrate specificity. Cruzain was localized to the lysosomes, flagellar pocket, and surface of different *T. cruzi* stages (Bontempi *et al* 1989; Souto-Padron *et al* 1990). Epimastigotes (the insect stage) contained much greater protease activity than amastigotes or trypomastigotes (Bonaldo *et al* 1991).

Multiple copies of the cruzain gene (Cazzulo *et al* 1989; Eakin *et al* 1990; Murta *et al* 1990) are present in tandem arrays on more than one chromosome (Campetella *et al* 1992; Eakin *et al* 1992). The gene encodes a papain-family cysteine protease with a typical pro-sequence and a carboxy-terminal extension similar to that of *T. brucei* and *Leishmania* cysteine proteases (Eakin *et al* 1992). The cruzain pro-region includes sequences that mediate intracellular targeting (Huete-Perez *et al* 1999). The carboxy-terminal extension differs from that of *T. brucei* in that it is rich in threonine rather than proline; it is not required for correct protein folding or enzymatic activity (Eakin *et al* 1992). Both the pro-region and the carboxy-terminal extension of cruzain are cleaved by autohydrolysis (Hellman *et al* 1991; Eakin *et al* 1992).

The function of cruzain is unknown. A number of protease inhibitors blocked adherence to and invasion of human cells (Franke de Cazzulo *et al* 1994; Piras *et al* 1985), and anti-cruzain antibody blocked invasion of macrophages (Souto-Padron *et al* 1990). Peptide cysteine protease inhibitors blocked parasite development (Bonaldo *et al* 1991; Meirelles *et al* 1992; Harth *et al* 1993; Franke de Cazzulo *et al* 1994), and *T. cruzi* engineered to overexpress cruzain had relative resistance to the deleterious effects of a diazomethyl ketone cysteine proteinase inhibitor (Tomas *et al* 1997). These results suggest that cruzain plays a role in parasite invasion of host cells, in intracellular protein degradation, and in parasite remodeling during transformation between parasite stages.

The antiparasitic effects of multiple inhibitors of cruzain suggest that it may be an appropriate target for new drugs to treat Chagas disease. Cruzain is the only protozoan cysteine protease for which a structure has been determined (McGrath *et al* 1995). This structure should be valuable in the prediction of inhibitors of the enzyme (Gillmor *et al* 1997). Extensive efforts to develop peptidyl (Harth *et al* 1993; Roush *et al* 1998; Scheidt *et al* 1998) and nonpeptide (Li *et al* 1996; Scheidt *et al* 1998) inhibitors of cruzain are currently ongoing. Concentrations of fluoromethyl ketones that interrupted the *T. cruzi* life cycle were not toxic to cultured mammalian cells (Harth *et al* 1993). Fluorescence microscopy showed that a biotinylated fluoromethyl ketone was selectively taken up by intracellular parasites and not host lysosomes (McGrath *et al* 1995). Fluoromethyl ketone and vinyl sulfone inhibitors of cruzain markedly altered the morphology of the *T. cruzi* Golgi apparatus, inhibited the transport of cruzain to lysosomes, and arrested parasite growth (Engel *et al* 1998a,b). These compounds also cured murine *T. cruzi* infections (Engel *et al* 1998a,b).

18.4.2 Other proteases of *T. cruzi*

T. cruzi contains an alkaline peptidase that is analogous to the serine oligopeptidase of *T. brucei* and may mediate the calcium signaling in host cells that is required for cell entry (Burleigh *et al* 1997). An alkaline *T. cruzi* collagenase may play a role in host cell infection (Santana *et al* 1997). A proteasome of *T. cruzi* has recently been identified (Gonzalez *et al* 1996). The proteasome inhibitor lactacystin inhibited parasite development, suggesting a role for the proteasome in this process (Gonzalez *et al* 1996).

18.5 MALARIA PARASITES

Malaria is the most serious protozoan infection of humans. *Plasmodium falciparum*, the most virulent human malaria parasite, is responsible for over a million deaths per year. Malaria parasites are increasingly resistant to available chemotherapeutic agents. Thus, new antimalarial drugs are urgently needed. Proteases of all four major mechanistic classes have been characterized from erythrocytic malaria parasites. Key roles for plasmodial proteases are mediation of erythrocyte invasion by free merozoites, erythrocyte rupture by mature schizonts, and the degradation of hemoglobin by erythrocytic trophozoites. Existing evidence suggests that serine and cysteine proteases mediate erythrocyte rupture and invasion and that cysteine, aspartic, and probably additional proteases mediate hemoglobin degradation.

18.5.1 Plasmodial cysteine protease (falcipain)

Falcipain is a *P. falciparum* papain-family cysteine protease (Salas *et al* 1995; Rosenthal *et al* 1988). It has an acidic pH optimum and its activity is enhanced by reducing agents (Rosenthal *et al* 1989). Falcipain has a cathepsin L-like substrate specificity, but clear differences in the hydrolysis of peptide substrates by falcipain and host proteases have been identified (Shenai, B.R. and Rosenthal, P.J., unpublished data). Falcipain activity is prominent in trophozoites, the erythrocytic stage during which most hemoglobin degradation occurs (Rosenthal *et al* 1987) and has been localized to the parasite food vacuole, the site of hemoglobin degradation (Gluzman *et al* 1994). The single-copy gene that encodes falcipain predicts a fairly typical papain-family enzyme, but it has an unusually long pro-sequence (Rosenthal and Nelson 1992). The function of the amino-terminal portion of the pro-sequence, which has no homology with other papain-family proteases, is unknown. Partial or complete sequences of falcipain homolog from nine other plasmodial species, including all species that infect humans, are available, and the sequences of these proteases are highly conserved (Rosenthal 1996). The falcipain analog of the murine parasite *P. vinckei* has also been characterized biochemically (Rosenthal *et al* 1993). The enzyme is similar, but not identical to falcipain in its substrate and inhibitor specificity.

Falcipain appears to be required for the degradation of hemoglobin by erythrocytic parasites. Falcipain does not cleave native hemoglobin in a nonreducing environment (Gluzman *et al* 1994), but readily cleaves this substrate under conditions (acidic pH, 0.5–2 mM glutathione) that are predicted to be present in the food vacuole, the site of hemoglobin degradation (Shenai, B.R. and Rosenthal, P.J., unpublished data). Falcipain more rapidly cleaves denatured hemoglobin, suggesting that the enzyme plays a major role in hydrolyzing globin to small peptides. Supporting a necessary role for falcipain in globin hydrolysis, treatment of cultured *P. falciparum* parasites with falcipain inhibitors causes a block in hemoglobin degradation such that parasite food vacuoles fill with undegraded globin (Dluzewski *et al* 1986; Rosenthal *et al* 1988; Bailly *et al* 1992). Falcipain also appears to play a role in initial steps of hemoglobin degradation, as the cysteine protease inhibitor E-64 inhibited the dissociation of the hemoglobin tetramer (Gamboa de Domínguez and Rosenthal 1996), the release of heme (Gamboa de Domínguez and Rosenthal 1996), the initial processing of α- and β-globin (Kamchonwongpaisan *et al* 1997), and the formation of the hemoglobin breakdown product hemozoin (Asawamahasakda *et al* 1994).

Falcipain inhibitors have been shown to block *P. falciparum* hemoglobin degradation and development (Dluzewski *et al* 1986; Rosenthal *et al* 1988; Bailly *et al* 1992; Rosenthal 1995). Peptidyl fluoromethyl ketone (Rosenthal *et al* 1989; Rockett *et al* 1990; Rosenthal *et al* 1991) and vinyl sulfone (Rosenthal *et al* 1996; Olson *et al* 1999) inhibitors of falcipain blocked *P. falciparum* development at nanomolar concentrations. The degree of inhibition of falcipain correlated with the extent of biological effects (Rosenthal *et al* 1991). Both fluoromethyl ketones (Rosenthal *et al* 1993) and vinyl sulfones (Olson *et al* 1999) also inhibited the *P. vinckei* falcipain analog and cured mice infected with an otherwise lethal malaria infection. Screening of potential nonpeptide inhibitors identified by molecular modeling techniques identified a lead compound (Ring *et al* 1993) acting at a low micromolar concentration and subsequent synthesis and screening identified chalcones (Li *et al* 1995) and phenothiazines (Domínguez *et al* 1997) that inhibited falcipain and blocked the development of cultured parasites at nanomolar-low micromolar concentrations.

Cysteine and aspartic proteases appear to act cooperatively in hemoglobin degradation (Rosenthal and Meshnick 1996; Francis *et al* 1997). The two classes of proteases acted synergistically to degrade hemoglobin *in vitro* (Francis *et al* 1994). Inhibitors of the two classes also inhibited the metabolism and development of cultured parasites and the progression of murine malaria in a synergistic manner (Bailly *et al* 1992; Semenov *et al* 1998). Thus, it may be appropriate to use protease inhibitor combinations to treat malaria.

18.5.2 Plasmodial aspartic proteases (plasmepsins)

The two identified *P. falciparum* aspartic proteases are plasmepsin I and plasmepsin II (Goldberg *et al* 1991; Vander Jagt *et al* 1992; Dame *et al* 1994; Gluzman *et al* 1994; Hill *et al* 1994). Both proteases are located in the food vacuole, have acid pH optima, and share significant sequence homology with other aspartic proteases. The enzymes are biochemically similar, but not identical; in evaluations of globin cleavage, substrate preferences of the two enzymes were distinct (Gluzman *et al* 1994; Luker *et al* 1996). The plasmepsins are synthesized as proenzymes, which are integral membrane proteins that are cleaved to soluble forms under acidic conditions (Francis *et al* 1997). The synthesis and processing of both plasmepsins peaks in trophozoites, but plasmepsin I is also synthesized, processed, and presumably active in young ring-stage parasites (Francis *et al* 1997). Homologues of the *P. falciparum* plasmepsins have recently been identified in the human malaria parasites *P. vivax* and *P. malariae* (Westling *et al* 1997). Only one aspartic protease was identified in each of these species; the biochemical properties of the proteases were quite similar to those of the *P. falciparum* plasmepsins.

Plasmepsins I and II are capable of hydrolyzing native hemoglobin at multiple sites (Gluzman *et al* 1994). In *in vitro* studies, both aspartic proteases initially cleaved native hemoglobin at a peptide bond that is a hinge region of the molecule (Gluzman *et al* 1994).

This cleavage apparently alters the structure of the substrate to expose other sites for additional cleavages by the plasmepsins, falcipain, and possibly other proteases (Francis *et al* 1997). The substrate specificities of plasmepsin I and plasmepsin II differ somewhat (Gluzman *et al* 1994), and each appears to cleave different sites in globin after the initial cleavage of hemoglobin (Gluzman *et al* 1994). The full pathway of hemoglobin degradation remains uncertain, and it is not clear whether hemoglobin is degraded in an ordered pathway, or whether multiple proteases act on the hemoglobin substrate simultaneously.

Recent studies suggest that, in addition to its role in hemoglobin degradation, plasmepsin II contributes to the hydrolysis of erythrocyte cytoskeletal proteins that is probably required for erythrocyte rupture by mature schizont-stage parasites (Le Bonniec *et al* 1999). Plasmepsin II was shown to be present in parasite fractions enriched for spectrin-hydrolyzing activity and to localize to the periphery of mature parasites. In addition, recombinant plasmepsin II cleaved spectrin, actin, and protein 4.1 at neutral pH, and the cleavage site of spectrin for the recombinant enzyme was identical to that of the enriched parasite fractions.

Aspartic protease inhibitors are under study as potential antimalarials. Peptidomimetic inhibitors of plasmepsin I and II inhibited hemoglobin hydrolysis and blocked parasite metabolism at low micromolar-nanomolar concentrations (Francis *et al* 1994; Silva *et al* 1996; Moon *et al* 1997). The determination of the structure of plasmepsin II (Silva *et al* 1996) has expedited the discovery of inhibitors of this enzyme, and combinatorial approaches have recently identified potent and selective inhibitors (Carroll *et al* 1998a,b; Haque *et al* 1999).

18.5.3 Plasmodial serine proteases

Genes encoding two *P. falciparum* serine proteases have recently been identified. PfSUB-1 (Blackman *et al* 1998) and PfSUB-2 (Barale *et al* 1999) encode subtilisin-like serine proteases. Both proteases are expressed in erythrocytic-stage parasites and concentrated in secretory organelles of invasive merozoites, and PfSUB-1 is released from merozoites in a truncated, soluble form. These data suggest that the plasmodial serine proteases play key roles in parasite invasion of erythrocytes. This process is blocked by serine protease inhibitors (Hadley *et al* 1983; Braun-Breton *et al* 1992).

A likely role for the plasmodial serine proteases is the processing of parasite proteins required for erythrocyte invasion. Multiple proteins of mature schizonts and merozoites are processed immediately before or during erythrocyte rupture and invasion. In the best characterized case, a major surface protein of merozoites (MSP-1), is processed in a series of steps to a carboxy-terminal surface portion which appears to be necessary for successful erythrocyte invasion (Blackman *et al* 1991). This final processing event appears to be due to the action of a membrane-bound calcium-dependent serine protease (Blackman and Holder 1992). The location of PfSUB-1 and PfSUB-2 in merozoites and their secretion after merozoite release suggest that one or both of these enzymes are responsible for cleavages of MSP-1 that are required for successful erythocyte invasion. Thus, the serine proteases are a new potential target for chemotherapy.

18.5.4 Plasmodial metalloproteases

A *P. falciparum* food vacuole metalloprotease, falcilysin, was recently purified (Eggleson *et al* 1999). The sequence of falcilysin shows that it is a member of the M16 family of metallopeptidases. Falcilysin cannot cleave native hemoglobin or denatured globin, but it cleaves relatively small (up to 20 amino acid) hemoglobin fragments at polar residues. Thus, it appears to act downstream of the plasmepsins and falcipain in the hydrolysis of hemoglobin. A *P. vivax* heat shock protein was found to contain metalloprotease sequence motifs, and sequence comparisons showed it to be a member of the M1 family, which consists of aminopeptidases (Fakruddin *et al* 1997).

18.5.5 Plasmodial aminopeptidases

Parasite aminopeptidase activity has been identified by a number of groups (Vander Jagt *et al* 1984; Curley *et al* 1994; Kolakovich *et al* 1997; Nankya-Kitaka *et al* 1998). This neutral activity has been identified in parasite, but not food vacuole lysates, suggesting a cytosolic site of action (Curley *et al* 1994; Kolakovich *et al* 1997). Recent studies suggest that the ultimate steps in hemoglobin hydrolysis occur in the parasite cytoplasm, as the incubation of hemoglobin with *P. falciparum* food vacuole lysates generated multiple peptide fragments, but not free amino acids (Kolakovich *et al* 1997). It was hypothesized that cytosolic proteases generate free amino acids after globin peptides are transported from the food vacuole. One or more aminopeptidases likely plays a part in this process. The *P. vivax* metallo-aminopeptidase gene that was recently identified may represent a *P. vivax* homolog of these proteins (Fakruddin *et al* 1997). The aminopeptidase inhibitors bestatin and nitrobestatin blocked the growth of cultured *P. falciparum* parasites, suggesting that the aminopeptidase is another potential chemotherapeutic target (Nankya-Kitaka *et al* 1998).

18.5.6 Plasmodial cysteine protease-like proteins (SERA and homologs)

The serine-repeat antigen (Bzik *et al* 1988) is under study as a potential component of a malaria vaccine. Portions of SERA and its homolog SERPH (Knapp *et al* 1991) have limited sequence homology with cysteine proteases, particularly near highly conserved active site residues. Recent studies have identified eight contiguous SERA genes on *P. falciparum* chromosome 2 (Gardner *et al* 1998) and multiple SERA homolog from other plasmodial species (Gor *et al* 1998). SERAs can be subdivided into proteins that have replaced the canonical active site cysteine with a serine and others that have conservation of all active site cysteine protease residues (Gor *et al* 1998). It is likely that a subset of SERAs are unusual cysteine proteases of mature parasites. As cysteine protease inhibitors block erythrocyte rupture by mature parasites (Hadley *et al* 1983; Lyon and Haynes 1986) and the SERAs are located in the vacuole surrounding mature schizonts (Delplace *et al* 1987), these proteins may be responsible for proteolytic cleavages required for erythrocyte rupture or invasion.

18.5.7 Role of host protease activity in erythrocyte rupture

The circulating serine protease urokinase was recently shown to bind to the surface of *P. falciparum*-infected erythrocytes, and the depletion of urokinase from parasite culture medium inhibited erythrocyte rupture by mature schizonts (Roggwiller *et al* 1997). This inhibition was reversed by the addition of exogenous urokinase, suggesting that urokinase activity is required for erythrocyte rupture. However, murine malaria parasites infecting mice deficient in urokinase, tissue plasminogen activator, or plasminogen replicated as efficiently as those infecting wild-type mice, arguing against a role for host plasminogen activators in the parasite life cycle (Rosenthal *et al* 1998).

18.5.8 Plasmodial proteasome

A proteasome of malaria parasites has not yet been reported. However, the proteasome inhibitor lactacystin inhibited the development of cultured *P. falciparum* parasites. Lactacystin also inhibited the exoerythrocytic development of murine malaria parasites and reduced the parasitemias of *P. berghei*-infected rats, although the therapeutic index of this therapy was low (Gantt *et al* 1998).

18.6 *ENTAMOEBA HISTOLYTICA*

E. histolytica causes colitis, dysentery, and serious extraintestinal disease including liver abscesses. However, only a small fraction of those infected with amebic parasites develops clinical disease. Among potential amebic virulence factors are proteases, including well characterized cysteine proteases.

18.6.1 Cysteine protease of *E. histolytica*

Cysteine protease activities of *E. histolytica* with broad pH optima, named amoebapain or histolysin, were reported by a number of investigators (Lushbaugh *et al* 1985; Keene *et al* 1986; Luaces and Barrett 1988; Scholze and Tannich 1994). The activities were localized to both the ameba cell surface and to subcellular vesicles (Scholze *et al* 1992). Activity was cathepsin B-like, with preference for substrates with arginine at the P1 and P2 positions (Scholze and Tannich 1994), but molecular analyses have identified only cathepsin L-like sequences. Six distinct *E. histolytica* cysteine protease genes (*ehcp1–6*) have been identified (Tannich *et al* 1991; Reed *et al* 1993; Bruchhaus *et al* 1996). It has been hypothesized that three cysteine proteases account for nearly all *E. histolytica* cysteine protease activity, and that *ehcp1* encodes amoebapain, *ehcp2* encodes histolysin, and *ehcp5* encodes a third major cysteine protease that is a key virulence determinant (Bruchhaus *et al* 1996).

Purified *E. histolytica* cysteine proteases degraded extracellular matrix components (Schulte and Scholze 1989; Keene *et al* 1986), suggesting that they may mediate the invasion of tissues by trophozoites. Supporting this possibility, purified cysteine proteases produced a cytopathic effect in cultured mammalian cells (Lushbaugh *et al* 1985; Keene *et al* 1986; Luaces and Barrett 1988). The virulence of laboratory strains has also been correlated with the protease activity of *Entamoeba* isolates (Gadasi and Kobiler 1983; Keene *et al* 1986; Luaces and Barrett 1988; Keene *et al* 1990). In studies of clinical isolates, extracellular cysteine protease activity was significantly greater in patients with colitis or liver abscess than in those with asymptomatic carriage of *Entamoeba* (Reed *et al* 1989).

The species *E. dispar* now denotes organisms that are morphologically identical to *E. histolytica* and colonize humans, but are nonpathogenic, probably due to limited invasiveness (Diamond and Clark 1993). Older reports noted that pathogenic and nonpathogenic strains expressed different proteases (Tannich *et al* 1991; Reed *et al* 1993), but they disagreed as to which genes differed in expression (Reed *et al* 1993; Bruchhaus and

Tannich 1996; Mirelman *et al* 1996). The most detailed analysis of the six known *E. histolytica* cysteine protease genes has suggested that at least three of the *E. histolytica* proteases have close homologues in *E. dispar*, but that two highly expressed *E. histolytica* proteases, EHCP1 and EHCP5, are not expressed in *E. dispar* (Bruchhaus *et al* 1996). In addition, EHCP5 was shown to be membrane associated and to produce a cytopathic effect on cultured cells (Jacobs *et al* 1998). It is intriguing to speculate that unique proteases may mediate cellular invasion and be specific virulence determinants of *E. histolytica*, although our current understanding of the roles of different proteases in this process remains incomplete.

E. histolytica cysteine proteases can activate complement by degrading C3 (Reed *et al* 1989) and can degrade the anaphylatoxic complement components C3a and C5a (Reed *et al* 1995). Complement activation by this mechanism leads to the lysis of nonpathogenic, but not pathogenic parasite isolates (Reed and Gigli 1990). Thus, the protease activity appears to contribute to the host immune response against nonpathogenic strains, but may inhibit an effective response, via degradation of C3a and C5a, against pathogenic strains. The cysteine protease activity also degrades IgA (Kelsall and Ravdin 1993) and IgG (Tran *et al* 1998), possibly explaining the lack of protective antibody responses against *E. histolytica*.

The inhibition of *E. histolytica* cysteine proteases may be a promising modality of chemotherapy. Antisense inhibition of cysteine protease expression did not affect *E. histolytica* cytopathic activity, but did inhibit parasite phagocytosis (Ankri *et al* 1998) and liver abscess formation in hamsters (Ankri *et al* 1999). Cysteine protease inhibitors blocked the cytopathic effect caused by parasite extracts and purified *E. histolytica* proteases (Lushbaugh *et al* 1985; Keene *et al* 1990). In studies with infected severe combined immunodeficiency mice, antibodies to EHCP1 identified the protease in extracellular amebic abscess fluid, and treatment with the cysteine protease inhibitor E-64 markedly reduced liver abscess formation (Stanley *et al* 1995). In other studies, allicin, from garlic, blocked both the cysteine protease activity of cultured *E. histolytica* trophozoites and the ability of these parasites to destroy monolayers of mammalian cells (Ankri *et al* 1997).

18.6.2 Other proteases of *E. histolytica*

E. histolytica also expresses collagenase activity which is independent of the cysteine protease activities discussed above (Munoz *et al* 1982). The collagenase is contained in electron-dense granules that are released upon incubation of trophozoites with collagen (Leon *et al* 1997), but it has not yet been well characterized. Two high molecular weight proteases of *E. histolytica* were recently identified (Scholze *et al* 1996). One of the activities was attributed to an *E. histolytica* proteasome based on its electrophoretic mobility and reactivity with an anti-proteasome antibody. The other activity was due to a complex of six subunits that had unique proteolytic activity (Scholze *et al* 1996). An *E. histolytica* proteasome α-subunit gene has also been identified (Ramos *et al* 1997).

18.7 *GIARDIA LAMBLIA*

Giardia lamblia is the most common parasite identified as a cause of diarrhea in humans. Cysteine protease activity of *G. lamblia* has been identified (Hare *et al* 1989; Parenti 1989) and localized to lysosome-like cytoplasmic vacuoles that move to the parasite periphery at the time of excystation and then release their contents between the cyst wall and emerging trophozoites (Lindmark 1988). Recently, three *G. lamblia* cysteine protease genes, each with a cathepsin B-like sequence, were characterized, and one gene product was purified and characterized as a neutral papain-family protease (Ward *et al* 1997). At least one of the *Giardia* cysteine proteases appears to be required for excystation, as cysteine protease inhibitors blocked this process in cultured parasites (Ward *et al* 1997). The *G. lamblia* cysteine proteases may thus be appropriate chemotherapeutic targets. A *G. lamblia* 20S proteasome has also recently been characterized (Emmerlich *et al* 1999).

18.8 *CRYPTOSPORIDIUM PARVUM*

Cryptosporidium parvum is an important cause of severe diarrhea in patients with AIDS and a cause of water-borne outbreaks of diarrhea in immunocompetent individuals. No effective therapy for cryptosporidiosis is currently available. A number of *C. parvum* protease activities have been identified, but most have not been well characterized. A *C. parvum* cysteine protease gene has been identified (C. Petersen, personal communication). An integral membrane aminopeptidase was identified in freshly excysted *C. parvum* sporozoites (Okhuysen *et al* 1994). The peptidase cleaved amino-terminal arginines. Aminopeptidase inhibitors and a metal chelator, but not endopeptidase inhibitors, blocked aminopeptidase activity and inhibited excystation (Okhuysen *et al* 1996). This enzyme is a potential therapeutic target.

18.8.1 *Trichomonas vaginalis*

T. vaginalis is a flagellated protozoan that is a major cause of sexually-transmitted vaginitis and urethritis and can cause serious gynecological complications. Multiple trichomonal cysteine protease activities have been identified (North *et al* 1990). Similar protease activities were identified in flagellated forms of the parasite, which have been most studied, and in amoeboid forms, which probably interact with vaginal epithelial cells (Scott *et al* 1995a). Molecular analyses have identified genes encoding four cathepsin L-like cysteine proteases of *T. vaginalis* (Mallinson *et al* 1994). The functions of the trichomonal proteases are unknown. Proteases are released by cultured parasites, possibly from an endosomal compartment (Scott *et al* 1995b). The proteases may thus act extracellularly, perhaps to mediate attachment of parasites to vaginal epithelium. Studies with cysteine protease inhibitors suggest that not all of the trichomonal proteases are essential, but that the inhibition of certain proteases blocks parasite growth (Irvine *et al* 1997). Thus, inhibitors of trichomonal cysteine proteases may be effective chemotherapeutic agents.

18.9 *TOXOPLASMA GONDII*

Infection with *T. gondii* is usually asymptomatic, but it can cause an acute febrile illness, severe disease in immunocompromised individuals, and serious congenital disease. Two proteases were partially purified from *T. gondii*, an acid cysteine protease and a neutral ATP-dependent serine protease (Choi *et al* 1989). A neutral cathepsin B-like activity has also been identified in *T. gondii* lysates (S.L. Reed personal communication) and two

T. gondii cysteine protease genes have been identified (S.L. Reed personal communication). In other studies, serine protease inhibitors blocked the invasion of host cells by *T. gondii*, suggesting that a serine protease is required for host cell invasion (Conseil *et al* 1999).

NOTES

Recent advances, including the availability of extensive genomic sequence data for protozoan parasites, have identified multiple new putative proteases. Thus, additional enzymes are likely to contribute to key activities in a number of the organisms discussed in this review. In the case of malaria parasites, see Shenai *et al* (2000) for an updated appreciation of the roles of multiple cysteine proteases in parasite biology.

ACKNOWLEDGMENTS

Work in the author's laboratory is supported by the National Institutes of Health.

REFERENCES

Ankri, S., Miron, T., Rabinkov, A., Wilchek, M. and Mirelman, D. (1997) Allicin from garlic strongly inhibits cysteine proteinases and cytopathic effects of *Entamoeba histolytica*. *Antimicrobial Agents and Chemotherapy*, **41**, 2286–2288.

Ankri, S., Stolarsky, T. and Mirelman, D. (1998) Antisense inhibition of expression of cysteine proteinases does not affect *Entamoeba histolytica* cytopathic or haemolytic activity but inhibits phagocytosis. *Molecular Microbiology*, **28**, 777–785.

Ankri, S., Stolarsky, T., Bracha, R., Padilla-Vaca, F. and Mirelman, D. (1999) Antisense inhibition of expression of cysteine proteinases affects *Entamoeba histolytica*-induced formation of liver abscess in hamsters. *Infection and Immunity*, **67**, 421–422.

Asawamahasakda, W., Ittarat, I., Chang, C.-C., McElroy, P. and Meshnick, S.R. (1994) Effects of antimalarials and protease inhibitors on plasmodial hemozoin production. *Molecular and Biochemical Parasitology*, **67**, 183–191.

Bailly, E., Jambou, R., Savel, J. and Jaureguiberry, G. (1992) *Plasmodium falciparum*: differential sensitivity *in vitro* to E-64 (cysteine protease inhibitor) and pepstatin A (aspartyl protease inhibitor). *Journal of Protozoology*, **39**, 593–599.

Barale, J.C., Blisnick, T., Fujioka, H., Alzari, P.M., Aikawa, M., Braun-Breton, C. *et al* (1999) *Plasmodium falciparum* subtilisin-like protease 2, a merozoite candidate for the merozoite surface protein 1–42 maturase. *Proceedings of the National Academy of Sciences of the U.S.A.*, **96**, 6445–6450.

Bart, G., Frame, M.J., Carter, R., Coombs, G.H. and Mottram, J.C. (1997) Cathepsin B-like cysteine proteinase-deficient mutants of *Leishmania mexicana*. *Molecular and Biochemical Parasitology*, **88**, 53–61.

Blackman, M.J., Whittle, H. and Holder, A.A. (1991) Processing of the *Plasmodium falciparum* major merozoite surface protein-1: identification of a 33-kilodalton secondary processing product which is shed prior to erythrocyte invasion. *Molecular and Biochemical Parasitology* , **49**, 35–44.

Blackman, M.J. and Holder, A.A. (1992) Secondary processing of the *Plasmodium falciparum* merozoite surface protein-1 (MSP1) by a calcium-dependent membrane-bound serine protease: shedding of $MSP1_{33}$ as a noncovalently associated complex with other fragments of the MSP1. *Molecular and Biochemical Parasitology*, **50**, 307–316.

Blackman, M.J., Fujioka, H., Stafford, W.H., Sajid, M., Clough, B., Fleck, S.L. *et al* (1998) A subtilisin-like protein in secretory organelles of *Plasmodium falciparum* merozoites. *Journal of Biological Chemistry*, **273**, 23398–23409.

Bonaldo, M.C., d'Escoffier, L.N., Salles, J.M. and Goldenberg, S. (1991) Characterization and expression of proteases during *Trypanosoma cruzi* metacyclogenesis. *Experimental Parasitology*, **73**, 44–51.

Bontempi, E., Martinez, J. and Cazzulo, J.J. (1989) Subcellular localization of a cysteine proteinase from *Trypanosoma cruzi*. *Molecular and Biochemical Parasitology*, **33**, 43–47.

Bouvier, J., Etges, R.J. and Bordier, C. (1985) Identification and purification of membrane and soluble forms of the major surface protein of *Leishmania* promastigotes. *Journal of Biological Chemistry*, **260**, 15504–15509.

Bouvier, J., Bordier, C., Vogel, H., Reichelt, R. and Etges, R. (1989) Characterization of the promastigote surface protease of *Leishmania* as a membrane-bound zinc endopeptidase. *Molecular and Biochemical Parasitology*, **37**, 235–245.

Bouvier, J., Schneider, P. and Etges, R. (1995) Leishmanolysin: surface metalloproteinase of *Leishmania*. *Methods in Enzymology*, **248**, 614–633.

Braun-Breton, C., Blisnick, T., Jouin, H., Barale, J.C., Rabilloud, T., Langsley, G. and Pereira da Silva, L.H. (1992) *Plasmodium chabaudi* p68 serine protease activity required for merozoite entry into mouse erythrocytes. *Proceedings of the National Academy of Sciences of the U.S.A.*, **89**, 9647–9651.

Brittingham, A., Morrison, C.J., McMaster, W.R., McGwire, B.S., Chang, K.P. and Mosser, D.M. (1995) Role of the Leishmania surface protease gp63 in complement fixation, cell adhesion, and resistance to complement-mediated lysis. *Journal of Immunology*, **155**, 3102–3111.

Bruchhaus, I., Jacobs, T., Leippe, M. and Tannich, E. (1996) *Entamoeba histolytica* and *Entamoeba dispar*: differences in numbers and expression of cysteine proteinase genes. *Molecular Microbiology*, **22**, 255–263.

Bruchhaus, I. and Tannich, E. (1996) A gene highly homologous to ACP1 encoding cysteine proteinase 3 in *Entamoeba histolytica* is present and expressed in *E. dispar*. *Parasitology Research*, **82**, 189–192.

Burleigh, B.A., Caler, E.V., Webster, P. and Andrews, N.W. (1997) A cytosolic serine endopeptidase from *Trypanosoma cruzi* is required for the generation of Ca^{2+} signaling in mammalian cells. *Journal of Cell Biology*, **136**, 609–620.

Button, L.L., Wilson, G., Astell, C.R. and McMaster, W.R. (1993) Recombinant *Leishmania* surface glycoprotein GP63 is secreted in the baculovirus expression system as a latent metalloproteinase. *Gene*, **134**, 75–81.

Bzik, D.J., Li, W.-B., Horii, T. and Inselburg, J. (1988) Amino acid sequence of the serine-repeat antigen (SERA) of *Plasmodium falciparum* determined from cloned cDNA. *Molecular and Biochemical Parasitology* , **30**, 279–288.

Campetella, O., Henriksson, J., Aslund, L., Frasch, A.C., Pettersson, U. and Cazzulo, J.J. (1992) The major cysteine proteinase (cruzipain) from *Trypanosoma cruzi* is encoded by multiple polymorphic tandemly organized genes located on different chromosomes. *Molecular and Biochemical Parasitology*, **50**, 225–234.

Carroll, C.D., Patel, H., Johnson, T.O., Guo, T., Orlowski, M., He, Z.M. *et al* (1998a) Identification of potent inhibitors of *Plasmodium falciparum* plasmepsin II from an encoded statine combinatorial library. *Bioorganic & Medicinal Chemistry Letters*, **8**, 2315–2320.

Carroll, C.D., Johnson, T.O., Tao, S., Lauri, G., Orlowski, M., Gluzman, I.Y. *et al* (1998b) Evaluation of a structure-based statine cyclic diamino amide encoded combinatorial library against plasmepsin II and cathepsin D. *Bioorganic & Medicinal Chemistry Letters*, **8**, 3203–3206.

Cazzulo, J.J., Couso, R., Raimondi, A., Wernstedt, C. and Hellman, U. (1989) Further characterization and partial amino acid sequence of a cysteine proteinase from *Trypanosoma cruzi*. *Molecular and Biochemical Parasitology*, **33**, 33–41.

Chan, V.J., Selzer, P.M., McKerrow, J.H. and Sakanari, J.A. (1999) Expression and alteration of the S2 subsite of the leishmania major cathepsin B-like cysteine protease. *Biochemical Journal*, **340**, 113–117.

Chang, K.P., Chaudhuri, G. and Fong, D. (1990) Molecular determinants of *Leishmania* virulence. *Annual Review of Microbiology*, **44**, 499–529.

Chaudhuri, G. and Chang, K.P. (1988) Acid protease activity of a major surface membrane glycoprotein (gp63) from *Leishmania mexicana* promastigotes. *Molecular and Biochemical Parasitology*, **27**, 43–52.

Chaudhuri, G., Chaudhuri, M., Pan, A. and Chang, K.P. (1989) Surface acid proteinase (gp63) of *Leishmania mexicana*. A metalloenzyme capable of protecting liposome-encapsulated proteins from phagolysosomal degradation by macrophages. *Journal of Biological Chemistry*, **264**, 7483–7489.

Choi, W.Y., Nam, H.W. and Youn, J.H. (1989) Characterization of proteases of *Toxoplasma gondii*. *Kisaengchunghak Chapchi*, **27**, 161–170.

Conseil, V., Soete, M. and Dubremetz, J.F. (1999) Serine protease inhibitors block invasion of host cells by *Toxoplasma gondii*. *Antimicrobial Agents and Chemotherapy*, **43**, 1358–1361.

Coombs, G.H. and Baxter, J. (1984) Inhibition of *Leishmania* amastigote growth by antipain and leupeptin. *Annals of Tropical Medicine and Parasitology*, **78**, 21–24.

Curley, G.P., O'Donovan, S.M., McNally, J., Mullally, M., O'Hara, H., Troy, A. *et al* (1994) Aminopeptidases from *Plasmodium falciparum*, *Plasmodium chabaudi*, and *Plasmodium berghei*. *Journal of Eukaryotic Microbiology*, **41**, 119–123.

Dame, J.B., Reddy, G.R., Yowell, C.A., Dunn, B.M., Kay, J. and Berry, C. (1994) Sequence, expression and modeled structure of an aspartic proteinase from the human malaria parasite *Plasmodium falciparum*. *Molecular and Biochemical Parasitology* , **64**, 177–190.

de Andrade, A.S., Santoro, M.M., de Melo, M.N. and Mares-Guia, M. (1998) Leishmania (*Leishmania*) amazonensis: purification and enzymatic characterization of a soluble serine oligopeptidase from promastigotes. *Experimental Parasitology*, **89**, 153–160.

Delplace, P., Fortier, B., Tronchin, G., Dubremetz, J. and Vernes, A. (1987) Localization, biosynthesis, processing and isolation of a major 126 kDa antigen of the parasitophorous vacuole of *Plasmodium falciparum*. *Molecular and Biochemical Parasitology* , **23**, 193–201.

Diamond, L.S. and Clark, C.G. (1993) A redescription of *Entamoeba histolytica* Schaudinn, 1903 (Emended Walker, 1911) separating it from *Entamoeba dispar* Brumpt, 1925. *Eukaryotic Microbiology*, **40**, 340–344.

Dluzewski, A.R., Rangachari, K., Wilson, R.J.M. and Gratzer, W.B. (1986) *Plasmodium falciparum*: protease inhibitors and inhibition of erythrocyte invasion. *Experimental Parasitology*, **62**, 416–422.

Domínguez, J.N., López, S., Charris, J., Iarruso, L., Lobo, G., Semenov, A. *et al* (1997) Synthesis and antimalarial effects of phenothiazine inhibitors of a *Plasmodium falciparum* cysteine protease. *Journal of Medicinal Chemistry*, **40**, 2726–2732.

Duboise, S.M., Vannier-Santos, M.A., Costa-Pinto, D., Rivas, L., Pan, A.A., Traub-Cseko, Y. *et al* (1994) The biosynthesis, processing, and immunolocalization of *Leishmania pifanoi* amastigote cysteine proteinases. *Molecular and Biochemical Parasitology*, **68**, 119–132.

Eakin, A.E., Bouvier, J., Sakanari, J.A., Craik, C.S. and McKerrow, J.H. (1990) Amplification and sequencing of genomic DNA fragments encoding cysteine proteases from protozoan parasites. *Molecular and Biochemical Parasitology*, **39**, 1–8.

Eakin, A.E., Mills, A.A., Harth, G., McKerrow, J.H. and Craik, C.S. (1992) The sequence, organization, and expression of the major cysteine protease (cruzain) from *Trypanosoma cruzi*. *Journal of Biological Chemistry*, **267**, 7411–7420.

Eggleson, K.K., Duffin, K.L. and Goldberg, D.E. (1999) Identification and characterization of falcilysin, a metallopeptidase involved in hemoglobin catabolism within the malaria parasite *Plasmodium falciparum*. *Journal of Biological Chemistry*, **272**, 32411–32417.

El-Sayed, N.M.A. and Donelson, J.E. (1997) African trypanosomes have differentially expressed genes encoding homologues of the leishmania GP63 surface protease. *Journal of Biological Chemistry*, **272**, 26742–26748.

Emmerlich, V., Santarius, U., Bakker-Grunwald, T. and Scholze, H. (1999) Isolation and subunit composition of the 20S proteasome of *Giardia lamblia*. *Molecular and Biochemical Parasitology*, **100**, 131–134.

Engel, J.C., Doyle, P.S., Hsieh, I. and McKerrow, J.H. (1998a) Cysteine protease inhibitors cure an experimental *Trypanosoma cruzi* infection. *Journal of Experimental Medicine*, **188**, 725–734.

Engel, J.C., Doyle, P.S., Palmer, J., Hsieh, I., Bainton, D.F. and McKerrow, J.H. (1998b) Cysteine protease inhibitors alter Golgi complex ultrastructure and function in *Trypanosoma cruzi*. *Journal of Cell Science*, **111**, 597–606.

Fakruddin, J.M., Biswas, S. and Sharma, Y.D. (1997) Identification of a *Plasmodium vivax* heat-shock protein which contains a metalloprotease sequence motif. *Molecular and Biochemical Parasitology*, **90**, 387–390.

Francis, S.E., Gluzman, I.Y., Oksman, A., Knickerbocker, A., Mueller, R., Bryant, M.L. *et al* (1994) Molecular characterization and inhibition of a *Plasmodium falciparum* aspartic hemoglobinase. *EMBO Journal*, **13**, 306–317.

Francis, S.E., Banerjee, R. and Goldberg, D.E. (1997) Biosynthesis and maturation of the malaria aspartic hemoglobinases plasmepsins I and II. *Journal of Biological Chemistry*, **272**, 14961–14968.

Francis, S.E., Sullivan, D.J. and Goldberg, D.E. (1997) Hemoglobin metabolism in the malaria parasite *Plasmodium falciparum*. *Annual Review of Microbiology*, **51**, 97–123.

Franke de Cazzulo, B.M., Martinez, J., North, M.J., Coombs, G.H. and Cazzulo, J.J. (1994) Effects of proteinase inhibitors on the growth and differentiation of *Trypanosoma cruzi*. *FEMS Microbiological Letters*, **124**, 81–86.

Funk, V.A., Jardim, A. and Olafson, R.W. (1994) An investigation into the significance of the N-linked oligosaccharides of *Leishmania* gp63. *Molecular and Biochemical Parasitology*, **63**, 23–35.

Gadasi, H. and Kobiler, D. (1983) *Entamoeba histolytica*: correlation between virulence and content of proteolytic enzymes. *Experimental Parasitology*, **55**, 105–110.

Gamboa de Domínguez, N.D. and Rosenthal, P.J. (1996) Cysteine proteinase inhibitors block early steps in hemoglobin degradation by cultured malaria parasites. *Blood*, **87**, 4448–4454.

Gantt, S.M., Myung, J.M., Briones, M.R., Li, W.D., Corey, E.J., Omura, S. *et al* (1998) Proteasome inhibitors block development of *Plasmodium* spp. *Antimicrobial Agents and Chemotherapy*, **42**, 2731–2738.

Gardner, M.J., Tettelin, H., Carucci, D.J., Cummings, L.M., Aravind, L., Koonin, E.V. *et al* (1998) Chromosome 2 sequence of the human malaria parasite *Plasmodium falciparum*. *Science*, **282**, 1126–1132.

Gillmor, S.A., Craik, C.S. and Fletterick, R.J. (1997) Structural determinants of specificity in the cysteine protease cruzain. *Protein Science*, **6**, 1603–1611.

Gluzman, I.Y., Francis, S.E., Oksman, A., Smith, C.E., Duffin, K.L. and Goldberg, D.E. (1994) Order and specificity of the *Plasmodium falciparum* hemoglobin degradation pathway. *Journal of Clinical Investigation*, **93**, 1602–1608.

Goldberg, D.E., Slater, A.F.G., Beavis, R., Chait, B., Cerami, A. and Henderson, G.B. (1991) Hemoglobin degradation in the human malaria pathogen *Plasmodium falciparum*: a catabolic pathway initiated by a specific aspartic protease. *Journal of Experimental Medicine*, **173**, 961–969.

Gonzalez, J., Ramalho-Pinto, F.J., Frevert, U., Ghiso, J., Tomlinson, S., Scharfstein, J. *et al* (1996) Proteasome activity is required for the stage-specific transformation of a protozoan parasite. *Journal of Experimental Medicine*, **184**, 1909–1918.

Gor, D.O., Li, A.C., Wiser, M.F. and Rosenthal, P.J. (1998) Plasmodial serine repeat antigen homologues with properties of schizont cysteine proteases. *Molecular and Biochemical Parasitology*, **95**, 153–158.

Hadley, T., Aikawa, M. and Miller, L.H. (1983) *Plasmodium knowlesi*: studies on invasion of rhesus erythrocytes by merozoites in the presence of protease inhibitors. *Experimental Parasitology*, **55**, 306–311.

Haque, T.S., Skillman, A.G., Lee, C.E., Habashita, H., Gluzman, I.Y., Ewing, T.J. *et al* (1999) Potent, low-molecular-weight non-peptide inhibitors of malarial aspartyl protease plasmepsin II. *Journal of Medicinal Chemistry*, **42**, 1428–1440.

Hare, D.F., Jarroll, E.L. and Lindmark, D.G. (1989) *Giardia lamblia*: characterization of proteinase activity in trophozoites. *Experimental Parasitology*, **68**, 168–175.

Harth, G., Andrews, N., Mills, A.A., Engel, J.C., Smith, R. and McKerrow, J.H. (1993) Peptide-fluoromethyl ketones arrest intracellular replication and intercellular transmission of *Trypanosoma cruzi*. *Molecular and Biochemical Parasitology*, **58**, 17–24.

Hellman, U., Wernstedt, C. and Cazzulo, J.J. (1991) Self-proteolysis of the cysteine proteinase, cruzipain, from *Trypanosoma cruzi* gives a major fragment corresponding to its carboxy-terminal domain. *Molecular and Biochemical Parasitology*, **44**, 15–21.

Hill, J., Tyas, L., Phylip, L.H., Kay, J., Dunn, B.M. and Berry, C. (1994) High level expression and characterisation of plasmepsin II, an aspartic proteinase from *Plasmodium falciparum*. *FEBS Letters*, **352**, 155–158.

Hua, S., To, W.Y., Nguyen, T.T., Wong, M.L. and Wang, C.C. (1996) Purification and characterization of proteasomes from *Trypanosoma brucei*. *Molecular and Biochemical Parasitology*, **78**, 33–46.

Huete-Perez, J.A., Engel, J.C., Brinen, L.S., Mottram, J.C. and McKerrow, J.H. (1999) Protease trafficking in two primitive eukaryotes is mediated by a prodomain protein motif. *Journal of Biological Chemistry*, **274**, 16249–16256.

Ilg, T., Harbecke, D. and Overath, P. (1993) The lysosomal gp63-related protein in *Leishmania mexicana* amastigotes is a soluble metalloproteinase with an acidic pH optimum. *FEBS Letters*, **327**, 103–107.

Irvine, J.W., North, M.J. and Coombs, G.H. (1997) Use of inhibitors to identify essential cysteine proteinases of *Trichomonas vaginalis*. *FEMS Microbiology Letters*, **149**, 45–50.

Jacobs, T., Bruchhaus, I., Dandekar, T., Tannich, E. and Leippe, M. (1998) Isolation and molecular characterization of a surface-bound proteinase of *Entamoeba histolytica*. *Molecular Microbiology*, **27**, 269–276.

Joshi, P.B., Sacks, D.L., Modi, G. and McMaster, W.R. (1998) Targeted gene deletion of *Leishmania major* genes encoding developmental stage-specific leishmanolysin (GP63). *Molecular Microbiology*, **27**, 519–530.

Kamchonwongpaisan, S., Samoff, E. and Meshnick, S.R. (1997) Identification of hemoglobin degradation products in *Plasmodium falciparum*. *Molecular and Biochemical Parasitology*, **86**, 179–186.

Keene, W.E., Petitt, M.G., Allen, S. and McKerrow, J.H. (1986) The major neutral proteinase of *Entamoeba histolytica*. *Journal of Experimental Medicine*, **163**, 536–549.

Keene, W.E., Hidalgo, M.E., Orozco, E. and McKerrow, J.H. (1990) *Entamoeba histolytica*: correlation of the cytopathic effect of virulent trophozoites with secretion of a cysteine proteinase. *Experimental Parasitology*, **71**, 199–206.

Kelsall, B.L. and Ravdin, J.I. (1993) Degradation of human IgA by *Entamoeba histolytica*. *Journal of Infectious Diseases*, **168**, 1319–1322.

Knapp, B., Nau, U., Hundt, E. and Küpper, H.A. (1991) A new blood stage antigen of *Plasmodium falciparum* highly homologous to the serine-stretch protein SERP. *Molecular and Biochemical Parasitology*, **44**, 1–14.

Kolakovich, K.A., Gluzman, I.Y., Duffin, K.L. and Goldberg, D.E. (1997) Generation of hemoglobin peptides in the acidic digestive vacuole of *Plasmodium falciparum* implicates peptide transport in amino acid production. *Molecular and Biochemical Parasitology*, **87**, 123–135.

Le Bonniec, S., Deregnaucourt, C., Redeker, V., Banerjee, R., Grellier, P., Goldberg, D.E. and Schrevel, J. (1999) Plasmepsin II, an acidic hemoglobinase from the *Plasmodium falciparum* food vacuole, is active at neutral pH on the host erythrocyte membrane skeleton. *Journal of Biological Chemistry*, **274**, 14218–14223.

Leon, G., Fiori, C., Das, P., Moreno, M., Tovar, R., Sanchez-Salas, J.L. *et al* (1997) Electron probe analysis and biochemical characterization of electron-dense granules secreted by *Entamoeba histolytica*. *Molecular and Biochemical Parasitology*, **85**, 233–242.

Li, R., Kenyon, G.L., Cohen, F.E., Chen, X., Gong, B., Dominguez, J.N. *et al* (1995) *In vitro* antimalarial activity of chalcones and their derivatives. *Journal of Medicinal Chemistry*, **38**, 5031–5037.

Li, R., Chen, X., Gong, B., Selzer, P.M., Li, Z., Davidson, E. *et al* (1996) Structure-based design of parasitic protease inhibitors. *Bioorganic & Medicinal Chemistry*, **4**, 1421–1427.

Lindmark, D.G. (1988) *Giardia lamblia*: localization of hydrolase activities in lysosome-like organelles of trophozoites. *Experimental Parasitology*, **65**, 141–147.

Liu, X. and Chang, K.P. (1992) Extrachromosomal genetic complementation of surface metalloproteinase (gp63)-deficient *Leishmania* increases their binding to macrophages. *Proceedings of the National Academy of Sciences of the U.S.A.*, **89**, 4991–4995.

Luaces, A.L. and Barrett, A.J. (1988) Affinity purification and biochemical characterization of histolysin, the major cysteine proteinase of *Entamoeba histolytica*. *Biochemical Journal*, **250**, 903–909.

Luker, K.E., Francis, S.E., Gluzman, I.Y. and Goldberg, D.E. (1996) Kinetic analysis of plasmepsins I and II, aspartic proteases of the *Plasmodium falciparum* digestive vacuole. *Molecular and Biochemical Parasitology*, **79**, 71–78.

Lushbaugh, W.B., Hofbauer, A.F. and Pittman, F.E. (1985) *Entamoeba histolytica*: purification of cathepsin B. *Experimental Parasitology*, **59**, 328–336.

Lyon, J.A. and Haynes, J.D. (1986) *Plasmodium falciparum* antigens synthesized by schizonts and stabilized at the merozoite surface when schizonts mature in the presence of protease inhibitors. *Journal of Immunology*, **136**, 2245–2251.

Mallinson, D.J., Lockwood, B.C., Coombs, G.H. and North, M.J. (1994) Identification and molecular cloning of four cysteine proteinase genes from the pathogenic protozoon *Trichomonas vaginalis*. *Microbiology*, **140**, 2725–2735.

McGrath, M.E., Eakin, A.E., Engel, J.C., McKerrow, J.H., Craik, C.S. and Fletterick, R.J. (1995) The crystal structure of cruzain: a therapeutic target for Chagas' disease. *Journal of Molecular Biology*, **247**, 251–259.

McGwire, B. and Chang, K.P. (1994) Genetic rescue of surface metalloproteinase (gp63)-deficiency-in *Leishmania amazonensis* variants increases their infection of macrophages at the early phase. *Molecular and Biochemical Parasitology*, **66**, 345–347.

McGwire, B.S. and Chang, K.P. (1996) Posttranslational regulation of a *Leishmania* HEXXH metalloprotease (gp63). The effects of site-specific mutagenesis of catalytic, zinc binding, N-glycosylation, and glycosyl phosphatidylinositol addition sites on N-terminal end cleavage, intracellular stability, and extracellular exit. *Journal of Biological Chemistry*, **271**, 7903–7909.

Medina-Acosta, E., Karess, R.E., Schwartz, H. and Russell, D.G. (1989) The promastigote surface protease (gp63) of *Leishmania* is expressed but differentially processed and localized in the amastigote stage. *Molecular and Biochemical Parasitology*, **37**, 263–273.

Meirelles, M.N., Juliano, L., Carmona, E., Silva, S.G., Costa, E.M., Murta, A.C. *et al* (1992) Inhibitors of the major cysteinyl proteinase (GP57/51) impair host cell invasion and arrest the intracellular development of *Trypanosoma cruzi in vitro*. *Molecular and Biochemical Parasitology*, **52**, 175–184.

Mirelman, D., Nuchamowitz, Y., Bohm-Gloning, B. and Walderich, B. (1996) A homologue of the cysteine proteinase gene (ACP1 or Eh-CPp3) of pathogenic *Entamoeba histolytica* is present in non-pathogenic *E. dispar* strains. *Molecular and Biochemical Parasitology*, **78**, 47–54.

Moon, R.P., Tyas, L., Certa, U., Rupp, K., Bur, D., Jacquet, C. *et al* (1997) Expression and characterisation of plasmepsin I from *Plasmodium falciparum*. *European Journal of Biochemistry*, **244**, 552–560.

Morty, R.E., Troeberg, L., Pike, R.N., Jones, R., Nickel, P., Lonsdale-Eccles, J.D. *et al* (1998) A trypanosome oligopeptidase as a target for the trypanocidal agents pentamidine, diminazene and suramin. *FEBS Letters*, **433**, 251–256.

Mottram, J.C., North, M.J., Barry, J.D. and Coombs, G.H. (1989) A cysteine proteinase cDNA from *Trypanosoma brucei* predicts an enzyme with an unusual C-terminal extension. *FEBS Letters*, **258**, 211–215.

Mottram, J.C., Robertson, C.D., Coombs, G.H. and Barry, J.D. (1992) A developmentally regulated cysteine proteinase gene of *Leishmania mexicana*. *Molecular Microbiology*, **6**, 1925–1932.

Mottram, J.C., Souza, A.E., Hutchison, J.E., Carter, R., Frame, M.J. and Coombs, G.H. (1996) Evidence from disruption of the lmcpb gene array of *Leishmania mexicana* that cysteine proteinases are virulence factors. *Proceedings of the National Academy of Sciences of the U.S.A.*, **93**, 6008–6013.

Mottram, J.C., Frame, M.J., Brooks, D.R., Tetley, L., Hutchison, J.E., Souza, A.E. *et al* (1997) The multiple cpb cysteine proteinase genes of *Leishmania mexicana* encode isoenzymes that differ in their stage regulation and substrate preferences. *Journal of Biological Chemistry*, **272**, 14285–14293.

Mottram, J.C., Brooks, D.R. and Coombs, G.H. (1998) Roles of cysteine proteinases of trypanosomes and Leishmania in host–parasite interactions. *Current Opinion in Microbiology*, **1**, 455–460.

Munoz, M.L., Calderon, J. and Rojkind, M. (1982) The collagenase of *Entamoeba histolytica*. *Journal of Experimental Medicine*, **155**, 42–51.

Murta, A.C., Persechini, P.M., Padron, T.d.S., de Souza, W., Guimaraes, J.A. and Scharfstein, J. (1990) Structural and functional identification of GP57/51 antigen of *Trypanosoma cruzi* as a cysteine proteinase. *Molecular and Biochemical Parasitology*, **43**, 27–38.

Mutomba, M.C., To, W.Y., Hyun, W.C. and Wang, C.C. (1997) Inhibition of proteasome activity blocks cell cycle progression at specific phase boundaries in African trypanosomes. *Molecular and Biochemical Parasitology*, **90**, 491–504.

Nankya-Kitaka, M.F., Curley, G.P., Gavigan, C.S., Bell, A. and Dalton, J.P. (1998) *Plasmodium chabaudi chabaudi* and *P. falciparum*: inhibition of aminopeptidase and parasite growth by bestatin and nitrobestatin. *Parasitology Research*, **84**, 552–558.

North, M.J., Robertson, C.D. and Coombs, G.H. (1990) The specificity of trichomonad cysteine proteinases analysed using fluorogenic substrates and specific inhibitors. *Molecular and Biochemical Parasitology*, **39**, 183–193.

Okhuysen, P.C., DuPont, H.L., Sterling, C.R. and Chappell, C.L. (1994) Arginine aminopeptidase, an integral membrane protein of the *Cryptosporidium parvum* sporozoite. *Infection and Immunity*, **62**, 4667–4670.

Okhuysen, P.C., Chappell, C.L., Kettner, C. and Sterling, C.R. (1996) *Cryptosporidium parvum* metalloaminopeptidase inhibitors prevent *in vitro* excystation. *Antimicrobial Agents and Chemotherapy*, **40**, 2781–2784.

Olson, J.E., Lee, G.K., Semenov, A. and Rosenthal, P.J. (1999) Antimalarial effects in mice of orally administered peptidyl cysteine protease inhibitors. *Bioorganic & Medicinal Chemistry*, **7**, 633–638.

Pamer, E.G., So, M. and Davis, C.E. (1989) Identification of a developmentally regulated cysteine protease of *Trypanosoma brucei*. *Molecular and Biochemical Parasitology*, **33**, 27–32.

Pamer, E.G., Davis, C.E. and So, M. (1991) Expression and deletion analysis of the *Trypanosoma brucei rhodesiense* cysteine protease in *Escherichia coli*. *Infection and Immunity*, **59**, 1074–1078.

Parenti, D.M. (1989) Characterization of a thiol proteinase in *Giardia lamblia*. *Journal of Infectious Disease*, **160**, 1076–1080.

Piras, M.M., Henriquez, D. and Piras, R. (1985) The effect of proteolytic enzymes and protease inhibitors on the interaction *Trypanosoma cruzi*-fibroblasts. *Molecular and Biochemical Parasitology*, **14**, 151–163.

Pupkis, M.F., Tetley, L. and Coombs, G.H. (1986) *Leishmania mexicana*: amastigote hydrolases in unusual lysosomes. *Experimental Parasitology*, **62**, 29–39.

Ramos, M.A., Stock, R.P., Sanchez-Lopez, R., Olvera, F., Lizardi, P.M. and Alagon, A. (1997) The *Entamoeba histolytica* proteasome alpha-subunit gene. *Molecular and Biochemical Parasitology*, **84**, 131–135.

Reed, S.L., Keene, W.E. and McKerrow, J.H. (1989) Thiol proteinase expression and pathogenicity of *Entamoeba histolytica*. *Journal of Clinical Microbiology*, **27**, 2772–2777.

Reed, S.L., Keene, W.E., McKerrow, J.H. and Gigli, I. (1989) Cleavage of C3 by a neutral cysteine proteinase of *Entamoeba histolytica*. *Journal of Immunology*, **143**, 189–195.

Reed, S.L. and Gigli, I. (1990) Lysis of complement-sensitive *Entamoeba histolytica* by activated terminal complement components. Initiation of complement activation by an extracellular neutral cysteine proteinase. *Journal of Clinical Investigation*, **86**, 1815–1822.

Reed, S.L., Bouvier, J., Pollack, A.S., Engel, J.C., Brown, M., Hirata, K. *et al* (1993) Cloning of a virulence factor of *Entamoeba histolytica*. Pathogenic strains possess a unique cysteine proteinase gene. *Journal of Clinical Investigation*, **91**, 1532–1540.

Reed, S.L., Ember, J.A., Herdman, D.S., DiScipio, R.G., Hugli, T.E. and Gigli, I. (1995) The extracellular neutral cysteine proteinase of *Entamoeba histolytica* degrades anaphylatoxins C3a and C5a. *Journal of Immunology*, **155**, 266–274.

Ring, C.S., Sun, E., McKerrow, J.H., Lee, G.K., Rosenthal, P.J., Kuntz, I.D. *et al* (1993) Structure-based inhibitor design by using protein models for the development of antiparasitic agents. *Proceedings of the National Academy of Sciences of the U.S.A.*, **90**, 3583–3587.

Robertson, C.D. and Coombs, G.H. (1990) Characterisation of three groups of cysteine proteinases in the amastigotes of *Leishmania mexicana mexicana*. *Molecular and Biochemical Parasitology*, **42**, 269–276.

Robertson, C.D., North, M.J., Lockwood, B.C. and Coombs, G.H. (1990) Analysis of the proteinases of *Trypanosoma brucei*. *Journal of General Microbiology*, **136**, 921–925.

Robertson, C.D. and Coombs, G.H. (1994) Multiple high activity cysteine proteases of *Leishmania mexicana* are encoded by the lmcpb gene array. *Microbiology*, **140**, 417–424.

Rockett, K.A., Playfair, J.H.L., Ashall, F., Targett, G.A.T., Angliker, H. and Shaw, E. (1990) Inhibition of intraerythrocytic development of *Plasmodium falciparum* by proteinase inhibitors. *FEBS Letters*, **259**, 257–259.

Roggwiller, E., Fricaud, A.-C., Blisnick, T. and Braun-Breton, C. (1997) Host urokinase-type plasminogen activator participates in the release of malaria merozoites from infected erythrocytes. *Molecular and Biochemical Parasitology*, **86**, 49–59.

Rosenthal, P.J. (1995) *Plasmodium falciparum*: Effects of proteinase inhibitors on globin hydrolysis by cultured malaria parasites. *Experimental Parasitology*, **80**, 272–281.

Rosenthal, P.J. (1996) Conservation of key amino acids among the cysteine proteinases of multiple malarial species. *Molecular and Biochemical Parasitology*, **75**, 255–260.

Rosenthal, P.J. (1999) Proteases of protozoan parasites. *Advances in Parasitology*, **43**, 105–159.

Rosenthal, P.J., Kim, K., McKerrow, J.H. and Leech, J.H. (1987) Identification of three stage-specific proteinases of *Plasmodium falciparum*. *Journal of Experimental Medicine*, **166**, 816–821.

Rosenthal, P.J., McKerrow, J.H., Aikawa, M., Nagasawa, H. and Leech, J.H. (1988) A malarial cysteine proteinase is necessary for hemoglobin degradation by *Plasmodium falciparum*. *Journal of Clinical Investigation*, **82**, 1560–1566.

Rosenthal, P.J., McKerrow, J.H., Rasnick, D. and Leech, J.H. (1989) *Plasmodium falciparum*: inhibitors of lysosomal cysteine proteinases inhibit a trophozoite proteinase and block parasite development. *Molecular and Biochemical Parasitology*, **35**, 177–184.

Rosenthal, P.J., Wollish, W.S., Palmer, J.T. and Rasnick, D. (1991) Antimalarial effects of peptide inhibitors of a *Plasmodium falciparum* cysteine proteinase. *Journal of Clinical Investigation*, **88**, 1467–1472.

Rosenthal, P.J. and Nelson, R.G. (1992) Isolation and characterization of a cysteine proteinase gene of *Plasmodium falciparum*. *Molecular and Biochemical Parasitology*, **51**, 143–152.

Rosenthal, P.J., Lee, G.K. and Smith, R.E. (1993) Inhibition of a *Plasmodium vinckei* cysteine proteinase cures murine malaria. *Journal of Clinical Investigation*, **91**, 1052–1056.

Rosenthal, P.J. and Meshnick, S.R. (1996) Hemoglobin catabolism and iron utilization by malaria parasites. *Molecular and Biochemical Parasitology*, **83**, 131–139.

Rosenthal, P.J., Olson, J.E., Lee, G.K., Palmer, J.T., Klaus, J.L. and Rasnick, D. (1996) Antimalarial effects of vinyl sulfone cysteine proteinase inhibitors. *Antimicrobial Agents and Chemotherapy*, **40**, 1600–1603.

Rosenthal, P.J., Semenov, A., Ploplis, V.A. and Plow, E.F. (1998) Plasminogen activators are not required in the erythrocytic life cycle of malaria parasites. *Molecular and Biochemical Parasitology*, **97**, 253–257.

Roush, W.R., Gonzalez, F.V., McKerrow, J.H. and Hansell, E. (1998) Design and synthesis of dipeptidyl alpha′, beta′-epoxy ketones, potent irreversible inhibitors of the cysteine protease cruzain. *Bioorganic & Medicinal Chemistry Letters*, **8**, 2809–2812.

Russell, D.G. and Wilhelm, H. (1986) The involvement of the major surface glycoprotein (gp63) of *Leishmania* promastigotes in attachment to macrophages. *Journal of Immunology*, **136**, 2613–2620.

Russo, D.C., Grab, D.J., Lonsdale-Eccles, J.D., Shaw, M.K. and Williams, D.J. (1993) Directional movement of variable surface glycoprotein-antibody complexes in *Trypanosoma brucei*. *European Journal of Cell Biology*, **62**, 432–441.

Salas, F., Fichmann, J., Lee, G.K., Scott, M.D. and Rosenthal, P.J. (1995) Functional expression of falcipain, a *Plasmodium falciparum* cysteine proteinase, supports its role as a malarial hemoglobinase. *Infection and Immunity*, **63**, 2120–2125.

Santana, J.M., Grellier, P., Schrevel, J. and Teixeira, A.R. (1997) A *Trypanosoma cruzi*-secreted 80 kDa proteinase with specificity for human collagen types I and IV. *Biochemical Journal*, **325**, 129–137.

Scheidt, K.A., Roush, W.R., McKerrow, J.H., Selzer, P.M., Hansell, E. and Rosenthal, P.J. (1998) Structure-based design, synthesis and evaluation of conformationally constrained cysteine protease inhibitors. *Bioorganic & Medicinal Chemistry*, **6**, 2477–2494.

Schlagenhauf, E., Etges, R. and Metcalf, P. (1998) The crystal structure of the Leishmania major surface proteinase leishmanolysin (gp63). *Structure*, **6**, 1035–1046.

Scholze, H., Lohden-Bendinger, U., Muller, G. and Bakker-Grunwald, T. (1992) Subcellular distribution of amebapain, the major cysteine proteinase of *Entamoeba histolytica*. *Archives of Medical Research*, **23**, 105–108.

Scholze, H. and Tannich, E. (1994) Cysteine endopeptidases of *Entamoeba histolytica*. *Methods in Enzymology*, **244**, 512–523.

Scholze, H., Frey, S., Cejka, Z. and Bakker-Grunwald, T. (1996) Evidence for the existence of both proteasomes and a novel high molecular weight peptidase in *Entamoeba histolytica*. *Journal of Biological Chemistry*, **271**, 6212–6216.

Schulte, W. and Scholze, H. (1989) Action of the major protease from *Entamoeba histolytica* on proteins of the extracellular matrix. *Journal of Protozoology*, **36**, 538–543.

Scory, S., Caffrey, C.R., Stierhof, Y.D., Ruppel, A. and Steverding, D. (1999) *Trypanosoma brucei*: killing of bloodstream forms *in vitro* and *in vivo* by the cysteine proteinase inhibitor Z-phe-ala-CHN2. *Experimental Parasitology*, **91**, 327–333.

Scott, D.A., North, M.J. and Coombs, G.H. (1995a) *Trichomonas vaginalis*: amoeboid and flagellated forms synthesize similar proteinases. *Experimental Parasitology*, **80**, 345–348.

Scott, D.A., North, M.J. and Coombs, G.H. (1995b) The pathway of secretion of proteinases in *Trichomonas vaginalis*. *International Journal of Parasitology*, **25**, 657–666.

Seay, M.B., Heard, P.L. and Chaudhuri, G. (1996) Surface Zn-proteinase as a molecule for defense of *Leishmania mexicana amazonensis* promastigotes against cytolysis inside macrophage phagolysosomes. *Infection and Immunity*, **64**, 5129–5137.

Selzer, P.M., Chen, X., Chan, V.J., Cheng, M., Kenyon, G.L., Kuntz, I.D. *et al* (1997) *Leishmania major*: molecular modeling of cysteine proteases and prediction of new nonpeptide inhibitors. *Experimental Parasitology*, **87**, 212–221.

Selzer, P.M., Pingel, S., Hsieh, I., Ugele, B., Chan, V.J., Engel, J.C. *et al* (1999) Cysteine protease inhibitors as chemotherapy: lessons from a parasite target. *Proceedings of the National Academy of Sciences of the U.S.A.*, **96**, 11015–11022.

Semenov, A., Olson, J.E. and Rosenthal, P.J. (1998) Antimalarial synergy of cysteine and aspartic protease inhibitors. *Antimicrobial Agents and Chemotherapy*, **42**, 2254–2258.

Shenai, B.R., Sijwali, P.S., Singh, A. and Rosenthal, P.J. (2000) Characterization of native and recombinant falcipain-2, a principal trophozoite cysteine protease and essential hemoglobinase of *Plasmodium falciparum*. *Journal of Biological Chemistry*, **275**, 29000–29010.

Silva, A.M., Lee, A.Y., Gulnik, S.V., Majer, P., Collins, J., Bhat, T.N. *et al* (1996) Structure and inhibition of plasmepsin II, a hemoglobin-degrading enzyme from *Plasmodium falciparum*. *Proceedings of the National Academy of Sciences of the U.S.A.*, **93**, 10034–10039.

Souto-Padron, T., Campetella, O.E., Cazzulo, J.J. and de Souza, W. (1990) Cysteine proteinase in *Trypanosoma cruzi*: immunocytochemical localization and involvement in parasite-host cell interaction. *Journal of Cell Science*, **96**, 485–490.

Souza, A.E., Waugh, S., Coombs, G.H. and Mottram, J.C. (1992) Characterization of a multi-copy gene for a major stage-specific cysteine proteinase of *Leishmania mexicana*. *FEBS Letters*, **311**, 124–127.

Souza, A.E., Bates, P.A., Coombs, G.H. and Mottram, J.C. (1994) Null mutants for the lmcpa cysteine proteinase gene in *Leishmania mexicana*. *Molecular and Biochemical Parasitology*, **63**, 213–220.

Stanley, S.L.J., Zhang, T., Rubin, D. and Li, E. (1995) Role of the *Entamoeba histolytica* cysteine proteinase in amebic liver abscess formation in severe combined immunodeficient mice. *Infection and Immunity*, **63**, 1587–1590.

Tannich, E., Scholze, H., Nickel, R. and Horstmann, R.D. (1991) Homologous cysteine proteinases of pathogenic and nonpathogenic *Entamoeba histolytica*. Differences in structure and expression. *Journal of Biological Chemistry*, **266**, 4798–4803.

Tomas, A.M., Miles, M.A. and Kelly, J.M. (1997) Overexpression of cruzipain, the major cysteine proteinase of *Trypanosoma cruzi*, is associated with enhanced metacyclogenesis. *European Journal of Biochemistry*, **244**, 596–603.

Tran, V.Q., Herdman, D.S., Torian, B.E. and Reed, S.L. (1998) The neutral cysteine proteinase of *Entamoeba histolytica* degrades IgG and prevents its binding. *Journal of Infectious Diseases*, **177**, 508–511.

Troeberg, L., Pike, R.N., Morty, R.E., Berry, R.K., Coetzer, T.H. and Lonsdale-Eccles, J.D. (1996) Proteases from *Trypanosoma brucei brucei*. Purification, characterisation and interactions with host regulatory molecules. *European Journal of Biochemistry*, **238**, 728–736.

Troeberg, L., Morty, R.E., Pike, R.N., Lonsdale-Eccles, J.D., Palmer, J.T., McKerrow, J.H. *et al* (1999) Cysteine proteinase inhibitors kill cultured bloodstream forms of *Trypanosoma brucei brucei*. *Experimental Parasitology*, **91**, 349–355.

Vander Jagt, D.L., Baack, B.R. and Hunsaker, L.A. (1984) Purification and characterization of an aminopeptidase from *Plasmodium falciparum*. *Molecular and Biochemical Parasitology*, **10**, 45–54.

Vander Jagt, D.L., Hunsaker, L.A., Campos, N.M. and Scaletti, J.V. (1992) Localization and characterization of hemoglobin-degrading aspartic proteinases from the malarial parasite *Plasmodium falciparum*. *Biochimica et Biophysica Acta*, **1122**, 256–264.

Ward, W., Alvarado, L., Rawlings, N.D., Engel, J.C., Franklin, C. and McKerrow, J.H. (1997) A primitive enzyme for a primitive cell: the protease required for excystation of *Giardia*. *Cell*, **89**, 437–444.

Westling, J., Yowell, C.A., Majer, P., Erickson, J.W., Dame, J.B. and Dunn, B.M. (1997) *Plasmodium falciparum, P. vivax*, and *P. malariae*: a comparison of the active site properties of plasmepsins cloned and expressed from three different species of the malaria parasite. *Experimental Parasitology*, **87**, 185–193.

Index

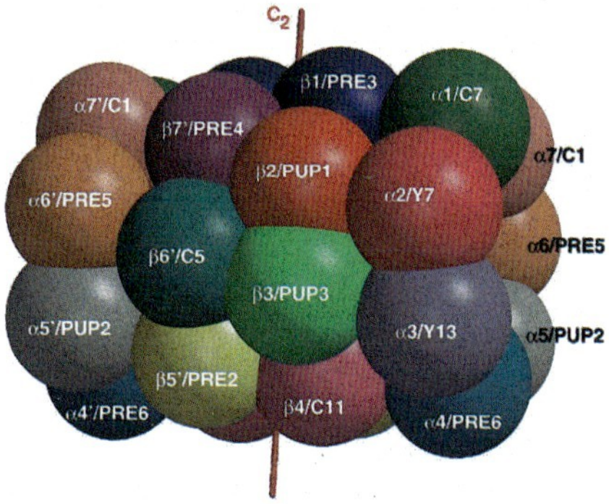

Color plate 1 Topology of the 28 subunits of the yeast 20S proteasome drawn as spheres. (*See page 65*)

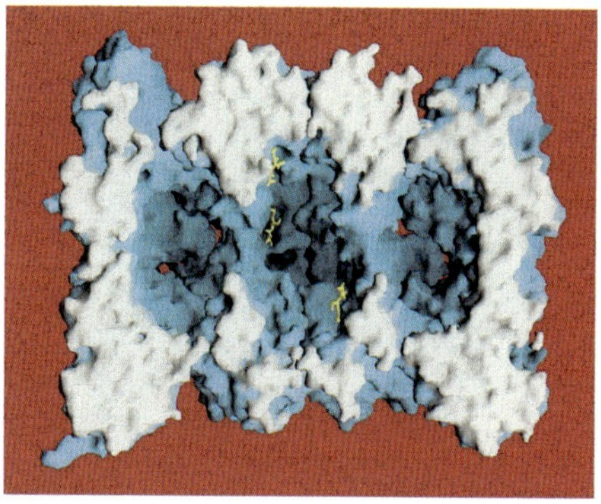

Color plate 2 Surface view of the yeast 20S proteasome crystallized in the presence of calpain inhibitor 1, clipped along the cylindrical axis. The inhibitor molecules are shown as space filling models in yellow. The sealed α-ports at both ends of the yeast proteasome and a few narrow side windows can be seen. (*See page 66*)

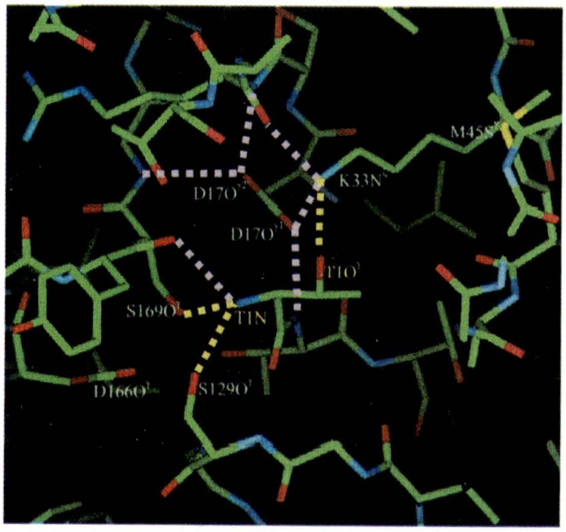

Color plate 3 The nucleophilicity of threonine 1 in the active site of the 20S proteasome shown for the subunit β5/Pre2. Hydrogen-bonds of the Thr1(Oγ) to Lys33(Nε) and of the N-terminus to Ser129(Oγ) and Ser169(Oγ) are shown as yellow dotted lines. (*See page 68*)

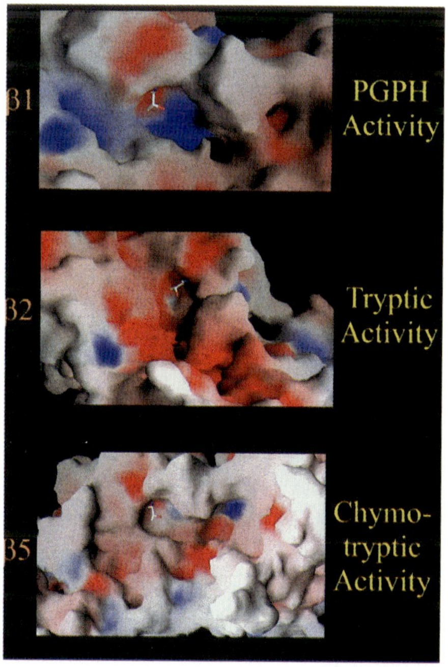

Color plate 4 Surface representation of the three active sites in the yeast 20S proteasome. Each picture shows the nucleophilic Th1 in sticks, the basic residues in blue, the acidic residues in red, and the hydrophobic residues in white. (*See page 69*)

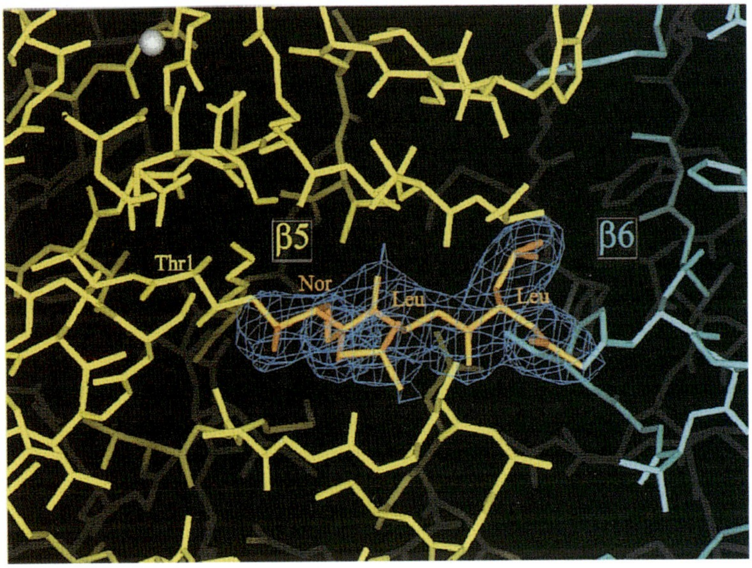

Color plate 5 Calpain inhibitor I binding and S1–S3 pocket of the subunit β5/Pre2. The inhibitor is shown with the electron density map (contoured from 1σ on) with 2Fo–Fc coefficients after two-fold averaging as orange sticks. The β5/Pre2 and β6/C5 subunit are shown as yellow and blue sticks respectively, the magnesium ion as a gray ball. (*See page 73*)

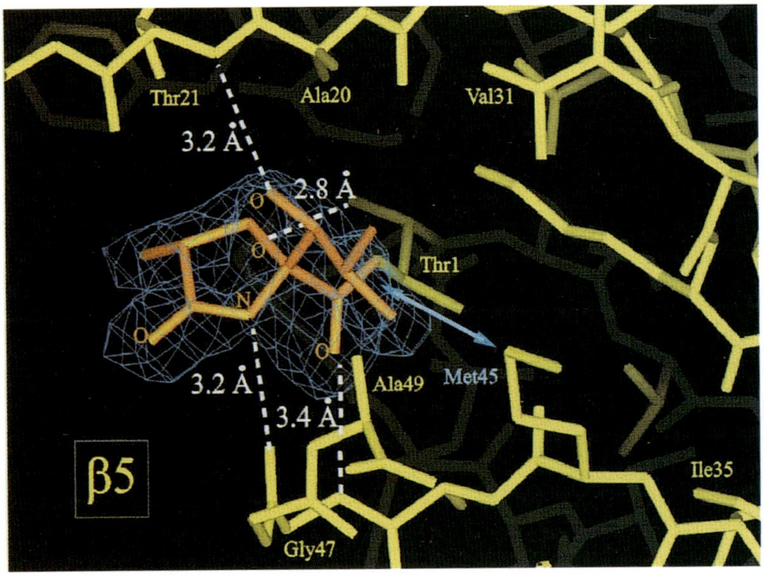

Color plate 6 β5/Pre2 with the covalently bound inhibitor lactacystin. The inhibitor is surrounded via the electron density map (contoured from 1σ on) with 2Fo–Fc coefficients after two-fold averaging. The ester bond between lactacystin and the active site threonine of β5/Pre2 is formed via an acetylation of the Thr1(Oγ) as a result of the β-lactone ring opening. Hydrogen-bonds to the β5-backbone are shown as dashed lines, the isopropyl-β5/Met45 interaction as a blue arrow. (*See page 74*)

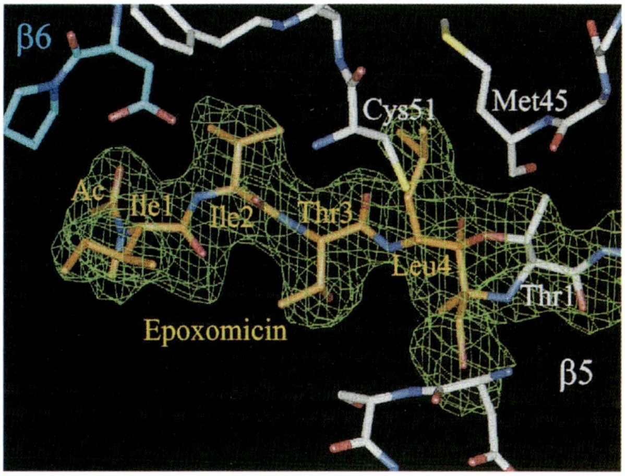

Color plate 7 View of the electron density map of the epoxomicin adduct at β5/Pre2. Epoxomicin is covalently bound to Thr1, resulting in the formation of a morpholino derivative, and the extended substrate binding site is composed of β5/Pre2 and β6. (*See page 75*)

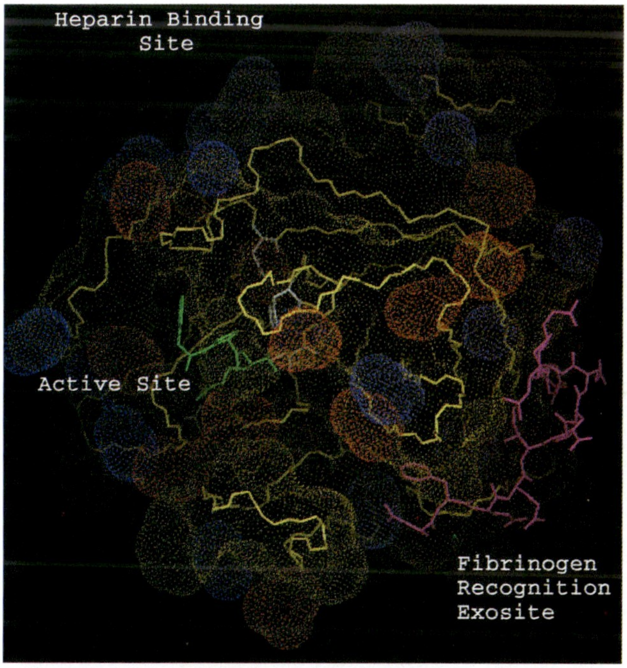

Color plate 8 Front view of the thrombin molecule (backbone in yellow) in complex with the active site-directed inhibitor PPACK (green) and the C-terminal hirudin tail (residues 55–65 in pink) bound to the "fibrinogen recognition exosite", generated from 1tmu.pdb (Priestle *et al* 1993). Thrombin is diplayed with a Connolly dot surface in blue, red, and yellow for basic, acidic or other residues, resp. The "heparin binding site" is located on the top of the thrombin molecule in this view defined as "standard" orientation (Stubbs and Bode 1993). (*See page 181*)

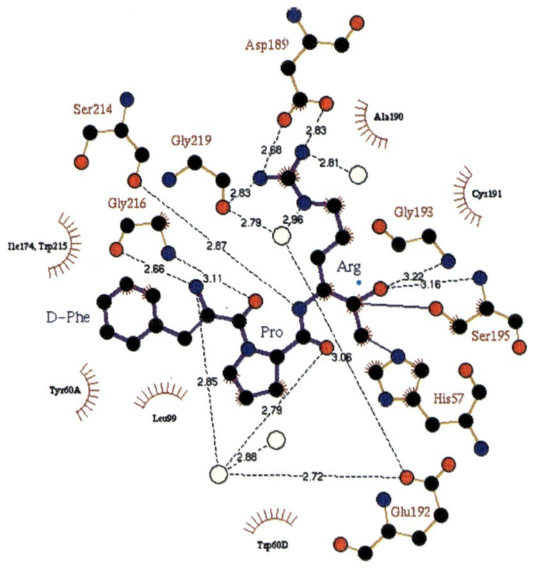

Color plate 9 Schematic diagram of the PPACK–thrombin complex (1ppb.pdb; Bode *et al* 1989) showing the key interactions, generated by LIGPLOT 4.0 (Wallace *et al* 1995). PPACK is covalently bound to Ser195 and His57. (The distances of the hydrogen bonds are given in Å.) (*See page 183*)

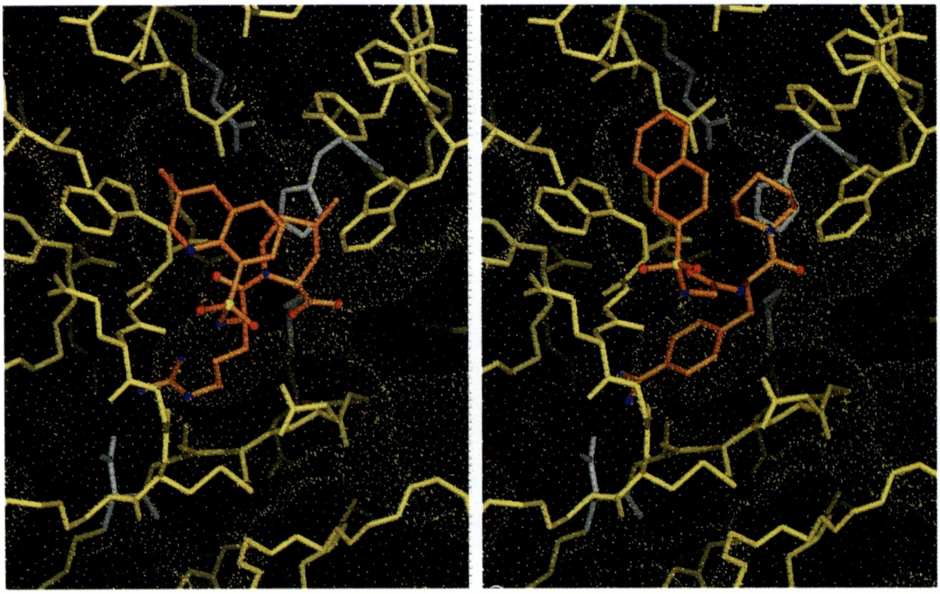

Color plate 10 Structure of the active site region of the complexes formed between thrombin (yellow) and the inhibitors (orange) argatroban (left, 1etr.pdb) and NAPAP (right, 1ets.pdb) (Brandstetter *et al* 1992). The amino acids of the catalytic triad (Ser195, His57, Asp102) and Asp189 in the S1-pocket are shown in light gray. (*See page 184*)

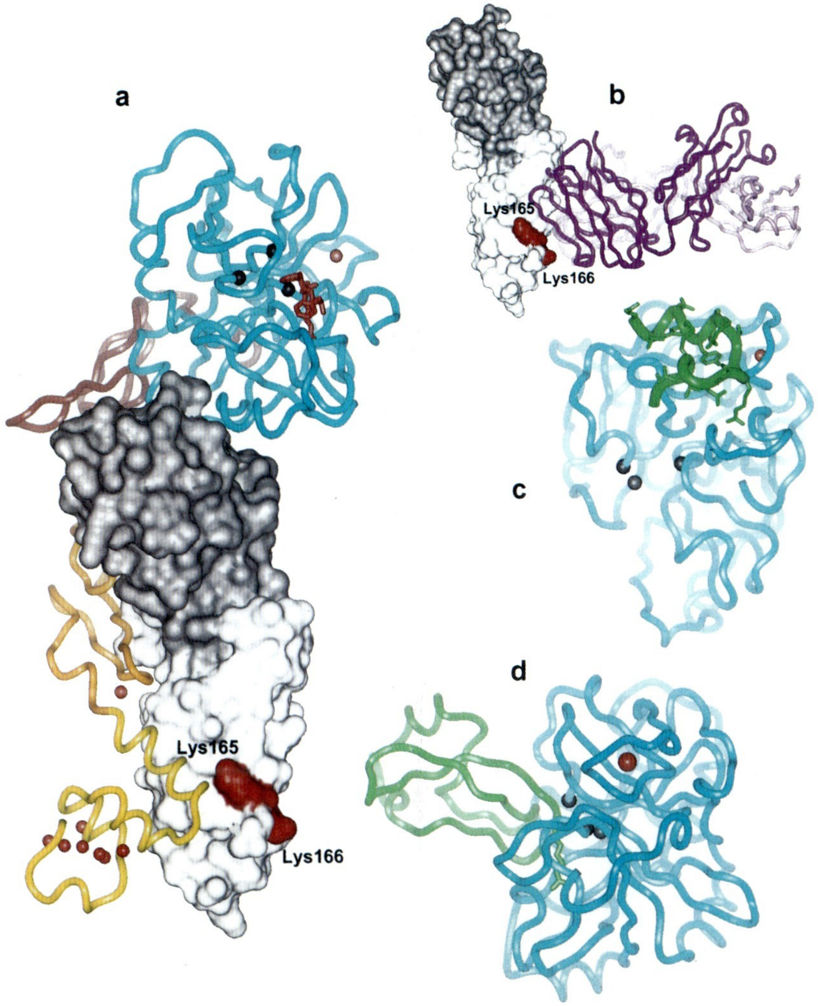

Color plate 11 Crystal structures of the tissue Factor⟨Factor VIIa (TF⟨FVIIa) complex and bound inhibitors. (**a**) The TF⟨FVIIa complex (Banner *et al* 1996). TF (gray) is shown in a solvent accessible representation with the N- and C-terminal fibronectin type III modules in dark and light gray, respectively. FVIIa light chain comprises the N-terminal γ-carboxyglutamic acid (Gla) domain (yellow) followed by the EGF-1 orange and the EGF-2 salmon domains. The catalytic serine protease domain (heavy chain) is shown in blue with the D-Phe-L-Phe-L-Arg chloromethyl ketone peptide (red) irreversibly bound in the active site. The positions of the catalytic triad residues His 57, Asp 102 and Ser 195 are indicated (black spheres). The calcium ions are shown in red. The substrate contact region in the C-terminal domain of TF is proximal to the phospholipid membrane and includes residues Lys 165 and Lys 166 (red). (**b**) F(ab) of anti-tissue factor antibody TF8–5G9 bound to tissue factor (Huang *et al* 1998). TF is represented as in Figure 9.4a with Lys 165 and Lys 166 in red. The epitope of the antibody F(ab) overlaps with the substrate interaction region of TF. The antibody heavy chain, comprising the variable and constant domains, is shown in dark purple and the corresponding domains of the light chain in light purple. (**c**) E-76 peptide bound to the FVIIa catalytic domain (Dennis *et al* 2000). The peptide (green) binds to an exosite proximal to the active site and the calcium (red)-binding site of FVIIa (blue). (**d**) BPTI variant 5L15 bound to FVIIa catalytic domain (Zhang, St. Charles and Tulinsky 1999). The inhibitor (green) binds in the active site of FVIIa (blue). Also depicted is the P1 residue (Arg) side chain of the inhibitor, which inserts into the S1 recognition pocket of the enzyme. (*See page 207*)

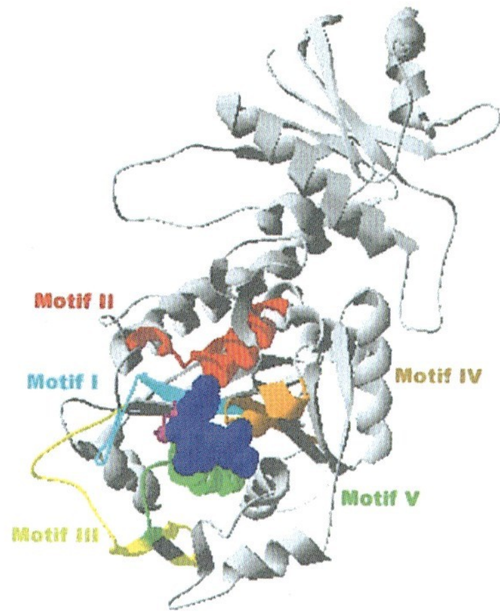

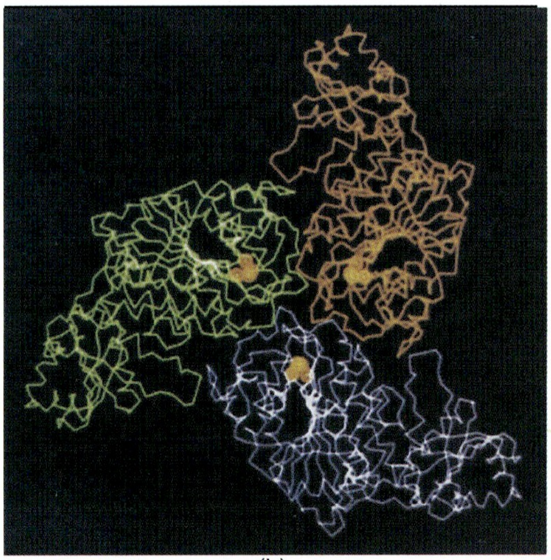

(b)

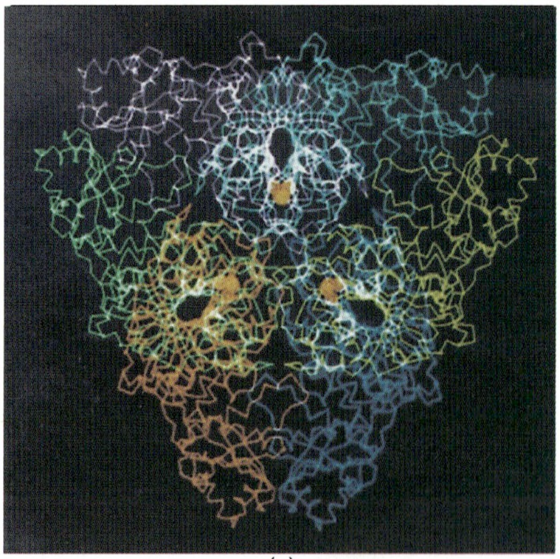

(c)

Color plate 12 (a) Schematic ribbon drawing of the LAP monomer. Blue spheres show the amastatin (or substrate) binding site. The five motifs which are common to many aminopeptidases are shown in color and are labeled I–V. The zinc ions are behind the amastatin and are shown in pink. (b) and (c) Tracing of the carbon backbone of the blLAP trimer (b) and hexamer (c) as viewed down the threefold axis. (*See page 312 and 313*)

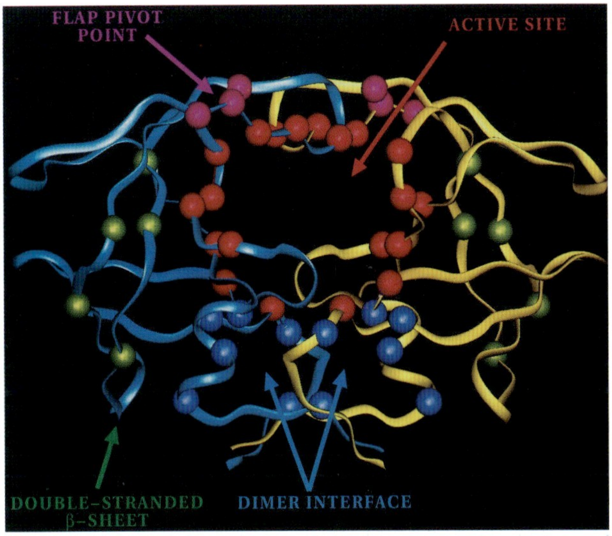

Color plate 13 Residues associated with HIV protease drug resistance. The color-coded spheres represent point mutations in different regions of the homodimer: red, active site; pink, flap pivot point; blue, dimer interface; and green, double-stranded β-sheet (Val56-Gly78). The C_α atoms of the monomers are drawn as blue and yellow ribbons. (*See page 368*)

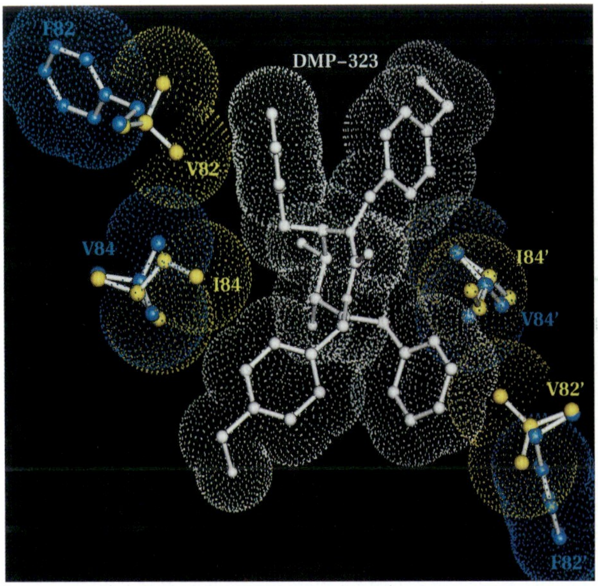

Color plate 14 Loss of vdw interactions. Conformations of Val82/82′ and Ile84/84′ in the wild type (yellow) and in the double mutant V82F/I84V (blue), when complexed to DMP-323 (white). A drawing of the vdw spheres (dotted spheres) of the atoms clearly shows that the two mutations enlarge the substrate binding pocket and reduce the extent of vdw interactions to the inhibitor; note that Val84 is too small to contact the inhibitor and the side-chain rotamer of Phe82 packs outside of the S1 subsite. (*See page 372*)

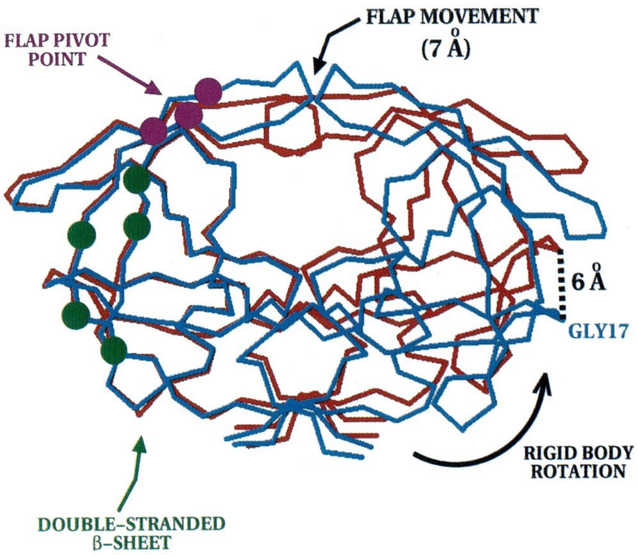

Color plate 15 Inhibitor-induced conformational changes in HIV protease. Overlay of the C_α atoms of the uncomplexed (blue) and DMP-323-complexed (red) structures of the protease. Note the contraction of the active site in the complexed state as the flaps close and the monomers rotate as two rigid bodies around the C-terminal dimer pivot point. (*See page 375*)

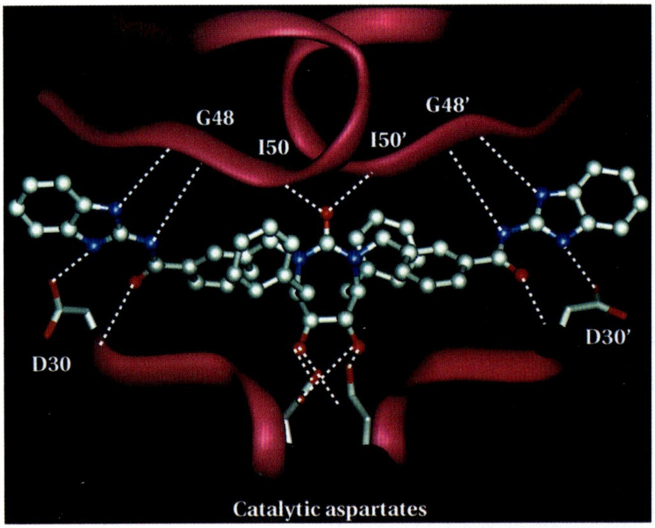

Color plate 16 Broad specificity of SD-146. This compound is a potent inhibitor of wild type and V82F/I84V because, (1) its diols form an extensive hydrogen bonding network (dashed lines) with the catalytic aspartates, (2) its urea oxygen hydrogen bonds directly to the backbone amides of Ile50 and 50′ of the flaps, (3) its P2 and P2′ substituents participate in six hydrogen bonds with the backbone atoms of G48, G48′, D30′, and D30′, and (4) it interacts with a total of 31 residues and participates in 177 vdw contacts. (*See page 376*)